◀ FIGURE EP.2

Fluorescence image from a DNA assay array

On the surface of this 12.8-mm square chip are localized thousands of different complements to DNA from the gene for human p53 tumor-suppressor and its possible mutated forms. From the pattern of binding of all the sites together, as probed by each individual site's level of fluorescence, the DNA and any mutations can be sequenced. This analysis system is manufactured by Affymetrix Corp. The colors code the levels of fluorescence.

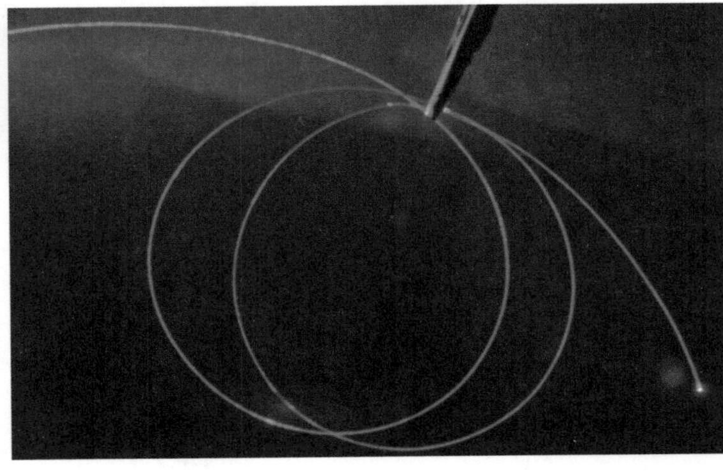

FIGURE EP.3 ▶

Fiber optic light propagation

Total internal reflection allows light to be propagated through optical fibers. As a result, the site of analysis need not be inside a spectrometer.

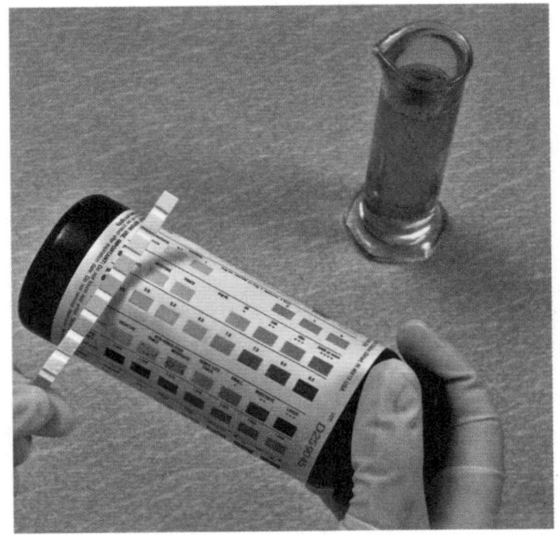

◀ FIGURE EP.4

A test strip for urine with multiple colorimetric measurements

Each pad on the test strip is a piece of porous material containing one or more reagents that can be stored dry and react specifically with some component in the urine. Some reagents are relatively simple, such as a pH indicator. Others are enzymes. Either the products of the enzyme reaction are colored or the products react with other reagents in the pad to produce a colored product. The colored spots on the container provide standards to compare with the strip's colors.

FIGURE EP.5 ▶

A laser beam for remote sensing

The beam from a laser in this truck is seen as it scans the atmosphere over Albuquerque. The light that returns can be analyzed to map a number of possible molecules or materials in the atmosphere. The map is made by determining the spectral properties at various positions. The angular position is determined by the direction of the beam, while the distance can be found from the time the pulses of light take to return to the detector after leaving the laser. On the cover of the book is a false-color map of particulate concentrations over Albuquerque obtained in this way.

Judith F. Rubinson

Kenneth A. Rubinson

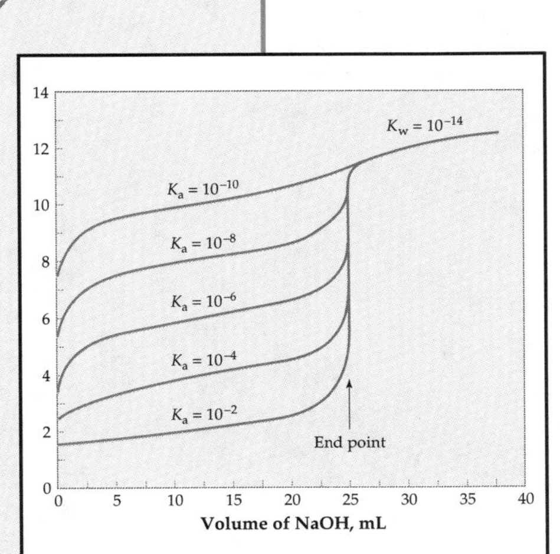

Contemporary Chemical Analysis

PRENTICE HALL
Upper Saddle River, NJ 07458

Library of Congress Cataloging-in-Publication Data

Rubinson, Judith F.
Contemporary chemical analysis / Judith F. Rubinson, Kenneth A
Rubinson. -- 1st ed.
p. cm.
Includes bibliographical references and index.
ISBN 0-13-519331-1
1. Chemistry, Analytic. I. Rubinson, Kenneth A. II. Title.
QD75.2.R825 1998
543--dc21 97-52137
CIP

Acquisitions Editor: *John Challice*
Editorial Assistant: *Betsy Williams*
Editor in Chief: *Paul F. Corey*
Assistant Vice President of Production and Manufacturing: *David W. Riccardi*
Executive Managing Editor: *Kathleen Schiaparelli*
Assistant Managing Editor: *Lisa Kinne*
Marketing Manager: *Linda Taft Mackinnon*
Manufacturing Manager: *Trudy Pisciotti*
Art Director: *Joseph Sengotta*
Text Designer: *Amy Rosen*
Art Editor: *Karen Branson*
Cover Designer: *Bruce Kenselaar*
Photo Researcher: *Yvonne Gerin*
Art Studio: *Wellington Studios*
Production Supervision/Composition: *Interactive Composition Corporation*
Cover Art: *Lidar map of air pollution showing the density of airborne particulates over Albuquerque, NM. The colors represent the concentration of particulates, from red (highest) to dark blue (lowest). The worst pollution is over a truck refueling station near the Interstate. The yellow lines are roads. Santa Fe Technologies Inc./Photo Researchers, Inc.*

Simon & Schuster/A Viacom Company
Upper Saddle River, New Jersey 07458

Printed in the United States of America
10 9 8 7 6 5 4 3 2 1

ISBN 0-13-519331-1

Prentice-Hall International (UK) Limited, *London*
Prentice-Hall of Australia Pty. Limited, *Sydney*
Prentice-Hall Canada Inc., *Toronto*
Prentice-Hall Hispanoamericana, S.A., *Mexico*
Prentice-Hall of India Private Limited, *New Delhi*
Prentice-Hall of Japan, Inc., *Tokyo*
Simon & Schuster Asia Pte. Ltd., *Singapore*
Editora Prentice-Hall do Brasil, Ltda., *Rio de Janeiro*

Brief Contents

Contents

Chapter 1

Preliminaries

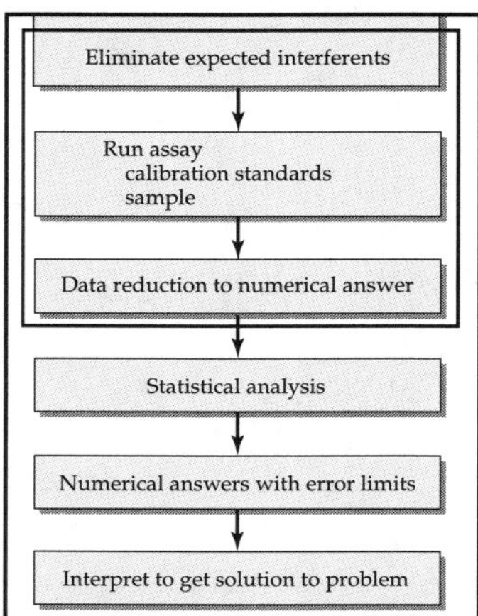

CHAPTER 2

Statistical Tests and Error Analysis

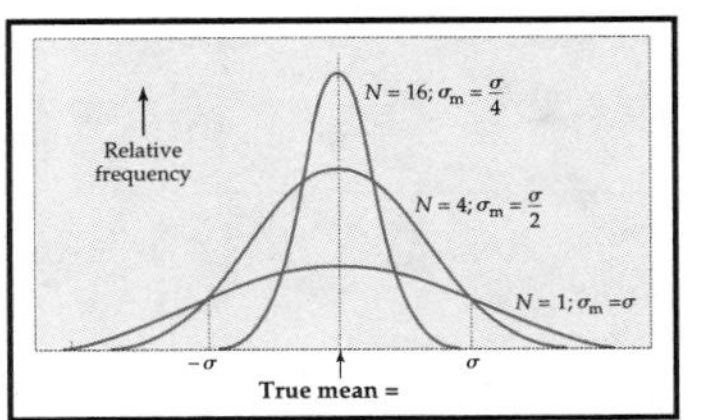

CHAPTER 3

Sampling

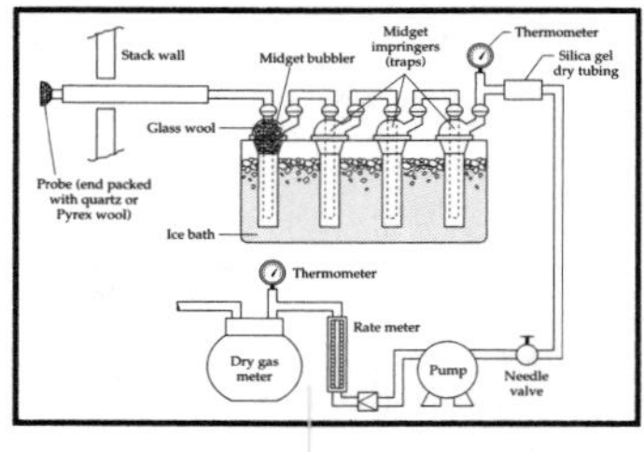

CHAPTER 4

Sample Treatments, Interferences, and Standards

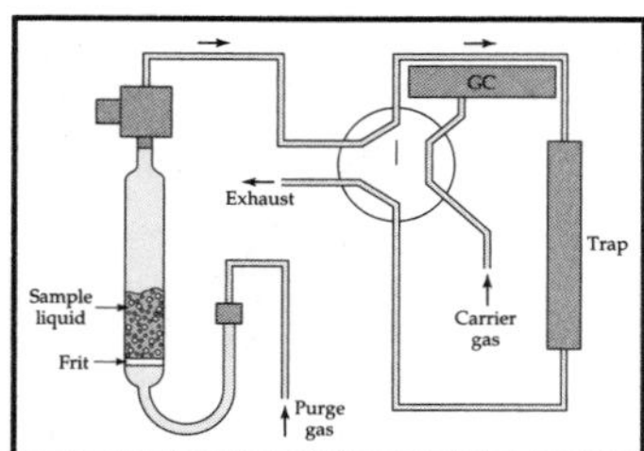

CHAPTER 5

Sample Size and Major, Minor, Trace, and Ultratrace Components

CHAPTER 6

Simple and Competitive Chemical Equilibria

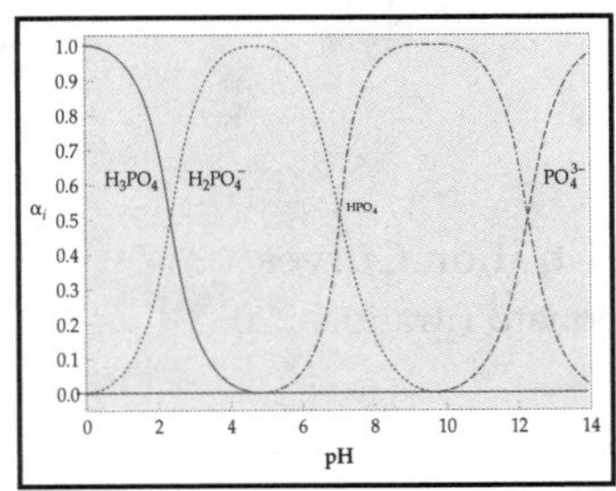

CHAPTER 7

General Introduction to Volumetric Titrations: Neutralization Titrations

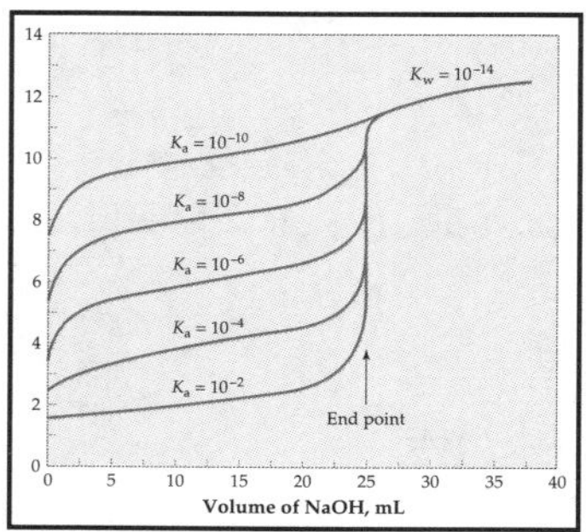

CHAPTER 8

Other Types of Equilibria

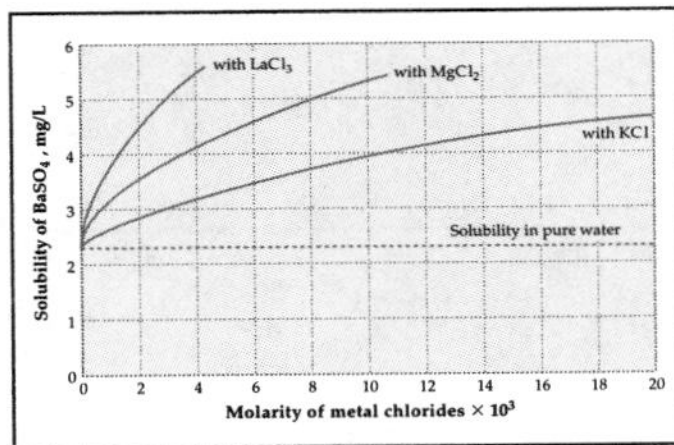

CHAPTER 9

Other Titrimetric Methods

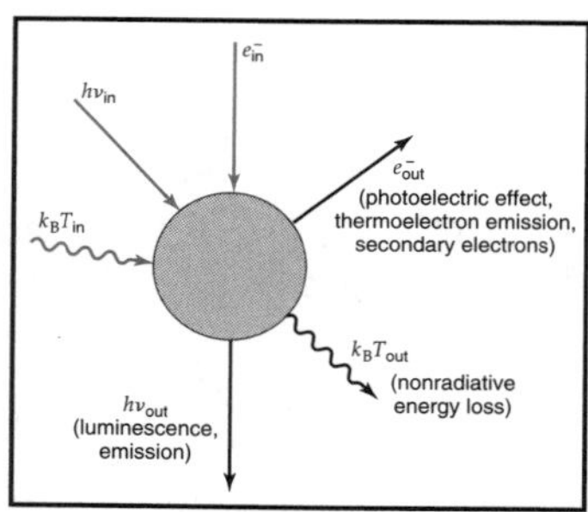
e^-_{in}
$h\nu_{in}$
k_BT_{in}
e^-_{out}
(photoelectric effect,
thermoelectron emission,
secondary electrons)
k_BT_{out}
(nonradiative
energy loss)
$h\nu_{out}$
(luminescence,
emission)

CHAPTER 12

Mass Spectrometry

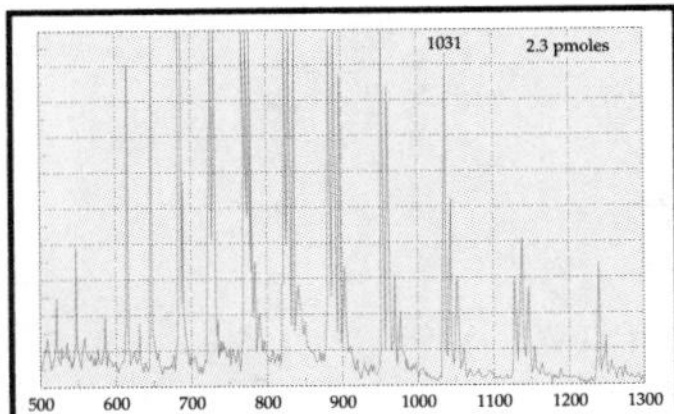

Chapter 13

Separations and Chromatography

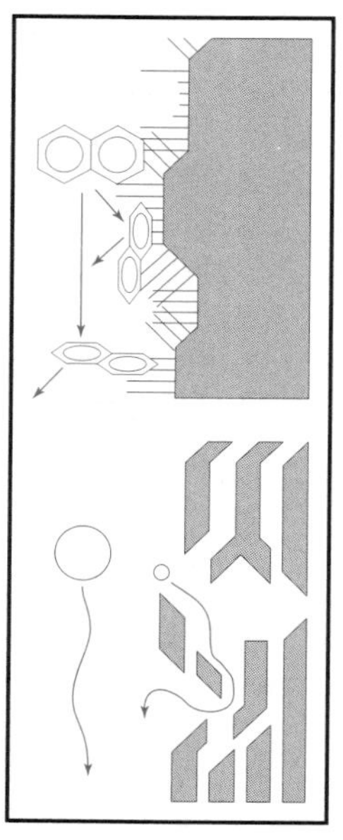

Chapter 14

Separations by Applied Voltage: Electroseparations

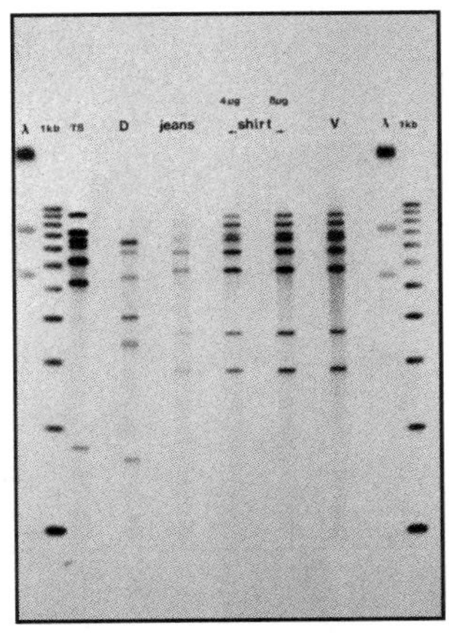

Chapter 15

Electrochemical Methods

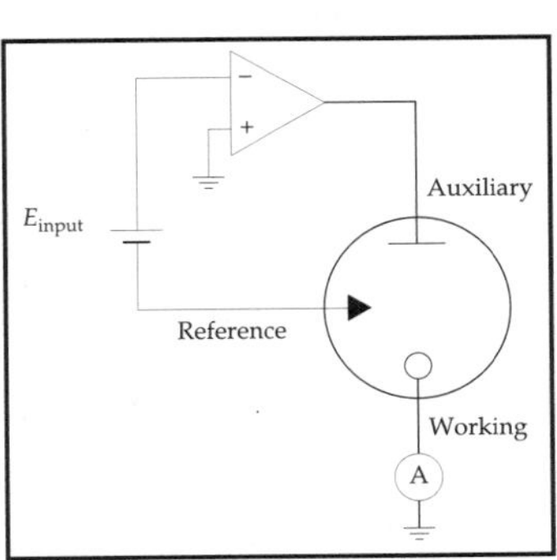

CHAPTER 16

Kinetic Methods

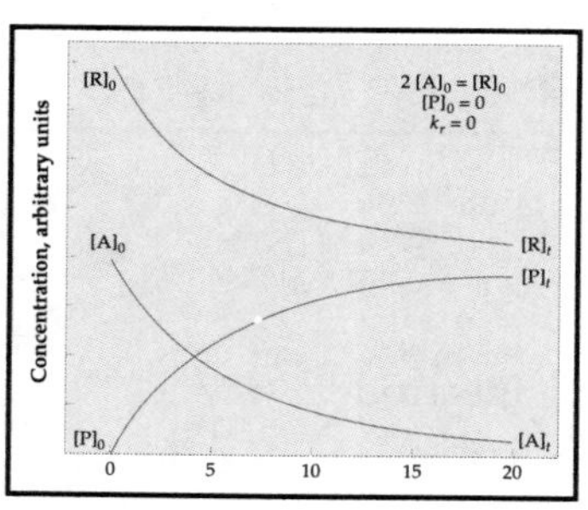

PREFACE

It is the mark of an educated mind to expect that amount of exactness which the nature of the particular subject admits. It is equally unreasonable to accept merely probable conclusions from a mathematician and to demand scientific proofs from an orator.

Aristotle, *Ethics*, Book 1, Chapter 3

Our first priority in this text is to convey the unchanging principles of analytical chemistry while indicating the range of applications of analytical chemical methods.

After an introduction/review chapter, the core concepts are presented: statistical treatment of experimental data; sampling; sample preparation and standards; the different approaches to analysis based on sample size and analyte level. We chose to include the discussion of statistics, sampling, and sample preparation (the Three S's) early in the book, since consequences of the "garbage-in, garbage-out" syndrome tend to propagate even more with the automated methods used today. Meaningless or suspect data (leading to erroneous conclusions) can be minimized by understanding these three subjects and improving their practice.

In order to allow class time for the much-expanded array of techniques in contemporary practice, we have organized the discussion of chemical equilibria more efficiently by exploiting the similarities between the various types of solution equilibria. The length of coverage of titration has been reduced by looking at pH titrations thoroughly and then showing the similarities of other types to them. We have eliminated a number of analytical techniques that now are primarily of historical interest. Treatment of gravimetric analysis has been reduced to the minimum necessary to understand how to work with solids and to understand the solubility equilibria that can be pertinent in sample preparations.

Our introduction to instrumental techniques begins with coverage of the principles of spectroscopy and chromatography. In the case of IR and NMR, we have limited the discussion to quantitation, since qualitative uses are often discussed in the organic sequence. On the other hand, in recognition of their ever-growing importance, mass spectrometry and electrophoresis have their own, separate chapters. We have also included a thorough discussion of modern electrochemical methods, with discussion of techniques such as electrogravimetry and classical polarography used only to introduce other, more contemporary methods. The book concludes with an introduction to the use of kinetics for chemical analysis.

Other strategies that we feel make this text effective include the following:

- In order to present a coherent, integrated coverage of major topics, we have avoided breaking each field down into artificially short chapters.
- In order to provide the maximum flexibility to the instructor, we have covered some topics in letter-labeled sections in each chapter (**A Deeper Look**) to be included at the instructor's discretion. These can be used to

bring more detail to specific topics as desired. The placement of these sections reflects only the fact that the number of variations in emphasis among the various topics in analytical chemistry is approximately equal to the number of people teaching the course.

- **Analytical Chemistry in Action** boxes provide interesting applications of the concepts of techniques introduced in the chapter. Considered together, these furnish a cross-section of the range of applications to which analytical chemistry methodology is applied.
- **Concept Review** questions at the end of each chapter provide a brief self-assessment of the principal points in the chapter. Answers for these questions appear at the end of the text.
- End-of-chapter **Exercises** draw extensively on the literature and present realistic applications of each analytical technique. The exercises are divided into two groups—Exercises and then Additional Exercises. Answers for the first set are included at the end of the text. Solutions to these are provided in a separate *Solutions Manual* (ISBN 0-13-779216-6).

All concept review answers, as well as solutions for both sets of problems, are provided in the *Instructor's Manual* (ISBN 0-13-080332-4). The *Instructor's Manual* also includes chapter outlines and suggestions for cooperative learning exercises.

In writing this book, we have had the benefit of expert comments on both technology and pedagogy from a large number of colleagues. We wish to thank them all publicly with a listing below (alphabetical—sorry to those in the second half of the alphabet . . . we know how you feel!). Other scientists around the world have supplied data, diagrams, photographs, and figures from their work, and we appreciate their efforts. Their thoughtfulness has greatly enlivened the book as well as illustrating contemporary capabilities. We are, of course, responsible for any errors that remain.

Robert Balahura	*University of Guelph*
E. Joseph Billo	*Boston College*
Doug Barofsky	*Oregon State University*
John Baur	*Illinois State University*
James A. Boiani	*SUNY-Geneseo*
John J. Breen	*Indiana University, Purdue University at Indianapolis*
Nancy Breen	*Indiana University, Purdue University at Indianapolis*
Thomas G. Chasteen	*Sam Houston State University*
Luis A. Colón	*SUNY-Buffalo*
John B. Cooper	*Old Dominion University*
Eric Grimsrud	*Montana State University*
Gene S. Hall	*Rutgers University*
John J. Harwood	*Tennessee Technological University*
Barbara Kebbekus	*New Jersey Institute of Technology*
William R. Lammela	*Nazareth College*
David A. Laude	*University of Texas Austin*
Charles H. Lochmuller	*Duke University*
Harry B. Mark, Jr.	*University of Cincinnati*
John Morawski	*Waters Instruments*
Gabor Patonay	*Georgia State University*
Edward H. Piepmeier	*Oregon State University*
Thomas H. Ridgway	*University of Cincinnati*
Marvin Rowe	*Texas A&M University*

Apryll M. Stalcup	*University of Cincinnati*
Peter C. Uden	*University of Massachusetts-Amherst*
Earl L. Wehry	*University of Tennessee*
George S. Wilson	*University of Kansas*

It also is a pleasure to acknowledge the support of the design and production team at Prentice Hall. Betsy Williams coordinated the process of reviewing and wore a multitude of other hats from the beginning. Karen Branson coordinated the complex art program, and Joseph Sengotta created the design. Yvonne Gerin obtained the photographs. Dave Wood at Wellington Studios produced the artwork. Production was by the team at Interactive Composition Corporation, with Judy Johnson coordinating and David Heath as compositor. Overseeing the whole process with great knowledge and insight was our editor John Challice, who must surely be the best in the business.

Judith F. Rubinson
rubinsjf@email.uc.edu
Kenneth A. Rubinson

A Guide to Using this Text

Analytical chemistry is definitely an undergraduate chemistry course where the applications are immediately obvious (even to that much maligned casual observer). Among other things, this means that there are interesting stories to tell about how analytical chemistry is important in day-to-day life. However, chemical analysis is also a science that requires problem-solving and numerical skills. So we have taken great care to provide opportunities to practice your new skills and test your understanding.

To get the most out of this book, we suggest you take a look at the Overview for each chapter before you begin. It provides a framework for the information you are about to encounter. Make sure you understand the worked examples. Work the exercises at the end of the chapter, including the concept review exercises. These will help improve your understanding of the science, and will identify areas where your understanding is still weak.

Below are highlighted some of the features of this book that have been designed to help you succeed in this course. Our aim has been not just to compile information on the principles of contemporary chemical analysis, but to help you learn to apply the principles after you have finished the course.

352 Chapter 11 Introduction to Spectrometry

ANALYTICAL CHEMISTRY IN ACTION
I'm a Wanderer, yes, I'm a Wanderer . . . on Mars

In July, 1997, a small robot carrying instruments was landed on the Martian surface to carry out photographic surveys and chemical analyses of the rocks. The chemical analysis was obtained with an alpha proton X-ray spectrometer (APXS). Designed specifically for the Mars Pathfinder mission, it provided a complete and detailed analysis of chemical elements in Martian soil and rocks near the landing site. The APXS instrument was carried on the Pathfinder microrover, which provided transportation to places of interest on the Martian surface. The spectrometer consisted of a sensor head mounted on a movable arm that enabled the head to be placed against soil and rock samples to perform chemical analyses in situ. The general layout of the sensor head is shown in the adjacent figure. You can see the sensor head in the photograph of the microrover, named Sojourner.

The APXS spectrometer works by bombarding the surface with α particles from a Curium source (^{244}Cm). When the particles hit the surface, three different types of interactions occur. The analysis of the spectra of these three provide the quantitative elemental analysis. Two of these interactions—Rutherford backscattering and X-ray fluorescence—are discussed briefly in this chapter. The third interaction, a nuclear reaction, results in the production of protons from the reaction between the α particles and the nuclei of light elements.

Measurement of the intensities and energy distributions of all three spectrometers together yields information on the abundance of chemical elements in the sample. The three spectrometries are partly redundant and partly complementary: Rutherford backscattering of the α-particles is superior for light elements (C, O), while the proton emission is mainly sensitive to Na, Mg, Al, Si, and S. X-ray emission is more sensitive to heavier elements (Na to Fe and beyond). A combination of all three measurements enables the quantitation of all elements with the exception of H and He when their concentrations are greater than about 0.2% (w/w).

[Ref: R. Rieder, H. Wänke, T. Economou, and A. Turkevich. 1997. "Determination of the chemical composition of Martian soil and rocks: The alpha proton X ray spectrometer," *Journal of Geophysical Research*. Vol. 102, No. E2, pp. 4027–4044, Feb. 25. © American Geophysical Union. Paper #96JE03918.]

PHOTO ACA11.2 ▲
Mars lander.

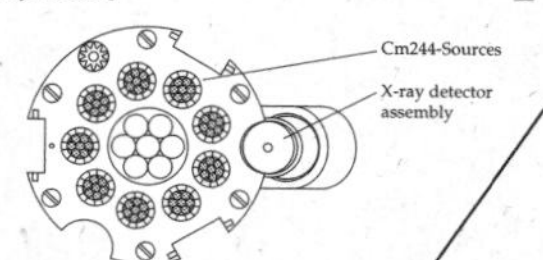

Working distance 39 mm

FIGURE ACA
APXS sensor

ANALYTICAL CHEMISTRY IN ACTION BOXES

Located throughout the text, these boxes describe recent situations where analytical chemistry was used to solve a real-world problem. These discussions involve many disciplines—forensic science, agriculture, biology, medicine, geology, space exploration, and environmental safety—and are not only interesting, but give you an example of how the information you are reading is being used today.

A DEEPER LOOK

The *Deeper Look* sections at the end of each chapter contain more detailed information about selected topics. Your instructor may ask you to read these, or you may just want more information about something you find interesting.

11A A Deeper Look: Waves, Wavelengths, and Electromagnetic Radiation

Waves

The description of light as a wave is one of the primary concepts of the action of the physical world. In chemical analysis, you will encounter the concepts of electromagnetic waves—radio, light, and X-rays, for example.

Waves carry energy from one location to another. They have a finite velocity. The "shape" of a wave is usually represented by a graph of height (for a water wave), or electric field strength (for light waves), or pressure (for a sound wave) vs. time or vs. distance. A graph of a general wave is shown in Figure 11A.1.

Much of the behavior of light can be explained by using the properties of sine waves. A graph of a sine wave is shown in Figure 11A.2. A sine wave can be described mathematically by

$$A = A_0 \sin \theta \qquad \textbf{(11A-1)}$$

where

A is the amplitude,
A_0 is the peak amplitude, and
θ is a continuous variable, the form of which is shown below.

EXAMPLE 4.1

During validation of an assay for polyaromatic hydrocarbons (PAHs) in soil in an old dump site, the relative standard deviation σ associated with each step was estimated for the 1–100 ppb range. The assay technique (gas chromatography) has $\sigma = \pm 1\%$. The sampling (core sampling and combining cores from the centers of four adjacent 10m × 10m sections of a grid) has an estimated $\sigma = \pm 20\%$ between sets, due mostly to the heterogeneity of the soil. The recovery during the sample preparation had been found by spiking soil samples to have $\sigma = \pm 7\%$ for PAHs in this type of soil. What is the total relative standard deviation? The analyte content is 3 ppb.

Solution

We use Equation 4-2.

$$\frac{\sigma_{tot}}{\text{analyte content}} = \text{R.S.D.} = \sqrt{0.2^2 + 0.07^2 + 0.01^2} = \sqrt{0.045} = 21\%$$

Notice that the value of 3 ppb is not needed since the *s*-values are relative deviations as given and are for samples of this type and PAH levels. Almost the entire error is from sampling, which might be expected from source such as a multiacre land area.

EXAMPLES

A multitude of in-text *Examples* shows you how to attack the kind of exercises you'll see on your exams. Make sure you read over these Examples carefully and understand each step. If not, go back and re-read the appropriate section of each chapter.

solution by a factor of two, you know that the ink concentration is now half what it was before dilution, yet you will not see any change. For this last dilution, you will be trying to run an assay of ink content outside of the usable range of your eyes. The bottom of the concentration range that can be seen is quantified by a number called the **limit of detection**.

Limits of detection are described in more detail in Section 5A.

Now think about the other end of the concentration range. Put enough black ink into a solution so that you cannot see any light through it at all. Double that concentration, and, again you see no light. This other extreme of concentration also is of no use to determine the changes in the amount of ink

MARGINAL ANNOTATIONS

Throughout the text, we use marginal cross-references to show additional connections between topics.

CONCEPT REVIEWS

The *Concept Review* section at the end of each chapter provides an opportunity to make sure you have understood the fundamental ideas of the chapter. Start your review here before you tackle the more mathematical Exercises and Additional Exercises.

Concept Review

1. What measured quantity provides information about analyte concentration in each of the following methods?
(a) coulometry
(b) pontentiometry
(c) amperometry
(d) conductimetry

2. What is the half-reaction that is the basis for each of the following reference cells?
(a) saturated calomel
(b) standard calomel

5. What three factors cause the current in an electrolytic cell to be lower than expected for a given E_{app}?

6. In a system which uses a three-electrode potentiostat, between which electrodes is the potential measured? Between which two is the current measured?

7. In an aqueous solution of 5% dextrose (a monosaccharide), what species would contribute to the conductivity of the solution?

8. What is the difference (if any) between the conductance

EXERCISES

Exercises at the end of each chapter allow you to practice your new skills. The answers to all quantitative exercises are located at the back of the book.

Exercises

15.1 A fluoride ion-selective electrode was used to analyze for fluoride in drinking water. The range for which the water treatment plant aims is 1.00 (±0.1). Each day three standards and six periodic samples were analyzed. The pH and ionic strength were buffered for all solutions before analysis. One day in the spring, the following values were obtained:

Solution	E (mV)
1.00×10^{-4} M F^-	40.5
5.00×10^{-5} M F^-	60.2
1.00×10^{-5} M F^-	100.0

(c) Is the average for the day within the range set by the water treatment plant?

15.2 A fluoride ion-selective electrode responds to both F^- and OH^- but not to HF. A measurement is made and interpreted assuming that all the fluoride ion is unassociated and no other species interferes with the ISE. Data: $K_a(HF) = 7.2 \times 10^{-4}$ at 25°C; $k_{FOH} = 0.06$ under the conditions of the experiments. Answer the following questions.
(a) Will the formation of HF tend to make the measured result low or high?
(b) Will the presence of OH^- tend to make the measured result low or high?

ADDITIONAL EXERCISES

Once you've had a crack at the *Exercises,* try these for more practice.

Additional Exercises

■ 7.23 (*A spreadsheet exercise*) Calculate and plot the expected titration curve for 50.00 mL of 0.0226 M monoprotic acid with $pK_a = 3.5$ for titration with KOH, 0.0100 M.

7.24 A 50.00 mL aliquot of 0.1000-M salicylic acid is titrated with 0.1000-M NaOH at 25°C in water.
(a) What equilibrium determines the pH values of the solution before the end point, at the end point, and after the end point?
(b) What is the volume of the solution and its pH at the midpoint between the beginning and the equivalence

that $a_{H^+} = [H^+]$ and T = 25°C. (Don't forget to include the added volume of HCl.)

7.27 What do you expect to be the volumes of added acid at the equivalence points in a titration of 10.000 mL of a solution that is 0.1000 M in Na_2CO_3 and 0.1000 M in $NaHCO_3^-$ and is titrated with 0.1077M-HCl?

7.28 You have available to you the following indicators: phenolphthalein, methyl red, methyl orange, and bromocresol green. Assume that for the calculation any indicator present is at a low enough concentration that the

COMPUTER EXERCISES

This icon indicates an exercise most easily solved using the spreadsheet of your choice.

SUGGESTIONS FOR FURTHER READING

Interested in learning more? Doing a project for your class? You may find these *Suggestions for Further Reading* helpful. We've annotated them to help you locate exactly what you're looking for.

Suggestions for Further Reading

Sawyer, D. T., Roberts, J. L., Jr. 1974. *Experimental Electrochemistry for Chemists*. New York: Wiley.
The best next place to go for more electrochemistry if you are going to do experiments despite its age.

Covington, A. K. 1979. *Ion-Selective Electrode Methodology*. Boca Raton, FL: CRC Press. Vols. I and II.

An elementary book of voltammetric methods.

Stock, J. T. 1965. *Amperometric Titrations*. New York: Interscience.
A complete book of amperometric titrations.

Heinze, J. 1984. "Cyclic Voltammetry." *Angewandte Chemie*, International Edn. 23, 831–847.

About the Authors

Faye Rubinson, currently a Visiting Scholar at the University of Cincinnati, has been an active member of the analytical chemistry community for over 20 years. She received her BS in Chemistry from the University of North Carolina at Chapel Hill and, after a brief period in industry, entered the University of Cincinnati, where she received her PhD in Analytical Chemistry. She then carried out postdoctoral work at the Charles F. Kettering Research Laboratory. In addition to her research and teaching responsibilities since that time, Faye has been active in the American Chemical Society, serving as an officer in both the Dayton and Cincinnati Sections and as a member of the Education Committee of its Analytical Division. She has been involved in outreach activities both through the ACS and through the Ohio Junior Academy of Science, and she has overseen or participated in a number of mentoring programs for students. Dr. Rubinson's present research efforts are directed toward the development and use of conducting polymer and polymer-modified electrodes for the investigation of redox properties of biological compounds.

Kenneth Rubinson received his undergraduate degree at Oberlin College and his PhD from the University of Michigan, and he carried out postdoctoral research for several years at Cambridge University. While in Britain, he was a recipient of fellowships from the NIH, NATO, and Britain's Science Research Council as well as an Honorary Ramsey Memorial Fellowship. Upon returning to the US, Dr. Rubinson taught General and Analytical Chemistry at the University of Cincinnati before founding the Five Oaks Research Institute, where he is currently the Director. The Institute is involved in basic research and provides research consulting services. Dr. Rubinson is the author of over fifty scientific publications as well as a prior analytical textbook. He is a member of the Education Committee of the Analytical Division of the ACS.

Contemporary Chemical Analysis

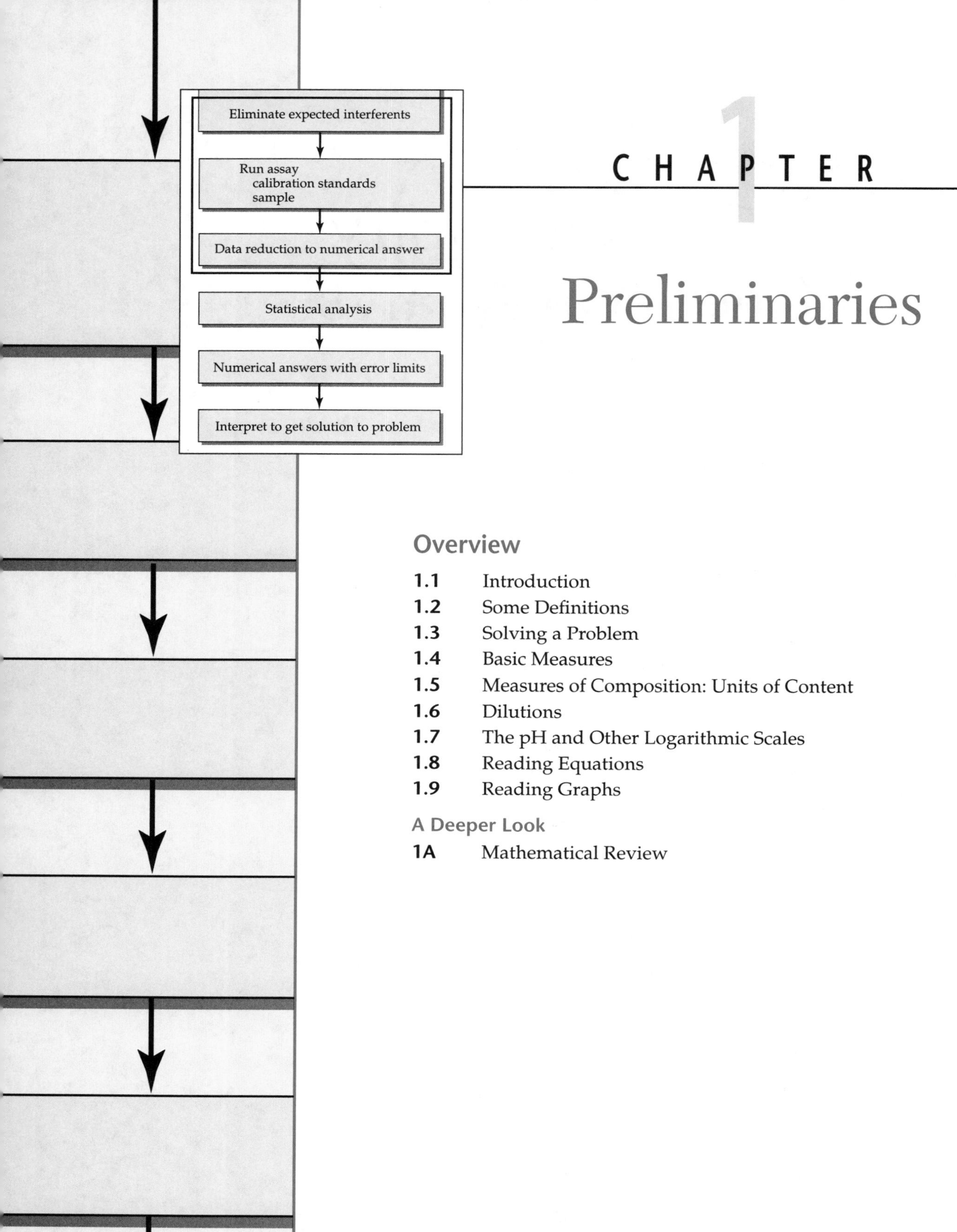

CHAPTER 1

Preliminaries

Overview

1.1 Introduction

What do archaeology, anthropology, botany, chemistry, engineering, forensic science, geology, materials science, medicine, molecular biology, pharmacology, and toxicology have in common? All of these disciplines depend on chemical analyses. Why? Because people working in these areas need answers to questions about the content of the materials they work with. Analytical chemistry is, then, the study of how we learn about materials: how to find their elemental and molecular compositions. Determining the compositions of solids, liquids, gases, solutions, glasses, flames, and other forms of matter enables us to understand their characteristics and to find their best uses or, perhaps, to discover where they came from.

Analytical chemists are interested not only in solving specific problems but also in discovering better methods to do so. This development work usually requires a deep knowledge of wide areas of chemistry and, perhaps, biology, statistics, and physics. Actually, carrying out chemical analyses requires significantly more knowledge of the details. Furthermore, analytical chemistry is very much a "hands-on" science that requires practice of the laboratory techniques.

To introduce some of the kinds of problems addressed by analytical chemistry, in each chapter of this book you will find brief examples of applications of the techniques described. These include: identifying the cause of sludge buildup in a chemical plant, identifying incendiaries in a suspected arson case, finding the chemical content of rocks on Mars, determining what vegetables early peoples in Great Britain ate, and running analyses for hazardous site remediation. In every case, the strategies and methods used were developed after a great deal of experimentation.

1.2 Some Definitions

The general principles and language of analytical **methodology**—that is, how analyses are carried out—are worth knowing. One important reason for this is the necessity of discussing your work with others and learning about the results of their efforts.

Inevitably, many technical words, phrases, abbreviations, and acronyms are used. These are useful since they are often more precise and can be used to communicate ideas more efficiently than less specialized language. In addition, knowing them will enable you to learn more easily from sources more advanced than this textbook.

As you learn the language, you likely will find that all analytical methods are straightforward in concept. Below are some basic definitions that are pervasive in the practice and literature of chemical analysis.

Identification, Determination, Analysis, Assay, Quantitation, and Analyte

Elements, ions, and compounds are **identified** in a sample. For example, we may say that an analyst has identified penicillin in an antibiotic. This is equivalent to saying that penicillin is present at or above some level. If the *amount* of penicillin in the sample is found, the word **quantitation** is often used to describe the process.

Only samples are **analyzed**. In the situation above, we would say that the sample is analyzed for penicillin. It would be incorrect to state that penicillin is analyzed unless we are referring to a sample of "pure" penicillin that is analyzed to determine whether other compounds are present.

If we intend to determine what fraction of a sample is some named material, the word **assay** or **determination** should be used. The named material is called the **analyte**. For instance, if the amount of a specific element, ion, or compound—say, carbon in iron—is to be determined, we can say that an assay for carbon is to be run. Carbon is the analyte.

Some confusion among these terms may occur when the sample is an element or compound of some unknown purity. Saying "an analysis of iron will be made" means an identification and quantitation of the impurities will be carried out for a sample of iron. However, if the iron content itself is to be determined, the correct word for the procedure is *assay*. Thus an *analysis of iron* conveys a different meaning from an *assay for iron*.

Validation

Most chemical analyses are developed through much trial and error. For each analytical procedure, it is extremely important to demonstrate that the procedure measures what the analyst says it measures on a specific type of sample. This is called **validation** of the analytical method. For instance, an analyst might develop a new procedure to measure the amount of cobalt in samples of animal livers. He or she must then validate the new method by showing that its results agree with those obtained using older, accepted procedures.

However, if a procedure is used with a different type of sample, unexpected problems might become obvious. Perhaps the sample contains some component that causes an incorrect result. When that happens, we say there is **interference** from that component. When developing chemical analyses, almost always you will find that problems arise from interfering species, or **interferents**. These interfering species cause the result to be greater or less than what it would be if they were absent. For example, when you use the new cobalt method for liver on a specimen of bone, high levels of calcium interfere with the assay. Calcium is, therefore, an interferent. As a result, the new procedure that is validated for use with animal liver is not validated for bone.

Methods, Protocols, and Techniques

A **method**, or **protocol**, is a fixed sequence of actions to be carried out. Often you will see references in the literature to standard methods approved by regulatory agencies such as the U.S. Environmental Protection Agency. These methods are specific for use with certain analytes in certain types of samples. They commonly define both the sample handling requirements and the analytical **techniques** that are to be used.

1.3 Solving a Problem

When approaching an analytical problem, there is a general sequence of steps that are followed in developing the protocols and deciding which analytical techniques to use. A general outline of these steps is shown in Figure 1.1. Some of the terms in the middle column may be unfamiliar to you now. Rest assured, you will become intimately acquainted with them soon.

Once *the problem is defined* the first step is to *decide on an appropriate method for the analyte of interest*. For example, say there are 120 tons of ore sitting outside the laboratory window to be analyzed. Before any method can be used

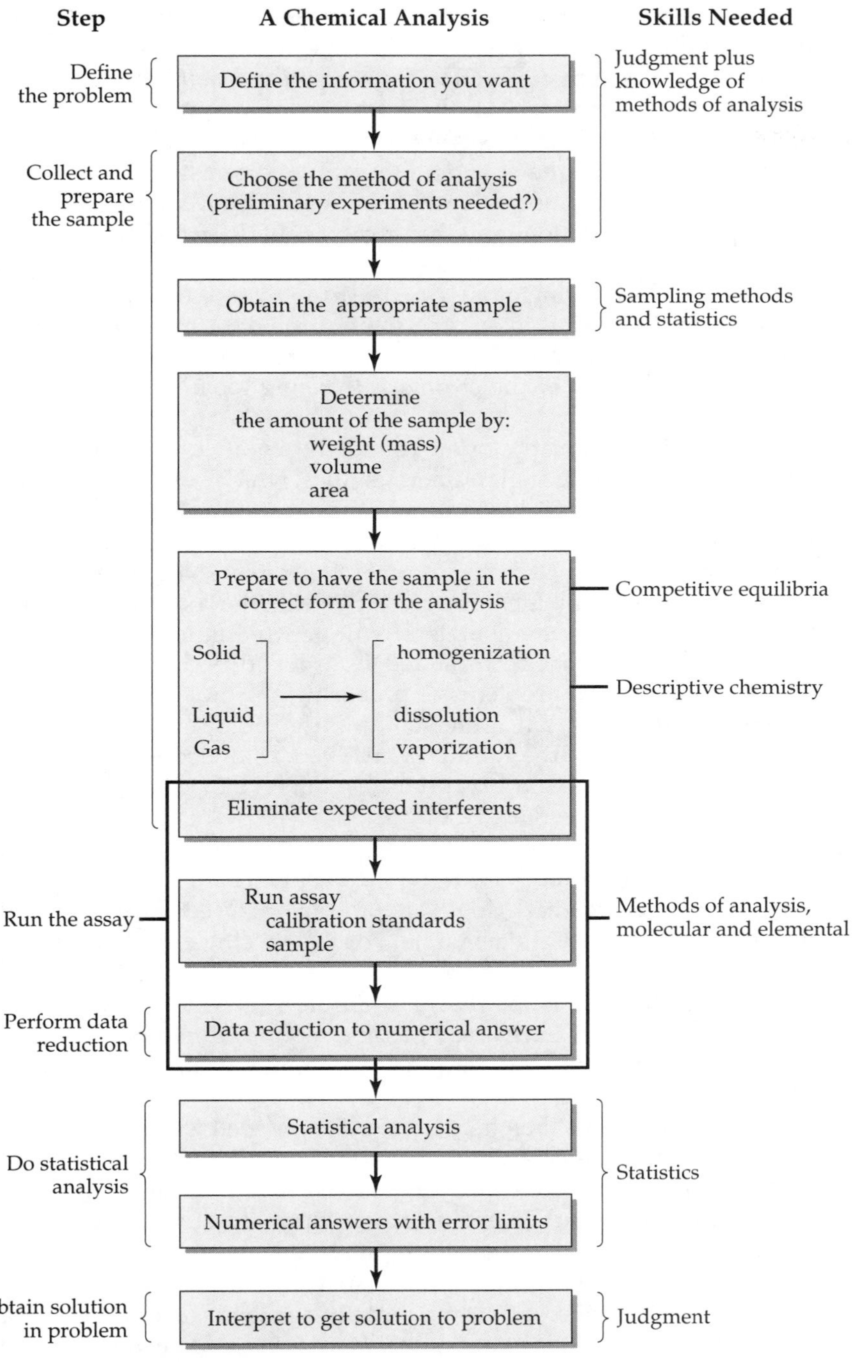

FIGURE 1.1 ◀
Diagram showing how the methodology of chemical analysis is used to solve a problem.
In the left column are the six general steps that comprise the methodology. In the middle column is a more detailed list of the stages involved. The top box delineates the steps of sample preparation. The lower box contains the steps involved in the assay. (Notice that the last part of the sample preparation is also part of the assay step.) The right column lists some of the less tangible skills involved in the process. Contemporary analysis is a combination of chemical principles, instruments and electronics, and practiced judgment. The "hardware" is an essential but relatively small part.

to do this, a *sample that represents the contents* of the pile must be taken. How can such a sample be obtained that will fit through the door?

Next, a decision must be made about what method can be used for *sample preparation*. The sample preparation method must be chosen with the characteristics of the assay method in mind. For instance, assume that the method we prefer can only measure some property of an aqueous solution, but we have 10 kg of solid chunks of ore in a bucket in the lab. What shall we do with it? Furthermore, if some interferents present in the prepared sample would be especially detrimental to performing the preferred assay, these interferents should be removed before running the assay, if possible. If some interferents cannot be removed, their effects must be accounted for by a correction to the results. Both these steps require some knowledge of the **descriptive chemistry** of the analyte. Descriptive chemistry is the body of knowledge that includes how elements and compounds react with each other under specific conditions and the properties of these reaction products, such as their volatilities or solubilities. Not all of the necessary descriptive chemistry may be fascinating or easy to unearth, but possessing this knowledge is enormously helpful in solving many problems involving chemical analysis, as well as in developing safe laboratory procedures. Descriptive chemistry is, in fact, one of the strongest thinking tools used in chemical analysis.

The descriptive chemistry often cannot be codified in mathematical terms but is understood through numerous rules, which have specific exceptions. Some of these rules apply regularly, others only occasionally. The organization of the periodic table is an example of descriptive chemistry. The elements that appear in any one column of the table share many similar chemical traits, which may include volatility, formation of stable oxides, or ease of oxidation. The fundamentals of **competitive equilibria**, a part of descriptive chemistry, must be mastered in order to understand chemical analysis. Not only do the problems of preparing a sample and eliminating interferents depend on it, but understanding the assay method itself may rely on your knowledge of the equilibria involved. Knowledge of descriptive chemistry—with its rules and exceptions—is one major characteristic that distinguishes an experienced analyst from a novice.

Basic concepts of chemical equilibria are presented in Chapters 6 and 8.

After we *perform the assays*, the next step must be to decide how to treat the data obtained. *Data reduction* is part of the assay. The response of an analytical instrument or method does *not* produce a direct measure of the amount of a given material in a sample. For instance, we might find that a certain amount of yellow light is absorbed by a sample, or that the sample solution conducts electricity, or that the sample's weight changes a given amount when it is heated. Each assay method requires different calculations to reduce the experimental data into the form required: the quantity of the analytes.

Usually, *statistical tests* then help us decide how well we really know the numerical answer to the analytical question. A statistical analysis of the results is required since each analysis and each sample analyzed is not exactly the same every time. To evaluate such variations, three or more samples of the same material are usually treated and assayed together. Then, quantitative measures can indicate how trustworthy the results are.

A computer will make only some parts of the analysis easier: the data acquisition, data reduction, statistical analysis, and record keeping. The rest still requires personal judgment and skill.

It is important to remember that developing analyses is not always a "straight-path" process. Often it is necessary to backtrack to an earlier step because of some unforseen difficulty. On the brighter side, an analysis may be changed because some new technique makes the task easier or cheaper, with the same or better quality of analysis.

This problem-solving process can be illustrated with a specific problem and its solution as found in the literature of chemical analysis. It comes from the oil industry and is adapted from an article in the journal *Analytical Chemistry* [*Anal. Chem. 48* **1976,** 972A]. The steps follow those of Figure 1.1.

Solids were building up in a reactor of a petrochemical processing plant, and the condition threatened to require a complete shutdown. This buildup was accompanied by relatively high corrosion rates of the metal reactors. The problem was thus defined: Could the solids be eliminated? From previous experience, the analysts believed the problem might be due to the presence of chlorine, but the presence of sulfur was also a possibility. The question was how much chlorine was in the system during the process.

The samples of the feed material (a liquid) that reacted in the synthetic process were obtained for an analysis of chlorine. In the company's laboratory, the chlorine content was determined by two different methods. The *total* chlorine in a sample was quantitated by a thermal method of analysis, in which changes in the temperature of a sample are measured as it is heated. The chlorine that originated in *organic* chlorine compounds was determined using an electrochemical method—one that depends on the oxidation or reduction of a substance. Samples were prepared differently for the two assay methods.

After the measurements of chlorine were made, the numerical results from the two methods were found to differ. The assay for organic chlorine compounds showed that there were none in the sample. The assay to determine the total chlorine content showed apparently high levels of chlorine.

By statistical analysis of results from a number of samples, the analysts determined that the different results could not be due to random experimental errors. A check for interferences was done. The assay using the thermal method was found to be influenced by a high level of sulfur that acted as an interferent in the assay. Due to this unanticipated problem, the original analysis was in error.

As a result, the analysts modified the analytical methodology by changing their technique for the sample preparation. Any sulfur present in the sample was removed before the thermal-method assay was run. After that modification, the results from both methods agreed. The chemists concluded that the problem in the reactor was not excessive chlorine.

The answer to the original question was that the problem of the sludge and corrosion lay in the high sulfur levels of the feedstock. There is more to the story, but this introduction serves to illustrate the sometimes circuitous route needed to solve a problem using chemical analyses.

1.4 Basic Measures

A **measure** labels the extent or size of something, especially as determined in comparison to a standard. For instance, the gram is defined as exactly 1/1000 of a standard mass kept near Paris at the International Bureau of Weights and Measures. It was selected to be identical with the mass of 1.0000 mL of water at the temperature of its maximum density, 3.98°C. Since originally defined, this value has been more accurately measured and differs from the original by about 0.03%. However, the standard gram has been kept the same and is close enough for all but the most exacting measurements. Similarly, the centimeter is defined by a standard. Finally, one degree on the Celsius and Kelvin temperature scales is defined as 1/100 of the difference between the normal freezing and boiling points of water. It is interesting to note that these three fundamental measures are arbitrarily defined.

Different measures are used for different physical states and for different ranges of concentration. These measures have been developed as the need arose over more than a century of endeavor. Overall, the units of these measures might seem to constitute quite a messy system. An attempt to standardize the units has been made: the Système International (SI). It is being adopted to varying degrees and, depending on the disciplinary areas and age of publications you read, the SI units may or may not be used. So you may have to convert various units of measure to those most useful to you.

Atomic Mass

The **atomic masses** are a set of relative numbers; their absolute values depend on a defined standard, the mass of the isotope ^{12}C. This is assigned mass 12 exactly, and 1/12 of this mass is called a **unified atomic mass unit** (u). This is now the recommended term, but the name atomic mass unit (AMU) still is commonly found. You will also see the unified atomic mass unit called the **dalton**, named after one of the key chemists in history. Atomic masses are commonly called **atomic weights**. It can be argued that the term *atomic weight* is less correct than *atomic mass*. However, on a day-to-day basis for chemical analyses, there is no essential difference.

Mole

One of the basic ideas of chemistry, the **mole** is defined as a specific *number* of units or a *weight* containing that number of units. A mole contains 6.0221367×10^{23} particles of any kind; this is known as **Avogadro's number**. A mole can be used in measuring atoms, ions, molecules, or units such as a chemical formula group.

As a mass, the mole is the mass formula weight (in u) expressed in grams. The **formula weight** is its common name. For example, 12 g exactly of ^{12}C is a mole of ^{12}C. This is the definition used on a day-to-day basis. In addition, there are two other commonly used specific definitions with which you probably are familiar. If the particles are atoms, the mole weight is called the **atomic weight** or **gram-atomic weight**. If the particle (or formula) is a molecule, the weight is called the **molecular weight** or **gram-molecular weight**. The three different names may be abbreviated f.w., a.w., and m.w., respectively.

Prefix Notation

The range of concentrations and the sizes of the samples that can be analyzed vary widely and benefit from a language to express this range. Primary measures such as meter or gram are given prefix names for various powers of ten (Table 1.1). For example, the prefix kilo- signifies 1000 times the measurement unit, as in kilogram. A larger measure is needed for other situations. For example, if the total amount of an element in a planet is desired, the mass of such a "sample" is in the range of 10^{15} teragrams. On the other end of the scale, analyses can be done to find small amounts of materials—in the femtogram or attomole range.

Table 1.1 Prefix Notation

Exponential	Prefix Name	Symbol
10^{12}	tera-	T-
10^{9}	giga-	G-
10^{6}	mega-	M-
10^{3}	kilo-	k-
10^{-1}	deci-	d-
10^{-2}	centi-	c-
10^{-3}	milli-	m-
10^{-6}	micro-	μ-
10^{-9}	nano-	n-
10^{-12}	pico-	p-
10^{-15}	femto-	f-
10^{-18}	atto-	a-
10^{-21}	zepto-	z-
10^{-24}	yocto-	y-

1.5 Measures of Composition: Units of Content

Four basic types of expressions are employed to characterize the compositions of materials that are not composed of a single pure element or a single pure compound. These are shown in Table 1.2 and described below.

Weight-to-Weight Measures

Weight-to-weight (w/w) measures express the ratio of the weight of one component to the weight of the whole. **Weight percent**, abbreviated wt% or % (w/w), is the ratio of the weight of a component as a part of the whole mixture expressed as a percentage, that is,

$$\% \text{ (w/w)} = \frac{\text{component mass}}{\text{sample mass}} \times 100$$

and the units of mass must be the same. Occasionally % (w/w) is called **parts per hundred**.

Table 1.2 Units of Content

Component Measured by	Total Sample Measured by	Nomenclature	Abbreviation
Weight*	Weight	Weight-to-weight	w/w
Weight*	Volume	Weight-to-volume	w/v
Volume	Volume	Volume-to-volume	v/v
Number	Volume	Molarity, Normality	M, *N*

**Mass* is more correct, but the term *weight* is almost always used in expressions such as weight-to-weight.

EXAMPLE 1.1

A sample weighs 1.2304 grams and contains 0.1012 g of iron. What is its iron content in wt%?

Solution

$$\frac{0.1012}{1.2304} = 0.0822$$

of the whole, and it has

$$0.0822 \times 100 = 8.22 \text{ wt\% iron in it.}$$

If a sample contains a smaller amount of some component than can be conveniently expressed as wt%, the next smaller unit you may see is **parts per thousand** (abbreviated ppt or ‰). This is analogous to parts per hundred.

EXAMPLE 1.2

A sample weighs 1.2304 grams and contains 0.0101 g of europium. What is its content of europium in parts per thousand?

Solution

$$\frac{0.0101\text{g}}{1.2304} = 0.0882$$

of the whole, and it has

$$0.0082 \times 1000 = 8.2 \text{ ‰ europium in it.}$$

Smaller fractions are described using **parts per million** (ppm).

$$\text{ppm} = \frac{\text{grams of analyte}}{\text{grams of sample}} \times 10^6$$

Parts per billion (ppb),[1] and **parts per trillion** (ppt, pptr) are defined similarly but the ratio is multiplied by 10^9 or 10^{12}, respectively. Note that there can be confusion between the abbreviations for parts per thousand and parts per trillion when both are abbreviated ppt. No confusion exists when pptr is used.

A weight-to-weight measure also can be expressed as a fraction. For example: The sample contains 33 μg/mg of ingredient X. Also, 33 μg/mg is equivalent to 33 parts-per-thousand. The following equivalencies also hold:

$$\text{parts per million} = \mu\text{g/g}$$
$$\text{parts per billion} = \text{ng/g}$$
$$\text{parts per trillion} = \text{pg/g}$$

To get an intuitive idea of the size of these measures, one ppm of 50 L (13 gallons) of water is one drop. One crystal of salt 0.5 mm on a side dissolved in

[1] Some confusion may occur because *billion* in Europe often means 10^{12}, whereas billion in the United States is always 10^9.

Table 1.3 More Common Units Used to Express Content

Name	Abbreviation	Units Used		
		w/w	w/v	v/v
parts per thousand	‰ or ppt	mg/g	mg/mL	mL/L
parts per million	ppm	μg/g	μg/mL	nL/mL
		mg/kg	mg/L	μL/L
parts per billion	ppb	ng/g	ng/mL	nL/L
		μg/kg	μg/L	

half a cup of liquid (250 g) is about one ppm in salt. Half of that crystal of salt inside your body amounts to about one ppb.

Weight-to-Volume Measures

In a weight-to-volume measure, a component's weight is compared to the total volume of the material. Common units are the weight in grams and the volume in mL. These and other common units are listed in Table 1.3.

EXAMPLE 1.3

What is the weight-to-volume ratio of 1.25 g of protein in 250 mL of solution, in g/mL?

Solution

$$1.25\ \text{g}/250\ \text{mL} = 0.005\ \text{g/mL}$$

We can extend the use of w/v measures to include percentage, written as % w/v, as well as ppm(w/v) and ppb(w/v). However, the measures of ppm and ppb can only be used with units converted into g with mL or kg with L. If these units were not agreed upon by convention, ppm and ppb would be ambiguous.

EXAMPLE 1.4

What is the weight-to-volume percentage of protein when 1.25 g of protein is dissolved in 250 mL of solution?

Solution

$$\%\ \text{w/v} = \frac{\text{solute mass}}{\text{sample volume}} \times 100$$

$$(1.25\ \text{g}/250\ \text{mL}) \times 100 = 0.5\%$$

Similar calculations are used for ppm and ppb.

EXAMPLE 1.5

What is the weight-to-volume concentration, in ppm(w/v), of sodium in a solution containing 2.500 mg of Na^+ in 500.0 mL of solution?

Solution

Once the mass has been converted into units of grams, we calculate ppm as defined above.

$$10^6 \times \left(\frac{2.500 \times 10^{-3}\ \text{g}}{500.0\ \text{mL}}\right) = 5.000\ \text{ppm(w/v)}$$

EXAMPLE 1.6

An analysis for cadmium in water gave a value of 1.20 ppb(w/v). What mass of cadmium is contained in 1.00 L of water?

Solution

The w/v units must be either g mL^{-1} or kg L^{-1}. Let us choose the first. So

$$1.20\ \text{ppb} = \frac{\text{g Cd}}{\text{mL } H_2O} \times 10^9$$

For the 1 L volume

$$1.20\ \text{ppb} = \frac{\text{g Cd}}{1.00 \times 10^3\ \text{mL}} \times 10^9$$

and

$$\text{g Cd} = \frac{1.20 \times 10^3}{10^9} = 1.20 \times 10^{-6}\ \text{g}$$

or 1.20 μg are present.

Since the densities of solvents vary with temperature, if you want to convert (w/w) to (w/v) measurements, the density of the solution is needed.

EXAMPLE 1.7

A stock solution of KI contains 107.6 g of KI per liter of solution. The solution's density at 20°C is 1.0781 g mL^{-1}. What is the concentration of the solution in % (w/v) and in % (w/w)?

Solution

To find the % (w/v), all you need to do is to convert the volume in L into mL.

$$\%\ \text{w/v} = \frac{\text{g solution}}{\text{mL solution}} \times 100$$

$$= \frac{107.6\ \text{g KI}}{10^3\ \text{mL}} \times 100 = 10.76\%\ \text{w/v}$$

To ensure that the conversion to % (w/w) is correct, let us keep an eye on the units of the numbers that are used. Here, the w/v measure of g KI mL^{-1} is converted to (w/w) by multiplying by mL g^{-1}, the reciprocal of the density.

$$\%\ \text{(w/w)} = \frac{\text{g solute}}{\text{mL solution}} \times \frac{\text{mL solution}}{\text{g solution}} \times 100 = \frac{\text{g solute}}{\text{g solution}} \times 100$$

$$= \frac{0.1076\ \text{g KI}}{\text{mL}} \times \frac{1\ \text{mL}}{1.0781\ \text{g}} \times 100 = \frac{0.0998\ \text{g KI}}{\text{g solution}} \times 100 = 9.98\%\ \text{(w/w)}$$

One useful consequence of water as a solvent is that for dilute solutions weight-to-weight and weight-to-volume measures are essentially equivalent. For instance, ppm(w/v) and ppm(w/w) can be considered the same because one mL of a dilute aqueous solution weighs almost exactly one gram. (At 3.98°C, one cm^3 of water weighs exactly one gram *by definition*.) This equivalence between w/w and w/v for aqueous solutions is *not* true for more concentrated solutions. Nor is it true if the solvent is not water.

Finally, two other common w/v measures that are probably familiar to you are **molarity** and **normality**. However, these are also **number-to-volume** measures.

Number-to-Volume Measures

Molarity (M) is defined as the number of moles of solute in 1 liter of solution. This is a number-to-volume measure. (Since a mole can be a weight, it is a weight-to-volume measure as well.)

EXAMPLE 1.8

If 2.354 g of KNO_3 is dissolved in exactly 250 mL of total solution, what is the molar concentration?

Solution

Since the molecular weight of KNO_3 is 101.1 g mol^{-1},

$$\frac{(2.354\text{ g}/101.1\text{ g mol}^{-1})}{0.250\text{ L}} = 0.0931\text{ M}$$

In acid-base and oxidation-reduction reactions, it is not necessary to know the chemistry of the sample in order to measure the number of moles of protons or electrons transferred per liter of solution. For proton transfer, the mass of reagent that can transfer one mole of protons is called an **equivalent**. Similarly, the mass of an oxidation-reduction reagent that can transfer one mole of electrons is also called an equivalent. Equivalents per mole is a w/w measure and is a small, whole number.

The number of equivalents contained in a liter volume is called the *normality*. For a reagent or analyte with a known stoichiometry, its normality can be related to its molarity by the following equation:

$$\text{normality} = \text{molarity} \times \text{number of equivalents mol}^{-1} \quad \textbf{(1-1a)}$$

For acid-base reactions,

$$\text{normality} = \text{molarity} \times \text{equivalents of protons mol}^{-1} \quad \textbf{(1-1b)}$$

EXAMPLE 1.9

Compare the normalities of a one-molar solution of acetic acid and a one-molar solution of sulfuric acid.

Solution

Acetic acid has the formula H_3CCOOH. Each molecule has four protons. However, the descriptive chemistry of acetic acid in water shows that only one of them is donated.

$$CH_3COOH + H_2O \rightleftharpoons H_3O^+ + CH_3COO^-$$

Since one mole of acetic acid has only one equivalent of protons to donate in water, the normality equals the molarity.

On the other hand, the strong acid H_2SO_4 can donate two protons to a base. A one-molar sulfuric acid solution contains two equivalents of protons and acts as a two-normal solution.

In oxidation-reduction reactions, the number of equivalents equals the number of moles of electrons that are actually donated or accepted in the reaction. The general equation for redox normality is

$$\text{normality} = \text{molarity} \times \text{equivalents of } e^- \text{ transferred mol}^{-1} \quad \textbf{(1-1c)}$$

EXAMPLE 1.10

Permanganate is used as a reagent in a number of analyses. It reacts by donating five electrons to form manganous ion:

$$5e^- + 8H^+ + MnO_4^- = Mn^{2+} + 4\,H_2O$$

What is the normality of 0.1 M $KMnO_4$?

Solution

Since five moles of electrons are donated per mole of permanganate ion, the normality is five times the value of the molarity. So a 0.1 M solution of permanganate is a 0.5 N solution under these acidic conditions.

One of the benefits of using normality as a measure of content is that the chemistry of the donating species is not needed. For example, in some very complex solutions, it is difficult to write equations for the pertinent reaction(s). But with normality you can still standardize the reagent to find the equivalents of protons or electrons in a given volume.

Volume-to-Volume Measures

A third type of content measure expresses the volume of a component as a part of the total volume of material. This **volume-to-volume** measure is expressed in much the same way as the other two types: volume percent (% v/v), as well as ppt(v/v), ppm(v/v), and ppb(v/v). It is most commonly used for liquid components of a liquid sample or gaseous components of a gaseous sample. Some of the expressions for volume-to-volume measures are collected in Table 1.3.

There is another volume-to-volume measure commonly used in conjunction with chemical analyses. It is written in the notation of ratios—for example, 1:2 methanol-water. This means that one volume of methanol is mixed with two times that volume of water to make a solution. Such measures are often used in instructions for mixed-solvent preparations.

1.6 Dilutions

It is often convenient to keep **stock solutions** of reagents in a laboratory. A stock solution is more concentrated than the solution that is used in some step of an analysis. Many reasons exist for using stock solutions, including ease and speed of preparing the less concentrated analytical reagents; stabilizing some materials that deteriorate when more dilute; and avoiding wasting hygroscopic reagents, which would pick up water once opened.

The amount of stock required to prepare the desired reagent solution is easy to calculate, since the total amount of solute does not change when we dilute it. That is, since

$$\text{amount of solute} = \text{concentration} \times \text{volume} \quad \textbf{(1-2)}$$

and the amount of solute remains fixed, we find a simpler form.

$$V_{\text{stock}} \times C_{\text{stock}} = V_{\text{desired}} \times C_{\text{desired}} \quad \textbf{(1-3)}$$

Both volume measures, V_{stock} and V_{desired}, have identical units, such as mL or L. Similarly, both of the concentrations, C_{stock} and C_{desired}, have the same units, such as M, mM, *N*, or ppm.

EXAMPLE 1.11

A stock solution of concentrated nitric acid is 10.0 M. How many mL of stock nitric acid is required to prepare 500 mL of a 0.50 M solution of nitric acid?

Solution

As for all dilutions of liquids, Equation 1-3 is used.

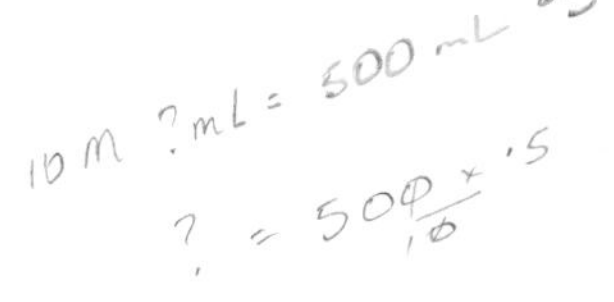

$$V_{\text{stock}} \times C_{\text{stock}} = V_{\text{desired}} \times C_{\text{desired}}$$

$$V_{\text{stock}} \times 10.0 \text{ M} = 500 \text{ mL} \times 0.500 \text{ M}$$

$$V_{\text{stock}} = 25.0 \text{ mL}$$

This approach simply involves conservation of mass.

EXAMPLE 1.12

As listed in the table in Appendix XII, concentrated hydrochloric acid is between 37% and 38% HCl (w/w) and its density is 1.18 g/mL in this narrow concentration range. How many mL of concentrated HCl is required to prepare 1.00 L of 0.100 M HCl? Assume 37.0% HCl and density 1.18 g/mL.

Solution

The units, as stated above, must be the same on each side, so we first convert 37.0% w/v to molarity. This can be checked by dimensional analysis.

$$\frac{37.0 \text{ g HCl}}{100 \text{ g reagent}} \times \frac{1.18 \text{ g reagent}}{\text{mL solution}} \times \frac{1 \text{ mol HCl}}{36.5 \text{ g HCl}} \times \frac{10^3 \text{ mL}}{\text{L}} = 12.0 \text{ mol} \cdot \text{L}^{-1} = 12.0 \text{ M}$$

Now that the molarity is known, Equation 1-3 applies directly.

$$V_{\text{stock}} \times 12.0 \text{ M} = 1000 \text{ mL} \times 0.100 \text{ M}$$

$$V_{\text{stock}} = 8.33 \text{ mL}$$

Preparing the HCl or HNO_3 reagents at the concentrations calculated in the above two examples is relatively easy. You could measure precisely 25.00 mL or 8.30 mL of the concentrated acids. But, say you need to make a 10-μM HCl solution. What volume would you measure to make 1.000 L of it? This concentration is 10,000 times less than the 0.100-M solution in the example above, so the amount of stock solution required is 10,000 times smaller: 0.83 μL. To measure such a volume precisely, or even to weigh it, is difficult, as would keeping it from evaporating while you work!

Doing **serial dilutions** is generally a better way to make the liter of this dilute acid solution. As its name implies, one dilution is followed by more dilutions. Assume you have made the 0.100 M solution of HCl already, and it is 10,000 times too concentrated. So, how would you proceed? Should you take 0.100 mL of the 0.100 M solution and dilute it further to 1.00 L? You could, but measuring exactly 0.100 mL (and not 0.102 mL or 0.098 mL) is moderately difficult. On the other hand, you could do two 100-fold dilutions. Take 1.00 mL of the 0.100 M solution and dilute it to make 100.0 mL of 1.00 mM solution, and then dilute 10.0 mL of that to make 1.00 L of the final solution.

EXAMPLE 1.13

Using the two 100-fold dilutions of the 0.1 M solution, convince yourself that you have 1.00 L of 10 μM acid.

Solution

Dilute 1.00 mL of stock solution into 100.00 mL for the first diluted solution. Then take 1.00 mL of this first dilution and, in turn, dilute it to 100.0 mL to obtain the final solution. The two steps are written

$$\frac{1 \text{ mL stock}}{100 \text{ mL 1st dilution}} \times \frac{1 \text{ mL 1st dilution}}{100 \text{ mL final solution}} = \frac{1 \text{ mL stock}}{10^4 \text{ mL final solution}}$$

So the two dilutions are equivalent to diluting 1 mL to 10^4 mL.

Of course, you could have used another series of dilutions, such as 1:200 followed by 1:50. All that is required is to end up with the volume and concentration you want. Clearly, some series of dilutions use more solvent than others, and the series you choose will depend on the cost of the solvent and the desired level of precision for the final solution.

A brief review of logarithms appears in Section 1A.

1.7 The pH and Other Logarithmic Scales

In 1909 the Danish biochemist Sven P. L. Sorensen invented a scale to make it easier to manipulate, calculate, and discuss the wide range of concentrations of hydrogen ions in solution. Instead of the molar concentration of protons, $[H^+]$, he decided to use the negative logarithm of the concentration, $-\log [H^+]$. This value was called the pH. Now, the effective concentration is used.

The concept of effective concentration is covered in Chapter 6 and its measurement in Chapter 15.

Consider the following examples. The hydrogen ion concentration of a 0.1-M solution of acetic acid in water at 25°C is 0.0013 M. The concentration of H^+ in pure water is 1.00×10^{-7} M. And a 0.1-M solution of NaOH has $[H^+] = 1.0 \times 10^{-13}$ M. It is a lot easier to write or discuss the equivalent pH values of 2.89, 7.00, and 13.0, respectively!

It is important to note that the above definition of pH is a theoretical one. The definition of pH as measured depends on comparing the analyte solution with a standard solution that possesses a stable, fixed pH.

In analogy to pH, a common practice is to use the negative logarithm of other quantities. The operation of taking the negative log of a number is signified by the letter p placed before the symbol. Examples are

$$-\log K = pK$$
$$-\log[Ca^{2+}] = pCa = pCa^{2+}$$

EXAMPLE 1.14

A neurophysiologist studying the effects of calcium on cytoplasm of a nerve axon makes up a solution that contains 1.5 μM free Ca^{2+}. What is the pCa_{free} for the solution?

Solution

$$pCa_{free} = -\log [Ca_{free}] = -\log (1.5 \times 10^{-6})$$
$$pCa_{free} = 5.82$$

1.8 Reading Equations

In any science, quantitative relationships are expressed mathematically in equations. Equations are a quantitative shorthand for sentences or paragraphs that describe our general knowledge about experimental relationships.

Take the ideal-gas law, for example. It describes relationships that have been generalized from experimental data. In words: The pressure of a gas multiplied by its volume is a constant, which is dependent on temperature. Written algebraically,

$$PV = nRT \tag{1-4}$$

where
- P is the pressure of the gas,
- V is the volume of the gas,
- T is the temperature (in K),
- R is a constant of proportionality, and
- n equals the number of moles of gas present.

This equation is a compact description of experimentally determined behavior. Mathematics is a powerful, symbolic way of describing it.

In addition to the algebra, you know the *meaning* of Equation 1-4, not just the expression. For instance, without looking at the equation you are well aware that the volume of a fixed amount of gas decreases as the pressure increases. You have a firm grasp of both the physical behavior of gases and the algebraic relationships that describe their behavior. That is the aim you should seek with all equations.

Equation 1-5 is, perhaps, less familiar. How do you approach this new equation? First, some of the terms may be unfamiliar, but the mathematical relationships should be clear. One requirement is to study each term carefully—after all, this is a shorthand notation.

$$\bar{x}(t) = \sqrt{2Dt} \tag{1-5}$$

where
- $\bar{x}$ is an average distance,
- D is the diffusion coefficient, and
- t the time.

Let us work our way back from the algebraic expression to the characteristics of the experiments that it describes. We first look for the terms we know.

Then we tackle the unfamiliar parameters and think of ways to find out more about them. Next, the information in the equation and some practical guesses might lead to an understanding of the equation without any further effort. For example, since time and distance can be measured experimentally, the equation might be used to determine D. Alternatively, if D is known, perhaps the relationship of Equation 1-5 might be used to predict $\overline{x}$.

Since D is the least understood variable, it might be helpful to determine its units. Since $\overline{x}$ is a distance, assume its units are centimeters. The time t is likely to be in seconds. By simple substitution, the dimensional analysis shows that D has units of $cm^2\ s^{-1}$. If you do not have any experience with this equation, these certainly may seem to be strange units. But from the description of the variable D, you do know that diffusion is involved. Looking into the literature of diffusion might be a worthwhile place to find more.

Understanding the behavior of the equation is worth a look. For example, it is often useful to look at simple substitutions. Let $\overline{x} = 1$ and $D = 1$. Then, $t = 1$. Let us double the value of $\overline{x}$: let $\overline{x} = 2$, and let D stay the same. Then, $t = 4$. In other words, the average distance increases at longer times, but not in a linear fashion. Perhaps it would be clearer if we let the time double. Then, the average distance increases only by about 40%. This is characteristic of the distances that particles diffuse. You have now discovered some useful concepts in this simple equation that will help you if you decide to look more deeply into the subject.

1.9 Reading Graphs

A number of different types of graphs are employed in association with chemical analyses. Selected aspects of these are examined briefly in this section.

Cartesian Graphs

Cartesian graphs usually consist of plots of assay results as they depend on the level of analyte in samples. One such graph appears in Figure 1.2. Each axis designates a linear scale, which should be labeled with the quantities

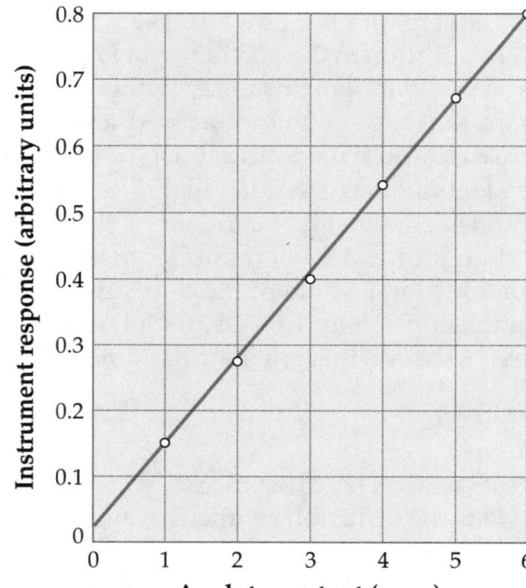

FIGURE 1.2 ▶
Cartesian graph with two linear scales.

Placing the units of the measure in parentheses is one of the common conventions. This graph is representative of numerous calibration plots found in instrumental chemical analysis.

that are plotted. In this figure, one labeling convention is shown—labels with the units inside parentheses. In this case, even though the instrument output is on a linear scale, the units are arbitrary. For instance, the output may simply be a voltage. The straight-line relationship between the data points (the circles) indicates that the instrument produces an output linear with the analyte content over the range shown.

Log-Log Graphs

A number of analytical methods can produce linear responses over a far larger range of concentrations of analyte than can be plotted clearly on a Cartesian graph. In such situations, it may be convenient to plot the results on a log-log graph, which consists of logarithmic scales on the perpendicular axes of the graph, as shown in Figure 1.3. Here the response is again linear with analyte content—for example, a ten-fold increase in analyte results in a ten-fold increase in counts per second. Notice that the units label on the vertical axis is (s^{-1}), since counts is a unitless value. The data points at the left in Figure 1.3 have vertical bars extending from them. These are called **error bars** and indicate the spread of measured values. The lengths of the error bars become too small to show when the number of counts increases. When such bars are shown, the caption of the graph should describe what statistical measurement they indicate.

Semilog Graphs

On a semilogarithmic graph, one axis has a logarithmic scale, and the axis perpendicular to it has a linear scale. Figure 1.4 shows an example of a semilog plot of concentration *vs.* time. A straight-line relationship on a semilog plot means that there is a constant percentage change in the quantity plotted along the logarithmic axis.

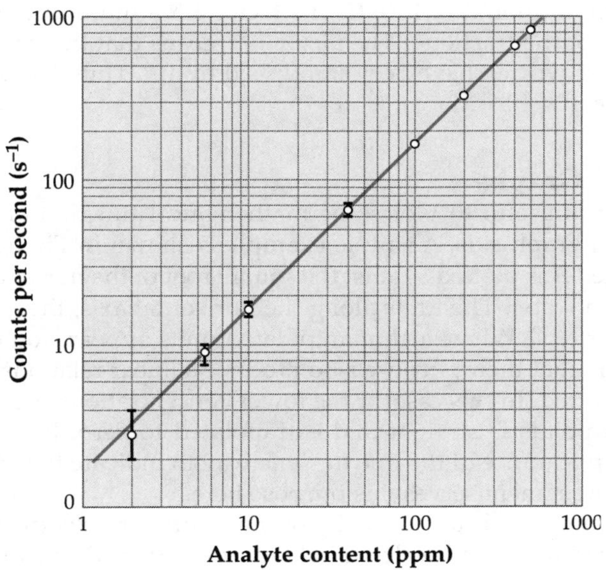

FIGURE 1.3 ▲
Three-cycle by three-cycle log-log graph with error bars.

This graph is representative of a calibration plot for a number of instrumental methods that have a wide range of linear response.

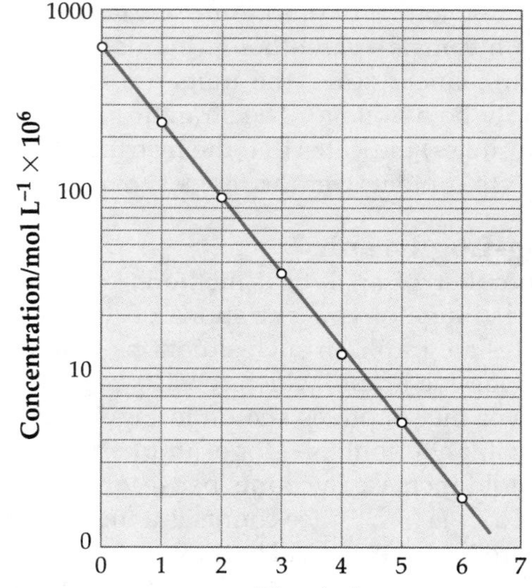

FIGURE 1.4 ▶
Semilogarithmic graph with three log cycles on the vertical axis.

The manner of labeling the units of the measure illustrates another of the common conventions. In this, each number label is unitless. This plot is representative of data from kinetic methods of analysis.

The labels on the axes of Figure 1.4 illustrate another labeling convention. The numerical values are to be unitless, so the quantities on each axis are written as quotients. The horizontal axis has the label (time/min), which means that the numerical values are the experiment times—which are in minutes—divided by one minute. This label conforms to the convention since (min/min) is unitless. The denominator of the label on the vertical axis is (mol $L^{-1} \times 10^6$), which is the style recommended. The negative exponent for liters avoids the confusion that would be possible if the label were written as a fraction (mol/L $\times 10^6$). (Is the factor 10^6 a multiplier or is it in the denominator of the fraction?) The factor 10^6 means that the units of concentration that are written have been multiplied by 10^6. Thus, the value of 1000 on the scale is 1×10^{-3} M.

Linearized Graphs

Sometimes it is useful to plot some mathematical function of a measured quantity on a graph axis. A simple example is shown in Figure 1.5. Here a quantity called A is plotted against the square root of the normality of a solution of HCl in water. The units along the horizontal axis, then, are (equivalents $L^{-1})^{1/2}$ or $N^{1/2}$. When such manipulated units appear, you can be moderately certain that their purpose is to produce a linear relationship between the values plotted. In this case, at the lower concentrations the line connecting the data points is straight and extrapolated to zero. It is clarifying to change the appearance of the line in some way to indicate the extrapolation. A dashed line segment serves this purpose here.

If you want to find the normality of the solution that corresponds to the value at the arrow, a simple calculation is needed. Since the factor of 100 multiplies the scale, the value of 4 corresponds to 0.04 $N^{1/2}$. The solution normality is then 0.0016 N at the arrow. Similarly, the data point at the right corresponds to a 0.005-N solution (0.071 $N^{1/2}$).

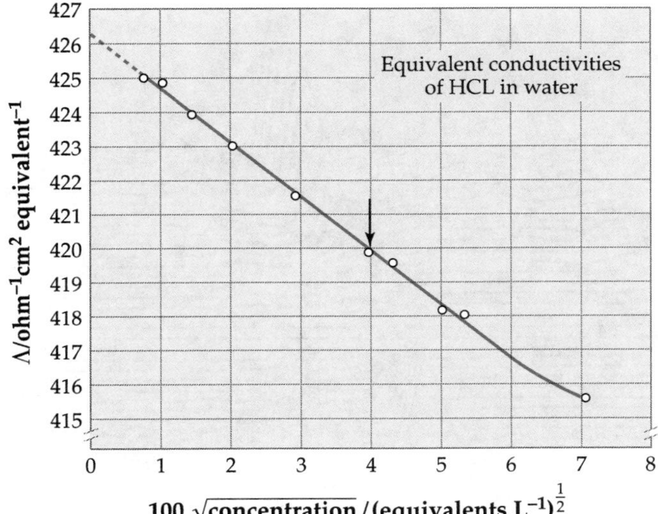

FIGURE 1.5 ▲
Plot on Cartesian axes but with a linearizing function plotted on one axis.

This plot shows the equivalent conductivities of hydrochloric acid in water at 25°C. This type of graph illustrates a fundamental theory in the chemistry of electrolyte solutions.

Three-Dimensional Graphs

In some cases, it is advantageous to plot data as a function of two different quantities. Two different displays of a three-dimensional format are shown in Figure 1.6. The upper graph shows a plot of an instrument output versus wavenumbers (a measure of energy) and time. It is drawn so as to have perspective. The bottom graph shows the same data drawn as a contour plot. Each contour represents a different specific level of instrument response. This type of graph can be displayed in "pseudo-color" with a color assigned to each of the various response levels.

Computerized instrumentation has made it possible to produce such 3-D graphs easily. They can be extremely useful in visualizing trends that might not be seen as easily with a series of two-dimensional graphs.

1A A Deeper Look: Mathematical Review

Logarithms

The logarithm of a number is the power to which 10 must be raised to give the number. Thus, for the number M and its logarithm a, the following relationship is true:

$$M = 10^a \qquad a = \log M \tag{1A-1}$$

The antilogarithm or antilog is the number itself (here M) that corresponds to the logarithm (here a).

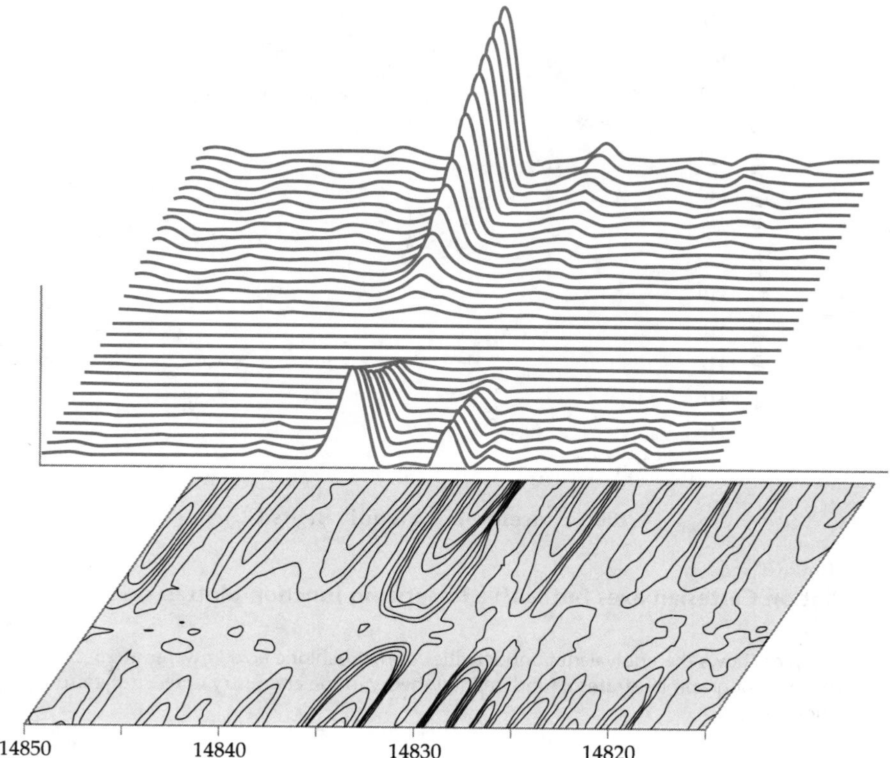

FIGURE 1.6 ▶
An example of three-dimensional plots.
Top: Instrument response versus both wavenumber (energy) and time. The experiment involves the measurement of light from a sample where, over time, one component decreases and another appears. (Bottom) A contour plot of the same data.
[Figure Courtesy of Bruker Instruments]

The logarithm of the reciprocal of a number M is $-\log M$. That is,

$$\frac{1}{M} = \frac{1}{10^a} = 10^{-a} \qquad -a = \log\frac{1}{M} = -\log M$$

With the advent of hand calculators, the use of logarithms for calculation has almost disappeared. However, logarithms are still used as a shorthand for writing a wide range of numbers and appear in many formulas in this text. Examples include expressing equilibrium constants and concentrations of species in a solution. For this reason, it is necessary to be able to find the logs of numbers and vice versa.

Since a logarithm is an exponent, it follows the rules of exponential mathematics. The rules for manipulating logs are simple. The log of a product is the sum of the logs of the multiplied numbers,

$$\log(ab) = \log a + \log b$$

while the log of the quotient of two numbers is given by

$$\log(a/b) = \log a + \log(1/b) = \log a - \log b$$

As for other exponents, multiplying two exponential numbers gives the same result as adding the exponents.

$$10^a \cdot 10^b = 10^{(a+b)}$$

and

$$\log(10^a \cdot 10^b) = \log(10^{a+b}) = a + b$$

EXAMPLE 1.15

Find log (3.45×10^{-4}).

Solution

$$\text{As for any product, } \log(3.45 \times 10^{-4}) = \log 3.45 + \log 10^{-4}$$
$$= 0.5378 + (-4)$$
$$= -3.462$$

In addition, you will see the abbreviation of logarithm to the base 10 as lg and $\log_{10}$. This notation differentiates it from the logarithm to the base e (=2.71828), which is written ln or $\log_e$. For all numbers,

$$\log_e M = 2.3026 \log_{10} M \qquad \textbf{(1A-2)}$$

EXAMPLE 1.16

Find ln 100.

Solution

Since

$$\log_{10} 100 = 2$$
$$\ln 100 = 2.3026 \times 2 = 4.6052$$

Quadratic Equations

In calculations that involve equilibria, you will inevitably encounter quadratic equations. These are equations in which the highest power of the unknown is the square. A general quadratic equation has the form

$$ax^2 + bx + c = 0 \qquad \textbf{(1A-3)}$$

Here a, b, and c are constants and can be positive or negative numbers or zero. The solution is

$$x = \frac{-b \pm \sqrt{b^2 - 4ac}}{2a} \qquad \textbf{(1A-4)}$$

We will find in our applications of Equation 1A-4, that one of the values ($\pm$ square root) for x can be eliminated because it makes no physical sense (for example, a negative concentration).

Equations of Higher Degree and Successive Approximations

Equations that contain an unknown to a power higher than two (for example, x^3) are called equations of higher degree. With a simple hand calculator, a successive approximation to answers is quite easy. With programmable calculators, even quite complicated equations can be solved in a few minutes by successive approximation. The method requires making an estimate of the value of the variable that is an exact solution to the equation, calculating the value of the function, and making a closer approximation until you are satisfied with the closeness of the exact and the approximate result.

EXAMPLE 1.17

Find the solution to the equation

$$(2 + x)^2 + 5x^3 = 8.33 = f(x)$$

Solution

If this is a description of a real physical system, then we may expect that x will be a positive number. We begin with $x = 1$. The trials and results are shown in the table below.

x Chosen	$f(x)$ Calculated	Comments
1	14	Too high. So try a lower value.
0.5	6.875	Too low.
0.7	9.005	Too high; x is between 0.5 and 0.7.
0.65	8.396	Slightly high.
0.63	8.167	Slightly low.
0.64	8.280	
0.645	8.338	
.	.	
.	.	
.	.	

If higher accuracy is needed, you may wish to use a spreadsheet or one of the many symbolic mathematics programs on the market. Both will eliminate the tedium.

Suggestions for Further Reading

TOWNSHEND, A. 1996. *Encyclopedia of Analytical Science*. London: Academic Press.

A recently published ten-volume encyclopedia.

POPPITI, J. A. 1994. *Practical Techniques for Laboratory Analysis*. Boca Raton: Lewis Publishers.

A handbook of analytical techniques.

MEITES, L., ed. 1982 reissue. *Handbook of Analytical Chemistry*. New York: McGraw-Hill.

A seasoned classic of the field, filled with condensed information about mostly wet analysis techniques.

MALUDZIŃSKA, G. 1990. *Dictionary of Analytical Chemistry*. Amsterdam: Elsevier.

A dictionary of terms used in analytical chemistry that you may find useful.

GRASSELLI, J. G. 1983. *The Analytical Approach*. Washington, D.C.: American Chemical Society.

Readings of case studies in problem solving using chemical analytical techniques.

KOLTHOFF, I. M., ELVING, P. J., eds. 1978. *Treatise on Analytical Chemistry*, 2nd ed. New York: Wiley.

A complete treatise on the methods and methodology of chemical analysis up to its publication. The treatise consists of three parts:

Part I: Theory and Practice

Part II: Analytical Chemistry of the Elements

Part III: Analysis of Industrial Products

Part I of the first edition is complete but now is significantly dated. (It first came out in 1959. The last volume appeared in 1976.) The second edition, still incomplete, began appearing in 1978. Part II is a list of the methods (with references) used for each element and is still useful. However, *Chemical Abstracts* contains references that are more up to date. The level of writing varies, but you should be able to make your way through it after your analytical course.

FREISER, H. and NANCOLLAS, G. H. 1987. *Compendium of Analytical Nomenclature: Definitive Rules 1987*. Oxford: Blackwell Scientific Publications.

A book of definitions of the terms used in chemical analysis. Despite the title, it is not the final word as to how the terms are now used. IUPAC (The International Union of Pure and Applied Chemistry) published it, but the reports in *Pure & Applied Chemistry*, which appear from time to time, are more up to date.

KUWANA, T., ed. 1978. *Physical Methods in Modern Chemical Analysis*. New York: Academic Press.

A multivolume set of descriptions of instrumental methods of analysis. These are at an intermediate level but are not "how-to" oriented.

Concept Review

1. What is the difference between an "analysis of iron" and an "assay for iron?"

2. What is meant by the "validation" of an analytical method?

3. What are the basic steps in the development of a new analytical method?

4. Under what conditions are weight/volume and weight/weight measures nearly equivalent?

Exercises

1.1 What is the molarity of K^+ in a solution that contains
(a) 75.6 pg $K_3Fe(CN)_6$ in 25μL solution?
(b) 62.4 ppm(w/v) of $K_3Fe(CN)_6$?

1.2 Calculate the following for 250 mL of a solution containing 0.250 g $MgCl_2 \cdot 2H_2O$:
(a) % (w/v)
(b) ppm(w/v)
(c) molarity

1.3 Calculate the following for a 0.100 M NaCl solution:
(a) mass NaCl in 5.00 mL
(b) % (w/v)

1.4 Concentrated H_2SO_4 has a specific gravity of 1.84 and contains 96.0% H_2SO_4 by weight. What is the molarity of the solution?

1.5 A 200-mg tablet of the drug ibuprofen (m.w. 306.3) contains 20 mg of the drug with the rest being filler.
(a) What is the w/w percentage of the tablet that is ibuprofen?
(b) What is the w/w percentage filler?
(c) How many ppm(w/w) of the tablet is active material?

1.6 Average human blood serum contains 2.2 mM K^+ and 77.5 mM Cl^-.
(a) What are their concentrations (w/v) in ppm?
(b) What are the %(w/w) concentrations of potassium and chloride if we assume that at body temperature, the density of serum is 1.005 g/mL?

1.7 One-half of a 1.00 liter container (0.500 L) of summer coolant is added to the 4.5 L already present in a car's radiator. What is the percentage (v/v) of coolant in the system?

1.8 How many mL of HNO_3 solution (density 1.3393 g mL^{-1}) which contains 55.00% HNO_3 by weight are required to prepare 1800 mL of 0.3100-N solution?

1.9 Water is considered "hard" when it contains 100 ppm(w/v) of $CaCO_3$.
(a) How many mg of Ca^{+2} and how many mg of CO_3^{2-} are in 0.5 L of a 100-ppm calcium carbonate solution?
(b) What is the molarity of the solution?

1.10 An impurity found in motor oil will form a sludge on the spark plugs when at or above 100 ppm(w/v) levels. How many mg of impurity would this be in a quart of oil? (1.06 quarts = 1 L)

1.11 In many states, a person is considered legally intoxicated if his or her blood contains 0.1% ethanol (v/v). What volume of ethanol in the blood is enough to produce legal intoxication in a 70-kg person with a total blood volume of 5.45 L?

1.12 A new metal composite has a density of 2.543 g/cm^3. Graphite fibers make up 14.4 parts per thousand of the composite (w/w). What weight of an airfoil weighing 35.00 kg and made of the composite is graphite?

1.13 Two shakes from a salt shaker dispense 150 mg of NaCl. This is added to a container holding 0.75 L of hot water on a stove.
(a) What is the content of Na^+ and Cl^- in the water in g/L and in mg/mL?
(b) What are the molar concentrations of Na^+, Cl^-, and NaCl?
(c) If a 10.0-mL sample of the solution is taken, what is the salt content in ppm(w/v) of the sample?

1.14 If 40 rat hairs weighing 0.15 mg each were found in a 1.00-kg sample of grain, what is the w/w content of rat hairs in mg/kg and ppm?

1.15 Calculate the grams of solute required to prepare 250 mL of the following solutions:
(a) 0.100 M $K_4Fe(CN)_6$
(b) 2.00 ppm(w/v) Na^+ (using NaCl)
(c) 2.50 ppb(w/v) Na^+ using Na_2SO_4

1.16 The compound 2,6-dichlorophenol indophenol (m.w. 256), abbreviated DCIP, is used in an analysis to determine the amount of vitamin C. Two electrons per mole of DCIP are transferred in the reaction. A 250-mL solution containing 15.0 mg of DCIP has what normality?

1.17 Given solutions with the following molarity value, what are their normality values:
(a) 0.01 M HCl
(b) 0.05 M H_2SO_4
(c) 0.04 M NaOH
(d) 0.02 M $Mg(OH)_2$

1.18 Iodate ion can be reduced as well as accept a proton in a reaction. The general reactions for these processes are

$$12\,H^+ + 2\,IO_3^- + 10\,e^- = I_2 + 6\,H_2O \qquad \text{redox}$$

$$HIO_3 = H^+ + IO_3^- \qquad \text{acid-base}$$

If 5.2 g of KIO_3 is contained in 1.00 L of solution,
(a) What is the molarity of the iodate solution?
(b) For an oxidation-reduction reaction, what is the solution normality?
(c) For an acid-base reaction, what is the normality?

1.19 Metal oxides react with H_2O to produce the corresponding hydroxide. If 1.50 g BaO is dissolved in water to produce 200 mL solution, what will be the normality of the basic solution?

1.20 An acid is 0.620 *N*. To what volume must 1.000 L be diluted to make it 0.500 *N*?

1.21 You are working in a poorly equipped laboratory that has only 10.00 mL pipettes and 100.0 mL volumetric flasks. Starting with a solution that is known to be 0.15 M in sodium, what steps of dilution would you use to get a 1.5×10^{-6} M solution?

1.22 Given the following [X], find the pX.
(a) $[H^+] = 0.010$ M
(b) $[H^+] = 1.00 \times 10^{-5}$ M
(c) $[K^+] = 2.5 \times 10^{-3}$ M
(d) $[OH^-] = 0.453$ M

1.23 Given the following pX values, find [X].
(a) pH = 6.00
(b) pH = 3.503
(c) pCa = 3.25
(d) pOH = 11.05

1.24 If 25.00 mL of an HCl solution reacts with 0.2178 g of pure $Na_2CO_3 \cdot 6H_2O$ replacing both sodiums, what is the normality of the acid?

1.25 What is the molarity of a 6.0% (w/w) NaCl solution if the specific gravity of the resulting solution is 1.0413?

1.26 A nitric acid solution containing 7.20% (w/v) of HNO_3 is needed. To what volume must 50.00 mL of concentrated HNO_3 solution—75.00% (w/w) HNO_3, specific gravity 1.4337—be diluted to obtain the 7.20% (w/v) solution?

1.27 When 50.02 mL of an aqueous solution containing only H_2SO_4 was treated with an excess $BaCl_2$ solution, a precipitate of $BaSO_4$ was obtained. This was filtered off and dried and found to weigh 1.2930 g. Assuming that the small amount of dissolved $BaSO_4$ can be neglected, what is the molarity of the original H_2SO_4 solution?

1.28 To make a special alkaline solution, 7.932 g of BaO, 3.976 g of NaOH, and 1.682 g of Na_2O were dissolved in enough water to make 1000.0 mL of the solution. Calculate the normality of the solution with respect to an acid-base titration.

1.29 Describe how you might prepare 100 mL of a 1.00 ppb solution of Cd^{2+} using deionized water, 10.00-mL pipettes, 100 mL volumetrics, and a mass of cadmium nitrate greater than 10 mg.

Section 1A Exercises

1.30 Find the logarithms of the following:
(a) 1.00×10^{-4}
(b) 2.00×10^{5}
(c) 0.0015

1.31 Find the antilogs of the following:
(a) −2.68
(b) 4.32
(c) −0.0015

1.32 A problem involving finding the H^+ concentration in an aqueous solution reduces to solving the following equation:

$$x^2 - 2.6 \times 10^{-6}\,x - 1.7 \times 10^{-11} = 0$$

where $x = [H^+]$. Determine $[H^+]$.

1.33 What is the pH of the solution in Problem 1.32?

1.34 To find the number of moles of a sparingly soluble salt that will dissolve in 1.00 L of solution reduces to solving the following equation:

$$x^3 + 2 \times 10^5\,x^2 + 250\,x - 5.08 \times 10^{-3} = 0$$

where S is the solubility in moles L^{-1}. Find S using the iterative method illustrated in Section 1A.

Additional Exercises

1.35 What is the molarity of Fe^{2+} in a solution that contains
(a) 75.6 pg $FeCl_2$ in 1.00 L of solution?
(b) 62.4 ppm(w/v) of $Fe(NH_4)_2SO_4$?

1.36 Calculate the following for 250 mL of a solution containing 0.250 g $Na_2B_4O_7 \cdot 10\,H_2O$:
(a) % (w/v)
(b) ppm(w/v)
(c) molarity

1.37 Calculate the following for a 0.100 M $K_4Fe(CN)_6$ solution:
(a) mass $K_4Fe(CN)_6$ in 5.00 mL
(b) mg Fe in 5.00 mL
(c) % (w/v)

1.38 Concentrated reagent ammonium hydroxide has a specific gravity of 0.90 and contains 56.6% NH_4OH by weight. What is the molarity of the solution?

1.39 An average human infant's blood serum contains 5–10 milliequivalents of potassium L^{-1}. What is the concentration (w/v) range of potassium in ppm?

1.40 The pH of the effluent from a factory is regulated so that it is between 6 and 7 to avoid killing the fish in their holding pond. To what minimum volume would you have to dilute 25 mL of 12 M HCl in order to adhere to the pH regulation?

1.41 How many mL of H_2SO_4 solution (density 1.1783 g mL^{-1}) which contains 25.0% HNO_3 by weight are required to prepare 1800 mL of 0.1000-N solution?

1.42 The local Department of Health guideline for chlorine in swimming pools is 1 ppm. At least how many grams should be present in a swimming pool that is 50 m × 30 m with an average depth of 1.3 m?

1.43 How many ppm(w/w) sodium are in one serving of the following products?

(a) ketchup, 180 mg of sodium/15 g serving
(b) salad dressing, 95 mg of sodium/15 g serving
(c) instant soup, 540 mg of sodium/180 mL serving (density = 1.001 g mL^{-1})

1.44 Calculate the grams of solute required to prepare 250 mL of the following solutions:
(a) 0.100 M potassium hydrogen phthalate
(b) 2.00 ppm(w/v) Na^+ (using Na_2CO_3)
(c) 2.50 ppb(w/v) Cr^{3+} using $Cr(NO_3)_3$

1.45 The compound 2,6-dichlorophenol indophenol (m.w. 256), abbreviated DCIP, is used in an analysis to determine the amount of vitamin C. Two electrons per mole of DCIP are transferred in the reaction. A 500-mL solution containing 27.5 mg of DCIP has what normality?

1.46 Given solutions with the following molarity values, what are their normality values:
(a) 0.01 M HCl
(b) 0.05 M H_2SO_4
(c) 0.04 M NaOH
(d) 0.02 M $Mg(OH)_2$

1.47 Metal oxides react with H_2O to produce the corresponding hydroxide. If 1.50 g Rb_2O were dissolved in water to produce 200 mL solution, what would be the normality of the basic solution?

1.48 A diluted sulfuric acid solution has a density of 1.068 g mL^{-1} and contains 10% (w/w) H_2SO_4. Can it be used as a stock solution to prepare 2.00 L of 0.5 M sulfuric acid? If so, how many mL of the 10% sulfuric acid solution would be needed?

1.49 Describe how you might prepare 100 mL of a 1.00 ppm solution of Ca^{2+} using deionized water, 1.00-mL pipettes, 100-mL volumetrics and a mass of calcium nitrate greater than 10 mg.

1.50 Given the following [X], find the pX.
(a) $[H^+]$ = 0.00100 M
(b) $[K^+] = 2.5 \times 10^{-3}$ M

1.51 Given the following pH values, find $[H^+]$.
(a) pH = 7.38 (average blood pH)
(b) pH = 1 (stomach acid)

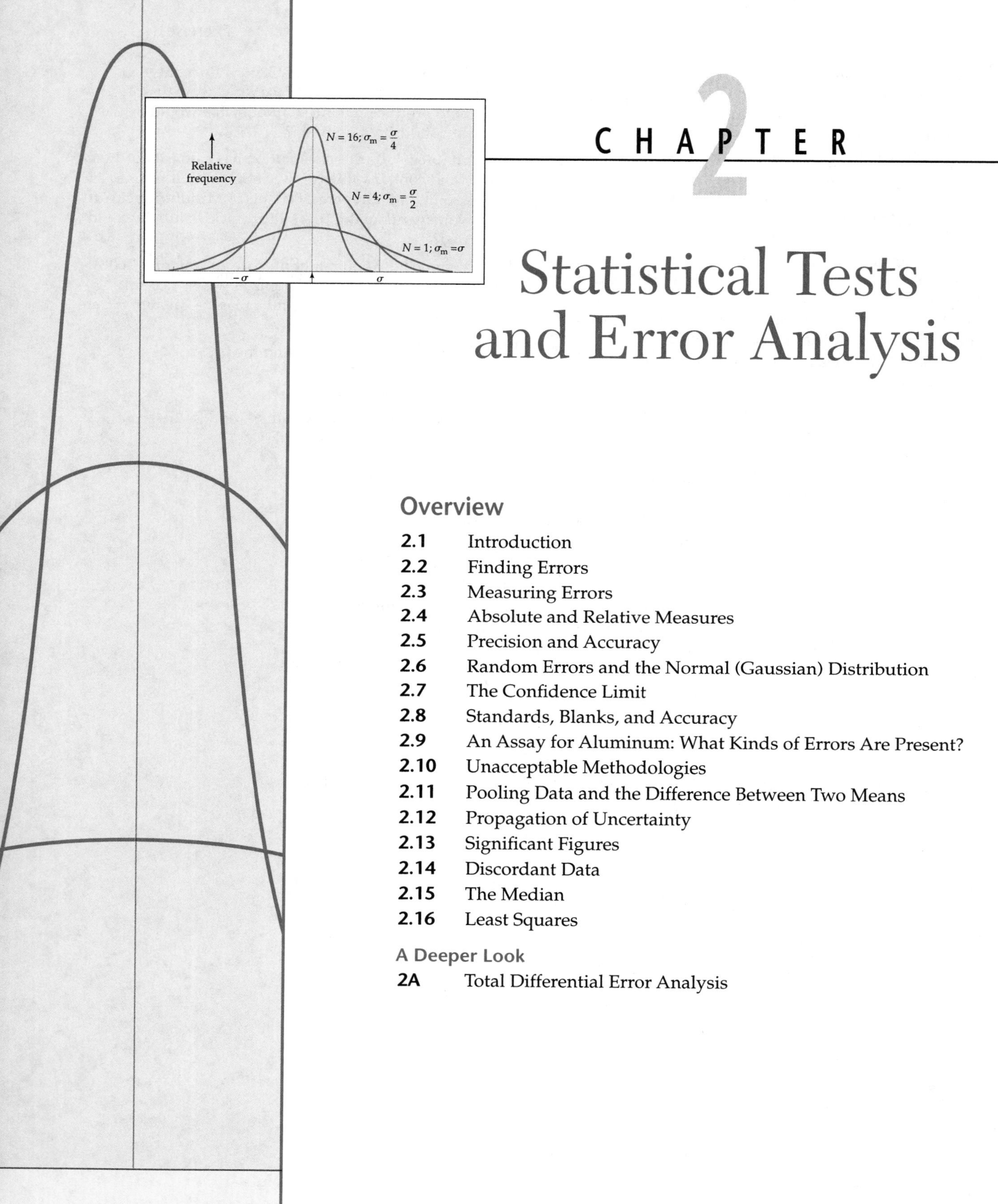

CHAPTER 2

Statistical Tests and Error Analysis

Overview

2.1 Introduction

When a conclusion is made, it is most useful to be able to say how certain it is. We all use phrases like "pretty sure," "very sure," "most likely," and "improbable" to express different degrees of certainty. However, in chemical analysis, more quantitative statements of certainty are desirable. Replacements for these rather vague phrases can be made by using mathematical statistical tests.

"Analytical chemistry is not spectrometers, polarographs, electron microprobes, etc., it is experimentation, observation, developing facts, and drawing deductions." [Ref: West, P. W. 1974. *Anal. Chem.* 46(9):784A.]

In this chapter, a number of these statistical methods for evaluation of experimental results are discussed. The two main questions that we want to answer are:

- How sure are you that the experimentally obtained value is close to the true value?
- How sure are you that the value you have obtained is the same (or different) from the value obtained on the same sample at a different time or by another person?

The first question asks the fundamental question of chemical analysis (and all experimental science). The second is quite important in the area of national or international regulation of the presence (or absence) of chemical substances. Information from numerous laboratories and analysts must be coordinated, and the analytical results must be reliable. If the results are unreliable, the regulation could be meaningless. What is desired is interchangeability of results among laboratories. To this end, significant effort is expended to *validate* an analytical methodology. To validate means to determine that the methodology developed to determine a specific analyte (such as vanadium) in a specific type of sample (such as beef liver) produces comparable results for the majority of analysts. Validation is extremely important for an individual analyst working alone, for a group within a single laboratory, and for groups of laboratories coordinating their analyses.

2.2 Finding Errors

Developing reliable chemical analyses entails identifying where errors can occur and being able to evaluate their magnitudes. To illustrate this skill, an assay is chosen that is likely to be familiar to you: a titration. (If you are not familiar with titrations, you may want to glance at the early sections of Chapter 7.)

A Titration Procedure

Some unknown powder was spilled in an accident. You are given a kilogram of the powder in a plastic bag. It is strongly acidic when tested with pH paper, and you are asked to find out approximately how much sodium carbonate will be needed to neutralize the material that remains after scooping up the bulk. The amount of carbonate required will be mixed with water and used to neutralize and wash away the residue. You can estimate the total amount remaining from the approximate area and depth. This estimate together with the results of your titration will provide the answer.

Normally, to obtain an accurate analysis of the acidity of a powder sample, the sample should be dried in an oven at 110°C to remove any associated water within the powder. The sample should be "dried" until it no longer

ANALYTICAL CHEMISTRY IN ACTION
Every Step Counts

So many different steps occur in even the simplest analysis that the variability in results between laboratories is usually greater than the variability within a single laboratory. The quote below describes the results from ten laboratories comparing the measurement for alcohol in blood. A term that may be unfamiliar to you is **outliers**. To do an analysis, the measurement is normally made a number of times on portions of the same sample. The results for all the portions are then averaged. If the result for one portion has a value significantly far from the others, it is called an outlier and may be rejected as invalid. The validity of outliers can be tested by using statistical methods.

> An irreducible difference exists between supposedly identical measurements made in different laboratories. This point was . . . demonstrated by a group of New Zealand government laboratories in attempting to minimize the discrepancies in values for blood alcohol between laboratories. The laboratories went to great pains to discover every source of error, even to the extreme of moving analysts from one laboratory to another. They found that an analyst increases his or her intra-analyst variability when moved to a different laboratory environment. They concluded that the only way to eliminate inter-laboratory variability was to conduct all analyses in a single laboratory and, presumably, by the same analyst. Under our legal system, this solution is impossible since defendants accused by laboratory evidence have a constitutional right to produce rebuttal evidence from any laboratory of their choice. Therefore, the important question to be answered in the evaluation of methods of analysis is how much allowance must be made for between-laboratory variability in interpreting the values produced by different laboratories. If the variability or error produced by the method is excessive—that is, it does not permit effective regulation as required by the statute—the method must be judged unacceptable for the intended purpose. . . .

[Reference: Horwitz, W. 1982. *Anal. Chem.* 54:67A–76A.] ■

loses weight by heating at 110°. In this case, however, drying may not be needed for two good reasons. First, the weight of the residue of the spilled material is not known accurately, and second, there may not be enough time to do the drying since the result is needed immediately.

Whether the sample is dried or not, a part of the sample (a few grams) is weighed accurately and added to water. (Let us assume that the powder easily dissolves in water.)

Next, a clean burette is filled with a solution of a known concentration of sodium hydroxide in water. The initial volume of liquid in the burette is measured by observing where the top of the liquid lies on the volume scale on the glass.

Acid-base colorimetric indicators are introduced in Section 7.9.

A small amount of acid-base indicator solution is then added to the solution of unknown acid in water. The hydroxide solution in the burette is added to the sample solution until the indicator just changes color. The volume of solution remaining in the burette is then read, and the amount of acid in the weighed sample is calculated.

ANALYTICAL CHEMISTRY IN ACTION
Not in My Chocolate Bar!

Interlaboratory variability apparently cannot be reduced below some minimum level. Interlaboratory tests over the years have shown that the relative error among laboratories increases as the analyte level decreases. One of these tests used a highly carcinogenic (cancer-causing) material called aflatoxin, which can be found as a natural contaminant on products such as cacao beans (used to make chocolate) and peanuts. Each of ten laboratories was sent a sample of uncontaminated cacao beans to which was added a known, fixed amount of the aflatoxin to test how well the laboratories' results compared. Each laboratory used the same analytical methodology and ran two separate analyses on its sample. The reported values are plotted in the adjacent Figure. It should be clear from the figure that the measured result of each analysis was not the quantity of aflatoxin that was originally added to the uncontaminated samples. In an update after a further decade of work, the following conclusion was made after similar tests were run hundreds of times.

> . . . The overall precision pattern . . . is almost identical with that presented 10 years ago. There has been no improvement in the among-laboratories precision of aflatoxin assays over the past 20 years despite the advances in technology. . . . All of the improvement has occurred in within-laboratory precision, indicating a need for laboratories to refer their measurements to common standards and to operate under the aegis of a strong external quality assurance program. . . .
>
> . . . The frequent reporting of contradictory test results at the 0–10 $\mu g/kg$ concentration for identical test materials, or even for the same extracts, suggests a random effect rather than traceable separation or faulty instrumental measurement techniques. The evidence suggests that the practical limit of measurement of aflatoxins is in the low single-digit $\mu g/kg$ concentrations. The limit cannot be specified with any greater exactness because of the high measurement uncertainty that exists at such low concentrations. . . .

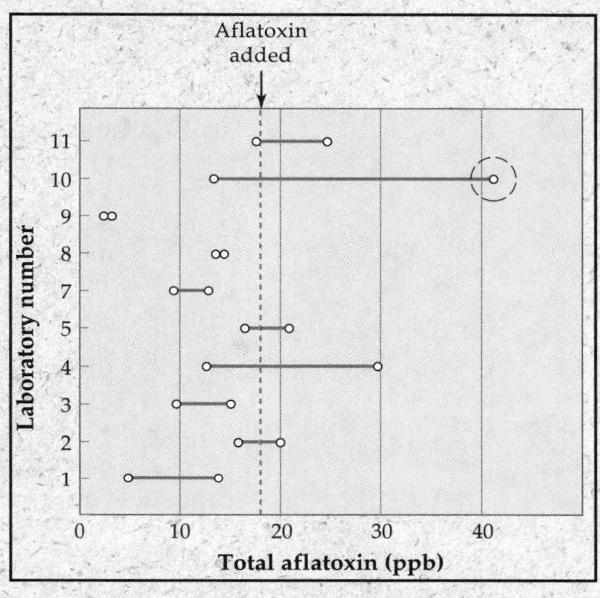

▲ **Graph of the results of the interlaboratory study of the determination of aflatoxin in cacao beans.**
The vertical (dashed) line indicates the amount of aflatoxin added to uncontaminated beans by the supervisors of the study. Open circles indicate the values obtained for the two duplicate determinations made by each of the 10 laboratories taking part in the study. The circled value was deemed an outlier. Note that laboratory 9 produced two results that were quite close to each other but not to the true value; the results were precise but not accurate.

[Figure Ref: Horwitz, W. 1982. *Anal. Chem.* 54:67A–76A.]
[Quote Ref: Horwitz, W., et al. 1993. *Journal of AOAC International* 76:461–491.] ■

In this procedure, as in *all* analyses, the possibility exists for making mistakes at every step. We call these possible mistakes **sources of error**. Possible sources of error in this analysis include incomplete drying of the sample (if it was dried), making an error in weighing the powder, misreading a volume measurement, and spilling or splashing some of the sample solution out of its container. Contamination of the reagents or sample with atmospheric components is also possible; for example, dissolved CO_2 increases the acidity of the solution.

Am I at Least in the Ballpark?

How can you find out if you have made a mistake? After all, no one knows the amount of acid in your sample. One way is to repeat the whole procedure for several **replicate samples**, or **replicates**. All the replicates are obtained from the same original sample. If the answers are similar for two replicate analyses of the acidic powder, it would appear that no errors like the ones mentioned above were made in either run. If you carry out the analysis on three replicates, you will be even more certain. Running a fourth replicate would make you yet more certain. The more experiments you perform, the more certain you will be that the next result is *going to be in the same range* as the previous ones. The statistical description of certainty—introduced in the next few sections—behaves in the same manner as this increase in your confidence in your average result.

2.3 Measuring Errors

It is important to recognize that a statistical measure of the quality of an assay cannot be obtained unless a series of tests is carried out. If only one sample is analyzed, we must take the measured value as the result. However, if a number of replicate runs are made, it is unlikely that two results will be exactly the same. You saw this in the ACA on page 31 that presented results of aflatoxin testing. If each run provides a different value, what value is to be reported as the result for the assay? One meaningful numerical result of a series of tests is the **arithmetic mean** of the individual results. The arithmetic mean of a series of values is found by dividing the sum of the results by the number of determinations in the series. This is expressed mathematically by

$$\overline{X} = \frac{\sum_i X_i}{N} \tag{2-1}$$

where

$\overline{X}$ represents the arithmetic mean,
X_i represents the numerical result of the ith run, and
N is the total number of runs.

EXAMPLE 2.1

A batch of iron-molybdenum protein in 0.1 M NaCl + pH 7.35 tris buffer was analyzed for protein content. The values found from five samples were 27.5, 28.3, 29.0, 28.5 and 28.2 mg protein mL^{-1}. What is the mean value of the protein content?

Solution

Based on Equation 2-1,

$$\overline{X} = \frac{\Sigma X_i}{N}$$

$$\overline{X} = \frac{(27.5 + 28.3 + 29.0 + 28.5 + 28.2)}{5}$$

$$= 28.3 \text{ mg mL}^{-1}$$

Often the arithmetic mean is simply called the **mean** or **average**. Calculation of the mean for a specific example is illustrated in Table 2.1.

Table 2.1 shows a statistical test of the quality of a technique for weighing a sample. To do the experiment, the sample was loaded onto the balance, weighed, and removed. This procedure was then repeated three more times. The results of the four weighings are shown in the first column. The sum of the four weights is 12.4115 g. The mean weight $\overline{X}$ is then 12.4115/4 = 3.1029 g, calculated according to Equation 2-1.

Some further definitions follow. The **deviation from the mean** d_i is defined for each individual measurement. It is the difference between each measured value X_i and the mean value for all the measurements, $\overline{X}$. Algebraically,

$$d_i = X_i - \overline{X} \tag{2-2}$$

(As written here, it is common to use upper-case letters to refer to quantities that are measured and lower-case letters to refer to errors or differences.) As an example, representative calculations are made for the data in Table 2.1. In the third column, the calculated values of the deviations from the mean of the individual weighing measurements are shown.

The **range** of the measurements is the difference in magnitude between the highest and lowest test results in a series.

$$\text{range} = w = X_{\text{highest}} - X_{\text{lowest}} \tag{2-3}$$

The range for the example shown in Table 2.1 is 0.0010 g, the difference between −0.0006 and +0.0004.

Table 2.1 Example of Statistics of Weight Determination

Weighing Number (n)		Observed Weight (X_i)	Deviation from the Mean d_i	d_i^2
1		3.1023 g	−0.0006 g	36×10^{-8} g^2
2		3.1032	+0.0003	9×10^{-8}
3		3.1027	−0.0002	4×10^{-8}
4		3.1033	+0.0004	16×10^{-8}
$N = 4$	Sum	12.4115		65×10^{-8}
Mean, $\overline{X} = 3.1029$ g		Av. dev., $\overline{d} = 0.00038$ g		Variance $= 22 \times 10^{-8}$ g^2 $s = 0.0004_6$* g

* The integer 6 written as a subscript means that the error in this digit is greater than ±1.

The most commonly used (and statistically meaningful) measure of the reproducibility of a set of measurements is the **standard deviation**. The standard deviation is the square root of the **variance**. The variance is defined by Equation 2-4.

$$\text{variance} = \frac{d_1^{\,2} + d_2^{\,2} + \cdots + d_N^2}{N - 1} \tag{2-4}$$

$$\text{standard deviation} = s = \sqrt{\frac{d_1^{\,2} + d_2^{\,2} + \cdots + d_N^2}{N - 1}} \tag{2-5}$$

Another statistical measure of error is the **relative standard deviation,** which is the standard deviation expressed as a fraction or percentage of the mean value. It is written algebraically as

$$\text{relative standard deviation} = \frac{s}{\bar{X}} \tag{2-6}$$

or

$$\text{percent relative standard deviation} = \frac{s}{\bar{X}} \times 100 \tag{2-7}$$

Occasionally the percent relative standard deviation is referred to as the **coefficient of variation** (a name no longer recommended for use; *Anal. Chem.* 1968, 40:2271).

EXAMPLE 2.2

Calculate the average deviation, the standard deviation, the relative standard deviation, and the range for the samples of Example 2.1.

Solution

The average deviation is calculated from the individual deviations, d_i.

$$d_1 = -0.8 \text{ mg mL}^{-1}$$

$$d_2 = 0.0$$

$$d_3 = +0.7$$

$$d_4 = +0.2$$

$$d_5 = -0.1$$

And $\bar{d} = \Sigma\, d_i / N$, so $\bar{d} = 0.0$ mg mL^{-1}

From Equation 2-5, the standard deviation is

$$s = \sqrt{\frac{\Sigma d_i^2}{N - 1}}$$

$$= \sqrt{\frac{(-0.8)^2 + 0 + (+0.7)^2 + (+0.2)^2 + (-0.1)^2}{5 - 1}}$$

$$= 0.5_4 \text{ mg mL}^{-1}$$

The relative standard deviation (RSD) is defined as

$$\text{RSD} = \frac{s}{\bar{X}} = \frac{0.5_4}{28.3} = 0.02 \quad \text{(unitless number)}$$

The range w is defined in Equation 2-3.

$$w = X_{\text{highest}} - X_{\text{lowest}} = 29.0 - 27.5 = 1.5 \text{ mg mL}^{-1}$$

2.4 Absolute and Relative Measures

In the previous section, we defined the standard deviation and the relative standard deviation. The standard deviation is a number: an **absolute measure**. Its units are the same as those of the mean value, such as g mL^{-1}. The relative deviation is a ratio of the standard deviation to the mean value. The relative deviation is a *unitless* measure.

A large number of quantities in analytical chemistry are such ratios. They are all called **relative** measures. When you see the label *relative*, it will indicate a ratio of two numbers that both have the same units. Therefore, relative measures are unitless.

2.5 Precision and Accuracy

Thus far, only the spread, or **scatter**, of the values of analytical results has been considered. When the scatter of the experimentally determined values is small, we say the **precision** is high. A quantitative measure of scatter is the standard deviation of a set of measurements: The standard deviation is small when the precision of the experiment is high.

However, we have not yet considered whether the mean value, which is calculated from a series of experimental runs, is close to the amount of the species of interest that is *actually present* in the sample. In other words, the precision of our data does *not* answer the question, How close to the true value is the mean value of the analyses? That is, how **accurate** is it?

The difference between precision and accuracy can be shown visually, as in Figure 2.1, where four different combinations of precision and accuracy are illustrated. The values from individual experiments are shown as dots along the x-axis. The first two plots show the results of precise determinations. However, the mean value of the second set of data is far from the "true value"; the mean value has a low accuracy, but the results have high precision. How can that be?

One possibility could be that in each sample some other component reacts in the same way as the assayed component: an interferent. This happened in the example in Chapter 1 where, in the method that was used to determine chlorine, the thiol in the samples reacted in the same way as the chlorine itself. Since each sample contained the same amount of thiol, the analysis produced a result that was always too high by a constant amount. No matter how precise the analytical methodology, the result was always inaccurate.

The difference between the "true value" of the analyte and the mean value of a series of analytical results is called the **mean error**, here abbreviated E_m. The mean error has units that are the same as the measurement, for instance, g L^{-1}. When the mean error is expressed as a percentage of the true result, it is called the **relative error**. As *relative* suggests, the relative error is a unitless number. (Do not confuse this with the relative standard deviation defined above in Equation 2-6.)

It is important to understand that from a purely statistical point of view, there is no such quantity as a mean error because the "true result" or "true

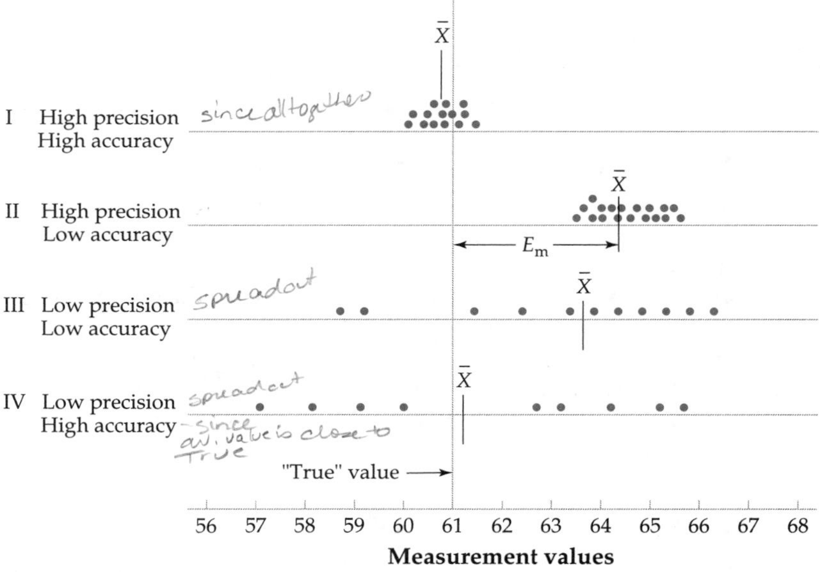

FIGURE 2.1 ▲
Accuracy and precision illustrated.

Four plots of the values of the results of four series of determinations. Dots indicate the values of results from individual experiments. Assume that a "true value" is known; this equals 61.

$\bar{X}$ is the mean value of each series of experiments I–IV.

E_m is called the mean error.

The relative mean error is $E_m/\bar{X}$ or, as a percentage, $(E_m/\bar{X}) \times 100$

One example of the nebulous nature of "true results" can be seen in McFarren, E. F., et al. 1970. *Anal. Chem.* 42:358–365.

value" is never known. However, we can approach the "truth" with cleverly designed validation experiments.

One approach could be to mix, completely and in the same ratios, highly purified materials to produce a known sample that is the same as the ones being analyzed. Known "artificial" samples can be tested in exactly the same way as the unknown ones until a mixture is found that gives the same results as the unknown sample. As you might imagine, such an approach is tedious, and easier experimental approaches are possible.

Incidentally, the mean error falls in the category called **determinate errors**. (Note that the word is not determinant but determin*ate*.) *Determinate* means to originate from a fixed cause. A determinate error in an analysis is either high every time or low every time the analysis is run. Determinate error is also called **systematic error**. On the other hand, the errors for which we use the standard deviation as a measure are assumed to be **random errors**, which means they originate in arbitrary or indeterminate processes. Random errors in an analysis produce a value that sometimes is high and sometimes is low.

2.6 Random Errors and the Normal (Gaussian) Distribution

When a number of events or errors occur at random, it means they are independent of each other. An example of random events is the appearance of heads or tails after flipping a coin. Half of the time the results are heads and

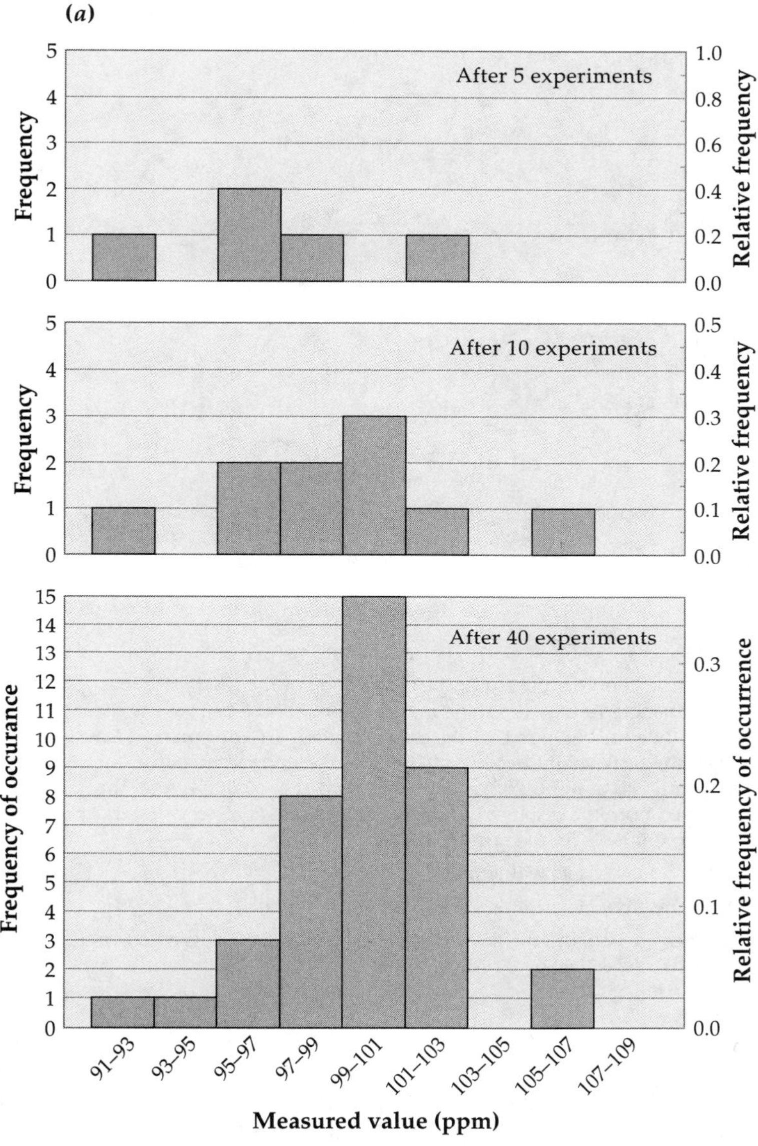

FIGURE 2.2 ◀ **Graphs of the results of a series of 40 experiments classified into sets of adjacent values.**

(a) Each bar represents a 2-ppm range. The height of each represents the number of results in each range. Two different ordinates are shown: the frequency of occurrence and the relative frequency of occurrence. The mean value is 100 ppm. As labeled on the histograms, the results are shown after 5, 10, and 40 experiments are run.

half of the time tails. The result of one flip will have no effect on the next: They are independent.

When the errors in a measurement are random, the values tend to be distributed in a characteristic manner on either side of the mean value. The following discussion of Figure 2.2 shows the statistical nature of this distribution.

First, a series of runs are made using replicate samples of the same material. Such a procedure is similar to weighing some mass repeatedly, as described earlier in Section 2.3. From the results obtained, the mean value can be calculated, and, following that, the deviations from the mean for each measurement can be determined. Let us assume for our illustration that the measurement was done 40 separate times. This number of runs might be done when validating a new method of analysis. However, for routine procedures, seldom are more than four replicate samples used.

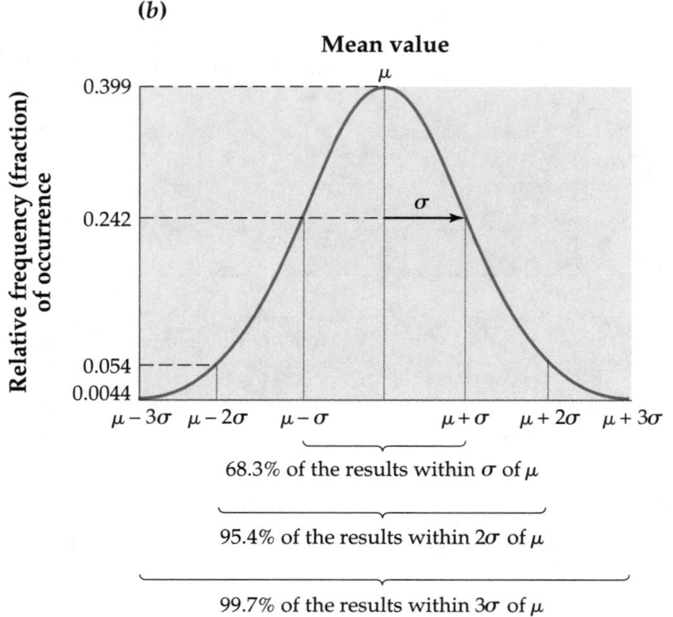

FIGURE 2.2 (cont.) ▲
(b) For a "very large" number of results, the distribution of the results approaches a Gaussian curve characterized by the standard deviation σ and a peak at the mean value. When we scale the curve so that the area under it is 1 (a **normalized** Gaussian curve), then the peak height becomes 0.399. For the normalized curve, the value of the vertical axis at σ is 0.242, at 2σ 0.054, and at 3σ 0.0044. After the curve is established from the results of a number of replicate analyses, we expect that for future experiments on the same material:

68.3% of the results will lie within σ and

95.4% of the results will lie within 2σ

These percentages are equal to the areas under the normalized Gaussian curve between $-\sigma$ and $+\sigma$, respectively.

As illustrated in Figure 2.2a, the data were collected and evaluated after the first 5, the first 10, and finally the full 40 replicate determinations. Next, each of the experimental values is classified into a group depending on how far it is from the mean value. Each group is equal in its range of error—the numerical difference between the highest and lowest values in the group. The data are then plotted as a bar graph. The graph shows the number of experiments in which the results fall into each of the ranges *vs.* the value of the experimental result. Such a plot is called a **histogram**. The mean value is 100 ppm, and each histogram bar is 2 ppm wide.

In the histogram, the height of each bar is proportional to the occurrence of values within its range, so the scale on the left can be labeled *number of experiments*. Since the bars are all equal in width, the area of each bar is also proportional to the number of experiments with results in that bar's range. As a result, the total area of all the bars is proportional to the total number of experiments: here 5, 10, and 40.

Another way to treat the data is to set the sum of the areas of all the bars equal to unity. The left scale then becomes the *proportion of experiments* run or *relative frequency of occurrence*. The area of each bar will then be proportional to the fraction of the experiments that have the error in the given range. (See the right-hand scale of the graphs.)

Another characteristic seen in these histograms is the *tendency* of the bar heights to decrease away from the mean. The number of errors in each group tends to decrease as the group's deviation from the mean increases (as the group lies further from $\overline{X}$). This is seen most clearly in the histogram for 40 replicate runs. However, the expected regular distribution is not necessarily observed after only a few results are measured, as seen in the histograms for 5 and 10 replicates.

The statistical treatment of random errors is only slightly different from plotting a histogram and calculating the relative frequencies of occurrence as they depend on the deviation. The main difference is that the statistical error is calculated *as if* there were an infinite number of experiments: Each range of error is infinitely narrow. In that case, the distribution of random errors about the mean value can be described quite well by a **Gaussian error** curve, as illustrated in Figure 2.2b. Other names for the curve include the **normal error** curve or **normal distribution** curve. In other words, the curve shown in Figure 2.2b summarizes the distribution of results from a large number of analyses.

The x-axis of the plot is a scale of experimental values. The mean value is at the peak of the curve. The height of the curve is a plot of the probability that an experimental result will have the value noted on the x-axis. Note that the errors can be both positive and negative from the mean value. The mean value for the Gaussian curve is labeled μ. It is the value of $\overline{X}$ in the limit of a very large number of trials.

Properties of Gaussian Distributions

The Gaussian curve is characterizable by a parameter σ. You will see that σ can be related to s, the standard deviation for a small number of runs (defined in Equation 2-6). The parameter σ is also called a standard deviation, but for a very large number of runs. The equation describing the Gaussian curve is

$$f(x) = \frac{1}{\sigma\sqrt{2\pi}} e^{-[(x-\mu)^2/2\sigma^2]} \qquad \textbf{(2-8)}$$

where

x is the experimental result value,
μ is the mean value and is located at the peak of the distribution,
σ is a parameter describing the relative width of the curve, and
$(x - \mu)$ is the deviation from the mean for a result.

In the statistical treatment of error, the height of the Gaussian curve itself is not as important as the *areas* under specific ranges along x. As pointed out for the histograms in Figure 2.2a, the total area under the Gaussian curve can be associated with the total number of experiments. Therefore, any *fraction* of the area under the curve will refer to a fraction of the total number of experiments.

The power of the statistical viewpoint of error is its ability to relate the standard deviation of a *series* of runs to the standard deviations of *individual* experiments in the series. Such a relationship is an important key in understanding chemical analysis results.

Now we can connect the standard deviation (Equation 2-6) to the statistical measurement of certainty. *For a "very large" number of identical experiments, the distribution of results will fall approximately on a Gaussian curve centered at the mean value.* We then expect that for *future* experiments:

68.3% of the results will lie within 1 standard deviation of the mean, and
95.4% of the results will lie within 2 standard deviations of the mean.

The fractional areas corresponding to these ranges are indicated in Figure 2.2b. Bear in mind that the relationships apply for a number of experiments far larger than the number we usually perform in a series of replicate samples.

2.7 The Confidence Limit

Earlier, an argument was presented that called upon your intuitive ideas of confidence: When more replicates are analyzed, you become more certain that the next result is going to be in the same range as the previous ones and more certain of the calculated mean value. However, this increase in assurance slows down beyond a certain number of replicates. That is, further repetition merely leads to a feeling that repeating the analysis is getting incredibly boring.

In other words, one determination alone leaves us clueless about the certainty of the result. Two determinations help. But only when three or more are made can you begin to be more sure of the result. The statistical calculation of the **confidence limit** makes quantitative these intuitive ideas. In addition, the statistical calculations allow us to determine how many replicates are needed to assure us that the mean value will remain within a specified range of values.

The essence of the calculation of the confidence limit lies in relating the standard deviation of a single result, σ, to the **standard deviation of the mean value**, σ_m. This relationship is illustrated in Figure 2.3. For an analysis, it is σ_m

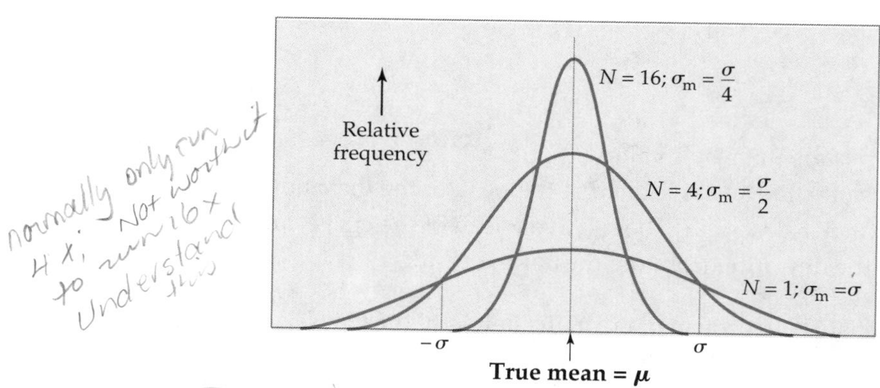

FIGURE 2.3 ▲
Illustration of how the standard deviation of the mean, σ_m, varies with the number of experiments run, *N*.

Each experiment has a random error of magnitude σ. The value of σ_m decreases as the number of trials increases but only as $N^{1/2}$. The reason that σ_m can be less than σ is that the average of the individual errors tends toward zero as more experiments are run.

that we wish to know; σ_m tells us how well we know the value found in the analysis. Another name for the standard deviation σ_m is the **error of the mean**. (Do not confuse this name with the *mean error* or *deviation from the mean*, which were defined earlier.) Unfortunately, both σ and σ_m are called standard deviations. We will retain the subscript to keep their differences clear.

Confidence Limits when σ Is Known

Although Figures 2.2 and 2.3 are both Gaussians, they are illustrating quite different concepts. Figure 2.2 reflects the data values that are collected. Figure 2.3 shows how the uncertainty in the mean value decreases with increasing numbers of trials.

In a simple case, let us *assume* that we know the standard deviation associated with each replicate. The relationship between the error of the mean and the σ associated with each measurement is

$$\sigma_m = \frac{\sigma}{\sqrt{N}} \tag{2-9}$$

where N is the number of measurements made. Simply stated, the intuitive ideas of increasing certainty with increasing number of replicate runs show up in the statistics as a square root of the number of experiments. So, according to statistics laws, you can be only twice as certain of the mean value if you run 16 replicates ($N = 16$) as if you run four replicates ($N = 4$). In addition, this square-root dependence on N also agrees with our intuition that eventually there are diminishing returns in additional precision as the number of runs increases, i.e., double the precision for four times the number of runs.

After studying Figure 2.3, you might reasonably inquire how the error of the mean σ_m can ever be less than σ, the average random error of each determination. In answer, recall that the result of each separate determination can be randomly high or low compared with the "true value." Therefore, as more replicates are run, the random errors tend to cancel; the average of the individual errors approaches zero. Equation 2-9 states that random errors tend to average to zero only as $N^{1/2}$, however.

From the simple idea expressed in Equation 2-9, we can make a quantitative declaration of confidence. As for any Gaussian, theory tells us that we can be 68.3% confident that the mean lies within σ_m. Also, we can be 95.4% sure that the mean lies within $2\sigma_m$ of the mean value that has been determined from the measurements already made. But overlay this confidence with the understanding that σ_m changes with the number of samples run.

The most common confidence limit, or **confidence interval**, stated is at the 95% level. This is written as 95% C.L. and spoken of as the **95% confidence limit**. The interval containing 95% of the area under the curve must lie within something less than $2\sigma_m$, since $2\sigma_m$ includes 95.4% of the results. In fact, the interval lies between plus and minus $1.96\sigma_m$. Therefore,

$$\text{mean} \pm 95\%\ \text{C.L. for } \mu = \overline{X} \pm 1.96\,(\sigma/\sqrt{N}) \quad \text{when } \sigma \text{ is known} \tag{2-10}$$

You need to recognize two characteristics of Equation 2-10. First, the parentheses contain the value σ_m, since $(\sigma/\sqrt{N}) = \sigma_m$. Also, the confidence limit provides a quantitative measure of the intuitive idea that if the encompassing range is wider, we can be more confident that the true value is within that range.

Table 2.2 Values of t* and $(t/\sqrt{N})$ to Use with Equation 2-11

Number of Measurements Used to Determine $\overline{X}$ N	Degrees of Freedom	t for 95% C.L.	$\frac{t}{\sqrt{N}}$
2	1	12.71	8.99
3	2	4.30	2.48
4	3	3.18	1.59
5	4	2.78	1.24
6	5	2.57	1.05
7	6	2.45	0.936
8	7	2.36	0.831
9	8	2.31	0.77
10	9	2.26	0.71
20	19	2.09	0.467
30	29	2.04	0.372
120	119	1.98	0.180
∞	∞	1.96	—

* For tables of t for other probability values of the C.L.—80%, 90%, and 99%—see Appendix VI.

Confidence Limits when σ Is Unknown

Equation 2-10 does not apply to the usual chemical analysis, because the value of the average random error for each experiment, σ, is seldom known. But Equation 2-10 can be modified so that it applies to the situation where σ_m and σ are *both* unknown. As you might suspect, we must be less certain in this case. To reflect this decreased certainty, a factor t is substituted for 1.96 in Equation 2-10. The value of t for the 95% C.L. is always greater than 1.96, since N is not infinite. Therefore, we must substitute s (Equation 2-6) for σ in Equation 2-10. So when σ is unknown, the following equation is used:

$$\text{mean} \pm 95\% \text{ C.L. for } \mu = \overline{X} \pm t(s/\sqrt{N}) = \overline{X} \pm (t/\sqrt{N}) \cdot s$$

$$\text{when } \sigma \text{ is unknown} \qquad \textbf{(2-11)}$$

Values of t and $(t/\sqrt{N})$ are shown in Table 2.2. Notice that t is quite large for small N values but approaches 1.96, as in Equation 2-10, as the number of trials N becomes very large. The trend in the values of t means that as N becomes larger, it does not matter that the magnitudes of σ are unknown, since the errors tend to cancel and tend to the same limit as with known σ-values. In this way, the standard deviation s (as defined in Equation 2-6) smoothly transforms to σ.

EXAMPLE 2.3

Calculate the 95% C.L. for the example of the four weighings listed in Table 2.1.

Solution

The value of s as shown in Table 2.1 is 0.0004_6. The value of $(t/\sqrt{N})$ is found in Table 2.2. Since the value of σ, the error, is unknown, we use Equation 2-11. Therefore,

$$95\% \text{ C.L.} = 0.0004_6 \cdot (t/\sqrt{N}) = 0.0004_6 \cdot 1.59 = 0.0007$$

and the result would be reported as

$$\text{weight} = 3.1029 \pm 0.0007 \text{ g (mean} \pm 95\% \text{ C.L.)}$$

One final point for this section: Remember, both the standard deviation of the mean and the confidence limit are measures of precision only and *not* accuracy.

2.8 Standards, Blanks, and Accuracy

So far, only the precision of a series of replicate measurements has been considered in detail. The level of precision of the series is determined by the magnitudes of the random errors associated with each of the measurements and the number of measurements made. In this section we consider the cause and prevention of determinate errors. Determinate errors are errors in the analytical method; their presence or absence and magnitude determine the accuracy of analytical measurements.

To say a result is *accurate* denotes that the "true value" and the mean value $\overline{X}$, which is experimentally determined, are the same within specified limits. *Poor accuracy* or *low accuracy* means that the value of the mean differs significantly from the "true value." As noted earlier, this difference is called the mean error E_m and is illustrated in Figure 2.2. Do not confuse it with the deviation *from* the mean, as defined in Equation 2-2.

Errors that cause inaccuracy must be eliminated in the experimental technique or at least accounted for by a correction in an analysis. Otherwise, the assay results will not correspond to the actual amount of the component being determined. This idea may seem elementary, and perhaps it is. In practice, however, it can be difficult to account for all the major causes of inaccuracies. The titration example at the beginning of this chapter illustrates some of the obstacles.

Determining the accuracy of a methodology is part of its validation. To develop any new analytical methodology, known samples of different amounts of analyte are assayed. Such samples, which contain *known* amounts of the species to be assayed, are the **standards**. If the analytical results from the standards agree with the known content, the accuracy of the methodology is confirmed. This procedure is called **calibrating** the method. Of course, the assay must be shown to work accurately with the standard samples before trusting the results with unknowns.

One other type of sample is essential to calibrate any method: the **blank**. A blank sample contains all the components in the sample *except* the assayed species. The blank sample is passed through all the steps of the procedure just as a regular sample. The results allow us to investigate interferences that affect the accuracy.

A more general term—one that may include standards, blanks, and samples with a well characterized content or well understood behavior—is **control**. Controls might be used to investigate the effect of different levels of specific interferences. In biochemical analysis, the word *control* is often used instead of *blank* since the substance or solution may be so complicated that the word *blank* seems inapplicable.

Through the analysis of control and blank samples, any causes of poor accuracy can be found. Then, with wise use of descriptive chemistry, changes in the technique can be explored to eliminate or reduce the errors. If necessary, appropriate corrections can be determined.

The following problem depicts how to examine experimental data to isolate the effects of determinate errors and to correct for or eliminate any errors by altering the procedure. In this case, the goal is to find an accurate analytical methodology to determine aluminum content.

This illustration is adapted from Benedetti-Pichler, A. A. 1936. *Ind. Eng. Chem., Anal. Ed.* (precursor to *Analytical Chemistry*) 8:373.

2.9 An Assay for Aluminum: What Kinds of Errors Are Present?

Gravimetric methods are the subject of Chapter 10.

The problem, which we shall study in detail, involves the development of a **gravimetric** assay for aluminum. Gravi/metric (weight/measure) methods are those in which the analyte is transformed into a solid that is weighed. The weight that is measured should be proportional to the amount of the analyte present. Currently, it is much more likely that an assay for aluminum would be done using one of a number of possible spectrometric methods. However, we shall examine the difficulties encountered in this gravimetric method to illustrate the concepts introduced in this chapter. The flowchart in Figure 2.4 illustrates the general steps in the procedure, and the results are listed in Table 2.3.

Two further definitions are needed before proceeding. Determinate errors, which cause inaccuracy, may be classified into two groups: **constant errors** and **proportional errors**. Constant errors are those that are the same magnitude even though the sample size changes. Proportional errors are those that have a magnitude directly proportional to the size of the sample; as the sample gets larger, so does the size of the error. As illustrated below, we can evaluate the proportional and constant errors by using different-sized control samples.

The purpose of the assay was to determine the amount of aluminum in a sample of alum, $AlK(SO_4)_2 \cdot 12\ H_2O$. First, samples were dissolved in a fixed volume of HCl solution. The aluminum was precipitated as a hydroxide,

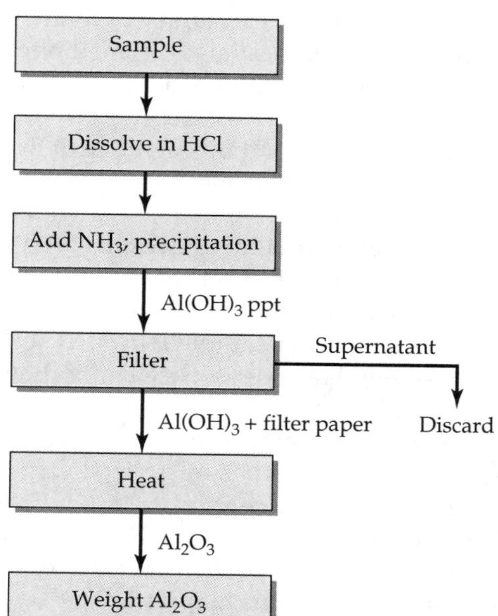

FIGURE 2.4 ▶ **Diagram of the steps in determining aluminum by a gravimetric method.**

Table 2.3 Results and Errors in the Determination of Aluminum as Al_2O_3

Col 1	Col 2	Col 3	Col 4	Col 5	Col 6	Col 7	Col 8	Col 9	Col 10
Size of Sample of $AlK(SO_4)_2 \cdot 12\,H_2O$ (g)	Expected Al_2O_3 weight (g)	Al_2O_3 Found Using Reagent Grade NH_3			Al_2O_3 Found Using Freshly Distilled NH_3			$Al_2O_3 \times 0.9903$	
		Weight (g)	Abs. Error (g)	Rel. Error (%)	Weight (g)	Abs. Error (g)	Rel. Error (%)	Weight (g)	Abs. Error (g)
1.0000	0.1077	0.1288	0.0211	16.4	0.1087	+0.0010	0.9	0.1076	−0.0001
2.0000	0.2154	0.2384	0.0230	9.7	0.2178	+0.0024	1.0	0.2154	+0.0000
3.0000	0.3231	0.3489	0.0258	7.4	0.3258	+0.0027	0.8	0.3226	−0.0005
4.0000	0.4308	0.4588	0.0280	6.1	0.4352	+0.0044	1.0	0.4310	+0.0002
									$s = 0.0003$

$Al(OH)_3$, by using an ammonia solution. The aluminum hydroxide was then filtered from the solution with filter paper. This paper was placed in a crucible that had previously been dried at high temperature, cooled, and weighed. The crucible with the filter paper and precipitate was heated until the aluminum hydroxide was converted to alumina, Al_2O_3. The filter paper produces an ash under these conditions, and the weight of this ash was reproducible. The weight of the ash together with the weight of the crucible were subtracted from the total weight to give the weight of the alumina that remained in the crucible. It may not be obvious but we could classify this operation as a correction.

The first column in Table 2.3 shows the weights of the control samples that were taken. The amount of aluminum in each is calculated as if it were all in the form of the final product Al_2O_3. This is 1.8895 times the weight of the aluminum itself. The weight of Al_2O_3 is shown in the second column. If the determination were exact, these numbers would correspond to the amount of Al_2O_3 that would be weighed at the end of the procedure.

The first set of experiments produced the weights shown in column 3. In order to consider the problems here, both the relative and the absolute errors are calculated. These are shown in columns 4 and 5, respectively. From perusal of these numbers, it appears that there is a relatively *constant* error associated with the assay. The error clearly is not a proportional error; if it were, the relative error would be about the same for all the samples. (Be sure you understand this before you continue.)

What might have caused such an error? The fact that the error is constant implies that some reagent that is added in a constant amount is at fault. The problem could be with the hydrochloric acid solution or with the ammonia. The amounts of both that are added remain nearly constant for all the samples.

But how could the ammonia solution cause a problem? The fact that the blank had no excess weight (beyond the filter-paper ash) was a key point. The root of the problem was that the basic ammonia solution had dissolved some of the silica of the bottle in which it had been stored. When the test solutions were neutralized, the silica precipitated as silicic acid (formula $SiO_2 \cdot (H_2O)_x$) along with the aluminum hydroxide. The descriptive chemistry of silica includes the fact that it will not precipitate out of solution as easily alone as with aluminum hydroxide present. Thus the silica acted as an interferent, which had to be removed. This discovery depended on a good knowledge of the chemistry of the reagents used.

The problem was overcome only by changing the analytical procedure. The ammonia was distilled immediately before using it. Since the silica is not volatile under the distilling conditions, the interference was removed. (Incidentally, these days, we would store the ammonia in containers made of more inert material, such as polyethylene.)

With the freshly distilled ammonia, the same series of experiments were rerun with the results shown in column 6. Again, the absolute and relative errors are calculated and are shown in columns 7 and 8. The absolute error values are plotted in Figure 2.5 as well. We can see that with the fresh NH_3 the absolute error seems to be about proportional to the sample size. This may show up more clearly in Figure 2.5a. Since a proportional error depends on the size of the sample itself, any relative measure of the proportional error is approximately constant.

With this error, can the assay now be used? There are a number of considerations to keep in mind when answering this question. It is possible that the assay as it stands could be used if the accuracy desired is no closer than ±1%.

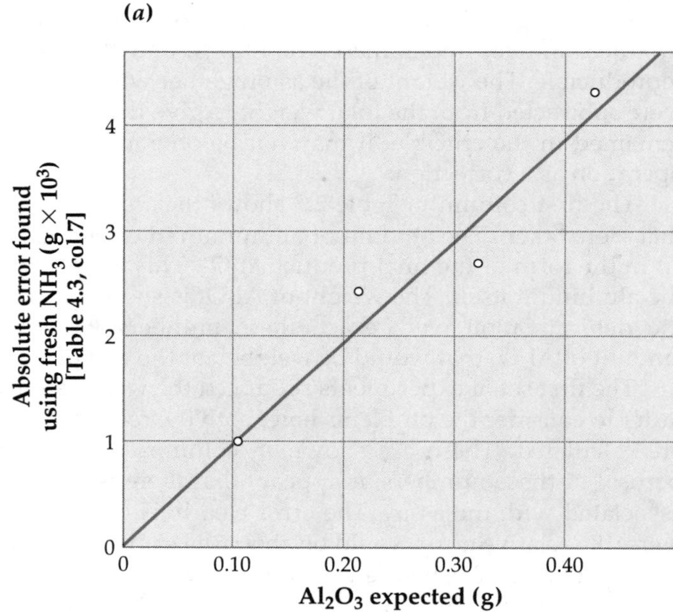

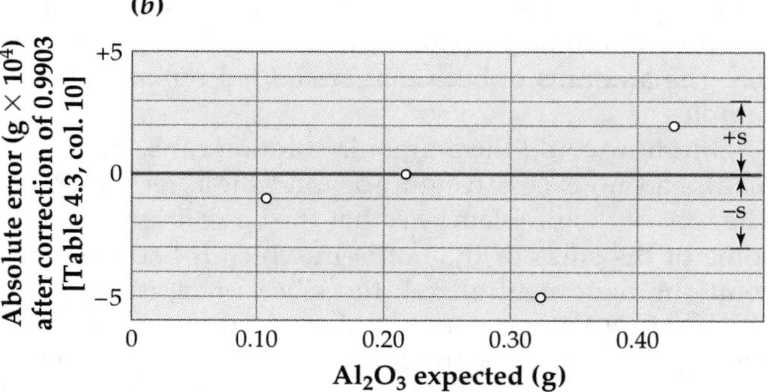

FIGURE 2.5 ▶
Illustration of the trends in errors found when developing a gravimetric assay for aluminum.

(a) Plot of the absolute error versus the expected weight of Al2O3 for the four samples tested. The trend of the points illustrates the effect of a proportional error associated with the determination. (Do not confuse the absolute error with an absolute value of a number.)
(b) Plot of the absolute error after a correction of 21.0% is applied to the results. (Each value was multiplied by 0.9903.) The error values are approximately randomly distributed about the zero-error (perfect accuracy) line. The magnitude of the standard deviation s, calculated from the four results, is indicated by the error bar at the right. Note that the final precision, after including the correction, is in the range of a few parts per thousand.

±1%. But greater accuracy can be obtained without further experimental work: a simple correction can be applied to the results.

Notice that the relative errors reported in column 8 are all positive. The slope of the line in Figure 2.5a indicates that there is still about 1% proportional error. A correction can be made by multiplying each of the results by 0.9903. The resulting errors are listed in column 10. It appears that some random error remains: there are as many positive as negative values of about the same magnitude. This random error can be used to calculate the precision of the method as quantified by the standard deviation.

Another way to visualize the randomness is to see the random distribution of points about the zero line in Figure 2.5b. The 1% correction that was made eliminates the **bias of the method**. The bias in the assay method described above is that the results are 1% too high.

In conclusion, this example shows the differences in the types of errors that affect the accuracy. If the absolute errors of the controls have constant values, a constant error is present. If the relative error values are the same, a proportional error is present. It is only through methods such as these—careful testing of procedures with blanks, calibration controls and, perhaps, a correction for inherent bias—that an effective and *accurate* assay can be created.

EXAMPLE 2.4

An assay for creatinine resulted in the following found *vs.* "true" values.

Actual creatinine level	Creatinine level found
0.00 mg dL^{-1}	0.07 mg dL^{-1}
0.10	0.16
0.20	0.26
0.50	0.57
1.00	1.04
2.00	2.06

Is the error in the method a proportional or a constant error? An adolescent with impaired renal function was found to have an uncorrected creatinine level of 1.35. What is the corrected value?

Solution

The errors for the found values are +0.07, +0.06, +0.06, +0.07, +0.04, and +0.06 mg dL^{-1}. All of the errors are positive and they are all about the same magnitude over the entire range of creatinine values. Therefore we conclude that a constant determinate error occurs in the method.

To obtain the creatinine value for the adolescent patient, we need to calculate the average error and then subtract this amount from the uncorrected value.

$$\text{average error} = (0.07 + 0.06 + 0.06 + 0.07 + 0.04 + 0.06)/6 = 0.06 \text{ mg dL}^{-1}$$

This means that a creatinine value of 1.35 mg dL^{-1} corrected for the determinate error of the assay corresponds to

$$(1.35 - 0.06) \text{ mg dL}^{-1} = 1.29 \text{ mg dL}^{-1}.$$

2.10 Unacceptable Methodologies

Let us now consider briefly the problems that occur when an analysis of a replicate series has poor precision. There are two cases, as shown in lines III and IV in Figure 2.2. From the results plotted, you now recognize that both determinations have wide confidence limits. No knowledge of the details or even the general test procedures is needed to make this evaluation.

It is also possible to evaluate the accuracy in the two cases. In Case III, the accuracy is low. In Case IV, the accuracy is high. However, such judgments can be made only if we know the "true value" of the result beforehand. Even for a known sample, we probably could not delineate constant, proportional, or random errors. The method certainly could not be used on unknown samples. The methodology of Case IV is clearly unacceptable; another must be found.

2.11 Pooling Data and the Difference between Two Means

In a great number of cases, an analytical method is developed to be used by a wide variety of people under some regulatory rules or as an often-used, standardized test for some component of a material. Additionally, to enhance meaningful communication and comparisons, standard testing methods are often developed voluntarily by various groups. Nevertheless, every analyst has certain habits and biases that have developed over years of practice. These are not "bad" or "good" habits but are usually convenient practices for the individual involved. This individuality inevitably shows up in the results from standardized methods of analysis. Disparities can result from differences in reagents, from instrument calibration and, in fact, from all the steps involved in an analysis.

As mentioned in the Analytical Chemistry in Action boxes (ACAs) early in this chapter, the range of results among laboratories, or individuals in the same laboratory, is inevitably larger than the range of results from a single analyst. Let us briefly consider the problem of defining the errors within collaborative studies between laboratories or within a single laboratory over time. Calculating these errors is essential in areas such as environmental regulation and quality control.

The key to manipulating data from multiple sources is to understand how to put the different sets of data together. This is called **pooling** the data. For example, we might want to find the mean value of the pooled data and its standard deviation. We assume that from each group we know three numbers: 1) the group's mean value, $\overline{X}$; 2) the standard deviation of the data, s; and 3) the number of determinations N that were involved in finding $\overline{X}$ and s. If all three of these values are not given, *the calculation cannot be done*. Let us consider only two groups' sets of data, Group A and Group B. The six values we obtain are $\overline{X}_A$, s_A, and N_A from Group A and $\overline{X}_B$, s_B, and N_B from Group B. The standard deviation is pooled by using the equation

impt

$$s_{\text{pooled}} = \sqrt{\frac{(N_A - 1)s_A^2 + (N_B - 1)s_B^2}{N_A + N_B - 2}} \tag{2-12}$$

Each group's standard deviation is multiplied by the number of points less one. This equation applies in most cases of pooling of analyses; that is, when the standard deviations are about equal from each group $s_A \approx s_B$.

To find the mean of the pooled data from two groups, it should be clear that the following equation holds:

$$\overline{X}_{\text{pooled}} = \frac{N_A\overline{X}_A + N_B\overline{X}_B}{N_A + N_B} \tag{2-13}$$

The denominator of the right hand side is the total number of data points. The numerator is the sum of the two means each weighted by the number of points that contribute to that mean. Alternately, if the data points from the two groups are available, the mean can be calculated using Equation 2-1.

EXAMPLE 2.5

Samples of a standard alloy were sent to labs A and B. In lab A, the mean value for six determinations was found to be (4.35 ± 0.07)% Ni ($\overline{X} \pm s$). In lab B, the mean value for eight determinations was found to be (4.47 ± 0.05)% Ni. What is the pooled mean and pooled standard deviation for the two labs?

Solution

After substituting the correct numbers into Equations 2-12 and 2-13, we can crank out the answer. For the pooled standard deviation,

$$s_{\text{pooled}} = \sqrt{\frac{(6-1)(0.07)^2 + (8-1)(0.05)^2}{6+8-2}}$$

$$s_{\text{pooled}} = \sqrt{\frac{0.0245 + 0.0175}{12}} = 0.059\%$$

For the pooled mean value,

$$\overline{X} = \frac{(6)(4.35) + (8)(4.47)}{6+8} = \frac{26.1 + 35.76}{14} = 4.42\%$$

Let us take a brief detour before continuing to comment on the denominator of Equation 2-12, ($N_A + N_B - 2$). This quantity corresponds to a number called the **degrees of freedom**. The number of degrees of freedom equals the total number of samples less the number of sets of samples that are used to calculate the mean. If one *set* of samples is present, the number of degrees of freedom is ($N - 1$). If two *sets* of samples are present, then the number of degrees of freedom is ($N_{\text{total}} - 2$), and so forth. The number of degrees of freedom is more fundamental in statistics, although it may be easier to think in terms of the number of samples analyzed. Table 2.2 on page 42 contains both degrees of freedom and number of measurements.

The Difference between Two Means

More often than pooling data, we are faced with the question of whether the mean values from two or more sets of data indeed do differ from each other. Suppose that analysts in two laboratories have each analyzed several replicates of some sample. The distribution of the results from each analyst is approximated by a Gaussian distribution as shown in Figure 2.6. Can we believe the apparent difference in the two sets of data or is assigning two separate sets simply our misinterpretation? In other words, are the "differences" only due to random error?

How do we know when the differences are due to random error? The calculation involves the difference between the two means: $\overline{X}_B - \overline{X}_A$. The key

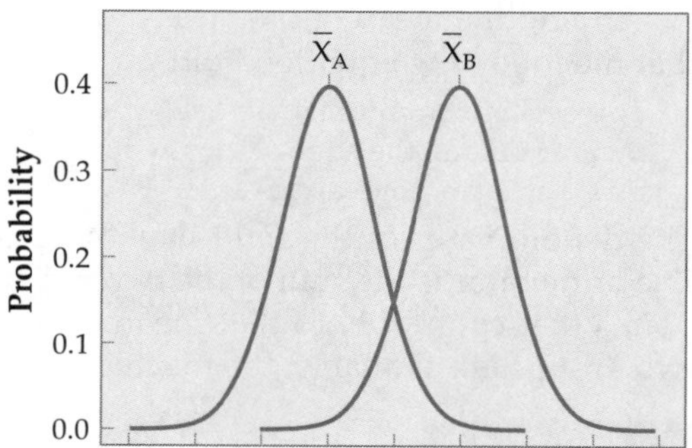

FIGURE 2.6 ▶
The results from two sets of laboratory data.
Set A and set B are approximated by Gaussian distributions that represent the probability (*y* axis) of obtaining a given number for the analysis (*x* axis). Calculation of a pooled *s* for the two sets of data allows us to evaluate the probability that the means are different (Equation 2-14).

point to recognize is that since both sets of data are uncertain, and since each mean has an *s* value—a standard deviation—associated with it, *the difference between the means is uncertain and can be characterized by an s value of that difference*. The *s* value of the *difference* is the same as given by Equation 2-12 above.

Since there is an *s* value for the difference, there must be a confidence limit for the difference as well. In this case, where each mean is uncertain, the confidence limit tells us how certain we can be that the two means *are* indeed different. However, the language used to describe the confidence limits differs from what you have learned in this chapter.

The confidence limit is given by the value of Student's *t*. But the value of *t* for the difference between the two means is calculated using the following equation:

$$t_{calc} = \frac{|\overline{X}_A - \overline{X}_B|}{s_{pooled}} \cdot \sqrt{\frac{N_1 N_2}{N_1 + N_2}} \qquad \textbf{(2-14)}$$

The calculation of the confidence limit cannot be as simple as before because t_{calc} itself depends on the difference of the two means. Also note that we cannot choose the value of the confidence limit—say, 95%—because t_{calc} (as well as *s*) is fixed by the experimental values of $\overline{X}_A$, $\overline{X}_B$, N_A, and N_B.

For the pooled data, we now compare t_{calc} to the values for $N - 2$ trials in a table of Student's *t*. If, for the desired confidence level and $N - 2$ degrees of freedom, $t_{calc} > t$, then you can conclude that the two means are indeed different *at that level*. For example, you could ask, Are the two means different at the 95% confidence level? (You could also ask whether the means differ for the 90% or 80% confidence level, etc. and answer either yes or no to each by comparing t_{calc} with *t* for 90% or *t* for 80% confidence levels.)

The behavior of the comparison—and the algebra of Equation 2-14—shows some general trends. It makes sense that the greater the difference between the two means, the greater the confidence that they are different. It is also worth noting, however, that either a small value for *s* or very large number of trials (reflected in σ_m) also results in a greater level of confidence that the two means are indeed different. Again, the comparison quantitates your intuitive understanding of confidence in the fact of a difference in two values being real.

Similar equations are used in the area called hypothesis testing, which is beyond the chosen scope of this text. More detailed information on hypothesis testing can be found in the sources listed in the bibliography of this chapter.

EXAMPLE 2.6

Two sets of chromium analyses were performed on water samples from a stream that runs along the edge of the property belonging to a plating plant, one before a major spill and one afterwards. The mean of the determinations before the spill was 0.95 ppb ($N = 5, s = 0.05$). After the spill, the mean was 1.10 ppb ($N = 6, s = 0.08$). Could we state with 95% confidence that there was an increase in the chromium content of water from the stream?

Solution

$$s_{\text{pooled}} = \sqrt{\frac{(5 - 1)(0.05)^2 + (6 - 1)(0.08)^2}{5 + 6 - 2}}$$

$$s_{\text{pooled}} = \sqrt{\frac{(4)(0.0025) + (5)(0.0064)}{9}} = 0.068$$

$$t_{\text{calc}} = \frac{|0.95 - 1.10|}{0.068} \cdot \sqrt{\frac{5 \cdot 6}{5 + 6}} = 6.02$$

The 95% confidence limit for $5 + 6 - 2 = 9$ degrees of freedom is $t = 2.26$ (from Table 2.2). So t_{calc} far exceeds the required value for 95% confidence. An increase in the chromium level *does* exist at the 95% confidence level.

Comparing an Experimental Mean with a True Value

Two specific situations call for using samples of a single material with a known amount of one or more components. One is when validating an analytical methodology and the other is for comparing the quality of analyses among different laboratories and among different workers. For these samples from the same batch of material, a "true value" is known. This procedure was illustrated at the beginning of this chapter where the Analytical Chemistry in Action described a comparison of results for aflatoxin at the ppb level in cacao beans.

We can illustrate the comparison of an experimental mean with a true value using Figure 2.7. Here, the vertical line represents the "known" mean, and the Gaussian curve illustrates the distribution of values of the mean. Note, s_m is *not* the standard deviation, σ or s.

The difference between the experimentally measured mean, $\overline{X}$, and the true value, is called the mean error. Recall that the mean error was defined in Figure 2.2 and Section 2.6 as $|X - X_{\text{true}}|$.

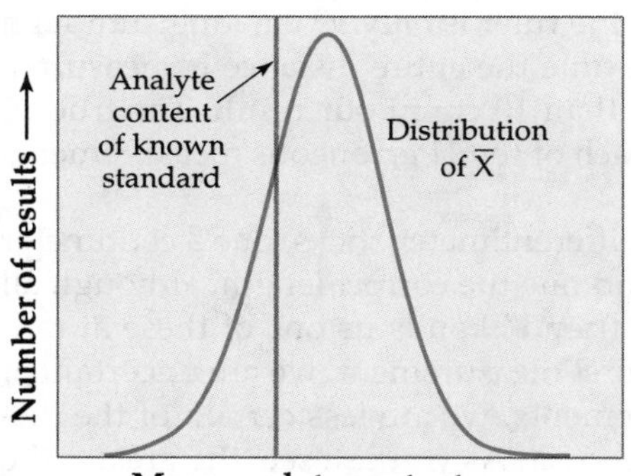

FIGURE 2.7
The distribution of the mean value calculated from $\overline{X}$ and s_m. The distribution in $\overline{X}$ results from uncertainty in all the results contributing to the analysis from interlaboratory and/or intralaboratory work. The separation of the known value and the peak of the $\overline{X}$-distribution is the mean error, and indicates the accuracy of the analysis.

In comparing the mean to the actual value, we want to know how many s_m's away from the true value our mean value lies. That is, when μ is known,

$$t_{calc} = \frac{E_m}{s_m} \qquad \text{known } \mu$$

This number will correspond to a Student t value and, therefore, to your level of confidence that the true value and the experimental mean are statistically the same.

EXAMPLE 2.7

Suppose for the nickel-containing alloy of Example 2.5 (used to illustrate the concept of a pooled mean) the true value was 4.44% Ni. Is the difference between the pooled mean for the two labs and the true value significant at the 95% confidence level? In Example 2.5, $s_{pooled} = 0.059\%$.

Solution

$$\text{Mean error} = 4.44 - 4.42 = 0.02\% \text{ Ni}$$

$$s_m = \frac{s_{pooled}}{\sqrt{N_{Tot}}} = \frac{0.059}{\sqrt{14}} = 0.016\%$$

$$t_{calc} = \frac{0.02}{0.016} = 1.25$$

The t value for 95% and $6 + 8 - 2 = 12$ degrees of freedom is 2.18. Since the value for t_{calc} is lower, this means that the true value is *within* the 95% confidence limits for our pooled mean. Another way to state the conclusion is the difference is not significant at the 95% confidence level.

2.12 Propagation of Uncertainty

So far in this chapter, you have read about the errors associated with the final results of a complete analysis, from sampling to numerical result. However, each analytical methodology consists of numerous steps. Each step has some random error associated with it. How do the random errors in each step contribute to the error in the final result? Can we calculate these contributions, and, if so, is this ability important? Following are answers to these questions.

Let us first relate a general picture of how a number of separate errors can affect a result in a particularly straightforward case. Consider that we wish to measure a 10-meter distance with a meter stick. We will mark the end position and slide the meter stick along to that end point and so forth the necessary number of times. However, say the ruler is only 99 cm long, although the scale appears to be normal. By the time the entire distance is measured, there will be an error of slightly more than 10 cm in our result. The true 10 meters appears to be 10.10 "meters." Each of the 11 erroneous measurements contributed to the larger total error.

Now, assume that there are three different meter sticks, one a centimeter too long, one a centimeter too short, and one the correct length, although all three have 100-cm scales on them. A "friend" hands us one of these at random for each measurement. After the first measurement, we are uncertain of the length by 1 cm. After two measurements, we are less certain of the distance. It may be 2 cm off, 1 cm off, or correct. As you can see, the greater the number of measurement steps we take, the less certain we become of the true measurement. Each of the errors contributes its part to a final, larger possible uncertainty. Since each of the errors is carried along to the final measurement, the process is called **propagation of error**. The mathematical treatment of this propagation might be more clearly and correctly named *propagation of uncertainty*.

Propagation of error is exceedingly important in deciding which of the possible methods should be used in an analysis. As the following sections will show, the final error can depend on certain steps far more than others in the analysis. As a result, making a single step more precise can be highly effective in creating a more precise analytical methodology. Such thinking plays a large part in planning the time and predicting the cost of an analysis, as well as contributing to planning experiments in all quantitative sciences.

Propagation of Absolute and Relative Random Error

Early in this chapter, two different measures of the error were defined: the absolute error and the relative error. The two are related by

$$\text{relative error} = \frac{\text{absolute error}}{\text{mean value}} \tag{2-15}$$

The absolute error could be converted to a relative error by dividing by the mean (or the true value if it is known), as in Equation 2-15. Conversely, the relative error can be converted to an absolute error measure by multiplying by either the mean or the true value.

Propagation-of-uncertainty calculations are straightforward even though the algebra may sometimes become messy. There are three simple rules to follow:

1. If the mathematical operation is addition or subtraction, the absolute error is used in calculations.
2. If the mathematical operation is multiplication or division, the relative error is used to make the calculations. The absolute error is then calculated by multiplication of the relative error by the value of the result.
3. All the terms of a sum must be of only one type: either absolute or relative error values. Absolute error values cannot be added to relative error values or vice versa.

The equations for determining the random error of a sum or difference of values are shown in Table 2.4. Note that if the error measure is a standard deviation, the result of the calculations will be a standard deviation.

EXAMPLE 2.8

As illustrated in Figure 2.8, a closed bottle containing a sample to be analyzed was weighed. The weight was 15.6784 g. A sample was poured from the bottle into a predried, weighed crucible, and the bottle with the remainder was reweighed. The second time, the weight was 15.5237 g. This technique is called **weighing by difference**.

Assume the weighings were done on the same balance, and the same person did both weighings. Further, assume the technique was the same as was done to obtain the data shown in Table 2.1. What is the error limit (standard deviation) that should be quoted along with the weight of the sample?

Solution

The weight of the sample is (bottle with material) − (bottle without material), or

$$\text{sample weight} = 15.6784 - 15.5237 = 0.1547 \text{ g}.$$

The random error associated with each weight (0.0004 g) is found in Table 2.1, since the conditions were the same. We could write

$$(15.6784 \pm 0.0004) - (15.5237 \pm 0.0004) = 0.1547 \pm \text{s.d.}$$

Since the result is a difference of two uncertain numbers, we use the absolute errors of both with the appropriate statistical formula from Table 2.4.

$$\text{s.d.} = (0.0004^2 + 0.0004^2)^{1/2} = 0.0005_6$$

The weight should be reported as

$$\text{weight} \pm \text{s.d.} = (0.1547 \pm 0.0006) \text{ g}$$

Another, shorter way to write this is

$$0.1547(6)$$

Table 2.4 Examples of Propagation-of-Uncertainty Calculations*

Sum and Differences (absolute error applies)	Products and Quotients (relative error applies)
For the equation:	
$R = A + B - C$	$R = \frac{AB}{C}$
Random errors:	
$s_R = \sqrt{s_A^2 + s_B^2 + s_C^2}$	$\frac{s_R}{R} = \sqrt{(s_A/A)^2 + (s_B/B)^2 + (s_C/C)^2}$
Determinate errors:	
Absolute	Relative
$r = a + b - c$	$r/R = a/A + b/B - c/C$

*R is the result to be calculated. Its absolute error is r. Each contributing factor A, B, and C has a determinate error a, b, and c, respectively. The s_A, s_B, and s_C are the standard deviations calculated from a finite number of trials, as in the example of Table 2.1.

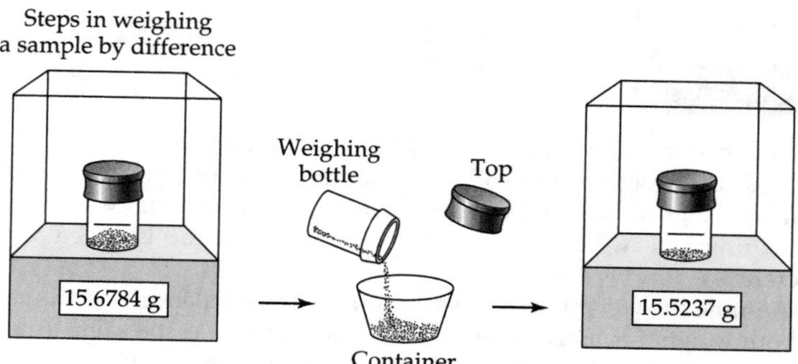

FIGURE 2.8 ▶ **Steps in weighing a sample by difference.**

Let us compare the statistically derived standard deviation, 0.0006 g in the above example, with the possible extreme values of the error. The extreme values occur when both errors are positive or both negative. For instance, if we simply added the two standard deviations—each being 0.0004 g or −0.0004 g—the weight and error would be

$$(0.1547 \pm 0.0008)\ \text{g}$$

On the other hand, the other extreme would be when the standard deviations were opposite in sign, one being positive and the other negative. The sum would be zero. In that instance, the answer might be written

$$(0.1547 + 0.0000)\ \text{g}$$

However, the statistical calculation indicates that it is unlikely that both weighings will be in error by the extremes (here measured by the standard deviations). The weight of the sample weighed by difference is unlikely to have an error of zero, and it is also unlikely to have an error of 0.0008 g. Statistically, the standard deviation is between the two possible extremes, 0.0006 g.

This is a general result: The propagated error is always less than the error found by assuming the worst case for each factor. It is not likely that all the random errors will have the extreme value of the random error *and* will have them all in the same sense. Both positive and negative deviations with different magnitudes are far more likely.

EXAMPLE 2.9

The 0.1547-g sample discussed in Example 2.8 is heated to drive off some water from crystallization. The sample is heated in a crucible, and, after heating, it is cooled and weighed. This procedure is repeated until, finally, two successive weighings are the same. This **constant weight** sample weighed 0.1234 g. What is the fraction of water in the sample and the error (standard deviation) in the measurement?

Solution

The fraction of water in the sample is simply

$$\frac{\text{weight of water}}{\text{weight of sample}} = \frac{0.1547 - 0.1234}{0.1547} = 0.202 \pm \text{s.d.} \qquad \textbf{(2-16)}$$

In the algebra of this calculation is the combination of a difference and a quotient. The three rules dictate that absolute errors are used for differences, relative errors are used for quotients, and the two types of error cannot be "mixed." So somewhere in the calculation, either the relative or absolute error will have to be transformed to the other type.

How can you approach this calculation? You could separate the quotient into two quotients (0.1547/0.1547) and (−0.1234/0.1547), calculate the relative errors for each, and then calculate the error of the difference. This will yield the correct answer. However, this path is slightly more complicated than if you calculate the error of the whole numerator first and then of the quotient. Such a calculation is done next.

The error of the numerator is calculated in the same manner as done in Example 2.8. However, as yet no error has been associated with 0.1547 and 0.1234. We must determine the values by taking into account the experimental details. Finding the weight (0.1547 g) of the original sample by difference required two weighings: the original sample in its container and then the weight of the remaining material. As a result, the associated error limit is $\pm 0.0005_6$ g as in Example 2.8.

In addition, it was implied in the description of the experiment that to find the final value (0.1234 g), the crucible was weighed alone and then with the dried sample. (Weighing of the crucible during the drying process has no effect on the results here since these intermediate results are used only to ascertain the progress of drying.) Therefore, again two weighings were required to determine the value 0.1234 g. As a result, the associated error limit is the same: $\pm 0.0005_6$ g. The numerator in Equation 2-16 is

$$0.0313 \pm \sqrt{2 \times (0.0005_6)^2} = 0.0313 \pm 0.0008$$

Equation 2-16 with errors then can be written

$$\text{fraction water} = \frac{0.0313 \pm 0.0008}{0.1547 \pm 0.0006}$$

For this simple quotient, manipulation of relative errors provides the overall error,

$$\sqrt{\left(\frac{0.0008}{0.0313}\right)^2 + \left(\frac{0.0006}{0.1547}\right)^2} = 0.026$$

The relative error of the quotient (0.0313/0.1547) is 2.6%. To obtain the absolute error, multiply the relative error by the value of the quotient.

$$\text{s.d.} = 0.026 \times 0.2023 = 0.005$$

The result (value ± s.d.) is then

$$\text{fraction that is water} = 0.202 \pm 0.005$$

Propagation of Determinate Error

Determinate errors also propagate through the steps of an analysis. The measurement with a meter stick *known* to be 1 cm short is an example. Accounting for the effects of determinate errors is relatively straightforward mathematically. Expressions for calculating determinate errors in two representative equations are listed in Table 2.4. In Section 2.9, you have already dealt with the essentials of problems with determinate errors through the discussion of a gravimetric assay for aluminum. The usual goal in analysis is not to make estimates of the propagation of determinate errors but to eliminate the errors altogether.

Logarithms and Exponentials

There are no simple rules for the propagation of uncertainty in logarithmic and exponential forms that appear in formulas of chemical analysis. To treat these, it is necessary to use the calculus form of the error propagation, the subject of Section 2A.

2.13 Significant Figures

Note in Example 2.9, although there are four digits in each of the numbers in the equation, only three digits were written in the final answer, 0.202. This is correctly written since the numerical result that is reported should communicate information about its error. The generally accepted convention is that the error of a measured value is indicated by error in the last digit *only*.

For instance, assume that an experimental result has a relative precision of 0.1%. We *could* write in a report, "The material contains 23.4522687% of ingredient Z." However, given the relative error of the value, 0.1%, only the first few digits of 23.4522687 are **significant**—the rest are insignificant. Also, to a certain extent the retention of all the digits is misleading. Since the relative error is 0.1%, the value should be meaningful to $0.001 \times 23.4 = \pm 0.02$, and the result should be written 23.46%, with or without a standard deviation listed.

There are other, simple conventions that are often used when there is an error greater than ±1 in the last digit. One convention indicates that the digit is less certain than ±1 by writing it as a subscript, for example, 1.98_7. This notation relates that the error is smaller than ±0.01 but greater than ±0.001. You have seen a number of examples of this earlier in this chapter.

Recall also a second convention shown earlier: Often, instead of writing 5.785 ± 0.003 (value ± s.d.), you could write 5.785(3). The digit enclosed in parentheses is the error in the last digit of the value.

Rules of Rounding Off

The rules for **rounding off** (or, simply, **rounding**) to the correct number of significant digits follow from the rules of error analysis and calculations of error propagation. The general rules for rounding are:

1. Retain no digits beyond the first uncertain one.
2. If the number in the place beyond the last significant one is less than 5, leave the figure as it is.
3. It the next digit is 6 or greater, add one to the last retained digit.
4. If the next digit is a 5, then round to the nearest even digit: 2, 4, 6, 8, or 0. This will cause a statistical 50:50 chance that you will round up or down. Thus no bias will be introduced by the rounding.

A slavish following of the rounding-off rules should be unnecessary once you are familiar with error calculations.

Significant Digits and Arithmetic Operations

Rule 1 above states that no digits beyond the first uncertain digit should be retained. A discussion of the rules for determining the cutoff points follows.

As for propagation of error, there are three different cases in determining the significant figures in arithmetic operations: those involving addition and/or subtraction, those involving multiplication and/or division, and those involving logarithms.

Addition and Subtraction

> Rule: In addition and subtraction, retain only as many decimal places as there are in the number with the fewest digits to the right of the decimal.

EXAMPLE 2.10

Find the sum of 21.2, 3.035, and 0.12, and round the answer to the correct number of decimal places.

Solution

$$\begin{array}{r} 21.2 \\ 3.035 \\ 0.12 \\ \hline 24.355 \end{array} \text{ becomes } 24.4$$

The last two digits have no meaning, since at least one of the numbers that were summed has no significant figures in those places. Compare this with the rules for the calculation of the absolute error for sums and differences.

Multiplication and Division

Rule: The factor containing the least number of significant figures determines the number of significant figures in the product or quotient.

This means that the *relative* error of the result cannot be less than the largest relative error of any factor.

The rules of significant figures for multiplication and division often cause confusion. For instance, the next example shows a situation where the rules of significant digits suggests an answer that disagrees with propagation-of-error calculations.

EXAMPLE 2.11

How should the result of the multiplication 2.0×43 be written so as not to be misleading?

Solution

By the rules of significant digits,

$$2.0 \times 43 = 86$$

The implication of this result is that the answer is 86 ± 1. However, we cannot be certain of the result better than to a relative error of 5%, i.e., $0.1/2.0 = 0.05$. But $0.05 \times 86 = 4.3$, and the answer should be written more correctly as

$$2.0 \times 43 = 86 \pm 4$$

Thus the rules for significant digits provide an answer that is misleading. It is technically more correct to write

$$86 \pm 4 \quad \text{or even} \quad 9 \times 10^1$$

than it is to write 86.

If in doubt, use the propagation-of-error calculations described in Section 2.12. The rules for significant digits were derived from them. Following are a few more examples of **truncation** (chopping off) and rounding of numerical results. Note that approximations of the uncertainties are used. In both examples, the relative error of the result is approximately the relative error of the *least* certain term.

EXAMPLE 2.12

Based on the implied uncertainty in the numbers 31.1, 0.063, and 98.9, find their product and round the answer to approximately the same relative error as that in the least certain factor.

Solution

$$\begin{array}{lccccc} & 31.1 \times & 0.063 \times & 98.9 & = 193.77477 \text{ rounded to } 1.9_0 \times 10^2 \\ \textit{Approximate} & \uparrow & \uparrow & \uparrow & \uparrow \\ \textit{uncertainty:} & 1 \text{ in } 300 & 1 \text{ in } 60 & 1 \text{ in } 100 & \text{round to } 1 \text{ in } 50 \end{array}$$

EXAMPLE 2.13

Calculate

$$\frac{961 \times 547}{0.053}$$

and round the answer to the correct number of significant digits.

Solution

1 in 1000 ↓ 2 in 1000 ↓

$$\frac{961 \times 547}{0.053} = 9{,}918{,}245.2 \quad \text{rounds to } 9.9 \times 10^6$$

↑ 1 part in 50 (2%) ↑ 1 in 100 (better than rounding to 1 in 10)

Choosing the last significant digit after multiplication and division becomes easy to do in your head with practice. The approximations (such as 1 in 50 ≅ 1 in 60) work because large changes in uncertainty must occur before an additional digit is dropped. A full **order of magnitude** (meaning a factor of 10) in the uncertainty must occur (for instance, 0.2% versus 2%) before you drop the next digit in the answer.

EXAMPLE 2.14

Calculate the product

$$15 \times 346 \times 165$$

and round to the correct number of significant digits.

Solution

$$15 \times 346 \times 165 = 856{,}350.0$$

The largest error is 1 part in 15 (or ±7%), thus the error in the answer must be ±7%. The answer becomes

$$8.6(6) \times 10^5 \quad \text{or} \quad 8.6 \times 10^5$$

The error would have to be around 12% (1 in 8, where 8 is the first digit of the product) before another digit would be dropped. On the other hand, in order to retain another digit, the relative error would have to be about l part in 856, or 0.1%.

Logarithms

Logarithms appear in numerous useful formulae. The rule for the number of significant digits is simple. When converting numbers to logarithms, use as many decimal places in the **mantissa**—the part of the logarithm following the decimal point—as there are significant digits in the number.

EXAMPLE 2.15

What is the value of log 12.35 with the correct number of decimal places?

Solution

There are four significant digits in the number and four decimal places in the mantissa of the log. Therefore,

$$\log 12.35 = 1.0917$$

EXAMPLE 2.16

Find the value of $[H^+]$ for a solution of pH 3.56 and report it with the correct number of significant digits.

Solution

The number of decimal places in the mantissa of the pH equals the number of significant digits in the concentration.

$$\text{pH} = 3.56$$

$$[H^+] = 10^{-3.56} = 2.8 \times 10^{-4}$$

2.14 Discordant Data

In the Analytical Chemistry in Action about aflatoxins, you saw a plot of data from analyses done at ten different laboratories. An analyst in each laboratory determined the content of aflatoxin in identical sample material. One of the resulting values was grossly different. The directors of the interlaboratory study rejected that single value from their sample.

The same kind of problem also arises in replicate determinations on a single sample done in a single location. Of a number of replicates, one result may seem to be quite far from the mean value and also far from all the other results. This is called a **discordant data point** or an **outlier**. Should it be kept as a part of the analysis or should it be ignored? There are two different approaches to answering this question: statistical and judgmental. Usually both skills are required.

The data in Table 2.5 illustrate this dual approach. The values in the table came from two sets of three determinations obtained in the order shown on the first line. The person doing the analyses acquired the second set since the value 7.19 *seemed* discordant. This judgment could have been based on extensive past experience with this analysis. For example, the value may be outside the range within which the random error usually lay. (Or it may have

Table 2.5 Results of Two Sets of Replicate Determinations

Results in order of determination:
7.06, 7.04, 7.19, 7.10, 7.02, 7.09
Results in order of value:
7.02, 7.04, 7.06, 7.09, 7.10, 7.19

Table 2.6 Critical Value for Q for Rejection of Outliers with 1-in-10 Probability of a Wrong Decision*

Number of Experiments (N)	Q Critical Value
3	0.94
4	0.76
5	0.64
6	0.56
7	0.51

*For a more complete table see the references cited here or the Youden and Steiner reference (their Table C.1) cited at the end of the chapter. With N larger, the criteria for Q and w change.
Sources: Dixon, W. J. 1953. *Biometrics* 9:74. Dean, R. B., Dixon, W. J. 1951. *Anal. Chem.* 23:636.

been due to those blue specks he remembered seeing in that sample. Strange . . . they were the same color as that paint on the ceiling!)

However, we ourselves do not have as much experience with the analysis. (In fact, we have not explained anything about it at all, except some numerical results.) So let us look at a statistical test to help us determine whether to reject the outlier. It is called **Dixon's test** or the **Q-test**. The following approach can be used to determine only if the extreme value is *statistically* subject to rejection. This is *not* a method to use mechanically but an aid in deciding whether to ignore the result of the specific run.

To see more clearly the distribution of the values obtained from the six replicates, they are rewritten in order of their values. Let d equal the difference between the possible outlier value and the result closest to it in value. Recall the definition of the range w from Equation 2-3, the difference between the highest and lowest measured values.

For three to seven replicates, Dixon's test involves comparing the value of Q,

$$Q = d/w \tag{2-16}$$

with an appropriate "critical value" of Q, which is listed in Table 2.6. The value in the table depends on the total number of measurements made, N. Then the conclusion can be made as follows: If the calculated value of Q exceeds the applicable critical value, then there is a 10% chance that rejection is a wrong decision. Restated from the opposite point of view, if Q exceeds the applicable critical value, you are 90% confident that it should be rejected.

EXAMPLE 2.17

Using the Dixon text on the results tabulated in Table 2.5, determine whether the value 7.19 should be rejected as an outlier.

Solution

For the results shown in Table 2.5,

$$d = 7.19 - 7.10 = 0.09$$

and

$$w = 7.19 - 7.02 = 0.17$$

so

$$Q = 0.09/0.17 = 0.53$$

According to Table 2.6, the critical value for six samples is 0.56. Thus, the Dixon test does not justify rejecting the result.

An analyst familiar with the assay may still wish to reject the 7.19-value. If the value is thrown out, then a new mean and standard deviation must be calculated for the five remaining results. The result will then be

$$5 \text{ points:} \quad \overline{X} = 7.06_2 \quad s = 0.03_3$$

This compares with the results from all the values

$$6 \text{ points:} \quad \overline{X} = 7.08_3 \quad s = 0.06_0$$

Notice that the relative change in the mean, about 0.3%, is much less than the relative change in the standard deviation, which is almost a factor of two. This is typical for the effects of outliers (or almost-outliers) on the results of replicate determinations. Note: If all values except one are identical, it is never permissible to discard the discordant value because the test is invalid in this situation.

There are more sophisticated methods to decide whether certain measurements are, indeed, discordant with the rest. They are used in carefully constructed interlaboratory comparisons and are beyond the level to which we wish to go in this text. References to these methods are included in the suggested readings of this chapter including those of Horowitz, of Kelly, and of Grubbs and Beck.

2.15 The Median

The **median** value of a set of numbers is the middle value. There are as many values above it as below. If the number of values is even, then the median value is the mean of the two middle values.

EXAMPLE 2.18

What is the median value of those in Table 2.5?

Solution

$$\frac{7.06 + 7.09}{2} = 7.07_5$$

The median is a less reliable estimate of the correct measured value than is the mean, but, on the other hand, the median is less sensitive to large variations in experimental values. Thus, when there is wide scatter in the data, the median may be better than the mean to estimate the experimental value. However, it is always preferable to run more samples and, even better, to isolate the cause of the scatter and correct the methodology.

2.16 Least Squares

For an analysis, a relationship is established between the amount of analyte present and the experimental measurement. Ideally, the analyte content and the result should exhibit a perfectly linear relationship, such as assumed in the graphs of Figure 2.9.

You have probably seen straight lines defined algebraically as

$$Y_i = mX_i + b \tag{2-17}$$

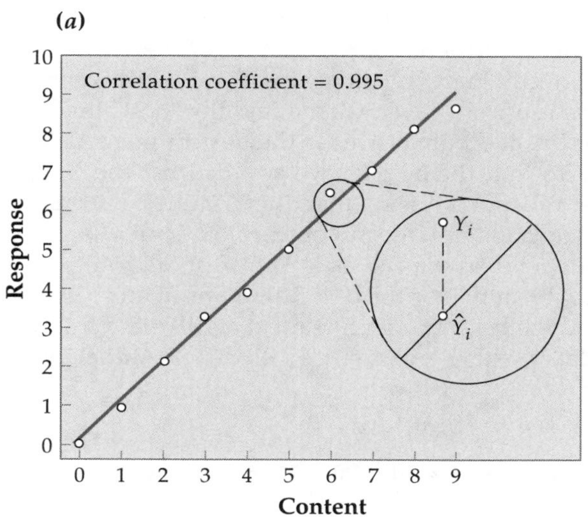

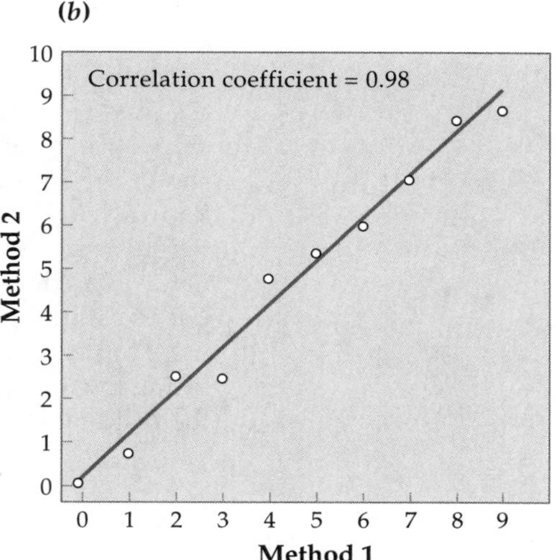

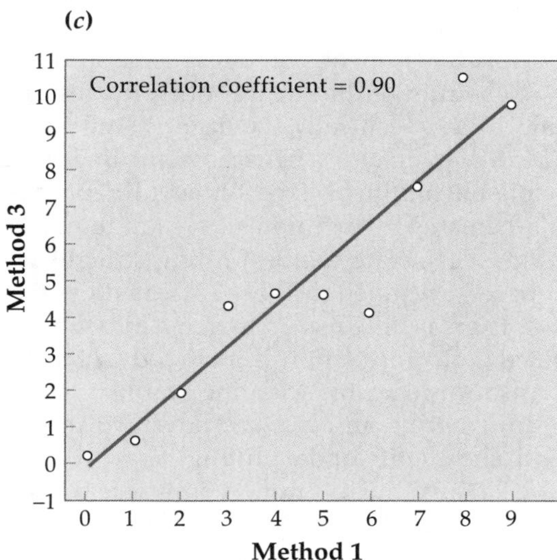

FIGURE 2.9 ▲
Example of linear regression.

(a) shows calibration data fitted with a least squares straight line. The correlation coefficient is 0.995. The magnifier shows the relationship between Y_i and $\hat{Y}_i$. They are at the same X_i-value. **(b)** shows a validation between two methods, but with a correlation coefficient of 0.98. **(c)** shows another validation with a much worse method that, for the data here, appears to be invalid in the middle of the range. Nevertheless, the correlation coefficient is 0.90.

where

Y_i is the analytical result,
X_i is the analyte content corresponding to that Y_i
m is the slope of a straight line (this is the sensitivity), and
b is a constant, which equals the value of Y_i when $X_i = 0$. The value of the blank b is also the **bias of the method**.

(A common, alternate form for Equation 2-17 is $Y_i = a_1 \cdot X_i + a_0$, where the subscript of each a corresponds to the power to which its X is raised. However, we shall use the form written as Equation 2-17.)

In reality, such perfectly linear relationships do not exist when real samples and standards are used. However, we *can* find the "best" linear relationship among the experimental points. This is the **best fit** line for the data, as shown in Figure 2.9. To find the best fit, we assume that the X_i values are known with far better relative precision than the Y_i values. Further, the individual Y_i measurements must not depend on each other. In other words, the value of Y_2 does not depend on the values of Y_1, Y_3, or of any other Y_i value. Let us call the values of m and b for the best fitting line $\overline{m}$ and $\overline{b}$.

The criteria to find the best straight line must involve the distance of the fitting line from the data points—a difference called the **residual**. For each X_i, the residual is

$$[Y_i - Y_{\text{line}}]$$

In the language of statistics, the residual for the point at X_i is written

$$[Y_i - \hat{Y}_i] \qquad \textbf{(2-18)}$$

where $\hat{Y}$ is the *predicted value* found from the equation for a straight line defined by some values of m and b (not necessarily $\overline{m}$ and $\overline{b}$). As you can see, the residual depends on the specific straight line chosen to fit the data.

The sum of the squares of all the residuals is used to determine the best line. More specifically, the *smallest* **sum of squares** characterizes the best line. (That, of course, is where the name linear least squares comes from. Another name for finding this best line is **linear regression**.) We express these ideas mathematically by saying we want to minimize $\Sigma_i [Y_i - \hat{Y}_i]^2$. The equations used to calculate the best fitting straight line (and minimum of the sum of squares) are given in Table 2.7, equations 2 and 3.

There is little reason to delve into the details of least-squares calculations since the computation is included in most scientific calculators, all general statistics programs, scientific graphing software, and most spreadsheet programs. Let us, instead, see how two important statistical parameters vary with the quality of data fitting.

The two most common parameters to assess the quality of the fit are the *residual sum of squares*, here signified by R, and the **correlation coefficient**, signified in the table by $\rho(X,Y)$.

To compare the quality of fit, a relative relationship must be used, and the correlation coefficient is such a measure. The correlation coefficient indicates how closely a change in the independent variable, X, is reflected in the dependent variable, Y. When $\rho(X,Y)$ is exactly one, it indicates a perfect correlation. The three graphs in Figure 2.9 illustrate how the correlation coefficient varies with a change in R. It is interesting to note how quickly the fit degrades as we move away from a ρ of 1.00.

Table 2.7 Equations of Linear Least Squares

$$\sum \equiv \sum_{i=1}^{N} \text{ in all equations}$$

Alternative ways to write some terms are

$$\text{Average value of } X_i = \overline{X} = \sum X_i/N$$

	Differences Format	Summation Format
1)	$R = \Sigma[Y_i(X_i) - \hat{Y}(X_i)]^2$ $= \Sigma[Y_i(X_i) - (\overline{m}X_i + \overline{b})]^2$	
2)	$b = \overline{Y} - m\overline{X}$	$b = \dfrac{\Sigma Y_i}{N} - m\dfrac{\Sigma X_i}{N}$
3)	$m = \dfrac{\Sigma(X_i - \overline{X})(Y_i - \overline{Y})}{\Sigma(X_i - \overline{X})^2}$	$m = \dfrac{\Sigma X_i Y_i - \dfrac{\Sigma X_i \Sigma Y_i}{N}}{\Sigma X_i^2 - \dfrac{(\Sigma X_i)^2}{N}}$
4)	$s_Y = \sqrt{\dfrac{\Sigma[Y_i - (\overline{m}X_i + \overline{b})]^2}{N - 2}}$	$s_Y = \sqrt{\Sigma Y_i^2 - \dfrac{(\Sigma Y_i)^2}{N} - \dfrac{\left(\Sigma X_i Y_i - \dfrac{\Sigma X_i \Sigma Y_i}{N}\right)^2}{\Sigma X_i^2 - \dfrac{(\Sigma X_i)^2}{N}}}$
5)	$s_{\text{slope}} = \sqrt{\dfrac{{s_Y}^2}{\Sigma(\overline{X} - X_i)^2}}$	$s_{\text{slope}} = \sqrt{\dfrac{{s_Y}^2}{\Sigma X_i^2 - \dfrac{(\Sigma X_i)^2}{N}}}$
6)	$s_{\text{intercept}} = s_Y \cdot \sqrt{\dfrac{\Sigma X_i^2}{N\Sigma(X_i - \overline{X})^2}}$	$s_{\text{intercept}} = s_Y \cdot \sqrt{\dfrac{\Sigma X_i^2}{N\Sigma X_i^2 - (\Sigma X_i)^2}}$
7)	$\rho(X,Y) = \dfrac{\Sigma(X_i - \overline{X})(Y_i - \overline{Y})}{\sqrt{\Sigma(X_i - \overline{X})^2 \Sigma(Y_i - \overline{Y})^2}}$	$\rho(X,Y) = \dfrac{N\Sigma X_i Y_i - \Sigma X_i \Sigma Y_i}{\sqrt{[N\Sigma X_i^2 - (\Sigma X_i)^2][N\Sigma Y_i^2 - (\Sigma Y_i)^2]}}$

As mentioned above, to find the best-fit line, only the residuals of Y are used. To use the method, we must assume that the relative uncertainties in the X-values are much smaller than the relative uncertainties in the Y-values. If this approximation cannot be made, a more complicated statistical calculation is required.

2A A Deeper Look: Total Differential Error Analysis

Determinate Errors

The equations of Table 2.4 can be derived from the equation expressing the total differential of any function. Call the function $f(a_1, a_2, \ldots, a_N)$. We shall also use the abbreviation F for the function. In the language of calculus, the total differential of a general function $f(a_1, a_2, \ldots, a_N)$ is given by

$$d[f(a_1, a_2, \ldots, a_n)] = \left(\frac{\partial F}{\partial a_1}\right)da_1 + \left(\frac{\partial F}{\partial a_2}\right)da_2 + \cdots + \left(\frac{\partial F}{\partial a_N}\right)da_N \qquad \textbf{(2A-1)}$$

The quantities in parentheses are partial differentials. This means that the derivative of the function is taken with respect to one parameter while all the rest are held constant. Take as an example the function $f(A, B, C) = A^2BC$.

$$d(A^2BC) = (2ABC)dA + (A^2C)dB + (A^2B)dC \qquad \textbf{(2A-2)}$$

In Equation 2A-2, dA, dB, and dC represent the estimated errors of the measurements of the quantities A, B, and C. In this notation, the relative errors would be written

$$dA/A, \quad dB/B, \quad \text{and} \quad dC/C$$

respectively, and$f(A, B, C)$ represents the result calculated from experiments.

Equation 2A-1 holds for the situation in which the changes in the parameters are infinitesimal. In order to use a total differential error calculation that is applicable to chemical analysis, the errors in the factors a_1, a_2, and so forth must be able to be finite. For a finite change, the equation to be used is

$$\Delta[f(a_1, a_2, \ldots, a_n)] = \left(\frac{\partial F}{\partial a_1}\right)\Delta a_1 + \left(\frac{\partial F}{\partial a_2}\right)\Delta a_2 + \cdots + \left(\frac{\partial F}{\partial a_N}\right)\Delta a_N \qquad \textbf{(2A-3)}$$

The value Δf is the now the error of a result calculated from an equation that has variables $a_1, a_2, \ldots, a_N$.

Random Errors

With the same nomenclature as used in Equation 2A-3, we can write the expression for the propagated random error of the result. This is

$$\Delta[f(a_1, a_2, \ldots, a_N)]_{\text{random}} = \sqrt{\left(\frac{\partial F}{\partial a_1}\right)^2 \Delta a_1^{\,2} + \left(\frac{\partial F}{\partial a_2}\right)^2 \Delta a_2^{\,2} + \cdots + \left(\frac{\partial F}{\partial a_N}\right)^2 \Delta a_N^{\,2}}$$

Suggestions for Further Reading

MILLER, J. C., MILLER, J. N. 1993. *Statistics for Analytical Chemistry*. Chichester: Ellis Horwood.
An introductory textbook that describes more advanced topics than covered here but does not cover them in sufficient detail to enable readers to use the techniques in their work.

GRUBBS, F. E., BECK, G. 1972. "Extension of Sample Sizes and Percentage Points for Significance Tests of Outlying Observations." *Technometrics* 14:847–852.

KELLY, P. C. 1990. "Outlier Detection in Collaborative Studies." *Anal. Chem.* 73:58–64.
Discussions of the best ways to decide whether some data points are really outliers.

MANDEL, J., LINNIG, F. J. 1957. "Study of Accuracy in Chemical Analysis Using Linear Calibration Curves." *Anal. Chem.* 29: 743–749.
This article covers confidence limits for least squares fits. Descriptive at first with a mathematical second half.

CAULCUTT, R., BODDY, R. 1983. *Statistics for Analytical Chemists*. London: Chapman and Hall.
An easy to read, profusely illustrated book covering most aspects well. It includes worked solutions to chapter problems.

HAMILTON, W. C. 1964. *Statistics in Physical Science*. New York: Ronald Press.
Rigorous mathematical treatment of estimation, least squares, and testing of hypotheses.

YOUDEN, W. J. 1951. *Statistical Methods for Chemists*. New York: Wiley.
A readable monograph with both mathematical derivations and examples and suggestions for experimental design.

GREEN, J. R., MARGERISON, D. 1968. *Statistical Treatment of Experimental Data*. Amsterdam: Elsevier.
Completely mathematical treatment of title subject.

MANDEL, J. 1978. In *Treatise on Analytical Chemistry*, Pt. 1, Vol. I, 2nd ed. I. M. Kolthoff and P. J. Elving, eds. New York: Wiley Interscience; Chap. 5.
A short, practical review of evaluating accuracy and precision in experimental results.

CURRIE, L. A. 1978. In *Treatise on Analytical Chemistry*, Pt. 1, Vol. I, 2nd ed. I. M. Kolthoff and P. J. Elving, eds. New York: Wiley Interscience; Chap. 4.
A chapter relating the testing of error to practical problems.

LITEANU, C., RICA, L. 1980. *Statistical Theory and Methodology of Trace Analysis*. Chichester: Ellis Horwood.
An exhaustive analysis of the title subject.

MORITZ, P. 1981. In *Comprehensive Analytical Chemistry*. Vol. 9. G. Svehla, ed. Amsterdam: Elsevier; Chap. 1.
Derivation of the statistical treatments used in analysis along with listings of computer programs used to calculate some of them. Examples with references for numerous different analytical methods are included.

YOUNG, H. D. 1962. *Statistical Treatment of Experimental Data*. New York: McGraw-Hill.
An introductory-level monograph on probability and statistics.

YOUDEN, W. J., STEINER, E. H. 1975. *Statistical Manual of the AOAC*. Washington, D.C.: Association of Official Analytical Chemists.
A compilation of statistical methods including interlaboratory comparison methods; complete and detailed.

Concept Review

1. What is the difference between:
(a) determinate and indeterminate errors?
(b) precision and accuracy?
(c) mean error and error of the mean?

2. An analytical protocol exhibits a 95% confidence interval of ±0.06. If a 90% confidence limit of ±0.06 is required by regulations, could the protocol still be used?

3. In order to decrease σ_m for an assay by a factor of two, by what factor must the number of trials increase (or decrease)?

4. In what case(s) might you want to report a median value instead of a mean value for a set of data?

Exercises

2.1 For a single observation, $\sigma = 0.01$ cm. How many observations would be required to report a mean with $\sigma_m = 0.001$ cm?

2.2 The results for seven determinations of the percentage of Cu in a sample are 39.3, 41.2, 40.4, 40.0, 41.1, 39.9, and 40.9 ppm Cu.
(a) Compute the mean and its standard deviation.
(b) What is the 95% confidence limit?

2.3 The overall random error associated with an analytical method depends on the random errors associated with sampling, sample preparation, and the measurement itself. The relationship is

$$\sigma^2_{\text{overall}} = \sigma^2_{\text{sampling}} + \sigma^2_{\text{preparation}} + \sigma^2_{\text{measurement}}$$

The sampling step of a protocol produces a random error of 0.2%, the sampling step 0.01%, and the instrumental measurement 0.1%. What is the overall random error?

2.4 The results of a set of determinations (in weight %) are

21.25, 21.27, 21.30, 21.23, 21.21

Find the mean, standard deviation, the relative standard deviation, and 95% C.L. of the mean.

2.5 The inscription on a 5-mL pipette says that it will deliver ("to deliver" is noted as TD on the pipette) 5.000 mL of water at 20°C. A set of experiments is made to determine the errors of the operation. The water and pipette are held at 20°, and the values of the volume were found by weighing the water delivered. The weights could be measured to 0.0002 g, so any error due to weighing can be ignored. Replicate measurements for a 5.000-mL TD pipette (in mL) gave the following results.

4.985, 4.981, 4.989, 4.970, 4.974, 4.981, 4.976, 4.988, 4.993, 4.973, 4.970, 4.985, 4.988, 4.982, 4.977, 4.982, 4.974, 4.988, 4.979, 4.985

Calculate the mean volume, the standard deviation, the 95% confidence limit, and the range of the measurements.

2.6 The Nernst equation describes a relationship between a voltage and a chemical concentration expressed as an activity, a_i.

$$E = E^\circ + \left(\frac{RT}{nF}\right)\ln a_i$$

For $n = 1$, what is the relative error in a_i for a 1-mV change in E at 25°C?

2.7 A standard alloy sample from the National Institute of Standards and Technology contains 57.85 wt% of chromium. Two 0.1000-g samples were prepared by acid dissolution followed by dilution to 500.0 mL. Each sample was assayed eleven times. The following results were obtained.

Sample No. 1: 57.64, 58.07, 57.88, 57.79, 57.67, 57.79, 57.67, 57.73, 57.59, 57.81, 57.76

Sample No. 2: 57.88, 58.00, 57.61, 57.80, 57.56, 57.61, 57.51, 57.77, 57.55, 57.40, 57.67

(a) Calculate the mean and relative s.d. for each sample.
(b) For Sample 2, calculate the 95% C.L., find the relative measure, and compare it to the relative standard deviation.

2.8 A method for the determination of chloride tested on pure NaCl gave the following results.

NaCl in the Sample (g)		Error in Analysis	
Used	Found	(in gm)	(in %)
none	0.0022		
0.1000	0.1023		
0.3927	0.3960		
0.8295	0.8311		
1.2976	1.3005		

(a) Complete the table.
(b) What type of correction should be applied, and what is the value of it?
(c) After the correction is made, what is the nature of the residual error, and what is its *average* value?
(d) How large should the samples be to keep this residual error less than 0.1% of the true value after making the correction made in part (b)?

2.9 Several beers and wines were tested for ethanol content by two methods. One used an instrument that utilizes an enzyme-electrochemical method, which involves injecting a 25-μL sample into the sample chamber and reading the (precalibrated) ethanol content on a digital display within a minute or so. The other involves distilling a large volume of sample and measuring the density of the resulting distillate. The following results %(w/w) were found. [Ref: Application note 110. Yellow Springs, OH: YSI, Inc.]

Sample	By Electrochemical Analyzer	By Distillation and Density
Beer A	3.80	3.82
Beer B	4.32	4.35
Beer C	3.48	3.47
Wine A	10.60	10.72
Wine B	5.90	5.83
Wine C	8.49	8.58

(a) For each of the six measurements, find the relative deviation between the two methods. (Does it matter to the results which set is considered to be "true?")
(b) Calculate the average relative deviation between the two methods. Is there a significant (greater than ± 0.005) *relative* bias in the method?

2.10 You want to analyze a sample for aluminum. The relative error inherent in the analyses for all components is 0.1%. The impurities found are Be, oxygen, and Sb, each 0.03 wt% of the total sample. The mean aluminum content is calculated at 99.09 wt% from an aluminum assay. Is the amount of aluminum found more precisely from the aluminum determination or from the impurity determinations? [Ref: Benedetti-Pichler, A. 1936. *Anal. Chem.* 8:373.]

2.11 The following questions refer to Figure 2.11.1. The figure shows results from a flow injection analysis (see Chap. 4) in which the height of a peak from the baseline is directly proportional to the sample content. Shown are the results from measurements of a neurotransmitter at an electrode coated with a special polymer based on 18-crown-6, a cyclic polyether. [Data courtesy of Suzanne Lunsford.]

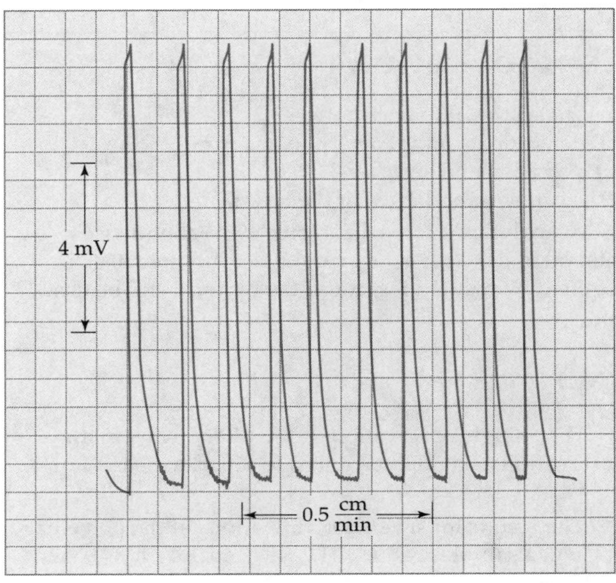

FIGURE 2.11.1 ▲

(a) What is the relative standard deviation of the method for the results from the ten samples? Assume no baseline fluctuation occurs.
(b) Estimate the relative standard deviation due to errors in reading the chart including the measurement and the baseline extrapolated under the peak. What, then, is the relative standard deviation due to the instrumental method, excluding the chart reading?

2.12 Calculate the propagated uncertainty in the calculation

$$7.07(\pm 0.03) + 6.5(\pm 0.4) = 13.57\ (\pm \quad)$$

Write the correct number of digits for the result and the standard deviation.

2.13 Calculate the propagated undertainty in the calculation

$$\frac{456.57(\pm 0.06)}{15.472(\pm 0.004)} = 29.5094\ (\pm \quad)$$

Write the correct number of digits for the result and the standard deviation.

2.14 Calculate the propagated uncertainty in the calculation

$$\frac{81.32(\pm 0.09) \cdot 0.1399(\pm 0.0002)}{-3.21(\pm 0.01)} - 22.3323(\pm 0.0001)$$
$$= -25.8764 \ (\pm \quad)$$

Write the correct number of digits for the result and the standard deviation.

2.15 In the titration done as described in Section 2.2, a solution of base was used that was assumed to be 0.1019 N exactly. The result from the assay is found using the product

$$\text{normality} \times \text{other terms} = \text{result} = 21.25 \pm 0.04$$

In fact, the correct description of the normality was (0.1019 ± 0.0003) N. Thus the standard deviation of the result must be less certain than was stated. The result, 21.25 ± 0.04, must be modified to account for the extra uncertainty.
(a) What is the propagated error in the result?
(b) What would be the propagated error if the normality were 0.1019 ± 0.00005?

2.16 Six replicates of the determination of zinc in an individual's hair gave the following results (in ppm):

2.67 2.75 2.82 3.01 2.94 2.87

(a) Determine the mean and standard deviation for the above results.
(b) After taking zinc supplements for two months, the study was repeated. The mean value was (3.03 ± 0.22) ppm (value ± s) for 5 replicate samples. Is the difference between the two sets of data significant at the 95% confidence level?

2.17 Boron nitride is used by semiconductor manufacturers for growth of semiconductor crystals. An old, time-consuming, but very accurate and precise analytical method for the determination of iron was to be replaced by a new, less labor-intensive method. The value of 0.150 ppm (w/w) obtained by the old method is accepted as the "true value." The new method gives a value of 0.146 ± 0.002 when used to analyze eight samples identical to that used with the old method.
(a) Calculate the mean error of the new method.
(b) Calculate the standard deviation of the mean (s_m) for the analysis by the new method.
(c) Is the new method verified against the old one within its 95% confidence limits?

2.18 In problems 2.12 through 2.14 the estimated errors of the results were calculated by the rules of propagation of error. Estimate the errors based on the rules for significant figures and compare them with those previously calculated.

2.19 Given the general rules for determining significant figures, how should the value of the following quantity be reported?

$$\log_{10}(1.125 \times 10^{13})$$

2.20 A small lecture class had the following lecture exam grades out of 100 points:

98 97 84 80 60 40 29

(a) What is the median grade?
(b) Calculate the mean, standard deviation, and 95% confidence limit for the grades.

■ **2.21** A set of calibration standards for Pb^{2+} yielded the following results during methods development. The analytical technique used produced a response that should have been linear with lead content.

μg/L Pb	Instrument Response
0.00	1
1.00	125
2.00	246
5.00	619
10.00	1250

(a) Graph the results with response on the y-axis.
(b) Find the best straight line through the point with your spreadsheet program and report the correlation coefficient, slope, intercept and any other statistical information given.
(c) If an unknown sample run under the same conditions produced a response of 1019, what is the concentration of the unknown?

■ **2.22** Using the data in Problem 2.21 and the equations in Table 2.7, calculate the intercept for the calibration and compare these with the values obtained using your spreadsheet graphics program. A helpful hint: This is a much less tedious process if you construct a table with values in each column for a given summation. For example:

X_i	Y_i	X_i^2	X_iY_i
. . .	. . .	. . .	. . .
ΣX_i	ΣY_i	$\Sigma(X_i^2)$	$\Sigma(X_iY_i)$

(Exercises 2.23–2.27 require calculus.)

2.23 For the function $R = A - B + C$ and the formula for the total differential of R, write the algebraic expressions for the propagated determinate error (*not* random) with errors a, b, and c.

2.24 For the function $R = AB/C$, find the algebraic expressions for the absolute and relative determinate errors if a, b, and c are the errors in each factor.

2.25 Derive the expression for the relative error and absolute error for R in terms of A.

(a) $R = \ln A$

(b) $R = e^A$

(c) $R = 2.193/e^A$

(d) $R = \log A$

2.26 What is the random absolute error of a computed result if $R = X + Y - Z$? Use the nomenclature of Table 2.4.

2.27 For random errors R,

$$r^2 = (\partial R/\partial X)^2 x^2 + (\partial R/\partial Y)^2 y^2 + \cdots$$

where x and y are the absolute standard deviations of X and Y, etc. Show that the relative error of a computed result is

$$(r/R)^2 = (x/X)^2 + (y/Y)^2 + (z/Z)^2$$

if

$$R = XY/Z$$

Note: The equation for the relative error is a general result.

Additional Exercises

2.28 Ten molasses samples were tested for "total sugar as invert" by two methods. One is a relatively fast instrumental method, the other a more tedious wet-chemical one. The following results were obtained. [Ref: Application note 100. Yellow Springs Instruments Co., Inc., Yellow Springs, OH.]

Sample	By Instrumental Analyzer	By Wet-Chemical Method
A	43.1	49.8
B	43.6	49.9
C	45.1	50.3
D	50.6	56.3
E	47.5	54.4
F	47.1	54.0
G	51.6	57.3
H	50.8	57.5
I	50.9	57.0
J	44.1	49.8

(a) Assume the "true" value is that found by the wet-chemical method. For each of the samples A–J, calculate the relative deviation between the two methods.

(b) Calculate the average relative deviation. Is there a significant bias (>0.5%) to the instrumental method?

(c) Is enough information available to decide whether a relative or absolute adjustment should be made for any necessary correction?

(d) If there is a significant bias, what is the value of the correction you can make to the instrumental results? Express the answer as

instrumental result (+ or ×) value

= true (wet-chemical) result

2.29 The following table contains data from five different research groups, each determining the content of a rock sample from the moon. The sample was collected by the Apollo 17 crew and was labeled as sample Mare Basalt 70215. [Reprinted with permission from Palme, C., Jagoutz, E. 1977. *Anal. Chem.* 49:717. Copyright 1977, American Chemical Society.]

Major Element Data (wt%)

Group	Mg	Al	Si	Ca	Ti	Fe
1	4.82	4.69	17.72	7.70	7.84	15.51
2	5.14	4.64	17.39	7.45	7.88	15.25
3	4.77	4.77	17.98	7.82	7.48	15.08
4	5.63	4.65	17.58	7.73	7.91	14.94
5	4.98	4.57	17.90	—	7.70	15.70
6	5.03	4.64	17.82	7.62	7.66	15.83

Minor Element Data (ppm)

Group	P	S	K	Cr	Mn
1	500	1880	340	2950	2044
2	390	1800	330	2870	2170
3	430	1700	415	2670	2250
4	300	—	660	2800	2090
5	—	—	340	2680	1940
6	440	1620	407	2820	2020

(a) Calculate the mean and standard deviation of the content for each of the elements.

(b) Calculate the relative standard deviation (in %) for each of the elements.

(c) Is the result for K from group 4 an outlier?

(d) Which of the sets of minor element data has the largest range? Which has the largest relative range?

2.30 A titration is done on a regular basis. The samples, all about 0.1 g, are weighed to the nearest 0.1 mg. Approximately 40 mL of titrant (liquid added from a burette) is needed to titrate each sample to the end point. The molarity of the titrant is known to ±0.1%. The titrant volume is measured by reading a burette before adding any liquid and again after all the liquid has been added. The error in reading the burette is estimated to be a random error of 0.02 mL. The titration is finished when an indicator color change is seen. This color change occurs over a 0.02-mL volume.

(a) Estimate the relative error of the result.

(b) If a 0.1000-g sample contained 22.2% of element Z, what is the absolute error in the content of Z?

2.31 For a single observation, $\sigma = 0.035$ cm. How many observations would be required to report a mean with $\sigma_m = 0.01$ cm? (Remember, it is more than a little difficult to make a fractional observation.)

2.32 The results for five determinations of the % (w/w) chromium in a sample of stainless steel were 17.3, 17.4, 18.0, 17.7, 17.9, and 17.6 ppm Cr.

(a) Compute the mean and its standard deviation.

(b) What is the 95% confidence limit for the value of the mean being the true value?

2.33 The sampling step of a protocol produces a determinate error of 0.2%. The actual analysis produces a determinate error of 0.1%. What is the overall relative determinate error?

2.34 The results of a set of determinations of mg of sodium in a 28 g serving of sliced turkey were

21.25, 21.27, 21.30, 21.23, 21.21

Find the mean, standard deviation, and 95% C.L. of the mean.

2.35 In calibrating a ten-mL pipette with distilled water at 20°C, the following masses were found for four replicates.

9.9886, 9.9843, 9.9881, 9.9973

(a) Calculate the mean volume, the standard deviation, the 95% confidence limit, and the range of the measurements.

(b) Is the second replicate an outlier? If so, recalculate the quantities in part (a).

2.36 What is the propagated error in the following calculation? The estimated error for each value is in parentheses.

$$17.6(\pm 0.2) - \frac{0.750(\pm 0.008)}{0.0411(\pm 0.002)}$$

2.37 After running a series of standards that contained Cr^{3+}, a regression line was calculated that related the molar concentration of the standard, *C*, to the instrument response, *R*.

$$R = 1.5 \times 10^{-4}\, C + 0.004$$

If three replicates give responses of 0.556, 0.552, and 0.555,

(a) Calculate the concentrations that correspond to each of the instrument responses.

(b) Calculate the mean and standard deviation for the chromium content of the unknown.

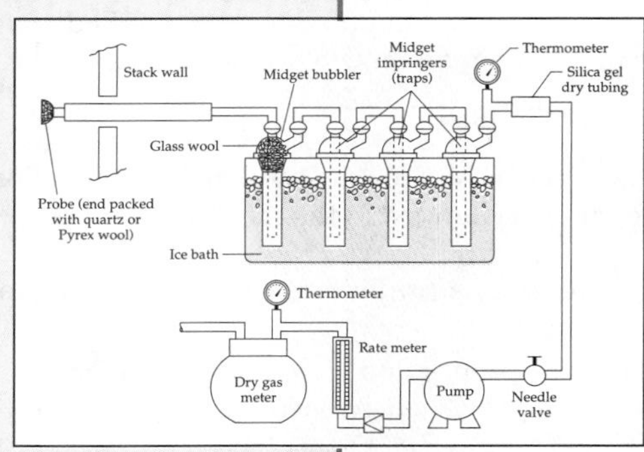

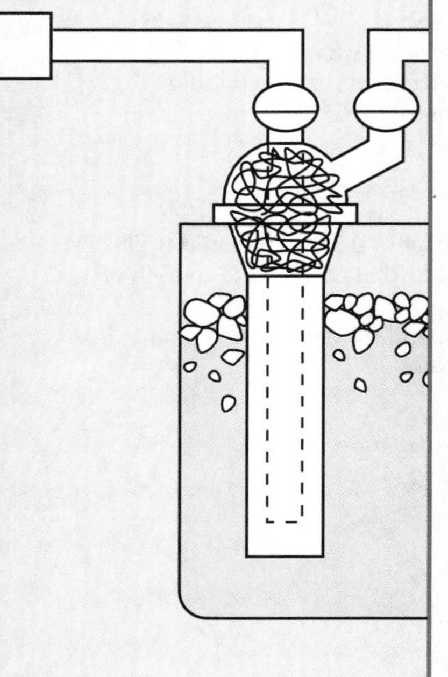

CHAPTER 3

Sampling

Overview

A Deeper Look

HW 1,3,4

3.1 Factors Involved in Effective Sampling

Suppose that a shipload of ore is delivered to a plant, and its average metal content is needed to plan the refining process. When you look at the delivered material, you see different-colored pieces in a large range of sizes. The larger pieces are striated, indicating that they are **heterogeneous**; they differ from one region to another. In addition, the larger rocks have rolled down to the bottom of the pile—at least on the outside where they can be seen. How do you sample the shipment so that you can ascertain the average metal content?

What if you needed to map a pollutant such as mercury in a system of local streams? You probably would want to know more than just how much mercury is in a sample scooped from a stream. You might need to know where and how the mercury moves. Is it dissolved in the water? Is it in or on small particles that can settle out of the liquid? Is it contained in or on the surfaces of particles so small that they remain suspended, essentially indefinitely? Is the mercury bound in a volatile organomercury compound? (Such compounds form when mercury interacts with bacteria.)

Or, consider the case of a few microscopic, smooth spheres that were found embedded in the threads of a sweater belonging to a person suspected of firing a gun in a murder. The spheres could be bullet propellant, but the suspect asserts that he is a welder, so they could have come from the sputtering of a welding rod. Where *did* they come from? How do you find out?

These are just three typical problems that involve the principles and methods of sampling of materials for analysis. If the laboratory samples do not represent the true content of the material, or fluctuate widely in content, this error will overwhelm errors from other parts of the analysis, and the determination could be meaningless.

This chapter will present some of the fundamentals of sampling methods. The details are too diverse and involved to discuss more than a few of them. As we hope you come to appreciate, the choice of sampling method often depends on the descriptive chemistry of the substances being determined and the other elements and compounds present in the sample.

Deciding how to obtain a sample for analysis depends on:

1. The *size* of the bulk to be sampled, which can vary from a shipload of ore to a subcellular organelle a fraction of a micron in diameter;
2. The *physical state* of the fraction to be analyzed, for instance, crystalline solid, glassy solid, liquid, or gas;
3. The *chemistry* of the material to be assayed. (If a specific species is sought, nothing can be done that would destroy or alter the identity or quantity of the analyte before the assay itself is run.)

Thus, the method of sampling a substance and preparing it for testing is, inevitably, linked to the assay method(s) that could be appropriate. The sampling method is a significant and integral part of a chemical analysis.

Sampling of the progress of a chemical reaction, whether it occurs in a flask in a laboratory, in a reactor in a process plant, in a river or open field, or in a biological organism, can be done either intermittently or continuously. If the sampling is continuous, the "laboratory" must somehow be connected directly to the reaction. While this idea may seem trivial at first, it is not so simple if the sample is a stack gas at 600°C or molten metal at 2000°C in a smelter. For many questions in biology, often a limited amount of sample can be taken from an organism without harming it irreparably. Also, the size of a sensor may be limited by its location (for instance, a sensor inside a needle in an arm vein). In a river or field, it can be quite complicated to decide when, where, and with what equipment to take the samples needed to answer an environmental question. These represent some of the challenges of sampling in chemical analysis.

3.2 Good Samples: Representative and Homogeneous

The two goals of sampling methods are to ensure that samples are **representative** of the material being analyzed, and that samples analyzed in the laboratory are **homogeneous**. The term *representative sample* has the same meaning in this context as in common usage: The content of the sample is the same overall as the material from which it is taken. *Homogeneous* means that the sample is the same throughout. The more representative and homogeneous the samples are, the smaller the sampling error will be. As can be appreciated from the discussion of propagation of error in Chapter 2, an analysis cannot be more precise than its least precise operation, and quite often, the sampling is the limiting factor of both the accuracy and the precision of an analysis.

Making the Sample Representative

If a sample is not representative of the material to be analyzed, the results of the analysis cannot accurately reflect the content of the material. Almost all materials to be sampled, with the exception of smaller volumes of liquids or gases, are **heterogeneous** in composition. For example, a heterogeneous solid might be composed of a mixture containing particles of various sizes and, within individual particles, varying compositions in different regions. Stone is one such heterogeneous solid; it is an agglomeration of small crystals of varying compositions. Another heterogeneous material might be a field of grass that is to be tested for natural nutrients such as phosphorus or for pollutants such as polychlorinated biphenyls (PCBs).

The chemical mix in any sample should be a miniature replica of the contents of the whole from which it came. One should take thoroughly mixed samples and reduce them to a homogeneous fraction that accurately represents the whole.

Consider again the shipload of ore. How would an analytical sample be taken when the huge load is composed of large and small heterogeneous particles and chunks? One way is to remove portions of the ore from the conveyor belt that carries it into the plant. However, when the ore with chunks of disparate sizes is shaken by vibration of the belt, the particles—from the size of dust to pieces weighing a few kilograms—usually separate into different locations. In addition, it is likely that the different-sized particles have different compositions. In fact, differences in composition may be the reason the material fell apart during transport. So, to take a representative sample from the conveyor belt, we must sample all the way through a cross section of the material. These periodic samples are sometimes called **increments**. These may be analyzed individually or collected to make a representative sample of the whole.

The next question is, how often are such samples taken? Are they obtained at regular or irregular times? Figure 3.1 demonstrates sampling

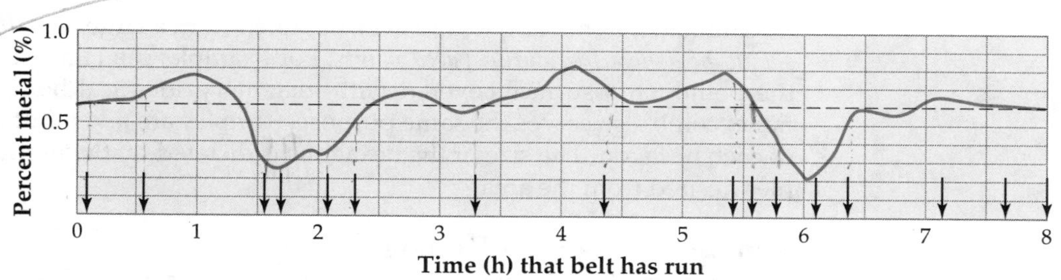

FIGURE 3.1 ▲

Plot of the true metal content in a cross section of ore versus the time at which the belt is sampled.

The content fluctuates about a mean value of 0.6%. The arrows indicate 16 random times at which the ore could be sampled; the ore thus sampled will yield an assay value significantly below the true average value. Sixteen regularly spaced sampling times half an hour apart will produce a more accurate indication of the true average content.

from a conveyor belt over the eight-hour period required to unload the entire shipload of ore. The solid line indicates a continuous measure of the metal content at one fixed point along the conveyor belt.

What might we find if the sampling were done randomly and far less often, such as at the points shown by the arrows? In all, 16 collections are made and analyzed. Just by chance, the samples collected at short intervals coincide with the points where the ore is relatively low in the metal: hours 1.5–2.5 and 5.5–6.5. The overall value—the average of the 16 measurements taken—will be significantly low. If, however, the samples were taken at regular intervals of half an hour, the average would approach that of the true average of the pile.

Notice that to obtain a representative sample in time, it seems reasonable (and can be justified by mathematical statistics) to take regular, frequent samples. But even then, it will still be possible to obtain an imperfect representation. Perfect sampling would require nearly infinite effort.

The general rule for sampling a large quantity of material on a conveyor belt (or any continuous stream) is: *The sampling should be taken at regular intervals and should be fixed in its form or method.* Recall that a whole cross section should be taken for each sample.

A similar general rule holds for a pile of material: To obtain a random sample, the sampling should be fixed in its method. One such method is **coning and quartering**, in which a pile of material is divided into quarters (if not in reality then in imagination, since it is difficult to divide a 10,000-kg pile). Samples are taken from each quarter of the pile (north, south, east, and west). These samples are crushed and formed into a smaller conical pile. This smaller pile is flattened and cut into equal quarters, and two opposite quarters are chosen at random, such as by flipping a coin. These quarters are crushed further, mixed thoroughly, and repiled. This fixed procedure (cutting, randomly choosing opposite quarters, mixing and grinding, and repiling) is continued until a sample of the size needed for replicate samples in the laboratory is obtained. This technique minimizes bias in the sampling.

If the original pile is so large that sampling cannot be done on the total—since the core of the pile is not really available—you might expect that our regular sampling from the conveyor belt, where the whole amount is accessible, produces a more representative sample than the preceding method. This has, in fact, been shown to be true.

Similarly, if the samples are to be obtained from a large area, *the sampling should be done at regularly spaced points*. For example, see Figure 3.2. Spacing the points on a surface covered with imaginary, identical hexagons like a honeycomb also serves the same purpose. Samples are taken from the center of each hexagon. The size of the hexagons is dictated by the number of samples desired from the area.

Making the Sample Homogeneous

One of the treatments for a solid sample is to crush it, pulverize it, grind it, or otherwise render it into a thoroughly mixed powder. The particular methods used to reduce particle sizes are too diverse to consider here due to the diversity of materials that are analyzed. On any given day, muscle tissue, rock salt, leather, wood, feathers, fish scales, jet-turbine blades, limestone, seaweed, granite, tomatoes, hair, calcium supplements, bean pods, iron ore, rubber, cardboard, butter, sheet steel, and many other materials are being anlyzed somewhere.

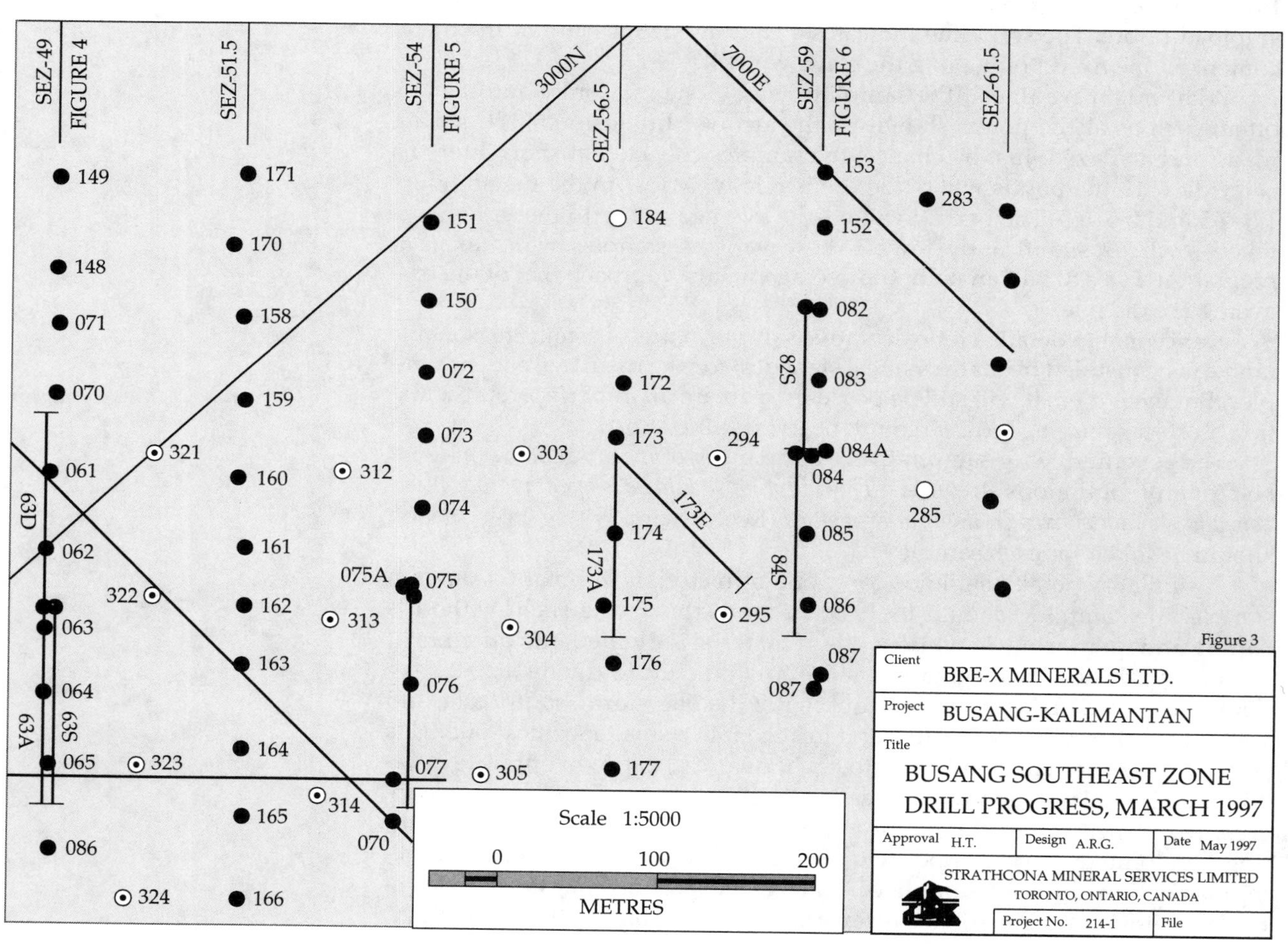

FIGURE 3.2 ▲

A part of a map of the regularly spaced locations of drilling in a promising mining region.

[Ref: Interim Report on the Busang Project, courtesy of Strathcona Mineral Services, Ltd.]

ANALYTICAL CHEMISTRY IN ACTION
There's Gold in Them Thar Hills?
Part 1: Sampling

In 1997, a spectacular gold discovery in Indonesia made by a small Canadian company called Bre-X Minerals, Ltd., was found not to exist. The stockholders of Bre-X lost billions overnight after the fraud was discovered. The sampling, sample preparation, and assay methodology performed by Strathcona Mineral Services in Toronto was exemplary of outstanding chemical analysis. The first quote from the report of this work describes the sampling for the analysis to check the Bre-X "find," which was also being checked by a Bre-X business partner, Freeport McMoRan Copper & Gold, Inc.

> . . . [A set of] six holes was laid out totalling 1500 metres, with the average hole depth being 250 metres. The latter depth was chosen as it was considered sufficient to meet our objectives given that Freeport had also drilled to a depth of 250 metres and it was more important to get broad area coverage rather than drilling to a depth of 400 metres as Bre-X did with most of their holes. All the holes had the same general orientation as the majority of the Bre-X holes. . .
>
> [As an example of the location for samples,] two holes, 63S and 202S, were selected to twin holes that Freeport had drilled, which in turn were twins of holes that Bre-X had drilled so there would be three holes all collared within a few metres of each other. [The 63S set is shown at the left on the map of Fig. 3.2.]. . .
>
> The same core size had been used by both Bre-X and Freeport in their drilling programs. The core was placed in metal boxes, covered with a wooden lid, enclosed with steel strapping, and then painted to indicate if any attempts were made to open the boxes. The core boxes were transported under armed guard to a locked container which in turn was also under observation by the military police. . . .
>
> We have followed to the maximum extent possible the same procedures used by both Bre-X and Freeport in sample preparation and assaying. This included taking samples at two-metre intervals . . . [However,] we split the core longitudinally leaving half the core in the core boxes that would be available for review by others for additional check assaying if necessary. Because of the large diameter of the core [63 mm], this still meant that the half core that was assayed weighed about 7 kg per sample, which is a sizeable sample by industry standards.
>
> In order to expedite delivery of our interim report and hopefully arrive at a resolution of the indicated discrepancies between the Freeport and Bre-X assays, we made the decision to select 350 metres of core to be done in the initial phase of assaying, which we expected to clearly indicate the likely direction of subsequent assays. This initial group of core samples included 100 metres of core from two separate 50-metre sections from hole 198, which . . . had indicated exceptional grades in the Bre-X analysis, and 50 metres in continuous intervals from each of the five other holes. The intervals chosen were simply those where Bre-X had obtained the highest grades and therefore presumably offered the best opportunity of finding measurable gold.

(To be continued . . .)

[Ref: Interim Report on the Busang Project, courtesy of Strathcona Mineral Services, Ltd.] ■

Subdividing and mixing a material is done to increase the homogeneity of the sample. In fact, the smaller the particle size, the lower the error in analyzing a given weight of material. For example, consider a sample consisting of a number of small, spherical particles of uniform size and density. Some of the particles contain pure Examplium (an uncommon material very seldom encountered outside of this text); the rest have none. The particles are well-mixed and are thus random in distribution, as illustrated in Figure 3.3. Let us assume that the particles of Examplium are 20% of the total number.

Also, assume we have an assay method for Examplium that is so precise we can ignore any random error in it. Then we can consider two extreme methods of sampling. At one extreme, we can take the whole sample and use it for the assay. Of course, the sample will be perfectly representative, and we will find 20% Examplium in it. However, because there can be only one sample, a statistical analysis of the data cannot be obtained. At the other extreme, we take only one particle at a time, and run an assay. In the majority of the assays, we shall find no Examplium at all. In others, we shall find 100% Examplium. These analyses would give a very large random error, which could be overcome only by running many more samples and finding the statistical mean value. Eventually, we would find the mean value of Examplium to be 20%. After that many runs, however, most analysts would be too exhausted to care.

From the difficulties associated with these two extreme methods, you might believe that there must be a better "middle way." Take a moderate number of individual samples, each made up of a number of randomly mixed particles. Of course, if the particles are large, such a sample might weigh hundreds of kilograms. Small particles enable us to maximize the number of particles in each reasonably sized sample, while assuring they would mix well, and, thus, minimizing variation in content between samples. In general, small particles produce better samples.

You might then ask, Why not always grind the particles as finely as possible? The answer is that there is a trade-off for such grinding. Very fine powders are more likely to be contaminated from the large amount of handling. They also can be hard to handle and transfer, since they behave like dust.

To review: Samples should contain large numbers of particles for two reasons:

1. Variation in content between individual samples is minimized, and
2. Each sample should be more representative of the material.

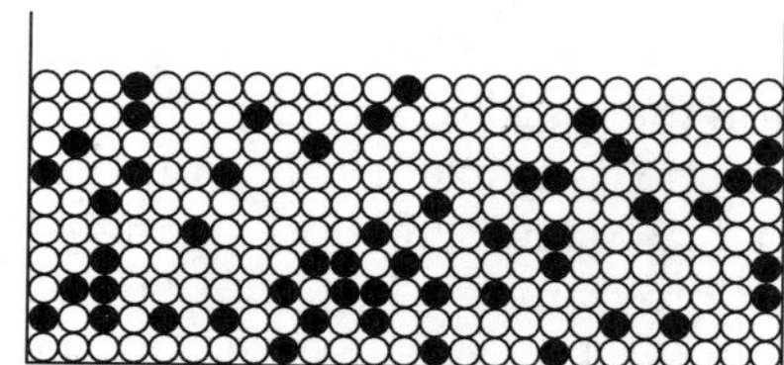

FIGURE 3.3 ▲

A material made up of particles of either pure Examplium (black) or filler (white).

Of the total mass, 20% is Examplium. In the accompanying text, two extreme cases of sampling the material are considered: a single assay of the entire sample and assaying each particle one at a time.

In the next section, these conclusions are justified in more detail mathematically.

3.3 A Pictorial Model

Let us consider a model for a finely ground, well-mixed sample. This model is still oversimplified, but it will serve to illustrate the trends in the results of further subdividing a sample (as by grinding). You can believe the results of the calculation because it has been shown to reflect the true results of experiments for more than two centuries, ever since first done by Bernoulli. (After all, not all calculations can be believed. Equations can agree internally while not reflecting the real world.)

Imagine three simple samples similar to the one in Figure 3.3. As before, each bottle contains balls of the same density. Twenty percent of the balls are pure Examplium, and 80% are pure filler. However, as illustrated in Figure 3.4, the balls in bottle A are twice the size of those in bottle B and four times larger than those in bottle C. Samples of equal mass are taken, one sample from each bottle. However, because of the different masses of the balls, the number of particles in each sample varies.

The average concentration of Examplium in all three samples is 20%. However, not every sample will contain exactly 20% Examplium; some will contain more, some less. A plot of the probability that a sample will have a given concentration of Examplium appears in Figure 3.5.

The calculation of the curves of Figure 3.5 is described in Section 3A.

Having smaller particles in a sample, then, has the same effect as doing a greater number of individual experiments. Both tend to make the measured mean value more precise. The main difference is that it is usually easier to grind the particles smaller.

The quantitative improvement in precision that occurs with reduction in particle size parallels that obtained by running more replicates. As you found

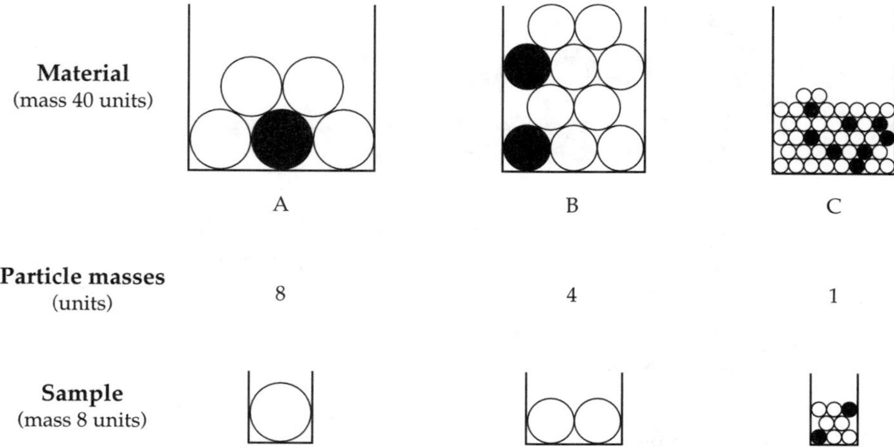

FIGURE 3.4 ▲
Three imaginary experiments to demonstrate the effect of particle size on samples.

The three materials A, B, and C contain particles with masses 8, 4, and 1, respectively. Black particles are Examplium; white particles are inert filler. All three materials contain 20% Examplium.

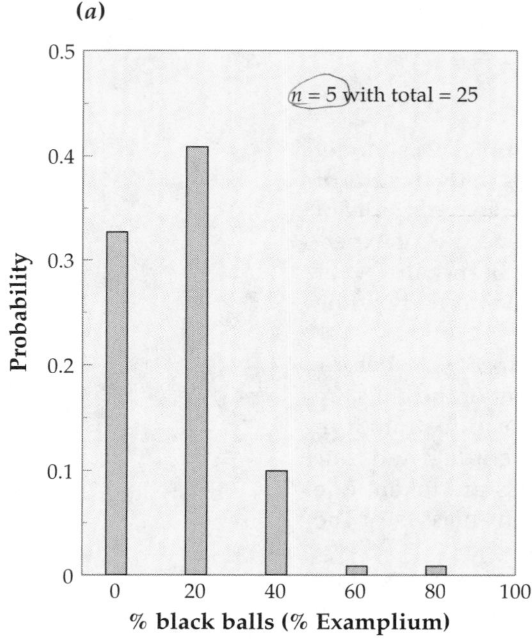

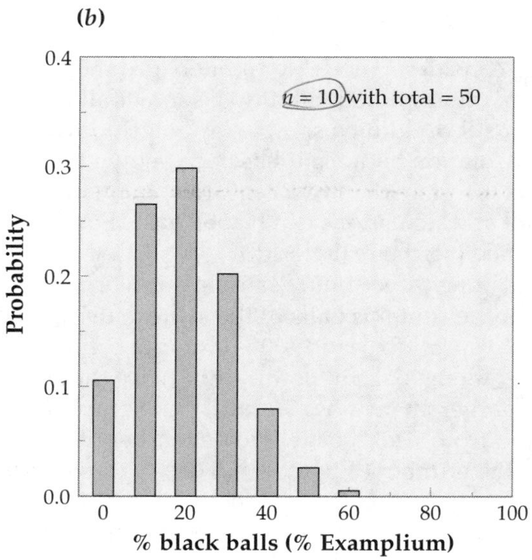

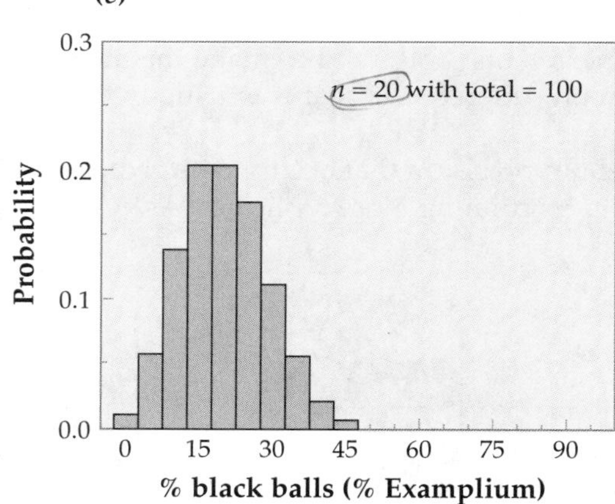

FIGURE 3.5 ▲
Probability that a sample will have a given percentage of black balls in it.
The material mass is constant at 100 units, and the masses of the samples are constant at 20 units. Three cases are shown: A, B, and C.

	A	B	C
Total number of particles in material	25	50	100
Particle mass	4	2	1
Number of particles in each sample	5	10	20

Notice that the standard deviation decreases with smaller particles. The derivation of these histograms can be found in Section 3A.

in the last chapter, when more replicates are analyzed, the error of the mean will be reduced as the square root of the number of samples run: $\sigma_m \propto 1/\sqrt{N}$. The same rule applies to reduction in particle size. The error expected from sampling decreases by a factor of 10 if we reduce the average particle mass by a factor of 100 while keeping the total masses of the analytical samples constant.

3.4 Samples of Mixtures

It may seem reasonable at first that obtaining homogeneity in samples of liquids and gases should present no problem. However, experience proves otherwise for samples that are larger than those that can be put in a small container in a laboratory. Consider again the sampling for mercury in a flowing stream. Say that a bottle of the stream water is taken and stored for a short time. Once in the bottle, the water is no longer mixed by the stream's motion, so the particulate matter settles to the bottom, with the heavier and larger particles doing so more quickly. If the solids are to be included in the analysis, the bottle contents must be thoroughly mixed before a sample, or **aliquot**, is taken out to be run. (A known fraction of a liquid sample, such as a volume taken out with a pipette, is called an *aliquot*, from the Latin *some* or *several*.) If the particles are small enough, samples taken from a well-stirred solution will be homogeneous. If only a few, large particles are present, homogeneity problems can arise, since one aliquot may have more or fewer of the particles than another. Such variations in particulate content will lead to less precise analyses. Finally, all the effects described for solid particles in liquids also occur in mixtures of liquids and solids suspended in gases.

3.5 Sample Integrity

Samples should not have any components added or lost between the time of sampling and the time an assay is run. This idea may seem obvious, but insuring that samples remain unchanged often requires a significant understanding of the chemistries of all the expected components and of the environment in which they will be stored. For instance, if an analysis is to be run to determine the quantity of residual insecticide, the stability of the molecule needs to be considered. If the insecticide is one of the more sophisticated ones, it is formulated to decompose relatively rapidly. In addition to time, perhaps the temperature, humidity level, sample acidity, oxygen content, and exposure to light may be important factors.

Also, the container must neither contribute interferents nor adsorb or absorb enough of the analytes to change the measurement. For instance, if the analyte can pass through or absorb into polyethylene and be lost, perhaps a glass container would be better. On the other hand, the surface of glass is relatively active and could catalyze the analyte decomposition or adsorb it strongly. No problem is foreseen if the sample is soil and the analyte is not volatile. However, if the sample is thoroughly wet—say lake water or swamp sediment—then migration to the container surface may be easy. All these problems of storage are more likely to occur the lower the levels of analytes in the sample. For example, adsorption might reduce a more concentrated analyte by 1%, but the same amount adsorbed from a less concentrated solution could virtually deplete the solution.

Contamination with other materials also must be minimized. For example, when a sample of deep-sea water is taken, the container is sealed at the sample depth so that no water at shallower levels can contaminate it as it is retrieved for analysis.

All these problems are significantly reduced when analyses are done immediately and at the site of the sampling—**in situ sampling**—such as with portable instruments (for example with x-ray fluorescence for lead in paint) or with **on-line instruments** that draw their samples from a process stream directly for analysis.

3.6 Physical Separations in Sampling

When sampling a heterogeneous material (solid, liquid, or gas) for analysis, the aim is to make it homogeneous. On the other hand, when one wants to analyze only part of a complex physical mixture, and it is necessary to separate the heterogeneous parts, the separation itself may introduce error. For example, what if you want to analyze the dust in the air near a copper smelter? The dust can be separated from air with a paper filter, but depending on the qualities of the paper, you might discover that 1% of all particles larger than 1 μm escape. The recovery would be high if all the particles were 1 μm or larger, but the sampling could be grossly deficient if particles smaller than 1 μm form a significant fraction of the solid.

Chemical separations are discussed in Chapter 13.

The separation of dust from the gases in which it is suspended provides an example of a **physical separation**, in which the separation of sample components is accomplished through their physical differences. A variety of methods for physical separation are available; some examples are included in Table 3.1. However, distinguishing a physical separation from a **chemical separation** is not necessarily simple. Certainly, collecting a solid precipitate from a gas or liquid with filter paper is a physical separation. However, there are now filters available with pores so small that they separate molecules by their molecular size. At this level, the distinction between physical and chemical separations disappears.

As an example, consider a production method that uses bacteria or immobilized enzymes (enzymes chemically anchored to a solid support) to synthesize the desired product. Assume the molecule that is to be assayed is

Table 3.1 Examples of Quantitative Physical Separations Used in Sampling and Analyses

	Separate Material that at Ambient Temperature Is a . . .		
	Solid	**Liquid**	**Gas**
From a Solid	By density-flotation By solubility differences	By heat and trap (e.g.,water from $CaCO_3$)	By melting solid and purging gas (e.g., radon from rock)
From a Liquid	By filtration (e.g., precipitate removal)	By distillation By decantation of immiscible liquids	By purge and trap (bubble inert gas through liquid and trap gases at low temperature) By gas-permeable membranes
From a Gas	By filtration (suspended particle removal)	By filtration (liquid aerosol removal; often includes solid nucleus)	By differential diffusion (used in GC/MS couplers) By selective freezing

a sugar with a molecular mass about 300. Also assume that there is a large amount of protein in the solution. (A good example of a solution of sugars and proteins is beer.) The protein can be removed and prevented from interfering in the analysis for sugars by using a membrane filter that does not pass molecules above molecular mass 2000.

A membrane separation is used in some gas-selective sensors described in Section 15.3.

Similarly, it is difficult to decide whether the separation of gaseous oxygen from water is a physical or a chemical separation. In this case, the gas can diffuse through a polymer membrane that will not allow the other components of a test solution to pass.

As can be inferred from the preceding information, a physical/chemical separation is often made immediately before the assay is run and may even be done by part of the assay apparatus itself. This close relationship makes many instruments far more selective than the assay methods on which they are based.

Sampling Gases and Volatile Species

Gases we might wish to analyze include vaporized volatile species as well as the lower boiling point gases such as CO_2 or oxygen. In other words, by gases we mean species that are gaseous at the temperature and pressure at which they are sampled.

Gas sampling can be as simple as opening and filling a plastic bag with ambient air, sealing the bag, and conveying it to the laboratory for an assay that can use the gas directly, such as mass spectrometry. It is more common to collect gaseous species in smaller volumes such as by condensing them to liquids by cooling, by **trapping** them in liquids through chemical reactions such as

Chapter 14 covers mass spectrometry techniques.

$$CO_2(g) + 2\,NaOH \rightarrow Na_2CO_3 + H_2O$$

or by **adsorbing** (note: not absorbing) them on the surfaces of solids. A vast number of different methods are possible, and the specific one chosen depends on the analyte, the **sample matrix,** and the assay method to be used.

Preconcentration often is part of the sample-collection stage; the materials of interest end up in a smaller, more concentrated sample. The analytes are removed from large amounts of the sample matrix during collection. Sample matrix is another name for the part of the material that is not of interest in an analysis. For example, the air matrix is eliminated when suspended particulate matter (for instance, smoke) is caught on a filter; simultaneously, the process concentrates the particulate matter.

Preconcentration in the sampling steps is commonly carried out for samples to be analyzed for trace and ultra-trace analytes, where the analyte is present at less than about 10 ppm.

Trace analyses are discussed in Chapter 5.

Figure 3.6 illustrates an apparatus for trapping a gas in a liquid. The gas is pulled through the apparatus at about 1 L/min by the pump, and the glass wool traps particulate matter (ash). The device pictured does three different processes simultaneously: 1) The trapped gas is concentrated; the SO_2 in 20 L of smokestack emissions is collected into about 100 mL of solution in less than an hour; 2) The gas is converted to a chemical form that will not be lost during storage; the SO_2 is converted into water-soluble sulfate ions by oxidizing the SO_2 with hydrogen peroxide; 3) Other gases are removed; gases such as O_2 and N_2 do not undergo the same reaction, so the SO_2 is separated from them as they pass through. The gas is dried in the fourth impinger,

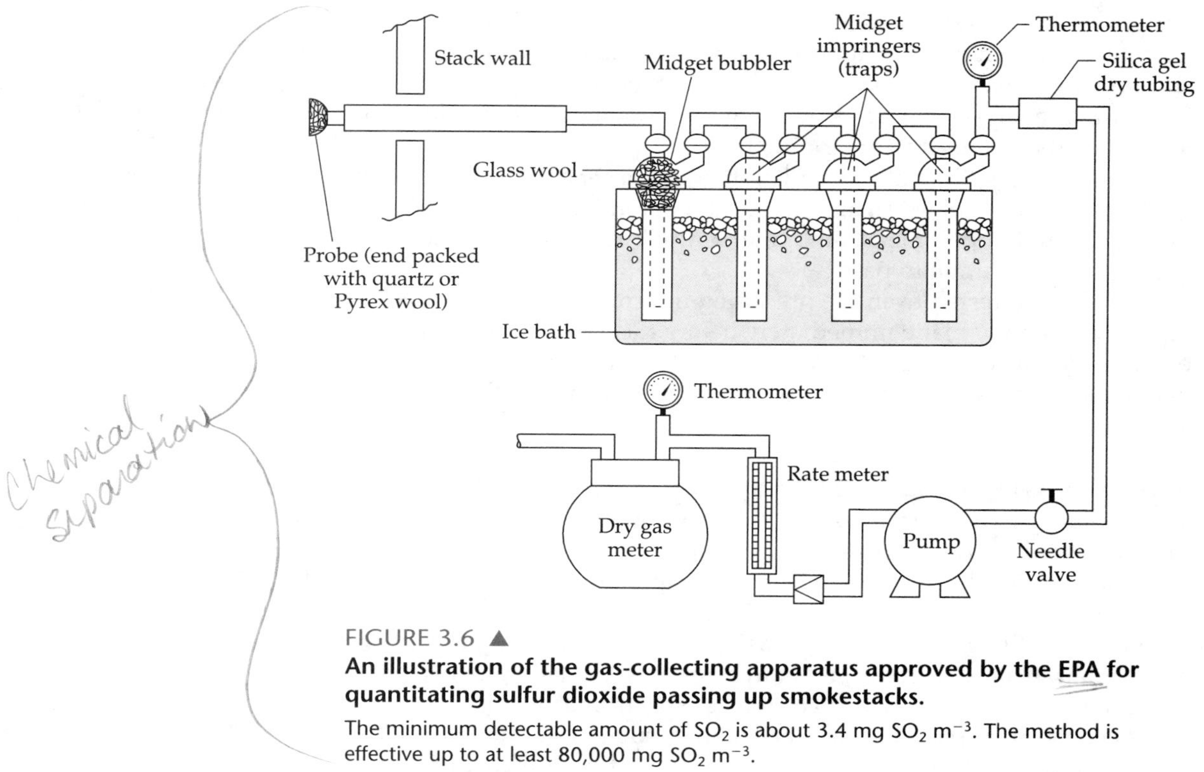

FIGURE 3.6 ▲
An illustration of the gas-collecting apparatus approved by the EPA for quantitating sulfur dioxide passing up smokestacks.
The minimum detectable amount of SO_2 is about 3.4 mg SO_2 m^{-3}. The method is effective up to at least 80,000 mg SO_2 m^{-3}.

which contains silica gel, and a second drying tube at the top right. The volume of the dried gas sample is measured by a "dry gas" meter. The thermometers enable the volume of the gas to be converted to standard temperature.

3.7 How Many Samples Do I Need?

In this section we turn the sampling question around. Assume you decide that you must be quite certain (say, 95%) that the relative error of the mean of an analysis does not exceed a *specific limit*. As you can infer from what you read in Chapter 2, if the upper limit is chosen and the confidence level is chosen, then the number of samples must be adjusted to satisfy both criteria simultaneously. How many samples are needed?

One question immediately arises: Do you know the random error in individual analyses? If not, it will be necessary to do some trial analyses to establish a estimate of that error. Then, look at Equation 2.11.

$$\mu = \overline{X} + t(\sigma/\sqrt{N}) \quad \text{when } \sigma \text{ is unknown}$$

Our task is to find an N so that with the method's s, the confidence limit is some fraction of $\overline{X}$. Now, let the maximum allowable relative error be represented by R. Then

$$t\frac{s}{\sqrt{N}} = \mathrm{R}\overline{X}$$

ANALYTICAL CHEMISTRY IN ACTION

Where There's Smoke, There's Fire. Where There's Fire, There's . . .

Having to create a representative sample from a large amount of a single material is one type of sampling problem. Another is described in the following brief account of sampling to find possible accelerants used in arson. The sampling occurs in an extremely heterogeneous environment.

> Fire investigators look at a fire in its totality and may probe into profit motive, opportunity, and similar factors. Reconstruction of the events that led to the fire and determination of its origin and cause involve the physical examination of materials. In many but not all cases, chemical analysis corroborates with the physical evidence and an accelerant is found where it is expected. . . . When physical and chemical analysis are done independently and if there is no communication between the investigator and chemist, bias is automatically eliminated. An investigator who detects a clear burn pattern may be more inclined to interpret a borderline chromatographic pattern to contain an accelerant than a chemist who has not been at the fire scene. . . .
>
> Although solid chemical incendiaries such as highway flares, thermite or improvised mixtures such as sugar and chlorates have been used by arsonists on occasion (<3%), liquid accelerants, in particular petroleum-based fluids, are usually used in arson-related crimes. . . .
>
> . . . A multi-step analytical procedure is only as strong as its weakest link and it is necessary to look at fire debris analysis in its entirety. The following stages can be defined: Selection of samples in the field; packaging and shipment to the laboratory; physical inspection of the sample in the laboratory; preparation of the sample for analysis by an instrumental method; instrumental analysis, usually by gas chromatography; and interpretation of data, reporting and documentation. . . .
>
> . . . Samples collected at the fire scene often consist of charred materials that are usually taken from the borderline of burn patterns. Experienced fire investigators can locate specimens from areas where interfacial adsorption and capillary action may have preserved traces of accelerants. . . . Adsorptive materials, in particular carpet, are obviously preferred. . . .
>
> . . . Accelerant components range from low-molecular-weight alcohols and ketones to substances that are solids under standard conditions. The ideal sample preparation method should thus be capable of recovering substances with a boiling point range that spans well over 1000°C. Since accelerants may be [either] polar [or] non-polar, it also should be effective for both. Even though quantification is usually not necessary, an indication of the approximate amount of accelerant present is desirable. ...
>
> . . . The experienced fire investigator is usually aware of potential interferences from materials such as tar paper, contact cements, roofing shingles, and other petroleum-based matrices. Control samples are sometimes used to rule out potential interferences. A control sample consists of material having a composition that is comparable to the matrix under investigation. Ideally, it should be taken from a location that is physically removed from the location of the fire. . . .

[Ref: Bertsch, W., Zhang, Q.-W. 1990. "Sample preparation for the chemical analysis of debris in suspect arson cases." *Analytica Chimica Acta* 236:183–195.] ■

From this formula, we find,

$$N = \frac{t^2 s^2}{R^2 \overline{X}^2} \tag{3-1}$$

Solving for N is not quite so straightforward, however, because the value of t itself depends on N, the number of samples. As a result, the equation is solved by iteration. That means the answer from a trial is placed back into the equation repeatedly until the answer does not change. Here, the most efficient way to work is to use the t for the 95% C.L. but for an infinite number of samples: that is, $t = 1.96$. N is found, rounded to the next highest integer, and substituted into the equation. A new N is found, and so forth.

EXAMPLE 3.1

Suppose a method for the determination of boric acid in a solution for production of eyedrops gives $3.0 \pm 0.1_7$ mg mL^{-1} boric acid. That is, $s = 0.17$ mg mL^{-1}. If the process requires that the solution be within 5% of the stated concentrations, how many trials should be done on each batch to ensure that the prescribed limits are attained with a 95% confidence level?

Solution

From the data, $R = 0.05$, $s = 0.17$, $\overline{X} = 3.00$.

Begin with t for an infinite number of samples, which for the 95% C.L. is 1.96

$$N = t^2 \cdot \frac{s^2}{R^2 \overline{X}^2} = (1.96)^2 \cdot \frac{(0.17)^2}{(0.05)^2(3.00)^2} = 2.8 \cdot 1.28 = 4.9 \quad \text{rounded to 5 samples}$$

For $N = 5$, $t = 2.78$ at the 95% C.L. This value of t is substituted and

$$N = (2.78)^2 \cdot 1.28 = 7.7 \cdot 1.28 = 9.9 \quad \text{rounded to 10 samples}$$

For $N = 10$, $t = 2.26$, which gives $N = 7$. The next iteration has $N = 8$, followed by another iteration with $N = 7$.

To obtain the result with the desired confidence, eight samples should be run. Clearly, if this is to be a routine analysis, some time should be spent improving the precision of the analytical methodology.

A Postscript

Incidentally, the analysis of the particles from the sweater mentioned on the first page of this chapter showed that they were, in fact, from a welding rod. The man was innocent. The sampling was not difficult in this case, since individual particles could be analyzed by microanalytical techniques. The particles were collected by lifting them from the sweater's surface with sticky tape.

Microanalytical techniques: see Chapter 5.

3A A Deeper Look: The Binomial Distribution

This section shows the mathematics used to calculate the probabilities of finding different levels of Examplium in samples ground to different particle sizes as graphed in Figure 3.4. The mathematics is based on the **binomial distribution**. The binomial distribution deals with yes-no, on-off, heads-tails, types of situations where a probability can be assigned to each result, and the events are independent.

For example, for a coin tossed four times, we can ask "What would be the probability of getting exactly three tails?" The outcome for each toss is independent of prior tosses. We would get the following possibilities for the four tosses (H=heads, T=tails, listed in the order of the toss left to right):

HHHH	HHHT	HTTH	TTHT
THHH	TTHH	HTHT	THTT
HTHH	THTH	HHTT	HTTT
HHTH	THHT	TTTH	TTTT

These are the different possible outcomes of the "experiment." We can reorganize these sixteen groups into five different sets.

4H	3H, 1T	2H, 2T	1H, 3T	4T
HHHH	THHH	TTHH	TTTH	TTTT
	HTHH	THTH	TTHT	
	HHTH	THHT	THTT	
	HHHT	HTTH	HTTT	
		HTHT		
		HHTT		

Now, let us ask, "What is the proportion of all the experiments where exactly three tails show up?" The fourth column above indicates that four experiments of the sixteen have exactly three occurrences of tails. The proportion, then, is $4/16 = 0.25$.

The statistics of the table of sets can be reproduced by the **combinatorial formula**. This formula is used so much, a special symbolism is employed: ${}_nC_r$. Here n is the number of samples (in the coin-toss case, $n = 4$; the experiment was four tosses), and r is the number of times a specific outcome occurs ($r = 3$ above; that is, three tails). The formula uses **factorials**, which are written as $n!$. Recall, for example,

$$4! = 4 \cdot 3 \cdot 2 \cdot 1 = 24$$

To be consistent, $0! = 1$. The general combinatorial formula is

$${}_nC_r = \frac{n!}{r!(n - r)!}$$

For our coin toss,

$${}_4C_3 = \frac{4!}{3!(4 - 3)!} = \frac{4!}{3!1!} = \frac{4 \cdot 3 \cdot 2 \cdot 1}{(3 \cdot 2 \cdot 1)(1)} = \frac{24}{6} = 4$$

This agrees with our table above: four possible outcomes have exactly three tails. (Note that to have exactly three *heads* also shows four outcomes. The combinatorial formula agrees!)

The proportion of heads is found with the **probability equation** with a slight change of nomenclature. We change our outlook to label tails as a "desired outcome" and heads as "not getting the desired outcome." Then, the proportion that will be calculated is called a probability, and the equation describing the coin toss is

$$\Pr[3 \text{ tails}] = {}_nC_r\, p^r q^{n - r}$$

where

p is the fraction of times that we will get the desired outcome in an individual toss (called a **trial**),

q is the fraction of times that we will *not* get the desired outcome in an individual toss, and

r and n are the same as above.

Let's apply the formula to the coin toss example. Half the time we expect heads to show, and half the time we expect tails. So, for our case, $n = 4$, $r = 3$, $p = 0.5$, $q = 0.5$, and

$$\Pr[3 \text{ tails}] = \frac{4!}{3!1!}(0.5)^3(0.5)^{4-3}$$

$$\Pr[3 \text{ tails}] = \frac{24}{6}(0.125)(0.5) = 0.25$$

which agrees with our count of 4 out of 16. Clearly, for a larger number of samples, it is far more efficient (and less tedious) to use the probability formula.

EXAMPLE 3.2

Using the sample described in Figure 3.5b, calculate the probability of finding exactly 4 black balls out of the 10 in the sample from the 50-ball total.

Solution

The overall probabilities are $p = 0.2$ and $q = 0.8$. We are looking for $r = 4$ black balls out of $n = 10$ samples. This is represented by the height of the bar at the 40% mark.

$$\Pr[4 \text{ blacks}] = \frac{10!}{6!4!}(0.2)^4(0.8)^6 = (210)(0.0016)(0.2621) = 0.088$$

We see that this does indeed correspond to a result of 40% black balls for $n = N = 10$ as shown in Figure 3.5b.

Suggestions for Further Reading

BICKING, C. A. 1978. In *Treatise on Analytical Chemistry*, Pt. 1, Vol. I, 2nd ed. I. M. Kolthoff and P. J. Elving, eds. New York: Wiley Interscience. Chap. 6.
A practical guide to the sampling of bulk liquids and solids with further references.

SMITH, R., JAMES, G. V. 1981. *The Sampling of Bulk Materials*. London: Royal Society of Chemistry.
The best single, readily available book on the subject of sampling. It is equally divided between theory and practice. The theory is developed in mathematical language.

Concept Review

1. What are the three major considerations that must be taken into account when deciding on a sampling procedure?
2. It is important that samples for analysis be *representative* and *homogeneous*. What does this mean?
3. What general rule with regard to sampling times or locations increases the likelihood that samples will be representative?
4. Give two reasons why it is a good idea to grind a bulk sample to produce small particles and mix it well before taking a smaller sample for analysis.
5. Using Table 3.1, suggest how you might separate:
(a) a gaseous analyte that forms an insoluble sulfate with BaO from a water sample.
(b) a liquid analyte that adsorbs strongly on glass wool from an air sample.
(c) a paint smudge (which is highly soluble in hexane) from the cotton sweater of a robbery suspect.
(d) two liquids that mix intimately and have boiling points that differ by 40°C.

Exercises

3.1 Use the data of Figure 3.1 for this exercise.
(a) Calculate, to the nearest 0.1%, the average concentration of the 16 samples taken at the random times indicated by the arrows.
(b) Repeat the calculations done in part (a) for 16 evenly spaced samples at regular half-hour intervals, beginning sometime in the first half hour. Repeat using evenly spaced intervals, but beginning at a different time in the first half hour.
(c) Compare the means of the trials with regularly spaced sampling times to each other and with the mean of the random-time samples.

3.2 A new instrumental method to determine mercury in water is being validated against an older one. The new method consists of reducing ionic mercury to its atomic form and then removing it as mercury vapor from solution. The mercury vapor is purged from the water with air and passed into an instrument especially constructed to measure Hg in air. The results were as follows (triplicate determinations); 50 ng of Hg metal was added for each run. [Ref: Murphy, P. J. 1979. *Anal. Chem.* 51:1599.]

Aeration and Collection Time (s)	Hg Measured (ng)
5	6
15	13
30	22
60	35
120	50
300	52
600	51

(a) What is the minimum time that aeration is needed to obtain the most precise results?
(b) Does additional time harm the results?
(c) Assume that the collection times were 60 s exactly and that the fraction of Hg collected is completely reproducible. By what factor would you have to multiply the measured results to obtain the correct ones?

3.3 An analysis had to be done within an hour on a sample that clearly had a significant amount of water associated with it. Therefore, part of the sample was dried for an hour, and part of the undried sample was assayed during the hour. The results showed that the wet sample was 32.4% analyte. The dried sample had a wet weight of 0.1362 g and a dried weight of 0.1128 g. What percentage of the sample is the assayed material, reported on a *dry basis*? (This means *as if the sample were dry*.)

3.4 A sample of gas was passed through a sampling train such as shown in Figure 3.6. A volume of 20.00 L was collected. The sample gas was at a temperature of 544 K and a pressure of 751 mm Hg.
(a) What is the volume of the sample at STP (273 K, 760 mm Hg) assuming ideal gas behavior?
(b) If the sample was found to contain 32.02 mg of SO_2, what was the SO_2 concentration (in g m^{-3}) in the original sample and at STP?
(c) The original sample gas was subsequently found to contain 20 volume % water. (Volume % = % v/v.) Correct the content of SO_2 to mg of SO_2 per m^3 of dry gas at STP.

3.5 An assay method involves injecting a liquid sample into the injection port of an instrument, either manually, using a small syringe, or with an automatic injector. The following results were obtained. Assume that the precision of the method depends entirely on the injection process.

Manual Injection		Automatic Injection	
Injection No.	**Instr. Response**	**Injection No.**	**Instr. Response**
1	96	1	169.6
2	101.5	2	168.5
3	100	3	166.5
4	100.5	4	164.8
5	98.5	5	165.8
6	101	6	167.8
7	100	7	167.2
8	97.5	8	166.3
9	98.5	9	167.3
10	101.5	10	167.7
11	100		
12	101		

(a) Calculate the relative standard deviations for experiments run under both conditions.
(b) Is the method more precise with manual or automatic injection?
(c) Assume that the relative standard deviation of a final result using this method is 10.8% regardless of the injection method. Further assume that the total error is due entirely to sampling error and injection error. What is the relative error due to sampling alone in each of these two cases?
(d) Assume that you obtain the error found in part (c) (10.8%) using manual injection. If more precision were wanted, would you invest first in buying an automatic injector or in developing a better sampling method?

3.6 A drum of chemical waste was analyzed for arsenic. The contents consisted of a top layer of water, a bottom layer of a water-immiscible liquid, and some metal-containing solid lumps at the bottom. The total volume of the contents was 55 U.S. gallons (1 U.S. gal = 3.785 L). The volume of the solid was 1.44 gal as measured by displacement. A 5.14-mL (wet) portion of the solid weighed 9.364 g when dried. The volume of the water was 62% of the liquid volume. The arsenic content of each part was determined in the same way with a method requiring a solid sample. A 10-mL aliquot of the water was placed in a foil cup, and the water was evaporated at low temperature. Subsequently, the remaining solids were dried at 110°; these weighed 83.2 mg. Similarly, 100 mL of the immiscible layer was evaporated, and the solids were redissolved in a small volume of acid and dried. Their weight was 55 mg. Each bulk solid was ground to 5 mm and smaller particles. A portion of each was dried and reground to a fine powder, which was used directly. The following results were obtained. The solids from the water sample were found to contain 220 ppm As. Those from the other liquid layer had 43 ppm As. The solid consisted of 157 ppm As.

(a) What are the volumes of the aqueous and the other liquid layer?

(b) What is the total mass of arsenic in each of the liquid layers?

(c) What is the total dry mass of nonvolatile solids in the barrel?

(d) What is the total mass of arsenic in the nonvolatile solids?

(e) An alternate method of analysis was also tested. The result was that the water contained 2.2 ppm (w/v) As. Were there any volatile (under the drying conditions) arsenic compounds present in the water, and, if so, what percentage were they of the total?

3.7 The relative standard deviation (in percent) of the binomial distribution can be described mathematically as

$$\sigma_{\text{rel}} = 100\sqrt{\frac{(1-p)}{np}}$$

To apply this equation to a mixed sample such as that illustrated in Figure 3.3, p is the fraction of particles composed of pure assayed component, and n is the total number of particles in the sample. For an analysis, the relative deviation should be less than 0.1%. Assume that the sample is composed of perfect spheres of density 3 g cm^{-3}, 10% of which are pure Examplium and 90% are inert filler.

(a) What is the minimum number of particles needed in each sample to achieve a relative standard deviation of less than 0.1%?

(b) What minimum weight (of sample in grams) will be needed to ensure that $\sigma_{\text{rel}} = 0.1\%$ if the diameters of the particles are 1 μm? 10 μm? 100 μm?

(c) If the balance used for weighing the sample measures up to 160 g with an accuracy of 0.1 mg, which sample particle sizes would be usable?

3.8 Will the answers of problem 3.8 differ if the particles are cubes with edges equal to the spheres' diameters? (This is equivalent to increasing the mass of each particle.)

3.9 A method for determination of PA6-5000 (an experimental antibiotic) has been tested for screening samples containing 7.5 mg of the drug in a 100 mg tablet. The method produced a result of 7.50 mg with $s = 0.3$ mg. It is now being adopted for evaluation of the compounded mixture for manufacturing the tablets. How many samples will be needed to make sure that the method assures (95% confidence level) that the content of an antibiotic capsule is 7.50 ± 0.04 mg in 100 mg samples of the compounded mixture?

*3.10 Atmospheric nitrogen monoxide (NO) and nitrogen dioxide (NO_2) were collected by pumping air through an absorbing solution. The solution contained a cobalt coordination compound—written Co(ligand)—which reacts with the NO in an equilibrium reaction. The reaction that occurs is

$$\text{Co(ligand)} + \text{NO} \rightleftharpoons \text{Co(ligand)NO} \qquad \textbf{(3.10-1)}$$

The equilibrium constant for this reaction is a mixed equilibrium constant.

$$K_p = \frac{[\text{Co(ligand)NO}]}{[\text{Co(ligand)}] \cdot P_{\text{NO}}} \qquad \textbf{(3.10-2)}$$

$= 4.5 \times 10^7$ atm^{-1} at 35°C in *o*-dichlorobenzene solvent. (Note from the units of K_p that P_{NO} must be in atmospheres.) Assuming that an equilibrium occurs between the gas and the coordination compound, from Eqs. 3.10-1 and 3.10-2,

$$P_{\text{NO}} = \frac{[\text{Co(ligand)NO}]}{[\text{Co(ligand)}]_{\text{initial}} - [\text{Co(ligand)NO}]} \cdot \left(\frac{1}{K_p}\right) \qquad \textbf{(3.10-3)}$$

Assume that the outlet pressure of NO after the gas passes through the solution is given by Equation 3.10-3. The pressure (in atmospheres) in the outlet stream is describable by the product of the atmospheric pressure, $P_{\text{atmospheric}}$, and the relative concentration of NO in the effluent stream, C_{outlet} in ppm (v/v). Algebraically,

$$P_{\text{NO}} = C_{\text{outlet}} P_{\text{atmospheric}}$$

Assume that $P_{\text{atmospheric}}$ is 1.00 atm. The efficiency of the collection is defined by

$$\frac{\%\ \text{collection efficiency}}{100} = 1 - \left(\frac{C_{\text{outlet}}}{C_{\text{inlet}}}\right)$$

Assume that $[\text{Co(ligand)}]_{\text{initial}} = 1$ mM in 5 mL of solution, and that the gas entering the trap is 9.8 ppm (v/v) NO. The rate of flow was 130 mL/min with the trap temperature held at 35°C. [Ref: Ishii, K., Aoki, K. 1983. *Anal. Chem.* 55:604.]

(a) If a 1% change in the volume of the gas can be ignored, what is the efficiency of gas collection as the gas flows in initially? ([Co(ligand)NO] is vanishingly small.)

*Indicates more involved problem.

(b) How many moles of NO gas can be absorbed in the collector and still have the efficiency remain above 98%?
(c) What volume of NO gas at STP can pass through the collector and still have the efficiency remain above 98%?
(d) What volume of analyte gas at STP can pass through the collector and still have the efficiency remain above 98%?
(e) Under the conditions of part d, how long will the 5-mL volume of absorber last?

Additional Exercises

3.11 A 10.45-m^3 sample of gas (310 K, 745 mm Hg pressure) was passed through a series of traps that removed the water from the gas. The gas that came out was found to be at 780 mm Hg pressure and 273 K. Its volume was 8.73 m^3. What was the percentage of water vapor (v/v) in the original sample of gas? Assume ideal behavior for all gaseous components.

3.12 An analysis was to be carried out on a sample of bauxite to determine the aluminum content as Al_2O_3. A truckload of ore was sampled at evenly spaced intervals by taking a one-foot-long portion as it passed by during unloading onto a conveyor belt. Each portion was bagged and transported to the central laboratory, where it was crushed and mixed thoroughly before four samples (with the weights given below) were taken. The remainder of the large sample was then placed into another bag for possible later use. Each of the smaller samples was dissolved in 5.00 mL of sodium hydroxide. The solution was diluted to 50.00 mL with distilled water and carried through a standard protocol resulting in another 10-fold dilution. The concentrations of the solutions were then determined to be:

Sample #	Sample weight (g)	ppm (w/v) Al in solution
1	0.6500	50.0
2	0.6450	48.0
3	0.6623	52.3
4	0.6557	55.3

(a) Calculate the mass of aluminum in the original small solid samples.
(b) What is the average content and 95% C.L. of Al(% w/w) as Al_2O_3 in the ore?

3.13 An alloy used for a magnetron sputtering target contains 7.0% vanadium. A method has been developed that is accurate and for which $s = 0.2$. The target manufacturer is bidding on a contract which specifies that the content must be 7.00 (±0.05)%. How many samples should the company analyze in order to make sure that their product is within the stated specifications?

3.14 The stack gases from a factory had been a source of annoyance to neighbors for years. Using a sample volume of 20 L collected one foot down from the top of the stack, an independent laboratory was commissioned to analyze the stack gas and found that the concentration of SO_2 was almost 30 ppm. A more responsible management team was brought in that installed a device that would chemically trap the SO_2. They then wanted to know the new concentration of SO_2. This time the independent laboratory knew they would need a larger volume sample. The stack gas was once again sampled at one foot below the top of the stack. Using a pump, stack gas was drawn through a series of traps containing pellets of a substance that reacts with SO_2 in the ratio of 100.0 mmole SO_2/100 g solid, but would not react with the other components of the stack gas. Each of the four traps contained 20.000 g of the pellets. After pumping at a rate of 100 L/min for 24 hours, the traps were emptied and change in mass determined. The results were as follows:

Trap #	Weight of pellets (g)
1	20.100
2	20.060
3	20.035
4	20.000

(a) Calculate the number of moles of SO_2 that were trapped, assuming that the change in mass is due to the SO_2 trapped.
(b) If the stack gas was at 250°C and 0.98 atmospheres, how many L of SO_2 were trapped?
(c) What was the concentration of SO_2 in ppm after installing the scrubbers?
(d) What was the efficiency of the trapping process?

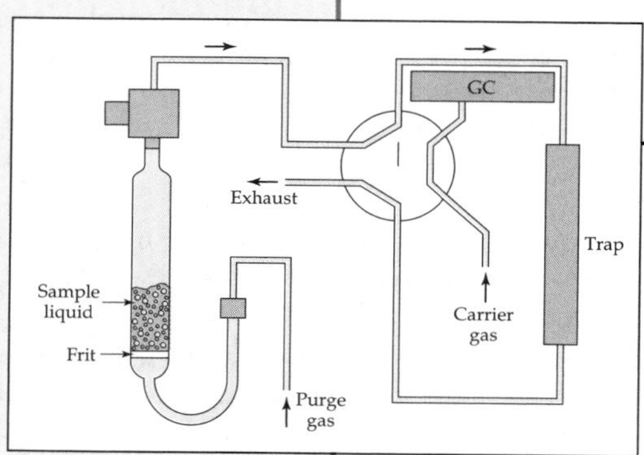

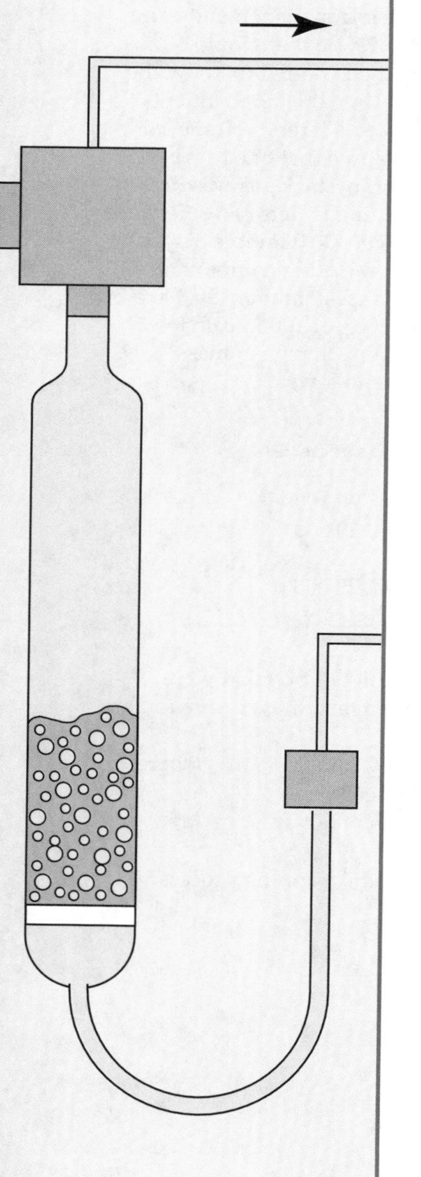

CHAPTER 4

Sample Treatments, Interferences, and Standards

Overview

A Deeper Look

4.1 Sample Preparation

The term **sample preparation** means to bring the sample into the correct size and form to be used with the chosen assay method. Other names for this process are **sample treatment** or **sample pretreatment**. Sample preparation is now the most time-consuming, most error-prone, and most labor-intensive step of an analysis. In addition, preparing a sample for an assay requires by far the most expertise in chemistry of any part of an analysis. No matter how simple or complex the procedures appear, the same five general principles apply to all sample preparations.

1. **The sample preparation should be done without losing any of the analyte(s).** Material should not be lost in the preparation, but if it is, some manner must be found to discover how much is being lost. The quantitative measure for the amount of the analyte that remains for the assay from the original sampling is called the **recovery** of the procedure.
2. **The sample preparation should include bringing the analyte(s) into the best chemical form for the assay method to be used.** Some assay methods require solids, some require liquids, and some require gaseous forms. If the sample must be transformed, sometimes this must be done by the analyst (such as dissolution of a salt into a solution). Sometimes the transformation is an intimate part of the analytical instrument and is not generally separated from the technique itself, such as when a hot flame is used to break up the sample matrix. It is not uncommon for the transformation step to be a major contribution to analytical error. In addition, the specific chemical form can be crucial. For instance, perhaps the atomic form of an element is necessary as opposed to the ionic form.
3. **The sample preparation should, perhaps, include removing some interferents in the matrix.** As noted in Chapter 1, all assay methods respond to a lesser or greater degree to species of molecules, atoms, or ions other than the one(s) being assayed. These other materials are said to interfere in the assay. An assay method that is less prone to interference is said to possess greater **specificity**. When the specificity of an assay is improved, the sample preparation steps can often be significantly simplified. Thus the properties of the assay method and the appropriate sample preparation method are closely linked.
4. **The sample preparation should be done without adding any new interferents.** The conditions should be chosen to prevent adding any new interferents from the reagents or reaction vessels. However, the most common problem in this area is **cross contamination**, which occurs when some material from one sample becomes included in another sample. Cross contamination will result if any part of a previous sample remains in a region when a subsequent sample arrives. Cross contamination becomes more difficult to eliminate when the levels of an analyte vary greatly between samples: for a given amount of material transferred, the more the concentrations differ, the greater the possibility for interference in the more dilute sample. While cross contamination is mentioned here first, it can occur in any stage of an analysis.

5. **The sample preparation should include, if necessary, diluting or concentrating the analytes to bring their concentrations into the best range for the assay method used.** It is essential that you recognize that every assay method has an optimum concentration range. The analysis will not work if the concentration (solid, liquid, or gas) is either too high or too low. Few methods can be used to determine content with the same precision over more than about three orders of magnitude in concentration—where the high end is 1,000 times the concentration of the low end. Here again the sample preparation and the assay method are closely linked.

These five requirements often involve tradeoffs, for example, a treatment that might tend to produce a better recovery might also add interferents. In addition, there are two more themes in contemporary analysis that also influence the sample preparation. The first is the speed of contemporary instruments: They are becoming faster all the time. If, in the space of one minute, a complete assay can be carried out, the data posted to the **laboratory information management system** (LIMS) through a computerized interface, and the next sample loaded from an automatic loader of some type, then in order to utilize the instrument fully, someone has to pretreat 60 samples per hour. Not long ago, when only wet chemistry assay methods were available, and all assays were done by hand—such as by manual titration—an efficient person might do ten to fifteen assays in an hour. If more than one analyte was titrated separately, perhaps one sample an hour would need to be prepared. So sample **throughput** has been increased by two orders of magnitude.

A second theme is the rapid increase in costs associated with purchase and disposal of solvents (and chemicals in general), not to mention the desire for less waste at all levels of chemical manipulation. The response to this trend has been to use smaller amounts of both organic solvents and acids or to find substitutes for the entire process. As an example, it has been common practice to **extract** organic compounds from water samples by shaking the water in a separatory funnel with an organic solvent such as ether and then using the ether phase for further processing. (For example, the ether could be evaporated to concentrate the extract.) The ether extraction is now commonly replaced by passing the water sample through a tube filled with a solid that has a non-polar surface. The organics adsorb on the non-polar surface—they are **stripped** out of the sample. A small amount of organic solvent such as methanol can then be used to recover the adsorbed analyte, or the solid can be heated and the analyte vaporized from the surface and transferred directly to the assay instrument.

Because of these factors, as well as the cost of sample preparation, a great deal of research is being done to develop better methods for sample preparation. The following sections, 4.2–4.5, explain how requirements 1–5 above are fulfilled in chemical analysis.

4.2 Maximize Recovery

Central to the idea of recovery is that an analyte exists first in some matrix from which it is to be recovered. Within the requirements of the assay, in order to choose the preparation method(s) you must understand

1. The chemistry of the analytes,
2. The chemistry of the matrix, and

3. The chemistry of the interactions between the matrix and the analytes.

That is, the chemistry to be understood here is relatively complicated. The desired outcome is to recover all of the analytes in the form desired while separating them from the matrix components that cause interference. We would like to be able to state some rules about this process, but, given the range of possibilities noted in Table 4.1, generalizations are almost impossible.

An analyst's dream might be to have a solid matrix that dissolves completely in water; an assay method that requires an aqueous solution of the (soluble) analyte; and no interference from molecules or atoms of the dissolved matrix. Then, only a simple dissolution of a known amount of the sample would be required for the preparation, and there would be no problem with loss of analyte. (Occasionally you too may find yourself in such a situation, but don't hold your breath!)

Loss of Analyte

How, then, can analyte be lost? One obvious way is through adsorption, absorption, or penetration of the analyte into its storage container between the time a sample is collected and the time it is run. For instance, metal ions can *ad*sorb on the surface of glass. Therefore, samples that are to be analyzed for trace metal levels require clean plastic containers (often polyethylene). However, organic materials can be *ab*sorbed into plastic and lost. Eventually the material can penetrate through the container wall to the outside. So clean glass containers would be better in that case. However, even for glass, the type of seal has to be considered so the analytes do not escape through the caps. Of course, sample losses caused in these ways are unlikely to be a problem when an analysis is done on-line or *in situ*.

Analytes are also lost when they decompose. Organics can decompose in a number of ways. Some may be oxidized by oxygen. So either oxygen has to be excluded or antioxidants added. Others materials may decompose with light, and so must be kept in the dark. A common problem with aqueous samples and biologicals is that bacteria can grow and decompose parts of samples, so bactericides or bacteriostats may be added to keep the sample's integrity. In addition, biological tissues contain numerous enzymes that can degrade a wide range of analytes.

The ACA on the next page about amyloid proteins shows some other ways to avoid degradation.

Table 4.1 Analyte/matrix combinations

Analyte	Matrix	Examples
Solid	Solid	Molybdenum in steel; polymer in polymer, fat in fish
	Liquid	Suspended particulates in lake water; ketchup
	Gas	Smoke
Liquid	Solid	Oil in sandstone
	Liquid	Volatile organic compounds in water; mayonnaise
	Gas	Exhaled breath
Gas	Solid	Argon in basalt
	Liquid	Acid rain
	Gas	Ozone in atmosphere; exhaled breath

ANALYTICAL CHEMISTRY IN ACTION
Breakin' Up Is Easy to Do

Polypeptides and proteins are relatively fragile molecules, and conditions must be chosen to extract and purify them while avoiding breaking of the intramolecular bonds that form the chain. During an extraction from tissue, the protocol must minimize digestion of the protein by enzymes (proteases) that are inherently present there. To stop protein digestion, a protease inhibitor must be included. Here, the inibitor phenylmethylsulfonyl fluoride (PMSF) is added to the mix. Another precaution to prevent decomposition includes refrigeration during most of the preparation time, including during centrifugation.

The scheme drawn here shows the steps in the isolation of one or more polypeptides in deposits of amyloid β-protein (abbreviated Aβ), which is found in brain tissue in Alzheimer's disease. Alzheimer's is a progressive dementia. The purpose of the study was to examine the amino acid order (the primary structure) and size of the soluble polypeptides that may subsequently form the deposits. The peptides isolated have molecular weights of about 4,000 Daltons (4 kDa) and are being prepared to be analyzed by mass spectrometry.*

[Ref: Näslund, J. et al. 1994. "Relative abundance of Alzheimer Aβ amyloid peptide variants in Alzheimer disease and normal aging." *Proc. Natl. Acad. Sci. USA* 91: 8378–8382.]

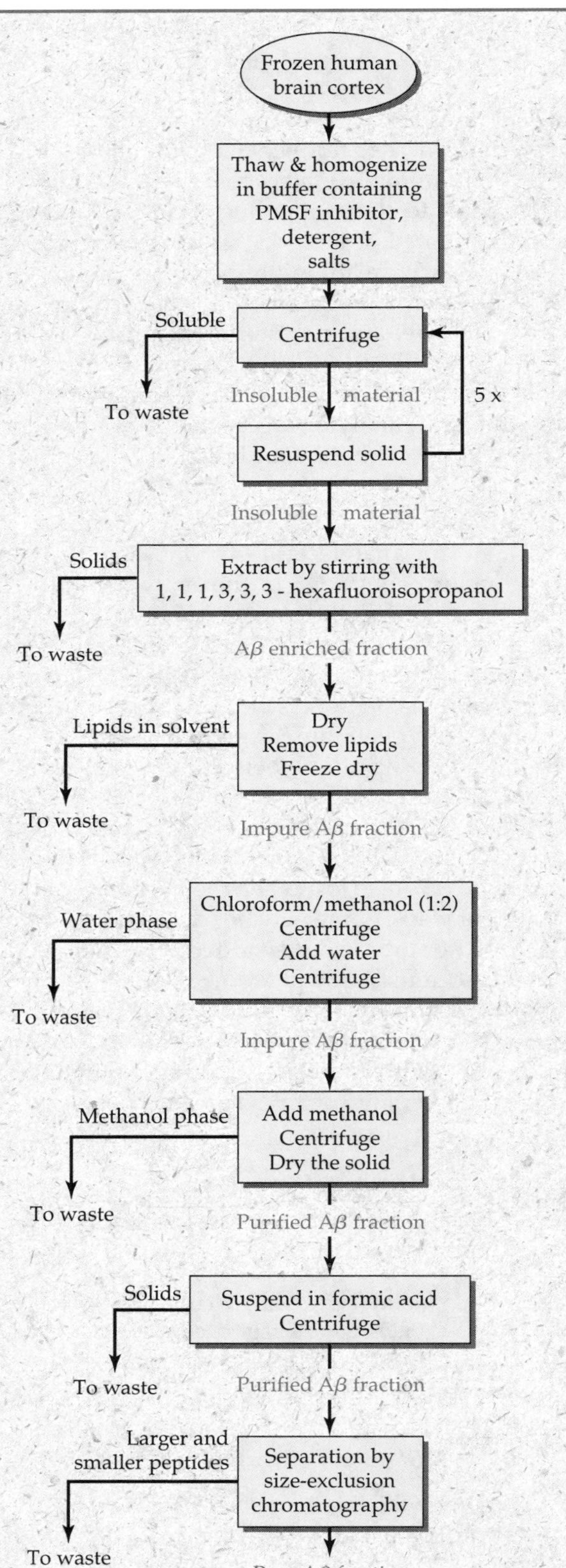

Amyloid β-protein isolation. ▶

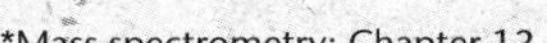

*Mass spectrometry: Chapter 12.

ANALYTICAL CHEMISTRY IN ACTION
There's Gold in Them Thar Hills?
Part 2: Sample Preparation

"Sample preparation involves a number of steps in which the original sample mass, for Busang core, about 14 kilograms, is reduced to such smaller mass as is required for the assaying procedure itself. So that this final aliquot is truly representative in its metal content of the original sample, the original sample is stepwise reduced in grain size by crushing and grinding. Only after such a reduction in grain size can a smaller sample safely be split off the original sample. . . . The sample preparation steps for the Busang core consist of the following:

- Those core intervals determined by the geologist to be favourable for the occurrence of gold are sub-divided and marked in one-metre intervals;
- From each one-metre interval, ten centimetres of core, usually at a particular position within the interval, is removed, sawed into halves, and one half returned to the core box, the other labelled and retained as a permanent sample. . . .
- In most cases, two one-metre intervals are combined to form one sample for subsequent assaying. A unique sample number is assigned to each such interval, the core removed from the core box and placed into plastic bags after breaking it into pieces. Two bags are required to hold one sample interval.
- Two samples, i.e. four plastic bags, are then placed in a larger fibreglass bag for shipment to Indo Assay Laboratories at Balikpapan. . . .
- After drying at 120° to 130° and primary crushing of the entire sample to minus 6 millimetres [meaning that it passes through a screen with 6-mm holes], a one-kilogram subsample is removed and returned to Bre-X as a permanent witness sample. This is also known as the coarse crusher reject sample;
- After secondary crushing to minus 2 mm, a one-kilogram sub sample is split out and pulverized to 95% passing 106 microns and 80% passing 73 microns. During each stage of crushing and pulverizing, the crushers and pulverisers are cleaned after each sample by passing inert material through it.

From the pulverized sample, 750 grams are taken . . . the sample is subjected to bottle leaching in a 1 ppm NaCN solution at pH 10.5 to 11 for 36 hours. The dissolved gold [leached as the soluble cyanide complex from the rock] is recovered in an organic component (DIBK) . . . As is well known, cyanide leaching does not produce an assay of the total gold in a sample, just that part that is cyanide extractable. The ratio of what can be . . . extracted . . . compared to the total gold in a sample can vary considerably, from 20% to close to 100%, and is often typical for a property. There are a number of . . . assays that have been done [that show] . . . based on nearly 1200 assay pairs, the ratio for Busang "ore" is around 95%, *i.e.* the under reporting by the cyanide leach process is very small. On this basis, the choice of [sample preparation] method is a good one, in our opinion, as the leaching which uses a 750 gram charge measures the gold content of a much larger sample than ordinary or even screen fire assay, which methods typically use charges of 50 and 250 grams, respectively. . . ."

[Ref: Interim Report on the Busang Project, courtesy of Strathcona Mineral Services, Ltd.] ■

Finally, although laboratories try to minimize human errors, some inevitably occur. Losses and apparent losses of analyte can result from putting the wrong solution into a sample storage vessel; forgetting to purge a solution of oxygen; weighing errors; and spills.

To review: some causes for the losses are:

- Adsorption of analyte onto container surfaces,
- Evaporation of volatile samples,
- Losses in unanticipated chemical side reactions,
- Leaks in transfer systems for gases or liquids, and
- Accidents and carelessness.

All these potential losses prevent **quantitative material transfer**. The quantitative part is usually expressed as a *percent recovery*. This value is simply a comparison of the concentration of analyte found from the assay with the concentration that is, in fact, present.

$$\%\ \text{recovery} = 100 \times \frac{\text{concentration of analyte from assay}}{\text{concentration of analyte in sample}} \tag{4-1a}$$

or

$$\%\ \text{recovery} = 100 \times \frac{\text{weight of analyte from assay}}{\text{weight of analyte in sample}} \tag{4-1b}$$

An example and discussion of how percent recoveries are determined will be given in Section 4.7.

Correction for Loss of Analyte

A relatively low recovery need not condemn a method outright. If the recovery is constant, we can compensate for it just as we would for any constant determinate error. For example, assume that a sample preparation method yields an 87% recovery of the analyte every time. It would only be necessary, then, to multiply the result by 1.149 to correct for the effect ($1.149 \times 0.87 = 1.00$).

Nevertheless, methods with inherently low recoveries are usually less precise. A simple comparison illustrates this point. Assume that a sample preparation procedure generally produces 99% recoveries. From the opposite viewpoint, there is a 1% loss. If the loss doubles or halves, the recovery varies from 98% to 99.5%, a relative range of about 1.5%.

In comparison, consider a sample preparation method with poor recoveries—say, 60%. With this method, 40% of the analyte is lost. Again, if the loss doubles or halves randomly, the recoveries will vary from 40% to 80%, a relative range of about 66%. Thus there is an inherently larger possible imprecision associated with low recoveries. Conclusions drawn from such analyses are, consequently, less certain.

It is best to aim for the highest analyte recovery possible, but many analytical methodologies are officially accepted for government regulation with reproducible recoveries as low as 50% for some of the analytes. As long as sample preparations are reproducible, reproducible recoveries can be obtained. Whether a treatment with a low recovery is usable can be decided by setting some maximum level of its statistical error, such as setting a maximum relative standard deviation of the recovery. The recovery error adds to the other statistical errors in the usual way:

$$\frac{\sigma_{\text{tot}}}{\text{analyte content}} = \text{R.S.D.} = \frac{\sqrt{\sigma^2_{\text{sampling}} + \sigma^2_{\text{recovery}} + \sigma^2_{\text{assay}} + \cdots}}{\text{analyte content}} \tag{4-2}$$

EXAMPLE 4.1

During validation of an assay for polyaromatic hydrocarbons (PAHs) in soil in an old dump site, the relative standard deviation σ associated with each step was estimated for the 1–100 ppb range. The assay technique (gas chromatography) has $\sigma = \pm 1\%$. The sampling (core sampling and combining cores from the centers of four adjacent 10m × 10m sections of a grid) has an estimated $\sigma = \pm 20\%$ between sets, due mostly to the heterogeneity of the soil. The recovery during the sample preparation had been found by spiking soil samples to have $\sigma = \pm 7\%$ for PAHs in this type of soil. What is the total relative standard deviation? The analyte content is 3 ppb.

Solution

We use Equation 4-2.

$$\frac{\sigma_{\text{tot}}}{\text{analyte content}} = \text{R.S.D.} = \sqrt{0.2^2 + 0.07^2 + 0.01^2} = \sqrt{0.045} = 21\%$$

Notice that the value of 3 ppb is not needed since the *s*-values are relative deviations as given and are for samples of this type and PAH levels. Almost the entire error is from sampling, which might be expected from such a heterogeneous source such as a multiacre land area.

4.3 Optimize the Chemical Form

Often, an analyte in its matrix is not in the form necessary or suitable for the chosen assay method. Bringing the analytes into the best chemical form for the assay is not limited to changes in the analytes. It may also demand a transfer of some analytes from one matrix to another. In Table 4.2, some terms that describe these transformations are briefly defined, and diagrams summarizing the types of transformations appear in Figure 4.1.

The remainder of this section describes some of the specific methods used to optimize the chemical form of analytes. There is no sharp division between inorganic and organic compounds, but there is a reasonable division between elemental and molecular analysis. Very powerful digestion methods can be used when elemental analyses are to be run, since we want to break up the matrix as completely as possible into its elements.

Working Directly with Solid Samples

In the previous chapter, it was argued that sample homogeneity could be improved by reducing solids into smaller particles or powders. In certain assays, this is a sufficient sample treatment since the assay method can use such a powder directly. Examples of such methods are thermal analysis, X-ray fluorescence, neutron activation, and a number of surface assay methods. In fact, one of the major strengths of these assay methods is that they can be used for solid samples.

X-ray fluorescence, neutron activation, and surface methods, Chapter 11.

Adding or Subtracting Heat

Sometimes, when the analyte needs only to be changed in physical form, this may be done simply by changing the temperature. A heterogeneous solid might be melted into a homogeneous liquid or a liquid might be vaporized to a gas. More esoteric methods of adding heat include firing high energy atoms or molecules at the surface of a material or using a high-powered laser

Table 4.2 Types of transformations of chemical form

Term	Meaning
Dissolution	To break down into component parts in a liquid driven by the binding of solutes by the solvent.
Digestion	Driven by externally applied energy to break down into component parts that dissolve in a liquid.
Decomposition	To break down into smaller component parts.
Solubilization	To break up the structure so that solution contained inside structures is free to mix with a surrounding solvent. Usually used in biology.
Desorption	To remove from a surface.
Trapping	To hold in any of a number of different ways: freezing, adsorption to a solid, or chemical reaction in solution.
Evaporation	To make into a gas at relatively low temperatures.
Ablation	To break up into small particles or into a gas by rapidly heating a surface.
Vaporization	Generically, to make into a gas; usually implies by strong heating.
Volatilization	To evaporate.
Atomization	To break up a collection of atoms into individual atoms.
Ionization	To cause atoms to become ions.
Neutralization	To cause acidic or basic solutions to become neutral. To cause gas phase ions to revert to atoms.
Condensation	To convert by cooling from a gas to a liquid.
Liquification	To convert a gas to a liquid either by cooling or by increasing pressure.
Melting	To convert a solid to a liquid by adding heat.
Extraction	To bring analytes from one matrix into another. The first matrix is a solid or liquid, the second a liquid.
Adsorption	To cause to bind to a surface.
Oxidation	To cause the oxidation state of an element or compound to become more positive.
Reduction	To cause the oxidation state of an element or compound to become more negative.

flash. Both of these methods can vaporize small volumes of solids or liquids where they hit.

Dissolution

If a liquid is required for an assay and a melted solid is not adequate or obtainable at a sufficiently low temperature, then a homogenous solution must be made. The simplest way to form a homogeneous solution is, of course, to dissolve the solid in an appropriate solvent. Examples are the dissolution of sodium bicarbonate or sodium chloride in water, or the dissolution of organic materials such as naphthalene or polymers such as polystyrene in toluene or other organic solvents. Dissolution can usually be hastened by shaking or gentle heating to warm the solution.

Often a better way to mix samples utilizes sound waves at a frequency far above that we can hear—**ultrasound**. Ultrasound assisted dissolution is described in more detail in Section 4A.

Digestion

If most or all of a sample will not dissolve in a solution, as is common, then more vigorous conditions are required. The process is called **digestion** and involves bringing the solid into solution with the aid of an acid, base, oxidizing agent, or enzymes. Some digestion mixtures are listed in Table 4.3.

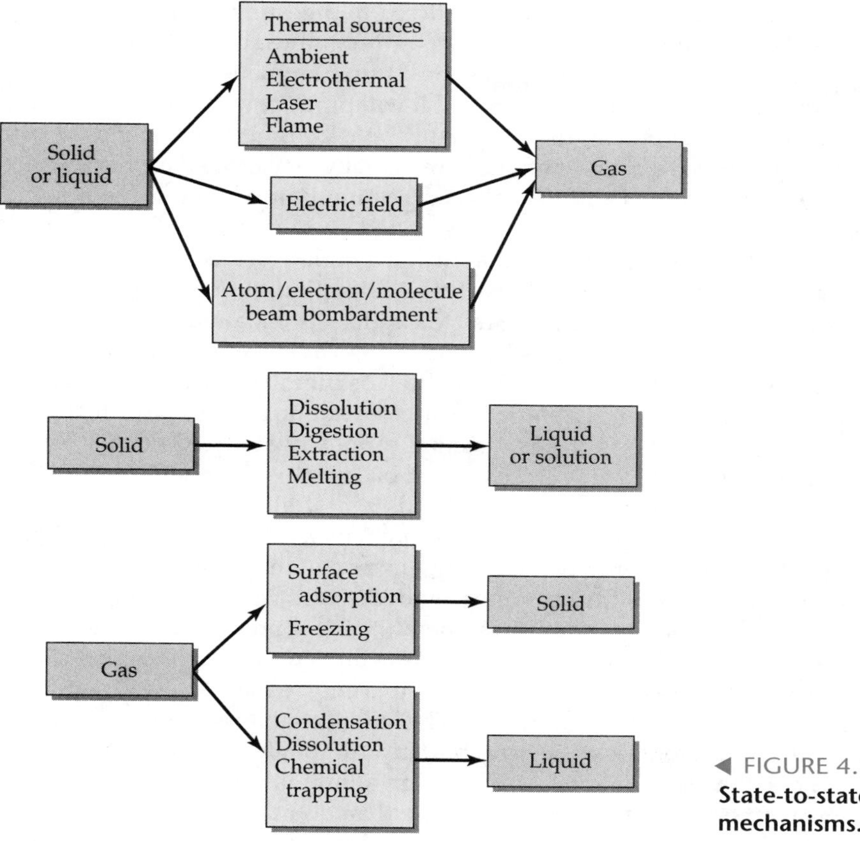

◀ FIGURE 4.1
State-to-state change mechanisms.

Table 4.3 Reagents used for digestion

Solution (concentrated)	Properties	Use
Acid		
Hydrochloric (38% w/w)	Mildly reducing	Metals more easily oxidized than hydrogen
Nitric (70% w/w)	Oxidizing	Metals not reacting with HCl; oxidizes organic matter
Sulfuric (98.3% w/w)	High boiling (~340°C)	Metals; destroys most organics
Perchloric (70% w/w)	Strongly oxidizing	Metals; NOT TO BE USED WITH ANY ORGANICS OR WITH INORGANIC REDUCING AGENTS
Hydrofluoric (49% w/w)	Forms stable fluorides	Silica and silicates; used along with other acids as well
Base		
Sodium hydroxide	Strong base; oxidizing when concentrated	Aluminum and amphoteric oxides of Sn, Pb, Zn, Cr

A complication that must be considered in digesting a sample for analysis is that one or more of the components to be assayed could be transformed into a volatile product by the digestion. If that happens, the component(s) could be lost as a gas, and any assay for it will produce a low result. If volatility is a potential problem, the material should be digested in a closed container to trap the gas. Such containers are usually constructed of thick-walled (2–5 cm) stainless steel and are called **digestion bombs**. The name is somewhat misleading, since they are constructed so that they will not explode even when the pressure rises to that inside a high-powered rifle when fired.

Another digestion method involves elimination of organic components of the matrix while leaving inorganic ones. This is usually done by destruction of the organic compounds by oxidation, leaving the inorganic components in the resulting ash. This ash can then be digested further by one of the methods described above. Ashing is commonly done in an electrically heated furnace, but, with special inserts, ashing can also be carried out with good control in microwave ovens, as discussed briefly next.

Microwave-assisted digestion affords a new alternative to aid sample dissolution and digestion. A number of such ovens are specifically designed for the laboratory, having heating compartments lined with relatively inert materials. In addition, they can be provided with a means to remove corrosive vapors and monitor sample temperatures and pressures. The advent of microwave-assisted digestion has greatly facilitated digestions in sealed containers, which is obviously a superior way to ensure that trace and ultratrace analyses will be accurate and precise. The loss of analyte is limited, and introduction of interferents is easier to avoid. In addition, for either open or closed containers, the use of microwave digestion allows far better control over the sample temperature and speed of heating than do hotplates or other heating methods. There are a number of drawbacks, however, which have prevented microwave digestion from completely replacing conventional heating, and these are elaborated in Section 4B.

If solution digestion cannot break down the solid, a still more vigorous method is **fusion**. This involves heating the solid sample with about 10 to 20 times its weight of a **flux**. Fluxes are solids at room temperature. A solid flux is a strong oxidizing agent, or acid or base, that will form a homogeneous solution (a **melt**) with the sample material when they are heated together. Upon cooling, the melt hardens to a homogeneous solid. Often, after fusion, the cooled melt can be dissolved in appropriate solvents. A number of fluxes are listed in Table 4.4.

Sometimes both acid digestion and fusion are needed in sample preparation; sometimes only one is necessary. Choosing the method requires knowledge of the descriptive chemistry of the materials being used, but making a good first choice for a specific material comes only with experience.

There are a number of reasons why fusion may work when acid or oxidative digestion cannot:

1. The temperature of fusion is usually higher (300–1000°C) than the boiling points of concentrated aqueous acid solutions (about 100–200°C).
2. The concentrations of the reagent flux (on the order of 10–20 M) are usually higher than in an acid solution (0.1–10 M).
3. In an aqueous solution, the water limits the maximum oxidizing power and the maximum acidity and basicity that can be obtained. In a flux, this restriction is not present.

Table 4.4 Reagents for Flux Digestion (Fusion)

Flux for fusion	Properties	Use
Sodium carbonate	Basic	Silicates, insoluble sulfates
Sodium hydroxide	Basic	Oxides and sulfides; Ores of Sn, Zn, Cr, Zr
Sodium peroxide	Alkaline and oxidizing	Ores of, e.g., Sb, Sn, Mo, V, Cr; Chrome steel
Potassium pyrosulfate	Acidic	Oxides of Fe, Co, Ni, Cr, Ti, W, Al; Steels
Potassium nitrate Potassium chlorate	Used with Na_2CO_3 for alkaline and oxidizing fluxes	Acid-insoluble metal sulfides

There are limitations, of course, with any digestion method. The principal one is that as the digestion conditions become stronger and harsher, it can be difficult to find a container for the mixture that will not itself be dissolved. (This brings up the old question, If you had a universal solvent, what would you keep it in?) One way to bypass this problem is to use containers composed of elements that are not to be determined and also do not act as interferents if corroded. One commonly used material for such crucibles is purified zirconium.

Another consideration in digestion is safety. You must be careful not to produce explosive products or create circumstances that might endanger your health.

Vaporization

A number of instruments used for assays require that the sample be introduced as a vapor. If the sample is a solid or liquid, the material first must be vaporized into a gas—either atomic, molecular, or ionized (a plasma).

Some instruments operate under high vacuum. Thus, if the sample is volatile enough, the vaporization might be done simply by introducing the sample into the vacuum. Or, if necessary, the material can be heated. Heating to vaporize the sample will be necessary if the instrument operates at atmospheric pressures or above. A more esoteric method of vaporizing materials is to add the necessary energy by bombarding them with high-energy (in effect, high-temperature) particles such as xenon atoms.

Vaporization is used to convert samples to the correct (gaseous) form inside the instruments for gas chromatography, mass spectrometry, and the atomic spectrometries. Details about vaporization in these instruments, including the use of electric sparks and high-intensity pulsed laser light, are described in the relevant chapters. Such vaporizations might also be classified as part of the assay or sample preparation process, but here we will keep this step separate for consistency.

Chapters 13, 12, and 11, include, respectively, gas chromatography, mass spectrometry, and atomic spectrometries.

A summary of the sample forms compatible with various assay methods appears in Table 4.5. A general conclusion can be made, however: the less done to convert a sample to a form that can be precisely and accurately analyzed, the better.

Table 4.5 Analyte introduction forms in various assay techniques

Form	Common assay methods accepting this sample form
Gas	Infrared spectrometry Raman (long path) UV-vis spectrometry Atomic absorption, emission Chemiluminescence Microwave spectrometry Gas chromatography Mass spectrometry Radiochemical methods
Solution/liquid	Liquid chromatographies Gas chromatography Supercritical fluid chromatography Titration Nuclear magnetic resonance Infrared spectrometry Raman spectrometry UV-vis spectrometry Electron paramagnetic resonance Gravimetry Electrochemical methods Atomic absorption, emission Mass spectrometry Radiofrequency methods Radiochemical methods Polymer light scattering
Solid	X-ray fluorescence X-ray crystallography Infrared spectrometry Raman spectrometry UV-visible reflectance Neutron activation Radiochemical methods Thermal analysis methods Nuclear magnetic resonance (polymers) Surface sensitive methods (ESCA, Auger, scanning tunneling microscopy, atomic force microscopy, Rutherford backscattering)

4.4 Minimize the Interferences

Optimizing the chemical form, minimizing the interferences, and maximizing the recovery are, clearly, intimately related through the requirements and limitations set by the nature of the sample and the chosen assay method. Sometimes, to minimize the interferences, it may be desirable to exchange the matrix surrounding the analytes. For instance, one exchange is to take the analytes from being associated with a solid alumino silicate matrix (clay) to a matrix of liquid acetone.

In essence, exchanging the matrix first requires that we perform a chemical separation of the analytes from the original matrix components. In order

Table 4.6 Extraction methods

Soxhlet
Separatory funnel and liquid-liquid extraction
Purge-and-trap
Supercritical fluid extraction (SFE)
High pressure/temperature
Accelerated Solvent Extraction (ASE™)
Microwave assisted
Ultrasonic assisted
Solid phase extraction (SPE)

to do such a separation, at some time we must move the analytes into a phase that is physically separable from the original matrix phase. (Here, by phase we mean solids, liquids, and gases.) Numerous methods exist, and some newer techniques have shown themselves to be better and/or faster than the more common ones. Table 4.6 lists a number of such extraction methods.

Let us briefly consider quantitative **solid-liquid extractions**. Experimentally, the idea is quite simple: A weighed solid is placed in a closable container, and some solvent is added. The solid and liquid are mixed well, and the liquid is separated from the solid. This procedure involves using a liquid to dissolve the analytes that are part of the solid (but not covalently bound within it). This distinction between the "material" that will dissolve and the "solid" is only useful for this discussion. After all, if you weighed the solid before extraction and after, you would say some of the *solid* was dissolved—and you would, of course, be right. But the interaction of the material that dissolved is stronger with the solvent than with the remaining solid, and so it transfers into the solution. We describe this preference for one phase over another with a **distribution coefficient**, K_D, based on the mass/volume of the analyte $i (m_i/V)$ in the two phases.

$$K_D = \frac{(m_i/V)_{\text{solvent}}}{(m_i/V)_{\text{solid}}} \quad \textbf{(4-3)}$$

You can understand from this equation that the weaker the interaction of the analyte with the solid matrix, the easier the separation into the solvent. This easier separation translates into a larger K_D value.

Let us look at a single analyte with $K_D = 2.0$. This means that twice the concentration of the material is found in the solvent as is found in the solid. This is not a large ratio, and it was chosen to illustrate a point about quantitative extractions: How do you best obtain a quantitative extraction when the extraction is relatively inefficient? (After all, in the extraction, if we use equal volumes of solvent and solid, approximately 33% of the analyte is left in the solid.) One obvious way would be to choose a better solvent, but let us assume this is the best, or the safest, or the most cost-efficient one available. The next suggestion might be to note that there is 66% recovery, and correct for that lower recovery as a determinate error. This tactic would probably work, but remember, the amount left in the solid depends on the volume of extractant. Half as much (in mass) will remain in the solid if we double the amount of solvent. Then, you might say, "Well, use ten times as much solvent, shake it and then concentrate the stuff!" This would result in twenty times as much analyte in the solvent as in the solid. But you need to hope that there is no loss of analyte during the solvent evaporation.

There is a simpler way, and that is to use *less* solvent and do a number of sequential extractions.

EXAMPLE 4.2

You want to extract 1 g of analyte from 10 mL of solid. (Note, we normally measure by weight for a solid, not volume.) For the extraction,

$$K_D = \frac{(m_i/V)_{\text{solvent}}}{(m_i/V)_{\text{solid}}} = 2$$

Calculate the percentage of analyte extracted with a single extraction of 100 mL compared to 5 sequential extractions of 20 mL each.

Solution

For the single extraction of 1 g, we would find $(1 - x)$ g of analyte in 10 mL of solid and x g in 100 mL of solvent. That is,

$$K_D = \frac{x/100}{(1 - x)/10} = 2$$

Carrying out the arithmetic, we have

$$21x = 20$$

$$x = 0.95$$

Now, let us do the calculation for 20 mL of solvent and then, subsequently, for four more 20 mL extractions. The first calculation is the same as the one above, but with 20 mL in place of the 100 mL volume. That is,

$$K_D = \frac{x/20}{(1 - x)/10} = 2$$

The solution found is $x = 0.8$.

The second extraction calculation is done the same way, but there is only 0.20 g of material left to distribute between the solid and liquid. So,

$$K_D = \frac{x/20}{(0.2 - x)/10} = 2$$

and carrying this through, we find $x = 0.16$. The second batch of solvent (containing 0.16 g of analyte) is added to the first fraction, giving us 40 mL with 0.96 g of material or 96% extracted. A comparison of the single 100-mL extraction with the multiple 20-mL extractions is shown in the table below. Clearly, multiple extractions are superior, and it appears that we get a slightly higher percentage extracted with only two 20-mL extractions than we did with one 100-mL extraction. A real-world limitation lies in retrieving the solvent that contains the analyte when the volume is too small. For example, it may be hard to decant 1 mL of solvent from 10 mL of solid. However, since the same principles and calculations hold for liquid-liquid extraction, smaller volumes of the extracting solvent *can* be used when extracting from another liquid.

Volume/mL	Amount extracted/g	Total % extracted
Single extraction		
100	0.95	95
Multiple extraction		
20	0.80	80
20	0.16	96.0
20	0.032	99.2
20	0.0064	99.84
20	0.0013	99.97

The benefits of multiple extraction elucidated in the above example explain the advantages of **soxhlet extraction**. (A soxhlet extraction apparatus is shown in Figure4.2.) By distillation and condensation of the solvent, the apparatus allows multiple extractions to be done repeatedly using the same (smaller) volume of solvent. The solvent is redistilled numerous times into a fiber filter cup that contains the sample. The extracted mass ends up dissolved in the lower flask. Soxhlet extraction has been a standard method for many decades, and it is often the method against which other extraction methods are measured and verified.

When we talk of a solvent in the above solvent extraction, where some material is removed from a solid matrix, what usually comes to mind as the "solvent" is water, or some organic solvent such as diethyl ether or ethanol, or, perhaps, a pure hydrocarbon such as hexane or toluene. These are all liquids at ambient temperature (around 20-25°C) and ambient pressure (about 1 atmosphere). In certain cases, however, there are benefits in using the much higher pressures of **supercritical fluid extraction** (SFE), which is described in Section 4C. A common supercritical fluid is supercritical carbon dioxide. The advantages of SFE include the possible use of fluid CO_2 that is essentially nontoxic and presents minimal difficulties with disposal. A drawback is that it requires special dedicated equipment capable of providing the conditions needed.

A relatively new technique for extraction utilizes equipment that holds the sample in a sealed-high pressure environment to allow higher temperatures to be used for conventional solvents. This method is called accelerated solvent extraction (ASE™). By allowing a higher temperature without boiling, ASE allows smaller volumes of solvent to be used in a single-step extraction. The extraction kinetics are also faster because of the raised temperature, so the entire process is far faster than soxhlet extraction. After the heating, the cell is allowed to cool to below the solvent's normal boiling point. High pressure is then applied to the cell, which forces the solvent and extracted materials out through a filter.

Solid-liquid extraction is also hastened by the use of microwave heating in both open and closed containers (see Section 4B). Solid-liquid extraction also can be accelerated and changed in effectiveness by using ultrasound (see Section 4A).

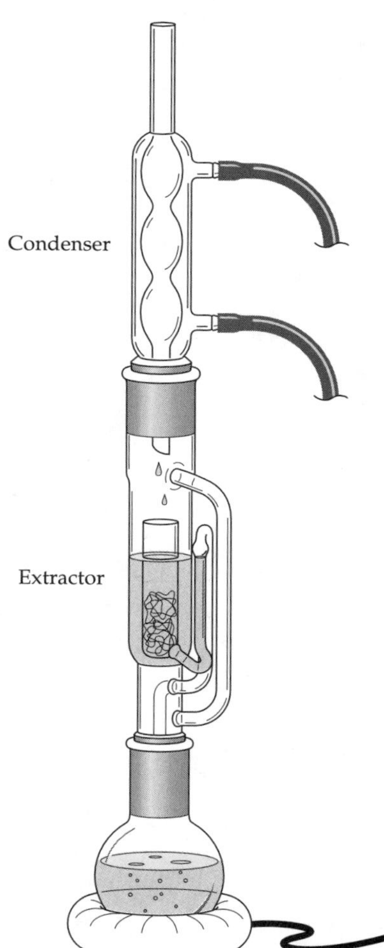

FIGURE 4.2 ▲
A soxhlet extraction apparatus.
The solvent boils from the bottom through the outer bypass tube and is condensed into a reservoir that holds a paper filter thimble. When the solvent nears the top, it is siphoned through the inner, u-shaped tube back to the bottom.

Liquid-liquid extraction is a standard and common method for changing matrices surrounding dissolved analytes. Most of you have already done liquid-liquid extraction in a separatory funnel. The two liquids are, at most, slightly soluble in each other; they are **immiscible**. (The antonym, **miscible**, means that two solvents dissolve in each other in any proportion.) Immiscible liquids, after being intimately mixed—such as by shaking—separate into two layers. The two layers are then segregated by draining one of them off until the interface is reached. Liquid-liquid extraction is often used to extract organic analytes from aqueous solutions. The analytes are concentrated, and any inorganic salts present are left behind in the aqueous layer. As mentioned above, by liquid-liquid extraction the analyte can be extracted into a much smaller volume and, thus, concentrated as well as isolated.

Nevertheless, current trends in regulation are causing the use of organic solvents to be minimized, and liquid-liquid extraction is being supplanted in many cases by **solid-phase extraction** (SPE). Solid-phase extraction is done by passing a solution containing analyte(s) over a solid phase that adsorbs the analytes specifically. Often this adsorption is called stripping the analytes from the solution. Examples of commonly available solid phases and their uses are listed in Table 4.7, along with some solvents used to remove the analytes from those phases. The solid phase is usually much smaller in volume than the solution containing the analyte: the solid phase can be the surface of a chemically modified filter-paper disk a few cm in diameter up to a small (1 to 35 mL) packed column; a tubular cartridge the shape of a hypodermic syringe; or a small funnel partly packed with small adsorbent particles. After adsorption, the analytes are removed with a small amount of solvent with which the analytes interact more strongly than with the solid phase. As a result, the solid-phase extraction not only changes the solution matrix, but also simultaneously reduces the sample volume; the analytes become more concentrated. It should be clear that a larger surface is needed to retain a larger amount of analyte, so the amount of the solid-phase extraction medium required depends on both the size of the sample and the concentration of the analytes present. As can be seen from Table 4.7, solid-phase surfaces can be classified with a convenient set of chemical attributes: polar or nonpolar; acidic, neutral, or basic; hydrophilic or hydrophobic; and cation-binding or anion-binding. These different surfaces are used to isolate different classes of compounds. Finally, two of the benefits of solid-phase extraction over liquid-liquid extraction are the significant decrease in the volumes of solvents needed for sample preparation and the ability to automate the extraction process.

Difference in solubility is not the only property that can be exploited to change a liquid matrix. By using **dialysis**, we can separate solutes by size and also change the solution composition. Dialysis is done with a membrane that has molecule-size holes in it, a **dialysis membrane**. Molecules larger than the holes cannot penetrate while those smaller can pass through with ease. There are two general uses for dialysis, the first being to change the small-molecule composition in a solution of macromolecules—usually a protein or other biopolymer. In this case, the polymer solution is placed in a dialysis container (a container of which at least part of the surface consists of a dialysis membrane). The container is brought in contact with a large reservoir (say, 100 times the volume of the sample) that contains a solution with the composition desired for the macromolecule. Eventually, the solution inside the dialysis container is nearly that of the larger reservoir, and the

Table 4.7 Solid phase extraction solids

Name	Surface identity	Surface characteristics	Binding type, conditions, and examples
Octadecyl	$(—C_{18}H_{37})$	Hydrophobic Non-polar	Hydrophobic species from aqueous solutions e.g., organics, peptides, oligonucleotides
Octyl	$(—C_8H_{17})$	Hydrophobic	Hydrophobic species from aqueous solutions (less strongly held than C_{18})
Silica	OH OH | | —O-Si-O-Si-O— | |	Hydrophilic Polar Neutral	Low to moderate polarity species from nonaqueous solutions e.g., fat-soluble vitamins
Florisil	Magnesium silicate	Hydrophilic Polar Slightly basic	Low to moderate polarity species from nonaqueous solutions e.g., fats, polychlorinated biphenyls
Alumina A (Acidic alumina)	OH OH | | —O-Al-O-Al-O—	Hydrophilic Polar Acidic Cation binding	Hydrophilic species in nonaqueous solutions e.g., antibiotics, caffeine
Alumina N (Neutral alumina)	OH O^- | | —O-Al-O-Al-O—	Hydrophilic, Polar Neutral	Hydrophilic species in nonaqueous solutions e.g., petroleum
Alumina B (Basic alumina)	O^- O^- | | —O-Al-O-Al-O—	Hydrophilic Polar Basic Cation binding	Hydrophilic species in nonaqueous solutions e.g., steroids, pesticides
Aminopropyl	$—C_3H_6NH_2$	Hydrophilic Moderately polar Slightly basic Anion binding	Analytes in aqueous or organic solvents e.g., phenols, petroleum, saccharides
Cyanopropyl	$—C_3H_6CN$	Hydrophobic Nearly non-polar Neutral	Analytes in aqueous or organic solvents e.g., hydrophobic peptides, pesticides
Diol	OH OH | | $—CH_2—CH_2—$	Hydrophobic Nearly non-polar Neutral	Trace elements in water; proteins and peptides in aqueous or organic solvents
Styrene divinylbenzene		Hydrophobic Neutral	Organics in water, e.g., polyaromatic hydrocarbons, Vitamin B_{12}
Anion binding (Strong anion exchange)	$-CH_2CH_2-\overset{+}{N}(CH_3)_3$	Hydrophilic Anion binding	Anions in water or aqueous mixed solvents e.g., Cl^-, SO_4^{2-}, PO_4^{3-}

larger reservoir has a small concentration of the solutes that were in the dialysis container initially and could pass through the membrane. Second, dialysis can be used to do an assay for small molecules while excluding larger ones. In this case, you might use a miniature dialysis tube that allows small neurotransmitter molecules (the messengers between nerve cells) to diffuse in from brain fluid but excludes the proteins.

We should also mention the classical method of changing liquid matrices: precipitation. Precipitate the analyte, filter or decant the solvent, and redissolve the analyte in a new solution. This method, however, tends not to be one used for trace analysis.

A less well-known method of matrix separation of particulates in a liquid is to isolate them by bubbling a gas, such as air, through the solution. The particles are removed from the liquid onto the surface of the gas bubbles. The separation and isolation may be aided by the addition of surfactants or coprecipitators such as Fe(III) hydroxide. The iron-hydroxide precipitate traps many other heavy metals and acts as a **carrier** for the low-concentration metal-ion analytes. The name for such methods is **flotation**.

If the analytes are volatile enough, the method of choice to isolate them is **purge-and-trap**. *Purging* means to eliminate from the original matrix. For solid samples, this is done by heating the solid while passing a gas through it to carry the volatile components to a trap. The volatile components are usually called **volatiles**, and VOC is a common abbreviation for **volatile organic compounds**. Purging of a liquid is done by passing a non-reacting gas through the liquid, and the volatile components are carried away. The traps halt and hold the volatiles either by cooling them or by adsorbing them on a small volume of powdered solid contained in the trap. Alternatively, both can be done: the volatile can be absorbed on a cooled solid. More information on methods of purge-and-trap is presented in Section 4D.

From these brief descriptions, you may see some similarity between purge-and-trap, flotation, and solid phase extraction: all concentrate the analytes from a larger volume. Concentration tends to improve the quality of the assay, especially those for trace or ultratrace levels of analytes. This brings us to the topic of the next section, that of optimizing the concentrations of analytes.

4.5 Optimize the Concentration

The final requirement to consider when developing a sample preparation method is to optimize the concentration of the analytes. The importance of this requirement cannot be emphasized enough. *Every* assay method has a range of concentrations within which it will give results with the greatest precision and accuracy. As a qualitative example, consider a test tube filled with a solution of black ink. As you look through it, you can see the ink's color in the solution. However, as you dilute the solution, at some point you will no longer be able to see the ink's color. If you then dilute the colorless solution by a factor of two, you know that the ink concentration is now half what it was before dilution, yet you will not see any change. For this last dilution, you will be trying to run an assay of ink content outside of the usable range of your eyes. The bottom of the concentration range that can be seen is quantified by a number called the **limit of detection**.

Limits of detection are described in more detail in Section 5A.

Now think about the other end of the concentration range. Put enough black ink into a solution so that you cannot see any light through it at all. Double that concentration, and, again you see no light. This other extreme of concentration also is of no use to determine the changes in the amount of ink present. So there is an upper limit to the useful concentration as well. The range of concentrations between the upper, useless range and the limit of detection is called the **working concentration range**. Different assay methods have different working concentration ranges, and, in most cases, the range is specific for the analyte.

Somewhere within the working concentration range, there is an optimum range in which the precision and accuracy will be best for the assay. As you might guess, the optimum range is somewhat removed from both the upper and lower extremes.

So how would you bring an analyte concentration into the optimum range if it were not there already? It depends on whether the assay requires a solid, liquid, or gaseous form and whether the analyte is below or above the optimum range.

For solids, **preconcentration**—that is, increasing the analytes' concentration before running the assay—would be difficult. It would likely involve dissolution, separation of the analytes from the matrix, and reprecipitation—either with an inert material or by driving off the solvent. Dilution is easier. This is usually done by grinding the solid and diluting it with another solid material that does not contribute any significant interferents.

For liquids, preconcentration usually requires exploiting differences in the chemistries of the analytes compared to the matrix components. A straightforward application for elements in solution is electrodeposition, as illustrated in Figure 4.3. Many elements in their metallic form dissolve in liquid mercury metal, forming solutions called **amalgams**. Among these metals are Ag, Au, Cd, Co, Pb, Sn, and Zn. If the elements are in solution as their ions (Ag^+, Au^+, and so on), they can be converted to their metallic forms by chemical reduction. This is easily done in an electrochemical cell using mercury metal as the cathode at which the ions are reduced. Once the ions are converted to their metallic forms, then they can dissolve in the adjacent mercury. However, when the liquid mercury metal is in contact with water, a number of ionic species are not reduced, and remain dissolved in the water. These include Ca^{2+}, Cs^+, Na^+, Rb^+, and Sc^{2+}. The different chemistries of the above two sets of metal ions allow their electrochemical separation using mercury as an electrode. A clever trick is to let the volume of the mercury be much smaller than the volume of the water containing the dissolved sample. Then,

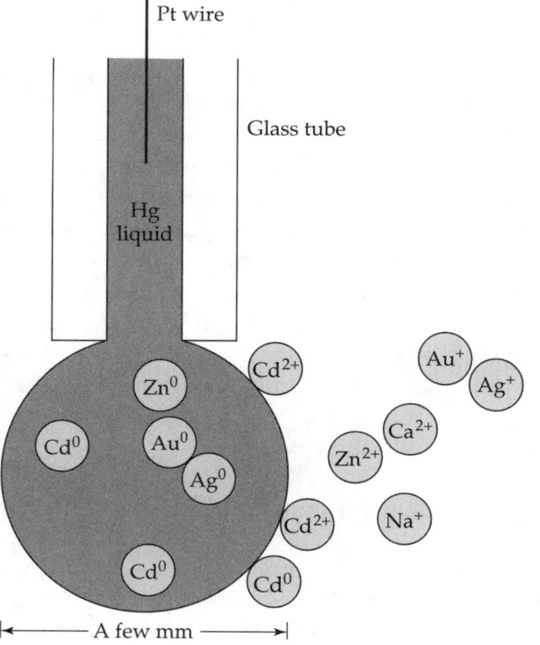

◀ FIGURE 4.3
Diagram of the process involved in concentration by electrodeposition.
Metal ions are reduced at the surface of the mercury-drop electrode and then dissolve in the mercury to form an amalgam. The maximum possible preconcentration depends on the volume of the mercury drop relative to the volume of the ionic solution. Ions that cannot be reduced to the metallic state are not taken up.

if the electrochemical reduction proceeds for a long enough time, most of the mass of the metals that amalgamate (form amalgams) will dissolve in the drop of mercury and become highly concentrated there.

Thus, in one operation—electrochemical reduction at mercury—components can be both separated and concentrated. Such a method is used quite often when Ca^{2+}, Na^{+}, K^{+}, and Mg^{2+} are present in large excess and elements such as Cd and Zn are being determined. Two substances for which this preconcentration method is used are seawater and blood. Of course, this method can be used in reverse, that is, to remove a matrix of metals such as Zn and Cd when analyzing for trace levels of Ca and Mg.

Other methods of preconcentration of analytes in solution have been mentioned when discussing changing the background matrices: purge-and-trap, solid-phase extraction, liquid-liquid extraction, flotation, evaporation, and precipitation with redissolution. Dilution is simple; add more of the solvent. While it may seem trivial simply to add solvent, when you do so, you dilute all the solutes including the interferents. This may have added benefits if the magnitude of the interference effects is reduced by the dilution. For example, say a homogeneous solution is needed, but an interferent precipitates the analyte at the original solution concentration. However, assume that by diluting by threefold, the precipitate will not form. Incorporating the threefold dilution into the protocol allows the assay to continue with the sample in a homogeneous solution as desired.

A purge and trap apparatus was shown in Figure 3.6 on page 84.

Most of these methods of preparing a sample in the optimum concentration range can be automated. Automating the dilution and preconcentration of solutions is the most developed since it is the most commonly done. Some of these methods are covered in Section 4E (flow injection analysis) and Section 4D (purge-and-trap method).

4.6 Calibration and Standards

In Chapter 2, you read how standard samples were used to test a method for the determination of aluminum. That set of experiments tested the reagents and chemistry of a relatively simple sample, such as might be used in an introductory lab course. Most realistic samples are far more complicated, and the use of standards in their analyses is the topic of Section 4.7. But before discussing some of the different standardization procedures, let us digress briefly on the nature of analytical measurement and types of standards.

In chemical analysis, a **direct measurement** (such as mass, time, volume, and voltage) seldom provides a measure of content. Chemical concentrations are found by **indirect measurements**, which are based on direct measurements together with **calibration**. More specifically, in the context of a chemical analysis, to calibrate means to ascertain the relationship between the content of the sample and the response of the assay method. Do not confuse calibration with validation: the latter term denotes ascertaining whether a specified analytical methodology is satisfactory to use both by a single analyst and among a number of laboratories and analysts.

Chemical standards are, then, standards of chemical content. This is in contrast to physical standards, which are used as standards of the characteristics of materials: for example, physical standards involve elasticity, color, and viscosity, as well as the more basic mass, time, length, and voltage. Chemical standards are pure substances, mixtures, solutions, gases, or materials such as alloys or biological substances that are used to calibrate and validate all or part of the methodology of a chemical analysis. Some of the types of standards available are described in Section 4F.

4.7 The Types of Standards

Standards of chemical content are used to determine preparation recoveries and to calibrate assays. The methods of standardization can be separated into two types: **external standards** and standards that are added to each sample—let us call them **added standards**.

An external standard is one that is analyzed separately from the replicate unknowns being tested. A series of such standards contains varying amounts of the analyte, together with a matrix that is similar or identical to that of the samples. For a very well-behaved assay, it may be possible to use only a blank and one standard to establish the calibration.

The added standards are further divided into two categories. They both involve adding a known amount of standard, a **spike**, to each sample being analyzed.

The first category, the **internal standard** technique, involves adding a single spike to each replicate sample and to the blank. The standard is sufficiently different chemically from the analyte such that, while it is detected in the same experiment, it does not interfere with the assay. Nevertheless, the standard is sufficiently similar such that its recovery reflects the recovery of the analyte and any losses can be compensated. Before an internal standard is used, calibration of the assay is needed for both the standard and the analyte. The internal standard can be added before the sample preparation or before the assay. In the first case, the calibration and recovery measurement includes both the preparation and assay steps; in the latter case, only the assay alone.

The second category is called a **standard addition**. Standard additions are fixed amounts of analyte that are added to each sample after an initial measurement; the assay measurement is then repeated after each addition. Additions and remeasurements usually are done one or two times. By extrapolation (described below), the amount of analyte originally in the sample can be found. Spikes for the standard addition technique are added after the sample preparation is completed; the calibration pertains to the assay step alone.

External Standards

External standards can be used to calibrate an assay method when the components of the matrix—including any reagents required for the preparation—do not cause an interference. External standards can also be used if the analyst has enough control of the conditions that the interferents' contribution to the measurement can be kept constant; a correction for the interferents' determinate error then can be made.

Let us look at a practical example in which the concentration of platinum in an automotive catalytic converter was determined. This illustrates the use of an external standard—or, more correctly, an external standard *series*. The manner in which a calibration can be done is shown in the following example. Here, *calibration* refers to relating the instrument output to the analyte concentration.

The external calibration of the instrument response is made by using standards containing different concentrations of analyte in solutions that contain the same matrix components as are in unknowns. Ideally, only the concentration(s) of the assayed species should change in a series of standard solutions. To emphasize the point: It is important to keep constant the concentrations of all the other species that could influence the assay measurement.

Atomic absorption makes its primary appearance in Chapter 11.

EXAMPLE 4.3

An atomic absorption assay was done on catalyst samples prepared by drying and grinding 300 g samples of the catalyst material (extremely stable alumina, Al_2O_3), dissolving it in an H_2SO_4-H_3PO_4 mixture, fusing insolubles from that step with $K_2S_2O_7$, and combining the solutions. Radioactive tracers were used to evaluate the completeness of a tellurium-stannous chloride precipitation of the platinum and palladium from the solution of dissolved catalyst material. No significant amounts of platinum were lost.

A set of platinum standards was made. The concentrations and the instrument response to each standard are shown in Table 4.8. The instrument response is given in *arbitrary units*. We need not care at this time what the units are. The trend in the output is seen more clearly in a graph of the instrument response versus the concentration of the standards. This graph is shown in Figure 4.4. The response is clearly linear—that is, directly proportional—with concentration. This calibration enables us to relate the instrument response to the content of platinum in the sample.

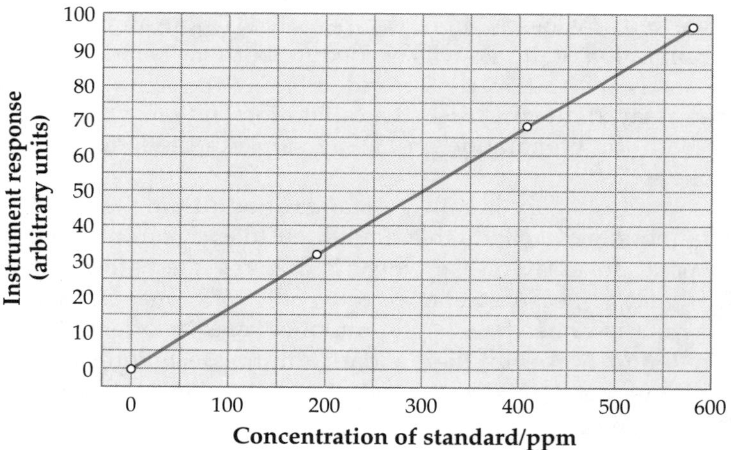

▲ FIGURE 4.4
Plot of the calibration data for platinum.

Table 4.8 Result of assay on a series of external standards

Concentration of standard (ppm)	Instrument response (arbitrary units)
0	0.0
190	31.7
410	68.3
580	96.6

Data from Potter, N. M., Lange, W. H. 1981. *Am. Lab.* 13:81–91. Potter, N. M. 1976. *Anal. Chem.* 48:531–534.

Table 4.9 Results of assay on replicate samples: instrument response

Sample number	Instrument response
1	64.3
2	63.5
3	64.5
4	63.3

Immediately after the instrument was calibrated using the standards, and *without changing the procedure used for the assay*, four samples were run and the output recorded. (See Table 4.9.)

The meaning of the linear response can be expressed as an equation:

$$\text{instrument output} = \text{constant} \times \text{concentration} \tag{4-4}$$

Find the value of the calibration constant and the platinum content of the catalyst.

Solution

Using the calibration data, the value of the calibration constant is determined. This can be done with a least-squares calculation, which is a built-in function of many hand calculators. (If you want to check the result in this example, do not neglect the point at ppm 0 with instrument response 0.)

Thus, from the calibration data of Table 4.8, Equation 4-4 becomes

$$\text{response (arbitrary units)} = 0.167 \times \text{concentration of Pt (ppm)}$$

and using the results listed in Table 4.9, the solution concentrations can be found. The results for the replicates are listed in Table 4.10.

Algebraic details of least squares regressions were presented in Section 2.16.

Table 4.10 Results of assay on replicate samples: concentration

Sample number	Solution concentration/ppm
1	386
2	381
3	387
4	380

The constant in Equation 4-4 will be specific for the following factors:

1. The instrument,
2. The conditions of the instrument at the time of the determinations, and
3. The other materials present in the sample matrix after the sample treatment.

Often, not all interferents can be eliminated during sample preparations because of time or cost constraints. However, carefully formulated standards make it possible to ascertain the effects of any interferents and allow the results to be corrected for those effects. Otherwise, a calibration could not be made. If any of the conditions or preparation steps are changed, the instrument must be recalibrated and, perhaps, new (external) standards prepared.

We have looked in detail at only two examples of how standards are used to calibrate an assay: those describing the gravimetric assay for aluminum (Section 2.9) and platinum in automobile catalysts (this section). However, the strategy is the same for all assay methods in daily use. The standards occupy the same place in the analyses: ascertaining the content and assuring the accuracies of the methodologies.

Added Standards

There are three situations in which standardization is done by adding the standard material to the samples themselves:

- When the solid or liquid matrix of a sample is either unknown or so complex that an external standard cannot be used with confidence;
- When the chemistry of the preparation or of the assay method is complex or highly variable; and
- When the assay depends on highly precise instrumental conditions that are difficult to control.

In the worst case, all three problems can occur during an analysis. Such a case might occur if the sample is complicated, the preparation is difficult, and the assay involves injecting the sample into a flame. The chemistries of all three stages are quite complicated, and flame conditions are sometimes difficult to regulate with the desired precision.

The use of added standards of various types are illustrated by a number of examples below. The first example involves the addition of a spike of standard to a blank to assess the percent recovery during a method development.

Chromatographic methods are the subjects of Chapter 13.

EXAMPLE 4.4

A pharmaceutical assay for a new polypeptide drug is being developed. The formulation is set with the pills having a particular multilayer filler. The laboratory wants to have recoveries between 95% and 105% in the preparation to ensure precise quantitation by a chromatographic method. A standard solution of polypeptide is used to spike a small batch of placebos such that each pill has 3.20 mg of the polypeptide in it. To dissolve the contents fully, the pill has to be stirred in an acidic solvent. When the dissolution is complete, the volume is brought to 10.00 mL in a volumetric flask. The output is found to be 15.42 units when 10 μL of the 10 mL solution is injected into the instrument. The instrument response to a blank is zero, and the response is known to be linear with concentration.

A standard solution was made consisting of 40.0 mg of the polypeptide spike in 100 mL of water and was then kept frozen at -70°C until just before use. A 10 μL sample of the standard solution at the same pH had a response of 18.25 units. Does the recovery fulfill the laboratory requirements?

Solution

We calculate that the concentration w/v of the standard is 0.400 mg mL^{-1}. That concentration in 10 μL gave 18.25 units response. For the spike, 15.42 units for a linear response represents

$$\frac{15.42}{18.25} \times 0.400 \text{ mg mL}^{-1} = 0.338 \text{ mg mL}^{-1}$$

The 10 mL sample has, then,

$$10 \text{ mL} \times 0.338 \text{ mg mL}^{-1} = 3.38 \text{ mg standard,}$$

and, the recovery is

$$(3.38 \text{ mg}/3.20 \text{ mg}) = 1.06$$

The method has 106% recovery. Another sample preparation needs to be devised, since the requirement was for 95% to 105% recovery. How can you get greater than 100% recovery? The answer is, there are interferences present.

The results of spiking with a standard are most certain when done with the analyte itself, such as is shown in the example above—there, a blank is spiked with the analyte. However, it is sometimes useful to spike a sample with a standard material that is chemically different from any analyte. In order to use such an approach, the assay method must produce this different output response for both the analyte in the sample and for the spiked sample. In this case, the best material to use is the analyte, but one that is isotopically substituted at some point in the molecule. The isotope may be radioactive (detected and quantitated with radioactive counting techniques) when the analyte is not radioactive. Examples are to use radioactive ^{57}Co to trace the natural ^{59}Co and tritium substituted for hydrogen in organics. Alternately, the isotope can be stable but of different mass (quantitated by mass spectrometric techniques). An example is to use stable ^{13}C for the more abundant ^{12}C; the confidence in the results is excellent.

See mass spectrometric quantitation in Chapter 12.

An alternate method is to spike with a compound that has a chemistry closely related to the analyte's but sufficiently different that the spike and analytes can be quantified separately. The measure of content is then calculated from the relative response for the analytes and for the separate standard. You may want to use this type of spike if:

1. A blank cannot be obtained;
2. You want to use an assay technique that does not require expensive, pure isotopes or radioactivity;
3. There is significant, unavoidable variability in some part of the analytical methodology; or
4. High precision is desired.

By carrying the spike through the assay, we can use the ratio of analyte to spike to correct for the losses and/or instrument response changes. In this way it is possible to use the output response to the standard material to compensate for any variations in the steps of the determination. For instance, if the standard is added to the original sample and only half of the standard is found at the end of the analysis, then we can assume that half of the closely-related analyte is lost as well. In essence, the percent recovery is established with every sample run.

Examples of spikes that differ from analytes are 2-pentanone as an internal standard spike for 2-hexanone, palladium with platinum, and bromochloromethane for tetrachloroethylene. For the polypeptide of Example 4.4, a spike might be made with a polypeptide having one amino acid different.

Internal standards that differ from analytes are used with almost all the methods presented in this text. Exceptions are a number of titrimetric methods and kinetic methods.

Titrimetric methods in Chapters 7 and 9. Kinetics in 16.

The third way that spikes of standards are used is to add them to each sample as a *standard addition*. The standard addition method can only be used when the assay response is selective enough so that response arises from the analyte alone.

The standard addition method is carried out as follows:

1. An assay measurement is made on the sample being tested.
2. A known amount of the analyte is added (the standard addition).
3. The sample, modified by spiking, is then assayed again under the same conditions.

4. Further spiking and remeasuring—steps 2 and 3—may be repeated a few times.
5. The amount of the original constituent is calculated from the data points by extrapolation.

An example of this procedure for a copper determination follows.

EXAMPLE 4.5

Three aliquots of 5 mL each are taken from the same original sample solution. Nothing is added to the first aliquot, and 5 μL of a 100 ppm solution of copper chloride standard is added to the second. 10 μL of the solution is added to the third. This procedure is outlined in Table 4.11, and the experimental results are illustrated in Figure 4.5. Ignoring the small change in volume for now, calculate the concentration of the copper in the sample.

Table 4.11 Outline of a copper assay using standard addition

5-mL aliquot number	Standard 100-pm Cu^{2+} added (μL)	Final concentration of Cu^{2+}added (ppm)
1	0	—
2	5	0.10
3	10	0.20

Solution

The spiked copper amounts to 0.10 ppm for the second sample and 0.20 ppm for the third. (The spike is diluted 1000-fold and 500-fold for aliquots 2 and 3, respectively.) In Figure 4.5, the instrument response (in arbitrary units) is plotted versus the amount of copper added. Each data point corresponds to a measurement. The "zero" (on the horizontal axis) is the instrument response when the original solution was measured. The other points are the measurements on the spiked aliquots. The higher responses are due to the sum of the copper in the original solution and the copper of the spike.

In this case, a linear extrapolation can be made, and the absolute value of the intercept on the horizontal axis is the calculated copper content of the original sample. Here, the concentration of copper is 0.17 ppm. More details are presented in the figure caption.

One drawback of the standard addition method is the number of separate analyses required. As mentioned above, the method is unable to compensate for a non-zero blank.

The benefit from standard additions is that the procedure allows compensation for quite complex effects. However, the concentration range over which the instrument response remains linear tends to be quite narrow—less than an order of magnitude (factor of 10). Thus the preparation must include bringing all samples into this narrow working range. At least part of the reason for this narrow linear range is not the inherent instrument limitations but the complicated chemical effects that necessitated the spiking in the first place. It always seems that an easy win is just out of reach!

When the instrument response is nonlinear, calculating the concentration in the original sample by using standard additions becomes more compli-

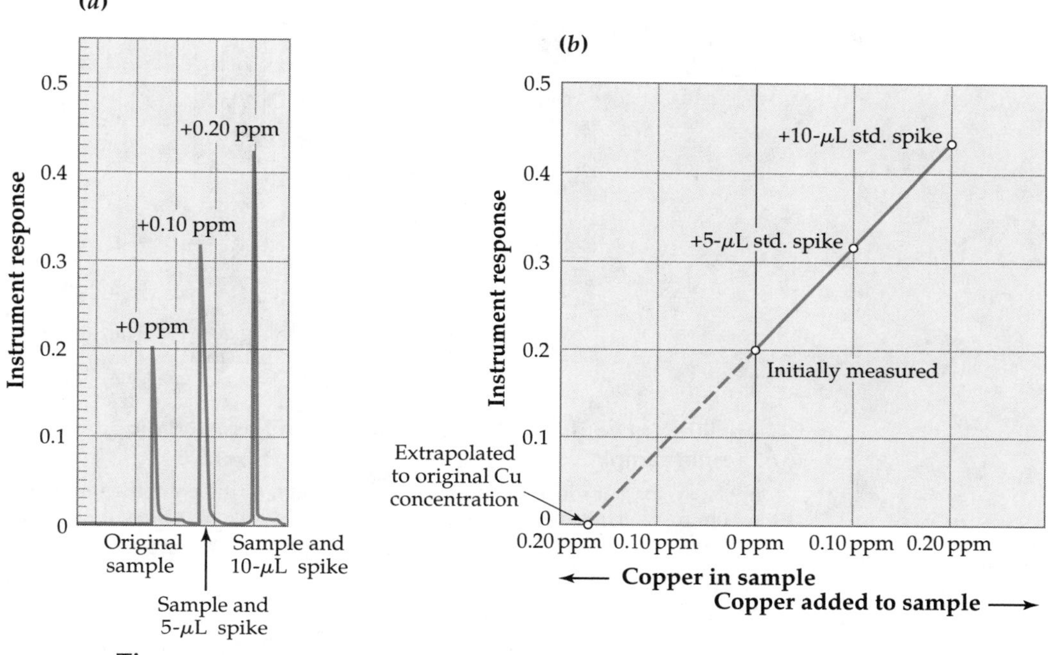

FIGURE 4.5 ▲

Data from a set of measurements of a sample and two standard additions.

(a) The data as they appear on a chart recorder. The peak heights are used as the measure of the instrument response; **(b)** plot of the instrument response versus concentration. The sample concentration is unknown and is plotted at "0." The two spikes add 0.1 ppm and 0.2 ppm of Cu, respectively. A linear extrapolation back to zero-response yields the content of the sample. Among the assumptions of this method are that the instrument response is linear over the entire range tested and that the instrument response reads zero when no Cu is present in the sample matrix.

cated than a simple extrapolation, and the result is significantly less certain. Standard addition is most commonly used with electrochemical methods, gas chromatography, mass spectrometry, gas-chromatography/mass spectrometry, and emission spectrometries such as atomic emission.

See (chapters): electrochemistry (15), GC (13), MS and GC-MS (12), and atomic emission (11).

4A A Deeper Look: The Basis of Ultrasound Assisted Dissolution and Digestion

Dissolving a solid, like any other chemical reaction, involves forward and reverse reactions that depend on concentrations. As the solid dissolves, material can move away from the surface only by diffusion if there is no stirring. Since diffusion is a relatively slow process, dissolution is usually accelerated if the solution is stirred. A more general term for stirring is "enhancing mass transfer." You are, of course, aware of the stirring done by a stirring bar on a magnetic stirrer or by hand. But stirring can also be done by using ultrasound. Ultrasound refers to sound vibrations that occur above the normal range of hearing. The vibrations stir a solution by generating microscopic bubbles that expand and contract. These bubbles can grow until a critical size is reached whereupon they collapse, producing a transient (<μs)

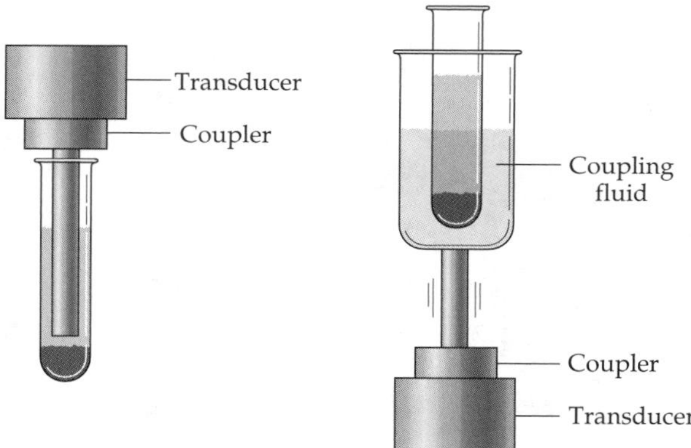

FIGURE 4A.1 ▲
Two different configurations of an ultrasound-assisted digestion or extraction of a solid sample.

On the left, the transducer is immersed in the solvent. At the right, the transducer attaches to the bottom of a container. The container is filled with liquid to couple the ultrasound power into the sample. It is possible to have different efficiencies by changing the geometry and the shape of the tube with the sample.

local production of high temperatures (~5000 K) and high pressures >1000 atm). In addition to stirring the solution, bubbles can break up a solid surface. They can form under surface films and detach the film from the surface. Bubbles also form in microscopic cracks, and their expansion then helps break up the solid.

As an example of the power of ultrasound, the following calculation has been made. At 20 kHz the surface of an ultrasound generator that displaces 50 μm in its vibration reaches a velocity of 10 m s^{-1} and has an acceleration approximately 100,000 × gravity. (The acceleration of a cannon is about 20,000 × g.) Thus, we can characterize ultrasonic dissolution assistance as providing a small displacement, a modest velocity of stirring, and incredibly rapid accelerations.

The wavelength of the sound is short enough that it can be on the order of the size of the containers. As a result, the sound waves can be focused and bent at surfaces where the velocity of the sound changes, just as light is bent and focused by changes in the index of refraction at a surface. This property means that there can be significant differences in effectiveness of sonication depending on the geometry of the apparatus, such as the sizes and shapes of the digestion tubes and placement and shape of the sonicating transducer. Two different geometries are shown in Figure 4A.1.

4B A Deeper Look: Microwave Assisted Sample Preparation

Most analytical techniques for the elements themselves use wet acid digestions. Microwave ovens offer an unprecedented level of control and speed to sample digestions and dissolutions that use such acidic aqueous solutions.

Samples for microwave digestion are held in open or closed containers most often made of a type of Teflon™, (perfluoroalkoxy)ethylene (Teflon PFA). (See Figure 4B.1.) One reason for using a polymeric material is that it is relatively transparent to microwave radiation at the commonly used fre-

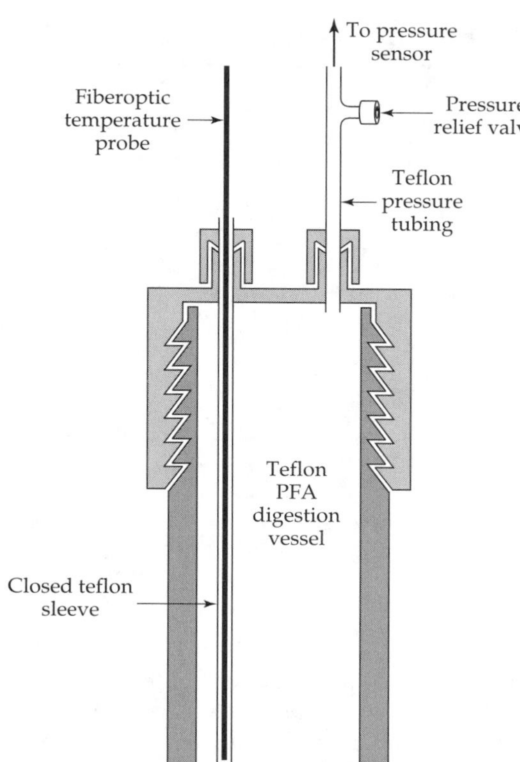

◀ FIGURE 4B.1
A microwave digestion vessel for carrying out digestions in a closed volume.

Such vessels often are composed of a polymer Teflon PFA. This material has a melting point of 306°C, significantly higher than other Teflons. Teflon PFA has a limit of temperature for repeated use of about 250°C. Regardless of their different designs, all such vessels have relief valves to prevent the gas pressure from rising too high unless continual temperature monitoring is used for all vessels.

quency of 2450 MHz. As a result, the microwave energy penetrates into the sample and heats it directly. Fused silica (often misnamed quartz) or glass containers can also be used; they have the same microwave properties.

Microwave heating results from the varying electric field of the microwaves. The electric field increases and decreases at the microwave frequency, and tends to cause molecules such as water to rotate, the ions in ionic solution to migrate, and the electrons in metallic materials to move. The rotational motions of the water molecules tend to be stopped by collision with the surrounding molecules, and these collisions heat the liquid. Similarly, the moving ions tend to be stopped by the liquid, and their energy is transferred throughout the solution as heat. In metals, the resistance to electron flow tends to heat them rapidly. (This effect limits the use somewhat for metal digestion, as will be described below.) None of these heating mechanisms is likely to occur in solids such as ores, rocks and dry soils; these solids are heated predominately by contact with the surrounding liquid.

In predominately aqueous samples, the microwaves are absorbed within a few cm of the surface, and the dropoff in heating is exponential. As a result of these absorption properties, a 10 mL sample is heated nearly the same throughout its volume, while a 100 mL sample will be slightly hotter at its outer surface than at its center. The exact difference depends on the sample's shape. But the center of the larger sample will still be heated to a given temperature from 10 to 100 times as fast as by conventional, thermal heating. That is, the samples can be brought to the desired digestion temperature in a microwave oven in only a few minutes. In addition, in closed containers,

Table 4B.1 Normal boiling points *versus* closed-vessel (175 psig†)

Closed-vessel		
Solvent	**Boiling point/°C**	**Temperature/°C at 175 psig**
Dichloromethane	39.8	140
Acetone	56.2	164
Methanol	64.7	151
Ethanol	78.3	164
Acetonitrile	81.6	194
2-Propanol	82.4	145
Acetone:hexane[a] 1:1 v/v	52	156
Acetone:cyclohexane[b] 1:1 v/v	52	160
Acetone:petroleum ether[c] 1:1 v/v	39	147

a. Hexane B.P. 68.7°C but it is not heated by microwaves
b. Cyclohexane B.P. 80.7°C but it is not heated by microwaves
c. Petroleum ether B.P. 35–52°C but it is not heated by microwaves
† psig = per square inch pressure of the gas
Source: Am. Lab. 1994. 26 (August):34–40.

digestion is accelerated since the boiling point rises above the normal boiling point of the acid as the pressure rises above atmospheric pressure. As an example, concentrated (constant-boiling) nitric acid boils in an open container (at 1 atmosphere pressure) at 120.5°C, but in a microwave pressure vessel, the temperature can be held at about 185°C. Table 4B.1 shows boiling points applicable to microwave digestions.

The limits of microwave digestion arise from the pressure and temperature limits of the containers and the electrical and chemical properties of the samples. For example, Teflon containers cannot be used for phosphoric or sulfuric acid, since the boiling points lie above the melting point of the container. Silica containers are usable for these, but silica cannot be used with hydrofluoric acid, which can attack it.

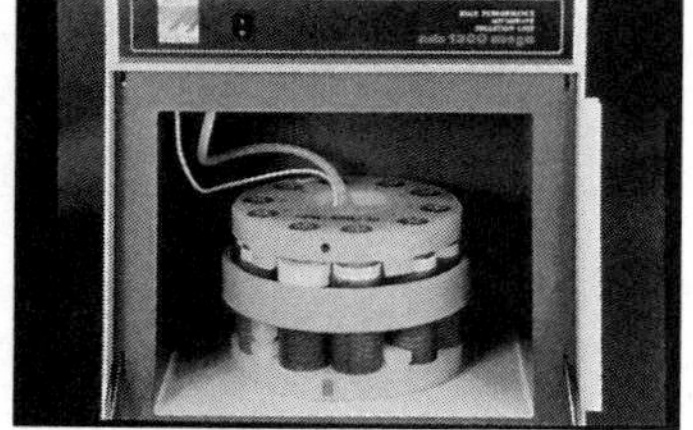

Microwave digestion oven.
Courtesy of Milestone, Inc.

Microwave digestion of metals and alloys presents a unique set of problems because of their high electrical conductivities. The sample particle can become so hot, even when immersed in an aqueous solution, that they can melt the polymer container. Again, a silica container can solve the heat problem, but other difficulties remain. One is the production of hydrogen by some metals when they react with acid. In closed containers, a relief valve will release the vapors if the container's limiting pressure is reached. This is not a problem as long as only gases are vented. However, sample material will be lost if droplets of liquid escape with the gases. Another difficulty arises since metals may actually produce sparks between pieces. If any oxygen is present with the hydrogen, an explosion can be ignited by these sparks. As a result, in both open and closed containers, oxygen must be excluded to avoid an explosion hazard, and the metal should be allowed to react thoroughly before closing the container and heating the sample. The oxygen can be excluded by purging the container with nitrogen or other gases that are inert under the digestion conditions. A third type of difficulty is the production of explosive mixtures if organic substances are present whether as solvent or as a sample, such as oil, to be digested. Their lower boiling point components can form explosive mixtures with oxygen. Finally, organic digestions can produce CO_2, and, as in the case of H_2 buildup, the CO_2 will be vented and some of the sample may be lost with liquid droplets.

Digestions that require taking a sample to dryness cannot be done in polymer containers. However, high-temperature ashing, digestion, and fusion, as well as taking samples to dryness, can be done in microwave ovens with the use of special inserts made of materials that are rapidly heated by microwave energy, such as SiC and ferrites. These inserts allow an increase in the rate of heating and improved temperature control compared to conventional high-temperature heating methods such as with flames or electric muffle furnaces. The microwaves do not heat the samples directly in this case. The heat is transferred by thermal conduction from the inserts to the samples in their containers.

Microwave ovens can also be used any time when rapid heating is useful. Examples are rapid drying of soil and wood samples, rapid sample concentration (as long as the analyte is not volatile), and heating samples during flow injection analysis.

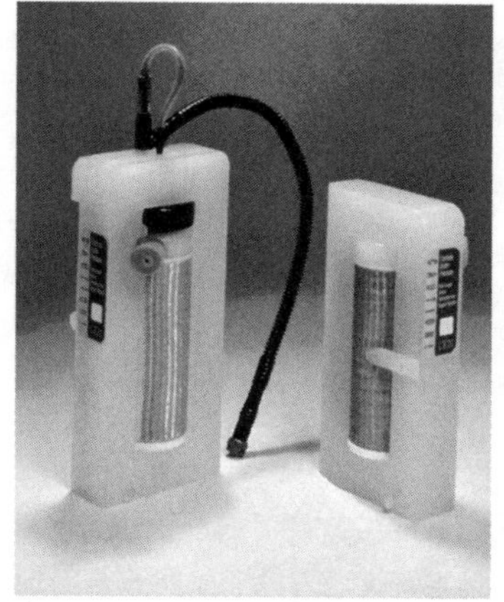

Microwave digestion vessels.
Courtesy of CEM Corp.

4C A Deeper Look: Supercritical Fluid Extraction

Just what is a supercritical fluid? Its properties are both gaslike and liquidlike. Say you have a closed container that has liquid CO_2 in it, and the volume above the liquid is filled by CO_2 gas. Now, increase the temperature and pressure in the container. The increase in temperature causes the liquid to expand and become less dense. Meanwhile, the gas becomes denser as the pressure rises. If the pressure and temperature are raised in just the right way, at a point called the **critical point**, the densities of the gas and the liquid phases become the same. At that point, the distinction between the gas and the liquid vanishes, the interface between the two disappears, and only a single *supercritical* phase remains. The critical point, where the gas and liquid just become indistinguishable, occurs at a specific temperature called the **critical temperature**, T_c, and a specific pressure called the **critical pressure**, P_c. The supercritical phase is neither a gas nor a liquid, so it is simply called a fluid. The values of T_c and P_c are characteristic of a compound, and some values of materials that are useful for supercritical fluid extraction are listed in Table 4C.1. The most commonly used supercritical fluid is CO_2, which is

Flow injection analysis is introduced in Section 4E.

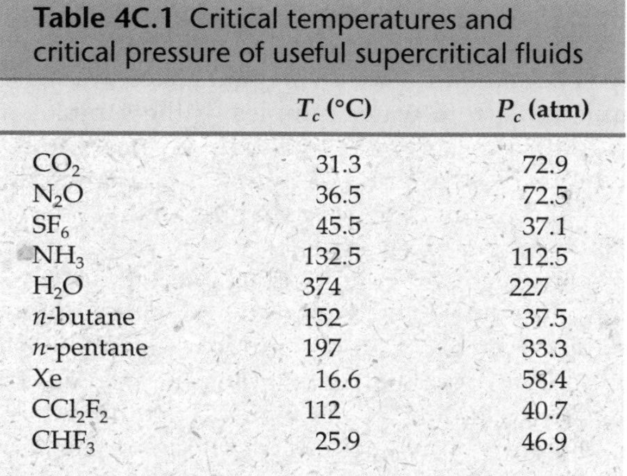

Table 4C.1 Critical temperatures and critical pressure of useful supercritical fluids

	T_c (°C)	P_c (atm)
CO_2	31.3	72.9
N_2O	36.5	72.5
SF_6	45.5	37.1
NH_3	132.5	112.5
H_2O	374	227
n-butane	152	37.5
n-pentane	197	33.3
Xe	16.6	58.4
CCl_2F_2	112	40.7
CHF_3	25.9	46.9

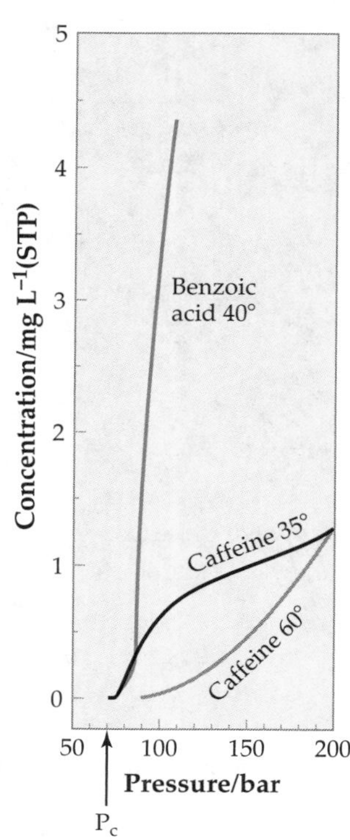

FIGURE 4C.1 ▲
Dependence of the solubilities of analytes in supercritical CO_2 on the pressure and temperature.

Shown here are the solubilities of pure caffeine at two temperatures, 35° and 60°C, and of pure benzoic acid at 40°C. The values of the solubilities of these pure materials will not be the same as when they are being desorbed from a matrix, but the trends remain the same. [Data from: Stahl, E., Schilz, W. 1979. "Mikroanalytische Untersuchungen zur Löslichkeit von Naturstoffen in Überkritischem Kohlendioxid." *Talanta* 26:675–679.]

essentially a non-polar solvent. When more polar molecules are to be extracted, a few percent of **modifiers** such as methanol or ethanol may be added to make a mixed fluid.

At temperatures and pressures higher than T_c and P_c, there is still only a single phase, and its fluid and chemical properties change relatively rapidly with P and T. Figure 4C.1 shows how easily changes occur in the solubilities of two representative compounds, caffeine and benzoic acid.

The general rule for supercritical fluid solubilization is that the equilibrium amount of solute increases with the density of the supercritical fluid (changed by increasing the pressure at a constant temperature), and with increasing temperature at a constant density. The ease of change in solubility has the benefit of allowing supercritical extraction to be "fine tuned" to dissolve some molecular species better than others. This fine tuning means that the selectivity improves by changing the temperature and pressure.

Compounds removed from their matrix by supercritical fluid extraction may be recovered in four different ways. All of them are done after lowering the pressure, which lowers the density of the fluid:

a. Lowering the pressure of the fluid alone in a collection container;
b. Trapping at low temperature;
c. Trapping on an adsorbent (see purge and trap, Section 4D) followed by elution; or
d. Trapping by bubbling through a liquid solvent.

Methods (c) and (d) are more commonly used, because it is difficult to recover the solutes reproducibly or well on a surface since the gas flow becomes so fast after dropping the pressure. For instance, supercritical CO_2 expands about 500-fold to form the one-atmosphere gas: so 1 mL min^{-1} of supercritical CO_2 becomes about 500 mL min^{-1} of gas. In cases (a) and (b), it is probable that volatile analytes would be lost. However, nonvolatile species can be recovered with any of the four techniques as long as the gas flow does not blow analyte particulates out of the collection container.

Of the compounds listed in Table 4C.1, CO_2 is the most widely used. This is because it has a convenient critical temperature, it is neither flammable nor toxic, and it is inexpensive.

4D A Deeper Look: Purge and Trap

Purge-and-trap is a method for removing and concentrating small amounts of organic compounds from water samples. If the samples must be transported and stored before analysis, it is usually necessary to refrigerate them to prevent any loss of volatile components. One configuration of the equipment for purging and trapping volatile analytes is illustrated in Figure 4D.1. The technique involves passing a gas, such as helium, through a solution that contains volatile organic compounds in low concentrations. As the helium bubbles through the solution, it picks up the organic compounds. The gas, which is called the **carrier gas**, is said to purge the solution. After its passage through the solution, the gas, now containing the purged organics, is then directed into a tube containing a solid material that has a large surface area for its volume. The purged organics adsorb on the surface. Commonly used

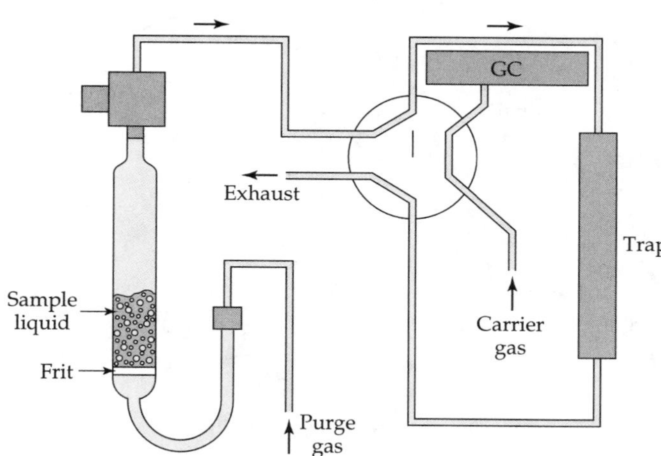

◀ FIGURE 4D.1
One design of a purge-and-trap apparatus.
The liquid sample is added to the purge compartment from the top while the purge gas is flowing. A frit is a thin, porous plug that lets gases flow through it easily. The purge gas carries the volatile components from the ~5 mL sample to the trap over a period of minutes. The six-port switch is then rotated counterclockwise by 60°. This rotation connects the purge solution to the exhaust, and the carrier gas then goes through the trap, though the switch again, and into the gas chromatograph (GC).

solid materials packed in the tube are charcoal, silica gel, and Tenax™. (Tenax™ is a porous polymer based on 1,6-diphenyl-*p*-phenylene oxide.)

After a time determined by previous testing (on the order of minutes), the gas flow is stopped. Some fraction of the total of each volatile organic has now been carried to and concentrated in the trap. Next, the trap is heated as rapidly as is convenient, and the volatile organics are driven out into the instrument being used for the assay. (The binding to the surface in the trap must be weak enough so the purged organics can be vaporized this way.) The instrument is usually a gas chromatograph, which for now is another "black box."

The GC black box is opened in Chapter 13.

Not all the organics are totally recovered, and the recovery of each may vary. Some factors affecting the recoveries are the analyte concentrations and the identities and amounts of other materials in the matrix. How the purging and trapping properties effect recoveries are illustrated in the graph of Figure 4D.2. There you can see that the final recovery is dependent on time. It should be clear that for a given mixture, developing and validating an analysis using purge-and-trap in the sample preparation will likely require spiking a number of internal standards to ascertain recoveries: at least one for each class of organics such as simple hydrocarbons, polyaromatics, alcohols, etc.

An important point to understand is that the trap has a fixed amount of surface area. The surface area has a limited **capacity**; it can hold only a finite amount of material. If some amount of some of the analytes passes through the trap and is lost, this is called **breakthrough**. Breakthrough can be defined as occurring when some fraction of the mass of a material is lost or, alternately, when the ratio of output concentration to input concentration reaches a specified level. In either case, the volume of carrier that has passed through the trap when the defined breakthrough occurs is called the **breakthrough volume**.

One requirement in driving off the trapped components is that all of them must be driven off the collection surface (**desorbed**) simultaneously, as quickly as possible, and quantitatively. The effect of the heating rate on this process is illustrated in Figure 4D.3.

The use of solid adsorbents at approximately ambient temperature is the most common purge-and-trap technique. In some cases, though, the trap is a tube held at low temperatures such as that of dry ice or liquid nitrogen. This

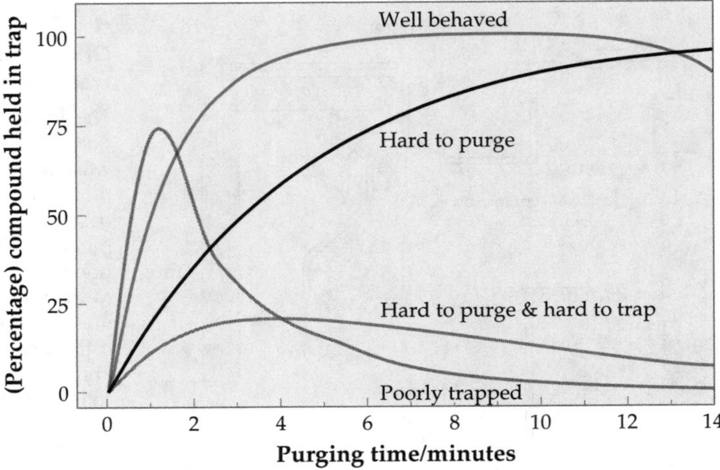

FIGURE 4D.2 ▲

Graph of the general behavior of the volatile, purged components from a mixture.

The components are divided into those hard or easy to purge and hard or easy to trap. The graph illustrates the fraction of each component that is held in the trap. If a component is easy to purge, it arrives quickly; hard-to-purge components are slow to arrive. Easily trapped components remain in the trap for minutes, while poorly trapped compounds bleed out and are lost. Thus, there are four types of behavior expected. If the desorption from the trap for all compounds is 100%, then this graph indicates the recovery *versus* time profile for each of the four general classes of compounds.

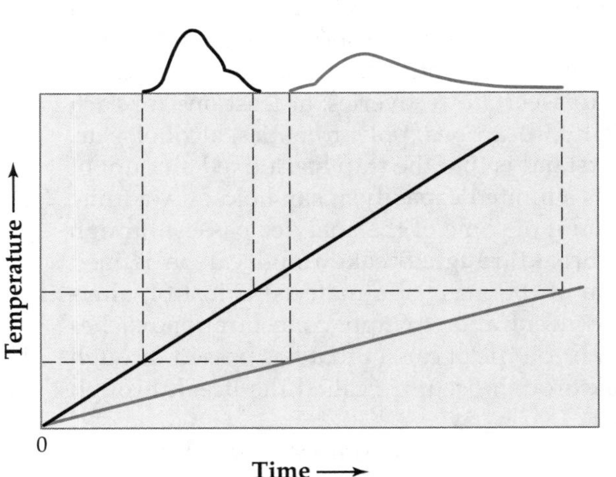

◀ FIGURE 4D.3

Change in desorption properties with rate of temperature rise.

Shown is the hypothetical concentration versus time for desorption of components. Time zero is the time heating is initiated. A slower temperature rise causes the compounds to be released beginning at a later time and to desorb over a longer time. The ripples in the profile indicate a specific substance that has a higher concentration than others released near the same temperature. A faster temperature rise produces an earlier desorption and shorter range of time. In general, a faster desorption is better.

is called a **cold trap**. The cold trap allows the carrier gas to pass through but retains the more easily condensed (or frozen) organic compounds. Since the sample is purged over a period of minutes, trapped on a cold surface, and desorbed over a short time (on the order of seconds), the name **cryofocussing** ("cryo" means *cold*) is also given to this and other, similar methods.

4E A Deeper Look: Flow Injection Analysis

Since sample preparation is so labor-intensive, much effort is directed toward automating the process. For assays that require or that can use solutions of analytes, once the sample is in the liquid form, later steps often involve adding reagents, waiting for reactions to occur, and heating or cooling the solution. These processes are particularly amenable to automation. One way to automate these steps is to pump a sample solution through a network of tubes and add reagents automatically as it moves along the network toward an assay section. Timing can be regulated by varying the lengths of the tubes, the flow rate, and the direction of flow. The temperatures can be changed by heating or cooling individual sections of the tube through which the sample passes. The general class of methods that involve this form of automatic sample preparation before the assay are called, reasonably, **flow injection analysis** (FIA). Devices for flow injection analysis fall in the class of instruments called automatic analyzers. Perhaps, because sampling and digestion procedures often come first and most often are done manually, the classification might more correctly be called automatic solution-sample preparers/analyzers.

In the simple flow injection analyzer shown in Figure 4E.1a, the box labeled P represents the pump that delivers both the carrier fluid and reagent solution. The sample is injected into the line at an injector into the carrier fluid and subsequently the reagent stream joins them. In this setup, the reagent to be mixed with the sample is pumped through a separate tube to mix with the carrier at the merging point and then with the sample after it is injected. The reagent and sample are allowed to mix thoroughly and react in the delay coil. The liquid sample, now reacted with the reagent, passes through the detector, labeled D, and out of the device. The carrier and reagent streams both are flowing when the sample is added.

As shown in 4E.1b, the sample volume begins flowing as a plug (or bolus) between carrier-sectors. As it flows, the reagent intermixes, initiating the desired reaction as shown in the lower tube. In some instruments intermixing of a bolus of sample and the reagent with the carrier is prevented by placing a gas bubble ahead of and behind the sample; this alternate method is called **segmented flow**.

The output for a series of samples and standards will usually resemble that shown in Figure 4E.1c. Either the heights of the peaks or the areas might be used as quantitative measures of the content of each, although for FIA the heights are usually a sufficiently precise measure. Note that the shape of the peaks is not square—with an instantaneous rise, a flat plateau at the top, followed by an instantaneous drop back to the baseline—as you might expect if the sample passed down through the tube in the same compact plug shown in the top tube in Figure 4E1.b. The peaks actually rise and fall more slowly due to two effects. First, the fluid at the center of the tube moves faster than that at the outside edge (as we saw in Figure 4E1.b), and second, the injected material diffuses from regions of higher concentration to regions of lower concentration. These effects always broaden the sample region and lengthen the time it takes to pass through the detector. In other words, the sample is always diluted.

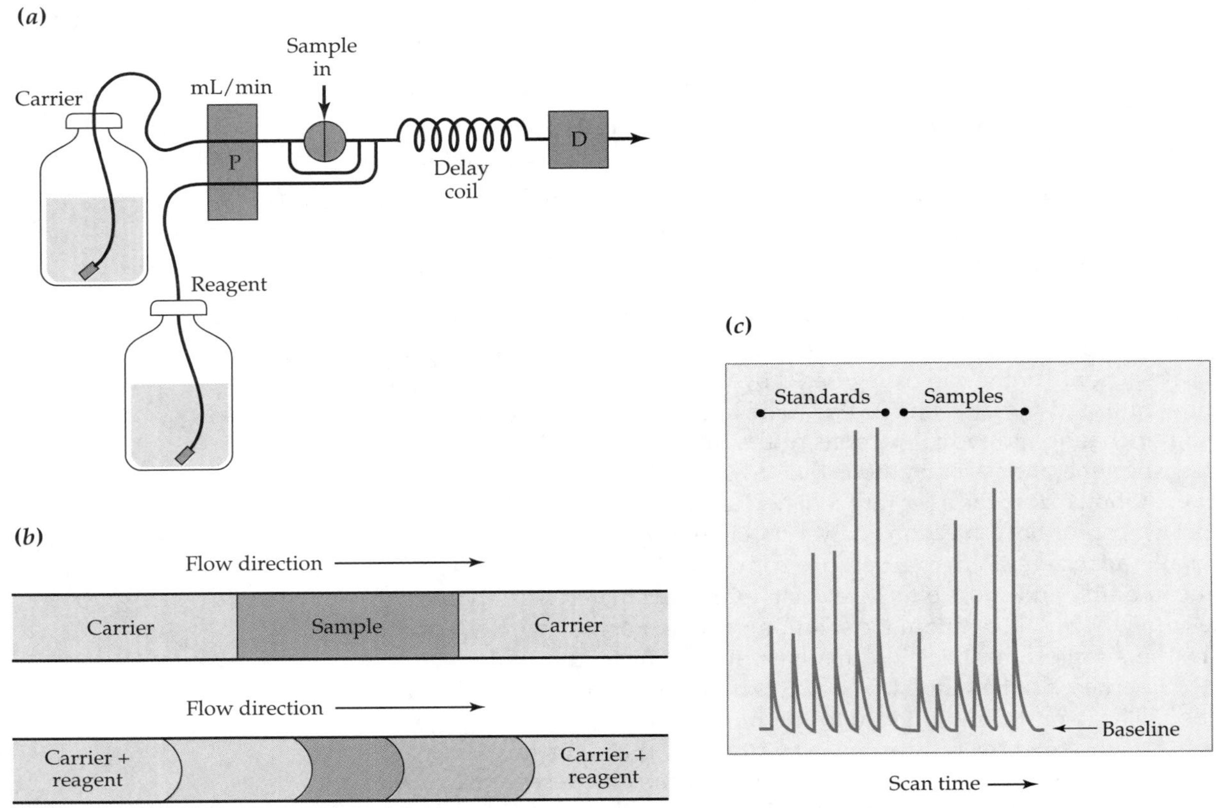

FIGURE 4E.1 ▲
Principles of flow injection analysis.
(a) A representation of a simple flow injection analyzer; **(b)** sample flow in tubes containing a plug (**bolus**) of sample; **(c)** typical response over time for a series of samples and standards.

The mixing of reagents and analytes in the lines occurs for two reasons. One was illustrated in Figure 4E.1b: the penetration of the middles of the sectors into the adjacent sections of the flow stream. An additional contribution to mixing is due to the curved tubes. The shorter path along the inside of a curve compared to the outside sets up mixing eddies across the tubing. That is why a coiled tube or sinuous channel is used in FIA mixing lines. Some analysts enhance this coiling by tying the tubing into knots, which produces even sharper turns.

In Figure 4E.1a, the sample injector is representational. One of the most common injectors is a six-port injector that can place a plug of liquid into a moving stream without interrupting the flow. Some details of its operation are shown in Figure 4E.2.

Another mode of use is illustrated in Figure 4E.3. Here, the flow of liquid is caused by a vacuum on the instrument end. The analyte is added by flow injection controlled at various rates. By adjusting the analyte rates of flow, a calibration set can be obtained with a single standard solution put in.

Figure 4E.4 on page 130 shows a diagram of a much more complex FIA train that is used to analyze for trace levels of molybdenum in digested agricultural plant specimens. The chemical basis of the assay is the conversion of

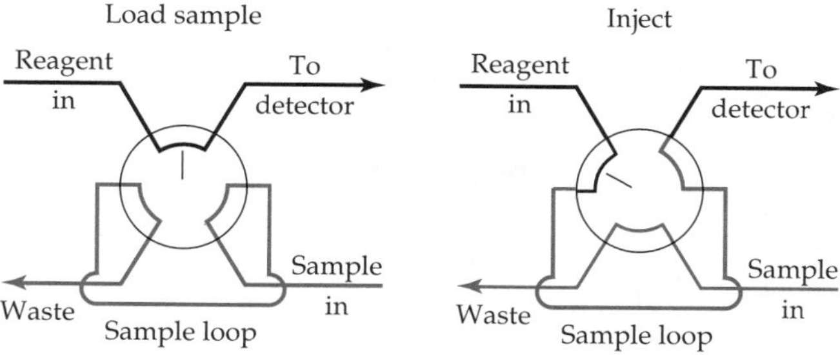

FIGURE 4E.2 ▲

Operation of a typical six-port valve used to inject a fixed sample volume into a flowing stream.

The valve has connections between sets of two ports. The sample is loaded while the valve is in the position shown on the left. Enough of the sample is allowed to flow through the sample loop to wash out the previously held material and ensure that the composition of the solution in the loop is the same as the sample. The sample is injected into the reaction path by turning the valve 60° so that the connections are changed to those at the right. For samples of tens of microliters, the relative standard deviation of the injected volume is generally lower than 0.5% for a series of injections. Electronically driven switching is even more reproducible.

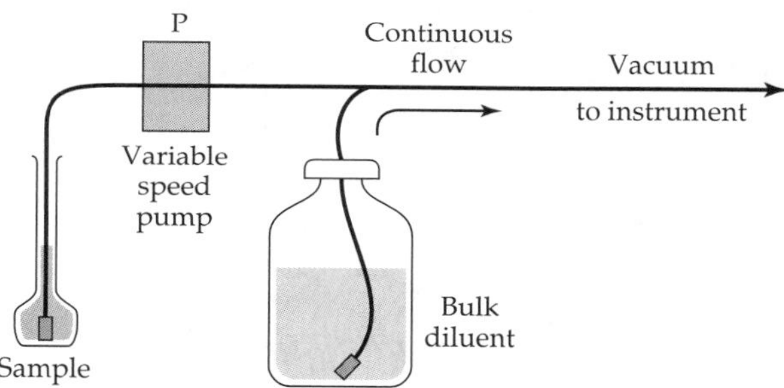

FIGURE 4E.3 ▲

Diagram of the general structure of a flow injection system where the main flow is pulled through the tube into a partial vacuum and the sample is metered into the line by a pump.

By varying the speed of the pump, a series of standards can be run from a single solution.

iodide to iodine by hydrogen peroxide. Molybdenum acts as a catalyst in the reaction

$$2\,I^- + H_2O_2 + 2\,H^+ \xrightarrow{\text{catalyst}} I_2 + 2\,H_2O$$

The amount of iodine produced depends on the concentration of molybdenum (the catalyst). In the assay, the color intensity of the iodine is measured at a fixed time after the catalyzed oxidation is begun.

4F A Deeper Look: Standard Reference Materials

The following is from a catalog published by what is now the U. S. National Institute of Standards and Technology (NIST). It describes different type of reference materials. NBS is the abbreviation for the Institute when it was

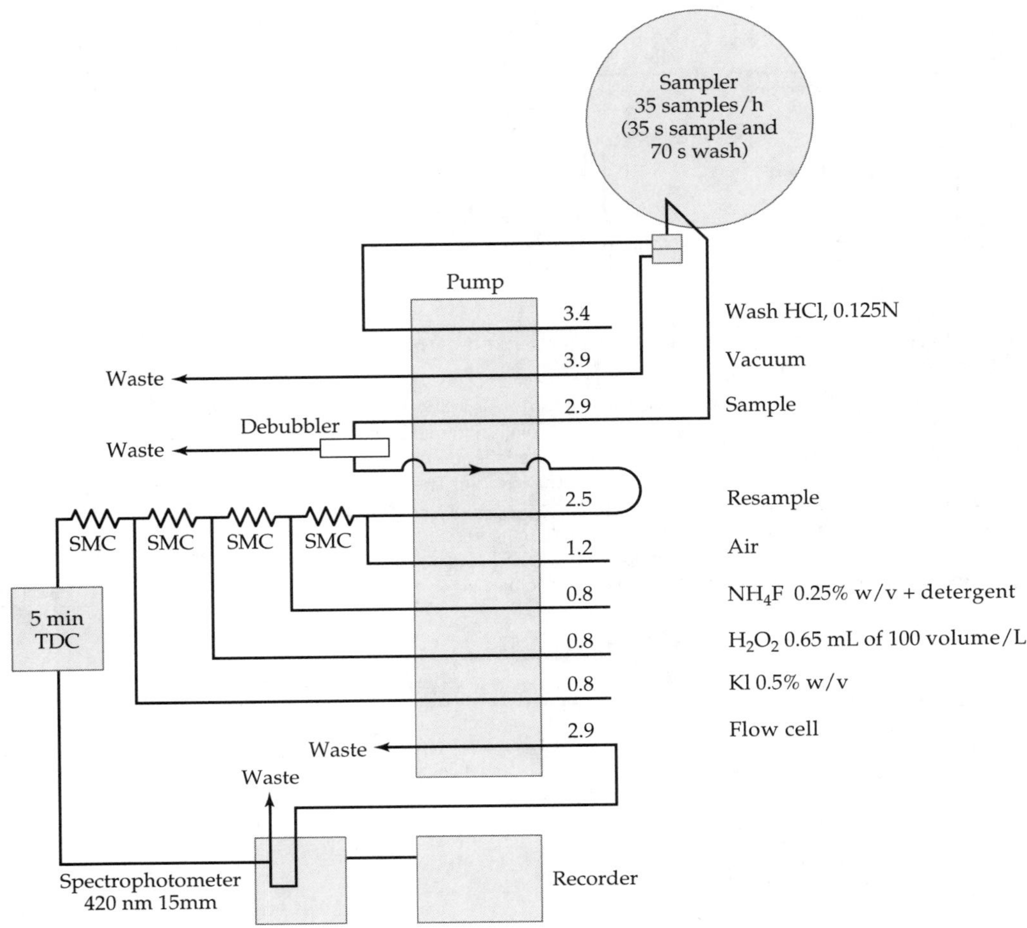

FIGURE 4E.4 ▲

Schematic diagram of the flow injection analyzer set up to determine molybdenum content by catalytic oxidation of iodide.

The pump simultaneously pumps nine different streams, each adding the material noted on the right. The numbers on the lines are the pumping rates in each line in mL min^{-1}. A debubbler removes gas bubbles that might have entered the line up to that point. Here the single-mixing coils (SMC) are symbolized by a jagged line, which is another symbol for a coil in FIA. As in Figure 4E.1, the coils allow each reagent to mix thoroughly before the next one is added. The time-delay coil (TDC) requires 5 min for the sample to travel through it. This is the fixed time allowed for the oxidation of iodide (the last reagent added) to occur. The numbers associated with the spectrophotometer (see Chapter 11) are the wavelength of light and length of liquid through which the light passes. Notice that only a part of the total volume is pulled through the spectrophotometer.

Fluoride is added to the solution in one stage to reduce interference from iron. Iron ions also catalyze the assay reaction and, thus, act as an interference in the molybdenum assay. The added fluoride coordinates the iron strongly (but not the molybdenum), and the iron coordination complex is not active in the reaction. Because the fluoride acts to mask the effects of iron, fluoride is called a masking agent for iron. [Ref: Quin, B. F., Woods, P. H. 1979. *Analyst* 104:552–559.]

called the U.S. National Bureau of Standards. [Ref: NBS Special Publication 260. NBS Standard Reference Materials Catalog, U. S. Dept. of Commerce, 1981.]

The National Bureau of Standards issues over 1000 different materials through its Standard Reference Materials Program. These materials are primarily Standard Reference Materials (SRMs) certified for their chemical composition, chemical property, or physical property, but also include Research Materials (RMs) and Special Reference Materials (GMs). All SRMs, RMs, and GMs bear distinguishing names and numbers by which they are permanently identified. Thus each SRM, RM, or GM bearing a given description is identical (within the required or intended limits) to every other sample bearing the same designation with the exception of individually certified items, which are further identified by serial number.

The first materials issued by NBS were called Standard Samples and consisted of a group of ores, irons, and steels certified for their chemical composition. Since the mid-1960s these materials have been issued as Standard Reference Materials, and cover a wide range of chemical and physical properties and an equally wide range of measurement interests.

Definitions

The different terms, SRM, RM, or GM, are used to indicate differences in the types of information supplied and in the purposes for which the material is intended.

Standard reference materials have been characterized by the National Bureau of Standards for some chemical or physical property and are issued with a Certificate that gives the results of the characterization. These results are obtained by one of the three established routes of certification, i.e., measurement of the property using: (1) a previously validated reference method, (2) two or more independent, reliable measurement methods, or (3) an *ad hoc* network of cooperating laboratories, technically competent, and thoroughly knowledgeable with the material being tested. These routes are described in detail in "The Role of Standard Reference Materials in Measurement Systems," NBS Monograph 148, 54 pages (Jan., 1975). SRMs are defined as being well-characterized and certified materials produced in quantity to improve measurement science. They are prepared and used for three main purposes: to help develop accurate methods of analysis (reference methods); to calibrate measurement systems used to: (a) facilitate the exchange of goods, (b) institute quality control, (c) determine performance characteristics, or (d) measure some property at the limit of the state-of-the-art; to assure the long term adequacy and integrity of quality-control processes. In these ways, SRMs help ensure the compatibility and accuracy of measurements in many facets of national life—from science and technology to trade and commerce.

Research materials, unlike SRMs, are not certified. Instead of a Certificate, RMs are issued with a "Report of Investigation," the sole authority of which is the NBS staff member who authored the report. An RM is intended primarily to further scientific or technical research on that particular material. The principal consideration in issuing an RM is to provide a homogeneous material so that investigators in different laboratories are assured that they are investigating the same material.

Special reference materials, differ from both SRMs and RMs in that NBS does not participate in the characterization of these materials. GMs are reference materials produced and certified or guaranteed by other government agencies, standards bodies, or other non-profit organizations. When deemed to be in the public interest and when alternate methods of national distribution do not exist, NBS acts as the distributor for such materials. This service is available to all organizations that qualify and have reference materials that would help solve a national measurement problem.

Suggestions for Further Reading

ALFASSI, Z. B., WAI, C. M., eds. 1992. *Preconcentration Techniques for Trace Elements*. Boca Raton, FL: CRC Press.

The book contains chapters on each of the usual methods for preconcentration for elemental and organic analysis. Good source for literature references for specific problems. A useful reference.

BAJO, S. 1992. In *Preconcentration Techniques,* Z. B. Alfassi and C. M. Wai, eds. Boca Raton, FL: CRC Press.

A good chapter on the chemistry of conventional sample preparation for elemental analysis.

COLEMAN, P. B., ed. 1991. *Practical Sampling Techniques for Infrared Analysis*. Boca Raton, FL: CRC Press.

This book is a mixture of general chapters full of useful tips about sample preparation for infrared spectrometry and any needed adjustments that depend on the nature of the sample and instrument. Other chapters are essentially specialist notes about specific types of infrared spectrometry. Outstanding if you are planning to do infrared spectral analysis.

KINGSTON, H. M., JASSIE, L. B., eds. 1986. *Introduction to Microwave Sample Preparation*. Washington, D.C.: American Chemical Society.

A broad treatment of microwave digestion methods and a good entry to the literature to the time of its publication. Chapter 3 by S. A. Matthes is a good starting point for the practice of microwave dissolution of solids. Microwave Kjeldahl digestions are also covered in detail.

MIZUIKE, A. 1983. *Enrichment Techniques for Inorganic Trace Analysis*. Berlin: Springer-Verlag.

A tremendously useful book on the subject with the broad array of techniques covered. Figures of the apparatus, references to the literature, and a large number of useful tables are included to allow one to harness the prior knowledge in the field up to 1982.

DE CASTRO, M. D. L., TENA, M. T. 1995. "Hyphenated Flow Injection Systems and High Discrimination Instruments." *Talanta* 42: 151–169.

A review of flow injection analysis and how it can reduce the sample size and reagent consumption when used to introduce samples into atomic emission, infrared, nuclear magnetic resonance, and mass spectrometries. A useful entry to the literature of a set of "hyphenated" methods starting with FIA.

KARLBERG, B., PACEY, G. E. 1989. *Flow Injection Analysis, A Practical Guide*. Amsterdam: Elsevier.

A book half with the fundamentals of FIA and useful hints on setting up an FIA system and half with applications and a extensive bibliography.

SARZANINI, C., MENTASTI, E. 1991. "Metal Ion Preconcentration Techniques with Particular Regard to Waters and Seawater." *Annali di Chimica* 81:185–232.

A highly useful review with an extensive bibliography that provides a fine entry to the literature of preconcentration techniques.

SEKULIC, S., SEASHOLTZ, M. B., KOWALSKI, B. R., LEE, S.E., HOLT, B. R. 1993. "Nonlinear Multivariate Calibration Methods in Analytical Chemistry." *Anal. Chem.* 65:835A–845A.

A difficult, jargon-filled review article that, nevertheless, provides an entry into some more recent literature of nonlinear calibration in analytical chemistry.

BEUKELMAN, T. E., LORD, S. S., JR. 1960. *Appl. Spec.* 14:12–17.

A clear exposition of the characteristics of nonlinear calibration.

WENCLAWIAK, B., ed. 1992. *Analysis with Supercritical Fluids: Extraction and Chromatography*. Berlin: Springer-Verlag.

Presents all the practical details to proceed toward actually doing SFE after you understand the properties of supercritical fluids described here. Chapter 2 presents a formidable physical chemical description of supercritical fluids. Some of the text requires a background in physical chemistry. Others will be clear after you understand this chapter.

WESTWOOD, S.A. 1993. *Supercritical Fluid Extraction and its Use in Chromatographic Sample Preparation*. Boca Raton, FL: CRC Press.

A book containing much of what you want to read about the title subject.

CHALMERS, R. A. 1975. In *Comprehensive Analytical Chemistry*, Vol. 3. G. Svehla, ed. Amsterdam: Elsevier. Chap. 2.

An older guide to standards for all aspects of analysis: with references.

BOGEN, D. C. 1982. In *Treatise on Analytical Chemistry*, Pt. I, Vol. 5, 2nd ed. I. M. Kolthoff and P. J. Elving, eds. New York: Wiley-Interscience. Sect. F, Chap. 1.

A chapter on the methods of sample preparation for inorganic compounds.

DUNLOP, E. C., GINNARD, C. R. 1982. In *Treatise on Analytical Chemistry*, Pt. I, Vol. 5, 2nd ed. I. M. Kolthoff and P. J. Elving, eds. New York: Wiley-Interscience. Sect. F, Chap. 2.

A chapter on both dry and wet methods of preparing samples composed of organic compounds.

SULCEK, Z., DOLEZAL, P. P. J. 1977. "Decomposition Procedures in Inorganic Analysis." *Critical Reviews in Analytical Chemistry* 6: 255–323.

URIANO, G. A., GRAVATT, C. C. 1977. "The Role of Reference Materials and Reference Methods in Chemical Analysis." *Critical Reviews in Analytical Chemistry* 6:361–411.

Two review articles in which expert authors critically survey the methods noted in the titles.

ALVAREZ, R., RASBERRY, S. D., URIANO, G. A. 1982. "NBS Standard Reference Materials: Update 1982." *Anal. Chem.* 54:1225A–1244A.

TAYLOR, J. K., OPPERMANN, H. V. 1985. *Handbook for the Quality Assurance of Metrological Measurements*. Gaithersburg, MD: U.S. National Bureau of Standards.
A short article and a handbook describing the U. S. National Bureau of Standards (now NIST) reference materials program and the uses of standards.

OBENAUF, R. H., et al., eds. 1994. *SPEX Handbook of Sample Preparation and Handling*. Metuchen, NJ: SPEX Corp.
This is mostly a catalog but contains useful practical information on sample preparation techniques for solids and the containers in which they are prepared.

Concept Review

1. What five general considerations should be taken into account when developing a protocol for sample preparation?
2. What is the difference between a method that is "specific" for calmodulin and one that is "selective" for calmodulin?
3. When a simple digestion by heating a sample with acid or base in the hood does not dissolve all of the sample that is required for analysis, a digestion with a flux or microwave digestion in a closed container may solve the problem. What are the advantages of each of these techniques over simple acid digestion?
4. What is the difference between an external standard and an added standard?
5. What are the three situations in which added standards might be used?

Exercises

The first three exercises refer to the data in Tables 4.9, 4.10, and 4.11.

4.1 Using the values found in Table 4.10, calculate the mean, standard deviation, and 95% confidence limit of the concentration of platinum in the sample.

4.2 What would the concentration of platinum be if the instrument output were 54.7 on the same scale that was determined from the standards?

4.3 Upon review of the laboratory notebook recording the experiment, it was found that the standards were made up incorrectly. The actual values of the concentrations of the standards are 32.43% lower than reported. Calculate the correct concentration of the platinum found in the sample.

4.4 Suppose a sample preparation method has an average loss of 25% (average recovery of 75%). The technician is still practicing the techniques, and the *relative* uncertainty in the average loss of material is ±0.16. What is the absolute uncertainty in the average loss?

4.5 Carry out the same calculations as in 4.4 for an average *recovery* of 98%.

4.6 A tablet was crushed and suspended in 50 mL of 0.1 M NaCl. After filtering to remove the insoluble solids, the aqueous solution was transferred to a separatory funnel. Addition of 1 mL of 0.1 M HCl converted the interferent to be removed to its less soluble acid form. The acid form has a K_D for an ether: water extraction equilibrium of 7.5. How many 10 mL portions of ether must be used to ensure that at least 99.5% of the interferent is removed from the aqueous solution? (Remember, the aqueous layer volume is 50 + 1 or 51 mL.)

4.7 The following values were found for the palladium standards when an instrumental assay similar to that used for platinum in Example 4.3.

Concentration of Pd in standard (ppm)	Instrument reading (arbitrary units)
0	14.1
80	27.7
205	48.4
325	68.6

The instrument reading for a sample with unknown Pd was 56.0 units. Plot the calibration curve and find the concentration of the sample.

4.8 This exercise involves deriving a mathematical expression to calculate the concentration of analyte using the method of standard additions. The machine response is found to be linear with concentration.

If the machine response is A_j, for the concentrations C_j, then

$$A_{\text{sample}} = k\, C_{\text{sample}}$$

where k is the factor that relates the concentration to the instrument response. Similarly, after a spike,

$$A_{\text{sample + spike}} = k\,(C_{\text{sample}} + C_{\text{spike}})$$

Using these two equations, solve for C_{sample} in terms of C_{spike}, A_{sample}, and $A_{\text{sample + spike}}$.

4.9 This problem refers to the copper determination described in Section 4.7. The volume of a 5-μL spike added to the 5.000-mL sample was not taken into account in calculating the copper concentration. (Not accounting for this volume change introduces a negligible determinate proportional error.)

(a) The spike was not exactly 0.10 ppm; calculate the true concentration of the spike accounting for the volume change. This is most easily done by calculating a new slope and intercept with the corrected concentrations.

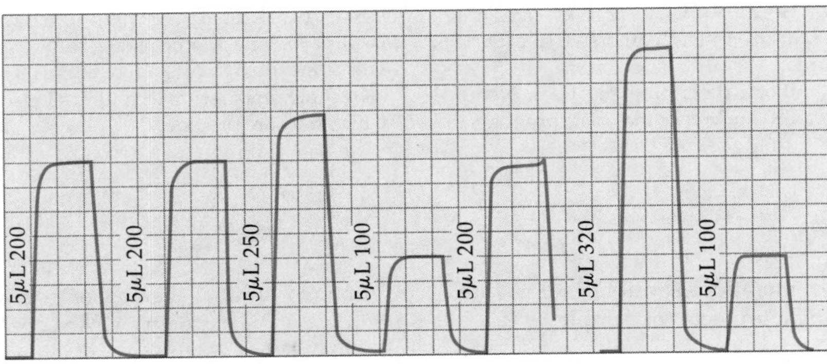

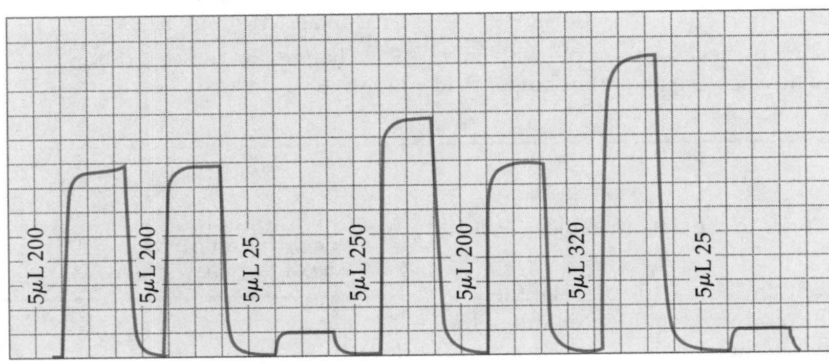

FIGURE 4.10.1 ▲

(b) The random error of the instrument is found to be 0.5% relative error. What is the ratio of the random relative error and the relative determinate error arising from ignoring the volume change?

4.10 Figure 4.10.1 illustrates the data for a linearity check of a membrane used in an electrochemical method to determine alcohols directly. Product lot 9208 is being tested. The output from the instrument is plotted on the vertical axis with time on the horizontal axis. The chart moves at 1 in. min^{-1}. The output of the chart recorder is set so that each of the 20 divisions equals 25 mV. The instrument output is set to have 1 mV = 1 mg dL^{-1} for the 5-μL injection of sample. The run starts from the left. The zero of the instrument is set on the bottom line. A 200-mg dL^{-1} sample is injected, and the instrument is allowed to respond fully. The plateau is taken as the instrument reading. The sample is washed out and another injected until the full series is run. The volumes and concentrations of each sample are noted on the chart. Answer the following questions. [Data courtesy of YSI, Inc., Yellow Springs, OH.]

(a) For the third and fourth 200-mg dL^{-1} samples, there is a slight hop in the plateau voltage. How will you treat these values?

(b) For the six 200-mg dL^{-1} samples, what is the mean (in mV) and standard deviation of the measurements?

(c) Is the response linear over the entire range tested (25–320 mg dL^{-1})?

(d) If the range is not linear, at what concentrations does it deviate by more than one relative standard deviation (as determined from part b) from the values expected by extrapolating the linear portion of the plot?

4.11 A determination of silver in waste water was done with a method involving spiking silver into the sample. A sample originally containing no silver was run six times with the following protocol. Each sample of 100.0 mL had added to it 200 μg of silver and was well mixed. For each run, a 5.00-mL aliquot was used. This was digested, and the following results were obtained. [Ref: Fu-sheng, W., Fang, Y. 1983. *Talanta* 30:190.]

μg Ag found: 8.8, 9.2, 9.2, 9.1, 9.1, 8.9

(a) How many μg Ag would you have expected to find?

(b) What is the mean recovery of the method?

(c) After correction for the recovery, what is the relative standard deviation of the analysis?

(d) Without correction for the recovery, what is the relative standard deviation of the analysis?

(e) If the assay for silver has a relative standard deviation of 0.1%, what is the relative standard deviation due to the recovery?

(f) If the assay for silver has a relative standard deviation of 5%, what is the relative standard deviation of the sample preparation procedure?

4.12 The following data were found in a determination of chromium in water. The protocol was to run a blank, the sample, and then the sample with two standard- addition spikes. Three replicate runs were made for each sample. Incidentally, this was all done automatically using a computer-controlled instrument. The significantly different responses to the spikes were caused by the other matrix components.

Solution	Conc. Cr (ng mL)	Instrument readings #1	#2	#3
Blank	0.000	0.005	0.009	0.005
Sample 1	—	0.054	0.058	0.058
Addition 1	+2.000	0.179	0.191	0.188
Addition 2	+4.000	0.305	0.310	0.318
Blank	0.000	0.000	0.000	0.000
Sample 2	—	0.537	0.548	0.531
Addition 1	+2.000	0.666	0.659	0.667
Addition 2	+4.000	0.790	0.784	0.805
Blank	0.000	0.001	0.003	0.000
Sample 3	—	0.086	0.083	0.078
Addition 1	+2.000	0.214	0.213	0.212
Addition 2	+4.000	0.341	0.350	0.337

Calculate the concentrations of Cr in the three samples, based on the means of the measured values for each sample and the standard additions. Make the blank correction if necessary. [Ref: Liddell, P. R. March 1983. *Am. Lab.* 15:111.]

4.13 A 0.750-g sample of corn syrup was assayed for dextrose with an instrument that specifically measures dextrose using an enzyme assay. The sample was diluted in a volumetric flask to 100.0 mL. A 25-μL sample was injected into the instrument, which read 373 mg dL^{-1} for the sample. The instrument had previously been calibrated with a 200-mg dL^{-1} standard using the same 25-μL sample injector. What is the content as %(w/w) dextrose in the corn syrup? [Ref: Application note 101. Yellow Springs, OH: YSI, Inc.]

4.14 A 1.596-g sample of a flavoring agent was assayed for dextrose with an instrument that specifically measures dextrose and records the result on a digital readout panel. The instrument was calibrated with a 200-mg dL^{-1} standard using a 25-μL sample injector. The sample was diluted in a volumetric flask to 50.0 mL. With the same sample injector as for the standard solution, a 25-μL sample was introduced into the instrument. The output read 46 mg dL^{-1} for the sample. What is the content as %(w/w) dextrose in the flavoring agent? [Ref: Application note 105. Yellow Springs, OH: YSI, Inc.]

4.15 A method was developed to determine oxygen dissolved in water. The method was tested to quantify any interferences that might be present. The following data were obtained from solutions which all contained 3.5 mg L^{-1} dissolved oxygen. Each result is the average value of between two and four replicates. [Ref: Data reprinted from Gilbert, W., Behymer, T. D., Casteneda, H. B. March 1982. *Am. Lab.*]

Potential interferent	Interferent conc. (mg L^{-1})	Avg. oxygen found (mg L^{-1})
Chromate (CrO_4^{2-})	0.1	3.4
	1.0	3.9
	10	6.4
Hypochlorite (OCl^-)	1.0	3.5
	10	4.9
	100	7.8
Thiosulfate ($S_2O_3^{2-}$)	1.0	3.4
	10	3.5
	100.0	3.4

(a) Which of the three compounds show an interference? Is it a positive or negative interference (raising or lowering the result)?
(b) Plot the result as it changes with a possible interferent. That is, plot O_2 concentration versus the interferent concentration. Calculate the best value for a correction factor that should be applied as F_c in the equation

$$[O_2]_{true} = [O_2]_{apparent} - F_c[\text{interferent}]$$

Do this for each interferent using its concentration in mg L^{-1}.
(c) What is the value of F_c for a species that does not interfere?

4.16 Table sugar (sucrose) molecules can be enzymatically broken down to produce fructose and dextrose (also called glucose). The reaction can be written

$$\text{sucrose} \xrightarrow{\text{invertase enzyme}} \text{dextrose} + \text{fructose}$$

The following procedure was followed to determine the sucrose content of sweet potatoes. A 78.7-g sample of sweet potato was reduced to juice, and the juice was collected in a beaker. The juicing machine was washed three times with 100-mL portions of sodium phosphate buffer with 2 to 3 min between rinsings. The juice and washings were poured into a 500-mL volumetric flask, and the containers were rinsed with several 10-mL portions of buffer, which were added to the flask. Buffer was then added "to the mark," which indicates the point where the volume is 500.0 mL. The sample was then refrigerated for 1 hour. After an hour, a 3-mL aliquot of the sample solution was taken, and the invertase enzyme was added to it. The sample that was not enzymatically decomposed was assayed for free dextrose. The content was found to be

208 mg dL^{-1}. After 20 min, the sample with invertase was assayed and found to contain 458 mg dL^{-1} dextrose. In each case, a 25-μL sample is used. [Ref: Application note 102. Yellow Springs, OH: YSI, Inc.]

(a) How many milligrams of dextrose are due to the decomposition of sucrose in the sweet potato?

(b) Given the molecular masses of dextrose (180.16) and sucrose (342.3), what is the percent sucrose (w/w) in the potatoes?

4.17 With the procedures described in problem 4.14 and a sample of 1.596 g of flavoring agent dissolved in 100 mL buffer, the following results were obtained. Dextrose in untreated sample, 23 mg dL^{-1}. Dextrose from sucrose in sample treated with invertase enzyme, 287 mg dL^{-1}.

(a) What is the content of sucrose in the sample expressed as %(w/w)?

(b) What is the total mass of sucrose in 1000 kg of the agent?

4.18 A new instrumental method to determine mercury in water is being validated against an older one. The new method consists of first reducing ionic mercury to its atomic form and then removing it as mercury vapor from the aqueous solution. The mercury vapor is purged from the water with air and passed into an instrument especially constructed to measure Hg in air. Possible interferents were investigated. The following results were obtained on water solutions of Hg containing the interferents. The figures in parentheses are the percentages by weight of the reagents. [Data reprinted from Murphy, P. J. 1979. *Anal. Chem.* 51:1599.]

	Hg measured (ng)		
Interferent	Hg added (ng)	Old method	New method
Acetone (20%)	100	>1000	95
Sodium sulfite (10%)	100	220	100
Sodium thiosulfate (10%)	100	750	102
Pyridine (5%)	100	>1000	91
Ammonium hydroxide (20%)	100	>1000	96

(a) Assume that the inherent precision of the assay methods is ±5% at this concentration level. For each of the potential interferents, and for the old method, what is the nature of the interference—positive or negative (causing the result to appear too high or too low, respectively)?

(b) For the new method, is there any interference from any of these compounds?

(c) Can you tell from the information given whether any interference in the new method results from the sample preparation step?

4.19 The Environmental Protection Agency lists a number of cancer-causing agents for which reliable analysis is important. These are called *polycyclic organic matter*, or POM. Among these are benz(a)anthracene, chrysene, benzo(b or k)fluoranthene, benzo(a)pyrene, and dibenz(a,h)anthracene. The data in Table 4.19.1 were found when these compounds were added to wastewater samples and extracted into methylene chloride. The volumes refer to the water phase. [Ref: EPA-600/7-/79-191.]

(a) Calculate the mean percent recovered of the amount added for each of the four compounds.

(b) Calculate the relative standard deviations of the percentage recovered.

(c) If a minimum relative error in the determination of each compound must be below 50% to be useful, which, if any, compound(s) cannot be determined with this sampling and extraction method as a part of the analysis?

4.20 In the manufacture of pharmaceuticals, it is necessary to make sure that no residual solvent remains in the finished product. Purge-and-trap is a technique whereby gas is passed through a sample and it carries away volatile components which are then trapped on some adsorbent. The trap is then heated so that the volatiles including any residual solvent come off in a much smaller volume and are analyzed. The following data summarizes the signals obtained from an analysis for benzene after the purge times indicated. [Washall, J. W., Wampler, T. P. December 1993. "A dedicated purge-and-trap/GC system for residual solvent analysis of pharmaceutical samples." *Am. Lab.* 20C.]

Purge time (minutes)	Detector response
0.2	10100
1	15000
2	22300
3	25300
4	24000
5	26300
6	25200
7	24800
8	27500
9	27300
10	23100
11	25000

(a) What is the minimum time that purging is needed to obtain the most accurate results?

(b) Does additional time harm the results?

(c) Assume that the purging times were 60 s exactly and that the fraction of benzene collected is completely reproducible. By what factor would you have to multiply the measured results to obtain the correct ones?

4.21 With the method of standard additions, sometimes it is useful to use a logarithmic extrapolation. For instance, sometimes the instrument response is

$$I = K\,c^{+m}$$

Table 4.19.1

Compound added	Amount added ($\mu g\ m^{-3}$)	Amounts measured in trials ($\mu g\ m^{-3}$)						
Chrysene	1.0	0.64	0.70	0.45	0.72	0.84	0.75	0.61
		0.92	0.80	0.59	0.63	0.86	0.58	0.77
Benzofluoranthenes	1.0	0.48	0.96	0.60	0.69	0.08	0.20	0.30
		0.85	0.90	0.69	0.88	0.76	0.33	0.39
Benzo(a) pyrene	10	6.9	5.7	7.8	5.2	8.0	5.4	4.7
		5.0	6.5	4.0	7.3	7.6	5.2	8.5
Dibenz(a,h)anthracene	50	32	22	37	39	40	30	17.5
		44	22	24	27	34	37	46

where I is the response, c the concentration, and K and m constants. A more useful form of the equation is

$$\log I = m \log c + \log K$$

If $\log I$ is plotted versus $\log c$, a straight line results with slope m and ordinate intercept $\log K$. The values of m and K will depend on the experimental conditions. Assume that two solutions with spiked concentrations $(c + \Delta_1)$ and $(c + \Delta_2)$ are run. The instrument responses are I_1 and I_2, respectively. Now define

$$A = \log(I_2/I)/\log(I_1/I)$$

[Ref: Beukelman, T.E.; Lord, S.S., Jr. *Appl. Spec.* 1960, 14:12.]
(a) Show that A can also be expressed as

$$A = \log[1 + (\Delta_2/c)]/\log[1 + (\Delta_1/c)]$$

(In other words, setting the two expressions for A equal to one another, we can find c from the three measured values I, I_1, and I_2 alone.)
(b) An assay for calcium was run with the following results:

Sample	Response (corrected for background)
c	22.5 chart divisions
c + 20 ppm Ca^{2+}	44.1 chart divisions
c + 80 ppm Ca^{2+}	106.8 chart divisions

Calculate c with both the usual linear extrapolation and with iterative substitution into the logarithmic equation. Compare the results.

4.22 A brand new technician was developing a method to determine the amount of iron present in a vitamin syrup using a placebo (syrup containing everything except the iron) plus a 5 mL spike of a known amount of iron. He found, when he compared the amount of iron recovered to that added initially, that he had only a 45% recovery. His laboratory notebook is excerpted below. If you find any sources of error, make the necessary correction(s) and report the correct recovery.

Preparation of standard: Five (5.00) mL of a standard solution of iron was treated with hydroxylamine hydrochloride and *o*-phenanthroline to yield Fe(*o*-phen)$_3$ and diluted to a final volume of 50.00 mL. The absorbance of the sample was read at 512 nm.

Preparation of vitamin sample: A 30.00 mL aliquot of placebo was transferred to a 100 mL microwave digestion vessel. Five (5.00) mL of the same standard solution of iron was added. In addition, 5 mL each of concentrated HNO_3 and concentrated H_2SO_4 were added to the container. The container was sealed and the sample digested. The resulting solution was transferred to a beaker and heated on a hot plate in the hood until no fumes were given off. At the end of this step some corrosion of the stainless steel tray of the hood was noted. The solution was transferred to a 100.0 mL volumetric flask, treated with hydroxylamine hydrochloride and *o*-phenanthroline and diluted to volume. The absorbance of the sample was read at 512 nm.

4.23 Microwave assisted extraction (MAE) of PAH's (polyaromatic hydrocarbons) with dichloromethane (DCM) and with acetone (ACE) was compared with soxhlet extraction, yielding the following data for naphthalene and chrysene:

Compound	Amount extracted (Microwave/soxhlet)			
	DCM	% RSD	ACE	%RSD
naphthalene	1.40	11.60	2.86	5.40
chrysene	0.86	16.30	1.03	2.50

(a) For which compound(s) and solvent(s) was the soxhlet extraction the better method?

(b) For naphthalene extracted into acetone, assume MAE gives 100% recovery. What is the percent recovery of the soxhlet method? (Note: The MAE method gives much better results for many other compounds!) [Data courtesy of CEM corp.]

4.24 Carrier precipitation is especially useful in cases in which the analytes are at low levels. For example, a relatively large 10-g replicate sample with 0.05% (w/w) platinum contains only 5 mg of platinum. So if the platinum is coprecipitated with 100 mg of tellurium as the chloride, a slight loss of precipitate causes less error. On the other hand, if the 5 mg of Pt metal could be collected and formed into a cube, the cube would be only half a millimeter on a side—the size of a grain of table salt. What relative error in the assay of a 10-g replicate would result if a single dust-sized cube 100 μm on a side of pure Pt (ρ = 21.45) were lost during a transfer?

4.25 The efficiency of extraction for soxhlet, sonic probe, and supercritical extraction for a set of organophosphorus pesticides was compared. As seen in Table 4.25.1 below, the percent recoveries, as well as the relative standard deviation of the recoveries, varied widely for the three methods. Based on the criteria given in the text for selecting a sample preparation method, which method would you immediately reject? [Ref: Snyder, J. L., Grob, R. L., McNally, M. E., Oostdyk, T. S. 1992. "Comparison of Supercritical Fluid Extraction with Classical Sonication and Soxhlet Extractions of Selected Pesticides." *Anal. Chem.* 92:1940.]

Additional Exercises

4.26 Using the data in Table 2.3 (page 45), calculate the average recovery of the original aluminum method.

4.27 Suppose a sample preparation method has an average loss of 15% (or average recovery of 85%). The technician is still practicing the techniques, and the *relative* uncertainty in the average loss of material is ±0.20. What is the range of loss (in %) to which this corresponds?

4.28 Carry out the same calculations as in 4.27 for an average *recovery* of 95%.

4.29 A caplet was crushed and extracted in 100 mL of 0.1 M NaCl. After filtering, the aqueous solution was transferred and the water solution extracted with ether. The ether-soluble ingredient has a K_D for an ether:water extraction equilibrium of 10. How many 10 mL portions of

Table 4.25.1 Comparison of extraction methods organophosphate pesticides: soxhlet vs. sonic probe vs supercritical fluid extractions

Compound	Amount added, ng g	Method[a]	Trials, n	Found (av), ng g	Std. dev, s	RSD %	% recovered (av)
dichlorvos	520	SX	8	333	28	8.4	64
		SP	6	375	42	11	72
		SFE	5	318	13	4.0	61
diazinon	515	SX	8	479	24	5.1	93
		SP	6	493	27	5.4	96
		SFE	9	433	13	3.0	84
ronnel	25	SX	8	26	1.4	5.2	104
		SP	6	27	2.4	9.1	106
		SFE	9	25	1.2	5.0	98
parathion (ethyl)	78	SX	8	22	20	92	28
		SP	6	77	1.3	1.6	99
		SFE	9	73	0.88	1.2	94
methidathion	100	SX	8	108	5.7	5.3	108
		SP	6	100	4.8	4.8	100
		SFE	9	106	3.4	3.3	106
tetrachlorvinphos	36	SX	8	42	4.7	11	117
		SP	6	41	1.3	3.2	113
		SFE	8	39	2.5	6.3	109

[a]SX = soxhlet extraction, SP = sonic probe extraction, SFE = supercritical fluid extraction.

ether must be used and combined to ensure that at least 99.5% of the active ingredient is transferred out of the water?

4.30 In order to check to see that the percent recovery for a new sample preparation method was constant over the range expected for pesticide residues in corn stalks (normally 1.0-5.0 ppb, depending on the time since application), a series of 20 g samples of untreated corn stalks were spiked with aliquots of a highly diluted solution of the pure pesticide. The spiked cornstalks together with 25 mL of acetonitrile were placed in a powerful blender and agitated for 45 seconds. The acetonitrile extract was filtered and then analyzed directly by gas chromatography. Comparison of the residue samples with standard samples produced the results in the table adjacent on the right.

(a) Calculate the concentration (in ppb) that would be expected in each of the solutions analyzed for the samples.

(b) Plot the response *vs.* concentration curve for the standards.

(c) Based on the graph from part b, what was the concentration in each of the solutions from the residue samples?

(d) Calculate the recovery for each sample. Is the recovery constant over the range tested?

Sample	Instrument response (arbitrary units × 10^{-5})
Standards	
1.00 ppb	2.15
2.00 ppb	4.25
3.00 ppb	6.52
4.00 ppb	8.57
5.00 ppb	10.76
Samples 200 g corn stalks + x mL of standard containing 2.5 µg/100 mL	
1.00 mL spike	2.10
2.00 mL spike	4.15
3.00 mL spike	6.28
4.00 mL spike	8.43
5.00 mL spike	10.28

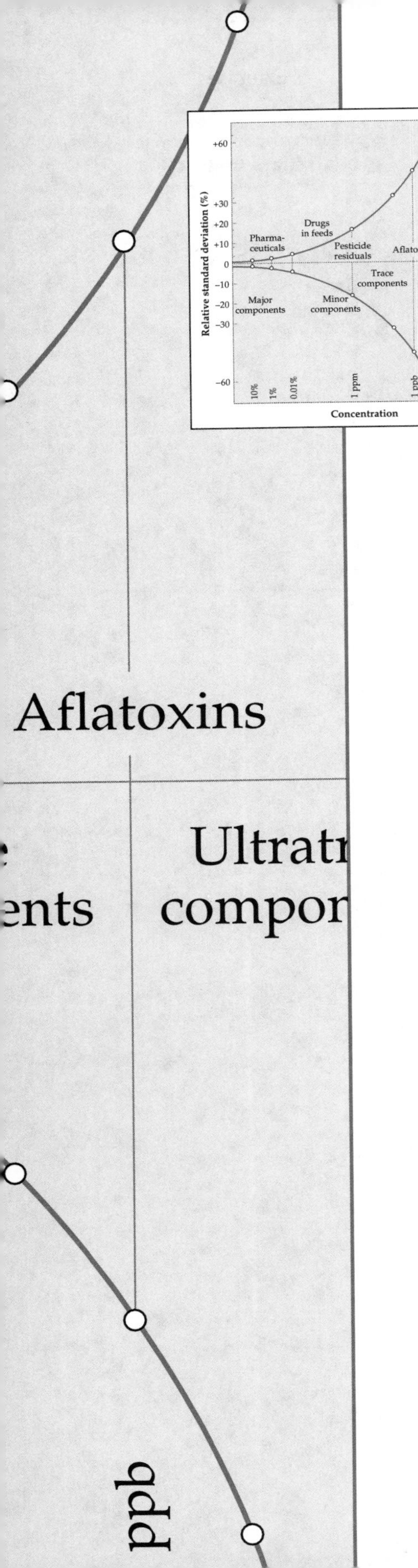

CHAPTER 5

Sample Size and Major, Minor, Trace, and Ultratrace Components

Overview

A Deeper Look

5.1 Sample Classification by Size and Analyte Level

Objects to be analyzed come in all sizes. On the large end of the range are the surfaces of the Moon and Mars, samples of which have been subjected to extensive elemental analysis. From the Moon, relatively large samples (~kg) could be obtained. A major problem was choosing the rock samples to bring back to earth to be analyzed. From these collected rocks, laboratory samples on the order of 0.1–1 g—**macro** or **normal** samples—could be taken for analysis. These samples are limited, and the demand for them is great, not only for chemical analysis.

Papers describing the Moon and Martian rock analyses are cited in the Suggested Readings for this chapter.

It is not hard to find an object to be sampled that is significantly smaller than an astronomical body—for instance, a single, small grain of powder found in the pocket of a suspected bank robber. The grain might be from some powder that was formulated for use as a marker for bank notes and would have been sprinkled on the money before it was handed over by a teller. The size and mass of the particle would place it in the class of **micro** samples, which are analyzed with the techniques of **microanalysis**.

The size of a laboratory sample should not be confused with the concentration level of analyte in a sample. A descriptive classification scheme is presented in Figure 5.1. The horizontal axis shows the sample size on a logarithmic scale, and the sample sizes are divided into macro- and microanalytical regions. On the vertical axis is a logarithmic plot of the percentage (w/w) of a constituent in a sample. The sample content is classified by the relative amount of an element or compound in it. A component can range from **major** (1–100%), **minor** (0.01–1%), or **trace** (less than 0.01% = 100 ppm) to **ultratrace** (in the range of parts per billion or less). The total or absolute amount of a component present depends on the size of the sample and the component's fraction of the whole. As a result, lines of constant analyte mass run diagonally across the graph.

However, chemical analysis is a dynamic area, and the term *trace analysis* has changed in meaning and may change again in the future. Such changes are reflected in the following comments by A. H. Hayes, the former commissioner of the U. S. Food and Drug Administration.

> The food safety law is really not that at all. It really is a series of laws, some of them old, some of them new. . . . When some of these laws were written, . . . scientist[s] would tell you, for a particular chemical, that they could find one part per thousand. . . . [This] means that zero, when you're talking about zero risk . . . [was] one part per thousand and one. Now, for some of these chemicals, 20 years after, . . . we can find one part per billion or one part per trillion. . . . If you can find one part per trillion, what is the new definition of zero risk, one part per trillion and one? . . . Do we really believe that the risk of zero being one part per thousand is the same as one part per trillion? [Quoted in *Chem. & Eng. News*, Jan. 18, 1982.]

How much is a part per billion (ppb)? How much is a part per trillion (pptr)? These concepts may be easier to understand by considering some concrete examples.

A gallon (U.S.) is equal to 3.8 L; so a 55-gallon oil drum contains about 200 L. The mass of water in a full drum weighs about 200 kg. If one grain

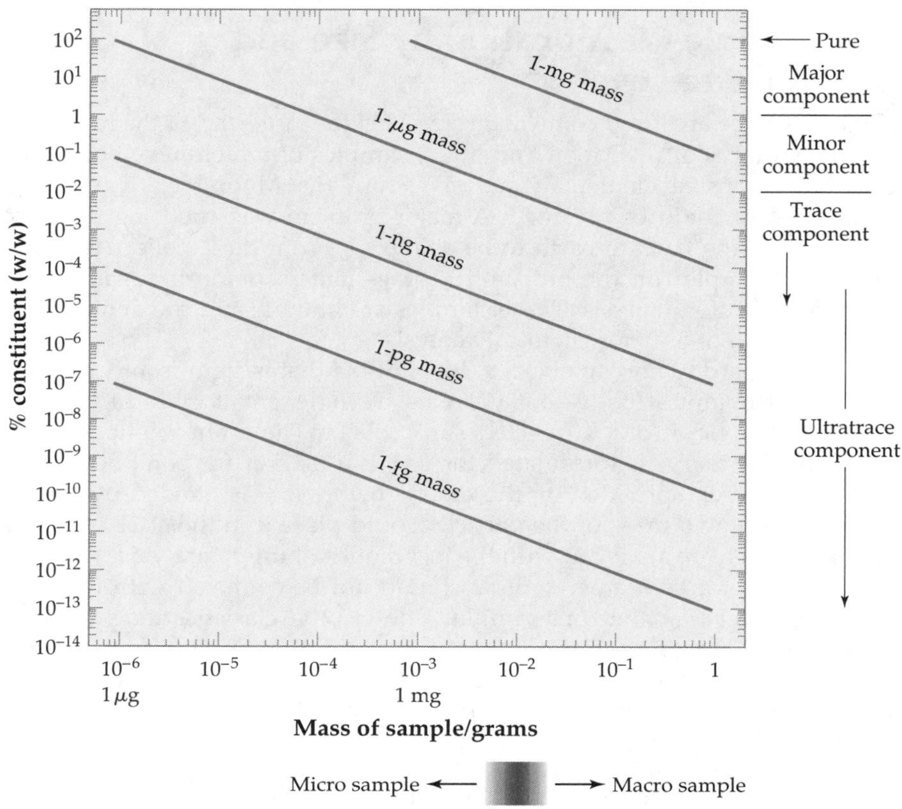

FIGURE 5.1 ▲
Illustration showing the names given to the ranges of sample size and of analyte content.

[Modified from Sandell, E. B. 1959. *Colorimetric Determination of Trace Metals*. 3rd ed. New York: Interscience.]

of table salt (mass about 0.2 mg) were dissolved in the barrel of water, the salt concentration would be about 1 ppb.

To dilute 1 g of salt (or any other substance) to a part per trillion (w/v) water solution would require about 10^9 L of water. This quantity of water would require a cube 100 m on a side to contain it.

We think that you will have to agree that the ability to quantitate a few parts per billion or parts per trillion of some substance must rank among the most impressive achievements of chemical technology.

5.2 Limits of Trace Analysis

Detection of aflatoxins: Analytical Chemistry in Action in Chapter 2.

As the content of a component decreases into the ppm and ppb range, eliminating interferences from the sample matrix becomes more and more difficult. Recall the difficulties of determining trace levels of aflatoxins in peanuts. Such difficulties become more acute and important when governments want to regulate concentrations of substances even into the ppb range. Disparities in results among laboratories may make meaningful exchanges of information impossible. An illustration of the trend in interlaboratory errors is shown in Figure 5.2. Attempts to reduce the limits indicated by this diagram have not been successful, especially at the lowest levels of content.

Another problem in carrying out trace analyses is contamination during the sample preparation. Every reagent, such as an acid used for digestion, must contribute as little interference as possible. Even containers made of glass or plastic can contaminate solution samples. Interfering impurities set a lower limit on our ability to ascertain trace amounts of a substance.

Tables 5.1a and 5.1b list the levels of some trace impurities in acids used for sample preparation. The amounts are stated in ng/mL, which is equivalent to ppb (w/v). An impurity concentration of 100 ppb in an acid may not seem very high. However, if the quantity of acid needed for digestion is 100 times the sample weight, 100 ppb of a species in the acid would be equivalent to 10 ppm in the analytical sample.

Analyses at the lowest concentration levels require that the areas where samples are handled be clean and that the air be filtered. In other words, the instrument used for the low-level assay is often not the limiting factor, as the example presented in ACA 5.1 demonstrates.

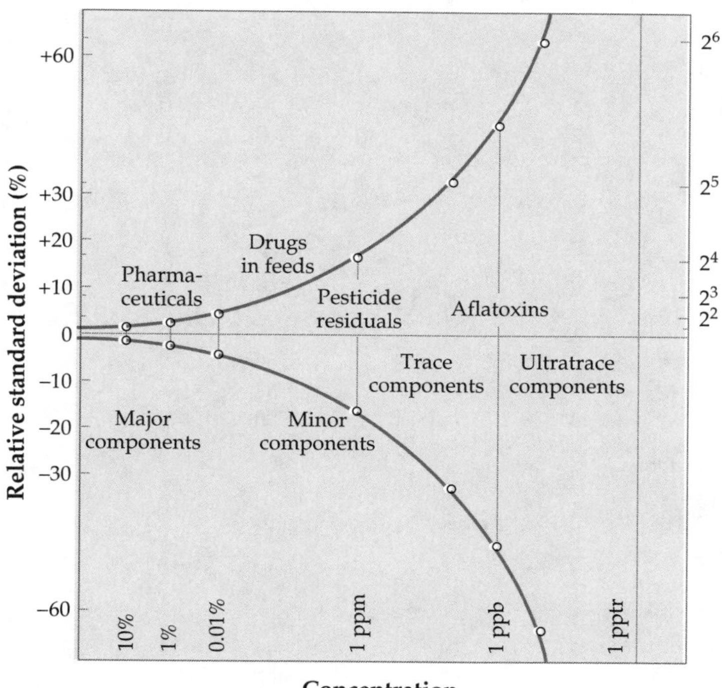

FIGURE 5.2 ▲

A plot illustrating the finding that in general an increase in interlaboratory errors occurs with decreasing concentrations of analyte.

The horizontal axis is a logarithmic plot of concentration; the vertical axis is the percent relative standard deviation of the results. This plot should be interpreted as showing a trend, rather than hard, fast conclusions. Note that in the ppb range, the 95% confidence limit approaches the value of the mean. [From Horowitz, W. 1982. *Analytical Chemistry* 54:67A–76A.]

Table 5.1a Concentrations of impurities found in acids

Element	Sample	Supplier → 1	2	3	4
		Concentration [ng/mL = ppb(w/v)]			
In nitric acids					
Mn	1	10.2	0.9	29.7	
	2	6.0	7.3	49.0	
	3	8.1	2.2	16.5	
	4*	0.6	—	—	
Cl	1	65	53	259	
	2	79	66	330	
	3	55	59	—	
	4*	50	500	80	
Na	1	66	109	1465	
	2	50	103	1400	
	3	98	124	—	
	4*	30	500	100	
Cu	1	19	13	355	
	2	12	—	—	
	3	13	—	—	
	4*	13	5	0.7	
In hydrofluoric acids					
Mn	1	0.3	0.2	0.3	1.0
	2	0.3	0.5	0.3	1.1
	3	0.6	—	0.2	0.9
	4*	0.5	—	—	—
Na	1	12	11	12	129
	2	23	15	27	140
	3	16	14	17	—
	4*	20	—	—	100
Cl	1	175	270	800	281
	2	180	472	805	287
	3	182	—	795	242
	4*	1000	100	—	80
In acetic acids					
Mn	1	7.0	0.5	2.4	2.4
	2	13.3	0.5	2.7	1.5
	3	—	—	3.0	2.0
	4*	1	5.0	—	—
Na	1	903	90	359	35
	2	916	24	359	39
	3*	400	50	—	—
Cu	1	66	4.8	3	—
	2*	30	9	100	5
Cl	1	655	101	187	133
	2	799	377	159	66
	3	—	—	—	58
	4*	300	300	—	500

*Concentration reported by supplier.
Source: Data from Mitchell, J. W., Luke, C. L., Northover, W. R. 1973. *Anal. Chem.* 45:1503–1506.

Table 5.1b Representative inorganic impurity levels in common concentrated reagent acids (Manufacturers' analyses in ng/g. Zeros are place holders only.)

	Impurity level (ng/g)		
Impurity	**Reagent grade**	**Better grade**	**Ultrahigh-purity grade**
In hydrochloric acid			
Ag	—	50	0.10
Al	—	50	10
As	20	5	1
B	—	100	30
Ba	—	50	<10
Br	50,000	500	400
Ca	—	50	1
Cd	50	5	<1
Co	50	5	<0.1
Cr	50	50	0.2
Cu	200	5	0.2
Fe	200	50	2
K	—	50	<5
Pb	—	5	0.5
Mg	—	10	1
Mn	50	5	0.1
Hg	—	5	0.2
Na	—	500	20
Ni	200	10	0.2
Si	—	100	200
Sn	—	50	<0.5
Sr	—	50	<1
Zn	500	5	<1
Ammonia	3000	—	1000
Phosphate	—	50	20
Sulfate	800	500	300
Sulfite	1000	800	200
In nitric acid			
Ag	—	10	<0.05
Al	—	50	2
As	5	5	<1
B	—	50	2
Ba	—	50	<10
Ca	5000	50	2
Cd	10	50	<1
Cl	200	100	100
Co	—	5	<0.1
Cr	100	50	0.3
Cu	50	5	0.1
Fe	200	20	0.5
K	—	50	<5
Pb	50	5	0.5
Mg	—	50	0.3
Mn	—	5	0.05
Hg	—	5	<0.1
In nitric acid (continued)			
Na	—	500	30
Ni	50	10	0.2
Si	—	1000	10
Sn	—	50	<0.5
Sr	—	50	<1
Zn	100	10	<1
Phosphate	200	100	<50
Sulfate	500	500	<100
In sulfuric acid			
Al	—	30	3
As	10	5	<1
B	—	50	3
Ba	—	50	<10
Ca	—	50	10
Cd	100	5	<1
Cl	200	100	40
Co	100	5	0.2
Cr	100	10	0.5
Cu	500	5	1
Fe	200	50	10
K	—	50	<5
Pb	100	10	0.5
Mg	—	10	3
Mn	100	50	0.3
Hg	5	5	1
Na	—	500	200
Ni	100	5	0.5
Si	—	500	20
Sn	—	10	0.5
Sr	—	10	<1
Zn	100	5	2
Ammonia	2000	—	200
Phosphate	—	500	10
Nitrate	1000	200	40

ANALYTICAL CHEMISTRY IN ACTION
Can Instruments Get Hay Fever?

The following example will provide you with some idea of the capability of contemporary analytical technology and the difficulty in defining the detection limit in practical situations. One term needs to be understood: absorbance units. Most instruments that measure absorbance produce an output within the range 0 to 2 absorbance units. If your eyes could register in absorbance units, a gray-colored solution registering 3 absorbance units would look black; a solution registering 0.01 units would appear colorless.*

> It's a beautiful late summer day, warm and with a gentle breeze. Any lab window or door that can be propped open is. Mother Nature has begun her work to ensure an abundance of next year's plant life. That means pollen—a completely innocuous substance to a human without hay fever. A typical grain of pollen may be about 10 μm in diameter, and have a density of 0.8 [g/cm^3], and contain about 5% magnesium. A rough calculation gives a total magnesium content of 7×10^{-11} g. The sensitivity† [of our method] for magnesium is about 10^{-14} g. This means that 10^{-14} g of magnesium produces 0.0044 absorbance units. Thus, if a single pollen grain were to get into [the sample chamber of the instrument], it alone would produce a magnesium signal of 31 absorbance units. That translates into a very off-scale reading on any instrument ever conceived.

Obviously it makes no sense to do a magnesium analysis during hay-fever season unless the work is done in an atmosphere filtered to remove particulate matter such as pollen. Actually, such filtering is required all year, since similar effects, though not quite as great, are produced by fungal spores or bacteria that are always present.

[Reprinted from Morgenthaler, L. 1977. In *Spectroscopy*. Series 11, Vol. 11. Fairfield, CT: International Scientific Communications. pp. 105–120.]

*A more detailed definition of absorbance is presented in Chapter 11.

†Here, sensitivity has the meaning of characteristic mass.

Two terms closely associated with measurements in trace and micro analyses are the **sensitivity** and the **detection limit**. Their definitions in a chemical context are similar to those in everyday language. If an instrument (or assay method) is more sensitive, it is able to measure smaller *changes* in content or concentration. The preferred measure of sensitivity is the slope of the **working curve** (the graph of output measured *vs*. concentration) at the concentration of interest. It is now the recommended definition (by the IUPAC) for all analytical methods. With this definition, a larger number (in the same units) means a method is more sensitive.

Unfortunately, in the past, *sensitivity* had a different, specific meaning: the concentration or mass of analyte needed to produce a signal that absorbed 1% of the light passing through it. It is now suggested that the amounts needed to produce that effect be called the **characteristic concentration** or **characteristic mass**, depending on which is applicable. For a characteristic concentration, a smaller number means the method is better in detecting small quantities. So when reading about reported sensitivities, you should consider the context of the word.

The detection limit is the limit below which a particular instrument or analytical method cannot report reliably the presence or absence of an analyte in a sample. The magnitude of the detection limit depends on the following factors:

1. the identity of the element or molecular species to be determined,
2. the identities of the other components in the sample,
3. the quality of the reagents required for the sample preparation, and
4. the instrument or assay method used.

The dependence of detection limits on the specific assay method and on the element assayed are illustrated in Appendix VIII; the numbers are associated with the lowest levels of each element that can be detected with each method. Notice in the table that for an individual technique, the concentration range for the elements can vary by well over a factor of a thousand; and, for a given element, the concentration limit can vary among techniques by over a factor of a million. However, as you may already appreciate, the selection of an assay method depends on many criteria in addition to its detection limit or characteristic concentration.

5.3 Trace, Ultratrace, and Microanalysis Compared

Earlier in this chapter, you read that if a single grain of sodium chloride is dissolved in a barrel of (ultrapure) water, the solution would contain about 1 ppb sodium chloride. It should be clear that both the original salt grain and the barrel of salt solution contain exactly the same amount of sodium and chlorine. Chemical analysis should show the same total sodium and chlorine content in either case. However, the approaches to the analysis of these two samples will be different, and each will present its own problems. Analyzing the grain requires a methodology that can be used for small, concentrated samples—the techniques of microanalysis. Determining the sodium and chlorine in the barrel of water requires a different methodology, that of trace analysis.

Within the classification scheme of Figure 5.1, trace analysis can be defined as analysis for components present at levels less than 1 ppm. Ultratrace analysis may be somewhat arbitrarily defined as being for total component concentrations below approximately 1 ppb.

In trace analysis and microanalysis, the assay methodology must be able to quantify small amounts of material. This may seem obvious only for microanalysis; but the same criterion is necessary for trace analysis so that the samples do not become unmanageably large. Instruments that can measure quantities in the subpicogram range ($<10^{-12}$ g) range are now relatively common.

Picogram methods: Anodic stripping (Chapter 15); atomic absorption (Chapter 11); X-ray fluorescence (Chapter 11); mass spectrometry (Chapter 12).

Among the techniques used for elemental analysis are anodic stripping voltammetry, atomic absorption spectrometry, neutron activation analysis, X-ray fluorescence, and a number of different surface-sensitive spectrometries. Also included is mass spectrometry combined with an internal standard of a stable isotope or mass spectrometry with a high-temperature (white heat) source.

To determine concentrations of organic compounds present at trace levels, gas chromatography in conjunction with mass spectrometry is the method of choice. This powerful analytical combination is almost always abbreviated GC/MS.

For more on gas chromatography see Chapter 13, and for GC/MS Section 12.6.

Preconcentration of Trace Analytes

Because of limitations of analytical sample size, for trace analysis it is necessary to concentrate, and at least partially separate, the component that is to be assayed from the interferents in the bulk matrix. A large number of different preconcentration methods are used and are carried out as part of the sample preparation steps.

A number of techniques that can be used to separate and concentrate trace elements from matrix materials are covered in Chapter 4.

5.4 Four Cases

Next in this chapter are four illustrative examples of the sorts of problems solved with trace and microanalytical methods.

1. A sample of a few grams of a high-purity material for which the purity is to be assayed,
2. A small (mg) solid sample to be analyzed for its elemental composition using the techniques of microanalysis,
3. A micro sample to be analyzed for organic components, and
4. A solid for which the distribution of elements in its structure is desired.

Case 1: A Sample of High Purity

To carry out a determination with a relative error of less than 0.1% (one part per thousand) is an exceedingly difficult task. How, then, can you answer the following question: Is this silicon for use in making microcircuit chips 99.99997% or 99.99999% pure? If we were to test for silicon at 100% using a precise and accurate assay, with enough different replicates to obtain good statistics, the result might be stated: This silicon is $(100.0 \pm 0.1)\%$ pure. But that result, as good as it is, does not answer the original question. In fact, it is about 10,000 times too crude to be useful.

To determine the purity of a highly pure material requires a different approach. What is done is to find the concentrations of all the trace impurities present: that is, all of the sample that is *not* silicon. (It is certain that they are present only in trace amounts because the total concentration of all of them is about a few hundred ppb.) Then, the total concentration of the impurities is subtracted from 100% to obtain the level of purity of the main component, silicon. Interestingly, the determination of high purity requires the use of trace element methodology.

To do this kind of analysis economically, it is necessary to determine simultaneously the levels of a number of trace elements. Some methods that are used for such analyses are mass spectrometry, emission spectrometry, instrumental neutron activation analysis, and X-ray fluorescence spectrometry.

Mass spectrometry (Chapter 12); emission spectrometry (Chapter 11); X-ray fluorescence spectrometry (Chapter 11).

Case 2: A Milligram Sample to be Analyzed for Its Elemental Composition

Another consideration in any kind of analysis is whether or not you want the original sample back. (Would you want to digest a part of a priceless Egyptian statuette?) If you want the sample to remain intact, then it is necessary to use **nondestructive** analytical methods, which cause no visible change in the sample. It may be modified on the atomic or molecular level, but these changes are not easily detectable. To cause as small a change as possible in the original sample, nondestructive tests are carried out with assay methods that are also used for micro or trace analysis. One method that is useful for both micro and trace analyses is neutron activation analysis. Two other microanalytical techniques are the electron microprobe and X-ray photoelectron spectroscopy.

Further information on microanalytical techniques can be found in the references at the end of this chapter and in Chapter 11.

A larger number of assay techniques require destruction of the sample. Any methodology that requires the digestion or fusion of samples must be classified as a **destructive** analytical methodology. For example, ACA 5.1 describes the measurement of magnesium in a single pollen grain. The assay method required that the sample be vaporized by rapid heating. The sample was destroyed.

When the sample is in the micro range, you might guess that an analysis would be easier if the sample is not digested or diluted in any way. After all, why add interferents from handling and pretreatments with an impure acid? With the capabilities of contemporary instrumentation, micro determinations (while often expensive) generally exhibit better precision and accuracy than trace determinations.

Samples for microchemical analysis arise in, among other areas, forensic analysis. (Forensic means suitable for a use in a law court.) Forensic analysts not only must run careful analyses, but also must take into account the rules of legal evidence by keeping track of the material that is analyzed. It is convenient to use nondestructive analytical techniques so the evidence can be kept if any legal questions arise. The use of a nondestructive microanalytical technique for elemental analysis in a forensic laboratory is shown in the following quotation. The method is based on the recording of X-rays emitted by a sample that is bombarded with electrons.

> . . . $119,000 worth of selenium was stolen in an armed robbery. Clothing from a person believed to have been involved in the attack was examined in the laboratory and a small fragment about 0.2 × 0.2 mm found on one of the shoes. [Micro]analysis showed it to be pure selenium. [Ref: Williams, R. L. 1973. *Anal. Chem.* 45:1076A–1089A.]

As usual, to achieve a precise and accurate quantitation, the analytical methodology used here requires that proper standards be used. Although it may seem odd, when the sampled mass is quite small relative to the total sample, care must be taken to ensure that the sampling of even a small fragment be representative. That is, the same statistical problems of ensuring a representative sample arise in micro analyses as in sampling a shipload of metal ore. However, in the case of the selenium fragment, it may be more difficult to find the original material to be analyzed. Notice, though, that no matter how large or small the sample, the methodology as shown in Figure 1.1 is followed to define and solve the problem.

Some of the more powerful analytical techniques, such as that used in the example above, can be used to quantitate composition with good precision for samples of less than 10^{-10} g. This number means that a one-milligram sample contains more than ten million times the amount of material needed for analysis. To get an idea of this scale, consider that a one-milligram sample with a specific gravity of four would fill a cube about 0.6 mm on a side. Therefore, a 10^{-10} g sample would be less than 1/1000 as long (about 600 nm) on each side.

Case 3: Micro-Organic Analysis

For micro, ultramicro, and trace chemical analysis of organic compounds, both man-made and biological, the methods of choice are mass spectrometry and gas chromatography, or the combination of the two methods—gas chromatography/mass spectrometry.

Our third case is, again, from forensic science and involves the analysis of the organic components of a sample of paint. The assay method used is gas chromatography alone. Here, the sample preparation method involves vaporizing 10–20 μg of paint rapidly and reproducibly. At high temperatures, the organic-polymer components of the paint break down into smaller molecular fragments: This breakdown is called pyrolysis (literally, cutting by fire or heat.) The results from two samples are shown in Figure 5.3. Some experimental details are described in the figure caption. The example of the use of these **pyrograms** involves a murder.

> . . . the pyrogram was obtained of a black speck of material found embedded in the skull of a 38-year-old woman who had died from head injuries. It turned out to be that of a styrenated alkyd, a coating used for tools; it matched similar pyrograms from paint on the tool kit of an abandoned car from which the [jack] was missing.

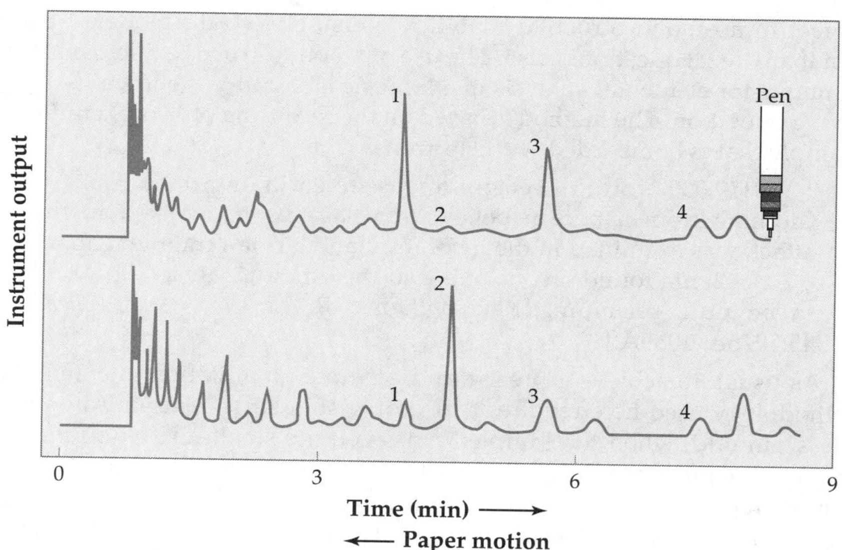

FIGURE 5.3 ▲

Pyrograms from heating solid organic polymers.

These are plots of the instrument output (vertical axis) versus time (horizontal axis). The instrument output is proportional to the amount of each component that passes through a detector. Each organic species arrives at the detector at a time that is characteristic for that species. The numbers identify the peaks that result from the same chemical species. The two samples are polymers of paint that are composed of the same two types of monomers, the only difference being the ratio of the monomers in the original mix. The difference in composition results in the different heights of the corresponding peaks. In this way, unknown polymer resins can be characterized even down to their specific compositions if the experiments are carefully done with good standards. [Reprinted with permission from Williams, R. L. 1973. *Anal. Chem.* 45:1076A–1089A.]

There is no need to point out the significance! [Ref: Williams, R. L. 1973. *Anal. Chem.* 45:1076A–1089A.]

Case 4: A Solid and the Distribution of Elements in Its Structure

In Case 3, you read about analyzing exceedingly small samples (less-than-nanogram amounts of microgram samples). It may now seem reasonable to ask: If such small samples can be analyzed, is it possible to do the same type of microanalyses at different places on a larger sample and determine the locations of various elements? A study was carried out in which small areas in a regular pattern over the surface of a material were analyzed and the results were plotted as a function of the position on the sample as illustrated in Figure 5.4. The method was tested to make sure it was reliable by following the sequence of steps in developing the methodology of an analysis.

Development of the methodology of analyses is outlined in Figure 1.1.

> Approximately 2 mm thick transverse sections cut from the femur of two autopsy cases were used for the study. Both persons had been working in different occupations in a metal smeltery in Sweden. The reference worker had been employed for 27 years during which time he worked with roasting and refining of arsenic and copper. He had not been exposed to excessive amounts of lead. He had retired from his work 8 years before his death. The lead exposed case had been working in the lead refinery of the same metal smeltery . . . [for] at least 21 years. The patient had experienced clinical symptoms of lead intoxication in periods, the first time about 20 years before his death. He had been away from lead exposure for more than 2 months before his death. The sections of femur were [frozen and then while still frozen] dried in a vacuum. . . . [Ref: Lindh, U., Brune, D., Nordberg, G. 1978. *Sci. Tot. Environ.* 10:31–57.]

The localized sampling of the bone was done by focusing a beam of protons from a particle accelerator onto spots about 1/20 mm apart. The X-ray emission characteristic of lead was separated from the X-rays due to most of the other elements present. After running a blank on a similar piece of bone, the results of the determination were plotted as shown in Figure 5.4. The magnitude of the corrected signal did, indeed, depend on the amount of lead present as found from another accepted method.

5.5 Signals, Noise, and Detection Limits

All analytical methods rely on a **signal** for the analysis. The signal may be a change in an electrical voltage that is being monitored. **Noise** is an unwanted random or nearly random time-dependent change in the output signal. However, do not confuse noise with a constant output level, which we would call a **background** or **baseline**. Also, the level of the background may change slowly with time, an effect that is called **drift**. However, the time over which drift occurs usually means that correcting for a drifting baseline is quite easy.

Not all signals are electrical ones. A signal can also be, for example, a rapid, visible change in color that tells you that a reagent has just used up all the analyte (the endpoint of a titration, for those of you familiar with them). The limit to seeing the color change arises from the limitations of your eyes, which will cease to recognize a color change while an instrument will still

(*a*)

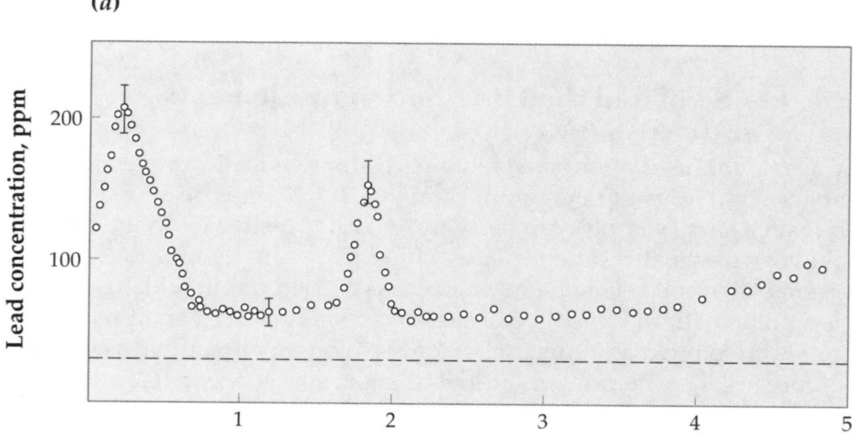

(*b*)

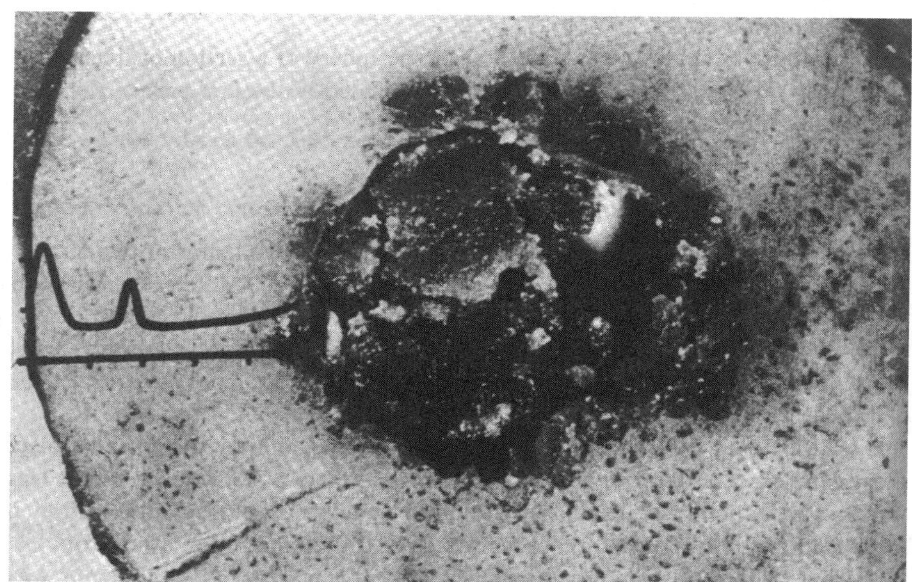

FIGURE 5.4 ▲

Lead distribution in a human femur.

(a) The concentration of lead is plotted as a function of the distance from the periphery of the bone. The dashed line is the mean of the concentration in a reference bone. The dotted curve displays the distribution in the bone of the poisoned individual. **(b)** The same plot as in Figure 5.5a but superimposed on the bone to show the distribution relative to the cross-section structure. The central dark area is the bone marrow cavity.

easily do so. It is the limits imposed by noise on the instrument's detection that concern us here.

To illustrate the effect of noise, let us look at the example shown in Figure 5.5a. The signal, which is the difference in the voltages of the blank and

(a)

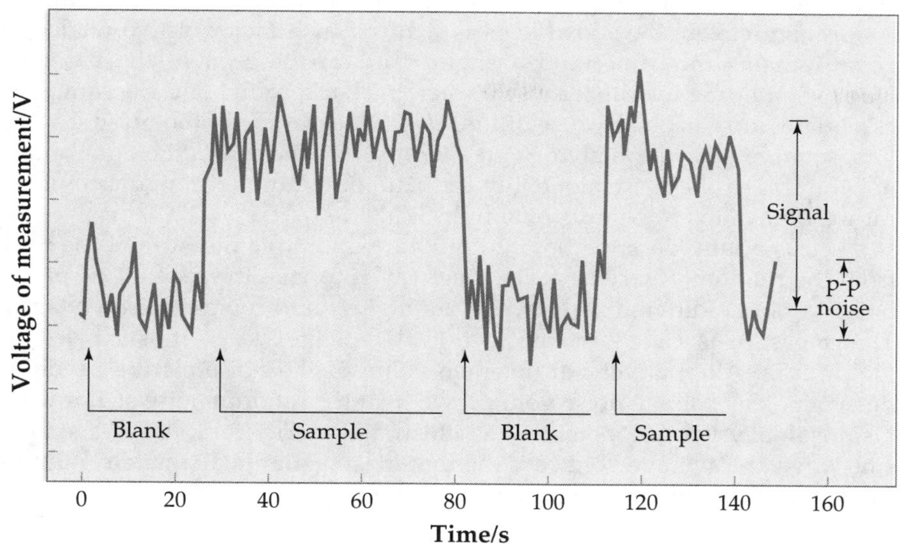

(b)

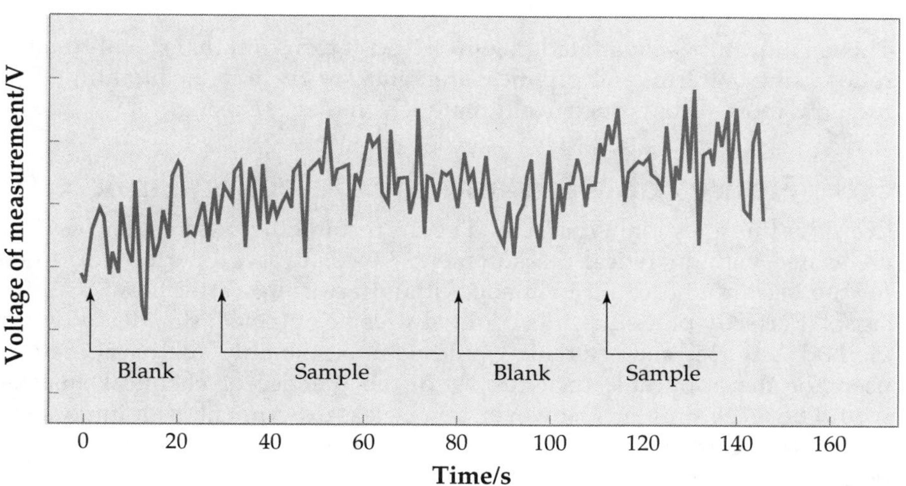

(c)

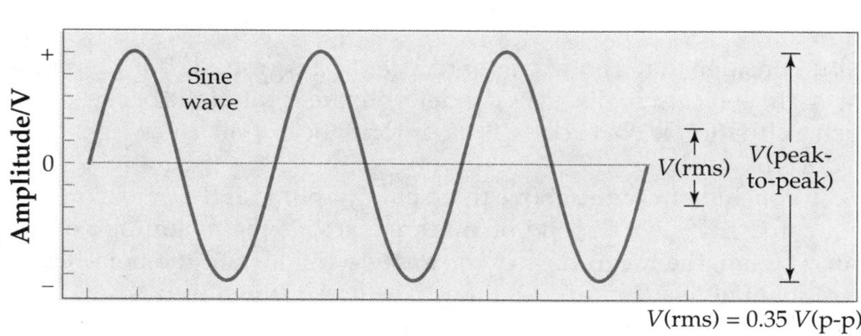

◀ FIGURE 5.5
Illustration of the concepts of signal-to-noise ratio and detection limits.

Plot **(a)** shows an instrument output when the noisy signal changes due to a sample and blank being interchanged. The signal and average peak-to-peak noise levels are labeled on the right. In **(b)**, the sample is now one tenth the concentration, and no obvious difference between it and the blank can be seen. There are a few spikes in the noise and some baseline drift present (the baseline is not strictly horizontal). The concentration of the sample is below the detection limit of the instrument under the conditions of the experiment. In **(c)**, if the noise is treated as if it were a sine wave, then a root-mean-square (rms) noise can be defined.

sample, is easy to label when the two are exchanged. Here, as noted on the right, the signal voltage is measured as the amount between the plateaus.

In Figure 5.5b, the signal level is reduced by a factor of five while the noise remains the same, and a slight drift can be seen in the baseline. However, finding the difference between the background and the sample is essentially impossible because of the *relative* magnitude of the noise. So, the limit to measuring the analyte in the sample is not the size of the signal alone nor the size of the noise alone, but the ratio of the two: the **signal-to-noise ratio** (abbreviated S/N ratio, or simply S/N).

Let us briefly describe how to form a quantitative measure of the S/N from the randomly varying level. One way is to take the magnitude of the noise to be the differences between the highest and lowest noise voltages. This measure is called the **peak-to-peak** voltage ($V_{\text{p-p}}$). It is labeled in Figure 5.5a. Alternatively, an average value of the noise can be used. To determine the average noise voltage, we treat the random noise as if it were a sinusoidally varying potential, as illustrated in Figure 5.5c. For a simple sine wave, the average voltage is the **root mean square voltage**. The numerical value of the root mean square voltage (V_{rms}) is

$$V_{\text{rms}} = \frac{0.707}{2}\text{(sine wave peak-to-peak voltage)}$$

$$= 0.35\ V_{\text{p-p}} \qquad \textbf{(5-1)}$$

This relationship is illustrated in Figure 5.5c. It turns out that signal-to-noise ratios using both rms and p-p measurements are given in the literature. Fortunately, most authors specifically state which they are using.

5.6 Figures of Merit and the Detection Limit

Consider this question: What would be the result if there were no noise at all associated with analytical measurements? In that case, a perfect correction for the background could be made without error no matter how small the signal. Perfectly precise results could always be obtained. (Incidentally, any method with this characteristic would soon be the only analytical method used for the applicable analytes. Also, the practice of chemical analysis would be *much* simpler!) However, a noiseless experiment is an impossibility. It must be, then, that the noise level is one of the limitations to finding a perfect method of analysis.

Similarly, if the signal measured from the background is not reproducible, the quality of the analysis suffers. Since you would not be sure of the value of the blank, you would not know what to subtract to correct for the background.

Finally, consider that you expect a constant background, but the noise level is so high, you cannot discover the determinate error of the background at all. Such a situation leaves you with no information at all unless the noise can be reduced.

You can conclude from the above three paragraphs that the precision and accuracy of an assay must depend on all three factors: the magnitude of the background signal, the magnitude of the analyte signal, and the noise level.

These qualitative ideas are of little use in reporting results, and we would like to provide a quantitative measure of such comparisons. You have already encountered a number of these quantitative **figures of merit**. Three are the signal-to-noise ratio, sensitivity, and detection limit. A figure of merit is a number *derived from experimental measurements* that aids us in evaluating an instrument or analytical technique.

Figures of merit are useful by themselves in evaluating the quality of instruments. However, for chemical analyses, the experimentally determined detection limit is probably the best figure of merit to use as a general guide. The judgment of "best" is based on the following three criteria:

1. the detection limit is a single number,
2. it can be used to compare data from different laboratories, and
3. when calculated with caution, it can be used to compare nearly all the instrumental techniques employed in chemical analysis.

A special committee of the American Chemical Society has defined the limit of detection thus:[1]

> The limit of detection is the lowest concentration of an analyte that the analytical process can reliably detect.

A key word in this definition is *reliably*, which suggests, correctly, that the detection limit is based on a statistical calculation. In reading published reports, you should be aware that there is still some debate about the calculation of detection limits from the experimental data.

The detection limit is a number that is either a measure of weight—such as nanograms of mercury—or of a concentration—such as 2×10^{-7}-M cadmium. This number corresponds to the amount of material producing a *signal* that is *2 to 3 times the rms noise level* (about equal to the p-p noise level). In general, the detection limit is an extrapolated value and is calculated for the best possible set of conditions with minimal interferences. The lowest limit that can *actually be measured* is typically 3 to 5 times the detection limit. That is,

> . . . when scientific researchers measure a detection limit, they give themselves the best possible opportunity to obtain the *lowest* value for which their instrument is capable, i.e., the simplest matrix (usually a simple, high purity salt in dilute, aqueous solution) and optimum experimental conditions for the analyte under consideration.
>
> When considering detection limits, several points must be kept solidly in mind:
>
> a. Quantitative determinations cannot be made at the detection limit. Relative precision at the detection limit will be ±30% to ±50% depending on the definition used.
> b. In simple cases, concentrations of 5 to 10 times the limits of detection are required for quantitative determinations.
> c. In samples with complex matrices, factors of 20 to 100 times the limits of detection are often required. [Ref: Parsons, M. L., Major, S., Forster, A. R. 1983. *Appl. Spec.* 37:411–418.]

---

[2]Committee on Environmental Improvement. 1980. *Anal. Chem.* 52:2242–2249.

Comparing Detection Limits

When comparing published detection limits, it is often necessary to discern whether the limit cited refers only to the assay or to the entire analysis. Incidentally, the detection limits for the instrument alone—which are found by using pure analytes without any interferences from sample preparation—are called, reasonably, the **instrumental detection limits**. Instrumental detection limits can be extraordinary. For example, in inductively coupled plasma mass spectrometry (ICP-MS), *instrumental* detection limits for some elements are in the parts per quadrillion (per 10^{18}) range. However, for samples that require preparation by, say, digestion with ultrapure acid, a more realistic detection limit is a ppb. That is still outstanding, but far from that of the instrument alone.

It is not uncommon to include the effects of preconcentration steps in detection limits. For instance, if an analyte is concentrated by a factor of 150 in the sample preparation, the instrumental detection limit that is found experimentally may have been divided by 150 and applied to the original sample. Without a clear statement of what part(s) of the analysis are accounted for in the calculated detection limit, direct comparison between methods and equipment is difficult. If no details of the detection limit calculation are included in a report, comparisons are useless.

Caution is also needed when comparing limits of detection for trace as opposed to microanalyses. The problem you encounter relates to comparisons between concentrations and quantities of material. For instance, in an aqueous, trace sample, the detection limit might be, for instance, 1 ng/mL. However, for a solid, micro sample, the minimum detectable quantity might be stated as 1 ng. These are not directly comparable.

Finally, you should understand that the detection limit is closely related to the standard deviation of a series of measurements, including blanks. The statistical relationships are described in greater detail in the next section.

5A A Deeper Look: The Limit of Detection in More Detail

A more precise definition and calculation of the detection limit is demonstrated in this section. For clarity, it is necessary to state most of the definitions algebraically. By understanding the derivation shown in this section, you should have a clearer picture of the assumptions underlying the concept of the detection limit.

As noted earlier, a statistical approach is required into which three important factors enter: background effects, instrument sensitivity to the analyte, and signal-to-noise (S/N) ratio. A rigorous derivation is somewhat beyond the level of expertise being pursued here. A less difficult demonstration is, therefore, presented.

The Gaussian distribution is described in Section 2.6.

In the statistical treatment of the data, the distribution of error is assumed to be Gaussian. Thus, all the rules you learned in Chapter 2 about the standard deviation, expressed as s (for a finite number of experiments) and σ (for a very large number of experiments), hold here. As with any statistical measure, by running more independent experiments, you can be more certain of the value of the detection limit *for a specific set of experimental conditions*. Different detection limits will be found for different experimental conditions (for example, note points b and c in the quote on page 155).

The following abbreviations are used in the development below:

S_t represents the total signal that is obtained from the instrument output.

S_X represents the part of the total signal that is due to analyte X. This is the net signal.

S_b is the part of the total signal that is due to the blank (baseline).

S_L is the signal due to the analyte alone at the detection limit.

C_X is the concentration of the analyte.

σ_b is the standard deviation of the blank measurements for a finite number of experiments. (Here, σ_b is used to avoid the confusion between s_b, which would be proper nomenclature, and S_b.)

c_L is the concentration of analyte at the limit of detection, or, simply, the detection limit.

The experimentally determined data are *assumed* to be well fit by a straight line. Such a fit is illustrated in Figure 5A.1a. The linear calibration line has a slope that equals the analytical sensitivity. Let us label the slope (sensitivity) by the letter g. Algebraically, the sensitivity relates simply to the analyte's concentration. That is,

$$S_X = gC_X \qquad \textbf{(5A-1)}$$

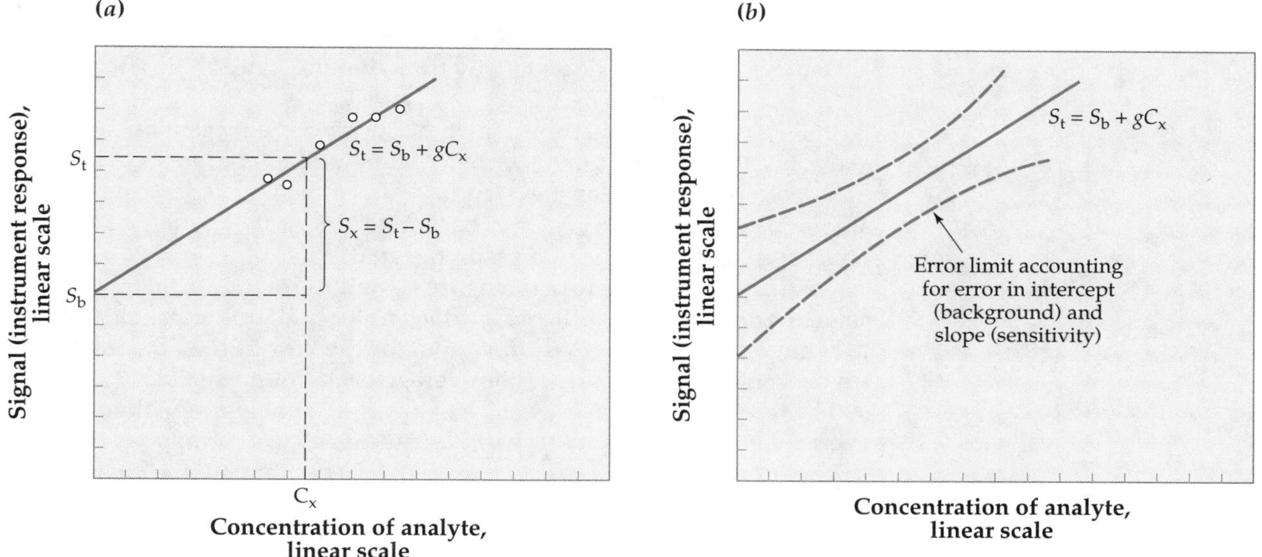

FIGURE 5A.1 ▲

Plot of the signal output of an instrument versus the analyte concentration.

(a) The value S_b is the signal due to the background alone as extrapolated from the data. A straight-line relationship is assumed. **(b)** The solid line is the same plot as above. The dashed lines indicate the error limits at some confidence level. The curvature is intentional and due to the fact that the ends of the straight line will move further than the middle with a change of slope in the line.

We now assume that the signal that is measured can be expressed as a simple sum of the background and analyte signals. Algebraically, this is

$$S_t = S_b + S_X \tag{5A-2}$$

Combining these two ideas—and two equations—we obtain an expression for the total signal,

$$S_t = S_b + gC_X \tag{5A-3}$$

This is the equation for the line in Figure 5A.1a. Note that this equation takes into account two of the three factors involved in determining the limit of detection—the instrument sensitivity and the background effects.

You should understand two points about Equation 5A-3. First, it is a correct description *only* if there are no determinate errors such as incorrect standards or blanks with $S_t \neq S_b$. Second, it is true only if the background is exactly constant for all samples and standards.

It should be clear that if both of the above points were true, any amount of analyte, no matter how small, would be detectable. As you know, there are always some errors in determining the background and the analyte signals for real instruments and samples. And that is the crucial point. The random errors can be determined and described by a confidence limit (C.L.).

Figure 5A.1b shows the same graph but with confidence limits indicated by dashed lines. This *could* be a 95% C.L., but the specific value is not important here. What is significant is that the intercept at zero analyte concentration is uncertain due to errors in the measurement. Note that this intercept determines the magnitude of the background signal. As a result, the background level is uncertain. We now make a reasonable assumption: the magnitudes of the errors occur with a normal (gaussian) distribution. The results then can be shown in a different way from that in Figure 5A.1. This alternate graph appears in Figure 5A.2.

In Figure 5A.2, the distribution of the background measurements is plotted as the gaussian curve (a histogram of the results). The gaussian curve can be characterized by a mean and standard deviation, which describe the mean value and statistical error of the background signal. The left scale of the graph represents the scale of measured signal.

On the right-hand side of the figure is the scale of the signal that is due to analyte *alone*. It is zero at the *mean value* of the background signal, which is the behavior that is expected. In words, for the best-fit line, the intercept at zero concentration S_b occurs at the average background signal.

The question that we now ask is, How much analyte signal do we need to be *certain* that we can observe it? Here the statistical certainty is expressed as a confidence limit, and *by definition*, the detection limit of the analyte *signal* is at the 99.7% C.L. This is three times the standard deviation of the background s_b. The magnitude of the signal at the detection limit ($S_b + 3\sigma_b$) is indicated in Figure 5A.2.

The ideas in the paragraph above can be expressed using the following algebraic expression:

$$\text{signal for analyte at limit of detection} = S_L = S_t - S_b \geq 3\sigma_b \tag{5A-4}$$

However, this expression concerns only the signal itself and not the concentration of analyte. In order to calculate the detection limit in terms of concen-

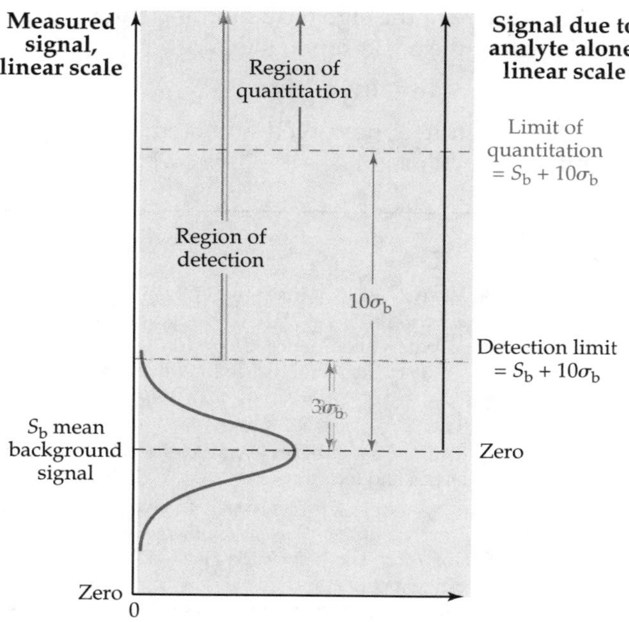

FIGURE 5A.2 ▲

Illustration of the relationship between the distribution of error in the background measurement, the detection limit, and the limit of quantitation.

The term detection limit is more widely used than quantitation limit. The total measured signal is plotted on the left scale, while the signal due to the analyte alone is plotted on the right.

tration c_L, the sensitivity must be included. Recall from equation 5A-1, the sensitivity g relates the signal and concentrations, so

$$c_L = \frac{S_L}{g} \geq \frac{3\sigma_b}{g} \qquad \textbf{(5A-5)}$$

where the value of S_L came from Equation 5A-4. Equation 5A-5 holds for the detection limit under the specific conditions during the data collection.

Recall that the value of a limit of detection depends on three factors: background effects, instrument sensitivity to the analyte, and signal-to-noise (S/N) ratio. However, the S/N ratio does not seem to appear in Equation 5A-5. What happened to it? The answer is that the effects of the noise appear in the value of σ_b, the standard deviation.

As you saw earlier in the chapter, when the noise level increases, the value of the background becomes more difficult to determine. In other words, as the noise level increases, the value of σ_b must increase. The detection limit will then increase proportionally.

Under fixed experimental conditions, it is more difficult to determine the concentration of an analyte than merely to detect its presence or absence. Thus, relatively recently, a new term has been suggested to delineate the lower end of the region where we can measure the quantity of analyte present. As shown in Figure 5A.2, the **region of quantitation** is defined when the signal is above the limit of quantitation. The **limit of quantitation** or

quantitation limit lies where the signal due to the analyte alone is 10 times the standard deviation of the background signal.

$$S_X = 10\sigma_b \quad \textbf{(5A-6)}$$

The term **quantitation limit** is now well accepted, although not used as widely as the limit of detection.

Suggestions for Further Reading

ALFASSI, Z. B., WAI, C. M. 1992. *Preconcentration Techniques for Trace Elements.* Boca Raton, Florida: CRC Press.

Contains chapters on each of the usual methods of preconcentration for elemental and organic analysis. A good source of literature references for specific problems. A useful reference.

MIZUIKE, A. 1983. *Enrichment Techniques for Inorganic Trace Analysis.* Berlin: Springer-Verlag.

A tremendously useful book with a broad array of techniques covered. Figures of the apparatus, references to the literature, and a large number of useful tables are included to allow one to access the knowledge in the field, unfortunately only up to 1982.

VANDECASTEELE, C., BLOCK, C. B. 1993. *Modern Methods for Trace Element Determination.* Chichester: Wiley.

Parallels much of this text but useful for its focus on spectrometric methods for trace analysis. Easily readable after this text. The mass spectrometry chapter is particularly thorough at its level.

BEYERMANN, K. 1984. *Organic Trace Analysis.* Chichester: E. Horwood, and New York: Halsted Press.

A survey of the different techniques usable for trace organic analysis. Some experimental methods of sample handling are included.

SKOGERBOE, R. K., MORRISON, G. H. 1971. In *Treatise on Analytical Chemistry*. Pt. 1, Vol. 9. I. M. KOLTHOFF and P. J. ELVING, eds. New York: Wiley-Interscience. Chap. 100.

A guide to the methods used in trace analysis.

LAWRENCE, J. F., ed. 1980. *Trace Analysis*. Vol. I; New York: Academic Press.

An annual series of review articles on organic and inorganic trace analysis and techniques.

MINCZEWSKI, J., CHWASTOWSKA, J., DYBCZYNSKI, R. 1982. *Separation and Preconcentration Methods in Inorganic Trace Analysis*. Chichester: E. Horwood, and New York: Halsted Press.

Despite its title, this book is mostly an exhaustive coverage of liquid-liquid extraction and ion-exchange methods for inorganic preconcentration.

BOWMAN, R. L., VUREK, G. G. 1984. "Analysis of Nanoliter Biological Samples." *Anal. Chem.* 56:291A–305A.

A fascinating, short review on how to handle liquid samples of nanoliter size.

MORRISON, G. H. 1971. *Anal. Chem.* 43(7):23A–31A.

A brief review of the Moon-rock analysis.

TREIMAN, A. 1996. *Science.* 272:1447–8.

The Martian rock analysis is described.

Concept Review

1. What is the difference between sensitivity and the detection limit?
2. Differentiate between detection limit, quantitation limit, and instrumental detection limit.
3. What is the relationship between peak-to-peak noise and rms noise?

Exercises 2, 4, 6

5.1 Classify the following according to the definitions in Figure 5.1.

(a) Lead as lead oxide in a 1 cm × 1 cm × 0.5 mm paint chip from an old house. (These paints were sometimes more than 50% lead by weight.)

(b) Micromolar levels of dopamine in a 10 mg sample of rat brain.

(c) Sodium levels in 2 mL of blood. (Blood is normally 140 mM in Na^+.)

(d) Nerve toxin level in a coffee residue at the bottom of a cup beside a murder victim's bed. (Some toxins are lethal at less than 100 μg per kg of body weight.)

5.2 Suppose an aliquot of a solution was to analyzed for manganese, and the organic impurities were to be removed by treating the sample with 10.0 mL of nitric acid and then diluting to 100.0 mL with deionized water. What is the minimum concentration of Mn (in ppm) in a 100.0 mL sample that could be reliably detected (i.e., Mn >10 times amount Mn in background due to acid) if we used acid sample 1 from supplier 3 in Table 5.1a?

5.3 Assume you are analyzing a 2-g biological tissue sample for the presence or absence of Ni in the 0.1 ppb (w/w) range. The tissue was obtained with a metal scalpel that is 10% Ni.

(a) What mass of the scalpel rubbing off on the sample would contribute a quantity of nickel to give a positive (0.1 ppb) result?

(b) Assume that the density of the steel is 7.0 g cm^{-3} and that the scalpel lost the material in a single, perfectly cubic chip. What is the length of the cube edge?

5.4 Dust picked up on a filter in your laboratory is analyzed. The following results are obtained for the metal content.

Element	Content (wt%)
Ca	10
Si	5
Fe	3
Al	1.5
Ni	1.5
K	1
Mg	1
Cu	0.5
Mn	0.5
Others	Traces

A dust particle that is a perfect sphere with a diameter of 100 μm and a density of 2 g cm^{-3} lands in your 10-mL acidified sample and dissolves.
(a) If you were to determine the sample's iron content in the range of 10 ppm, would you see the effects of the presence of this single dust particle?
(b) If you were to assay a sample for trace Si at the 1-ppm level, what size spherical dust particle would contain an amount of Si equal to that contained in your sample?
(c) Could you effectively analyze a 1-g sample for Mn in the ppb range if the sample were collected with one spherical 100 μm dust particle (with the composition listed in the table) on it? (In other words, would the Mn content be within 5% of the true value)

5.5 Activated charcoal can be used to concentrate trace metals. The metals are adsorbed in basic solution and desorbed with acid. The solution used to desorb the ions then is analyzed. Under the experimental conditions, the charcoal was found to have the recoveries shown in the table. In addition, some impurities wash out from the charcoal, the amounts of which depend on the quantity of charcoal used. The impurity levels in the charcoal are listed below.

Element adsorbed	% recovery	Impurities in charcoal (μg/g)
Zn	85	<1
Cu	96	18
Ni	98	19
Mn	98	150
Ag	92	<0.2
Cd	96	<0.1
Pb	92	2.3

0.50 g of charcoal is used per sample. The charcoal was acid-washed so that 10-mL blanks showed a background that was no greater than 1% of the total of each impurity present in the charcoal. (This means that no more than 1% of each impurity will be leached from the charcoal into the acid solution during the analysis.) Final solution volumes are made to 10.0 mL. The original solutions, which were passed through a bed of the charcoal, were 500 mL. The metals are adsorbed from this.
(a) If the recoveries are 100%, how much more concentrated is the eluted solution than the sample?
(b) What are the apparent background concentrations (in ppb w/v) in the 500-mL blank if 1% of the charcoal impurity elements wash out with the 10-mL desorbing solution?
(c) Samples with 10 ppb of each of the elements were used as standards. The analytical method is considered to be usable if the samples have five times the concentrations of the background from the charcoal. Can all the metals be analyzed at the 10-ppb range?

5.6 Assume that a trace analytical method has an inherent relative standard deviation of 2% for random errors. You have a sample that has a low content of the component to be assayed. You measure a number of blanks and a number of samples and find that the sample measurement is only 25% above the blank level. In other words, when a sample content is assayed, the measurement (on average) consists of a background value that is 80% of the measure, whereas the sample content is 20% of the whole.
(a) What is the relative standard deviation of the sample content? (Note that the 2% RSD applies to the background measurements alone as well as to those with the sample.)
(b) What is the RSD of the sample content alone if the sample measurement is equal to the blank?

5.7 Table 5.7.1 on the next page shows the results of determinations of the inorganic content of commonly used organic solvents. [Data from Jacobs, F. S., Ekambaram, V., Filby, R. H. 1982. *Anal. Chem.* 54:1240.]

Inorganic trace elements in oil from oil sand are to be determined. The method involves extracting with an organic solvent followed by a multielement quantitation. The elements of most importance are Co, Fe, K, Na, V, and Zn.
(a) Assume that the extraction of the oil is done by heating and stirring the oil sand with a fixed volume of solvent and that the extraction is complete. Assume that the instrumental assay is done directly on the solution. If there is no interference in the assay for Zn, which would be the best solvent to use for determining only Zn?
(b) If there is a large interference from Na in the assay method, which solvent(s) would you choose as being the best?
(c) If the interference from Na is fully correctable, which solvent would be the best to use?
(d) Assume that the solubilities of the organic components of the sample are about ten times greater in methanol than in any other solvent. As a result, only one-tenth as large a solvent volume is needed for the extraction. How would you answer parts (b) and (c) in this case?

Table 5.7.1 Upper limits found of trace-element concentrations in representative organic solvents (in ng/L) (Zeros are place holders only.)

Element	Hexane	Toluene	Chloroform	Methanol	Ether
As	39	200	16	4,200	9.1
Br	10	230	13	140	50
Ce	200	120	200	200	200
Cl	20,000	10,000	40,000	1,800	30,000
Co	90	72	90	90	90
Cr	170	6,700	70	490	140
Cs	30	30	30	30	30
Eu	20	20	20	20	20
Fe	7,000	71,000	10,000	7,000	5,000
Ga	30	300	40	200	50
Hf	20	20	20	20	20
Hg	0.7	59	25	0.7	0.3
K	3,000	250,000	10,000	9,000	1,000
La	0.6	100	40	50	4
Mn	500	4,200	700	700	40
Mo	400	2,000	400	1,000	400
Na	4,800	390,000	2,500	99,000	580
Sb	4	4	8	40	40
Sc	2.6	22	2.2	1.1	0.5
Se	10	10	60	7	80
Sm	2	7.6	1.9	5	3
Tb	5	6	6	5	5
Th	20	35	30	20	20
V	9,000	8,000	20,000	10,000	20,000
Zn	40	17,000	200	2,700	580
Zr	200	600	1,000	700	2,000
Total inorganics (μg/L)	45.3	759	85.0	137	37.1

(e) Assume you are doing analyses for all six important elements. All six occur in approximately the same range of concentration, and the same volume of each solvent would be used for extraction. Are there any of the solvents in the table you would *not* use because of significant background contributions?

(f) Can you decide on the best solvent to use without determining the general range of each of the six most important elements?

5.8 The bitumen (organic phase) was extracted from ten grams of Athabasca oil sand with 1000 mL of toluene using a soxhlet extractor (such as pictured in Section 4.4). The solvent was removed with a rotary evaporator, and the bitumen was analyzed for trace elements with neutron activation analysis. The values measured for nine elements are shown in the table to the right, along with the standard deviations of the measurements.

(a) Assume that the upper-limit numbers in Table 5.7.1 are true and exact. Correct the bitumen data for background.

Element	Observed concentration (ng/g)
Br	276 ± 20
Co	239 ± 28
Fe	242,000 ± 17,000
Mn	4,760 ± 210
Hg	31 ± 5
Cr	854 ± 70
K	44,300 ± 2900
Na	21,200 ± 1600
Zn	2,730 ± 140

(b) Are any of the corrected values outside the $\pm 2\sigma$ range for the observed values?

(c) Does the value for the Na content of the sample have any meaning using a toluene extraction?

Additional Exercises

5.9 An instrument manufacturer lists the detection limit of a detector as 1 pg for a specific analyte (MW = 100 *u*) under optimal conditions. If the detector has a volume of 2.5 μl, what is the detection limit in terms of

(a) molarity?

(b) ppb (w/v)?

5.10 A sample of 600 L of air at 25°C and 1.00 atm has pumped through a trap containing a solid phase, which will trap the VOCs in the air. The analytes are desorbed at 200°C over a 12-second time period into helium carrier gas which had been saturated with hexane vapor and maintained at a flow rate of 10 mL/min. The helium was then passed through 10.0 mL of hexane, which dissolved the VOCs. By what factor were VOCs preconcentrated?

5.11 (Requires information from Section 5A.) If we assume that σ_b for an instrumental analysis is dominated by the noise of the instrument, that the peak-to-peak noise of the instrument is 0.20 mV, and that the response factor for the analyte is 1.25 mV/μM, what is the concentration limit of detection? (Hint: Which of the S/N definitions reflects the average value of the *random* noise?)

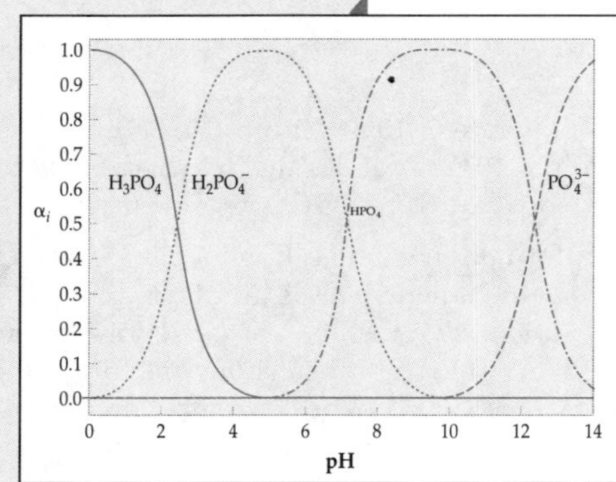

CHAPTER 6

Simple and Competitive Chemical Equilibria

HPO_4

Overview

6.1 Properties of Chemical Equilibria
6.2 Introduction to Simultaneous Equilibria
6.3 Calculating the pH for Solutions of Weak Acids and Bases
6.4 When Water Ionization Becomes a Significant Source of Protons
6.5 Estimating pH for Polyprotic Acid Solutions
6.6 Fractional Composition: What Is the Predominant Species?
6.7 Finding the Fraction of Each Species: α-Values
6.8 pH Buffers
6.9 Activities and Activity Coefficients

6.1 Properties of Chemical Equilibria

Knowledge of the principles of chemical equilibria and the ability to carry out calculations about competitive equilibria are essential to understanding the fundamentals of chemical analysis. These principles underlie a number of different stages of an analysis. In sample preparation, a knowledge of chemical equilibria aids in making the sample homogeneous and removing the parts of the sample that will interfere; chemical equilibria can often be manipulated to minimize interferences. You will more easily understand the methods presented in this text if you are conversant with the ideas and calculations of chemical equilibria.

The Concept of Chemical Equilibrium

If nitrogen gas, N_2, and hydrogen gas, H_2, are placed together in a container at one atmosphere pressure and 400°C (a dull-red heat) and left for a week, an analysis of the contents of the vessel should show the presence of hydrogen, nitrogen, and ammonia. The reaction that occurs can be written

$$N_2(g) + 3H_2(g) \rightleftharpoons 2NH_3(g) \qquad \textbf{(6-1)}$$

Since all three chemical species are present, the reaction did not go to completion. If it had, N_2 or H_2 would not be detected in the container.

If ammonia gas alone is placed in a container under the same conditions, and the contents are analyzed a week later, the analysis again should show the presence of the same three components: hydrogen, nitrogen, and ammonia. The reaction can be written

$$2NH_3(g) \rightleftharpoons N_2(g) + 3H_2(g) \qquad \textbf{(6-2)}$$

The conclusion drawn is that this reaction (6-2) does not go to completion either.

One possible explanation of these observations is that neither reaction had time to progress all the way to its ultimate state; perhaps the reaction would have gone to completion if it were allowed to react further. This hypothesis can be tested by running both reactions for a few more weeks, followed by another analysis of the contents. If the analytical results remain the same, we can say that the system of ammonia, nitrogen, and hydrogen has reached its **equilibrium** condition. At equilibrium, the identities of reactants and products are the same regardless of the side from which the equilibrium was started (in this case, nitrogen and hydrogen alone or ammonia alone). When we talk about the position of the equilibrium, we say it lies *toward the left* (very little of the products are present compared to reactants) or *toward the right* (very little of the reactants are present compared to products).

It has been found from experiments that the position of the equilibrium for the reaction that involves nitrogen, hydrogen, and ammonia can be described by a constant, the **equilibrium constant**, which is dependent on temperature and pressure. For the ammonia reaction

$$2NH_3(g) \rightleftharpoons N_2(g) + 3H_2(g) \qquad \textbf{(6-2)}$$

the equilibrium constant K is

$$K = \frac{[N_2][H_2]^3}{[NH_3]^2} \qquad \textbf{(6-3)}$$

where the square brackets indicate that molar concentration units are to be used. Each concentration is raised to the power of its stoichiometric coefficient.

An alternate means of expressing the equilibrium constants of gas phase reactions uses the pressures of the species. For example, we can define another equilibrium constant for reaction 6-2

$$K_p = \frac{P_{N_2}P^3_{H_2}}{P^2_{NH_3}} \quad \textbf{(6-4)}$$

The subscript p in K_p is used to denote this type of equilibrium constant, since the value of K_p would have pressure units while K was calculated from quantities that were in molar concentration units. For the present, we shall continue to use only the molar concentrations in the equilibrium expressions.

Given the general reaction

$$a\text{A} + b\text{B} \rightleftharpoons c\text{C} + d\text{D} \quad \textbf{(6-5)}$$

the concentrations of the reactants and products at equilibrium are related by the equation

$$K = \frac{[\text{C}]^c[\text{D}]^d}{[\text{A}]^a[\text{B}]^b} \quad \textbf{(6-6)}$$

K is a constant, the value of which depends on the temperature and the identities of the products and reactants. Equation 6-6 expresses the **law of mass action**. In other words, a quantitative description of the equilibrium follows the law of mass action. Equation 6-6 also can be called a **mass-action expression**.

As can be seen in Equations 6-3 and 6-6, the equilibrium constant expressions are written with the reaction products in the numerator and the reactants in the denominator. Also, as mentioned above, the concentration of each species is raised to a power equal to its stoichiometric coefficient in the chemical equation.

If the mass-action expressions describe the behavior exactly, the system is said to follow ideal behavior. For real equilibria, this ideal behavior is nearly true for gases at less than a few atmospheres pressure and for solutions that are dilute. Absolute rules defining when "ideal" behavior describes the system cannot be stated since the limit depends on the specific chemical system. For example, ionic solutions are ideal only up to about 1 mM, while organic solutes in organic solvents can behave as ideal systems over a much wider range. These limits are pursued in more detail in Section 6.9.

Conventions

The chemical equations and mass-action expressions reflect only the *overall* reactions that occur. For instance, consider the equilibrium constant for the dissociation of acetic acid. The chemical equilibrium can be represented by the equation

$$CH_3COOH + H_2O \rightleftharpoons CH_3COO^- + H_3O^+ \quad \textbf{(6-7a)}$$

However, the exact formula of the species formed by water interacting with the proton from the acid is uncertain. Therefore, although we can write H_3O^+ as above, some workers maintain that three waters associate with the proton to form $H_7O_3^+$, and others suggest that $H_9O_4^+$ is formed by association with four waters. Equation 6-7a might then be written

$$CH_3COOH + 3H_2O \rightleftharpoons CH_3COO^- + H_7O_3^+ \quad \textbf{(6-7b)}$$

In other words, the number of associated water molecules to be specified in the equation is questionable. Actually, both the acetic acid molecules and the acetate ions are themselves strongly associated with waters of hydration. These waters have been ignored as well in Equations 6-7a–d. We have also ignored the molecular structures of the species in these simple equations (for example, $H_9O_4^+$ is thought to be tetrahedral). Instead, chemists find that sufficient information is conveyed when the equation is written

$$CH_3COOH \rightleftharpoons CH_3COO^- + H^+ \quad \textbf{(6-7c)}$$

Such simplified equations are useful to represent the observed behavior of reactions that we know are highly complicated in their chemical details.

Besides the selective simplification of the chemistry, other conventions are used to make chemical equations easier to write. As long as the abbreviations are understood, there is no confusion. For instance, acetate ion CH_3COO^- can be abbreviated as OAc^- or OAc. The form OAc^-, used in this text because it is short, indicates the oxygen to which the proton is bound and retains a reminder of the charge to help avoid errors in balancing the charges in stoichiometric equations. Therefore, the simple equation representing ionization of acetic acid in water can be written

$$HOAc \rightleftharpoons H^+ + OAc^- \quad \textbf{(6-7d)}$$

with the mass-action expression

$$K = \frac{[H^+][OAc^-]}{[HOAc]} \quad \textbf{(6-8)}$$

Often, for acid dissociation reactions like that for acetic acid, the equilibrium constant is written with a subscript as K_a, where the subscript a stands for acid. You may also see the equilibrium constant written K_d in certain circumstances, where the d indicates dissociation.

Properties of Equilibrium Constants

The equilibrium reaction 6-7d can be reversed and written

$$H^+ + OAc^- \rightleftharpoons HOAc \quad \textbf{(6-9)}$$

Then the mass-action expression 6-8 becomes

$$K' = \frac{[HOAc]}{[H^+][OAc^-]} \quad \textbf{(6-10)}$$

The equilibrium constant written with a prime K' differs from K, but comparison of the two equilibrium expressions, Equations 6-8 and 6-10, shows that

$$K' = \frac{1}{K} \quad \textbf{(6-11)}$$

K' is called an **association equilibrium constant** and the algebraic relationship in Equation 6-11 illustrates a general rule:

> *When a stoichiometric equation is reversed, the equilibrium constant for the reversed reaction is the reciprocal of that for the forward reaction.*

If we add Equation 6-7d to itself

$$\begin{array}{r} HOAc \rightleftharpoons H^+ + OAc^- \\ + \; HOAc \rightleftharpoons H^+ + OAc^- \\ \hline 2\,HOAc \rightleftharpoons 2\,H^+ + 2\,OAc^- \end{array} \quad \textbf{(6-12)}$$

then, according to the rules for writing mass-action expressions, we write for reaction 6-12

$$K = \frac{[H^+]^2[^-OAc]^2}{[HOAc]^2} = K_a(HOAc) \cdot K_a(HOAc)$$

This example of adding the chemical equations in 6-12 illustrates another general rule:

When stoichiometric equations are added to give an overall equilibrium equation, the associated equilibrium constants must be multiplied together to obtain the equilibrium constant for the overall equation.

Most equilibrium expressions relating to real systems can be evaluated with the operations of addition and reversal of chemical equations and with the appropriate manipulation of the mass-action expressions. To obtain numerical results, the stoichiometry of the chemical reaction and the value of the equilibrium constant (at the temperature of the experiment) must be available.

6.2 Introduction to Simultaneous Equilibria

Sometimes a chemical system can be described by a single mass-action expression and a single stoichiometric equation. It appears as if only one chemical reaction occurs. However, competitive equilibria are more typical; that is, when two or more different reactions occur that have at least one species in common. For example, if two bases, A^- and B, compete for protons in solution by the reactions

$$H^+ + A^- \rightleftharpoons HA \qquad \textbf{(6-13a)}$$

$$H^+ + B \rightleftharpoons HB^+ \qquad \textbf{(6-13b)}$$

then H^+ is the species in common.

In the competitive equilibria represented by Equations 6-13, a proton may associate almost exclusively with only one of the two competing bases. Or it may associate with both by spending part of the time in association with one and part of the time with the other. This seems straightforward enough. However, some problems arise in quantitative calculations of concentrations of the species present: here [HA], $[HB^+]$, $[A^-]$, [B], and $[H^+]$. The algebra often becomes involved, but with the computing power that is available now, reasonable approximations can be made.

The essential algebra for even the most complicated equilibrium calculations has been demonstrated in Section 6.1. No matter how involved the algebra of a problem appears, you will only be adding and reversing chemical equations and manipulating the equilibrium constants using the two general rules given above. Nevertheless, until you work through a number of problems of each type, you should not expect to have a clear understanding of the techniques.

The rules of manipulating equilibrium constants and chemical equilibrium equations also allow us to calculate quantitatively the concentrations of chemical species at equilibrium under conditions where the same information might be difficult to measure by direct experiment or may even be unobtainable. However, you should understand that these extrapolations are dependent on the quality of the calculations, and many imperfections are possible. Some of these difficulties are noted in the discussion of the nature of activities and activity coefficients that appears in Section 6.9. For the present such difficulties are ignored.

Weak and Strong Acids and Bases

Let us write the general formula HA for an acid consisting of a proton H^+ bound to an anionic base A^-. The dissociation of this general acid is written

$$HA \rightleftharpoons H^+ + A^- \qquad \textbf{(6-14)}$$

In other words, the acid HA dissociates to produce a proton and the acid's **conjugate base**, A^-.

As represented by their pK_a-values, in most solvents (including water) most acids fall into one of two categories: **weak acids** and **strong acids**. From Figure 6.1 you can see that when weak acid HA is added to water to produced a concentration C_{HA}, only a small fraction of it dissociates. For example, the equilibrium of the acetic acid dissociation reaction, represented by Equation 6-8, lies far to the left. Acetic acid is an example of a weak acid. Incidentally, the concentration C_{HA} represents the number of moles of HA dissolved per liter of the solution. This is the **analytical concentration** of the acid.

On the other hand, a strong acid is one for which the equilibrium 6-14 lies far to the right. Only a small fraction of C_{HA} of the strong acid HA remains undissociated. Hydrochloric acid (HCl) is an example of a strong acid. This is the nomenclature of elementary Brønsted acid-base theory.

The fact that an acid is weak means that the conjugate base holds on to the proton relatively strongly; the conjugate base is strong. When an acid is strong, the base holds the proton relatively poorly; the conjugate base is weak. Put simply:

The weaker an acid, the stronger its conjugate base.

Few intermediate-strength acids exist. As a result, by studying demonstration calculations for weak acids, most of the characteristics of simultaneous equilibrium calculations can be understood. The next section begins with an illustration for a typical weak acid, acetic acid.

6.3 Calculating the pH for Solutions of Weak Acids and Bases

How might you estimate the expected pH of an aqueous solution of a weak acid at 25°C? For illustration, we use a 0.1 M solution of acetic acid.

From tables such as those in Appendix II, the equilibrium constant for acetic acid dissociation is found to be

$$K_a = 1.76 \times 10^{-5}\ \text{M}; \qquad pK_a = 4.75 \qquad \textbf{(6-15)}$$

This K_a applies to the mass action expression in Equation 6-8.

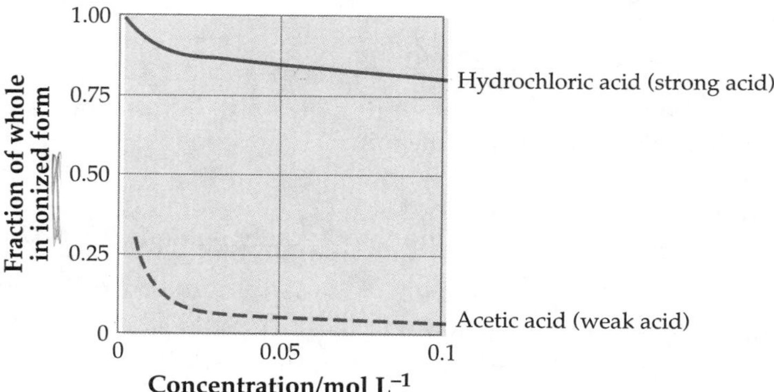

◀ FIGURE 6.1 **Graph of the fraction of molecules dissociated as a function of concentration for a strong acid (HCl) and a weak acid (acetic) in water.** The strong acid is fully dissociated at low concentrations, and the fraction dissociated slowly decreases with concentration. The weak acid is less dissociated, but the extent of dissociation depends strongly on concentration in the low range.

To make some algebraic expressions easier to write, we let $X = [H^+]$. The stoichiometric equation for the acid dissociation (and the concentrations that would be expected) are:

$$\begin{array}{ccccc} HOAc & \rightleftharpoons & H^+ & + & OAc^- \\ (C_{HOAc} - X) & & (X) & & (X) \end{array} \qquad \textbf{(6-16)}$$

This equation says that for each proton produced by dissociation, one OAc^- ion forms, and one HOAc molecule vanishes. For the 0.1 M solution of acid, the information in Equation 6-16, can be combined with the mass action expression to give Equation 6-17a.

$$K_a = \frac{[H^+][OAc^-]}{[HOAc]} = \frac{X^2}{C_{HOAc} - X} = \frac{X^2}{0.1 - X} = 1.76 \times 10^{-5} \qquad \textbf{(6-17a)}$$

Once this equation is set up, we can solve it exactly, or we can simplify it by approximation. Do not confuse the method of solving the equation, which is strictly algebra, with the chemical behavior being described. The stoichiometry as well as the K_a value were *found through experimentation.* This chemistry is expressed in Equations 6-15 and 6-16.

In the next two subsections, Equation 6-17a will be solved both ways: first exactly and then after making an approximation. As you become more familiar with the approximation methods, you will be more confident in their use.

Exact Solution

First, the calculation is done without approximations in the algebra. We write Equation 6-17a as a quadratic function:

$$X^2 = (1.76 \times 10^{-5})(0.1 - X) \qquad \textbf{(6-17b)}$$

After multiplication and rearrangement, the equation becomes

$$X^2 + (1.76 \times 10^{-5})X - 1.76 \times 10^{-6} = 0 \qquad \textbf{(6-17c)}$$

We then solve for X using the quadratic formula (Section 1A).

$$X = \frac{-1.76 \times 10^{-5} \pm \sqrt{3.10 \times 10^{-10} + 7.04 \times 10^{-6}}}{2}$$

$$X = \frac{-1.76 \times 10^{-5} \pm 2.653 \times 10^{-3}}{2} = \frac{2.64 \times 10^{-3}}{2} = 1.3_2 \times 10^{-3}$$

That is,

$$[H^+] = 1.3_2 \times 10^{-3}$$

and the pH of the solution is

$$pH = -\log(1.32 \times 10^{-3}) = 2.88.$$

Note that the negative-value solution, $X = -1.34 \times 10^{-3}$, was not used because a negative concentration has no meaning.

A Further Algebraic Approximation

The pH for an acetic acid solution was found through solving Equation 6-17a,

$$K_a = X^2/(C_{HOAc} - X) \qquad \textbf{(6-17a)}$$

but this equation can be simplified further. The key test whether this further approximation might be used is to observe whether the analytical concentration C_{HOAc} is much larger than K_a for the acid. If so, you can ignore the value of X in the denominator of Equation 6-17a. In the acetic acid calculation, "much larger than" means a factor of 10^4. In the algebra, this means that $(C_{HOAc} - X) \approx C_{HOAc}$. On a practical basis, the error in the calculation is less than a typical difference between a calculated value and the corresponding measured value.

The equation simplified from Equation 6-17 is, then,

$$K_a = \frac{X^2}{C_{HOAc}} \tag{6-18}$$

or

$$X = \sqrt{K_a C_{HOAc}} \tag{6-19}$$

Substituting the values for C_{HOAc} and K_a, we find

$$X = \sqrt{(1.76 \times 10^{-5})(0.1)} = 1.33 \times 10^{-3} = 1.33 \text{ mM}$$

The answer, 1.33×10^{-3} M, is within 1% of 1.32×10^{-3} M, which was the answer calculated without mathematical approximations.

You might ask, Why should we solve such a simple equation approximately at all? There is a good reason. Simplifications such as this often can make our perception of a problem clearer, enabling us to see what trends of behavior are expected without going through an entire calculation. This kind of broad overview can help indicate whether we might be pursuing a wrong direction when attempting to solve a problem. Another reason for simplifying the algebra occurs in more complicated equilibrium problems. As the calculations become messier, the need to perceive a trend may be much more profound—a need that a computer may not fulfill.

What About the Water?

Up to now, we have ignored another other source of protons in the solution: the water itself. Why could we do that? To answer that question, let us calculate the value expected for $[H^+]_{Water}$ in pure water. For the dissociation of water

$$H_2O \rightleftharpoons H^+ + OH^- \tag{6-20}$$

the equilibrium constant is designated K_w and

$$K_w = [H^+][OH^-] = 1.00 \times 10^{-14}; \quad 25°C \tag{6-21}$$

(Note that water does not appear in the denominator. The reason behind this will become clear in Section 6.9.)

As Equation 6-21 indicates, the maximum proton concentration that can be contributed by water alone equals the concentration of hydroxide, so

$$X^2 = 1.00 \times 10^{-14} \qquad \text{and} \qquad [H^+] = [OH^-] = 10^{-7}$$

In a 0.1 M acetic acid solution, we found that $[H^+]_{HOAc} = 1.34 \times 10^{-3}$, more than ten thousand times the amount that might come from the water alone.

Two conclusions can be drawn from these numbers. First, acetic acid is a much stronger acid than water. And second, the large difference in the contributions of protons allows us to assume that $[H^+]_{Total} \approx [H^+]_{HOAc}$, the protons produced from the ionization of acetic acid alone. It is reasonable to ignore this small, extra H^+ contribution.

This simplification cannot be used, however, when an acid is only very weakly dissociated or the added acid's concentration is only in the range of

10^{-7} M or less. In either situation, the approximation that the acid remains the only source of protons may become suspect.

The Salt of a Weak Acid: Hydrolysis

The next situation presents the question, What happens when, instead of adding acetic acid to the water, we add the conjugate strong base, acetate ion? Experimentally, this can be done by making a solution of a salt such as potassium acetate. The known chemistry of this compound indicates that it is completely dissociated in the solution to yield hydrated potassium ions and acetate ions. Some of the acetate ions react with the water with the following chemistry:

$$OAc^- + H_2O \rightleftharpoons HOAc + OH^- \qquad \textbf{(6-22)}$$

This type of reaction is called **hydrolysis** (meaning splitting water).

Can we calculate the pH of this acetate salt solution? We have enough information to solve the problem by using the rules for manipulating chemical equations and some data we already have used. The first pieces of information needed are contained in

$$HOAc \rightleftharpoons H^+ + OAc^-; \quad K_a = 1.76 \times 10^{-5}\ \text{M at } 25°\text{C} \qquad \textbf{(6-7d)} \text{ and } \textbf{(6-15)}$$

The other equation needed links the water equilibrium to the acetate equilibrium. The term "links" means that a **common ion**, here OH^-, appears in both equilibria. This other equation describes the dissociation of water.

$$H_2O \rightleftharpoons H^+ + OH^-; \quad K_w = 1.0 \times 10^{-14} \text{ at } 25°\text{C} \qquad \textbf{(6-20)} \text{ and } \textbf{(6-21)}$$

The hydrolysis reaction, Equation 6-22, can be obtained by manipulating the two chemical equations. Since acetate occurs on the left, the first of the two equations must be reversed before adding them and the equilibrium constants must be treated by the usual rules (Section 6.1).

Thus

$$\begin{array}{lll} & H^+ + OAc^- \rightleftharpoons HOAc & K' = 1/K_a \\ + & H_2O \rightleftharpoons H^+ + OH^- & K_w = 10^{-14} \\ \hline & \not{H}^+ + OAc^- + H_2O \rightleftharpoons \not{H}^+ + OH^- + HOAc & K_h = (1/K_a)K_w = K_w/K_a \end{array}$$

where the subscript h indicates that it corresponds to a hydrolysis reaction. The equilibrium constant K_h is called a **hydrolysis constant**—in this case for the hydrolysis of acetate. However, K_h is not a special kind of equilibrium constant. On the contrary, it is like any other equilibrium constant, but has been labeled to indicate the type of chemical reaction to which it applies.

If we now cancel H^+ from both sides of our sum reaction and substitute the values for K_a and K_w, we have the information we need.

$$OAc^- + H_2O \rightleftharpoons HOAc + OH^- \qquad \textbf{(6-22)}$$

$$K_h = K_w/K_a = (1.00 \times 10^{14})/(1.76 \times 10^{-5}) = 5.68 \times 10^{-10}$$

The mass action expression for the hydrolysis is, in turn,

$$K_h = \frac{[HOAc][OH^-]}{[OAc^-]} = 5.68 \times 10^{-10}\ \text{M}$$

We can now calculate the pH of an acetate solution in much the same way as we did for acetic acid.

EXAMPLE 6.1 

What is the expected pH of a 0.1 M solution of potassium acetate in water at 25°C?

Solution

Potassium acetate dissociates completely in water to give acetate and potassium ions. Therefore $C_{acetate} = 0.1$ M. Our hydrolysis constant that we calculated for acetate indicates that acetate is a weak base but is still much stronger than water. So, we can make assumptions similar to those that we made earlier. There, relatively accurate calculations could be made with the simplest form of the equation. That is, $[OH^-]_{Tot} \approx [OH^-]_{acetate}$ and that $[OAc^-] \approx C_{acetate}$.

$$[OH^-] = \sqrt{K_h C_{acetate}}$$
$$= \sqrt{(5.68 \times 10^{-10}) \cdot (0.1)}$$
$$= 7.53 \times 10^{-6} \text{ M}$$

These examples illustrate some of the power of the methods of calculation involving simultaneous equilibria. For instance, the calculation shows the relationship between the properties of an acid and its conjugate base. If we measure the K_a for an acid, we can calculate the K_b for its conjugate base. On the other hand, if the base does not yield the "expected" value of pH in an experiment, you might expect that either the experiments measuring the properties of the acid or base were not done correctly, or perhaps one of the materials might have been impure. In other words, separate experiments on an acid and its conjugate base can be used to check each other through the use of these calculational methods.

The Common Ion Effect

When two different sources of a single species exist, the equilibria of both may shift concentrations as a result. When ions are involved, this shift is called the **common ion effect**.

EXAMPLE 6.2 

Suppose that enough concentrated HCl (a strong acid) is added to our 0.1 M acetic acid solution to give a total $[H^+]$ of 0.1 M. What would be the new concentration of acetate ion in the solution?

Solution

For HOAc alone, we calculated $[H^+]$ to be 1.34×10^{-3} M. HCl is a strong acid, and, thus, practically all the H^+ contribution comes from the added HCl. Therefore, make the assumptions that $[HOAc] \approx 0.1$ M (just as we did before), but this time the total proton concentration $[H^+]_{Tot} = 0.1$ M. Substituting these values in the mass action expression for acetic acid,

$$K_a = \frac{[H^+][OAc^-]}{HOAc}$$

$$1.76 \times 10^{-5} = \frac{(0.1)[OAc^-]}{0.1}$$

$$[OAc^-] = \frac{(0.1)(1.76 \times 10^{-5})}{0.1} = 1.76 \times 10^{-5} \text{ M}$$

Henri Louis Le Chatelier (1850–1936) while a student at Ecole Polytechnique. He developed this principle in the 1880s.

The value we found for $[OAc^-]$ is also the concentration of H^+ *due to the acetic acid alone*. It is so much smaller than the 0.1 M contribution from the HCl, our initial assumption that $[H^+]_{Tot} \approx [H^+]_{HCl}$ was correct. Furthermore, the H^+ from acetic acid is decreased compared to a solution containing only acetic acid. The large $[H^+]$ from the HCl has suppressed the acetic acid ionization, *i.e.*, the equilibrium has shifted toward undissociated HOAc.

You may be familiar with the shift in equilibrium as **Le Chatelier's principle**. Le Chatelier's principle states that if an equilibrium is disturbed, the reaction will proceed in a manner to re-establish its equilibrium. In other words, either the forward or back reaction will predominate until the concentrations are back to the levels dictated by the equilibrium constant. The disturbance may be a change in temperature, a change in pressure (for reactions involving gases), or the removal or addition of one of the chemical components, as for the HCl-HOAc system.

6.4 When Water Ionization Becomes a Significant Source of Protons

So far in this chapter, it has been possible to calculate the expected pH of some solutions based on a single equilibrium. In addition, the calculations have been relatively easy because a number of arithmetic approximations are valid for these calculations. In this section, you will see what happens when the water ionization no longer can be ignored: when two equilibria both contribute similar concentrations of a single species.

Very Dilute or Very Weak Acids and Bases

We have seen from our examples above that two assumptions can often be made when working with weak acids in aqueous solution. First, the H^+ in solution comes primarily from the acid as opposed to the water itself. Second, since little dissociation occurs, undissociated acid remains nearly at the level initially added. There are, however, some instances where one or the other of these assumptions cannot be made.

One is when the concentration of the acid is very low. For example, even for a strong acid such as HCl, as the concentration approaches 10^{-7} M, the contribution from the water equilibrium becomes an appreciable part of the total $[H^+]$. Call the total proton concentration $[H^+]_{Tot}$.

EXAMPLE 6.3

What is the expected pH of a 10^{-7} M solution of HCl?

Solution

The HCl ionization goes to completion, so its contribution will be 10^{-7} M. However, at this low concentration of acid, we cannot neglect the contribution from water dissociation. It will *not* be 10^{-7} as it would be in pure water. However, there is no change in the *equilibrium constant* for water,

$$K_w = 10^{-14} = [H^+][OH^-] \quad \textbf{(6-21)}$$

When the HCl is added, the water equilibrium is shifted back slightly toward undissociated water as the solution $[OH^-]$ decreases in response to an increase in $[H^+]$. But the *decrease* of $[OH^-]$ to form water is accompanied by a *decrease* in $[H^+]_{Tot}$:

$$OH^- + H^+ \rightleftharpoons H_2O$$

So the final proton concentration at equilibrium is less than the sum of the initial values $[H^+]_{acid}$ and $[H^+]_{water}$. This chemical reaction occurs in less than one microsecond.

What happens to the chloride ions? Since Cl^- is such a weak base, it does not reprotonate; the HCl remains completely dissociated. Changes in the proton concentration do not shift the acid equilibrium back toward undissociated HCl. Therefore $[Cl^-] = C_{HCl}$.

To make a quantitative estimate, we write the **charge balance equation** for the solution (total anion charges equals total cation charges). The charge-balance equation for the solution of HCl in water is

$$[H^+]_{Tot} = [Cl^-] + [OH^-] \qquad \textbf{(6-23)}$$

as derived from the stoichiometric equations. Using Equation 6-21 and the fact that $[Cl^-] = C_{HCl}$, we find

$$[H^+]_{Tot} = C_{HCl} + K_w/[H^+]_{Tot}$$

which is a quadratic equation in $[H^+]_{Tot}$. Upon multiplying both sides by $[H^+]_{Tot}$ and rearranging, we get

$$[H^+]^2_{Tot} - C_{HCl}[H^+]_{Tot} - K_w = 0$$

If we then substitute numerical values for C_{HCl} (= 10^{-7} M) and K_w(= 10^{-14}), we can use the quadratic formula to solve for $[H^+]_{Tot}$. The formula gives

$$[H^+]_{Tot} = 1.62 \times 10^{-7}\ \text{M}$$

and the final pH of the solution is expected to be

$$-\log\,[H^+]_{Tot} = -\log\,(1.62 \times 10^{-7}) = 6.79$$

What would have happened if the water equilibrium did not shift? To answer that question, consider the change in $[H^+]_{Tot}$ upon adding 10^{-7} M acid if no shift in the water's equilibrium had occurred. Then

$$[H^+]_{acid} + [H^+]_{water} = 2 \times 10^{-7}\ \text{M}\,; \quad \text{pH} = 6.70$$

The true pH, 6.79, is higher because the added acid shifts the water equilibrium slightly toward water, and, thus, the pH shifts toward neutrality. (Le Chatelier strikes again!)

A Very Weak Acid

Another instance where our simplifying assumptions cannot be made occurs for very weak acids. A representative acid is hydrogen peroxide. The K_a for hydrogen peroxide is 2.2×10^{-12}, so it is not significantly stronger as an acid than water.

EXAMPLE 6.4

Estimate the pH of a 0.01 M aqueous solution of hydrogen peroxide.

Solution

Initially, we might write

$$K_a = \frac{[H^+][HO_2^-]}{[H_2O_2]} = \frac{[H^+]_{Tot}[HO_2^-]}{C_{H_2O_2} - [HO_2^-]}$$

As you can see, the dissociation of the peroxide is taken into account by subtracting $[HO_2^-]$ from the analytical concentration since $[H^+]$ contains a contribution from water. In this case $C_{HA} >> K_a$, so we can use the approximation that the concentration of undissociated hydrogen peroxide remains constant at $C_{H_2O_2}$. On the other hand, it is not clear if the proton contribution of water can be ignored. Therefore, our mass-action expression becomes

$$K_a = \frac{([H^+]_{water} + [H^+]_{H_2O_2})[HO_2^-]}{C_{H_2O_2}}$$

Of course we cannot measure a concentration for each "type" of hydrogen ion, but we can think about them in this way. There are three variables along with the two constants, K_a and $C_{H_2O_2}$. However, by using the mass action expressions for the dissociation of water and of hydrogen peroxide, we find

$$[H^+]_{H_2O} = [OH^-] \quad \text{or} \quad [H^+]_{H_2O} = \frac{K_w}{[H^+]_{Tot}}$$

and

$$[H^+]_{H_2O_2} = [HO_2^-] \quad \text{or} \quad [H^+]_{H_2O_2} = \frac{K_a C_{H_2O_2}}{[H^+]_{Tot}}$$

Substituting these expressions into our simplified equation, we get

$$K_a = \frac{\left(\frac{K_w}{[H^+]_{Tot}} + \frac{K_a C_{H_2O_2}}{[H^+]_{Tot}}\right)\frac{K_a C_{H_2O_2}}{[H^+]_{Tot}}}{C_{H_2O_2}}$$

This equation certainly looks awful, but upon examination consists almost exclusively of constants that we know. So, after substituting the constants, including the initial concentration of hydrogen peroxide, we will find a quadratic equation in $[H^+]_{Tot}$. Solving the quadratic equation yields a value of $[H^+]_{Tot} = 1.79 \times 10^{-7}$ M. Therefore the peroxide solution has a pH of 6.75. (If we had used both of our usual simplifying assumptions, we would have calculated $[H^+] = 1.48 \times 10^{-7}$ or an error of over 17%.)

6.5 Estimating pH for Polyprotic Acid Solutions

The chemical world contains many important polyprotic acids, those which contain a base that can bind more than one dissociable proton. These include H_3PO_4, H_2CO_3, and H_2SO_4, as well as amino acids and organic acids such as *o*-phthalic acid (which you will probably remember as its infamous potassium salt, KHP). In this section, you will see how to calculate the pH of solutions of a diprotic acid and its salts. It turns out that calculations for triprotic acids and higher are no more complicated. As before, the calculations are important for understanding (and, possibly, predicting) the behavior of the chemical species present.

A Diprotic Acid: *o*-Phthalic Acid

In many cases, the use of the appropriate mass action expressions and stoichiometric equations remains similar to what you studied for the monoprotic acid, acetic acid. For example, for the diacid in water, the calculation remains the same as for acetic acid. For the dianion salt, the calculation remains the same as for the hydrolysis reaction of acetate. However, for the monoprotonated salt solution, calculating the pH is far more complicated.

We will use the *o*-phthalic acid system to illustrate these calculations. The mathematical techniques used for the *o*-phthalate equilibrium are general, so it is worthwhile to understand the approach. From experiments, the following stoichiometries and equilibrium constants have been determined (phth = the *o*-phthalate dianion).

1st ionization $H_2phth \rightleftharpoons H^+ + Hphth^-$; $K_1 = 1.12 \times 10^{-3}$ M; $pK_1 = 2.95$ **(6-24)**

2nd ionization $Hphth^- \rightleftharpoons H^+ + phth^{2-}$; $K_2 = 3.91 \times 10^{-6}$ M; $pK_2 = 5.40$ **(6-25)**

The subscript nomenclature of the *K*s in Equations 6-24 and 6-25 is general. The steps of the ionization are labeled starting from the fully protonated form of the acid as 1. Connected with each ionization is a K_a value that replaces the subscript *a* with the same number as the ionization step label: K_1 and K_2. The same convention extends to polyprotic acids; so the third ionization has associated with it a K_3, and so forth.

The pH of Solutions of Diprotic Acids and Their Anions

The strategy for these estimates has been laid out, so let the calculations begin.

EXAMPLE 6.5

Suppose that you prepare a 35 mM solution of *o*-phthalic acid. Calculate the pH you expect.

Solution

To decide whether the water equilibrium needs to be included, compare K_1 and K_2 to K_w. K_w is more than seven orders of magnitude smaller, so you can ignore any proton contribution from the water. Similarly, the second ionization constant is more than two orders of magnitude smaller than the first, so it too will contribute little to $[H^+]$. As a result, we can calculate the pH based on the first ionization equilibrium alone—the same method used for acetic acid. Let $X = [H^+] = [Hphth^-]$. Then,

$$K_1 = \frac{[H^+][Hphth^-]}{[H_2phth]}$$

$$1.12 \times 10^{-3} = \frac{X^2}{0.035 - X}$$

Rearrangement produces a quadratic equation in *X*, and its solution is

$$X = [H^+] = [Hphth^-] = 5.73 \times 10^{-3}\ \text{M}$$

We estimate the pH of the solution to be 2.24.

Let us now switch from the acid dissociation to hydrolysis of the dianion. The calculation for the hydrolysis of phthalate parallels that of acetate. The predominant equilibrium is

$$phth^{2-} + H_2O \rightleftharpoons Hphth^- + OH^-$$

The numerical value of K_h for the phthalate reaction with water is found from the related acid reaction K_2.

$$K_h = \frac{K_w}{K_2} = \frac{1.00 \times 10^{-14}}{3.91 \times 10^{-6}} = 2.56 \times 10^{-9}$$

You may reasonably ask, Doesn't a second hydrolysis occur? The answer is yes, but the amount is so small that it can be ignored. An estimate of this ratio is easily made. The reaction for the second hydrolysis has a hydrolysis constant $(K_w/K_1) = 8.9 \times 10^{-12}$. At a basic pH, the relative amount of hydrolysis equals the ratio of the first and second hydrolysis constants, that is, $2.56 \times 10^{-9}/8.9 \times 10^{-12}$. The first hydrolysis produces approximately 300 times more hydroxide than the second. So we ignore the second.

EXAMPLE 6.6

What is the pH expected in a 10 mM solution of K_2phth?

Solution

The applicable mass action expression is

$$2.56 \times 10^{-9} = \frac{[\text{Hphth}^-][\text{OH}^-]}{[\text{phth}^{2-}]}$$

The same approximations can be made for hydrolysis as were made for the acetate (or any other equilibrium calculation). For example, since K_h (order of 10^{-9}) is small relative to the concentration of the added salt (10^{-2} M), we can make the usual assumption that the concentration of the added salt remains constant. Is the hydroxide from water a factor? As you can see, K_h (order of 10^{-9}) is still large relative to K_w (10^{-14}), so any contribution by water to $[\text{OH}^-]$ can be ignored. As a result, we can let $[\text{OH}^-] = [\text{Hphth}^-] = X$.

$$2.56 \times 10^{-9} = \frac{X^2}{0.01}$$

$$X = [\text{OH}^-] = 5.06 \times 10^{-6}\text{ M}$$

From the hydroxide concentration, the pH can be found.

$$[\text{H}^+] = \frac{K_w}{[\text{OH}^-]}$$

$$[\text{H}^+] = 1.98 \times 10^{-9}\text{ M}$$

$$\text{pH} = 8.70$$

The monoprotonated salt KHphth can either pick up a proton by hydrolysis or it can deprotonate further. Molecules that behave this way are **amphiprotic** or, equivalently, **amphoteric**. The algebra is a lot more complicated than for monoprotic acids, especially if water dissociation must be taken into account. However, setting up the calculation is made far easier using the method shown next.

An efficient calculational tool for this type of problem in effect extends the practice of using other species to account for the proton concentration. For example, when the proton contribution from acetic acid was needed, $[\text{OAc}^-]$ was used as a measure. However, for the amphoteric species, where protons can both be generated and taken up, both protonation and dissociation reactions must be included.

In a solution containing potassium hydrogen phthalate, KHphth, in water, the following equation can be written.

$$[\text{H}^+] = [\text{phth}^{2-}] - [\text{H}_2\text{phth}] + [\text{OH}^-]$$

That is, term by term, a hydrogen ion appears for each phthalate ion formed by dissociation, a hydrogen ion is removed (hence, the subtraction) for each phthalic acid molecule formed, and a hydrogen ion appears for each hydroxide from water. This equation is the basis for the calculation of pH for the KHphth salt, and similar equations provide a general method for amphoteric species. Let us call such equations **allocation equations**.

EXAMPLE 6.7

What is the expected pH of a 0.01 M solution of KHphth?

Solution

Begin by writing the allocation equation.

$$[H^+] = [phth^{2-}] - [H_2phth] + [OH^-]$$

Now, express each concentration on the right in terms of the initial salt ion Hphal$^-$ and proton concentrations by using the mass action expressions. These are K_2, K_1, and K_w, respectively, for the three terms. After this substitution,

$$[H^+] = \frac{K_2[Hphth^-]}{[H^+]} - \frac{[H^+][Hphth^-]}{K_1} + \frac{K_w}{[H^+]}$$

Some algebraic gymnastics give

$$[H^+]^2 = \frac{K_1(K_2[Hphth^-] + K_w)}{K_1 + [Hphth^-]}$$

It is at this stage that we look for simplifications in the algebra. Recall that $K_2 = 3.91 \times 10^{-6} >> K_w = 10^{-14}$. This means that the numerator can be simplified by dropping K_w because $K_2[Hphth^-] >> K_w$. The equation then can be written

$$[H^+] = \sqrt{\frac{K_1K_2[Hphth^-]}{K_1 + [Hphth^-]}}$$

$$[H^+] = 3.93 \times 10^{-7}\ M$$

The equation derived above (before simplication) for Hphth$^-$ can actually be used for the amphiprotic salt of any acid. For the general acid HA$^-$, if K_1 is the equilibrium constant for the deprotonation reaction that produces the salt and K_2 is the equilibrium constant for the deprotonation to form A^{2-}, we can write the general equation

$$[H^+] = \sqrt{\frac{K_1(K_2[HA^-] + K_w)}{K_1 + [HA^-]}} \quad \textbf{(6-26)}$$

In some very specific situations, we can simplify Equation 6-26 greatly. First, as in Example 6.7, the numerator can be simplified if $K_2[HA^-] >> K_w$, which allows dropping K_w. Further, if the concentration of the general amphiprotic salt, $[HA^-] >> K_1$, then K_1 in the denominator can be dropped to yield the simple equation

$$[H^+] = \sqrt{K_1K_2} \quad \textbf{(6-27)}$$

Equation 6-27 can be used to estimate an approximate pH. It is interesting to note that under the conditions that produced Equation 6-27, the pH does not depend on the salt concentration.

What about Triprotic Acids?

For most triprotic acids and the few tetraprotic acids, *exactly* the same approach can be used as for the diprotic acids, since the pH is controlled by one dissociation and/or one hydrolysis equilibrium. For example, for $H_2PO_4^-$ added to water, the equilibria that determine the pH are those involving the loss or uptake of a proton by $H_2PO_4^-$. That is,

$$H_2PO_4^- \rightleftharpoons H^+ + HPO_4^{2-}$$

$$H_2PO_4^- + H_2O \rightleftharpoons H_3PO_4 + OH^-$$

The use of an allocation equation to set up the calculation followed by the substitutions and simplifications is, thus, widely applicable.

Overview of pH Calculations

At this point, you are equipped to take on even the most complicated calculations of pH. The first step of the approach is to write all the mass-action expressions and, if needed, the allocation equation. After the controlling mass-action expression is written, look at the values of the analytical concentrations, such as C_{HA}, compared to K_a or K_h to decide what approximations might be useful during the calculation. In other words, decide what terms can be eliminated to make the calculation simpler.

Although a number of different, simplified equations were developed, these tend to be applicable only under narrowly defined conditions. An example is $[H^+] = \sqrt{K_1K_2}$, Equation 6-27. Whenever you use a simplified equation, be sure to check carefully that the approximations made in deriving them are valid for your specific problem.

Finally, check whether your answer makes sense. For example, it would not make sense if the calculated $[H^+] < 10^{-7}$ M in water or if $[H^+] > 20$ M.

6.6 Fractional Composition: What Is the Predominant Species?

Although the types of problems we have described so far predict how acidic or basic solutions might be, it is often the case that we want to understand what species predominate in solutions of a given pH.

To illustrate how to approach such estimates, let us switch from phthalate to carbonate equilibria. From experiments, the following stoichiometries and equilibrium constants have been determined.

$$\text{1st ionization } H_2CO_3 \rightleftharpoons H^+ + HCO_3^-; \quad K_1 = 4.2 \times 10^{-7}\,\text{M}; \quad pK_1 = 6.38 \qquad \textbf{(6-28)}$$

$$\text{2nd ionization } HCO_3^- \rightleftharpoons H^+ + CO_3^{2-}; \quad K_2 = 4.8 \times 10^{-11}\,\text{M}; \quad pK_2 = 10.32 \qquad \textbf{(6-29)}$$

The carbonate equilibria are of special interest because they are involved in significant, fundamental chemical processes. For instance, the carbonate/bicarbonate (CO_3^{2-}/HCO_3^-) equilibrium is the main determinant of the pH of your blood and also important in the following diverse phenomena: the formation of sea shells, which are calcium carbonate; the formation of carbonate scale in boilers; and the balance of carbon dioxide in the atmosphere.

Given the information included in Equations 6-28 and 6-29, we shall be calculating the expected values of such quantities as $[H_2CO_3]$, $[HCO_3^-]$, and $[CO_3^{2-}]$. A primary question is whether Equations 6-28 and 6-29 include enough information to make such calculations. One way to view this question is strictly mathematically. There are four *possible* unknown values in the equations: $[H^+]$, $[H_2CO_3]$, $[HCO_3^-]$, and $[CO_3^{2-}]$. There also are four algebraic equations interrelating the four quantities: the equations we use are the two stoichiometric equations and the two mass-action expressions (not written out above). With four equations and four unknowns, in principle the *algebraic* problem must be solvable uniquely.

To obtain numerical values, the pK_a values and at least one concentration, such as that of added HCO_3^- are needed. An exact solution of the equations can be awkward, and approximations can simplify the mathematics significantly while still providing estimates with sufficient precision. One way of simplifying the mathematics is to use the chemical principle of conservation of mass to construct a **conservation equation** or **conservation relation**, also called the **mass-balance equation**.

$$C_{Tot} = \text{(The total concentration of all carbonate species)} \quad \textbf{(6-30a)}$$

For the carbonates, a conservation equation can be derived from the stoichiometric equations 6-28 and 6-29. As expressed in Equation 6-30b below, the total carbonate concentration remains constant. That is,

$$C_{Tot} = [H_2CO_3] + [HCO_3^-] + [CO_3^{2-}] \quad \textbf{(6-30b)}$$

It is important to recognize that the conservation equation 6-30b is equivalent to both stoichiometry equations. As a result, the information needed to calculate the carbonate species' concentrations in solutions is included in the conservation relation, the two mass-action expressions, and the numerical values of C_{Tot}, pK_1, and pK_2.

Sometimes a charge balance equation may be needed (defined in Example 6.3 on page 175). In this case, the following relationship has to be true for the charge to balance:

$$[H^+] = [HCO_3^-] + 2[CO_3^{2-}] + [OH^-]$$

The dianion term has a factor of 2 to reflect the fact that it has a −2 charge.

Carbonate at pH 6.38

Let us calculate the concentrations of H_2CO_3, HCO_3^-, and CO_3^{2-} expected in a solution with 0.10 M total carbonate when the pH is 6.38. (This aqueous solution might be made by adding 0.10 mol of $NaHCO_3$ to 950 mL of water. Then the pH can be brought to 6.38 with concentrated HCl, and the solution diluted to 1.000 liter. In this way, the addition of HCl does not produce a diluted solution.)

Notice that pH 6.38 is equal to the pK_1 for carbonate. This special pH value is used here to show how approximations can simplify the calculations.

The statement of the problem has given us two different pieces of information. First, the total carbonate concentration C_{Tot} is 0.10 M. Second, having the solution pH = 6.38 means that $[H^+] = 4.2 \times 10^{-7}$. Let us substitute the $[H^+]$ and the K_a values into the mass-action expressions:

$$K_1 = \frac{[H^+][HCO_3^-]}{[H_2CO_3]} = \frac{(4.2 \times 10^{-7})[HCO_3^-]}{[H_2CO_3]} = 4.2 \times 10^{-7} \quad \textbf{(6-31)}$$

$$K_2 = \frac{[H^+][CO_3^{2-}]}{[HCO_3^-]} = \frac{(4.2 \times 10^{-7})[CO_3^{2-}]}{[HCO_3^-]} = 4.8 \times 10^{-11} \quad \textbf{(6-32)}$$

Upon rearrangement, ratios of the concentrations of the carbonate species can be calculated.

From K_1: $4.2 \times 10^{-7}/4.2 \times 10^{-7} = 1 = [HCO_3^-]/[H_2CO_3]$ **(6-31)**

From K_2: $4.8 \times 10^{-11}/4.2 \times 10^{-7} = 1.14 \times 10^{-4} = [CO_3^{2-}]/[HCO_3^-]$ **(6-32)**

The third equation, the conservation equation for a solution with a total carbonate concentration of 0.10 M, is

$$0.10 \text{ M} = [H_2CO_3] + [HCO_3^-] + [CO_3^{2-}] \quad \textbf{(6-30b)}$$

All the information we have is now included in the three equations: 6-31, 6-32, and 6-30b.

The rest of the calculation is straightforward but messy algebra. However, let us see if an approximation might make the calculation even easier.

An Approximation

Look at the carbonate species ratios that were calculated above.

$$1 = [HCO_3^-]/[H_2CO_3] \quad \textbf{(6-31)}$$

$$1.14 \times 10^{-4} = [CO_3^{2-}]/[HCO_3^-] \quad \textbf{(6-32)}$$

The carbonic acid concentration $[H_2CO_3]$ and the bicarbonate ion concentration $[HCO_3^-]$ are the same. Their ratio is one. On the other hand, compared with the bicarbonate ion concentration $[HCO_3^-]$, the carbonate ion concentration $[CO_3^{2-}]$ is quite small—only about one ten-thousandth the amount. In the conservation equation, Equation 6-30b, the value of $[CO_3^{2-}]$ hardly contributes to the sum. We shall lose little accuracy if we ignore $[CO_3^{2-}]$, so we can simplify the conservation equation. It becomes

$$0.10 \text{ M} = [H_2CO_3] + [HCO_3^-] \quad \textbf{(6-30c)}$$

Being able to ignore the presence of CO_3^{2-} causes a second, significant simplification in the algebra. Since we can assume that CO_3^{2-} is not present, then the mass-action expression containing $[CO_3^{2-}]$ *is no longer needed or applicable* at this pH. It is *most* important that you understand this point. We ignore the concentration CO_3^{2-} and the connected mass action expression. The reason is that here the concentration of other carbonate species are far larger than $[CO_3^{2-}]$.

As a result there are now only two equations with two unknowns to solve. These are the mass-action expression for K_1 at pH 6.38

$$1 = [HCO_3^-]/[H_2CO_3] \quad \textbf{(6-31)}$$

and the modified conservation relation

$$0.10 \text{ M} = [H_2CO_3] + [HCO_3^-] \quad \textbf{(6-30c)}$$

Equation 6-31 says

$$[H_2CO_3] = [HCO_3^-]$$

And, by using the conservation relation, we obtain

$$[H_2CO_3] = [HCO_3^-] = 0.05 \text{ M}$$

Thus, the answer to the original question is that at pH 6.38

$$[H_2CO_3] = 0.05 \text{ M}$$

$$[HCO_3^-] = 0.05 \text{ M}$$

and, from Equation 6-32,

$$[CO_3^{2-}] = [HCO_3^-] \times 1.14 \times 10^{-4} = 1.14 \times 10^{-4} \times 0.05\ M = 5.7 \times 10^{-6}\ M$$

and all the carbonate species are accounted for.

6.7 Finding the Fraction of Each Species: α-Values

Carrying out calculations to mimic and explain simultaneous equilibrium systems more complicated than carbonate use other algebraic methods. These allow calculations to be made systematically so as to account for the important chemical effects. One of the most useful of these involves manipulating ratios such as $[HCO_3^-]/[CO_3^{2-}]$ and $[H_2CO_3]/[HCO_3^-]$ in a methodical way. The method, which involves calculating the *fraction of each of the species*, becomes especially useful for acids with three or more protons and is the subject of this section.

To illustrate the method, we will use the phosphoric acid-phosphate equilibria. The mathematics shown below remains *exactly equivalent* to solving the simultaneous equations using the mass action expressions and the conservation equation—just as you have seen above for simpler simultaneous equilibria. At first glance, the mathematics on the next few pages may appear to be extraordinarily complicated. However, the algebra is highly systematic and only requires careful bookkeeping to yield useful equations. Our compensation is a set of equations that explain and predict the behavior of multiequilibrium systems in a relatively clear way.

The total amount of phosphate present in all forms is called C_{Tot}. From the data in Appendix II, the following phosphate species are known to exist, and together they comprise the total amount.

$$C_{Tot} = [H_3PO_4] + [H_2PO_4^-] + [HPO_4^{2-}] + [PO_4^{3-}] \tag{6-33}$$

You will recognize this as the conservation equation for phosphate species. The following mass action expressions summarize the dissociations of protons.

$$K_1 = \frac{[H^+][H_2PO_4^-]}{[H_3PO_4]} = 7.11 \times 10^{-3} \qquad pK_1 = 2.148$$

$$K_2 = \frac{[H^+][HPO_4^{2-}]}{[H_2PO_4^-]} = 6.32 \times 10^{-8} \qquad pK_2 = 7.199 \tag{6-34a}$$

$$K_3 = \frac{[H^+][PO_4^{3-}]}{[HPO_4^{2-}]} = 4.5 \times 10^{-13} \qquad pK_3 = 12.35$$

These mass-action expressions only need to be *rearranged* to obtain expressions of concentrations of the phosphate species in terms of the K_i and of $[H^+]$.

$$[H_2PO_4^-] = K_1[H_3PO_4]/[H^+]$$

$$[HPO_4^{2-}] = K_2[H_2PO_4^-]/[H^+] \tag{6-34b}$$

$$[PO_4^{3-}] = K_3[HPO_4^{2-}]/[H^+]$$

Notice the repetition in Equations 6-34b: The left-hand side of each equation has the same phosphate term as the right-hand side of the one below, and the denominators are all the same, $[H^+]$. Therefore, by simple algebraic substitution,

$$[H_2PO_4^-] = K_1[H_3PO_4]/[H^+]$$
$$[HPO_4^{2-}] = K_1K_2[H_3PO_4]/[H^+]^2 \quad \textbf{(6-35)}$$
$$[PO_4^{3-}] = K_1K_2K_3[H_3PO_4]/[H^+]^3$$

Now all three equations have the factor $[H_3PO_4]$ on the right side. Substituting Equations 6-35 (derived from the mass-action expressions) into Equation 6-33 (the conservation relation) and then factoring out the common term leaves

$$C_{Tot} = [H_3PO_4]\left(1 + \frac{K_1}{[H^+]} + \frac{K_1K_2}{[H^+]^2} + \frac{K_1K_2K_3}{[H^+]^3}\right) \quad \textbf{(6-36)}$$

Here, we make a useful change in nomenclature and express the concentration of each species of phosphate as a fraction of the whole. This is done by dividing the concentration of each by C_{Tot}. The fractions are abbreviated as α-values.[1]

$$\alpha_0 = [H_3PO_4]/C_{Tot}$$
$$\alpha_1 = [H_2PO_4^-]/C_{Tot}$$
$$\alpha_2 = [HPO_4^{2-}]/C_{Tot}$$
$$\alpha_3 = [PO_4^{3-}]/C_{Tot} \quad \textbf{(6-37)}$$

The numerators of the α_i are the concentrations of specific species; the denominator is the total concentration as given by Equation 6-33 or 6-36.

Let us now derive the algebraic expression for the value of α_2, the fraction present as HPO_4^{2-}. To begin, from Equation 6-37

$$\alpha_2 = [HPO_4^{2-}]/C_{Tot}$$

and from Equation 6-35,

$$[HPO_4^{2-}] = K_1K_2[H_3PO_4]/[H^+]^2$$

Therefore

$$\alpha_2 = \frac{K_1K_2[H_3PO_4]/[H^+]^2}{[H_3PO_4]\left(1 + \frac{K_1}{[H^+]} + \frac{K_1K_2}{[H^+]^2} + \frac{K_1K_2K_3}{[H^+]^3}\right)}$$

After carrying out some simplifying algebra, we obtain a somewhat more convenient equation.

$$\alpha_2 = \frac{K_1K_2[H^+]}{[H^+]^3 + K_1[H^+]^2 + K_1K_2[H^+] + K_1K_2K_3} \quad \textbf{(6-38)}$$

Simply by substituting the appropriate values of $[H^+]$ into the equation, we can find the fraction of phosphate in the form HPO_4^{2-} at any pH. This is an excellent algebraic form to use for computation—only a single variable $[H^+]$ appears. By carrying out such calculations at a number of pH values, graphs of the fractional concentrations, α_i, versus pH can be constructed.

---

[1]Some workers use α_i to represent the inverse $1/\alpha_i$.

Incidentally, since the α_i values represent fractions of each phosphate species, the sum of all the α-values is unity. That is,

$$1 = \alpha_3 + \alpha_2 + \alpha_1 + \alpha_0$$

This equation is derived by dividing both sides of Equation 6-33 by C_{Tot} and comparing the terms on the right side with Equations 6-37. It is another statement of mass conservation.

Rules for Writing α-Expressions

As formidable as the α-expressions such as Equation 6-38 appear, they are quite regular, and it is not necessary to go through so much algebra to find the correct expression. The rules are:

1. All α in the system have identical denominators.
2. The denominator is a decreasing power series in $[H^+]$ with the following properties:
   a. The power series begins with $[H^+]^N$, where N is the total number of protons that can dissociate.
   b. The first term has no factor involving a K, and the factor for each successive term contains an additional stepwise dissociation constant, starting with K_1.
   c. There will be $(N + 1)$ terms, the last of which is $K_1K_2 \cdots K_N$.
3. Each term in the denominator corresponds to the numerator for one of the species present. Thus the first term forms the numerator of α_0, the second for α_1, and so on, finishing with the numerator for α_N being $K_1K_2 \cdots K_N$.

EXAMPLE 6.8

Write the α-term for PO_4^{3-} and find the PO_4^{3-} concentration in a 0.10-M solution of phosphate at pH 4.0. The pK_a values for H_3PO_4 are 2.23, 7.21, and 12.32.

Solution

Following the rules laid out above, we find that the denominator is

$$[H^+]^3 + K_1[H^+]^2 + K_1K_2[H^+] + K_1K_2K_3$$

The α_i ($i = 0, 1, 2,$ and 3) values refer, respectively, to H_3PO_4, $H_2PO_4^-$, HPO_4^{2-}, and PO_4^{3-}. Therefore

$$\alpha_3 = \frac{K_1K_2K_3}{[H^+]^3 + K_1[H^+]^2 + K_1K_2[H^+] + K_1K_2K_3}$$

The calculation is carried out simply by substituting the appropriate values into the equation. The result is

$$\alpha_3 = 2.80 \times 10^{-12} \text{ at pH 4.0}$$

Since the α-values are the fractions of each of species,

$$[PO_4^{3-}] = C_{Tot} \cdot \alpha_3 = 0.10 \times 2.80 \times 10^{-12} = 2.80 \times 10^{-13}\ \text{M}$$

This extremely low concentration is reasonable since the pH of the solution lies more than 8 pH units below the pK_a of the HPO_4^{2-}/PO_4^{3-} pair.

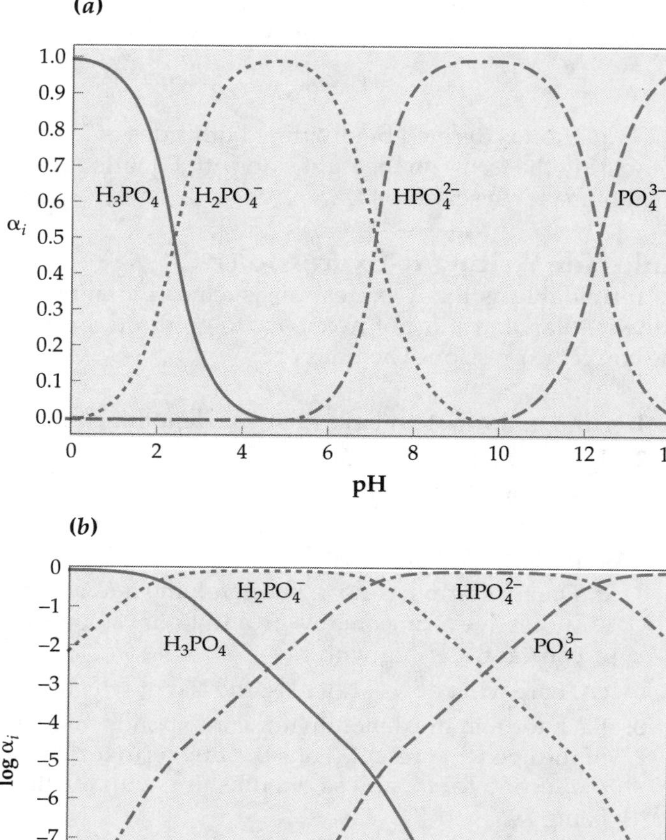

FIGURE 6.2 ▶ **Fractional composition of phosphate solutions as a function of pH.**

(a) α-values for phosphate species as a function of pH in water at 25°C. **(b)** The logarithm of α_i vs. pH. Both types of graphs are useful in visualizing the behavior of polyprotic acids. Graph (a) is easier to interpret when determining the major species while (b) provides a clearer picture of the concentrations of the minor species.

Figure 6.2a shows plots of the α_i values for all phosphate species as they change with pH. Figure 6.2b shows the same data plotted on a semilogarithmic graph. Some characteristics of these plots should be noted. First, for the phosphoric acid, as for most polyprotic acids, adjacent α plots cross at each pK_a. The slopes of the α plots are the steepest at the pK_as, which means that the largest changes in the ratios of two species occur with a given change in pH. Second, with the exception of half a pH unit to each side of the pK_a values, one species predominates. Both of these characteristics together indicate that changing the pH across the pK_a value changes the identity of the predominant species in the solution.

Overview of Species Calculations

For the equilibrium systems described in this chapter, given the specific conditions involved (for instance, pH), there is either a single principal equilibrium or two approximately equally influential principal equilibria. If only a single equilibrium is involved, calculation of the species present requires only straightforward algebra. If two principal equilibria are involved, it is useful to use a conservation equation and the mass action expressions together.

You have seen how it is possible to ignore the reactions having pK_a values that are three or more pH units away from the pH value of the solution. In that case, at least one of the species will be less than 1/1,000 of the other. Since many organic and inorganic acids have pK_1 and pK_2 values that differ by three pH units (K_1 and K_2 differing by three orders of magnitude), this simplification is often justified since the precision still allows a valid estimate to be obtained.

The general procedure for carrying out the algebra to calculate the species present is

1. Write the mass-action expressions and stoichiometric equations for all the applicable equilibria.
2. Using the stoichiometric equations, derive the conservation equations.
3. Use the equilibrium constant (mass-action) expressions to find the ratios of the species needed.
4. Decide if approximations can be made and do so if possible.

Do not lose sight of the fact that these calculations are mathematical models made to match the chemical system in the solution. For example, rejecting the need for the K_2 equilibrium constant and $[CO_3^{2-}]$ in the conservation equation for carbonate (Eqs. 6-30b) merely says that at the specified pH the reaction $HCO_3^- \rightleftharpoons H^+ + CO_3^{2-}$ plays no *significant* part in determining the solution composition.

6.8 pH Buffers

The properties of other buffers can be found in Section 8.7.

Solutions that contain a pH buffer resist change in pH with added base or acid. Such pH buffers present one example of the general property of buffering, which will be discussed later. The buffering action results from the equilibrium properties of an acid and its conjugate base. To explain why such combinations act as buffers, let us take another look at the acetic acid-acetate equilibrium and rearrange the mass-action expressions. Recall

$$K_a = \frac{[H^+][OAc^-]}{[HOAc]} \tag{6-8}$$

Next, take the negative logarithm of both sides:

$$-\log K_a = -\log[H^+] - \log \frac{[OAc^-]}{[HOAc]}$$

or

$$pK_a = pH - \log \frac{[OAc^-]}{[HOAc]}$$

A rearrangement of the equation gives

$$pH = pK_a + \log \frac{[OAc^-]}{[HOAc]}$$

Note that since this is merely the mass-action equation written in a different mathematical form, there is no new information in it, although it is convenient to describe buffers. This form of the mass-action expression is often called the **Henderson-Hasselbalch equation.** The general form of this equation is

$$\mathrm{pH} = \mathrm{p}K_a + \log\frac{[\text{conjugate base}]}{[\text{acid}]} \tag{6-39}$$

Since the value of pK_a is a constant, the pH changes in the same way as the logarithm of the ([base]/[acid]) concentration ratio changes. For instance, the pH changes by one pH unit for every factor-of-ten change of the ratio.

EXAMPLE 6.9

What pH is expected for a buffer made from 0.100 mol of acetic acid and 0.100 mol of sodium acetate dissolved in 1.00 L of water?

Solution

Use the Henderson-Hasselbalch equation

$$\mathrm{pH} = \mathrm{p}K_a + \log\frac{[\mathrm{OAc^-}]}{[\mathrm{HOAc}]}$$

Since any proton that leaves acetic acid protonates acetate forming another acetic acid, there is no change in the concentrations of either acetate or acetic acid. Therefore, simply substitute the analytical concentrations of both species along with the value of pK_a.

$$\mathrm{pH} = 4.75 + \log\frac{0.1}{0.1} = 4.75$$

EXAMPLE 6.10

What do you expect the pH of the buffer described in Example 6.9 to be after 1.0 mL of 12.0 M HCl is added to the solution? (Assume the change in volume is negligible.)

Solution

Since the acetate is a much stronger base than chloride or water, the protons from the added HCl react with acetate to form acetic acid. The acetic acid formed essentially equals the HCl added. So, first, calculate the moles of HCl added.

$$(12\ \mathrm{M})(0.001\ \mathrm{L}) = 0.012\ \mathrm{mol}$$

The initial solution had 0.100 moles of acetate and 0.100 moles of acetic acid. After the HCl is added, in concentration units

$$[\mathrm{OAc^-}] = 0.100 - 0.012 = 0.088\ \mathrm{M}$$

and

$$[\mathrm{HOAc}] = 0.100 + 0.012 = 0.112\ \mathrm{M}$$

Using these values in the Henderson-Hasselbalch equation,

$$\mathrm{pH} = 4.75 + \log\frac{0.088}{0.112} = 4.75 - 0.10_5 = 4.65$$

Just for comparison, a 0.012 M HCl solution in water is expected to have a pH of approximately 1.9.

It may seem obvious, but only rapidly exchanging protons are effective in buffering. This is clear for acetic acid, CH_3COOH. The methyl hydrogens of acetic acid do not participate in the buffering action since they do not exchange at an appreciable rate. Only the carboxyl proton is involved.

Buffer Capacity

In the example above, the pH change would have been larger if the concentration of the acid/conjugate base pair were lower. That is, the **buffer capacity** depends on the concentration of the buffer. The larger the concentration of the pair, the more the solution resists changes in pH. Here, we show how to calculate that concentration dependence.

The buffer capacity for the general weak acid HA in the presence of its salt MA is described by the derivative given in Equation 6-40.

$$\frac{db}{d\text{pH}} \approx 2.303\left(\frac{C_{\text{HA}} \cdot K_a \cdot [\text{H}^+]}{(K_a + [\text{H}^+])^2} + [\text{H}^+] + [\text{OH}^-]\right) \qquad \textbf{(6-40)}$$

Here, db (in equivalents L^{-1}) is the differential amount of strong base added to the buffer solution. The constant $C_{\text{HA}} = [\text{HA}] + [\text{A}^-]$, and K_a is the equilibrium constant for HA.

The buffer capacity is always a positive quantity. This property is obtained by defining the addition of a strong acid to the solution as $-db$. Since the pH decreases with added acid, the change in pH must be $-d$pH. As a result, the buffer capacity db/dpH remains positive.

Between pH 3 and pH 11 the terms $[\text{H}^+]$ and $[\text{OH}^-]$ in Equation 6-40 can be neglected. In that range of pH, the buffer capacity is well described by the first term in the parentheses. The buffer capacity is directly proportional to the buffer's concentration.

The maximum buffer capacity occurs when the solution pH $= \text{p}K_a$. At that point,

$$\left(\frac{db}{d\text{pH}}\right)_{\text{max}} \approx \left(\frac{2.303}{4}\right)C_{\text{HA}}; \qquad \text{pH} = \text{p}K_a \qquad \textbf{(6-41)}$$

However, the buffer capacity changes with pH. Let us explore how much with a rough calculation. When the solution pH $= \text{p}K_a$, the ratio $([\text{A}^-]/[\text{HA}]) = 1$. If $([\text{A}^-]/[\text{HA}])$ is decreased by a factor of 10—from unity to 1/10—the pH of the solution changes from $\text{p}K_a$ to $\text{p}K_a - 1$. However, when the solution pH $= \text{p}K_a - 1$, then $[\text{H}^+] = 10\,K_a$. In the denominator of Equation 6-40, we can approximate $(K_a + [\text{H}^+]) \approx [\text{H}^+]$. Then

$$\frac{db}{d\text{pH}} \approx 2.303\left(\frac{C_{\text{HA}}\,(0.1 \cdot [\text{H}^+])[\text{H}^+]}{[\text{H}^+]^2}\right)$$

and

$$\frac{db}{d\text{pH}} \approx \left(\frac{2.303}{10}\right)C_{\text{HA}}; \quad \text{pH} = \text{p}K_a - 1 \qquad \textbf{(6-42)}$$

By comparing Equation 6-42 with 6-41, you can see that the buffer capacity falls by a factor of about 2.5 (down to 40% of the maximum) when the pH lies one pH unit away from the $\text{p}K_a$.

Similarly, a change of pH in the opposite direction to $\text{p}K_a + 1$ occurs if the ratio $([\text{A}^-]/[\text{HA}])$ is increased by a factor of 10. (In other words, the calculation is symmetric about $\text{p}K_a$.) As a result, in practice it is best to use a buffer within ± 1 pH unit of its $\text{p}K_a$.

Some Practical Points

There are seven NIST primary-standard buffers, and their properties are listed in Table 6.1. Some other useful buffering substances with widely varying pK_a values are listed in Table 6.2. In Table 6.3 on page 193, the pK_a values of a number of buffers that are especially useful with biological analyses are listed. You might note that many of the buffers listed in Table 6.3 are **zwitterionic** buffers. They have both positively and negatively charged sites on an electrostatically neutral molecule. A number of these buffers have been found that function well in the physiological range (pH 6–8), and most have few biological effects such as inhibition of enzyme activity. In addition, they are highly water-soluble, which allows concentrated stock solutions to be made and minimizes buffer partitioning into biological phases such as membranes. More information on biological buffers can be found in most biochemistry handbooks.

One important point about the storage of buffers should be mentioned: over a wide range of pH, aqueous buffer solutions support the growth of microorganisms. Over time, the solution components are degraded, and metabolic byproducts are produced. As a result, buffer properties change over time due to the presence of such microorganisms. To impede their growth, it is wise to refrigerate buffer solutions, and growth inhibitors can be added if they do not affect the analysis at hand.

6.9 Activities and Activity Coefficients

Equations of the type

$$K = \frac{[C]^c[D]^d}{[A]^a[B]^b}$$

accurately describe the equilibrium concentrations of ideal systems, with a value of K that is constant. In reality, equilibria in solutions, especially ionic solutions, do not behave so simply.

Why do the concentrated solutions not behave as ideal solutions? The reason is that there are interactions between molecules and especially between ions. These interactions are exceedingly difficult to quantify. Consider a simple salt solution such as 0.1 M sodium chloride in water. Assume that the only other species present is water with a pH of 7. This allows us to ignore the very low concentrations of protons and hydroxide ions. Even with this simplification, there are interactions of

Na^+ with Na^+

Cl^- with Cl^-

Na^+ with H_2O

Cl^- with H_2O

H_2O with H_2O

Na^+ with Cl^-

The situation quickly becomes extremely complicated!

Suppose an equilibrium constant was determined experimentally by measuring the concentrations of the species involved. For example, to evaluate acetic acid dissociation, $[H^+]$, $[OAc^-]$, and $[HOAc]$ would be measured, and the equilibrium constant calculated from the concentrations (in mol L^{-1}).

Table 6.1 U.S. National Bureau of Standards Primary Standard Buffers (at 25°C)

Buffer Compositions

I.D.	Composition	Molality	Molarity	$g\ L^{-1}$ salt* at 25°C
A	Potassium tetroxalate	0.05	0.04962	12.61
B	Potassium hydrogen tartrate	0.0341	0.034	Saturated at 25°
C	Potassium hydrogen phthalate	0.05	0.04958	10.12
D	Potassium dihydrogen phosphate	0.025	0.02490	3.39
	Disodium hydrogen phosphate	0.025	0.02490	3.53
E	Potassium dihydrogen phosphate	0.008695	0.008665	1.179
	Disodium hydrogen phosphate	0.03043	0.03032	4.30
F	Borax	0.01	0.009971	3.80
G	Calcium hydroxide	0.0203	0.02025	Saturated at 25°

Buffer Properties

I.D.	A	B	C	D	E	F	G
Name	Tetraoxalate	Tartrate	Phthalate	Phosphate	Phosphate	Borax	Ca hydroxide
Formula	$KH_3(C_2O_4)_2 \cdot 2H_2O$	$KHC_4H_4O_4$	$KHC_8H_4O_4$	KH_2PO_4 Na_2HPO_4	KH_2PO_4 Na_2HPO_4	$Na_2B_4O_7 \cdot 10H_2O$	$Ca(OH)_2$
pH at 25°C	1.679	3.557	4.008	6.865	7.413	9.180	12.454
Dil. value, $\Delta pH_{1/2}$	+0.186	+0.049	+0.052	+0.080	+0.07	+0.01	−0.28
db/dpH (equiv./unit)	0.070	0.027	0.016	0.029	0.016	0.020	0.09
dpH/dT (units/°C)	+0.001	−0.0014	+0.0012	−0.0028	−0.0028	−0.0082	−0.033

* Weights in air near sea level. Tartrate, phthalate, and phosphates dried at 110°C for 1–2 h. Potassium tetroxalate dihydrate dried at less than 60°C. Borax not heated. CO_2-free distilled water should be used.

Source: Bates, R.J. 1992. *Journal of Research of the National Bureau of Standards* 66A:179–184.

Table 6.2 Approximate pK_a Values of Aqueous Buffering Systems (at 25°C)

pK_a	Proton*	Compound
1.27	1	Oxalic acid
1.97	1	Maleic acid
2.35	1	Glycine
2.95	1	Phthalic acid
3.13	1	Citric acid
3.75		Formic acid
4.21	1	Succinic acid
4.76		Acetic acid
4.76	2	Citric acid
5.41	2	Phthalic acid
6.04	2	Histidine
6.40	3	Citric acid
6.95		Imidazole
7.20	2	Phosphoric acid
7.67		N-Ethylmorpholine
8.08		Tris(hydroxymethyl)aminomethane
8.95	4	Pyrophosphoric acid
9.24		Boric acid
9.25		Ammonia
9.78		Glycine
10.33	2	Carbonic acid
10.72		Triethylamine
11.12		Piperidine
12.38	3	Phosphoric acid

* For polyprotic acids, this indicates to which proton of the stepwise removal of pK_a refers: pK_1, pK_2, etc. If only one proton ionizes, no number appears.

Further, suppose that this experiment was repeated for various concentrations of acid but at a fixed temperature. As the solution is made more concentrated, the measured equilibrium "constant" changes! However, rather than throw out the very useful concept of mass action, an effective concentration can be substituted in place of the measured ones. This allows the idea of an equilibrium constant to be useful over a wider range of conditions.

These effective concentrations are called **activities** (abbreviated a), and the activity of a salt tends to be lower than the concentration of the salt. The activities of various species under specific solution conditions are found through experimental measurements. This information is conveniently stored as **activity coefficients** which relate the *effective* concentration (that is, the activity) to the analytical concentration.

Activities can be related to the concentrations of solutes on three scales:

the molal scale, for which $$a_A(m) = \gamma_A m_A \quad \textbf{(6-43a)}$$

the molar scale, for which $$a_A(c) = y_A C_A \quad \textbf{(6-43b)}$$

the mole-fraction scale for which $$a_A(N) = f_A N_A \quad \textbf{(6-43c)}$$

Here m_A is the molality, C_A the molarity, and N_A the mole fraction of solute A.

The factors γ_A, y_A, and f_A are called, respectively, the molal, molar, and rational activity coefficients of species A. As can be seen in the representative

Table 6.3 Selected pH Buffers with Special Applicability to Biological Media

Approximate pK_a Values at 20°C	Abbreviation	Structure
6.1	MES	$O(CH_2CH_2)_2\overset{+}{N}H CH_2CH_2SO_3^-$ (morpholine ring)
6.8	*ADA	$H_2NCOCH_2N^+H(CH_2COO^-)(CH_2COONa)$
6.8, 2.8	PIPES	$^-O_3SCH_2CH_2N^+H(CH_2CH_2)_2NCH_2CH_2SO_3^-$ (piperazine ring)
7.2	MOPS	$O(CH_2CH_2)_2\overset{+}{N}H CH_2CH_2CH_2SO_3^-$ (morpholine ring)
7.5	HEPES	$HOCH_2CH_2N^+H(CH_2CH_2)_2NCH_2CH_2SO_3^-$ (piperazine ring)
8.0, 3.7	PIPPS	$^-O_3SCH_2CH_2CH_2N^+H(CH_2CH_2)_2NCH_2CH_2CH_2SO_3^-$ (piperazine ring)
8.1	*TRIS	$(HOH_2C)_3CNH_3^+$
8.4	*TAPS	$(HOH_2C)_3C\overset{+}{N}H_2CH_2CH_2CH_2SO_3^-$
8.7, 4.4	PIPBS	$^-O_3SCH_2CH_2CH_2CH_2N^+H(CH_2CH_2)_2NCH_2CH_2CH_2CH_2SO_3^-$ (piperazine ring)
8.8, 4.7	DEPP	$H_3CCH_2N^+H(CH_2CH_2)_2NCH_2CH_3$ (piperazine ring)
9.3	*CHES	C_6H_{11}—$\overset{+}{N}H_2CH_2CH_2SO_3^-$ (cyclohexyl)
9.4, 5.8	DESPEN	$(CH_3CH_2)(^-O_3SCH_2CH_2CH_2)N^+H$—$CH_2CH_2$—$N(CH_2CH_3)(CH_2CH_2CH_2SO_3^-)$
9.8, 6.5	TEEN	$(CH_3CH_2)_2N^+H$—CH_2CH_2—$N(CH_3CH_2)_2$
10.4	*CAPS	C_6H_{11}—$\overset{+}{N}H_2CH_2CH_2CH_2SO_3^-$ (cyclohexyl)
11.0, 1.3	TEMN	$(CH_3CH_2)_2N^+H$—CH_3—$N(CH_2CH_3)_2$

*Can form coordination complexes with metal ions.

Sources: Good, N. E., et al. 1966. *Biochemistry* 5:467–477; Ferguson, W. J., et al. 1980. *Analytical Biochemistry* 104:300–310; Yu, Q., Kandegedara, A., Xu, Y., Rorabacher, D. B. 1997. *Analytical Biochemistry* 253:50–56.

Table 6.4 Experimental Mean Activity Coefficients of Solute NaCl and Water in Solutions at 25°C

Molality	Molarity	$\gamma_{\pm}$	$y_{\pm}$	$f_{\pm}$	a_{H_2O}
0	0	1.000	1.000	1.000	1.000
0.001	0.000997	0.965	0.965	0.965	0.9999
0.01	0.009969	0.902	0.902	0.902	0.9997
0.1	0.09953	0.778	0.779	0.781	0.9966
0.2	0.1987	0.735	0.738	0.740	0.9933
1.0	0.9788	0.656	0.668	0.681	0.9669
2.0	1.921	0.668	0.693	0.716	0.9316
4.0	3.696	0.783	0.845	0.896	0.8515
6.0	5.305	0.986	1.111	1.199	0.7598

In the table the subscript ± means that the activity coefficient relates to both ions.
Source: Data from Robinson, R. A., Stokes, R. H. 1965. *Electrolyte Solutions.* 2nd ed. rev. London: Butterworths.

example in Table 6.4, activity coefficients are concentration dependent, but at concentrations less than 0.2 m, the three types of activity coefficients differ by less than 1% for sodium chloride solutions.

EXAMPLE 6.11

What is the activity of sodium chloride in an aqueous solution in which the concentration of added NaCl is 0.10 M, and the temperature is 25°C?

Solution

From the information in Table 6.4 and Equation 6-43b,

$$a_{NaCl}(c) = y_{\pm}C_{NaCl} = 0.779 \times 0.10 = 0.0779 \text{ M}$$

The *effective* concentration of sodium chloride is less than the concentration measured by, say, adding a known weight of NaCl to make a known volume of solution.

Ionic Strength

The activity of a solute that dissolves to form ions depends on the concentration and the charge of every type of ion present in the solution. For instance, a doubly charged ion has a greater effect on the activities than the same concentration of a singly charged ion. A single number that is used to characterize the total ionic concentration (and accounts for the charges on them) is the solution **ionic strength**. Ionic strength is defined by Equation 6-44. The concentrations C_j are given in mol L^{-1} for each ionic species j, and z_j is the charge on species j.

$$\mu = \frac{1}{2}\sum_j c_j z_j^2 \qquad \textbf{(6-44)}$$

Each type of negative and positive ion is included in the sum. Note that neutral molecules do not contribute to the ionic strength.

EXAMPLE 6.12

What is the ionic strength of a 0.01 M $CaCl_2$ solution? The salt is completely dissociated under these conditions.

Solution

In the solution, the calcium concentration is 0.01 M, and the chloride ions are present at 0.02 M. Therefore,

$$\mu = \frac{1}{2}[0.01 \times 2^2 + 0.02 \times (-1)^2] = \frac{1}{2}(0.04 + 0.02)$$
$$= 0.03\text{ M}$$

EXAMPLE 6.13

What is the ionic strength of a 0.01 M solution of optimal sulfate $OpSO_4$, which, under the conditions of the experiment is, remarkably, exactly 50% dissociated?

Solution

In the solution, both the Op^{2+} and SO_4^{2-} concentrations are 0.005 M. The undissociated, uncharged $OpSO_4$ makes no contribution to μ. Therefore

$$\mu = \frac{1}{2}[0.005 \times 2^2 + 0.005 \times (-2)^2] = \frac{1}{2}(0.04) = 0.02\text{ M}$$

Single-Ion Activities

A theoretical interpretation of the interactions of ions in solution provides an equation that can be used to assign activity coefficients to each species of ion in a solution. For aqueous solutions at 25°C,

$$\log \gamma_i = -0.51\, z_i^2 \sqrt{\mu} \qquad \textbf{(6-45)}$$

where z_i is the charge on the ion species i, γ_i the molal activity coefficient, and μ the ionic strength of the solution. The numerical value 0.51 is only slightly temperature dependent. This equation applies to about 3% accuracy for solutions with ionic strengths less than about 1 mM. Incidently, Equation 6-45 is called the **Debye-Hückel equation.**

EXAMPLE 6.14

Calculate the single-ion activity coefficients you would expect to find for Na^+ and Cl^- in an aqueous solution containing 0.0001 M NaCl and 0.0002 M K_2SO_4 at 25°C. Under these conditions, both electrolytes are completely dissociated.

Solution

The ionic strength of the electrolyte solution is calculated from Equation 6-45.

$$\mu = \frac{1}{2}([Na^+] + [Cl^-] + [K^+] + [SO_4^{2-}] \cdot 4)$$

Substituting the appropriate concentrations of the ions, this becomes

$$\mu = \frac{1}{2}(0.0001 + 0.0001 + 0.0004 + 0.0002 \cdot 4) = 0.0007\text{ M}$$

The activity coefficients of Na^+ and Cl^- are expected to be equal since both possess unit charge and can be calculated using Equation 6-46:

$$\log \gamma_{Cl^-} = \log \gamma_{Na^+} = -0.51 \cdot 1^2 \sqrt{0.0007}$$

Thus

$$\gamma_{Cl^-} = \gamma_{Na^+} = 0.97$$

From this calculation we find that the activity coefficients for both sodium and chloride ions are expected to be the same and to equal $\gamma_{\pm}$. This behavior is general for so-called **symmetric** strong electrolytes, in which the anion(s) and cation(s) are in the ratio 1:1 (e.g., KCl, $CaSO_4$, and $La[Fe(CN)_6]$) where the electrolyte is completely dissociated and the ionic strength is less than 1 mM.

An extension of the Debye-Hückel theory that allows predictions of activity coefficients up to about 0.1 M is called (most appropriately) **extended Debye-Hückel theory**. In this case, a term is added to the denominator. It takes into account the radius r_i of the hydrated ion involved. Equation 6-45 then becomes

$$\log \gamma_i = \frac{-0.51\, z_i^2 \sqrt{\mu}}{1 + 3.3\, r_i \sqrt{\mu}} \tag{6-46}$$

where r_i is in nanometers. (Values of r_i can be found in tabulations of hydrated ionic radii.)

As mentioned earlier, at less than about 0.2 molar ion concentrations, the molal and molar activity coefficients are essentially equal. As a result, γ_i can be substituted for y_i in Equation 6-43b in the applicable concentration range. For example, with protons as a representative ion,

$$\text{activity of protons} = a_{H^+} = \gamma_{H^+} \cdot [H^+] \tag{6-47}$$

The γ_i that represent *molal* activity coefficients are used with molar units of concentration. Why mix these concentration units? We do so mainly because molal activity coefficients are available in standard references, and, in our daily work, we use mostly molar concentrations.

EXAMPLE 6.15

What is the expected activity coefficient of H^+ in a solution that is 0.0001 M in HCl but is 0.03 M in K_2SO_4. (The r_i value for H^+ is 0.90.)

Solution

For this ion concentration, the extended Debye-Hückel equation, Equation 6-46, must be used. To do so, first we must calculate the ionic strength of the solution using Equation 6-44. The H^+ and Cl^- contributions to the ionic strength are very small compared to those from K^+ and SO_4^{2-}. Therefore the ionic strength is

$$\mu = \frac{1}{2}[(0.06)(1)^2 + (0.03)(-2)^2] = 0.09 \text{ M}$$

Then, use Equation 6-46 to calculate the activity coefficients.

$$\log \gamma_{H^+} = \frac{-0.51(1)^2\sqrt{0.09}}{1 + 3.3(0.9)\sqrt{0.09}}$$

$$\gamma_{H^+} = 0.83$$

(If Equation 6-45 were used alone, we would have calculated a value of 0.71, a difference of about 17%.) It is worth noting that even in what might seem to be a relatively dilute solution, the activity of H^+ is expected to decrease by almost 20%.

Activity coefficients have been measured and tabulated only for relatively simple solutions and almost exclusively for water as solvent. For more complex solutions, such as seawater, approximations can be made with the aid of a computer, a list of the molar concentrations of all the ionic species, and one of a number of different mathematical models that have been found to describe the activities reasonably well.

Mean Activity Coefficient

When the activity coefficient is measured for a dilute solution of electrolyte A_xB_z with $x \neq z$, we no longer can assume that the activity coefficients for the cation and anion are the same. A **mean activity coefficient** $\gamma_\pm$ can be measured, and the calculated activity coefficients for the individual ions are not expected to be equal to it. The mean activity coefficient is defined in relation to the individual-ion activity coefficients by

$$\gamma_\pm^{(x+z)} = \gamma_A^{\ x} \cdot \gamma_B^{\ z} \quad \text{for } A_xB_z \qquad \textbf{(6-48)}$$

As you can see from Equation 6-48, the mean activity coefficient is a geometric mean, not an algebraic mean.

EXAMPLE 6.16

What is the mean activity coefficient expected for NaCl in a solution of 0.0001 M NaCl and 0.0002 M K_2SO_4?

Solution

Using the ionic strength found in Example 6.14 on page 196,

$$\gamma_\pm^{(1+1)} = \gamma_{Na^+}^{\ 1} \cdot \gamma_{Cl^-}^{\ 1} = 0.970 \cdot 0.970$$

$$\gamma_\pm^{\ 2} = (0.970)^2$$

$$\gamma_\pm = 0.970$$

The sodium chloride provides an example of a symmetric electrolyte. What about a compound like $K_2Cr_2O_7$?

EXAMPLE 6.17

What is the mean activity coefficient expected for K_2CrO_4 in a solution of 0.0001 M HCl and 0.0001 M K_2CrO_4?

Solution

We first need to find the ionic strength. From Equation 6-44, $\mu = 0.0004$ M. Since the ionic strength is less than 10^{-3} M, we can use Equation 6-45 to find the activity coefficients for K^+ and CrO_4^{2-}.

$$\log \gamma_{K^+} = -0.51 z_{K^+}^2 \sqrt{\mu} \qquad \log \gamma_{CrO_4^{2-}} = -0.51 z_{CrO_4^{2-}}^2 \sqrt{\mu}$$

$$\log \gamma_{K^+} = -0.51(1)(0.02) \qquad \log \gamma_{CrO_4^{2-}} = -0.51(4)(0.02)$$

$$\gamma_{K^+} = 0.977 \qquad \gamma_{CrO_4^{2-}} = 0.910$$

Now $\gamma_{\pm}$ can be calculated.

$$\gamma_{\pm}^{(2+1)} = \gamma_{K^+}{}^2 \cdot \gamma_{CrO_4^{2-}}$$
$$\gamma_{\pm}{}^3 = (0.977)^2 \cdot (0.910)$$
$$\gamma_{\pm} = 0.954$$

In the practice of chemical analysis, the ability to calculate activities and activity coefficients is usually less critical than understanding that ionic equilibria change with ionic strength. When ionic equilibria are involved in an analysis, it is necessary to keep the ionic strength as constant as possible to obtain the most precise results.

Equilibrium Constants and Activity Coefficients

Using activities in place of concentrations allows us to extend the range in which an equilibrium is found to obey the law of mass action. For example, the complicated behavior of the equilibrium of acetic acid with concentration can be described by Equation 6-49 below. The complexity is incorporated into the behavior of the factor consisting of the ratio of activity coefficients:

$$K = \frac{[H^+][OAc^-]}{[HOAc]} \cdot \frac{\gamma_{H^+} \cdot \gamma_{OAc^-}}{\gamma_{HOAc}} \qquad \textbf{(6-49)}$$

Here, as before, *molal* activity coefficients are used with molar units of concentrations.

Standard States

The activity coefficients γ_i are usually regarded as dimensionless quantities. The use of a dimensionless coefficient in Equation 6-47 indicates that the activity has the same units as the concentrations. However, a more fundamental activity measure, the mole-fraction activity $a(N)$, is dimensionless. Recall, again using protons for the example, that a mole fraction is defined by

$$\text{Mole fraction } H^+ = N_{H^+} = \frac{\text{moles } H^+}{\text{Total moles of all components}}$$

The reason mole-fraction activities are unitless and the important consequences of this fact are described next.

The mole-fraction *activity* is defined as a ratio of the effective concentration of a species relative to its **standard state**. The standard state is *defined* in a different way for different parts of an equilibrium system. The standard state for

solids is the pure solid;
solvents is the pure solvent;
solutes is a 1 molar solution of that solute; and
gases is the concentration of a pure gas at standard temperature (298 K) and pressure (1 atm)

Note: 298 K and 1 atm is *not* STP.

The standard state is written as f°_A, and the mole-fraction activity is then written

$$\frac{f_A}{f^\circ_A} = a_A(N) \qquad \textbf{(6-15)}$$

The effective concentration is written as f_A in order not to confuse it with the molar concentration [A]. The value of f_A° is chosen as appropriate to the system: solids, solutes, and so forth. For instance, if the system is an aqueous NaCl solution, f° is 1 molar NaCl. The correct term for the effective concentration is **fugacity**, which is why the letter f is written.

The Activities of Solids

The concept of an activity as a dimensionless ratio unifies a great deal of equilibrium chemistry. We *should* use activities in all equilibrium constant expressions. For example, instead of writing

$$K_{\text{concentration}} = \frac{[H^+][OAc^-]}{[HOAc]} \tag{6-8}$$

it is more proper to write

$$K_{\text{thermodynamic}} = \frac{a_{H^+} \cdot a_{OAc^-}}{a_{HOAc}}$$

Although this type of equilibrium constant is not often used in day-to-day analyses, it serves to illustrate the origins of the rules for including or excluding a species in the mass-action expression. For instance, let us consider the equilibrium that includes a pure solid dissolving in a solvent. No matter how much solid dissolves, any solid that remains continues to have the same concentration—that of the pure solid. So the activity of the remaining solid *at all times* is

$$a_{\text{solid}} = \frac{f^\circ}{f^\circ} = 1$$

EXAMPLE 6.18

What is the thermodynamic equilibrium constant for a saturated solution of sodium chloride in contact with solid sodium chloride?

Solution

$$K_{\text{thermodynamic}} = \frac{a_{Na^+}a_{Cl^-}}{a_{NaCl(s)}} = (a_{Na^+})(a_{Cl^-})$$

In the example here, the equilibrium constant is called the **solubility product** and is designated K_{sp}. It is most important to understand that there is nothing inherently different between solubility products and the mass-action expressions that were described earlier. As a result, for chemical equilibria that include solids, the rules for the addition and reversal of the stoichiometric equations and allied mass-action expressions are the same as for other equilibria.

More about solubilities and the related solubility products is presented in Chapter 10.

The Activity of Water in Solutions

When we write the equilibrium constant for a reaction such as acetic acid dissociation (Equation 6-8), the water that is present does not appear in the equilibrium expression. Why not? Also, is there any time when the water activity must be left in?

Let us begin answering the questions by writing the equation for the reaction and include the water:

$$HOAc + n\ H_2O \rightleftharpoons H(H_2O)_n^+ + OAc^- \tag{6-51}$$

The mass-action expression is then

$$K_{thermodynamic} = \frac{a_{H(H_2O)_n^+} \cdot a_{OAc^-}}{a_{HOAc}\, a_{H_2O}^n}$$

The key to answer the questions is to find the value of a_{H_2O}. Representative values for NaCl solutions can be seen in the right column in Table 6.4 on page 194.

But what is the value of a_{H_2O} for a 0.1 M solution of acetic acid? Since $a_{H_2O} = f/f°$, we can make an approximate calculation in the following way. The concentration of pure water is around 55.5 M (1000 g/L and 18 g/mol). One liter of a solution that is 0.1 M in acetic acid requires adding 6.0 g (5.7 cm^3) of pure acetic acid to approximately 995 cm^3 of water: 1000 mL of solution contains 995 mL of water. The concentration of water itself is lower in the acid solution compared with pure water, but only by about 0.5%. That is, the activity of water is within a few parts-per-thousand of unity: $a_{H_2O} \approx 995/1000$, and so we let it equal one.

$$K_{thermodynamic} = \frac{a_{H(H_2O)_n^+} \cdot a_{OAc^-}}{a_{HOAc}\, a_{H_2O}^n} \approx \frac{a_{H(H_2O)_n^+} \cdot a_{OAc^-}}{a_{HOAc}} \tag{6-52}$$

Leaving the water activity out of the mass-action expression is really an approximation. In other words, Equation 6-52 is not exact without the water activity, but it is an approximation that is adequate for all but the most exacting work. If the solution is very concentrated, however, this approximation may no longer be reasonable.

Mixed Equilibrium Constants

Ideally, the equilibrium constants we use should always be thermodynamic equilibrium constants that utilize the activities of each of the species involved. In reality, analytical chemical problems (among others) involve extremely complicated solutions and mixtures, and it simply is not clear what values should be used for the activity coefficients in these complicated solutions. In fact, usually only a few of the components in these solutions are known. Thus, concentration equilibrium constants are used for calculations most of the time.

Electrochemical methods are the topics of Chapter 15.

However, with certain methods, notably the electrochemical ones, the activities of certain species are actually measured. Perhaps the most common case is with the measurement of the activity of protons, a_{H^+}, made by measuring the solution pH. This proton *activity* value a_{H^+} is often used in the equilibrium expression along with the concentrations of other species. When activities and concentrations occur in equilibrium expressions together, the equilibrium constant is said to be a *mixed equilibrium constant*. For instance, the mixed equilibrium constant for acetic acid is

$$K_{mixed} = \frac{a_{H^+} \cdot [OAc^-]}{[HOAc]} \tag{6-53}$$

The only problem with such mixed constants is that they cannot be manipulated using all the known rules of chemistry. However, they are adequate for calculations relating to a series of experiments in which the conditions remain the same for every experiment. This is one reason why standardized and reproducible conditions are so important in chemical analyses.

Ionic Strength Effects upon Dilution of pH Buffers

If a solution containing a pH buffer is diluted, the concentrations of both the weak acid and its conjugate base are diluted equally. According to the Henderson-Hasselbalch equation (Equation 6-39), the pH would remain the same. However, the equation as it is stated does not include changes in the ion activities upon dilution. Diluting the solution causes the ionic strength to decrease, and the pH *does* change. An indication of the effect of dilution is given by a function called the **dilution value** ($\Delta pH_{1/2}$). Its numerical value measures the change in pH found when a solution of initial concentration c_i is diluted by an equal volume of pure solvent so that the concentration becomes ($C_i/2$).

$$\Delta pH_{1/2} = (pH)_{c_i/2} - (pH)_{c_i} \quad \textbf{(6-54)}$$

Representative values are shown in Table 6.1 on page 191.

Increases in ionic strength also affect the pH-buffer equilibria. The results of these changes are indicated in Table 6.5. Such effects must be considered if pH changes of this magnitude are unacceptable (for reference, a ΔpH of 0.04 equals a 10% change in [H^+]).

One Final Reminder

Do not lose sight of the fact that all the equilibrium calculations described are made to match and explain experiments. The properties of the equilibrium constant (law of mass action) and its value merely codify the experimental information. In addition, the rules of addition and reversal of chemical-equilibrium equations allow us to calculate concentrations under conditions that may not be obtainable experimentally. But there are many imperfections in the calculations. Some of these difficulties were reflected in the discussion of activity coefficients.

Nevertheless, equilibrium calculations can be used to determine, if only approximately, equilibrium properties that might be too costly to determine experimentally for every case. More importantly, though, they help to explain the chemistry that is encountered in analyses and thus can help in optimizing the analytical conditions.

Table 6.5 The Effect of Added Salt on the pH of Three NBS Standard Buffers

I.D.	C	D	F
Buffer	Phthalate	Phosphate	Borax
Salt added	KCl	NaCl	NaCl
ΔpH 0.02-*m* salt	−0.019	−0.022	−0.014
ΔpH 0.05-*m* salt	−0.044	−0.051	−0.035

Source: Bates, R. G. 1964. *Determination of pH*. 2nd ed. New York: Wiley.

Suggestions for Further Reading

LAITINEN, H. A., HARRIS, W. E. 1975. *Chemical Analysis: An Advanced Text and Reference*. 2nd ed. New York: McGraw-Hill.
A rather old, but excellent, description of equilibrium calculations.

RAMETTE, R. W. 1981. *Chemical Equilibrium and Analysis*. Reading, MA: Addison-Wesley.
A carefully written textbook, mostly of methodical equilibrium calculations—ignoring, in general, activity effects.

ALBERT, A., SERJEANT, E. P. 1984. *The Determination of Ionization Constants: A Laboratory Manual*. 3rd ed. London and New York: Chapmann and Hall.
A short book that shows how ionization constants are determined and the care that is required.

AZAB, H. A. 1993. "Potentiometric Determination of the Second-Stage Dissociation Constants of Some Hydrogen Ion Buffers for Biological Research in Various Water + Organic Solvent Mixtures." *J. Chem. Eng. Data* 38:453–457.
The pK_a of the zwitterionic biological buffers (MES, PIPES, ACES, HEPES) in mixed solvents are measured and tabulated.

SERJEANT, E. P., DEMPSEY, B. 1979. *Ionisation Constants of Organic Acids in Aqueous Solution*. Oxford and New York: Pergamon Press.

PERRIN, D. D. 1982. *Ionisation Constants of Inorganic Acids and Bases in Aqueous Solution*. 2nd ed. Oxford and New York: Pergamon Press.
Extensive tables of pK_a values and references printed under the auspices of the IUPAC (International Union of Pure and Applied Chemistry).

MCBRYDE, W. A. E. 1974. "Spectrophotometric Determination of Equilibrium Constants in Solution." *Talanta* 21:979–1004.
A short, more advanced review article.

COVINGTON, A. K. 1985. "Potentiometric Titrations of Aqueous Carbonate Solutions." *Chem. Soc. Rev.* 14:265–281.
A review article illustrating that there is always more to learn about "simple" equilibria.

MARTELL, A. E., MOTEKAITIS, R. J. 1988. *The Determination and Use of Stability Constants*. New York: VCH Publishers.
This text shows how equilibrium constants are obtained experimentally with examples and methods to eliminate commonly made errors. A practical book.

SHAKHASHIRI, B. Z., SCHREINER, R. 1993. *Chemical Equilibrium*. Champaign, IL: Stipes Pub. Co.
A workbook-style text at about the level of this chapter on acid-base, solubility products, and complexation.

Concept Review

1. Given the following reactions,

$$A + B \rightleftharpoons C$$

$$C \rightleftharpoons A + B$$

$$2\,A + 2\,B \rightleftharpoons 2C$$

write the mass action expression for each equilibrium and describe how the three mass action expressions are related.

2. If we say that the acid HA is a weaker acid than HB, what does this imply about
(a) the K_a values for the two acids?
(b) the strength of the conjugate bases A^- and B^-?

3. What species would be present at equilibrium in an aqueous solution of
(a) the weak acid HOOC-CH_2-CH_2-COOH?
(b) the weak base H_2N-CH_2-CH_2-NH_2?

4. For a solution of the weak base $Na_2C_4H_4O_6$ (sodium tartrate), describe the trends in the concentrations of $C_4H_4O_6^{2-}$, $HC_4H_4O_6^-$, and $H_2C_4H_4O_6$ as the solution pH is varied from pH = 0 to pH = 14.

5. What do α values tell us about solutions of polyprotic acids and their salts?

6. A solution containing a mixture of a weak acid and its conjugate base resists a change in pH with addition of H^+ or OH^- to the solution. Explain this in terms of the Henderson-Hasselbalch equation.

7. How would the pK_b value for a weak base help you decide where a mixture of the base and its conjugate acid would function most effectively as a buffer?

8. Which would be expected to have a higher buffer capacity: an aqueous solution that is 0.5 M in NaOAc and 0.5 M in HOAc, or an aqueous solution that is 0.1 M in NaOAc and 0.1 M in HOAc?

9. For what effect do activity coefficients correct?

Exercises

6.1 Write the mass action expressions for the following reactions.
(a) $CH_3OH(g) \rightleftharpoons CO(g) + 2\,H_2(g)$
(b) $CO(g) + H_2O(g) \rightleftharpoons CO_2(g) + H_2(g)$
(c) glucose + $3\,OH^- + I_3^- \rightleftharpoons$ gluconate$^-$ + $3\,I^- + 2\,H_2O$

6.2 Write the mass action expression for the following reaction and compare it to 6.1a.

$$CO(g) + 2\,H_2(g) \rightleftharpoons CH_3OH(g)$$

6.3 Based on Le Chatelier's Principle, what would be the effect on the equilibrium of Exercise 6.2 if
(a) the pressure in the reaction vessel were increased?
(b) a small amount of CO were added to the reaction vessel (assume pressure remains constant)?

6.4 Given the following reactions and equilibrium constants

$$A + B \rightleftharpoons C \qquad K = 10^{-5}$$

$$2\,C \rightleftharpoons D \qquad K = 10^4$$

write the mass action expression and the equilibrium constant for

$$2\,A + 2\,B \rightleftharpoons D$$

6.5 Given the following reactions and equilibrium constants

$$A + 2\,B \rightleftharpoons C \qquad K = 10^4$$

$$2\,D \rightleftharpoons C \qquad K = 10^4$$

write the mass action expression and the equilibrium constant for

$$A + 2\,B \rightleftharpoons 2D$$

6.6 A table is found in a handbook that lists "Dissociation constants for water-soluble organic bases." (Actually, a base does not dissociate protons, but, on the contrary, takes them up.) The "base dissociation constant" for sodium barbiturate (NaBar) is given as $K = 1.08 \times 10^{-10}$. You look up the pK_a of the conjugate acid, and it is 4.035. Given that the reaction is not a dissociation of a base, what is the reaction that corresponds to K?

6.7 Determine the pH of a 0.15-M taurine solution made by adding pure taurine to water. Taurine takes up a proton in water with a pK_b of 5.26.

6.8 Some enzymatic assay procedures produce phenol as the product. Phenol is a monoprotic acid with $pK_a = 9.85$ at 25°C. What is the pH of a 0.01-M phenol solution?

6.9 Another term used to describe the dissociation of weak acids is **percent dissociation**. The number describes the percentage of the total acid in the solution that is in the dissociated (ionized) form. The total acid is the sum of both the dissociated and undissociated acid. The percent dissociation is, algebraically,

$$\%\ \text{dissociation} = 100 \times \frac{[\text{dissociated acid}]}{[\text{total acid added}]} = 100 \times \frac{\text{moles dissociated acid}}{\text{moles acid added}}$$

The percent dissociation is simply 100 times the fraction dissociated. What is the percent dissociation of the acids in the following aqueous solutions?
(a) 0.100 M acetic acid
(b) 0.200 M acetic acid, adjusted to a pH of 5.00
(c) 0.100 M picric acid

6.10 HNO_3 is a strong acid.
(a) Calculate the pH you expect in a 0.1-M nitric acid solution.
(b) Calculate the pH you expect in a 1×10^{-8} M nitric acid solution.

6.11 Ammonia reacts with water with the reaction

$$NH_3 + H_2O \rightleftharpoons NH_4^+ + OH^-; \qquad K_b = 1.8 \times 10^{-5}$$

(a) Calculate the pH you would expect from a 0.10-M solution of ammonia in water.
(b) What percentage difference is there between the exact algebraic solution and one calculated using the assumption that the concentration of NH_3 at equilibrium remains at 0.1 M?

6.12 Calculate the pH for the following aqueous solutions of hypothetical acid and its salts, where pK_1 and pK_2 for H_2hyp are 6.00 and 9.00, respectively.
(a) 0.1 M H_2(hyp)
(b) 0.1 M NaH(hyp)
(c) 0.1 M Na_2(hyp)

6.13 Solubilities of fatty acids, $CH_3(CH_2)_nCOOH$, in water increase when the acids are in their charged form, $CH_3(CH_2)_nCOO^-$. Would the solubilities increase, decrease, or remain the same as the pH of an aqueous solution is lowered?

6.14 Assume for the following question that the activity coefficients of the substance behave similarly to those shown in Table 6.4; the higher the ionic strength, the lower the $\gamma_\pm$. Also, assume each ion is affected equally.

Nonesuch acid dissolves in water with the following reaction,

$$HNs \rightleftharpoons H^+ + Ns^-; \qquad K_a = 0.01$$

(a) A nonreactive salt such as KNO_3 is added to a 0.01-M nonesuch acid solution. When the solution is 0.1 M in the salt, will the equilibrium shift to the left, shift to the right, or stay the same?
(b) Under the conditions of (a), will the pH increase, stay the same, or decrease compared to the acid alone?

6.15 Citric acid is a triprotic acid with pK_as of 3.13, 4.76, and 6.40 for the three successive dissociations.
(a) Write the mass action expressions for the three dissociations.
(b) Determine the following three ratios of the citrate species

$$A = [H_3Cit]/[H_2Cit^-]$$

$$B = [H_2Cit^-]/[HCit^{2-}]$$

$$C = [HCit^{2-}]/[Cit^{3-}]$$

at pH 1, at pH 4.76, at pH 7.00

6.16 Suppose 100 mL of three citrate solutions (pH 1.00, pH 4.76, and pH 7.00) each was diluted to 200 mL with a salt solution (to give essentially the same ionic strength in each solution). For the ratios A, B, and C of Exercise 6.15:
(a) Would the ratio A increase, decrease, or remain the same?
(b) Would the ratio B increase, decrease, or remain the same?
(c) Would the ratio C increase, decrease, or remain the same?
(d) For each of a, b, and c, would you expect the pH to change for any of the solutions? If so, in which direction?

6.17 Calculate the concentrations of H_2CO_3, HCO_3^- and CO_3^{2-} in a solution with 0.10 M total carbonate when the solution pH = 10.00.

6.18 Calculate the fraction (α_1) of lysine that is in the zwitterionic form at physiological temperature (37°C) for venous neonatal blood with pH 7.35. The pK_a values for lysine are 2.20, 8.90, and 10.28 at 37°C.

6.19 For boric acid,

(a) Write out the equations for α_0, α_1, etc. Then use a spreadsheet program to generate a table with the following columns for pH = 0 to 14.

pH	[H⁺]	α_0	α_1	...	α_N

(b) Graph the α values vs. pH for the boric acid.

(c) Use the values of α_i at pH 8.2 to estimate the salt concentrations needed to produce a borate buffer of pH 8.2 that is 0.1 M in total borate.

6.20 Salicylic acid (abbreviated Hsal; formula C_6H_5OCOOH; F.W. 138.12), at 25°C has $K_a = 1.07 \times 10^{-3}$.

(a) If 125.0 mg are dissolved in 50.0 mL water, what [H⁺] do you expect the solution to have?

(b) What percentage difference is there between the exact solution and a solution calculated with the usual approximation method? Express the answer both as percent [H⁺] and percent pH.

6.21 A phosphoric acid solution, H_3PO_4 in water, was neutralized with ammonia to a pH of 8.19. Using the K_a values for phosphoric acid from Appendix II and the K_b value for ammonia shown in Exercise 6.11,

(a) What is the value of the ratio $[NH_4^+]/[NH_3]$?

(b) Which of the following species make up more than 1% of the total phosphate at pH 8.19: H_3PO_4, $H_2PO_4^-$, HPO_4^{2-}, PO_4^{3-}?

6.22 Calculate the ionic strength for the following solutions.

(a) 0.1 M NaCl

(b) 0.2 M Na_2SO_4

(c) 0.1 M $K_4Fe(CN)_6$

6.23 What would be the activity coefficient for Fe^{2+} ($r_{Fe^{2+}}$ = 0.600 nm) in

(a) 0.1 M NaCl?

(b) 0.2 M Na_2SO_4?

6.24 The addition of an inert salt to a solution of a weak base results in a shift in the measured equilibrium constant, $K_{measured}$, due to activity effects. In the following calculations of these effects, sufficient accuracy will be obtained if you assume that $y_i = \gamma_i$.

(a) Calculate the ionic strength of a 0.0200 M KOAc solution in water.

(b) Calculate the ionic strength of an aqueous solution which is 0.0200 M in KOAc and 0.1 M with KCl.

(c) The following equation has been found to describe the empirical behavior of the mean activity coefficient of potassium acetate as a function of ionic strength, μ.

$$\log \gamma_{\pm} = \frac{-0.82\sqrt{\mu}}{1+\sqrt{\mu}} + 0.33\,\mu$$

What is the value of $\gamma_{\pm}$ for KOAc for the solutions in parts (a) and (b)?

(d) We measure the activity of protons using a pH electrode and define the measured equilibrium constant to be

$$K_{measured} = \frac{a_{H^+}[OAc^-]}{[HOAc]}$$

where we use the concentration values instead of the activities for HOAc and OAc^-. Since acetic acid is uncharged, its activity will be nearly unchanged by changes in ionic strength. Using the activity coefficient for acetate in 0.1 M KCl calculated in part (c), calculate the ration of $K_{measured}$ and $K_{thermodynamic}$.

Additional Exercises

6.25 Write the mass action expressions for the following reactions.

(a) $[Co(NH_3)_6]^{2+} + H_2O \rightleftharpoons [Co(NH_3)_5(OH)]^+ + NH_4^+$

(b) $N_2(g) + 4\,H_2O \rightleftharpoons 2\,NO_2(g) + 4\,H_2(g)$

(c) $CaCO_3(s) \rightleftharpoons CaO(s) + CO_2(g)$

(d) $NH_3(g) + H_2S(g) \rightleftharpoons NH_4HS(s)$

6.26 Write the mass action expressions for the following reactions and compare them to those for Problem 6.26a and 6.26b, respectively.

(a) $[Co(NH_3)_5(OH)]^+ + NH_4^+ \rightleftharpoons [Co(NH_3)_6]^{2+} + H_2O$

(b) $2\,NO_2(g) + 4\,H_2(g) \rightleftharpoons N_2 + 4\,H_2O$

6.27 Given the following reactions and equilibrium constants

$$A + B \rightleftharpoons 2C \qquad K = 10^{-7}$$

$$2C \rightleftharpoons D \qquad K = 10^{4}$$

write the mass action expression and the equilibrium constant for

$$D \rightleftharpoons A + B$$

6.28 For the reaction

$$2\text{ aniline} + 4\,Cr^{3+} \rightleftharpoons \text{azobenzene} + 4\,Cr^{2+} + 4\,H^+$$

what would be the effect on the equilibrium if additional acid were added to the solution?

6.29 Calculate and compare the percent dissociations expected from 10^{-1}, 10^{-3}, and 10^{-5} M acetic acid solutions.

6.30 Determine the pH of a 0.25-M ammonia solution made by bubbling pure ammonia into water. Ammonium has a pK_a of 9.244.

6.31 HCl is a strong acid.

(a) Calculate the pH you expect in a 0.2-M hydrochloric acid solution.

(b) Calculate the pH you expect in a 1×10^{-10} M hydrochloric acid solution.

6.32 What would be the trend(s) in the solubility of glycine in water as the pH varies from 1 to 12?

6.33 Calculate the relative concentrations of phenol and phenolate (its conjugate base) at
(a) pH 4
(b) pH 8
(c) pH 12

6.34 Calculate the concentrations of phenol and phenolate in a solution with 0.10 M total phenol when the solution pH = 9.98.

6.35 For lysine,
(a) Write out the equations for α_1, α_2, etc. Then use a spreadsheet program to generate a table with the following columns for pH = 0 to 14.

pH	$[H^+]$	α_1	α_2	...	α_N

(b) Graph the α values vs. pH.

6.36 If 125.0 mg of nitrous acid are dissolved in 50.0 mL water,
(a) What $[H^+]$ do you expect the solution to have?
(b) What percentage difference is there between the solution calculated using $[H^+]^2 = K_a C_{HNO_2}$ and a solution calculated without the approximation that $[HNO_2] = C_{HNO_2}$?
(c) What is the difference in the pHs found using the two calculations?

6.37 Calculate the expected pH of a 0.001 M solution of acetic acid, assuming that K_a at $\mu = 0$ applies. If the ionic strength is adjusted to 0.1 M with NaCl, will the pH increase, decrease or stay the same?

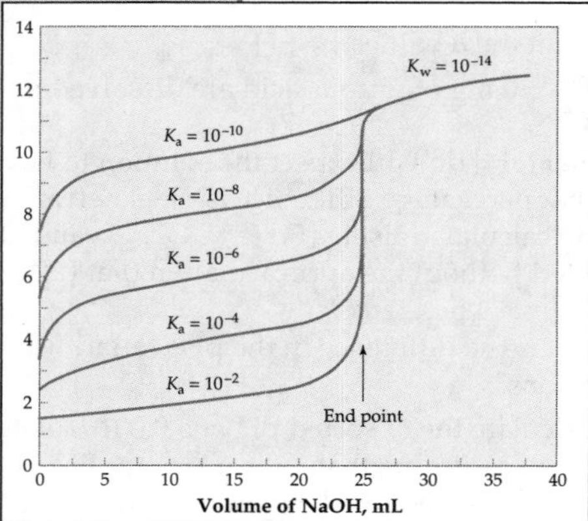

CHAPTER 7

General Introduction to Volumetric Titrations: Neutralization Titrations

Overview

7.1 Requirements for Volumetric Titrations

Titration is a method in which a volume of a standard(ized) solution is added to an unknown solution to ascertain the titer of some component of the unknown. Titer, a word now not widely used, means the weight of a substance equivalent to a 1-mL volume of this standard solution. As an equation,

$$\text{titer} = \frac{\text{mass of substance}}{\text{mL of standard solution}}$$

Since an unknown amount of some substance is calculated from a known *volume* of added solution, the method of titration falls under the classification of **volumetric** methods. Precision in the range of 0.1% is possible with moderate care.

The requirements for volumetric titrimetric assays include the following:

a. A **titrant** solution with a known concentration of reagent (a standardized reagent). It must react to completion with the analyte with a *reproducible stoichiometry* and with an adequate reaction rate, and it must be stable over the time required for the analyses.
b. A technique to measure the volume of the titrant solution to the desired precision and accuracy.
c. A technique to measure the sample weight or volume to the desired precision and accuracy.
d. A pretreatment to remove interferents if the reagent-analyte reaction is not sufficiently specific to eliminate effects of the matrix components.
e. A technique to measure when the titrant-analyte reaction has just used up all the analyte; the moles of titrant added are equivalent to the moles of analyte. This is the **equivalence point**. The volume of titrant added up to this point is the best quantity to use in calculating the analyte content. However, to discern the equivalence point, some physical change must be observed. The center of that change is the **end point** of the titration. Ideally, the end point and the equivalence point coincide. If they do not, the difference is called the **titration error**. In this chapter, where no explicit distinction is made it is assumed that the end point and equivalence point coincide.

One variation of volumetric titration is **back titration**. This technique involves, for acids, adding a known quantity of base that is greater than the amount required for neutralization. The acid reacts completely, and by titrating the "leftover" base with a standardized acid, the original amount of acid present can be determined. In other words, you go past the endpoint and then back. Back titration is especially useful when the initial acid-base reaction is sluggish. The excess base can be left to react for as long as needed to react completely, and then the solution titrated as usual. One example where a sluggish reaction will occur is when the acid being titrated resides on and/or in porous solids, where the base takes a relatively long time to reach and react with the acid groups.

7.2 The Chemistry of Titration

To understand the chemistry underlying titration, it may be helpful to describe a simple titration and analyze the results in detail. Before continuing, you should realize that titration is not a method used for trace analysis. In general, the lower limit for the use of titration to do precise determinations arises primarily from two requirements.

1. The reagent-analyte reaction must go to completion.

If the reagents get too dilute, the equilibrium shifts back toward unreacted forms. As a result, smaller amounts of analyte must be contained in smaller volumes, which leads to the second requirement.

2. The volume of the analyte solution must be large enough to handle—on the order of 0.1 mL.

The amount and precision of titrant addition is not a problem. Contemporary automatic titrators can add tens of microliters with precisions of less than a microliter.

Demonstration of a Simple Titration

A demonstration experiment was done for a class one day. A beaker contained 500 mL of a 0.020-M solution of acetic acid. This had been made by mixing pure acetic acid (>99.7% CH_3COOH) with distilled water. Since the acetic acid could be weighed accurately (0.600 ± 0.001 g; f.w. 60.05) and the water volume measured accurately (500.0 ± 0.2 mL), the acid concentration was known to 0.2%. This preparation satisfied requirements (c) and (d) of the list in Section 7.1. In this demonstration, the concentration of the sample solution is known.

Immersed in the solution was a sensor (an electrode to measure the hydrogen-ion activity in the presence of other ions and molecules). Its presence satisfied requirement (e). The electrode's response was calibrated immediately before the experiment using a standard solution of accurately known pH.

The solution was stirred constantly with a magnetic stirrer. This ensured that the solution was homogeneous and that the measured pH represented the entire solution.

To this solution was added 1.00-M KOH solution. The concentration of the KOH had been calibrated by titrations against replicate samples of a National Bureau of Standards standard acid, potassium acid phthalate.

The stoichiometry of the potassium hydroxide reaction with acetic acid is

$$K^+ + OH^- + HOAc \rightleftharpoons K^+ + H_2O + OAc^-$$

or, after omitting the K^+ spectator ion,

$$OH^- + HOAc \rightleftharpoons H_2O + OAc^-$$

The choice of KOH, a strong base, as the reagent, and the calibration of the concentration of the KOH solution satisfied requirement (a). The KOH was added to the sample solution 1.00 mL at a time. This level of precision satisfied requirement (b). Thus all the requirements for a titration were present.

The pH of the initial acetic acid solution was measured as pH 3.5. Next, 1 mL of the KOH titrant was added to the solution, followed by a delay of

Table 7.1 Data for Titration of Acetic Acid with KOH*

Added Titrant (mL)	pH (expl)	$[H^+]$ (expl)	$[OH^-]$ (calc)
0.0	3.5	3.2×10^{-4}	3.2×10^{-11}
1.0	4.0	1.0×10^{-4}	1.0×10^{-10}
2.0	4.2	6.3×10^{-5}	1.6×10^{-10}
3.0	4.4	4.0×10^{-5}	2.5×10^{-10}
4.0	4.5	3.2×10^{-5}	3.2×10^{-10}
5.0	4.7	2.0×10^{-5}	5.0×10^{-10}
6.0	4.9	1.3×10^{-5}	7.9×10^{-10}
7.0	5.1	8.0×10^{-6}	1.25×10^{-9}
8.0	5.4	4.0×10^{-6}	2.5×10^{-9}
9.0	6.1	8.0×10^{-7}	1.25×10^{-8}
10.0	8.5	3.2×10^{-9}	3.2×10^{-6}
11.0	11.3	5.0×10^{-12}	2.0×10^{-3}
12.0	11.5	3.1×10^{-12}	3.2×10^{-3}
13.0	11.7	2.0×10^{-12}	5.0×10^{-3}
14.0	11.8	1.3×10^{-12}	6.3×10^{-3}

*Acetic acid: 500 mL of 0.020 M; KOH: 1.00 M.

at least 10 s to allow the stirred solution to become homogeneous. Only then was the pH reading on the meter recorded. Further 1.00-mL additions were done the same way 13 times, and the pH values were noted. These are listed in Table 7.1. The data are also plotted as circles on a graph shown in Figure 7.1, with a continuous line drawn to join the data points. This is a **titration curve**. The table and titration curve show the same results in two different ways.

The titration curve in Figure 7.1 shows the pH behavior clearly. What we see is an initial, slow rise in pH with the addition of base. However, only after 7 or 8 mL of the KOH titrant have been added does the rise become rapid. The pH rises very rapidly between 9 and 11 mL of added KOH, and then the curve flattens again—meaning, of course, that the pH does not change much with addition of more base. Note that the slope of the graph looks even flatter on the top (above pH 11) than in the early flat part (pH 4 to 5). Let us now analyze this curve in detail using the law of mass action and the descriptive chemistry of the reactions. This information is reflected in the equilibrium constants and stoichiometries for the reactions that occur (the chemical equilibria involving acetic acid, KOH, and water).

7.3 Acetic Acid-KOH Titration: Equilibria and pH Changes

Before any base was added, the pH of the acetic acid solution was measured to be 3.5. Then, upon addition of KOH, the solution pH changed as plotted in Figure 7.1. Can we explain these empirical pH values? (*Empirical* means based solely on an experiment, without reference to theory.) We seek an explanation that tells us how the pH values are related to the amount of added KOH.

The procedure involves using our data (or someone else's) to calculate an equilibrium constant for the acid. Then, from the law of mass action, the entire titration curve can be explained. At the same time, the calculation tests

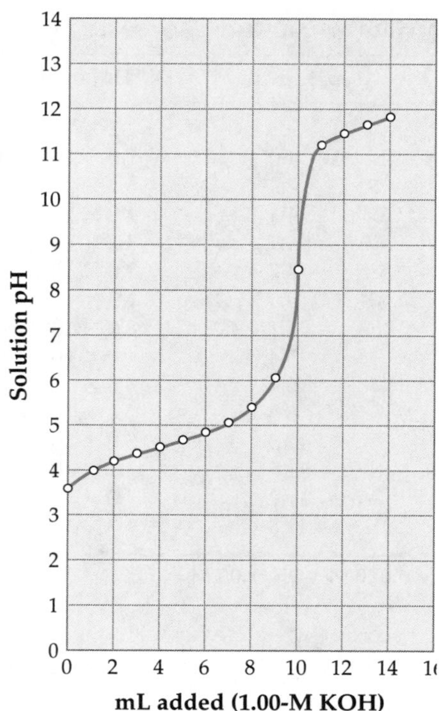

FIGURE 7.1 ▲
Results of the titration of 500 mL of 0.020-M acetic acid with 1.00 M KOH.

Graphed here are the measured pH-values of the solution vs. the total volume of KOH solution that had been added.

whether the law of mass action applies to the titrated solution and the ions that we believe are present.

It is important to realize that we can only explain the experiments with theories. If the data do not agree with the theory, then the theory is either wrong or incomplete. By incomplete we mean some chemical reaction or chemical species that affects the measured behavior has not been included in the calculation. The experiment is primary, and the explanation using chemical theory follows.

In the remainder of this section, we first ascertain whether the equilibrium of acetic acid/acetate can explain the experimental results. Following that, the behavior of various regions of the titration curve are explained. (These regions are the initial pH, the range around pH 4–5, the end point, and the region beyond the end point.)

Determining K_a

Here, we shall see how the equilibrium constant K_a for acetic acid is determined from the data. You will see that we do the same equilibrium calculation for each data point (see Table 7.2). What we especially want to know about the titration is how the concentrations of H^+, HOAc, and OAc^- change with the addition of OH^-. This is the key to understanding the titration curve.

Table 7.2 Result of Calculations of Species Concentrations

Added Titrant (mL)	[HOAc](M)	[$^-$OAc](M)	$\frac{[^-\text{OAc}]}{[\text{HOAc}]}$	K_a	pK_a
0.0	0:02	(3.2×10^{-4})*	0.016	5×10^{-6}	5.3
1.0	0.018	0.002	0.111	1×10^{-5}	5.0
4.0	0.012	0.008	0.667	2×10^{-5}	4.7
5.0	0.010	0.010	1.00	2×10^{-5}	4.7
6.0	0.008	0.012	1.50	2×10^{-5}	4.7
8.0	0.002	0.018	9.0	4×10^{-5}	4.4
9.0	0.001	0.019	19.0	2×10^{-5}	4.7

*The value is set equal to the proton concentration.

In order to calculate K_a we need to determine three concentrations—$[H^+]$, [HOAc], and $[OAc^-]$—to put into the mass-action expression

$$K_a = \frac{[H^+][OAc^-]}{[HOAc]}$$

During the titration, we measured a_{H+}, the H^+ activity. We will *assume* that the activity and concentrations are equal: $a_{H+} = [H^+]$. We also need to find [HOAc] and $[OAc^-]$. We know the original value of [HOAc] = 0.020 M. From this value and the amount of OH^- added, we can find the two needed concentrations [HOAc] and $[OAc^-]$ at any point. The method of this calculation is important and will be done step by step.

The reaction

$$OH^- + HOAc \rightleftharpoons H_2O + OAc^- \quad \textbf{(7-1)}$$

goes essentially to completion; the equilibrium lies far to the right. Therefore, for every mole of OH^- added, there is a *decrease* of one mole of HOAc and an *increase* of one mole of OAc^-. Through this stoichiometric relationship we can find the concentration of acetate and acetic acid. (The reaction's effect on the water activity is negligible.)

Equation 7-1 says

$$\text{moles } OH^- \text{ added} = \text{moles } OAc^- \text{ formed} = \text{decrease in moles HOAc}$$

It should be recognized that this information is the same as contained in a conservation equation. From the data of the titration, the conservation equation that applies here is

$$0.020 \text{ M} = [HOAc] + [OAc^-] \quad \textbf{(7-2)}$$

Conservation equations such as 7-2 were used in Sections 6.6 and 6.7 for solving simultaneous equilibrium problems.

Notice that the units of the quantities in Equation 7-2 are mol L^{-1}. Since the initial volume of the solution was 500 mL, Equation 7-2 could also be written with units of moles.

$$0.010 \text{ moles} = \text{moles HOAc} + \text{moles } OAc^- \quad \textbf{(7-3)}$$

EXAMPLE 7.1

Calculate the K_a and pK_a for the solution after 4.00 mL of 1.00-M KOH titrant has been added to 500 mL of a solution 0.020 M in acetic acid. This is the calculation done for the data of line 3 in Table 7.2.

Solution

From the experimental data in Table 7.1, we find that the pH of the solution was 4.5. Equating the activity of H^+ with its concentration gives

$$[H^+] = 10^{-pH} = 10^{-4.5} = 3.2 \times 10^{-5}\ M$$

We now must find the concentrations of acetic acid and acetate. This is done by calculating the changes from the numbers of moles present initially and then converting to concentrations of the species. That is, addition of 4.00 mL of 1.00-M KOH means that

$$\text{moles KOH} = 0.00400\ L \times 1.00\ mol\ L^{-1} = 0.00400\ \text{moles KOH added}$$

Then, using the information in the stoichiometric equation 7-1, we find that

$$\text{moles } OAc^- = 0.00400$$

after 4.00 mL of KOH titrant was added. From the conservation equation 7-3,

$$\text{moles HOAc} = 0.0100\ \text{moles} - 0.0040\ \text{moles} = 0.0060\ \text{moles}$$

Since the mass-action equation requires that the quantities of the species be in concentration units, mole quantities are now converted to mol L^{-1}. This requires knowing the solution volume. But what is the solution volume? The original volume was 500 mL, and to this were added 4.00 mL, giving a total volume of 504 mL. The final concentrations (to within ±1% relative error) are

$$[HOAc] = 0.0060\ mol/0.504\ L = 0.0120\ \text{molar}$$

and

$$[OAc^-] = 0.0040\ mol/0.504\ L = 0.0080\ \text{molar}$$

Substituting into the mass-action expression, we obtain

$$K_a = \frac{[H^+][OAc^-]}{[HOAc]} = \frac{(3.2 \times 10^{-5})(0.0080)}{0.0120} = 2.2 \times 10^{-5}$$

and

$$pK_a = 4.70$$

(Note: In addition to our approximation that $a_{H+} = [H^+]$, it was also assumed that the ionic strength effects are irrelevant throughout the addition of 4 mL of 1-M KOH.)

As shown in Table 7.2, from the data that were collected, the equilibrium constant is 2×10^{-5} in the concentration range and at the temperature of the solution. This compares with a "literature value" of 1.76×10^{-5} at 25°C ($pK_a = 4.75$). The equilibrium constant calculated from the demonstration experiment data agrees with the K_a value for acetic acid, which others have experimentally determined. More important, though, is the confirmation that the description of the system using the stoichiometric equation and the equilibrium constant expression is self-consistent. We can say we have a correct chemical model of the system.

The disagreement of about 10% between the average K_a value obtained in this experiment and the literature value can arise from a number of sources:

1. Experimental errors may exist. Among these are a loss of titrant or solution (by splashing), a surplus of titrant (by dripping), and inaccuracy of the pH electrode.
2. The temperature of the solution used in the demonstration may not have been 25°C.
3. The K_a values were calculated using a mixed equilibrium constant with concentrations of acetic acid and acetate but with the activity of hydrogen ions as measured by the electrode. That is,

Mixed equilibrium constants are described in Section 6.2.

$$K_a = \frac{a_{H^+}[OAc^-]}{[HOAc]}$$

   This approximation will become worse the further the concentrations differ from activities, and this difference increases with ionic strength.
4. The ionic strength changes throughout the titration.

To review: the equilibrium constant K_a for acetic acid was calculated from the data obtained using standardized solutions and calibrated measuring equipment to find the pH of the acid solution as a function of the amount of added base. Together with the descriptive chemistry of the base-neutralization reaction (confirmed by other methods), the data were used to find the equilibrium constant for the dissociation of acetic acid.

The pH of the Initial Sample Solution: 0.020-M Acetic Acid

From the regularity of K_a (as seen in Table 7.2), we have concluded that the equilibrium of acetic acid/acetate controls the pH of the solution. We can now use this information to calculate the pH expected for a 0.020-M acetic acid solution. This is the same problem described in detail in Section 6.3, and here we state the result only: The original 0.020-M acetic acid solution is expected to have a pH of 3.2. This value is consistent with the rest of the data, but it differs from the pH that was measured, pH 3.5. One possible cause of this discrepancy is the presence of some base in the water before the acid was added. However, it is more likely that there is an error in the pH measurement. Such an error is likely to be due to using only one pH-value to calibrate the electrode; it was calibrated at pH 7 with a standard. However, the hydrogen concentration at the measured pH is more than 6,000 times higher than at the calibration pH.

Description of the Region of pH 4–5: The Buffer Region

So far, we have found that the equilibrium

$$HOAc \rightleftharpoons H^+ + OAc^-$$

with its mass-action expression

$$K_a = \frac{[H^+]\,[OAc^-]}{[HOAc]} = 2 \times 10^{-5}$$

can be used to describe the titration curve of acetic acid in the pH range 3 to 6.2. The values of K_a in Table 7.2 show this. Let us now consider the slopes of the curve plotted in Figure 7.1.

Between 0 and 1 mL of base, the pH rises faster per milliliter of added KOH than it does further along in the titration (after a few milliliters have been put in). The rise is again faster after about 6 mL are added. In between, the pH change for a given amount of added base is the smallest: The slope of the curve is a minimum. In the pH range around 4.5 to 5, the solution is a buffer which, by definition, resists a change in pH when base or acid is added, relative to other regions of the titration. (See Figure 7.1.)

A most effective way to clarify such behavior is to look at extreme regions. Assume that a small, fixed amount of OH^- is added to the solution, and the pH change observed. Let us answer the following two questions.

1. At what pH will the change be smallest?
2. At what pH will the change be largest?

To answer the questions, we note the following: From the stoichiometry of the reaction, the magnitude of the changes in [acid] and [base] are equal and opposite. One increases when the other decreases, and the result depends on a change in a ratio of two numbers.

Mathematically, with a fixed hydroxide addition, the change in the ratio ([base]/[acid]) will be the smallest when the *relative* change in *both* concentrations is the smallest. But, recall that the *absolute* changes are the same. So the smallest relative change in *both* concentrations occurs when they are both the same size. In other words, for acetate/acetic acid the smallest changes in the ratio will occur when

$$1 = \frac{[OAc^-]}{[HOAc]}$$

which occurs in the center of the buffer region. The pH at the center of the range can be found using the Henderson-Hasselbalch equation.

$$pH = pK_a + \log 1 = pK_a$$

This is a general result for any acid-base pair: *The buffer region is centered around the* pK_a *of the acid*. We have answered question 1.

Conversely, when one of the concentrations—either $[OAc^-]$ or [HOAc]—is much smaller than the other, a fixed addition of base will cause a much larger relative change in the smaller concentration. The change in the ratio will then be larger than in the buffer region.

To answer question 2, there are two extreme points at which the pH change will be the largest. The first is before any base has been added. (Here, $[OAc^-]$ is at its smallest value.) The second is just before all the acid has reacted with the added base. (There, [HOAc] is at its smallest value.) The volume at which all the acid has reacted and at which the curve has its steepest slope is the end point of the titration. It is labeled in Figure 7.1. As the end point is approached, the volume of base needed to change the solution pH by large amounts becomes quite small. Comparisons between different regions of the titration are shown in Table 7.3.

The pH at the Equivalence Point

As the titration progresses, the pH changes have been explainable as resulting from the quantitative transformation of acetic acid to acetate due to its reaction with hydroxide. The pH depends on the ratio of acetate and acetic acid. The result is a titration curve with an initial, relatively rapid rise, then a

Table 7.3 Volume of 1-M KOH Needed to Change the pH by 1 Unit in a 0.02-M Acetic Acid/Acetate Solution

pH Range	Volume 1-M KOH (mL)
4.0 to 5.0	4.9
5.0 to 6.0	3.0
6.0 to 7.0	0.5
7.0 to 8.0	0.05

flattening as the buffer region is titrated, followed by an evermore rapidly rising slope as the equivalence point is approached.

Now, let us answer the following questions.

1. What species are in the solution at the equivalence point?
2. What is the pH there? (From Figure 7.1 you can see that the pH is not 7.)

At the equivalence point, the acetic acid has all been converted to acetate. Therefore, the solution contains acetate ions, and the concentration of K^+ in the solution equals that of acetate.

To answer the second question, note that the solution is *exactly* the same as the one we would obtain if we added 0.010 mol of KOAc, potassium acetate, to 510 mL of water. When potassium acetate is added to water, the salt dissociates, and the acetate hydrolyzes. We can explain the pH at the equivalence point in the same way as we explain a conjugate base hydrolysis. It follows that the calculation of the pH at the equivalence point is done as for any other hydrolysis problem.

The hydrolysis calculation for acetate was done in Section 6.4 and is not reproduced here.

To understand the equivalence-point equilibrium, it is necessary to know some descriptive chemistry of acetate salts. Sodium acetate, potassium acetate, rubidium acetate, and cesium acetate all are completely dissociated in water. This means that we can treat the potassium ion as a spectator and consider only the acetate-ion hydrolysis.

EXAMPLE 7.2

What is the pH at the equivalence point of the titration of 500 mL of 0.020-M acetic acid by 1.00-M KOH in water? 10.00 mL of titrant is required.

Solution

The chemical equation that describes the hydrolysis is

$$OAc^- + H_2O = HOAc + OH^-$$

At 25°C, the hydrolysis constant describing the reaction is

$$K_h = \frac{K_w}{K_a} = \frac{1.00 \times 10^{-14}}{1.76 \times 10^{-5}} = 5.68 \times 10^{-10}$$

$$K_h = 5.68 \times 10^{-10} = \frac{[OH^-][HOAc]}{[OAc^-]}$$

We begin by making our usual substitutions for the concentrations of H^+, OAc^-, and HOAc. In other words, $[H^+] = [OAc^-] = X$ and $[HOAc] = 0.0196 - X$. (0.0196-M is 0.010 mol in 510 mL.) Then,

$$5.68 \times 10^{-10} = \frac{X^2}{(0.0196 - X)}$$

So,

$$X = [OH^-] = 3.33 \times 10^{-6}\text{ M}, \quad \text{and}$$

$$[H^+] = K_w/[OH^-] = 3.0 \times 10^{-9}\text{ M}$$

At the equivalence point, the pH is expected to be 8.52.

The pH beyond the Equivalence Point

We are able to describe the pH at any volume along the titration curve to be a consequence of the acid-base interaction of acetic acid and acetate. The acid-base equilibrium controls the solution pH. The concentrations of acetic acid and acetate ions are determined by the initial concentration of the acid and the amount of KOH that is added and reacts completely with HOAc to produce OAc^-.

The pH continues to change with added KOH in the region beyond the equivalence point. Can the acetic acid-acetate pair control the pH here? Let us reason through the answer. The acetic acid has already been converted to acetate; so the acid form can have no effect. It might be that the acetate itself has some influence. However, a 0.02-M acetate solution can contribute, at most, only the hydroxide due to hydrolysis. The pH at the equivalence point, pH 8.53, indicates that the 0.02-M acetate can generate only about 10^{-5}-M OH^-. This is three orders of magnitude too small to bring the solution into the range of the pH measured after the equivalence point, pH 12 ($[OH^-] = 10^{-2}$ M). We must conclude that the acetic acid-acetate equilibrium is no longer determining the pH of the solution.

This changeover of controlling equilibrium is an important idea. The equivalence point marks where there is a change in the chemical equilibrium that determines the pH.

Therefore, after the equivalence point, some other acid-base equilibrium must be controlling the pH changes when KOH is added. The only other possible acid in this simple potassium acetate solution is water itself. The chemical equilibrium that determines the pH beyond the equivalence point must be, then,

$$^*OH^- + HOH = H^*OH + OH^-$$

The asterisk is used to label the hydroxide added as titrant. (Of course, in the solution itself, we cannot actually distinguish the hydroxide added from that due to hydrolysis.) So after the equivalence point, the equilibrium constant describing the pH in relation to the hydroxide concentration is

$$K_w = [H^+][OH^-] = 1.0 \times 10^{-14} \text{ at } 25°\text{C}$$

To see if this idea is correct, let us see whether the calculated pH agrees with that measured at one point (volume) in the titration.

EXAMPLE 7.3

What pH do we expect the solution to have after 3.00 mL of 1.00-M KOH has been added beyond the equivalence point?

Solution

This addition will produce 513 mL of 5.83-mM KOH solution:

$$5.83 \times 10^{-3}\ \text{M} = \frac{3.00\ \text{mL} \cdot 1.00\ \text{M}}{513\ \text{mL}} = [\text{OH}^-]$$

With $K_w = [\text{OH}^-][\text{H}^+] = 10^{-14}$ at 25°C, then

$$1.0 \times 10^{-14} = [\text{H}^+](5.8 \times 10^{-3})$$

and

$$[\text{H}^+] = 1.7 \times 10^{-12}\ \text{M, or the pH} = 11.70$$

This result is in agreement with the experimental data shown in Table 7.1. This agreement gives us confidence that the water is, in fact, the next acid system that is titrated after acetic acid.

7.4 Alternate Titration Curves

You might ask the question, Why are the data plotted as pH versus the volume of KOH added, as in Figure 7.1? The answer is that it clearly shows the rapid change in pH around the equivalence point compared to the remainder of the titration curve. However, the data can be plotted other ways that accentuate the equivalence point. Three of these can be found in Figure 7.2.

When a titration is used as the assay method in an analysis, the main piece of information desired is the volume of titrant added to the end point. Recall, the equivalence point and end point are not necessarily the same, but we set up the titration so that the difference is negligible. The calculated quantity of analyte in a sample is only as precise as the volume measurement.

For acid-base titrations, say, acetic acid with KOH, usually you will see plots of pH vs. titrant volume such as Figure 7.1. For many other titrations, such as those covered in Chapter 9, plots of the type shown in Figure 7.2a are used. This graph helps us find the end point by indicating the volume at which excess titrant first appears. Figure 7.2b, the plot of the slope of the pH curve, certainly shows the end point clearly. If the ΔpH data are encoded into an electrical signal, it is relatively easy to build a circuit to detect the end point. This is the basis of operation for many automatic titrators.

7.5 Assigning Equilibria to Different Parts of Titration Curves

An interesting and important relationship showed up in the analysis of the experimental titration curve of acetic acid: The pH of the solution is determined by different equilibria in different parts of the titration curve, and the end point marks the point of change in control from the acetic acid/acetate equilibrium to the water/hydroxide equilibrium. (See Figure 7.3a.)

How does one tell which equilibrium controls the pH in each region? This is not easy to determine in complicated solutions with many acids and bases present. However, in the simpler acid-base solutions that you will become acquainted with here, the applicable equilibrium can be decided from the numerical values of the pK_a of the acids or the pK_b of the bases present. Just as for acetic acid, the buffer regions lie centered at the pK_a values.

Some general descriptive chemistry also simplifies the assignment of the controlling equilibrium. In water, the strongest acid that can exist is H_3O^+,

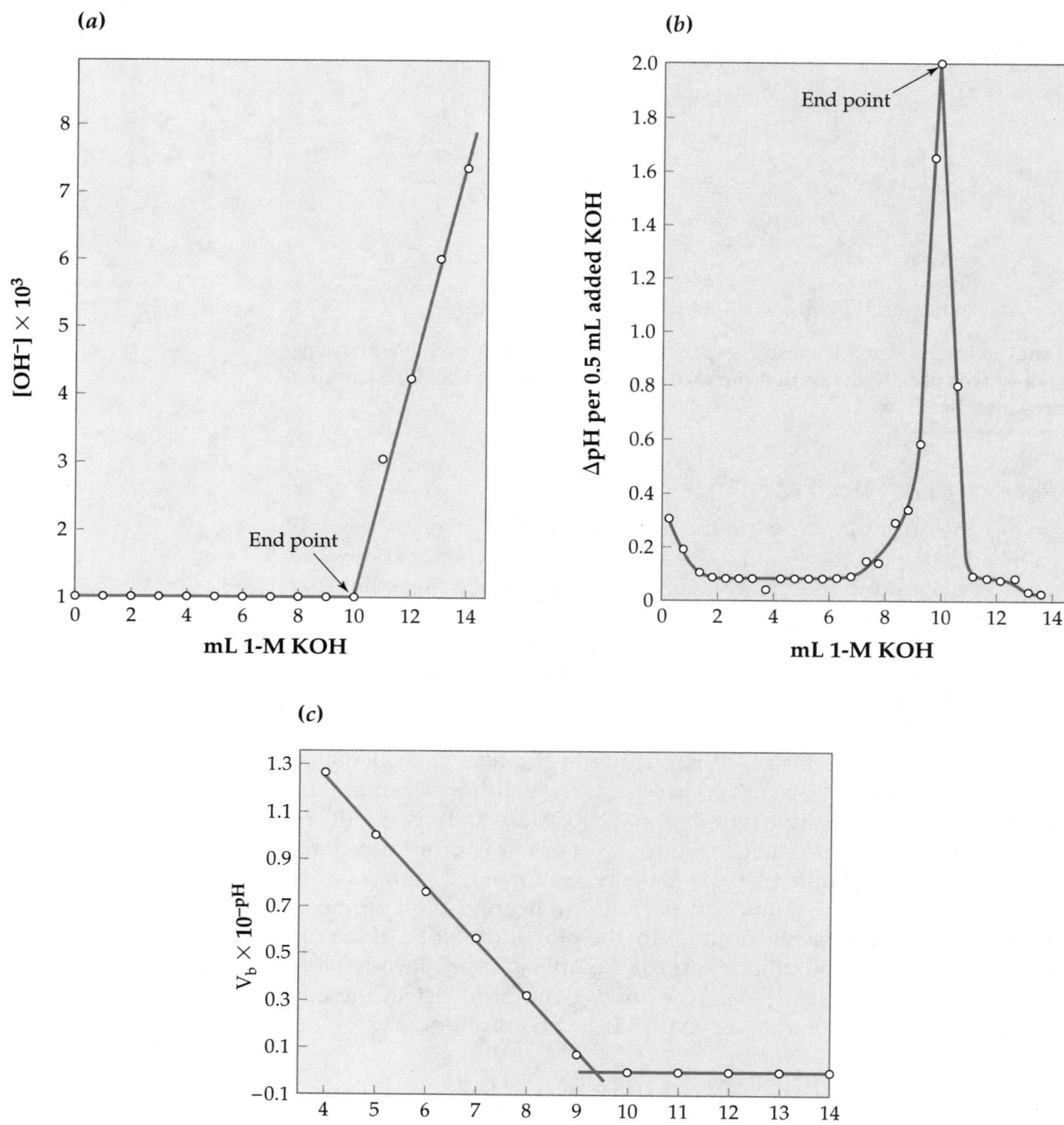

FIGURE 7.2 ▲

Results of the titration of 500 mL of 0.020-M acetic acid with 1.00-M KOH titrant plotted in three more ways.
The base titrant was added 1.00 mL at a time. The horizontal axis is the same in all three graphs: titrant volume. **(a)** Solution hydroxide concentration as a function of titrant added. The values used here are shown in Table 7.1. The plot of the concentration of OH- has an abrupt break at the end point. A general name for this graph might be titrant concentration vs. titrant volume. It is the most common type of plot for titrations that are not based on acid-base neutralizations. **(b)** Plot of the slope of the titration curve plotted in 7.1. The points of the graph were calculated by taking the pH differences over half-milliliter intervals along the titration curve. Their positions on the volume axis are plotted at the mid-points of each half-milliliter interval. A similar plot can be produced electronically if the titrant is added at a constant rate. The end point is then found at the peak value. This is one basis of operation of automatic titrators. **(c)** A **Gran Plot**, which shows the volume of base $\times$ 10^{-pH} vs. volume of base. This plot makes the typical titration curve become linear and allows the end point to be extrapolated from a few data points just before and just after the end point. Some auto titrators use this **linearizing** procedure to speed up titrations. Note that the result here is not accurate due to the relatively large increments of base added near the end point. The slope decreases as the end point is approached, and that region is not sampled by this data.

(*a*)

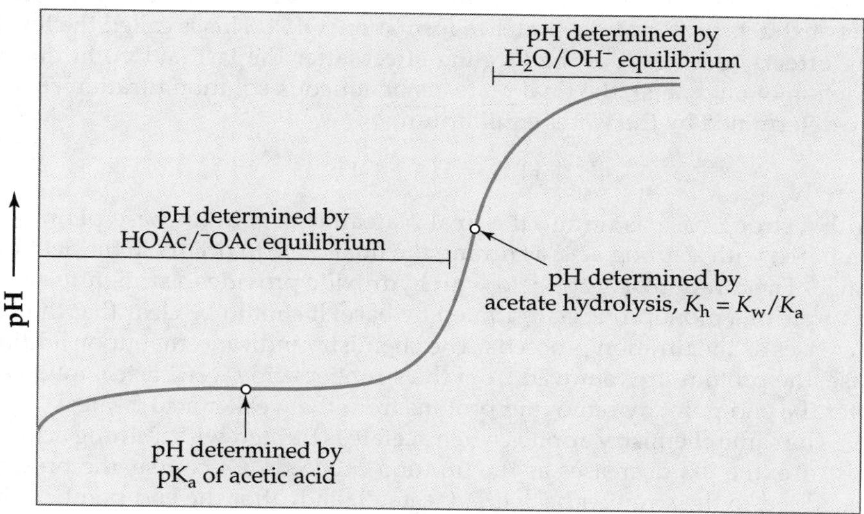

(*b*)

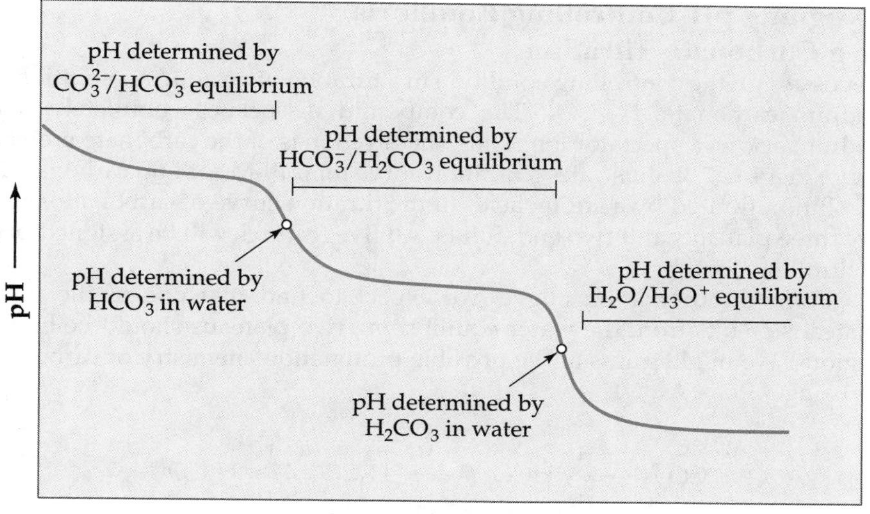

FIGURE 7.3 ▲

Comparison between a monoprotic neutralization titration (acetic acid) and a sequential diprotic titration (CO_3^{2-}).

The two sequential titrations are explained in the same manner as two individual monoprotic titrations. **(a)** Acetic acid titration. The end point occurs at the volume at which all the acetic acid has just been neutralized by the strong base. The pH is determined by hydrolysis of acetate. **(b)** CO_3^{2-} titration. The first end point is the volume at which all the CO_3^{2-} has just been neutralized by the strong acid. The pH at the end point is determined by HCO_3^- alone. The second end point is the volume at which all the HCO_3^- has just been neutralized by the strong acid. The pH at the end point is determined by H_2CO_3. The HCO_3^- neutralized in the second range may have two origins: (1) any HCO_3^- in the original sample, and (2) HCO_3^- from neutralization of CO_3^{2-} in the sample. The equivalents of titrant needed between the first and second end points equals the sum of the two sources of HCO_3^-.

and the strongest base that can exist is OH^-. Any acid more acidic than H_3O^+ donates its protons to water to form more H_3O^+. Any base more basic than OH^- extracts protons from water to form more OH^-. This is called the **leveling effect**. As a result of the leveling effect, after the last end point due to reaction of analyte(s), the final plateaus of aqueous-solution titration curves are determined by the water equilibrium,

$$H_2O \rightleftharpoons H^+ + OH^-$$

With a strong base as titrant, the final plateau occurs in the base pH range; similarly, with a strong acid as titrant, the final plateau occurs in the acid pH range. The titration of acetic acid with hydroxide provides a straightforward example of a monoprotic acid titrated by base. It should be clear that the pH increases as the titration proceeds. The chemistry indicates that upon adding base, the protons are removed from the stronger acid (acetic acid), followed after the end point by removing protons from the weaker acid (water).

The same chemistry applies when acetate is titrated with a strong acid. In this case, the pH decreases as the titration proceeds. Of course, the protons are added to the stronger base first (acetate), and, after the end point, to the weaker base (water). The results are only slightly different with a titration of a polyprotic acid, as shown next.

Assigning pH-Controlling Equilibria to a Carbonate Titration

Let us assign the controlling equilibria in a titration of an aqueous solution of sodium carbonate, Na_2CO_3. The compound dissociates completely, and sodium acts as a spectator ion. Thus, the titration is of the carbonate **moiety**, CO_3^{2-}. Figure 7.3b illustrates a titration curve for 0.10-M sodium carbonate in a solution titrated by a strong acid. In the titration curve of carbonate, there are three plateaus and two end points. All five features will be assigned pH-controlling equilibria.

From the discussion above, we expect to find that one of the three plateaus results from the water equilibrium; two plateaus should be buffer regions. We might guess at the possible protonation chemistry of carbonate as being

$$CO_3^{2-} \underset{-\,H^+}{\overset{+\,H^+}{\rightleftharpoons}} HCO_3^- \underset{-\,H^+}{\overset{+\,H^+}{\rightleftharpoons}} H_2CO_3 \underset{-\,H^+}{\overset{+\,H^+}{\rightleftharpoons}} H_3CO_3^+$$

However, the descriptive chemistry of carbonate reveals that H_2CO_3 is a weaker base than H_2O so the last step on the right need not be included to describe the titration. We are left with

$$CO_3^{2-} \underset{-\,H^+}{\overset{+\,H^+}{\rightleftharpoons}} HCO_3^- \underset{-\,H^+}{\overset{+\,H^+}{\rightleftharpoons}} H_2CO_3 \qquad \textbf{(7-4)}$$

Since this is a titration of a base with a strong acid, we expect the final plateau (the one with the greatest volume of titrant added) is due to the water equilibrium: $H_2O + H^+ \rightleftharpoons H_3O^+$. In other words, the pH after the second end point is determined by the concentration of protons from the titrant. This conclusion is illustrated in Figure 7.3b.

How do we make the assignments of the other two plateaus, the initial and middle ones? The two carbonate pH-determining equilibria are

$$CO_3^{2-} + H^+ \rightleftharpoons HCO_3^-$$

and

$$HCO_3^- + H^+ \rightleftharpoons H_2CO_3$$

To assign the controlling equilibria, it is necessary only to determine the order of protonation of CO_3^{2-} and HCO_3^-. The CO_3^{2-}-HCO_3^- equilibrium has the highest pK; pK_a = 10.22. Therefore, the initial plateau—the one at the highest pH—must be controlled by the carbonate-bicarbonate equilibrium. The middle plateau is then due to the HCO_3^--H_2CO_3 equilibrium; pK_a = 6.46. These are the assignments shown in the figure.

Assigning End Point Equilibria

The equilibria that control the pH of the two end points of the carbonate titration are relatively simple to discover once the equilibria controlling the plateaus have been assigned. The method is the same as in the example of acetic acid. All you need ask is, What are the major species in solution at each end point?

At the first end point (left-most in Figure 7.3b), all the carbonate has been converted to bicarbonate, HCO_3^-, so the equilibrium is the same as if a dissociated bicarbonate salt were added to water. Thus the pH is determined by the hydrolysis of bicarbonate,

$$HCO_3^- + H_2O \rightleftharpoons H_2CO_3 + OH^-; \quad HCO_3^- \rightleftharpoons H^+ + CO_3^{2-}$$

At the second end point, all the bicarbonate has been converted to carbonic acid, H_2CO_3. So the equilibrium is the same as if carbonic acid were added to water, such as can be done by bubbling gaseous CO_2 through the water. The controlling equilibrium is then

$$H_2CO_3 + H_2O \rightleftharpoons H_3O^+ + HCO_3^-$$

In the same manner, any other diprotic or any polyprotic acid titration or base titration can be analyzed. The calculations for any individual region of the curve are done as they are for monoprotic acids. All that is necessary is to separate the appropriate or applicable equilibrium from those that are possible. In this way, a complicated system can be analyzed logically—by splitting it into a number of simpler problems involving only one or two equilibria.

One final reminder: These calculations are used to match the experimental data. If the results agree with the experiment, then we have some confidence that we understand the underlying chemistry. After all, the chemistry is what we seek to understand first. Once the chemistry is clear, the technique can be used for analysis with confidence, and problems that arise in the analysis can be approached intelligently.

7.6 Volumetric Calculations for Titrations

In contrast to the moderately involved calculations required to explain the chemical changes during titrations, the calculation for obtaining sample content from the end-point volume is usually quite straightforward. There are three parts to such a calculation:

1. Relate the number of moles of titrant to the number of moles of material titrated.
2. Find the total mass or volume of the analyte.
3. Convert the result to the form desired; for example, milligrams, volume percent, or ppm.

The following are a few examples of such calculations.

EXAMPLE 7.4

A pound of dried, white powder is given to you, a sample from a carload of salicylic acid (formula weight 138.12, $pK_{a1} = 2.97$, $pK_{a2} = 13.40$). It is to be acetylated to make aspirin. Find out how pure (in % w/w) the material is if none of the impurities are acidic or basic. Let us consider only one replicate that weighs 0.4208 g. It was titrated with 21.92 mL of a 0.1354-M KOH solution to its first end point.

Solution

To solve the problem requires knowing the stoichiometry of the titration reaction. To the first end point, the reaction that occurs can be written

$$OH^- + H_2sal \rightleftharpoons H_2O + Hsal^-$$

The abbreviation sal^{2-} is used for salicylate. From the equation, the stoichiometry is found; each mole of base added reacts with one mole of salicylic acid. This is the first of the three parts needed for the calculation.

For the second part, we must find the total mass of salicylic acid that was titrated. This can be found by multiplying the number of moles times the molecular mass. It can be done in two stages. Initially,

$$\text{moles salicylic acid} = 1 \times \text{moles KOH} = 1 \cdot 0.1354\text{ M} \cdot 0.02192\text{ L}$$

$$= 2.968 \times 10^{-3}\text{ moles}$$

The factor 1 reflects the stoichiometry found in the first step. Then,

$$\text{mass of salicylic acid} = \text{moles salicylic acid} \times 138.12 = 0.4099\text{ g}$$

The third stage of the calculation is to convert this number into the desired form. Asking for the percentage purity is the same as asking us to calculate the percentage by weight of the sample that is salicylic acid. Thus,

$$\%\text{ w/w salicylic acid} = \frac{0.4099\text{ g}}{0.4208\text{ g}} \cdot 100 = 97.4\%$$

Notice that at no time was it necessary to know the volume of solution in which the salicylic acid was dissolved.

The calculations of content become even easier if the sample is a liquid. Say it is an acid, and you wish to find its normality. It will be titrated with a base. At the equivalence point of the titration,

$$\text{equivalents of base used} = \text{equivalents of acid (as } H^+\text{) reacted}$$

But the rest of the stages of a volumetric calculation need not be done because

$$\text{equivalents of base used} = \text{base normality} \times \text{base volume}$$

and

$$\text{equivalents of acid reacted} = \text{acid normality} \times \text{acid volume}$$

As a result, we can write the important, simple, general equation

$$\text{titrant normality} \times \text{volume} = \text{sample normality} \times \text{volume} \qquad \textbf{(7-5a)}$$

or

$$N_1 \times V_1 = N_2 \times V_2 \qquad \textbf{(7-5b)}$$

If it is desired to find the molarity of an acid or base, its formula and chemistry must be known so that the molarity can be related to the normality.

EXAMPLE 7.5

Assume that you have 500 L of an aqueous sulfuric acid solution for which you wish to find the molarity. You titrate a 50.00-mL aliquot. It requires 42.20 mL of 0.1354 M KOH to titrate both protons. What is the molarity of the sulfuric acid solution?

Solution

Since KOH can donate one hydroxide,

$$\text{normality}_{KOH} = \text{molarity}_{KOH}$$

Thus, letting $N_{H_2SO_4}$ = normality of sulfuric acid, from Equation 7-6b

$$0.1354 \text{ N} \cdot 42.20 \text{ mL} = N_{H_2SO_4} \cdot 50.00 \text{ mL}$$

$$N_{H_2SO_4} = 0.1143 \text{ N}$$

Then, since

$$2 \cdot \text{molarity}_{H_2SO_4} = \text{normality}_{H_2SO_4}$$

$$M_{H_2SO_4} = \frac{0.1143 \text{ N}}{2} = 0.0571 \text{ M}$$

7.7 pK_a Values and the Equivalence Point

The titration curve for acetic acid as titrated by the strong base KOH was explained using the equilibrium mass-action expressions and the stoichiometry of the base-acid interaction. The titration curve consists of two plateaus with a pH jump in between. The first plateau is centered at a volume halfway to the end point where the pH equals the pK_a of the acid. The pH values of the second plateau are determined by the equilibrium of water itself, pK_w = 14.0. (When a monoprotic acid is titrated, the second plateau always is that of the water equilibrium.) It follows that the lower the pK_a of the acid being titrated, the larger the jump of pH at the end point to the water plateau. This is illustrated in Figure 7.4. Since a jump in pH is needed to determine the end point, there is an upper limit to the pK_a of an acid that can be titrated in water.

The same ideas apply to polyprotic titrations such as that for carbonate seen in Figure 7.5. Note that the spread between pK_a values and the sharpness of each break are closely related. In general, a larger jump allows a more precise measurement of the end point as well as ensuring a minimum titration error: that is, minimizing the difference between the end point and the equivalence point.

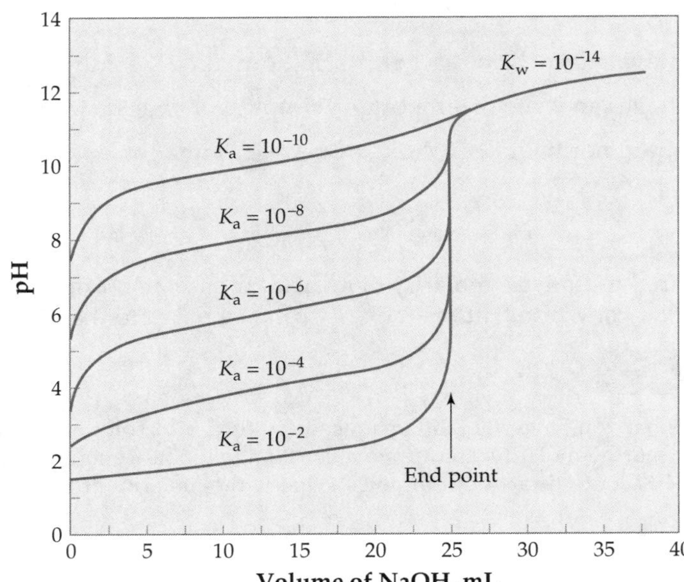

FIGURE 7.4 ▲
Calculated titration curves for weak acids with differing pK_a values titrated with an aqueous strong base.

Each curve is calculated as if 25.00 mL of 0.1000-N acid is titrated with 0.1000-N strong base. Since the curve in the region to the right of the end point is determined by the same equilibrium in all cases, the pH jump at the end point will be smaller when there are smaller differences between the pK_a and the pH range of the water-equilibrium plateau. The precision will be reduced. The pH jump for the weaker acids can be enhanced by using another solvent or solvent mixture.

Titrations in Different Solvents

The leveling effect of water sets the pH limits of titration, as you have seen. However, a wider range of pH is available by titrating in a different solvent. For example, as you have seen, during a titration of a weak acid by, say, hydroxide, the plateau level centers around the acid's pK_a. If the acid is weak, the jump in pH at the end point can be quite small; the end point is indistinct. However, by using a solvent or solvent mixture that is a weaker acid than water (higher pK_a), the second, solvent-controlled plateau can be raised. A clearer end point can be obtained since the pH jump is made larger.

Also, nonaqueous solvents may be considered simply for solubility reasons. For example, an analyte can be more soluble or, if desired, some interferent can be less soluble. The book by Huber listed at the end of this chapter presents a comprehensive view of that field.

7.8 Indicators

Titration methods can be useful in seeking answers to questions such as, What is the concentration of acetic acid in this sample of white vinegar?, or, How much iodine is there in this pharmaceutical preparation?, or, How many times its own weight in stomach acid will this little white pill neutralize? In other words, usually we are interested only in the volume of standardized titrant that is required to reach the end point. From this volume and the volume or weight of the sample, the content of the analyte sought can be

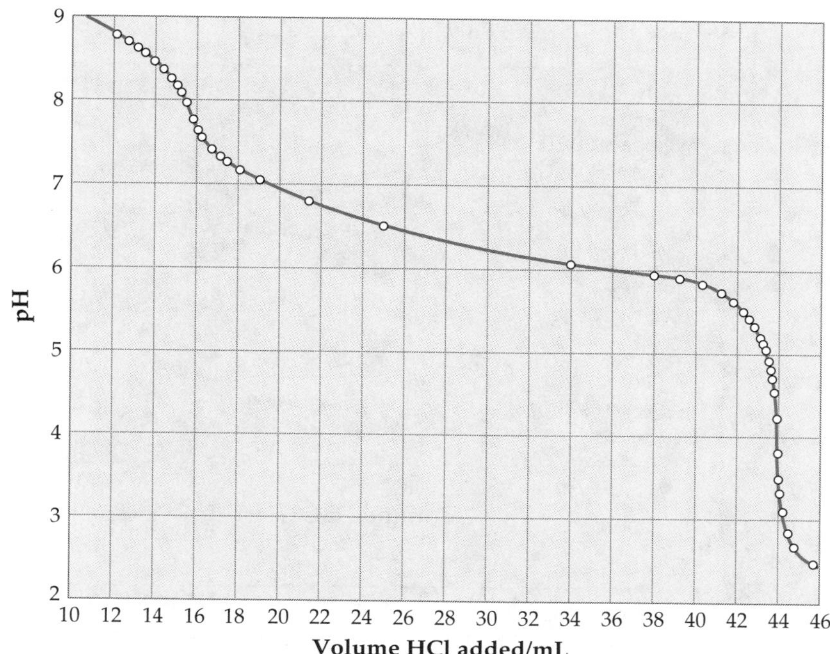

FIGURE 7.5 ▲
Part of a titration curve for a carbonate/bicarbonate mixture titrated with standardized HCl.

The early points are not shown, but both end points are included. Notice that where the pH jump is smaller, the end-point slope is shallower. (The end point is not as sharp.) The number of data points is small in the buffer region and greater in the end-point region. This saved time since the volumes at the end points comprise the desired data.

calculated. With this viewpoint, the pH values at other points along the titration curve are irrelevant.

As we have seen, near the end point of acid-base titrations there is a rapid change in the pH of the solution as the titrant reacts with the last remaining part of the titrated species. At the end point, a different chemical equilibrium begins to determine the pH. In the acetic acid titration of Section 7.3, a pH meter measured the jump in pH at the end point. However, this jump could also have been detected by a change in the color of a dye added to the solution. Such a dye is called, reasonably, a **colorimetric indicator**.

The way these indicators work can be explained as follows. Let us consider a titration of a solution containing two monoprotic acids with different pK_a values. The more acidic component will be a strong acid; It is totally dissociated in the water. The less acidic material has a pK_a of 7.6. Let us do this imaginary titration on 100 mL of solution containing 0.0100 moles of each acid. We shall titrate the solution with 0.200-M KOH. The titration curve will appear as shown in Figure 7.6a. The equilibria that determine the pH values of the three plateau regions are labeled on the figure.

Next, let us do a second imaginary titration. Everything is the same except that we reduce the concentration of the weak acid (pK_a 7.6). Thus, we now titrate 100 mL of solution that contains 0.100 M strong acid and 0.020 M weak acid. The titration curve will now appear as shown in Figure 7.6b. The

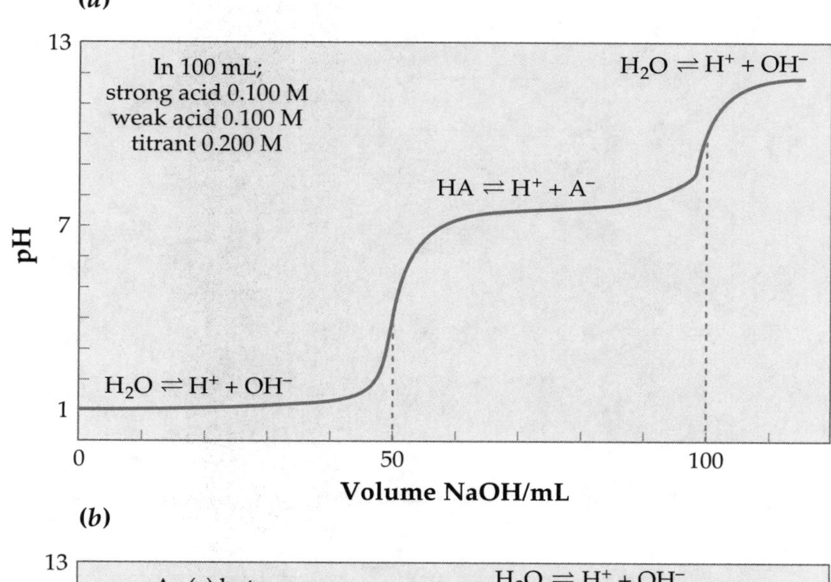

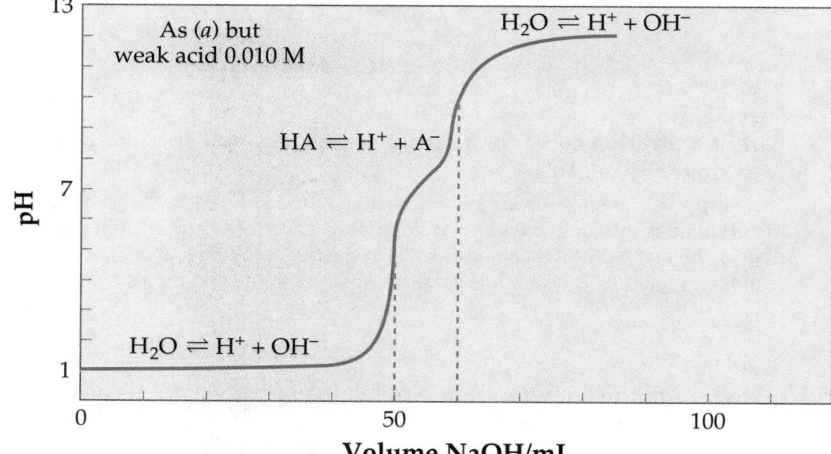

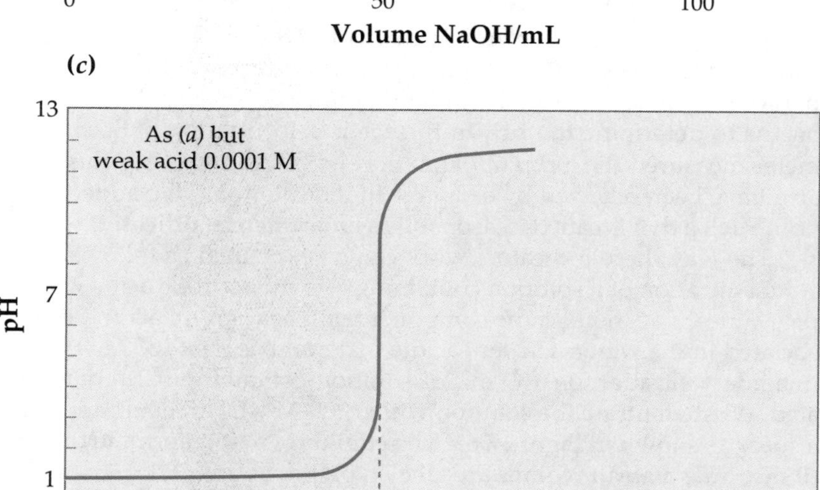

FIGURE 7.6 ▶
Titration of a solution composed of a mixture of a strong acid and a weak acid: some properties of neutralization indicators.

The middle plateau of the curve is due to titration of the weak acid. As the quantity of the weak acid is reduced (acid in Figure a > b > c), its influence on the titration curve is lessened. If the weak acid is highly colored and exhibits a color change with pH, then a dilute solution can be used as a colorimetric indicator to detect the end point of the titration of the strong acid. To ensure optimal precision in the assay, the pK_a of a neutralization titration indicator should have a value near the pH of the solution at the end point of the strong-acid titration.

ANALYTICAL CHEMISTRY IN ACTION
Just When You Thought It Was Safe to Go Back in the Water

Titrations have been declining in use for some time as they are supplanted by instrumental methods. Nevertheless, in some instances acid-base titrations still fulfill an important analytical need and do so efficiently. For example, titration can be used to determine the amounts both of bicarbonate and volatile fatty acids present during wastewater treatment. The titration gives results that are comparable to instrumental methods. The effort is also comparable, and the equipment is far less expensive.

Anaerobic bacteria are held in reactors to treat the wastewater and digest sludges at the last stage of sewage treatment. There, the remaining dissolved organic matter is converted by fermentation to volatile fatty acids. Then, the acids are broken down further into CO_2 and methane. However, the unwanted fatty acids can build up because of heavy metal contamination or a sharp increase in water feed rate. If the stress is removed without delay, the system restabilizes. But, if the volatile acids build up so far that all the bicarbonate present is neutralized, the pH can drop, and, after a short time, the system will fail, and the formation of the CO_2 and methane stops.

The onset of this failure cannot be detected by monitoring the pH alone, since the bicarbonate buffer dominates the pH control of the solution. However, by titration, both the bicarbonate and volatile acids can be determined through two separate end points. Each sample is titrated with standard sulfuric acid first to pH 5.2 and then from pH 5.1 to 3.5. Both the carbonate and volatile fatty acid levels can be monitored simply by using the two titration volumes together with the equilibrium constant for carbonic acid (under the digester's ionic strength conditions) with the combined dissociation constant for the C2 to C6 fatty acids.

[Ref: Anderson, G. K., Yang, G. 1992. "Determination of bicarbonate and total volatile acid concentration in anaerobic digesters using a simple titration." *Water Environment Research* 64(1): 53–59.] ■

second plateau becomes much narrower; the volume of titrant needed to neutralize the weak acid is smaller by a factor of 5. Nevertheless, the same neutralization reaction is occurring as it did when the concentration of the weak acid was higher, so the same pH range is being traversed from beginning to end.

Finally, let the weak acid be another 200 times more dilute, 0.0001 M, while keeping the strong acid the same as before (0.100 M). In this case, the titration curve will appear as in Figure 7.6c. Now we cannot see the effect of the weak acid on the pH of the solution although we know that there is a second end point between pH 7 and 8, and that the weak acid must be proceeding through its usual reaction. But the plateau and end-point pH change are occurring over a small relative volume addition: only 1/500 of that needed to titrate the strong acid to its end point. In other words, while 50.0 mL of the KOH are needed to neutralize the strong acid, all the weak acid present is titrated to its end point with only 0.05 mL.

The disappearance of the plateau due to the weak acid illustrates a limitation of the titration technique; only a relatively narrow range of concentrations can be titrated at one time. However, we can use this limitation beneficially if the weak acid is a highly colored material that changes color from its acid to its base form. First, because the color is intense, it can be seen even though the concentration is small. Second, the color change occurs without changing the titration of the stronger acid significantly. In the above example, the relative error was 0.05 mL in 50 mL, or 0.1%. Thus, we can use a highly colored, weak acid as a colorimetric indicator of the end point of the titration of the strong acid.

The above example is consistent with a solution in which the strong acid is HCl and the weak acid, which we will now call an indicator, is phenol red (or phenolsulfonphthalein, $pK_a = 7.6$). In its *acid form*, abbreviated HA or HIn, it is yellow. In its *base form*, A^- or In^-, it is red. As seen from the chemical structures in Figure 7.7, this indicator is closely related to both phenolphthalein and bromocresol green. Phenolphthalein is commonly used in manual volumetric acid-base titrations. It is colorless in acid and reddish-purple in base.

The Properties of Indicators

For an indicator to work as a marker of the end point of a neutralization titration of an acid or a base, ideally its pK_a (often more than one—see below) should be nearly the same as the end-point pH. The reason is as follows. As illustrated in Figure 7.7, the color change is due to a change in the chemical form of the indicator, from its acid (protonated) form to its basic form. The chemical transformation for which a color change would be observable is from about 90% in one form to 90% in the other. (An obvious color change can be seen by eye only when a major proportion of the indicator changes color.) This means that the ratio of acid-to-base forms of the indicator must change from approximately

$$\frac{[\text{HIn}]}{[\text{In}^-]} = 9$$

to

$$\frac{[\text{HIn}]}{[\text{In}^-]} = 0.11$$

or vice versa. Compare these ratios to the mass-action expression

$$K_a = \frac{[\text{H}^+][\text{In}^-]}{[\text{HIn}]}$$

Since K_a is a constant, to change the indication from 90% in one form to 90% in the other, the hydrogen-ion concentration must change by about a factor of 80. Thus, the color change will be observable over a range of pH values, not just at a single, specific pH. This range is called the **indicator range**, and for *all* color-changing indicators is at least 1.5 to 2 pH units.

Indicator ranges and colors are shown in Figure EP.1 on the inside front cover.

In addition, to give the best, most accurate results we want this pH change to occur over the minimum volume of added titrant. As you can see, the minimum volume will occur when the color change occurs at the most steeply rising part of the titration curve. When an incorrect indicator is used, the results tend to be significantly less precise, as illustrated in Figure 7.8. The difference between the end point and equivalence point is called the indicator error.

As a result of changing their form over a pH range, indicators are listed by their useful pH ranges for a color change rather than by their pK_a values. A second reason for listing a range is that the changes often involve two or more protons on an indicator molecule. (For example, see phenolphthalein in Figure 7.7.) Thus, a listing of pK_a values is not always easy to interpret in terms of deciding on a useful range. Again, listing an operational range is more useful. A third reason is that the indicators' color changes are all differ-

Methyl orange
range pH 3.1–4.4
red–yellow

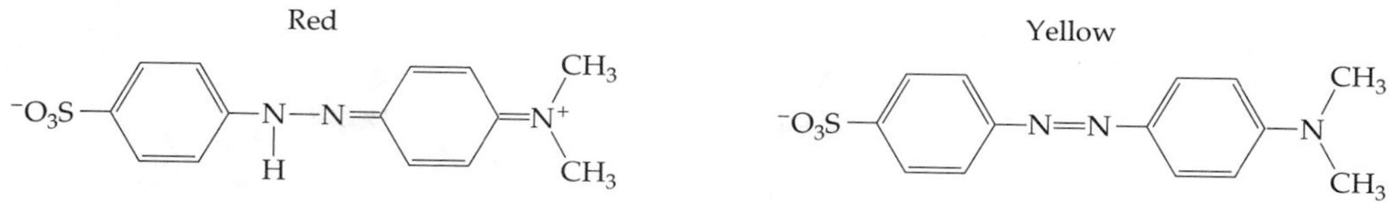

Red

Yellow

Bromocresol green
range pH 3.8–5.4
yellow–blue-green

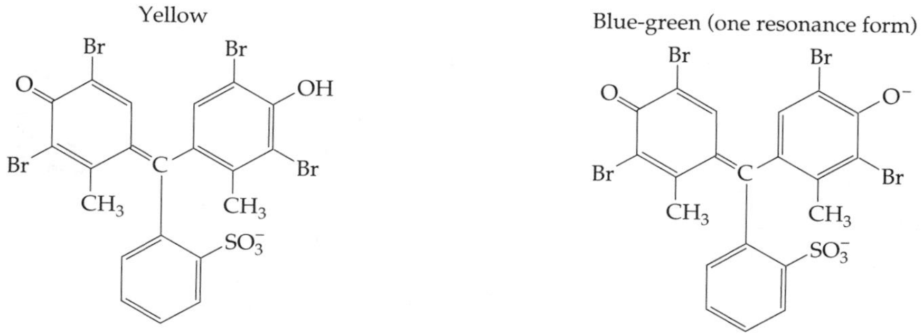

Yellow

Blue-green (one resonance form)

Methyl red
range pH 4.2–6.2
red–yellow

Red

Yellow

Phenol red
range pH 6.8–8.4
yellow–red

Yellow

Red (one resonance form)

Phenolpthalein
range pH 8.0–9.6
colorless–pink

Colorless (hydrated form)

Pink (one resonance form)

FIGURE 7.7 ▲

Chemical structures of five commonly used neutralization colorimetric indicators.

Each of these indicators has a different useful pH range. More data appear in Appendix VII and in Figure EP.1 on the inside front cover.

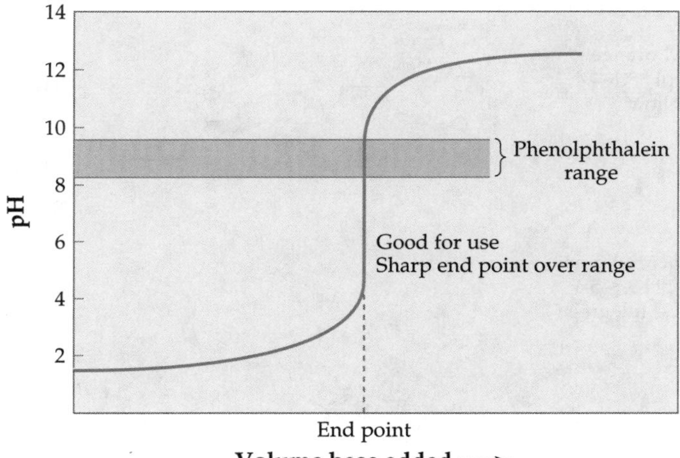

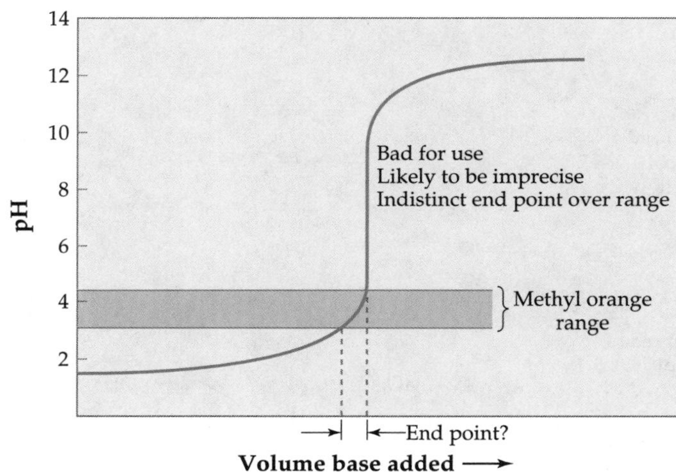

FIGURE 7.8 ▲
The use of a good and an inappropriate indicator to determine an end point.

The titration curve results from an acid with a pK_a around 2. The shaded regions illustrate the useful ranges of phenolphthalein and methyl orange. Phenolphthalein is a good indicator to use because the entire useful range is encompassed by the pH jump at the end point. On the other hand, the end point for methyl orange will be far less certain because the color change occurs over a relatively large volume of added base.

ent. For instance, it may be easier to see a change from colorless to pink or from yellow to blue than from purple to green. So, a useful pH range is a better guide, although some caution may still be needed because individuals differ in their abilities to perceive specific color changes.

Titrations with two end points require two different indicators if both end points are needed for the analysis. The reason can be seen by considering the titration curve of carbonate shown in Figure 7.5. If the end-point detection is done with colorineteric indicators, one indicator is required to change through the pH range of the first end point (around pH 9). A second is required to change around pH 4. In addition, the colors of the two indicators cannot interfere; otherwise, two separate titrations will be required (one to the first end point with the first indicator and another to the second end point with only the second indicator present). It is best to have a colored-to-colorless transition for the first and a colorless-to-colored transition for the second. Then, both indicators can be present together, and both end points can be detected. As an example, for the titration in Figure 7.5, phenolphthalein can be used at the first end point and methyl orange for the second one.

Finally, note that there is nothing special about the equilibrium properties of a colorimetric indicator. Its pK_a value(s) will depend on the ionic strength of the solution and the temperature just as any other equilibrium, and any changes in solvent will change the pK_a value(s). As a result, the useful range in water at temperatures around 20°C will not necessarily be the useful range in other solvents or under more extreme conditions.

Suggestions for Further Reading

SERJEANT, E. P. 1984. *Potentiometry and Potentiometric Titrations.* New York: Wiley.

An advanced textbook on potentiometric titrations—not much about how to do the work but loaded with cautions about the techniques.

JORDAN, J., ed. 1974. *New Developments in Titrimetry.* New York: Marcel Dekker.

WAGNER, W., HULL, C. J. 1971. *Inorganic Titrimetric Analysis: Contemporary Methods.* New York: Marcel Dekker.

Two books in a series on modern titrimetry.

KUCHARSKY, J., SAFARIK, L. 1965. *Titrations in Non-Aqueous Solvents.* Amsterdam: Elsevier.

HUBER, W. 1967. *Titrations in Nonaqueous Solvents.* New York: Academic Press.

Two books with similar, comprehensive coverage and a similar outlook on the principles and practice of non-aqueous titrations. More recent work has not apparently been collected in book form.

BARNARD, A.J., JR., et al. 1963. In *Handbook of Analytical Chemistry,* L. Meites, ed. Sec. 3. New York: McGraw-Hill.

A useful, practical guide to various titrimetric techniques, reagents, and indicators.

Numerous authors. 1974. In *Treatise on Analytical Chemistry.* I. M. KOLTHOFF and P. J. ELVING, eds. Pt. 1, Vol. 11, Sec. I–2. New York: Wiley-Interscience. pp. 6845–7225.

A comprehensive set of chapters on acid-base titration in aqueous and nonaqueous solutions, plus complexometric, precipitation, and redox titrations.

RAMETTE, R. W. 1981. *Chemical Equilibrium and Analysis.* Reading, MA: Addison-Wesley.

A carefully developed textbook approach to aspects of equilibria with titrations in some chapters.

FREISER, H. 1970. "Acid-Base Reaction Parameters." *J. Chemical Education* 47:809–811.

Only three pages, but worth reading for the discussion about equilibrium calculations. Written at an intermediate level.

Concept Review

1. What are the five basic requirements for volumetric titration analyses?

2. If a sample containing equimolar concentrations of three acids HA, HB, and HC (pK_as of 3.5, 10.0, and 6.7, respectively) are titrated with NaOH, which acid would be deprotonated first?

3. Sketch the titration curve that would result from the titration in Question 2, indicating the relative equivalence volumes on the x-axis and the pHs of the plateaus on the y-axis.

4. You wish to titrate a solution of HA, which has a pK_a of 12.00. Why might you want to use an nonaqueous solvent for the titration?

5. The endpoints during the titration of a diprotic acid with 0.1 M NaOH occur at pHs of 4.3 and 8.4. Describe the color changes that would occur if a mixed indicator containing bromocresol green and phenolphthalein were used for the titration.

6. Figure 7.2b shows a derivative plot (change in pH/change in volume added) for the titration of an acid with a base. Sketch the general appearance of such a plot for the titration of a base with acid.

Exercises

A table of K_a values appears in Appendix II.

7.1 Calculate the pH you would expect to observe in a 0.200 M acetic acid solution in water at 25°C.

7.2 Calculate the *molarity* of the unknown for each the following titrations:
(a) 50.00 mL of unknown HCl which required 12.25 mL of 0.05050 N NaOH
(b) 50.00 mL of unknown CH3COOH which required 14.23 mL of 0.1002 N KOH
(c) 50.00 mL of unknown phthalic acid which required 10.38 mL of 0.1075 N M NaOH to reach its second end point

■ 7.3 Calculate and plot the expected titration curve for 100.0 mL of 0.0150-M monoprotic acid with $pK_a = 3.5$ for titration with KOH, 0.0100 M.

7.4 As calculated, the pH of a 0.0200 M acetic acid solution is 3.23. However, in an experiment that was run on 500.0 mL of solution, the pH was measured and found to be 3.5. One hypothesis to explain the discrepancy is that there was some small amount of base on the glassware prior to the addition of acetic acid. If the base is NaOH, how many g would be needed in the 500 mL to bring the solution from pH 3.23 to 3.5?

7.5 For 500.0 mL of a 0.0200 M acetic acid solution,
(a) How much 0.0200 M KOH must be added to change the solution from pH 4.00 to pH 5.00? Assume that there is no change of volume. Assume that the activity of H^+ equals its concentration.
(b) Recalculate part (a) but include changes in the volume. What do you expect the total volume of the solution to be at pH 4.00 and at pH 5.00?

7.6 A 50.00-mL aliquot of 0.1000 M HCl is titrated with 0.1000 M NaOH. T = 25°C, in water.
(a) Which equilibrium determines the pH values of the solution before the end point, at the end point, and after the end point?
(b) What is the volume of the solution and its pH at the midpoint between the beginning of the titration and the equivalence point?
(c) What is the volume of the solution and its pH at the equivalence point?
(d) Repeat parts (b) and (c) for a titration of 50.00 mL of 1.00 mM HCl with 1.00 mM NaOH.

7.7 A back titration is being used to titrate a sample of a polyelectrolyte. A 1.056 g sample of the polyelectrolyte is suspended in 100.0 mL of 0.1021 N HCl and stirred for 10 minutes. The solution is filtered and 25.00 mL of the filtrate is transferred to a wide-mouth erlenmeyer flask. It required 30.45 mL of 0.04857 M NaOH to reach the equivalence point.
(a) How many mmoles of HCl were added to the polyelectrolyte?
(b) How many mmoles of NaOH were required to titrate the excess HCl?
(c) Based on your answers to parts (a) and (b), characterize the polyelectrolyte in mEq/g (mEq = milliequivalents).

7.8 What are the pH, pOH, and concentrations of CO_3^{2-} and HCO_3^- after 1.000 mL of 0.1500 M HCl is added to 100.0 mL of 0.0100-M sodium carbonate solution? Ignore the added 1% volume. Assume that $a_{H^+} = [H^+]$ and T = 25°C.

7.9 What do you expect to be the volumes of added acid at the equivalence points in a titration of 2.000 mL of a solution that is 0.4000 M in Na_2CO_3 and 0.3000 M in $NaHCO_3^-$ and is titrated with 0.1024 M HCl? Ignore activity effects. Do the calculation for two cases:
(a) The original sample is titrated directly.
(b) The original sample is titrated after having added it to 98.00 mL of water. Does the added water make a difference?

7.10 Determine the species present and their concentrations in solution at the following points in the titration of 500.0 mL of 0.020 M acetic acid with 1.000 M KOH. Ignore the change in volume due to the titrant addition.
(a) The initial acetic acid solution
(b) Where the pH = pK_a of acetic acid
(c) At the end point
(d) At a point 5.00 mL past the end point
(e) By what factor does the ionic strength increase between the beginning of the titration and 5 mL past the end point?

7.11 Calculate the ionic strength for the following solutions.
(a) A solution of 0.0200 M acetic acid in water
(b) The same solution at the point pK_a = pH after the solution is brought to that point by adding 1.000 M KOH. Ignore volume changes.
(c) The following formula gives the activity coefficients for potassium acetate versus ionic strength (I). Assume molality and molarity are equal.

$$\log \gamma_{\pm} = \frac{-0.82\sqrt{\mu}}{1 + \sqrt{\mu}} + 0.33\mu$$

What is the value of $\gamma_{\pm}$ for the solutions of parts (a) and (b)?

7.12 You have available to you the following indicators: phenolphthalein, methyl red, methyl orange, and bromocresol green. Assume that for the calculation any indicator present is at a low enough concentration that the actual titration of the indicator can be neglected. Assume that the indicator ranges are applicable at their exact values. 50.00 mL of 0.1000 M HCl is titrated with 0.1000 M NaOH.
(a) Calculate the expected titration curve in the region of the equivalence point: The pH at the equivalence point and at the equivalence point ±0.05-mL, ±0.10-mL, and ±0.50-mL NaOH should be sufficient.

(b) Which of the indicators would be satisfactory to determine the end point to moderate accuracy (within 1% volume)?

(c) Which indicators can be used for the titration if 0.1% accuracy is desired?

7.13 The seldom-encountered acid, nonesuch acid, abbreviated H_4Ns is one of the (exceedingly) rare tetraprotic acids. It has the following pK_a values:

H_4Ns, 2.0; H_3Ns^-, 4.5; H_2Ns^{2-}, 7.5; HNs^{3-}, 10.0

A 50.00-mL sample for a test titration is prepared from 10.00 mL each of five solutions: each of these is 0.0500 M in, respectively, H_4Ns, NaH_3Ns, Na_2H_2Ns, Na_3HNs, and Na_4Ns.

(a) How many equivalence points will there be in the titration?

(b) What volumes of 0.1000-M NaOH are required to reach these equivalence points for the sample?

■ 7.14 The following data were obtained by titrating 4.00 mL of an unknown carbonate/bicarbonate mixture with 0.1455 M HCl. What is the concentration of CO_3^{2-} and HCO_3^- in the original mixture?

mL HCl	pH	mL HCl	pH	mL HCl	pH
0.00	9.75	16.91	7.45	43.50	4.98
2.02	9.61	17.19	7.39	43.57	4.98
4.00	9.50	17.44	7.30	43.68	4.74
7.00	9.28	17.70	7.24	43.71	4.70
9.96	9.03	18.20	7.16	43.83	4.38
12.47	8.75	19.10	7.00	43.90	4.05
12.85	8.70	21.45	6.69	43.95	3.70
13.40	8.60	25.02	6.48	44.00	3.48
13.65	8.55	29.32	6.21	44.05	3.30
13.92	8.50	34.00	5.97	44.11	3.18
14.34	8.40	38.12	5.80	44.21	3.00
14.80	8.30	39.27	5.75	44.44	2.88
15.05	8.20	40.49	5.60	44.80	2.57
15.29	8.12	41.49	5.60	45.05	2.45
15.50	8.01	42.01	5.50	45.33	2.38
15.71	7.95	42.56	5.40	45.71	2.32
16.00	7.80	42.90	5.30	46.45	2.32
16.20	7.71	43.12	5.20	47.72	2.00
16.40	7.62	43.30	5.11		
16.78	7.55	43.40	5.05		

Source: Data courtesy of A. Borchers.

■ 7.15 Plot the data in problem 7.14 as

(a) A differential plot of $\Delta pH/\Delta$titrant vs. volume titrant as shown in Figure 7.2b.

(b) A second-differential plot. What part of the curve indicates the end point?

■ 7.16 Plot the five points on each side of the first estimated endpoint for the titration in Exercise 7.14 as a Gran plot (for a base titration with an acid, the graph is plotted as $V_{acid} \cdot 10^{pH}$ vs. volume acid).

7.17 A convenient manner to determine concentrations of alkyllithium reagents is by titration of pure 1,3-diphenyl-2-propanone tosylhydrazone (m.w. 378) in tetrahydrofuran (THF). The reactant undergoes the following reactions with alkyllithium reagents in THF.

$$\underset{\text{(colorless)}}{H_2\text{tosylhydrazone}} + RLi \rightleftharpoons \underset{\text{(colorless)}}{[H\ \text{tosylhydrazone}]^-} + RH + Li^+$$

$$\underset{\text{(colorless)}}{[H\ \text{tosylhydrazone}]^-} + RLi \rightleftharpoons \underset{\text{(orange)}}{[Li\ \text{tosylhydrazone}]^-} + RH$$

R is an alkyl group. A commercially available solution of methyllithium (CH_3Li) in an organic solvent was the titrant in this method.

Into an oven-dried 50-mL Erlenmeyer flask is weighed 0.2835 g of the tosylhydrazone solid. The flask is then covered with a serum cap (a tightly fitting rubber cap with a relatively thin section so that syringe needles can be inserted through it). Through an inlet and vent in the cap the flask is purged with pure nitrogen. Then, 10.0 mL of anhydrous THF is added, and the contents are stirred. The flask is cooled in an ice bath (to reduce reaction of RLi with THF), and the reagent liquid is added dropwise with a 1.00-mL syringe until the orange color persists. The color indicates the end point. (The product is a self-indicator.) The syringe volume could be read to the nearest 0.01 mL. In the titration, 0.45 mL of the CH_3Li solution was required to obtain the orange color. [Ref: M. F. Lipton, et al. *J. Organomet. Chem.* 1980, *186*, 155.]

(a) What is the molarity of the methyllithium solution?

(b) Is the volume of THF in the reaction mixture crucial to obtain a precise result?

(c) Could the same result be obtained with the methyllithium in the flask and with the tosylhydrazone reagent added? How would the color be used as an indicator?

7.18 An assay procedure for the enzyme papain is done by a titrimetric assay of the acid produced by the enzymatic hydrolysis of benzolyl-L-arginine ethyl ester (BAEE). A unit of the enzyme is defined as the amount of enzyme that hydrolyzes 1 μmol min^{-1} of BAEE at 25°C and pH 6.2 under the assay conditions. The enzyme is quite unreactive until it is activated by treatment with mild reducing agents such as cyanide or cysteine. An enzyme activating/diluent solution is prepared consisting of

10 mL 0.01 M edta
0.1 mL 0.06 M mercaptoethanol
10 mL 0.05 M cysteine-HCl prepared fresh daily
70 mL doubly distilled water

The substrate solution is made fresh daily by mixing the following

15.0 mL 0.058 M BAEE made fresh daily

0.8 mL 0.01 M edta

0.8 mL 0.05 M cysteine-HCl

This solution has the pH adjusted to 6.2 with HCl or NaOH and diluted to 21.0 mL with doubly distilled water.

The assay is done on an automatic titrator that operates by adding titrant NaOH to the reaction solution to keep the pH constant at 6.2 as the reaction progresses to form a carboxylic acid by the reaction

$$R-\overset{\overset{\Large O}{||}}{C}-O-C_2H_5 + H_2O \xrightarrow[\text{(cat)}]{\text{enzyme}} R-\overset{\overset{\Large O}{||}}{C}-OH + HOC_2H_5$$

An assay was run under the following conditions. The titrant was standardized 0.0164 *N* NaOH. The enzyme in solution was activated by reaction in the enzyme diluent solution for 30 min. While the enzyme was activating, into the titration vessel were placed 5.00 mL of substrate solution, 5.00 mL of 3.0 M NaCl, and 5.00 mL of doubly distilled water. The solution was allowed to equilibrate to 25°C. At zero time, 1.000 mL of the enzyme solution was added, and the titrator was turned on to keep the pH adjusted to pH 6.2. After a constant rate of NaOH addition was achieved, the titration was allowed to progress for 5.00 min. During the five minutes, 3.811 mL of titrant was used. How many units mL^{-1} are present in the enzyme solution?

■ *7.19 Calculate the titration curve (at enough points to construct a smooth curve) of 100.00 mL of a 0.0100 M $H_2Fe(CN)_6^{2-}$ solution as it is titrated with 0.2000 M KOH. The equilibria are

$$H_2Fe(CN)_6^{2-} \rightleftharpoons H^+ + HFe(CN)_6^{3-}; \quad pK_a = 3.0$$

and

$$HFe(CN)_6^{3-} \rightleftharpoons H^+ + Fe(CN)_6^{4-}; \quad pK_a = 4.25$$

(The two different equilibria cannot be separated as well as carbonic and phosphoric acids.) Assume the volume remains 100mL.

Additional Exercises

7.20 Calculate the pH you would expect to observe in a 0.200-M benzoic acid solution in water at 25°C.

7.21 Calculate the *molarity* of the unknown concentration for each the following titrations:
(a) 50.00 mL of unknown HCl that required 15.45 mL of 0.05050 N NaOH
(b) 50.00 mL of unknown CH_3COOH that required 27.26 mL of 0.1002 N KOH
(c) 50.00 mL of unknown salicylic acid that required 10.38 mL of 0.1075 N M NaOH to reach its first end point

7.22 An automatic titration instrument is being set up to find the % (w/w) NaOH in incoming batches of bulk NaOH. The endpoint volume is determined by the instrument and fed into a computer program which calculates the % (w/w) NaOH in the batch. Before the instrument is used, it is necessary to find the constant factor that relates the endpoint volume to the % (w/w) NaOH in the sample. Once that is done, the analysts will need to enter only the lot number of the NaOH, the mass of NaOH used for the titration and the normality of the acid in the titrant. Find the constant that relates these three variables to the % w/w NaOH in the samples.

■ 7.23 Calculate and plot the expected titration curve for 50.00 mL of 0.0226 M monoprotic acid with $pK_a = 3.5$ for titration with KOH, 0.0100 M.

7.24 A 50.00 mL aliquot of 0.1000-M salicylic acid is titrated with 0.1000-M NaOH at 25°C in water.
(a) What equilibrium determines the pH values of the solution before the end point, at the end point, and after the end point?
(b) What is the volume of the solution and its pH at the midpoint between the beginning and the equivalence point?
(c) What is the volume of the solution and its pH at the equivalence point?

7.25 A back titration is being used to titrate a sample of a polyelectrolyte. A 2.056 g sample of the polyelectrolyte is suspended in 100.0 mL of 0.1021 N HCl and stirred for 10 minutes. The solution is filtered and 25.00 mL of the filtrate is transferred to a wide-mouth erlenmeyer flask. It required 30.45 mL of 0.04857 M NaOH to reach the equivalence point.
(a) How many mmoles of HCl were added to the polyelectrolyte?
(b) How many mmoles of NaOH were required to titrate the excess HCl?
(c) Based on your answers to parts (a) and (b), characterize the polyelectrolyte in mEq/g (mEq = milliequivalents).

7.26 What are the pH, pOH, and concentrations of CO_3^{2-} and HCO_3^- after 5.000 mL of 0.1000 M HCl is added to 100.0 mL of 0.0100 M sodium carbonate solution? Assume that $a_{H^+} = [H^+]$ and T = 25°C. (Don't forget to include the added volume of HCl.)

7.27 What do you expect to be the volumes of added acid at the equivalence points in a titration of 10.000 mL of a solution that is 0.1000 M in Na_2CO_3 and 0.1000 M in $NaHCO_3^-$ and is titrated with 0.1077 M HCl?

*Indicates more involved exercise.

7.28 You have available to you the following indicators: phenolphthalein, methyl red, methyl orange, and bromocresol green. Assume that for the calculation any indicator present is at a low enough concentration that the actual titration of the indicator can be neglected. Assume that the indicator ranges are applicable at their exact values. 25.00 mL of 0.1000-M imidazolium chloride is titrated with 0.1000-M NaOH. Which of the indicators would yield the smallest indicator error?

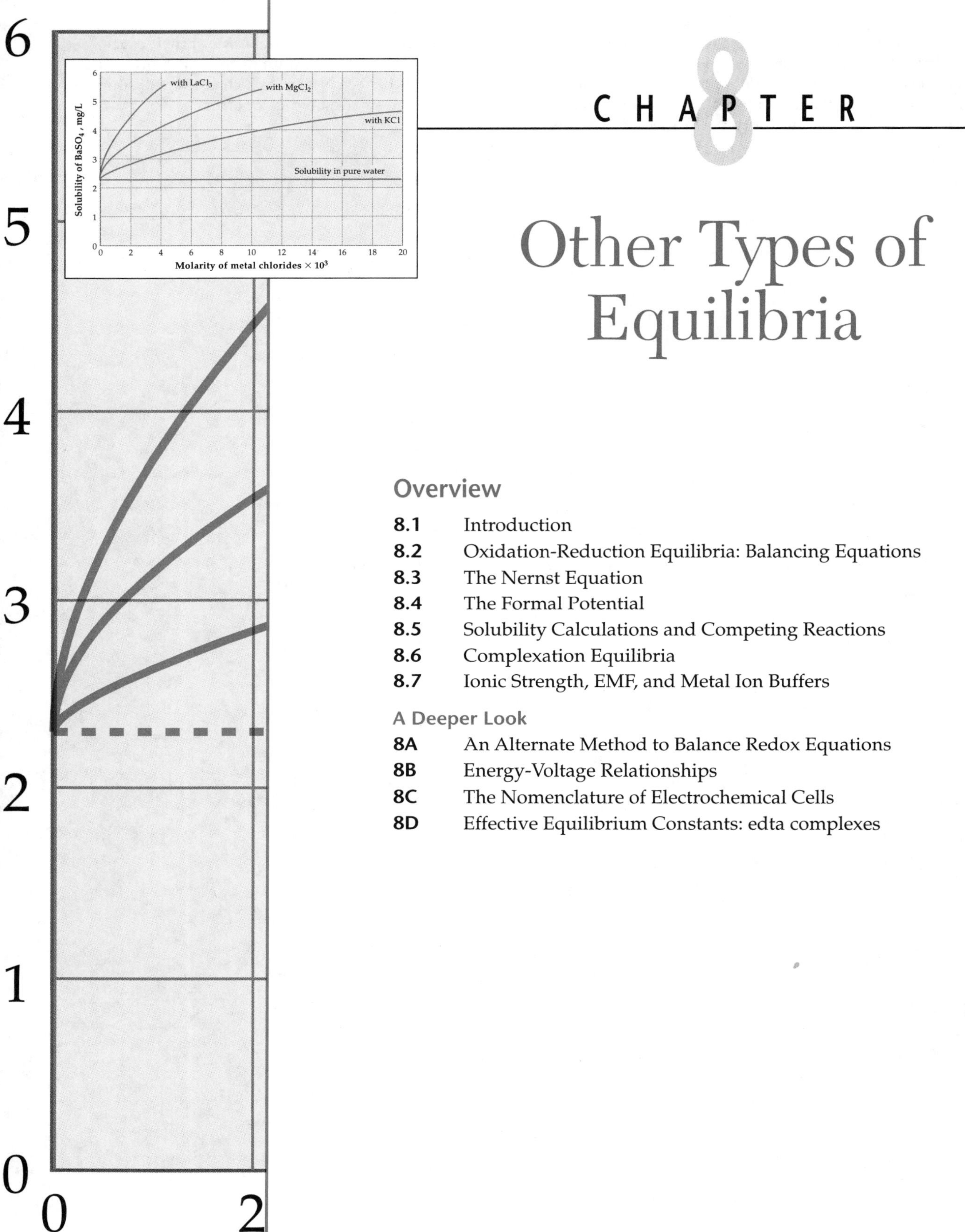

CHAPTER 8

Other Types of Equilibria

Overview

A Deeper Look

8.1 Introduction

You have studied acid-base equilibria and acid-base titrations in the last two chapters. Let us now turn our attention to equilibria with three other molecular mechanisms involved. First are **oxidation-reduction** or **redox** reactions, in which at least one electron is transferred between two reactants. A large part of this chapter is devoted to electron transfer reactions, because they form the basis both for a number of titrations as well as for a wide range of important **electrochemical methods** of analysis.

Electrochemical methods of analysis are the topics of Chapter 12.

Second, **complexation equilibria** involve the binding of two stable chemical reactants to form a stable product. The product is called a **coordination complex**, and such equilibria are the basis for complexometric titrations. Complexation also can be important in changing the expected behavior of the third type of equilibrium, the precipitation and dissolution of a solid in a solvent: **solubility equilibria**.

Recognize this important point: *these equilibria are governed by all of the general rules for equilibria that were presented in Chapter 6*. Only the classes of chemical species that are involved differ.

8.2 Oxidation-Reduction Equilibria: Balancing Equations

Oxidation-reduction reactions can be separated into two parts, one in which the reactant species is oxidized (loses electrons) and one in which the reactant species is reduced (gains electrons). These parts are called, reasonably, **half-reactions**. For example, the reaction

$$6\,I^- + O_2 + 4\,H^+ \rightleftharpoons 2\,I_3^- + 2\,H_2O$$

can be divided into its (unbalanced) half reactions:

The oxidation:	$I^- \rightleftharpoons I_3^-$
The reduction:	$O_2 \rightleftharpoons H_2O$

The balanced half reactions for these species can be found among those listed in Appendix III.

Let us emphasize that this first step, that of writing the oxidized and reduced species, is not done by magic. This is descriptive chemistry; someone has done experiments to show what species are present in the reaction.

A preliminary requirement in dealing with redox equilibria is to feel at ease balancing oxidation-reduction equations. A number of approaches are used to do this balancing. But since chemical analyses usually do not require knowing the formal oxidation state of some specific atom in a molecule, the equation-balancing method shown here usually suffices.

Once the products and reactants of the oxidation-reduction reaction are identified, the individual half-reactions are balanced independently and then added together to balance the overall reaction. The steps for balancing a redox equation are

Another method to balance oxidation-reduction equations is shown in Section 8A. The method requires assigning oxidation states to individual atoms.

1. Write the unbalanced chemical equation for the half-reactions involved. This requires knowing both the reacting species and the product species of the reaction.
2. Balance the atoms, beginning with the species other than those of water (H^+, OH^-, H_2O). Then, if the reaction is in aqueous solution,

add water and protons or water and hydroxide. (This is called material balance.) Two different situations are possible:

a. In acid solution, only water and protons are allowed.
b. In basic solution, water and hydroxide ions are allowed.

3. Make the charge balance. The half-reactions will require adding electrons to balance the charge.
4. Obtain the full reaction by multiplying (where necessary) and adding the half-reactions so that the electrons cancel.

EXAMPLE 8.1

Balancing a half-reaction
Balance the half-reaction for the reduction of permanganate, MnO_4^- to Mn^{2+} in an acidic, aqueous solution.

Solution

Step 1. The reactant is MnO_4^-, and the product is Mn^{2+}. This chemical equation (not balanced) is written

$$MnO_4^- \rightleftharpoons Mn^{2+}$$

Step 2. In acid, H_2O and H^+ are allowed to balance material. We add water to balance the oxygens of permanganate.

$$MnO_4^- \rightleftharpoons Mn^{2+} + 4\,H_2O$$

To match the hydrogens that are now on the right, we can add protons to the left.

$$8\,H^+ + MnO_4^- \rightleftharpoons Mn^{2+} + 4\,H_2O$$

Step 3. The left side has net 7+ charges, the right has net 2+. We need 5 e^- on the left.

$$5\,e^- + 8\,H^+ + MnO_4^- \rightleftharpoons Mn^{2+} + 4\,H_2O \qquad \text{(balanced)}$$

The equation is now balanced with respect to both charge and material balance.

EXAMPLE 8.2

Balancing a full redox reaction
An organic compound is oxidized with the inorganic oxidant permanganate. (The reaction is the basis for one method of analysis for alcohols.) The chemistry of the reaction of methanol with *basic* permanganate has been thoroughly investigated. Following from experimental data, the reaction is described by

$$MnO_4^- + CH_3OH \rightleftharpoons CO_2(g) + MnO_2(s) \qquad \text{(unbalanced)}$$

Balance the equation.

Solution

We split this into its two half-reactions. For permanganate:

Step 1. $$MnO_4^- \rightleftharpoons MnO_2(s)$$

But since OH^- and water are the relevant species in a basic, aqueous solution,

Step 2. $$2\,H_2O + MnO_4^- \rightleftharpoons MnO_2(s) + 4\,OH^-$$

Step 3. $3e^- + 2\,H_2O + MnO_4^- \rightleftharpoons MnO_2(s) + 4\,OH^-$

For methanol:

Step 1. $CH_3OH \rightleftharpoons CO_2$

Next, one *more* oxygen is needed on the left and more hydrogens on the right to complete the material balance. We can add hydroxide(s) on the left as long as we add one less water on the right to balance the oxygens. In addition, *at least* 4 OH^- are required to take up the hydrogens from the methanol since no free H^+ can show up in a solution of base. Therefore,

Step 2. $6\,OH^- + CH_3OH \rightleftharpoons CO_2 + 5\,H_2O$

(Another way to think of this step is that on the left side, there are 4 OH^- to take up the protons from the methanol, 1 OH^- to balance the oxygen, and 1 OH^- to pick up the proton from the oxygen-balancing OH^-, for the total of 6 OH^-. Alternately, the number of OH^- and H_2O needed may be solved by trial and error.)

Step 3. $6\,OH^- + CH_3OH \rightleftharpoons CO_2 + 5\,H_2O + 6\,e^-$

For permanganate plus methanol:

Step 4. To eliminate the electrons between the two half-reactions, we must have equal numbers of electrons on both sides of the equation. To do this, the permanganate equation must be multiplied by 2 before the half-reactions added.

$$\begin{array}{r} 2 \times [3\,e^- + 2\,H_2O + MnO_4^- \rightleftharpoons MnO_2(s) + 4\,OH^-] \\ + 6\,OH^- + CH_3OH \rightleftharpoons CO_2 + 5\,H_2O + 6\,e^- \\ \hline 6\,e^- + 4\,H_2O + 2\,MnO_4^- + 6\,OH^- + CH_3OH \rightleftharpoons 2\,MnO_2 + 8\,OH^- + CO_2 \\ + 5\,H_2O + 6\,e^- \end{array}$$

After canceling species that appear on both sides of the equality, the answer is

$$2\,MnO_4^- + CH_3OH \rightleftharpoons 2\,MnO_2(s) + 2\,OH^- + CO_2(g) + H_2O \qquad \text{(balanced)}$$

8.3 The Nernst Equation

In this section we review the relationships between electrical potentials and the chemical reactions that can cause them. Other names for the electrical potential are the electromotive force (EMF), electrochemical potential, and, of course, voltage. For a voltage to be generated by a chemical reaction, a chemical transformation *must* involve charged species. This may not be obvious from the stoichiometric equations for some reactions. For example:

$$3\,H_2 + 2\,N_2 \rightleftharpoons 2\,NH_3.$$

Walther Nernst (1864–1941)

The magnitude of the voltage associated with an equilibrium involving charge transfers is described quantitatively by the **Nernst equation**.

$$E = E^\circ - \frac{RT}{n\mathcal{F}} \ln K_{act} \qquad \textbf{(8-1)}$$

E is the electrochemical potential, and is related to ΔG for the reaction by the equation $\Delta G = -n\mathcal{F}E$. This relationship with ΔG is discussed in greater depth in Section 8B.

E° is the electrochemical potential for the reaction *under standard state conditions*; E° is called the **standard potential**.

Standard state conditions are described in Section 6.8.

R is the gas constant (8.3145 joules K^{-1} mol^{-1}).

T is the temperature in K.

n is the number of moles of charges transferred in the reaction per mole of reactant.

$\mathcal{F}$ is the Faraday constant, which represents the number of coulombs in a mole of electrons ($\mathcal{F}$ = 96,485 coulombs mol^{-1} = 96,485 joules V^{-1} mol^{-1}).

K_{act} is the thermodynamic equilibrium constant corresponding to the specific E of the equation. In other words, just as equilibrium constants depend on T and P, they also depend on E. So, K_{act} is constant for fixed T, P, and E. In addition, to provide a correct description of the equilibrium and its related electrochemical potential, all concentrations must be expressed as activities, which is why K_{act} is used. As before, we assume concentrations equal activities.

Activities are the subject of Section 6.9.

The Nernst equation applies to two types of chemical processes that involve the motion of electrical charge. One type is the oxidation-reduction reactions discussed earlier; in these, an electron is transferred. The other process involves transfer of ionic charge by **diffusion** of ions from a region where they are more concentrated to a region where they are more dilute. Diffusion is the process in which, by random motion, molecules eventually become equally distributed in the volume accessible to them. As will be reviewed below, equilibria associated with each of the processes—both redox reactions (transfer of e^-) and concentration changes (transfer of ionic charges)—can *always* be treated separately with the Nernst equation. When *both* redox and concentration changes occur simultaneously, the EMF of the total reaction equals the sum of the EMFs of for both the redox and concentration steps. Such calculations will be demonstrated later in this chapter.

Two special situations involving equilibria described by the Nernst equation are worth noting. First, when the concentration of each of the species present is equal to its respective standard state, then each activity in the mass-action expression equals 1. Under this special condition, all mass-action expressions have the same equilibrium constant, namely,

$$K_{act} = 1; \qquad \text{all species in standard states.}$$

And thus,

$$\log K_{act} = 0; \qquad \text{all species in standard states}$$

The Nernst equation then becomes

$$E = E°; \qquad \text{all species in standard states}$$

Recall $E°$ is called the standard potential.

The second special condition occurs when a chemical reaction has come to equilibrium. At that point, there can be no further *net* reaction and no further net transfer of charge. Charge transfer ceases only when the EMF equals zero, that is,

$$E = 0; \qquad \text{at equilibrium}$$

The voltage is zero at equilibrium. Therefore, when we measure a positive or negative potential, the system is not at equilibrium. How do these measured potentials relate to the chemistry? The general answer can be found by analyzing the descriptive model, the Nernst equation. We can write a general reduction reaction,

$$A^{2+} + e^- \rightleftharpoons B^+$$

which has K_{act}

$$K_{act} = \frac{a_{B^+}}{a_{A^{2+}}}$$

Let us keep this inquiry simple. Assume that the standard potential $E^\circ = 0$ volts, and molar concentrations can be substituted for activities. From the stoichiometry, you can see that one charge is transferred, and $n = 1$ in the Nernst equation.

At equilibrium, when $E = 0$,

$$E = 0 = 0 - \frac{RT}{n\mathcal{F}} \ln \frac{[B^+]}{[A^{2+}]}$$

This is true only when $[B^+]/[A^{2+}] = 1$, which means that

$$[A^{2+}] = [B^+]; \quad \text{at equilibrium.}$$

Now let us assume that we mix two more solutions of the same species A^{2+} and B^+ but with different concentration ratios and see how the voltage will change from its equilibrium value of zero. The first solution we mix has more A^{2+} than B^+; specifically, let $[A^{2+}] = 10[B^+]$. Then, E is expected to be

$$E = 0 - \frac{RT}{n\mathcal{F}} \ln \frac{[B^+]}{[A^{2+}]} = 0 - \frac{RT}{n\mathcal{F}} \ln \frac{1}{10}$$

$$= -\frac{RT}{n\mathcal{F}} \cdot (-2.303) = +2.303 \frac{RT}{n\mathcal{F}} \text{ volts}$$

Now, you know from Le Chatelier's principle that the reaction $A^{2+} + e^- \rightleftharpoons B^+$ where $[A^{2+}] = 10\,[B^+]$ will proceed to the right; it tries to go back to the equilibrium condition where $[A^{2+}] = [B^+]$. So a reaction that proceeds spontaneously to the right (from Le Chatelier) is expected to have a potential E that is positive (from Nernst).

The second solution has more B^+ than A^{2+}; let $[A^{2+}] = 0.1[B^+]$. Then E is expected to be

$$E = 0 - \frac{RT}{n\mathcal{F}} \ln \frac{[B^+]}{[A^{2+}]} = 0 - \frac{RT}{n\mathcal{F}} \ln \frac{1}{0.1} = -2.303 \frac{RT}{n\mathcal{F}} \text{ volts}$$

As you probably have guessed, a negative potential means the reaction proceeds spontaneously to the left, toward the reactant side.

In summary: when the reaction "goes" spontaneously to the product side, the potential is positive. When the reaction "goes" spontaneously to the reactant side, the potential is negative. This is a general result.

A chemical reaction proceeds spontaneously as written if the voltage associated with it is positive.

This statement means that the products are formed in preference to the reactants if the potential for the reaction is greater than zero. The reaction does not spontaneously proceed toward forming the products if the potential is negative. This also makes sense given that $\Delta G = -n\mathcal{F}E$.

In the Nernst equation, numerical substitutions for $T = 298$ K (25°C), R, and $\mathcal{F}$ can be made, along with a conversion from natural to base-10 logarithms. The potential (in volts) for an electrochemical *reduction* is, then, expressed as

$$E = E^\circ - \frac{0.059}{n} \log K_{act} \qquad \text{at } 298\ K \tag{8-2}$$

Redox Equilibria, Half-Reactions, and the Nernst Equation

In the same way that a redox reaction can be split into half-reactions, the Nernst equation can be used to express the concentration dependence of a half-reaction potential. In that case, the sign in front of the logarithmic term depends on the way the half-reaction is written. The sign is positive (+) if the half-reaction is written as an oxidation and negative (−) if written as a reduction.

The electrons that appear in half-reactions do not appear in the equilibrium constants for those half-reactions. The basis for this practice is that each half-reaction potential is thought of as being measured against a well-characterized, reference half-reaction with its reactants and products both in their standard states (having unit activity). For instance, we see the sum of two equations:

$$Fe^{3+} + e^- \rightleftharpoons Fe^{2+}$$
$$+ \text{[reference species(reduced, a = 1)} \rightleftharpoons \text{reference species(oxidized, a = 1)} + e^-]$$
$$\overline{Fe^{3+} + \text{reference(reduced form, a = 1)} \rightleftharpoons Fe^{2+} + \text{reference(oxidized form, a = 1)}}$$

The electrons for the implicit reference half-reaction cancel those of the "test" half-reaction. As a result, the iron half-reaction is described by

$$E_{Fe} = E_{Fe}{}^0 - (RT/n\mathcal{F}) \log ([Fe^{3+}][1]/[Fe^{2+}][1])$$

EXAMPLE 8.3

Write the equilibrium constants for the half-reactions associated with the redox reaction between thallium and iron.

Solution

The half-reactions and their sum are

$Tl^+ + e^- \rightleftharpoons Tl^0$	$K_{Tl} = 1/[Tl^+]$
$+\ Fe^{2+} \rightleftharpoons Fe^{3+} + e^-$	$K_{Fe} = [Fe^{3+}]/[Fe^{2+}]$
$Fe^{2+} + Tl^+ \rightleftharpoons Tl + Fe^{3+}$	$K = [Fe^{3+}]/([Tl^+][Fe^{2+}])$

The associated equilibrium constants are written at the right. No factor such as $[e^-]$ appears. The usual rules for addition of stoichiometric chemical reactions apply. As a result, the thermodynamic equilibrium constant is the product of the two half-reaction constants,

$$K = K_{Fe} \cdot K_{Tl}$$

In other words, the mass-action expression for the sum reaction is the product of those for the two half-reactions. Note that because the expression for K_{eq} is a thermodynamic equilibrium constant, the activity of Tl^0, metallic thallium, is unity, and is not written.

You should know one other term associated with electrochemical half-reactions. The pair of oxidized and reduced species that can interconvert, such as Fe^{3+}/Fe^{2+} or Tl^0/Tl^+, is called a **redox couple**. The symbol Fe^{3+}/Fe^{2+} is read as *the iron three-iron two redox couple.*

Cataloging Half-Cell Potentials

A selected list of reduction potentials appears in Appendix III.

Many reduction potentials for half reactions are conveniently stored as the potentials for half-reactions with all reactants in their standard states. Two examples of these reduction potentials are

$$Tl^+ + e^- \rightleftharpoons Tl^0 \qquad E° = -0.34\ V$$

and

$$O_2 + 2\ H_2O + 4\ e^- \rightleftharpoons 4\ OH^- \qquad E° = 0.401\ V$$

The equations are a shorthand for the more complete expressions

$$Tl^+(1\ M) + e^- \rightleftharpoons Tl^0 \qquad E° = -0.34\ V$$

and

$$O_2(1\ atm, 298°) + 2\ H_2O(pure) + 4\ e^- \rightleftharpoons 4\ OH^-(1\ M) \qquad E° = 0.401\ V$$

In other words, since the potentials are standard potentials, $E°$s, the species are understood to be in their respective standard states.

Experimentally, only the voltages associated with full cells can be measured. One half-cell can be the **reference** against which all other half-cells are measured. From the measurements of each cell against the reference, the differences in potential between all the others can be found. For instance, Table 8.1 lists the results of a set of ten hypothetical experiments and the calculated half-cell potentials. Ten cells require five half-cell potentials.

Table 8.1 Hypothetical Cell Potentials and Possible Half-Cell Potentials

The electrochemical half-cells are labeled A, B, C, D, and E.

There are an *infinite* number of *sets* of half-cell potentials that will produce the measured cell potentials. For instance,

Set I: A +1.22; B −0.34; C 0.98; D −1.06; E 0.51

Set II: A 0.86; B −0.70; C 0.62; D −1.42; E 0.15

Both would yield the set of measured potentials shown above. To make comparisons easier, one cell potential is arbitrarily defined as zero. Let that be Cell E. The scale then becomes

Set III: A 0.71; B −0.85; C 0.47; D −1.57; E 0.0 exactly.

Cell composed of	Measured cell potential (V)
A against B	1.56
A against C	0.24
A against D	2.28
A against E	0.71
B against C	−1.32
B against D	0.72
B against E	−0.85
C against D	2.04
C against E	0.47
D against E	−1.57

Although the half-cell potentials are fixed relative to one another, there are an infinite number of absolute half-cell potentials that can still be assigned, as indicated by the three different sets of half-cell potentials of Table 8.1. An arbitrary, fixed value must be given to one half-cell in order to assign a specific voltage to each of the other half-cells. In electrochemistry, the half-cell that has been chosen to be the reference and assigned a potential of zero is

$$2\,H^+(1\text{ M}) + 2\,e^- \rightleftharpoons H_2(1\text{ atm}); \qquad E° = 0.00\text{ V exactly, any } T$$

Two names have been given to this half-cell: the **Standard Hydrogen Electrode** (SHE) and the **Normal Hydrogen Electrode** (NHE).

EXAMPLE 8.4

On the scale based on NHE, the following half-cell potentials are found.

$$Fe^{3+} + e^- \rightleftharpoons Fe^{2+} \qquad E° = +0.771\text{ V}$$

$$AgCl(s) + e^- \rightleftharpoons Ag^0 + Cl^-(1\text{ M}) \qquad E° = +0.228\text{ V}$$

What would the potential of the iron half-cell be if based on a normal[1] silver-silver chloride electrode?

$$AgCl(s) + e^- \rightleftharpoons Ag^0 + Cl^-(1\text{ M}) \qquad E° = 0.00\text{ V exactly}$$

Solution

To visualize the answer, a simple figure with the three potentials involved can be drawn.

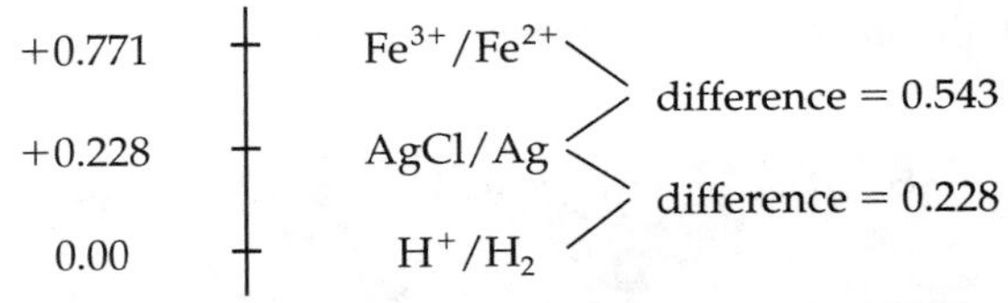

Potential	Couple	
+0.771	Fe^{3+}/Fe^{2+}	
		difference = 0.543
+0.228	$AgCl/Ag$	
		difference = 0.228
0.00	H^+/H_2	

The silver-silver chloride is now assigned zero, and this shift is done by subtracting 0.228 from the value relative to NHE. The other two potentials are similarly shifted. The diagram now appears as

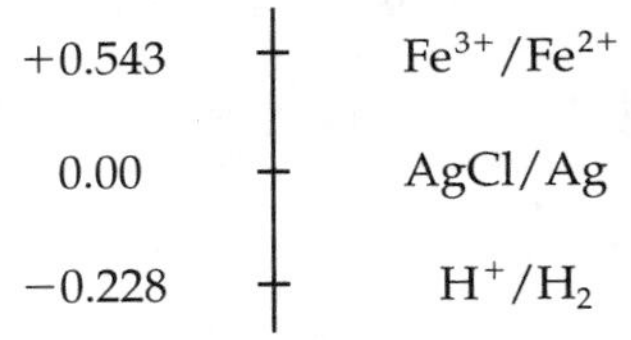

Potential	Couple
+0.543	Fe^{3+}/Fe^{2+}
0.00	$AgCl/Ag$
−0.228	H^+/H_2

Note that the *differences* in the potentials remain the same.

[1] *Normal* in this context means $a_{Cl} = 1.0$.

To summarize, half-cell potentials have the following properties:

1. The Standard Hydrogen Electrode (SHE) is assigned a potential of exactly zero volts at all temperatures.
2. When all reactions are written as reductions,

$$\text{oxidized form} + ne^- \rightleftharpoons \text{reduced form}$$

the reactions proceeding to the right more readily than H^+/H_2 are assigned positive voltages. Those proceeding with a smaller driving force are assigned negative half-cell voltages.

3. The half-cell potential is a quantitative measure of the tendency of the half-reaction to proceed from left to right.

Calculating Full-Cell Potentials

Since any two half-cells can be combined to form an electrochemical cell, the data for a large number of cells can be stored efficiently as half-cell potentials. For example, from 50 half-cell potentials it is possible to calculate the expected potentials for more than 1200 different pairs.

The cell potentials are calculated by adding two half-reactions and finding the expected potential from the sum. However, the voltages associated with half-reactions are handled differently from the equilibrium constants. The following rules are needed to calculate full-cell potentials from the tables of half-reactions.

The rules for handling equilibrium constants and their associated stoichiometric equations were stated in Section 6.1.

a. If the direction in which a half-reaction is written is reversed, the sign of its half-cell potential is reversed.
b. If the half-reaction is multiplied by a positive number, its voltage remains unchanged.

The foundation for the rules for calculating full cell potentials is described in Section 8B on energy-voltage relationships.

Two examples of calculating electrochemical cell potentials from half-reaction potentials are shown next.

EXAMPLE 8.5

Calculate the potential (EMF) of the electrochemical cell from the half-reactions for the cell illustrated in Figure 8.1 on the next page. The cell is at the standard state. If current were allowed to flow, in which half-cell do you expect oxidation to occur?

Solution

The cell reaction can be written

$$Fe^{2+} + Tl^+ \rightleftharpoons Tl + Fe^{3+}$$

From the table in Appendix III, the half-cell reactions and reduction potentials are

$$Tl^+ + e^- \rightleftharpoons Tl^0 \qquad E^\circ = -0.34 \text{ V}$$

$$Fe^{3+} + e^- \rightleftharpoons Fe^{2+} \qquad E^\circ = 0.77 \text{ V}$$

For the reaction in which thallium is reduced and iron is oxidized, the following sum is written. Note that the reaction for iron is reversed from the sense above. E_1 and E_2 are merely convenient labels.

$$\begin{array}{ll} Tl^+ + e^- \rightleftharpoons Tl & E_1 = E_{red}{}^\circ = -0.34 \text{ V} \\ Fe^{2+} \rightleftharpoons Fe^{3+} + e^- & E_2 = -E_{red}{}^\circ = -0.77 \text{ V} \\ \hline Fe^{2+} + Tl^+ \rightleftharpoons Tl + Fe^{3+} & E_{tot} = E_1 + E_2 = -1.11 \text{ V} \end{array}$$

With a negative potential, is this reaction spontaneous as written? With iron being oxidized and thallium reduced, the reaction is not spontaneous (to the right) as written. The reaction will proceed spontaneously to the left, and we expect the oxidation to occur in the Tl half-cell.

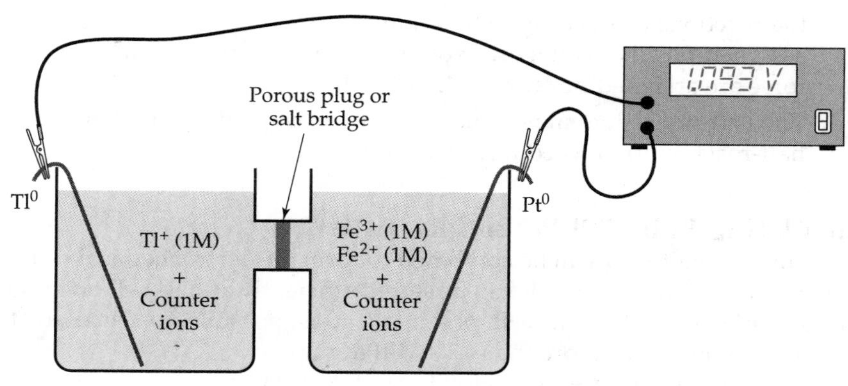

FIGURE 8.1 ▲

Diagram of an electrochemical cell composed of two half-cells.

Each solution with its solid electrode can be described by a redox half-reaction. The content of each solution is indicated in the figure. However, no counter ions are shown; the counter ions in this case might be NO_3^-, SO_4^{2-}, Cl^-, and so forth. The circuit is completed through a porous plug that allows current to flow through it but prevents the two solutions from mixing and reacting. Often this porous plug will contain a concentrated solution of KCl or similar simple salts. The connection then is called a **salt bridge** (also see Section 12A). The voltage of the electrochemical cell is easily measured using a digital voltmeter, as illustrated. Using such an instrument, we can measure the voltage without appreciable current flowing. As a result, no appreciable oxidation or reduction occurs. This means that the measured voltage is an *equilibrium* potential.

EXAMPLE 8.6

Find the potential of the electrochemical cell with the reaction

$Cu^0 + 2\,Fe^{3+} \rightleftharpoons 2\,Fe^{2+} + Cu^{2+}$ with all solutions 1M in the ionic species.

Solution

A summary of the nomenclature used to describe the construction of electrochemical cells is the subject of Section 8C.

This problem consists of adding half-cell reactions that have different n-values. The reaction can be run in an electrochemical cell,

$$Cu|Cu^{2+}(1\text{ M})\|Fe^{2+}(1\text{ M}),\ Fe^{3+}(1\text{ M})|Pt$$

The two half-reactions and their reduction potentials are

$$Cu^{2+} + 2\,e^- \rightleftharpoons Cu^0 \qquad E^\circ = 0.34\text{ V}$$

$$Fe^{3+} + e^- \rightleftharpoons Fe^{2+} \qquad E^\circ = 0.77\text{ V}$$

The algebraic equations that are added to obtain the required full-cell equation are

$$\begin{array}{ll} Cu^0 \rightleftharpoons Cu^{2+} + 2\,e^- & E_1 = -E^\circ = -0.34\text{ V} \\ \underline{2 \times [Fe^{3+} + e^- \rightleftharpoons Fe^{2+}]} & E_2 = E^\circ = 0.77\text{ V} \\ Cu^0 + 2\,Fe^{3+} \rightleftharpoons 2\,Fe^{2+} + Cu^{2+} & E_{cell} = E_1 + E_2 = 0.43\text{ V} \end{array}$$

By the rules: When the equation for copper is reversed, the potential changes sign. When the equation for iron is multiplied, the potential remains the same. This reaction will be spontaneous to the right as written since the voltage is positive.

Concentrations and the Nernst Equation

As shown earlier, the Nernst equation consists of two terms:

$$E = \quad E^\circ \quad - \frac{RT}{n\mathcal{F}} \ln K_{act} \tag{8-1}$$

Reaction under standard conditions — Adjustment for change from standard conditions

The value of E° derives from a reaction done under standard-state conditions and the second, logarithmic, term results from any changes (diluting or concentrating) from the standard-state conditions. In other words, an electric potential is generated by changes in concentrations of ionic species. That is, dilution causes an equivalent voltage. The adjustments apply both to full reactions and half-reactions.

EXAMPLE 8.7

Determine the E for the reaction

$$Sn^0 + 2\,Fe^{3+} \rightleftharpoons Sn^{2+} + 2\,Fe^{2+}$$

where all ion concentrations are 0.1 M.

Solution

The two half-reactions and their sum are

$$\begin{array}{rl} Sn^0 \rightleftharpoons Sn^{2+} + 2\,e^- & E^\circ = +0.136 \\ 2 \times [Fe^{3+} + e^- \rightleftharpoons Fe^{2+}] & E^\circ = +0.771 \\ \hline Sn^0 + 2\,Fe^{3+} \rightleftharpoons Sn^{2+} + 2\,Fe^{2+} & \end{array}$$

The number of moles of electrons transferred between half-cells per mole of reaction is two, which is the value of n. The Nernst equation for this reaction is

$$E_{cell} = E^\circ_{cell} - \frac{0.059}{2} \log \frac{[Fe^{2+}]^2[Sn^{2+}]}{[Fe^{3+}]^2}$$

Note that the tin reaction is written as an oxidation, and the potential is an oxidation potential. Therefore, to obtain E°_{cell} simply add the two half-cell potentials. (Recall, the value of E° does not change with multiplication of the reaction.) The result is

$$E^\circ_{cell} = +0.907 \text{ V.}$$

Now, put in the numerical values of the concentrations.

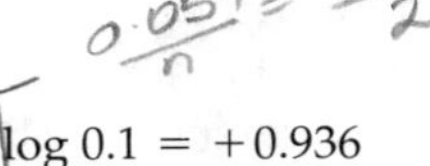

$$E_{cell} = +0.907 - 0.029 \log \frac{(0.1)^2(0.1)}{(0.1)^2} = 0.907 - 0.029 \log 0.1 = +0.936$$

The reaction is spontaneous as written.

Note: If one chooses n based on the reaction

$$\tfrac{1}{2}\,Sn^0 + Fe^{3+} \rightleftharpoons \tfrac{1}{2}\,Sn^{2+} + Fe^{2+}$$

the result of the calculation is the same.

Concentration Cells

The voltage associated with a concentration difference alone can be measured using an electrochemical cell, such as this one for copper:

$$Cu|Cu^{2+}(0.1\text{ M})\|Cu^{2+}(1\text{ M})|Cu$$

The only difference between the two half-cells is the concentration of Cu^{2+} in the solution phase. Such a cell is called a **concentration cell**. The voltage in this cell is equivalent to diluting a 1-M solution of Cu^{2+} ions to produce a 0.1-M solution. The concentration cell is drawn with the lower concentration on the left since oxidation occurs there during a spontaneous reaction. (To understand this last point, consider how $[Cu^{2+}]$ could be equalized in the two half-cells when the solutions cannot be mixed.)

The equation that describes the full-cell reaction is

$$Cu^{2+}(0.1\text{ M}) \rightleftharpoons Cu^{2+}(1\text{ M}) \qquad \textbf{(8-3)}$$

In this case, the value of n is equal to the number of electrons transferred in the half-reactions. In general, the value of n is the moles of charge transferred in a balanced full-cell reaction. The Nernst potential for the cell is

$$E_{cell} = E_{cell}{}^{\circ} - \frac{0.059}{2}\log\frac{1}{0.1} \qquad \textbf{(8-4a)}$$

As you can calculate, the logarithmic term has a value of -0.029 V. Thus,

$$E_{cell} = (E_{cell}{}^{\circ} - 0.029)\text{ V} \qquad \textbf{(8-4b)}$$

However, to find the cell potential, the value of E°_{cell} is required.

The value of the standard potential E°_{cell} in Equation 8-4 is for a cell

$$Cu|Cu(1\text{ M})||Cu(1\text{ M})|Cu$$

In this case, the standard cell is copper vs. copper. The result of the calculation for the potential for this cell is expected: $E_{cell}{}^{\circ} = E_{Cu}{}^{\circ} - E_{Cu}{}^{\circ} = 0$ V. So Equation 8-4b becomes

$$E_{cell} = (0 - 0.029)\text{ V} = -0.029\text{ V} \qquad \textbf{(8-4b)}$$

The entire potential results from the concentration difference alone.

pH and the Nernst Equation

As you have seen, there are, in fact, only a few rules needed to set up the Nernst-equation calculations associated with any equilibrium. These rules concern the reversal and addition of stoichiometric equations and how the equilibrium constant and voltage of the sum equation are related to those of the reactions added.

So far, acid-base and redox equilibria have been treated separately even though they are closely related. The final juncture between acid-base and redox chemistry is presented here, where we discuss how changing the pH alone can change the EMF of a redox reaction, even if the protons themselves are not reduced.

In other words, changes in the concentrations of species other than the redox couple can alter the EMF associated with it. We can use the Nernst equation to describe this effect. Let us consider a redox half-reaction in which protons must be included in the equilibrium constant.

A well-studied example of such a redox reaction involves vanadium. The following equation has been found to describe the reduction of vanadate, $VO_2{}^{+}$ in acid solution:

$$VO_2{}^{+} + 2\,H^{+} + e^{-} \rightleftharpoons VO^{2+} + H_2O \qquad \textbf{(8-5)}$$

(In this stoichiometric equation, the formulae for the vanadium species do not include a number of water molecules of hydration that are present. That is, there is ambiguity in the formulae similar to that which was discussed for hydrated protons. For instance, a possible form of VO_2^+ involves four additional waters, forming $[VO_2(H_2O)_4]^+$.)

How can the equilibrium be shifted by altering the solution pH, and why does it shift? To describe the expected shift qualitatively only requires reference to Le Chatelier's principle (yes, once again): an equilibrium shifts so as to remove the changes imposed on it. Thus, if the pH is lowered ($[H^+]$ raised), the extra protons in the solution will force the reaction 8-5 to the right. The vanadium is forced toward its reduced form, VO^{2+}. In other words, in a more acidic solution, the reduced form of vanadium is more stable.

The explanation of the above chemistry is that protons compete with both oxidized and reduced forms of vanadium for the oxygens. However, the reduced form of vanadium, VO^{2+}, holds its oxygens less strongly than VO_2^+. Thus, when an oxygen is removed from a vanadium by combining with the protons, the system is more stable if the weaker-binding VO^{2+} loses them.

The following calculation describes this phenomenon quantitatively. In the standard state, $a_{H^+} = 1$.

$$VO_2^+ + 2\,H^+ + e^- \rightleftharpoons VO^{2+} + H_2O; K = \frac{[VO^{2+}]}{[H^+]^2[VO_2^+]}; E° = 1.00\text{ V} \quad \textbf{(8-6)}$$

The Nernst equation for reaction 8-6 is

$$E = E° - \frac{0.059}{1}\log\frac{[VO^{2+}]}{[H^+]^2[VO_2^+]} \quad \text{at } 25°\text{C}$$

With the use of the rules of logarithms, this can be written equivalently as

Logarithms are reviewed in Section 1A.

$$E = E° - \frac{0.059}{1}\left(\log\frac{[VO^{2+}]}{[VO_2^+]} - \log\,[H^+]^2\right) \quad \textbf{(8-7a)}$$

If we assume $a_{H^+} = [H^+]$, then from the definition of pH $= a_{H^+}$, the last term in the curly brackets can be changed to give Equation 8-7b.

$$E = E° - \frac{0.059}{1}\left(\log\frac{[VO^{2+}]}{[VO_2^+]} + 2\text{ pH}\right) \quad \textbf{(8-7b)}$$

A little more algebra (that is, pulling the terms out of the parentheses) gives

$$E = E° - \frac{0.059}{1}\log\frac{[VO^{2+}]}{[VO_2^+]} - 0.118\text{ pH} \quad \textbf{(8-7c)}$$

How much does the potential change with a shift in pH? Since $E°$ is a constant, the entire shift appears in the last term of Equation 8-7c. Therefore the change in potential, ΔE, that results from a change in pH *alone*, ΔpH, should be

$$\Delta E = -0.118\,\Delta\text{pH}$$

This calculation indicates that the potential of the vanadium couple is expected to decrease 118 mV per pH unit. In other words, for each 10-fold increase in $[H^+]$—equivalent to ΔpH $= 1$—the potential of the half-cell will increase by 118 mV. An increase in potential means that the equilibrium 8-5 lies further to the right; it is *easier* to reduce the vanadium(V). This is the same conclusion obtained from considering Le Chatelier's principle, but we have calculated a quantitative value for the change with pH.

The term **per decade** means "a factor of 10"; thus, another way to state the change is to say that the potential changes 118 mV *per decade* change in

$[H^+]$. Note that *per decade* refers to both directions of change: both raising and lowering $[H^+]$.

8.4 The Formal Potential

The Nernst equation correctly describes voltage changes of electrochemical cells or half-cells only if thermodynamic equilibrium constants K_{act} are used. The EMF depends on the *activities* of the components of the redox couple.

Most often, however, an equilibrium potential is measured in complicated solutions in which numerous ionic species, known and/or unknown, are present and could influence the activities of the redox pair. A biological fluid presents such conditions. Biological solutions can be so complicated that it is not possible to understand fully all the effects of the thousands of species present. All these effects make it difficult or impossible to compare the measured potentials with the standard potentials stored in our lists. Even relatively simple solutions such as acid-digested samples must be considered complicated by this criterion.

To avoid these problems, an electrochemical equilibrium can be characterized by a **formal potential** $E°'$ relative to the hydrogen standard. The prime sign (′) indicates that not all the concentrations are at unit activity. When the general reaction is

$$\text{oxidized form} + ne^- \rightleftharpoons \text{reduced form}$$

then the formal potential is defined as the potential measured when

$$\frac{[\text{reduced form}]}{[\text{oxidized form}]} = 1 \qquad \textbf{(8-8)}$$

This is a ratio of concentrations, not activities. Further, the concentrations are not necessarily 1 M.

If the EMF published for a redox couple (half-cell) is a formal potential, the detailed conditions under which it was measured *must* accompany the measured value. Otherwise the data would be worthless, since the conditions are unknown to the readers. For a wide range of analytical methods, the formal potential is more useful than the standard potential because it is a quantity that can be measured directly. The variations in the formal potential for two representative half-reactions are shown in Table 8.2. The variations result from changes in the solution environment including effects of ionic strength, complex formation, and pH. The largest effects tend to be from pH changes. This point is illustrated in the following example, at the end of which is a *Note*, the first of a number of comments on the calculations involving ionic equilibria in complicated solutions. These illustrate the approximations that are made and the size of the errors expected due to approximations.

EXAMPLE 8.8

What is the expected value of $E°'$ for the vanadium(V)/vanadium(IV) couple at pH 3 in a solution in which the total vanadium concentration is 0.02 M? Ignore possible effects of ionic strength and complex formation.

Solution

Given that the reaction is described by Equation 8-6,

$$VO_2^+ + 2\,H^+ + e^- \rightleftharpoons VO^{2+} + H_2O; \quad K = \frac{[VO^{2+}]}{[H^+]^2[VO_2^+]}; \quad E° = 1.00 \text{ V} \qquad \textbf{(8-6)}$$

The formal potential is defined at the point where the ratio of the oxidized and reduced vanadium species is unity, Equation 8-8. Since the total concentration is 0.02 M, then the formal potential will be measured when $[VO_2^+] = [VO^{2+}] =$ 0.01 M. Thus, from Equation 8-7,

$$E^{\circ\prime} = E^\circ - (0.059/1)\log[(0.01)/(0.01)(10^{-3})^2]$$

$$= 1.00 - 0.059 \cdot (+6) = 0.65 \text{ V}$$

The formal potential of vanadium in an aqueous solution at pH 3 with 0.02 M total vanadium is expected to be about 0.65 V versus NHE. This compares with a standard potential of 1.00 V.

Note: We assumed that the total vanadium concentration has no effect on the $E^{\circ\prime}$ value calculated. This is true only if the vanadium species involved have the same stoichiometric coefficients, that is, their concentrations are raised to the same power. In addition, effects of ionic strength and complex formation were ignored. If the solution were made with pure water, a vanadium salt, and an acid, both the ionic strength and formation of coordination complexes would depend on the identity of the acid—effects such as are seen in Table 8.2.

One final point of nomenclature: If the electrochemical potential is measured in an experiment in which the ratio

$$[\text{reduced}]/[\text{oxidized}] \neq 1$$

the measured potential is usually written as E. Only when the data can be connected[2] to a standard potential E° or a formal potential $E^{\circ\prime}$ should these symbols be used. The lack of a superscript standard-state symbol (°) or the presence of a prime sign (′) should alert you to look for a listing of the specific conditions of the measurement.[2]

Table 8.2 Formal Potentials of Iron and Dichromate Half-reactions in Various Solutions vs. SHE

Acid and concentration/M	$E^{\circ\prime}_{Fe^{3+}/Fe^{2+}}$/V*	$E^{\circ\prime}_{Cr_2O_7^{2-}/Cr^{3+}}$/V*
HCl		
0.1	0.73	0.93
0.5	0.72	0.97
1	0.70	1.00
3	0.68	1.08
H_2SO_4		
0.1	0.68	0.92
0.5	0.68	1.08
4	0.68	1.15
$HClO_4$		
0.1	0.735	0.84
1	0.735	1.025
Standard potential	0.771	1.33

*These values include voltages that arise in parts of the measuring apparatus—the junction potentials. These are discussed in Chapter 15.
Source: Smith, G.F. 1951. *Anal. Chem.* 23:925.

[2]Connected means linked by extrapolation, interpolation, or direct measurement, and includes various possible corrections.

8.5 Solubility Calculations and Competing Reactions

Solubility and Ionic Strength

We now turn our attention to calculations of solubilities. Solubility equilibria are treated mathematically in the same way as any other equilibrium. The apparent change in form reflects the convention that the activity of the pure solid is always unity and, thus, does not appear in the activity equilibrium constant. As a result, the activity solubility product for a general solid A_mB_n is

$$K_{sp} = (a_A)^m(a_B)^n \tag{8-9}$$

In the calculations made here, the molar concentrations will be used instead of activities; for example,

$$K_{sp} = [A]^m[B]^n \tag{8-10}$$

As a result, the numerical values calculated will not conform precisely to the experimental values except for dissolution in pure solvent. For example, as can be seen in Figure 8.2, with the addition of salts that will precipitate with neither Ba^{2+} nor SO_4^{2-}, the solubility of $BaSO_4$ increases. Even in this range of inert electrolyte concentration—a few millimolar—the solubility of $BaSO_4$ changes by factors up to two or three.

Activities were discussed initially in Section 6.9.

These changes in solubility are due to activity effects. Mathematically, they can be accounted for by changes in activity coefficients of the precipitated species—here Ba^{2+} and SO_4^{2-}—with ionic strength. The mathematics will not be pursued further here. However, it should be noted that some margin of error must be allowed if calculations are made from data taken at ionic strengths different from that of the assay.

Solubility and Solubility Products

Equilibrium calculations of solubilities involve the solubility product and any simultaneous equilibria that might change the concentrations of the ionic species in the solubility product expression. Such calculations merely reverse the process of deriving equilibrium constants from the solubility data. Thus, a great deal of information can be stored in a table of solubility

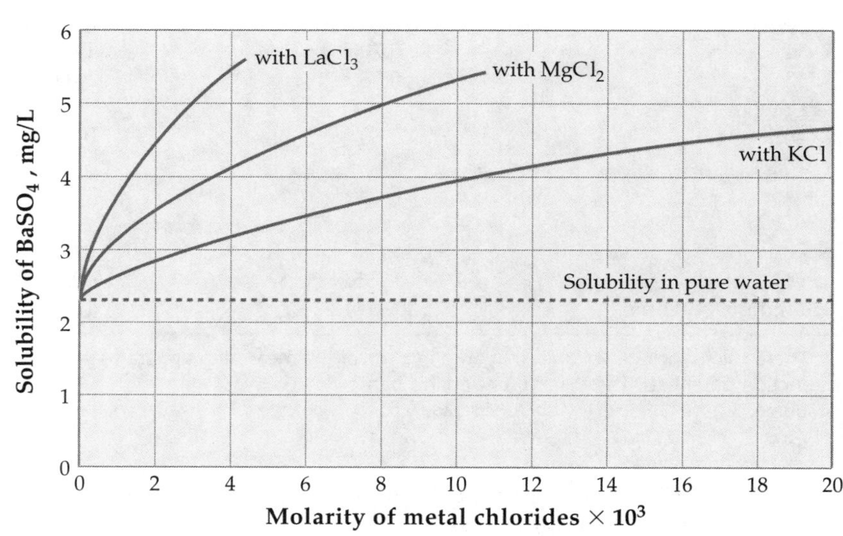

FIGURE 8.2 ▶ **The variation in equilibrium solubility of $BaSO_4$ at 25°C in solutions of inert electrolytes present at different concentrations.**

The nonspecific ion effects, as seen for these salts, will be similar for other salts of the same type: KCl for uni-univalent salts, $MgCl_2$ for uni-bivalent salts, and $LaCl_3$ for uni-trivalent salts. These are ionic strength effects. [Data from Linke, W. F. 1958. *Solubilities*. 4th ed., Vol. 1. Washington, D.C.: American Chemical Society. p. 392.]

products. Appendix I contains a short list of K_{sp} values for salts dissolved in pure water. The concentration units for the ions are mol L^{-1}.

EXAMPLE 8.9

Calculate the solubility product for barium sulfate, given that the solubility of $BaSO_4$ at 25°C in pure water is 2.23 mg L^{-1}.

Solution

A saturated solution of $BaSO_4$ contains

$$(2.23 \times 10^{-3}\ \text{g L}^{-1})/(233.40\ \text{g mol}^{-1}) = 9.554 \times 10^{-6}\ \text{mol L}^{-1}$$

The solubility product at 25°C is, then,

$$K_{sp} = [Ba^{2+}][SO_4^{\ 2-}] = (9.554 \times 10^{-6})^2 = 9.13 \times 10^{-11} \text{ for } BaSO_4 \text{ in pure water.}$$

EXAMPLE 8.10

The log of the solubility product at 25°C and zero ionic strength of lanthanum iodate, $La(IO_3)_3$, is −10.92 for the dissolution equilibrium.

$$La(IO_3)_3 \rightleftharpoons La^{3+} + 3\ IO_3^{\ -}$$

Assume that only La^{3+} and $IO_3^{\ -}$ are formed in the solution when the salt dissolves. What is the solubility of lanthanum iodate under these conditions? [Ref.: Firsching, F. H., Paul, T. R. 1966. *J. Inorg. Nuclear Chem.* 28: 2414.]

Solution

The formula weight of $La(IO_3)_3$ is 663.68. From the logarithm of K_{sp} we calculate

$$K_{sp} = 10^{-10.92} = 1.20 \times 10^{-11}$$

and

$$K_{sp} = [La^{3+}][IO_3^{\ -}]^3$$

The **molar solubility** of lanthanum iodate equals the concentration (in mol L^{-1}) of lanthanum since one lanthanum ion results for each formula unit dissolved. Call this molar solubility X for the calculation. Upon dissolution, for each lanthanum ion, three times as many iodate ions dissolve. Thus,

$$1.20 \times 10^{-11} = (X)(3X)^3 = 27X^4$$

yielding

$$X = 8.17 \times 10^{-4}\ \text{mol L}^{-1}$$

This is the molar concentration of lanthanum iodate in a saturated solution. An 8.17×10^{-4} molar solution contains 0.542 g L^{-1} lanthanum iodate since

$$8.17 \times 10^{-4}\ \text{mol L}^{-1} \times 663.63\ \text{g mol}^{-1} = 0.542\ \text{g L}^{-1}\ La(IO_3)_3$$

Note: It has been assumed that all of the lanthanum is in the form La^{3+}. However, in reality highly charged ions such as La^{3+} generally exist in a number of different hydrolyzed forms such as $La(OH)^{2+}$ and $La(OH)_2^+$. In other words, we have assumed that for

Correcting for hydrolysis is discussed in Section 8.6 below.

$$[La^{3+}]_{Tot} = [La^{3+}] + [La(OH)^{2+}] + [La(OH)_2^{\ +}] + \cdots$$

La^{3+} is the predominant form, and the concentrations of the others are negligible.

Predicting Precipitation Conditions

Under what conditions do we expect ions in a solution to begin to precipitate? The following example illustrates a calculation of this expectation. The key point is to find what ion concentrations are needed to exceed K_{sp}.

EXAMPLE 8.11

How many mL of 0.10 M K_2CrO_4 solution would have to be added to 50.0 mL of 1.0 mM $Cu(NO_3)_2$ to begin to precipitate the $CuCrO_4$?

Solution

The K_{sp} for $CuCrO_4$ is 3.6×10^{-6}. This means that anytime the **ion product**

$$[Cu^{2+}][CrO_4^{2-}] > 3.6 \times 10^{-6}$$

a precipitate should form such as to keep the ionic concentrations such that

$$[Cu^{2+}][CrO_4^{2-}] = 3.6 \times 10^{-6}$$

To determine the minimum amount of CrO_4^{2-} solution, we must find the minimum volume X of the chromate solution that would give an **ion product** equal to the solubility product. If we add X mL of CrO_4^{2-}, the concentrations of the ions will be

$$[Cu^{2+}] = 1.0 \times 10^{-3}\,\text{M}\left(\frac{50.0}{50.0 + X}\right) \qquad [CrO_4^{2-}] = 0.1\,\text{M}\left(\frac{X}{50.0 + X}\right)$$

Volumes are left in mL since they appear in unitless ratios. Next, substitute these concentrations into the solubility product equation.

$$\left[1.0 \times 10^{-3}\,\text{M}\left(\frac{50}{50 + X}\right)\right]\left[0.1\,\text{M}\left(\frac{X}{50 + X}\right)\right] = 3.6 \times 10^{-6}$$

The equation can be rearranged to give

$$3.6 \times 10^{-6}\,X^2 - 4.64 \times 10^{-3}\,X + 9.0 \times 10^{-3} = 0$$

The value of X is found from the quadratic formula, and gives

$$X = 1.94\ \text{mL}$$

If any volume less than this is added, some precipitate may form, but will redissolve when equilibrium is reestablished.

In reality, the initial formation of a precipitate requires concentrations in excess of this calculated limit. This condition, called **supersaturation**, is covered in Section 10.2.

Changing Solubility by Adding a Common Ion

The conditions of precipitation can also be affected by adding more of any of the precipitating ions. For example, if a solution is saturated with barium sulfate,

$$BaSO_4(s) \rightleftharpoons Ba^{2+} + SO_4^{2-}; \qquad K_{sp} = 9.12 \times 10^{-11}\ (25\ °C),$$

and then more sulfate is added to the solution (say, as $NaSO_4$) with a minimum volume change, the reaction is driven to the left. Similarly, if more Ba^{2+} were added (as, say, $BaCl_2$), the reaction would be forced to the left as well. This is the well-known **common ion effect**, which results from adding one of the precipitating ions to a solution already at equilibrium. Note that in both cases, some ionic strength effects will occur. However, the common ion effect is much larger than ionic strength effects.

EXAMPLE 8.12

What is the calculated solubility (in mg L^{-1}) of $BaSO_4$ in 1.0 mM Na_2SO_4 at 25°C? Ignore the effects of changes in ionic strength. At 25°C, 2.23 mg $BaSO_4$ dissolves in a liter of high-purity water.

Solution

As calculated earlier, $K_{sp} = 9.13 \times 10^{-11}$. In the original solution, $[SO_4^{2-}] = 9.55 \times 10^{-6}$ M. The sulfate originating from the precipitate can be ignored relative to that from the sodium sulfate. The calculation is done as follows.

$$K_{sp} = [Ba^{2+}][SO_4^{2-}] = 9.13 \times 10^{-11}$$

$$[Ba^{2+}](0.001) = 9.13 \times 10^{-11}$$

$$[Ba^{2+}] = 9.13 \times 10^{-8}\ \text{M}$$

$$9.13 \times 10^{-8}\ \text{mol L}^{-1} \cdot 233.40\ \text{g mol}^{-1} = 0.021\ \text{mg L}^{-1}$$

The solubility of $BaSO_4$ has been reduced by a factor greater than 100 by the addition of the sodium sulfate. The ionic strength effect of Na_2SO_4 will be approximately the same as that due to $MgCl_2$ as seen in Figure 8.2. Thus, the prediction made here will be in error by about 30–40%.

EXAMPLE 8.13

A liter of a solution contains 10.0 ppm SO_4^{2-}. What fraction of the original sulfate ion remains in the solution if 10.0 mL of 1.00 M $BaCl_2$ is added to the solution? The temperature of the solution is 25°C.

Solution

In this case, we ignore the volume change (1.0%) due to the addition of the barium solution and assume the solution specific gravity is unity.

10 ppm of SO_4^{2-} corresponds to

$$\frac{10 \times 10^{-6} \cdot 1000\ \text{g L}^{-1}}{96.06\ \text{g mol}^{-1}} = 1.04 \times 10^{-4}\ \text{M}\ SO_4^{2-}$$

If the $BaCl_2$ remained unreacted, its diluted concentration would be

$$[Ba^{2+}] = (10.0/1000) \cdot 1.00\ \text{M} = 0.010\ \text{M}$$

At most, only about 1% of this could react with the sulfate present, so 0.010 M may be considered the final concentration of Ba^{2+}. Thus,

$$K_{sp} = 9.12 \times 10^{-11} = [Ba^{2+}][SO_4^{2-}]$$

$$9.12 \times 10^{-11} = (0.010)[SO_4^{2-}]$$

$$[SO_4^{2-}] = 9.1 \times 10^{-9}\ \text{M}$$

Since the original sulfate contained in the solution was 1.04×10^{-4} M, only the fraction

$$\frac{9.1 \times 10^{-9}}{1.04 \times 10^{-4}} = 8.8 \times 10^{-5}$$

remains in solution. Quantitative precipitation—arbitrarily defined as greater than 99.99% precipitated—is expected.

Note: At 25°C the density of pure water is 0.99707 g cm^{-3}, which means that 1000.0 g of water will occupy 1.0029 L. The density of a 10-ppm ionic solution will differ negligibly from this value. However, the assumption of unity for the specific gravity at 25°C introduces a minimal error of approximately 0.3%. Nevertheless, this error is small relative to the error introduced into the calculation by ignoring ionic strength effects. The ionic strength effect of $BaCl_2$ will be nearly the same as that due to $MgCl_2$ shown in Figure 8.2. Thus, the calculation is expected to be in error by a factor of about 2 to 3.

Solubilities and Simultaneous Equilibria

The solubility product describes the relationship between a precipitate in contact with a solution containing its constituent ions. But consider the precipitate $Mg(OH)_2$. The concentration of hydroxide $[OH^-]$ depends on the pH of the solution, and so the precipitation of $Mg(OH)_2$ depends on the pH as well. Any calculation concerning the solubility of $Mg(OH)_2$ must involve competitive equilibria such as those you studied in Chapter 6.

EXAMPLE 8.14

$Mg(OH)_2$ has a $K_{sp} = 7.1 \times 10^{-12}$ at zero ionic strength and 25°C. Assuming equilibrium is reached, will a precipitate of $Mg(OH)_2$ form in a solution of pH 9.0 (buffered) containing 1.0-mM $Mg(NO_3)_2$? Assume that ionic strength effects can be neglected. Under these conditions, magnesium nitrate is completely dissociated in water.

Solution

At a pH of 9.0, the pOH is 5.0 and $[OH^-] = 1.0 \times 10^{-5}$ M. Then, under these conditions,

$$[Mg^{2+}][OH^-]^2 = [1.0 \times 10^{-3}][1.0 \times 10^{-5}]^2 = 1.0 \times 10^{-13}$$

The ion product is smaller than K_{sp}, and the precipitate will not form even under equilibrium conditions. The factor of 70 difference between the K_{sp} and the ion product ensures that this result will be correct even accounting for errors due to ionic strength effects.

Note: The calculation is made with the implicit assumption that the species used to buffer the solution does not bind to the Mg^{2+} ions.

8.6 Complexation Equilibria

When a molecule that is stable by itself in solution binds to a metal ion, the compound that is formed is called a coordination complex. The nomenclature of this area is reviewed in Figure 8.3. The strength of the molecule-metal-ion binding depends on the conditions under which the equilibrium exists. In addition to the ionic strength affecting the equilibrium somewhat, pH changes produce significant effects because the protons compete with the metal for the ligand.

It is worthwhile to go through a calculation of the simultaneous equilibria involved to get an idea of the factors producing changes in the equilibrium and how truly complicated such multiple equilibria are. Our example will show how the formation of coordination complexes with the ions in solution can shift a precipitation equilibrium toward dissolution. As you will see, when additional equilibria are accounted for, the calculated solubility becomes greater and more closely related to the solution chemistry.

For instance, if ammonia is added to a solution in equilibrium with silver chloride, the silver ions will react to form the diamino silver complex.

(*a*)

$$\left[Co(NH_3)_6\right]^{+3}$$

(*b*)

$NH_2{-}CH_2{-}CH_2{-}NH_2$ or N⌒N

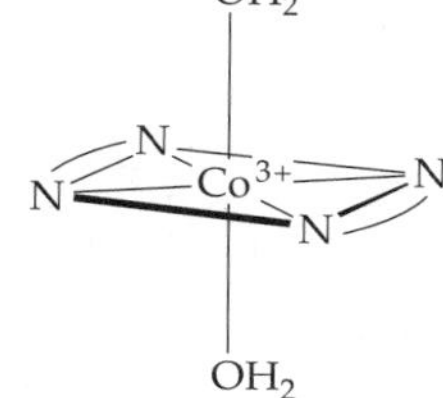

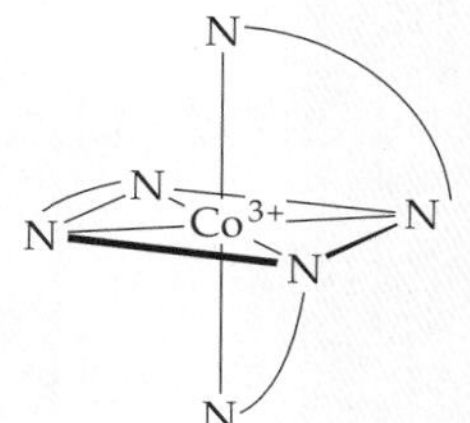

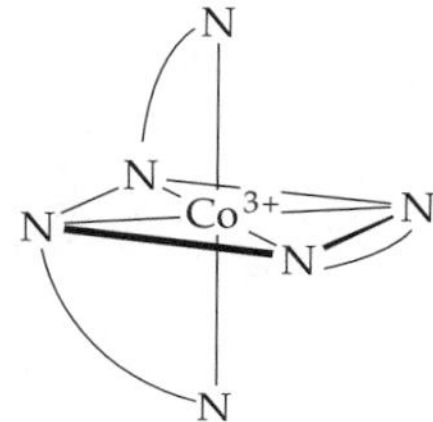

(*c*)

$$(HOOC{-}CH_2)_2N{-}CH_2{-}CH_2{-}N(CH_2{-}COOH)_2$$

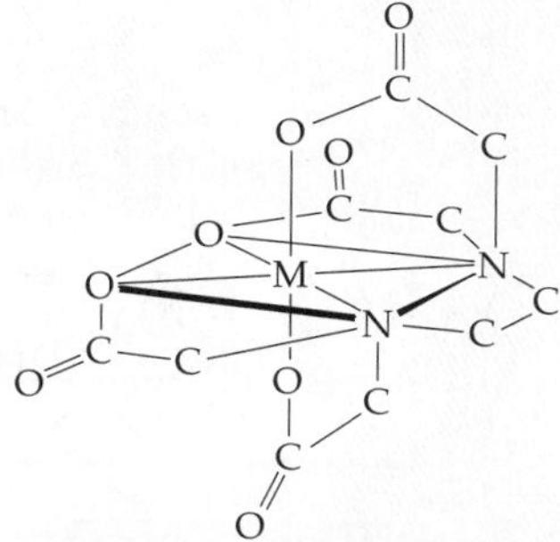

FIGURE 8.3 ▲
Examples of ligands and metal complexes.

The structure of a cobalt-ammonia coordination complex is shown in **(a)**. The compound ethylenediamine, in **(b)**, is a **chelating agent**. Ethylenediamine in water forms stable chelates with Co^{3+} with the structures shown, where each ethylenediamine binds at two sites on the cobalt ion. Figure **(c)** shows the chemical structure of H_4edta and an example of an M(edta) complex. A number of different metal ions bind with $edta^{4-}$ as shown. Some metal ions are bound at seven or eight coordination sites and, thus, bind all six edta sites and still bind other groups. (The structures of a few other chelating agents are shown in Figure 9.1.)

$$Ag^+ + 2\,NH_3 \rightleftharpoons Ag(NH_3)_2{}^+; \qquad K_f = 1.67 \times 10^7$$

K_f is called the **formation constant**. The ammonia competes for Ag^+ with the AgCl precipitate.

EXAMPLE 8.15

A precipitate AgCl(s) is placed in contact with a solution containing an initial concentration of 0.10-M NH_3. What is the final concentration of NH_3 in the solution? What is the solubility of AgCl (in mol L^{-1}) in the solution?

$K_{sp}(AgCl) = 1.7 \times 10^{-10}$

$K_f(Ag(NH_3)_2^+) = 1.67 \times 10^7$

and for $NH_3 + H_2O \rightleftharpoons NH_4^+ + OH^-$, $K_b = 1.8 \times 10^{-5}$

Solution

As for other simultaneous equilibria, it is necessary to write the total reaction for dissolution and formation of the silver-ammonia complex as the sum of the two competing reactions.

$$\begin{array}{rl} AgCl(s) \rightleftharpoons Ag^+ + Cl^- & K_{sp} = 1.7 \times 10^{-10} \\ Ag^+ + 2NH_3 \rightleftharpoons Ag(NH_3)_2^+ & K_f = 1.67 \times 10^7 \\ \hline AgCl(s) + 2NH_3 \rightleftharpoons Ag(NH_3)_2^+ + Cl^- & K = K_{sp} \cdot K_f \end{array}$$

Then, the mass-action expression for the sum equation is

$$\frac{[Ag(NH_3)_2^+][Cl^-]}{[NH_3]^2} = 2.8 \times 10^{-5}$$

Let X equal the molar concentration of $Ag(NH_3)_2^+$ (and Cl^-). So,

$$\frac{X^2}{(0.10 - 2X)^2} = 2.8 \times 10^{-3}$$

The 2 in the denominator arises because two ammonia molecules are needed for each complex. Taking the square root of both sides and solving for X gives

$$X = 4.8 \times 10^{-3}\text{ M}$$

which is the concentration of the diamino silver complex and Cl^-. This compares with a solubility in water alone for AgCl(s) of 1.3×10^{-5} M. The ammonia increases the solubility by a factor of nearly 400.

Note: Three assumptions were made in the above calculation. First, $[Ag^+]$ was negligible compared with $[Ag(NH_3)_2^+]$. Second, the monoamino complex $[Ag(NH_3)^+]$ was negligible. And third, NH_3 was the form of ammonia in the solution and, thus, $[NH_4^+]$ could be neglected. These assumptions will be shown to be reasonable in the next three examples.

EXAMPLE 8.16

Show that $[Ag^+]$ is small (less than 0.1%) relative to $[Ag(NH_3)_2^+]$ under the conditions.

Solution

Using the data

$$Ag(NH_3)_2^+ = Ag^+ + 2\,NH_3 \quad ; \quad 1/K_f = 6.0 \times 10^{-8}$$

then

$$\frac{[NH_3]^2[Ag^+]}{[Ag(NH_3)_2^+]} = 6.0 \times 10^{-8}$$

We can calculate that for a solution 0.10 M in NH_3,

$$\frac{[Ag^+]}{[Ag(NH_3)_2^+]} = 6.0 \times 10^{-6}$$

Essentially all the silver is tied up as the diamino silver complex, even allowing for moderate changes in the free-ammonia concentration.

EXAMPLE 8.17

Show that the monoamino complex $Ag(NH_3)^+$ can be neglected in this analog (to the equilibrium) calculation. The ammonia concentration remains 0.10 M. The stepwise equilibrium constant for the silver-ammonia complex is

$$Ag(NH_3)_2^+ = Ag(NH_3)^+ + NH_3 \quad ; \quad K_1 = 1.4 \times 10^{-4}$$

Solution

With the ammonia concentration of 0.10 M,

$$\frac{[Ag(NH_3)^+]}{[Ag(NH_3)_2^+]} = 1.4 \times 10^{-3}$$

From this ratio, it can be seen that $[Ag(NH_3)^+]$ is only about one part per thousand of the total complexed silver. It is negligible relative to the concentration of the diamino silver complex $Ag(NH_3)_2^+$.

EXAMPLE 8.18

Show that NH_3 is the predominant form of ammonia in the solution and, thus, $[NH_4^+]$ can be neglected under the conditions of the solution.

Solution

The data needed are

$$NH_3 + H_2O \rightleftharpoons NH_4^+ + OH^- \quad ; \quad K_b = 1.8 \times 10^{-5}$$

For a total ammonia concentration of 0.10 M,

$$0.10 = [NH_3] + [NH_4^+]$$

and

$$\frac{[NH_4^+][OH^-]}{[NH_3]} = 1.8 \times 10^{-5}$$

Let X be $[NH_4^+] = [OH^-]$, then

$$X^2/(0.10 - X) = 1.8 \times 10^{-5}$$

and $X = 0.0013 = [NH_4^+]$

$$[NH_4^+]/[NH_3] = 0.013$$

This ratio suggests that there is little competition for the ammonia from the protons in solution under the conditions.

The presence of a fraction of ammonia as NH_4^+ will affect the value of the solubility of AgCl which was calculated. The competition of H^+ with Ag^+ for the ammonia will tend to decrease the calculated AgCl solubility somewhat. However, within the limits of neglected ionic strength effects, making the correction will probably not be worthwhile.

However, another competitive reaction is suggested by Example 8.18. The concentration of hydroxide is equal to $[NH_4^+]$. The pH of the solution is expected to be about 11. Will any silver hydroxide form? A calculation suggesting the answer to this question is made next.

EXAMPLE 8.19

Will any silver hydroxide form in the 0.10 M ammonia solution with $[OH^-] = 0.0013$ M?

Solution

The data needed are

$$K_{sp}(AgOH) = 6.8 \times 10^{-9} \text{ in 1-M } NaClO_4 \text{ solution at 25°C.}$$

From the solubility calculation of Example 8.15, the total silver concentration in solution was found to be 4.8×10^{-3} M. We have found that almost all this silver is complexed with ammonia. Then, $Ag(NH_3)_2^+ = 4.8 \times 10^{-3}$ M. But from Example 8.16 we found

$$\frac{[Ag^+]}{[Ag(NH_3)_2^+]} = 6.0 \times 10^{-6}$$

Thus, substituting for the concentration of the diamino complex,

$$\frac{[Ag^+]}{(4.8 \times 10^{-3})} = 6.0 \times 10^{-6}$$

From this,

$$[Ag^+] = 2.9 \times 10^{-8} \text{ M}$$

Under the conditions of the experiment,

$$[Ag^+][OH^-] = [2.9 \times 10^{-8}][0.0013] = 3.8 \times 10^{-11}$$

which is two orders of magnitude less than the solubility product. The ammonia in the solution prevents the silver hydroxide precipitate from forming.

Note: The value of K_{sp} used here is probably the best one available for AgOH. [Ref: Sillen, L. G., Martell, A. E. 1971. "Stability Constants of Metal-ion Complexes." Supplement No. 1, Special Publication No. 25. London: The Chemical Society.] This K_{sp} was measured in a solution 1 M in $NaClO_4$, whereas the other K_{sp} values, which have been quoted from the same source, are extrapolated to zero ionic strength. K_{sp} data on similar compounds run at a range of ionic strengths suggest an error of about a factor of two will result. However, this is a smaller difference than the usual disagreement among various workers.

Up to this point, we have dealt only with ammonia as a ligand. There are a number of other compounds that can form coordinate covalent bonds with metal ions; among them are those shown in Figure 8.3. Ethylenediaminetetraacetic acid (edta) is by far one of the most often used for analytical purposes. You will notice that edta has more than one site than can bind to a metal—two nitrogen atoms as well as four carboxyl groups that can bind when it is deprotonated. We call compounds like edta **chelates**.

Effective Equilibrium Constants: edta Complexes

The chelate edta remains the reagent of choice for titrating calcium and magnesium in water. The sum $([Ca^{2+}] + [Mg^{2+}])$ is called the **hardness of water,** and tests of hardness are run in every water treatment plant on a regular basis. However, metals other than Ca and Mg are not determined this way because the equilibria are not as favorable for binding. The underlying cause

Table 8.3 Stoichiometric Equations and Equilibrium Constants for edta

$H_4edta \rightleftharpoons H^+ + H_3edta^-$	$\log K_1 = -2.0$
$H_3edta^- \rightleftharpoons H^+ + H_2edta^{2-}$	$\log K_2 = -2.67$
$H_2edta^{2-} \rightleftharpoons H^+ + Hedta^{3-}$	$\log K_3 = -6.16$
$Hedta^{3-} \rightleftharpoons H^+ + edta^{4-}$	$\log K_4 = -10.26$

Source: Sillen, L. B., Martell, A. E. 1971. "Stability Constants of Metal-Ion Complexes." Supplement 1, Spec. Publ. No. 25. London: The Chemical Society.

Table 8.4 α_4 for edta at Various pH Values in Water at 20°

pH	α_4 for edta
2.0	3.7×10^{-14}
4.0	3.6×10^{-9}
6.0	2.2×10^{-5}
8.0	5.4×10^{-3}
10.0	0.35
12.0	0.98

is that unlike Ca and Mg, at the high pHs where edta binds most effectively, the other metals form hydroxides that bind less strongly.

The formation of hydroxides by hydrolysis of the metals is modeled in Section 8D below.

The fully deprotonated form of edta binds most effectively to calcium and magnesium

$$M^{2+} + edta^{4-} \rightleftharpoons M(edta)^{2-} \quad ; \quad K_f = \frac{[M\,(edta)^{2-}]}{[edta^{4-}][M^{2+}]} \tag{8-11}$$

and, thus, the concentration of the $edta^{4-}$ form must be used to model the system correctly. However, the only measure of the edta concentration we have is the total of all the various protonated forms of the chelate together; call it [edta(all forms)]. We must relate the K_f above to an **effective equilibrium constant,** K_{eff}, for the reaction

$$\text{edta(all forms)} + M^{2+} \rightleftharpoons M(edta)^{2-} \quad ; \quad K_{eff} = \frac{[M\,(edta)^{2-}]}{[\text{edta(all forms)}][M^{2+}]}$$

To relate the $edta^{4-}$ to edta(all forms), requires calculating α_4, the fraction of the edta that is in the fully deprotonated form. The required K_a values of edta are collected in Table 8.3. A brief table of α_4 values appears in Table 8.4.

EXAMPLE 8.20

What is the fraction of edta that is in the form $edta^{4-}$ at pH 10.0 (the pH where titrations for hardness of water are generally done)? Assume $a_{H^+} = [H^+]$.

Solution

$$[H^+] \text{ at pH } 10.0 = 1.00 \times 10^{-10}$$

Using the K_a values in Table 8.3 and our rules for calculating α values, the fraction of edta(all forms) that is $edta^{4-}$ is

$$\alpha_4 = \frac{K_1K_2K_3K_4}{[H^+]^4 + [H^+]^3K_1 + [H^+]^2K_1K_2 + [H^+]K_1K_2K_3 + K_1K_2K_3K_4}$$

$$\alpha_4 = 0.35$$

But how can we use our α_4 value to relate K_{eff} and K_f? To find our answer to this we will go back to our definition of α. Since

$$\alpha_4 = \frac{[edta^{4-}]}{[\text{edta(all forms)}]}$$

we can express [edta(all forms)] in terms of [$edta^{4-}$] as

$$[\text{edta (all forms)}] = \frac{[\text{edta}^{4-}]}{\alpha_4}$$

and therefore,

$$K_{\text{eff}} = \frac{[\text{M(edta)}^{2-}]}{[\text{M}^{2+}][\text{edta}^{4-}]/\alpha_4}$$

$$K_{\text{eff}} = \alpha_4 \cdot K_{\text{f}}$$

EXAMPLE 8.21

What is K_{eff} for the complexation of Ca^{2+} by edta at pH 10.0?

Solution

With the use of the value of α_4 at pH 10.0 = 0.35 and the value of K_f for $Ca(edta)^{2-}$, we find

$$K_{\text{eff}} = \alpha_4 K_{\text{f}} = 0.35 \cdot 5.0 \times 10^{10} = 1.75 \times 10^{10}$$

We now have a numerical value for the relationship between the concentrations of free Ca^{2+}, $Ca(edta)^{2+}$ and the total concentration of all of the forms of uncomplexed edta at pH 10.0.

EXAMPLE 8.22

What is the free Ca^{2+} concentration in a solution that is 0.001 M in $Ca(edta)^{2-}$ and 0.001 M total edta at pH 10.0?

Solution

Using our K_{eff} that we found above and substituting the known values,

$$K_{\text{eff}} = \frac{[\text{Ca(edta)}^{2-}]}{[\text{Ca}^{2+}][\text{edta (all forms)}]}$$

$$[\text{Ca}^{2+}] = \frac{[\text{Ca(edta)}^{2-}]}{[\text{edta (all forms)}] \cdot K_{\text{eff}}}$$

$$= \frac{0.001}{0.001 \cdot (1.75 \times 10^{10})}$$

$$= 5.7 \times 10^{-11}\ \text{M}$$

8.7 Ionic Strength, EMF, and Metal Ion Buffers

To say a solution is buffered means that the solution resists changes in some property. Four types of buffers are most commonly found in analyses.

1. Ionic strength buffers, which reduce the relative change in ionic strength with changes in the concentrations of ionic species
2. pH buffers, which resist the change in pH with added acid or base (see Section 6.8.)
3. EMF buffers, which resist changes in the solution electrochemical potential with added oxidant or reductant

4. Metal-ion buffers, which resist the change in a specific metal-ion concentration when that ion is added or removed

The action of pH, EMF, and metal-ion buffers results from the equilibrium properties of at least one chemical reaction involving the buffered species. As for any equilibrium, these depend on the following solution properties:

a. Temperature
b. Ionic strength
c. pH (for EMF and metal-ion buffers); metal-ion content (for pH buffers)
d. Molds, yeast, and/or bacteria growing in the buffer solutions

The first two influences, (a) and (b), perhaps are obvious. Point (c) reflects the usual competition between protons and metals for ligands.

If side reactions could be eliminated, an ideal buffer might be made. That is, the components of an *ideal* pH buffer would not coordinate with metals while the components of an *ideal* EMF or metal-ion buffer would not bind protons or hydroxyl groups. Neither of these ideals is obtainable, although in a few cases, the difference in affinity for metals and protons is quite large.

One further practical point from Chapter 6 should be restated: over a wide range of pH and ionic strength, aqueous buffer solutions support the growth of microorganisms. Over time, the solution components are degraded, and metabolic byproducts are produced. The buffer properties change over time due to the presence of microorganisms. To impede their growth, it is wise to refrigerate buffer solutions, and growth inhibitors can be added if they do not affect the analysis at hand.

Ionic Strength Buffers

As was discussed in Chapter 6, the activities of ions in solution depend on that solution's ionic strength. Recall, the ionic strength is defined as a summation over all ions present:

$$\mu = \frac{1}{2} \sum_i c_i z_i^2 \tag{6-44}$$

The ionic strength of an analyte solution can change for a number of reasons such as adding ionic species in a titration or having different concentrations of ionic species in a series of samples. In order to minimize the effects of possible changes in ionic strength, it is common to make up the solutions with a constant high ionic strength: on the order of 0.1-M. Solutions possessing high ionic strengths cause any *absolute* changes in ionic strength to become small *relative* changes. This reduces significantly the changes in activity effects on ionic equilibria. Thus, the high-ionic strength acts as an ionic-strength buffer.

The high-ionic strength solutions are obtained by adding fully dissociated salts such as NaCl, KCl, and K_2SO_4 and $NaClO_4$. These are used because they can be obtained in highly pure forms at relatively low cost.

EMF Buffers

EMF buffers resist changes in the electrochemical potential upon addition of oxidizing or reducing equivalents. EMF buffers are compounded by mixing the two species of a redox couple—for example, Fe^{3+} and Fe^{2+}—together. In order for the electrochemical couple to control the EMF, they must dominate all the other couples in the solution. This dominance is a complicated function of concentration and the rate of electron transfer. In other words, a redox couple can be present in the solution, but the electron transfer rate might be so slow that it will not buffer the solution. Such a consideration is not explicitly stated with regard to pH buffers since the H^+ transfer rate is fast. One result of the requirement of a fast electron transfer is that only a few organic compounds will function as EMF buffers in aqueous solutions.

In analogy with pH buffers, the maximum buffering capacity for a given total concentration of the redox couple occurs when the concentrations of the oxidized and reduced components are equal. The potential when the concentrations of the two are equal is the formal potential of the half-cell, $E^{\circ\prime}$. The formal potential will change with the following:

1. The identity of the redox couple
2. The solution pH (due to binding with H^+ or OH^-)
3. Any chemical binding such as coordination that may occur
4. The effects of ionic strength
5. The temperature
6. The solvent

However, since the formal potential is relatively easy to measure, it is not difficult to adjust it to a chosen value experimentally by increasing or decreasing the mole ratio

$$[\text{oxidized form}]/[\text{reduced form}]$$

As with pH buffers, it is best to work with a buffer near its $E^{\circ\prime}$. In this case the equivalent of ± 1pH unit is $\pm(0.059/n)$ volts, where n is the number of electrons transferred in the redox reaction of the buffer.

Metal-Ion Buffers

Some chelates are shown in Figure 8.3.

Metal ions in solution can be bound so strongly with a chelate, or other strongly binding ligands, that the bound ions are made inert to many reactions. The ions, which are still present in the solution but unreactive, are said to be **sequestered**. For instance, heavy metals catalyze a number of oxidative reactions in foods. The chelate edta is often added in small amounts to sequester the metal ions, which eliminates the catalytic action and slows the degradation.

When a coordinating ligand is present in stoichiometric excess, it will buffer metal-ion concentrations because each free-ion concentration is established primarily by the identity of the ligand and by the pH of the solution. It is common to use M to designate a metal or metal ion, and M(ligand) represents the complex. The context should provide little ambiguity between abbreviations for metal as opposed to molarity.

The Henderson-Hasselbalch equation is Equation 6-39 on page 188.

Compare to the Henderson-Hasselbalch equation the last of the following three equations for a 1:1 metal chelate

$$K_f = \frac{[\text{M(chelate)}]}{[\text{M}][\text{chelate}]}$$

$$[\text{M}] = \frac{[\text{M(chelate)}]}{K_f \cdot [\text{chelate}]}$$

$$\log\,[\text{M}] = -\log K_f + \log \frac{[\text{M(chelate)}]}{[\text{chelate}]} \qquad \textbf{(8-15)}$$

Analogous to the pH of a hydrogen-ion buffer, the metal-ion-ligand system can be characterized by a pM.

Secondary contributions to the buffered free-ion concentration are made by the temperature and ionic strength, as usual. However, as mentioned before, the pH of the solution has a major effect on the buffer concentration since the metal ions and hydrogen ions compete for the ligand binding.

Calculations to find the expected pM of a metal-ion buffer rapidly become complicated and can be simplified by use of the concept of an effective formation constant. Briefly, the effective formation constant is found by calculating what fraction of the ligand in solution will be in the form L^{n-} to react as

Some details of the complexity of the proton-metal-ion competition for binding and pM calculations are described in Section 8D.

$$L^{n-} + M^{m+} \rightleftharpoons [LM]^{(m-n)}$$

when the ligand can also bind stepwise with q protons

$$L^{n-} + H^{+} \rightleftharpoons HL^{(1-n)}$$

$$\vdots$$

$$H_{(q-1)}L^{(q-1-n)} + H^{+} \rightleftharpoons H_qL^{(q-n)}$$

The calculation of the effective binding constant parallels the calculation of the pH of a polyprotic acid, such as was done for phosphate in Section 6.5.

EXAMPLE 8.23

A buffer solution for $Cu^{2+}(aq)$ is being made using edta as ligand. The pH of the solution must be 4.0 to prevent hydrolysis to the metal hydroxide. What concentration of Cu^{2+} will be present if the solution is made from 50.0 mL of 0.0040 M Na_2H_2edta plus 50.0 mL of 0.0020 M $CuNO_3$? The stoichiometry of the $Cu(edta)^{2-}$ reaction is 1:1. Assume that metal hydrolysis is negligible at this pH. Recall that to form a buffer, the bound-form of the ligand should be about half the material, just as the acid and base forms in a pH buffer. The formation constant for $Cu(edta)^{2-}$ is

$$Cu^{2+} + edta^{4-} \rightleftharpoons Cu(edta)^{2-}; \qquad K_f = 10^{18.9} = 7.9 \times 10^{18}$$

Solution

To use the data given to us, we want to know

$$K_{eff}(\text{pH } 4.0) = \frac{[Cu(edta)^{2-}]}{[edta^{4-}][Cu^{2+}]} \qquad \textbf{(8-16)}$$

where the following is true:

$$K_{eff} = K_f \cdot \alpha_4$$

As you have seen it used before, the value of α_4 is the fraction of the ligand in the form $edta^{4-}$, that is,

$$\alpha_4 = \frac{[edta^{4-}]}{C_L}$$

where C_L is the total concentration of edta in all its forms.

Therefore, with the value of α_4 from Table 8.4 on page 261 and the value of K_f for $Cu(edta)^{2-}$,

$$K_{eff}(\text{pH } 4.0) = 7.9 \times 10^{18} \cdot 3.6 \times 10^{-9} = 2.8 \times 10^{10}$$

The final volume is 100.0 mL, so before complexation, $C_{Cu} = 0.0010$ M and $C_L = 0.0020$ M. Because of the large effective formation constant, after the complexation reaction, half of the copper will be bound. As a result,

$$[Cu(edta)^{2-}] = 0.0010 \text{ M}$$

and

$$C_L = 0.0020\ \text{M} - 0.0010\ \text{M} = 0.0010\ \text{M}$$

The complex dissociates slightly to form $[Cu^{2+}] = X$. As a result,

$$C_L = (0.0010 + X)$$

Substituting into the equation for K_{eff}, we obtain

$$2.8 \times 10^{10} = \frac{(0.0010)}{(0.0010 + X)(X)}$$

and

$$X = 3.5 \times 10^{-11}\ \text{M} = [Cu^{2+}]$$

The pCu = 10.45 at pH 4.0.

EXAMPLE 8.24

To 100 mL of the above copper buffer solution (3.5×10^{-11} M in Cu^{2+}) was added 10 mL of 5.0-mM Cu^{2+} solution. The pH remained fixed. What do you expect the Cu^{2+} concentration to be in this solution?

Solution

The total volume of the solution will be 110 mL. If no complex were formed between Cu^{2+} and the ligand,

$$C_{Cu} = \frac{1 \times 10^{-3}\ \text{M} \cdot 100\ \text{mL} + 5 \times 10^{-3}\ \text{M} \cdot 10\ \text{mL}}{110\ \text{mL}}$$

$$= 1.36 \times 10^{-3}\ \text{M}$$

$$C_L = \frac{2 \times 10^{-3}\ \text{M} \cdot 100\ \text{mL}}{110\ \text{mL}} = 1.82 \times 10^{-3}\ \text{M}$$

Again assuming hydrolysis is negligible, let the complexation reaction take place, and

$$C_L = (1.82 - 1.36) \times 10^{-3}\ \text{M} = 0.46 \times 10^{-3}$$

Now, put the values of C_{Cu} and C_L into Equation 8-16. We find

$$2.8 \times 10^{10} = \frac{(0.00136)}{(0.00046 + X)(X)}$$

and

$$X = 1.05 \times 10^{-10}\ \text{M} = [Cu^{2+}]$$

The pCu = 9.97.

What does this numerical result tell us? Enough Cu^{2+} was added to increase its concentration by 4.5×10^{-4} M (a factor of about 3). However, as a result of the buffering action, the actual increase in free copper was only

$$(1.05 - 0.35) \times 10^{-10}\ \text{M} = 7.0 \times 10^{-11}\ \text{M}$$

This increase is only about 1 part in 6.4×10^6 of the amount of Cu^{2+} added.

STOP

8A A Deeper Look: An Alternate Method to Balance Redox Equations

One method of balancing redox equations was described in Section 8.2. An alternate procedure is presented here. This procedure requires that oxidation numbers be assigned to each atom in the stoichiometric equation. The assignments are carried out according to the following rules.

a. Elements are always zero.
b. Hydrogen is +1 unless it is in a metal hydride in which case it is −1.
c. Oxygen is −2 unless it is in a peroxide O_2^{2-} or superoxide, O_2^-.
d. Alkali metals (Na, K, and so forth) in compounds are assigned +1; alkaline earths (Be,Ca, and so forth) are assigned +2.
e. Halogens (F, Cl, Br, and I) are −1 unless combined with oxygen or another halogen. When combined with oxygen, the oxidation number is assigned using oxygen as −2. When combined with another halogen, the element closest to the top of the periodic table is assigned an oxidation number of −1. (Example: BrCl has Br +1 and Cl −1.) OF_2 is an exception: O is +2, F is −1.
f. All other values are assigned to make the charge on the compound add to zero or, in the case of a polyatomic ion, equal the total ionic charge.

There are six steps to follow when balancing a redox equation.

1. Assign oxidation states.
2. Write half-reactions.
3. Balance half-reactions with regard to the atoms that change oxidation state and then to e^-.
4. Balance the charge on each side of the half-reactions by adding H^+ in acid or OH^- in basic solution.
5. Balance the half-reactions with regard to hydrogen and water.
6. Multiply the half-reaction equations by the appropriate factors to obtain equal numbers of e^- in each and add.

EXAMPLE 8.25

Balance the following redox equation of the oxidation of oxalate by bromate in acid solution.

$$H_2C_2O_4(aq) + KBrO_3 \rightarrow CO_2 + KBr + H_2O \qquad \text{(unbalanced)}$$

Solution

Step 1. Assign oxidation states.

−6 (bracket from $KBrO_3$ to KBr)

$H_2C_2O_4$	+	$KBrO_3$	→	CO_2	+	KBr	+	H_2O
+1+3−2		+1+5−2		+4−2		+1−1		+1−2

+1 (bracket from $H_2C_2O_4$ to CO_2)

As shown, the reaction involves oxidation of carbon from +3 to +4 with concomitant reduction of bromine from +5 to −1.

Step 2. Write half-reactions.

$$H_2C_2O_4 \rightarrow CO_2$$

$$KBrO_3 \rightarrow KBr$$

Step 3. Balance with respect to e^- and the atom changing oxidation state.

$$H_2C_2O_4 \rightarrow 2CO_2 + 2\ e^-$$

$$6\ e^- + KBrO_3 \rightarrow KBr$$

Step 4. Balance charge with H^+ since this is in acid.

$$H_2C_2O_4 \rightarrow 2\ CO_2 + 2\ H^+ + 2\ e^-$$

$$6\ H^+ + 6\ e^- + KBrO_3 \rightarrow KBr$$

Step 5. Balance hydrogen and oxygen with water.

$$H_2C_2O_4 \rightarrow 2\ CO_2 + 2\ H^+ + 2\ e^-$$

$$6\ H^+ + 6\ e^- + KBrO_3 \rightarrow KBr + 3\ H_2O$$

Step 6. Multiply to obtain equal e^- and add.

$$\begin{array}{r} 3 \times [H_2C_2O_4 \rightarrow 2\ CO_2 + 2\ H^+ + 2\ e^-] \\ + 6\ H^+ + 6\ e^- + KBrO_3 \rightarrow KBr + 3\ H_2O \\ \hline 3\ H_2C_2O_4 + 6\ H^+ + 6\ e^- + KBrO_3 \rightarrow 6\ CO_2 + 6\ H^+ + 6\ e^- + KBr + 3\ H_2O \end{array}$$

After canceling the 6 H^+ and 6 e^- occurring on both sides, we find

$$3\ H_2C_2O_4 + KBrO_3 \rightarrow 6\ CO_2 + KBr + 3\ H_2O \qquad \text{(balanced)}$$

8B A Deeper Look: Energy-Voltage Relationships

Associated with a chemical reaction is its Gibbs free energy change ΔG. If the species involved are in their standard states, the associated free energy is denoted by $\Delta G°$. For an electrochemical reaction, which involves transfer of charge across a potential, the electrochemical potential is related to the molar free energy by

$$\Delta G = -n\mathcal{F}E \tag{8B-1}$$

With n the number of moles of electrons, $n\mathcal{F}$ denotes a quantity of electrical charge since $\mathcal{F}$ represents 96,486 coulomb mol^{-1}. If this negative charge is transferred across a voltage from 0 V to $+E$ V, the motion is spontaneous, and the magnitude of the work equals $|\Delta G|$. Spontaneous reactions have $\Delta G° < 0$. The minus sign in Equation 8B-1 keeps the convention consistent.

If the species are in their standard states, Equation 8B-1 becomes

$$\Delta G° = -n\mathcal{F}E° \tag{8B-2a}$$

or

$$E° = -\Delta G°/n\mathcal{F} \tag{8B-2b}$$

The manipulations of the voltages associated with reactions follows in a straightforward way from the relationship between ΔG and E. Recall the two rules:

a. If the direction in which a half-reaction is written is reversed, the sign of its half-cell potential is reversed.

b. If the half-reaction is multiplied by a positive number, its voltage remains *un*changed.

Reversal of a Reaction

As an example, the reduction of Fe(III) to Fe(II) under standard conditions is used.

$$Fe^{3+} + e^- \rightleftharpoons Fe^{2+} \qquad E^\circ = -\Delta G^\circ/\mathcal{F} \tag{8B-3}$$

Since $n = 1$, the voltage is related to the free energy as shown. If this reaction is reversed,

$$Fe^{2+} \rightleftharpoons Fe^{3+} + e^-$$

then the free energy changes sign and becomes $-\Delta G^\circ$. From Equation 8B-2, the sign of E° changes as well.

$$Fe^{2+} \rightleftharpoons Fe^{3+} + e^- \qquad -E^\circ = -(-\Delta G^\circ)/\mathcal{F}$$

Thus, when we reverse a reaction, the sign of the potential is reversed as well.

Multiplication of a Reaction

If the Fe(III)/Fe(II) half-reaction is multiplied by 2, it becomes

$$2\,Fe^{3+} + 2\,e^- \rightleftharpoons 2\,Fe^{2+} \tag{8B-4}$$

When two times the amount of reactants and products react, the free energy doubles—algebraically, $\Delta G_{2Fe}{}^\circ = 2\,\Delta G_{Fe}$. However, as reaction (8B-4) shows, $n = 2$. Therefore

$$E^\circ = -2\Delta G^\circ/2\mathcal{F} \qquad \text{for Equation 8B-4}$$

or

$$E^\circ = -\Delta G^\circ/\mathcal{F}$$

which is the same as for the original reaction, Equation 8B-3. The potential E° does not change simply by multiplying a reaction. It depends on the identities of the reactants and products only. Thus, when we multiply a reaction by a constant value, the potential remains the same.

Both rules hold even when the reactions are not done under standard conditions; the results here are general.

8C A Deeper Look: The Nomenclature of Electrochemical Cells

An electrochemical cell is constructed of two half-cells as represented by the experimental arrangement shown in Figure 8.1. This example consists of two half-cells: One forms a Tl(I)/Tl(0) couple and the other a Fe(III)/Fe(II) couple. The Tl(I)/TI(0) half-cell consists of a thallium metal wire or foil partly immersed in a 1-M Tl^+ solution. The metallic thallium and the thallous ion in solution comprise the electrochemical couple. The Fe(III)/Fe(II) half-cell contains a platinum wire or foil immersed in a solution with 1-M Fe^{3+} and 1-M Fe^{2+} present. Through the platinum wire, we can measure the potential of this electrochemical couple. The electrodes from the half-cells are connected through a voltage-measuring device. One other connection is needed to complete the circuit. This requirement is fulfilled by a salt bridge.

A typical electrochemical cell is shown in the color photograph EP.14 inside the back cover.

These types of electrochemical cells are so common that a shorthand has been developed to show the same information as is contained in the figure of

A number of different forms of salt bridges are described in Section 12A.

the apparatus. In this shorthand, a salt bridge is drawn as two parallel vertical lines ‖. An interface between a solid electrode and a solution (and any boundary between phases—solids, liquids, or gases) is a single vertical line |. If more than one relevant soluble species is present, the components' labels are separated by a comma. For instance,

$$\|Fe^{2+}(1\ M),\ Fe^{3+}(1\ M)|$$

Convention now requires that for an electrochemical cell, the reaction involving oxidation be written on the left. With the contemporary convention, the cell shown in Figure 8.1 can be represented compactly by

$$Tl|Tl^{+}\ (1\ M)\|Fe^{2+}(1\ M),\ Fe^{3+}(1\ M)|Pt$$

This is the nomenclature for full electrochemical cells now commonly used. Consistent with this convention, the cell potential is then

$$E_{cell} = E_{right} - E_{left} \qquad \textbf{(8C-1)}$$

when the potentials E_{right} and E_{left} are *both* reduction potentials.

EXAMPLE 8.26

Calculate the potential expected from the following cell, and determine which side undergoes reduction and which side oxidation.

$$Pt|Fe^{3+}(0.1\ M),\ Fe^{2+}(1\ M)\|Tl^{+}(1\ M)|Tl$$

Solution

The reaction for the diagrammed cell is written

$$Fe^{2+}(1\ M) + Tl^{+}(1\ M) \rightleftharpoons Fe^{3+}(0.1\ M) + Tl^{0}$$

Why is this stoichiometric equation written this way and not the reverse? Since thallium is on the right side of the cell, it must be written as the species reduced in the overall equation. Recall the convention, remembered as an **R** mnemonic: **R**ight (half-cell) = **R**eduction.

Step 1. Calculate the E for each half reaction.

For thallium

Written as a reduction, the reaction is

$$Tl^{+}(1\ M) + e^{-} \rightleftharpoons Tl^{0}$$

The half-cell is operating under standard conditions. So, from the data in Appendix II

$$E_{Tl} = E°_{Tl} = -0.34\ V$$

For iron

Written as a reduction, the reaction is

$$Fe^{3+}(0.1\ M) \rightleftharpoons Fe^{2+}(1\ M) + e^{-}$$

and the half cell potential expected is

$$\begin{aligned} E_{Fe} &= E^{\circ}_{Fe} - (0.059/n)\log([Fe^{2+}]/[Fe^{3+}]) \\ &= 0.771 - 0.059\log(1/0.1) \\ &= 0.712\ V \end{aligned}$$

Step 2. Calculate E for the overall reaction. By looking at the cell as written, we know to write $E_{cell} = E_{right} - E_{left}$

$$E = E_{Tl} - E_{Fe} = -0.34 - 0.712 = -1.05\ V$$

But this negative voltage says that the reaction runs spontaneously from right to left—that iron, not thallium, is reduced! Therefore, the cell was incorrectly written. The correct way to draw it is

$$Tl|Tl^+(1\ M)||Fe^{3+}(0.1\ M),\ Fe^{2+}(1\ M)|Pt$$

Therefore, the stoichiometric equation must be reversed from above, and the potential must become positive, which indicates the cell is drawn correctly.

8D A Deeper Look: Effective Equilibrium Constants: edta Complexes

The general structure of a metal-edta complex is shown in Figure 8.3 on page 257.

In this section, the binding of one of the most useful coordination agents, ethylenediaminetetraacetate (edta), is used to illustrate such calculations. The metal to which we shall consider it binding is vanadium, which in the presence of air can be in one of two different oxidation states. The applicable equilibria and equilibrium constants are listed in Table 8D.1. The profound effect of metal hydrolysis is included in this calculation for vanadium.

In a vanadium-edta solution there are free and bound (by ligand) vanadium and specific concentrations of free and bound (to vanadium) ligand. We attempt to describe these concentrations correctly as they depend on pH by accounting for all the important equilibria that simultaneously occur. This calculation, in effect, is reversing the course through which the equilibrium constants were derived from experimental data. You will see how *all* the applicable equilibria must be included in modeling simultaneous chemical equilibria.

Two reasons to use edta and vanadium for the example are that the equilibria are relatively well understood, and the necessary equilibrium constants are available. These two criteria often do not apply to samples for analysis, and such calculations can only be used for guidance. As an example, the quantities and chemistry denoted in Table 8D.1 were found under conditions of constant ionic strength. Usually this is done by running experiments in solvents containing 0.1 to 1 M inert 1:1 electrolyte such as sodium perchlorate or potassium chloride. However, after common sample pretreatments, it is unlikely that these are the solution conditions. As a result, it is questionable whether or not the literature values apply. Even if all the applicable equilibria are taken into account, the error in the calculation can range up to a factor of five due to ionic strength effects alone.

To simplify the multiple simultaneous equations describing the ligand and metal chemistry requires developing a single equilibrium equation that

Table 8D.1 Stoichiometric Equations and Equilibrium Constants for edta and Vanadium edta

Equation	Constant
$H_4edta \rightleftharpoons H^+ + H_3edta^-$	$\log K_1 = -2.0$
$H_3edta^- \rightleftharpoons H^+ + H_2edta^{2-}$	$\log K_2 = -2.67$
$H_2edta^{2-} \rightleftharpoons H^+ + Hedta^{3-}$	$\log K_3 = -6.16$
$Hedta^{3-} \rightleftharpoons H^+ + edta^{4-}$	$\log K_4 = -10.26$
$VO^{2+} + edta^{4-} \rightleftharpoons VOedta^{2-}$	$\log K(V^{4+}) = 18.0$
$VO_2^+ + edta^{4-} \rightleftharpoons VO_2edta^{3-}$	$\log K(V^{5+}) = 15.55$
$H_2O + VO_2^+ \rightleftharpoons VO_3^- + 2H^+$	$\log K_h = -7.0$

Source: Sillen, L. B., Martell, A. E. 1971. "Stability Constants of Metal-Ion Complexes." Supplement 1, Spec. Publ. No. 25. London: The Chemical Society.

has been corrected for all the other simultaneous equilibria. This correction is accomplished by multiplying the formation constant K_f by a numerical factor. In the case of $edta^{4-}$ and a metal, only the single equation

$$M^{+n} + edta^{4-} \rightleftharpoons [M(edta)]^{n-4} \qquad \textbf{(8D-1)}$$

is used directly. However, there are six other edta species present: H_6edta^{2+}, H_5edta^+, H_4edta, H_3edta^-, H_2edta^{2-}, and $Hedta^{3-}$. There are six competing equilibria involved.

$$H^+ + [H_{n-1}\,edta]^{n-5} \rightleftharpoons [H_nedta]^{n-4} \qquad n = 1 \text{ to } 6 \qquad \textbf{(8D-2)}$$

These can all be accounted for in the mass-action expression

$$K_4 = \frac{[H^+][edta^{4-}]}{[Hedta^{3-}]}$$

by finding the fraction of the total of all seven edta species that is in the form $edta^{4-}$. As done in Section 8.6, for the calculations done below, the formation of H_5edta^+, and H_6edta^{2+} will be ignored because with pH > 1 their concentrations are negligible.

The same procedure can be done for the metal-ion equilibrium to compensate for any forms other than M^{n+} that may be present. For instance, one possibility is the formation of hydrolysis products such as $M(OH)^{(n-1)+}$, where one proton has been removed from the waters of hydration. The effective concentration of M^{n+} is lowered when both the hydroxide and chelating agent compete for the metal. How to account for each of these effects—protonation of $edta^{4-}$ and hydrolysis of the metal—is shown in the sections following.

8D.1 Correcting for the Protonation of $edta^{4-}$

The final correction accounts for the protonation of $edta^{4-}$. The correction is equivalent to finding α_4 value at the experimental pH. At pH 7.4, $[H^+] = 3.98 \times 10^{-8}$, and $\alpha_4 = 1.3 \times 10^{-3}$ using the values of the K_i in Table 8D.1. By carrying out such calculations at a number of pH values, graphs of the fractional concentrations, α_4, versus pH can be constructed. Two different ways to plot α_i variation with pH are illustrated in Figure 8D.1. From such plots, you can see easily how α_4, as well as the other α_i, changes with pH.

8D.2 Determining the Effective Formation Constant

The most common application of edta is to quantify a metal by titration. However, only the total quantity of the chelating agent present is measured experimentally, such as with a buret. We now calculate the equilibrium based on the total chelate only. The total quantity of edta is signified below by the symbol edta(all forms). For a general metal M, the equilibrium that can be applied is

$$\text{edta(all forms)} + M \rightleftharpoons M(\text{edta}) \;;\; K_{eff} = \frac{[\text{Medta}]}{[\text{edta(all forms)}][M]}$$

In other words, we define an *effective* equilibrium constant K_{eff} that applies to experimentally measured values. For pentavalent vanadium (VO_2^+), the stoichiometric equation that represents the system is

$$VO_2^+ + \text{edta(all forms)} \rightleftharpoons VO_2edta^{3-};\; K_{eff}(VO_2^+) = \frac{[VO_2edta^{3-}]}{[\text{edta(all forms)}][VO_2^+]}$$

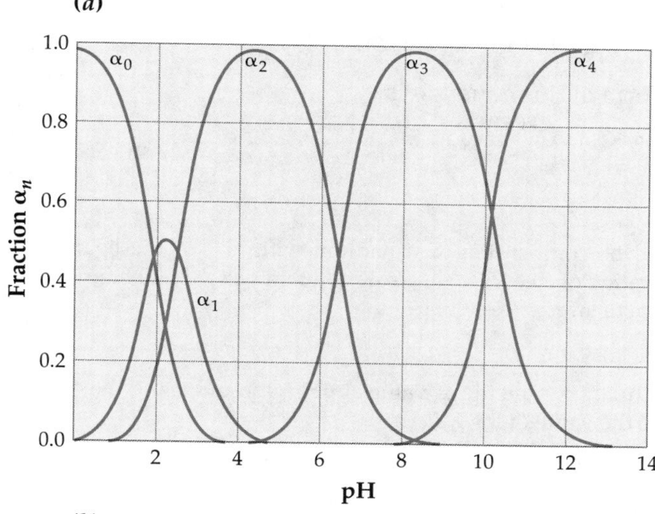

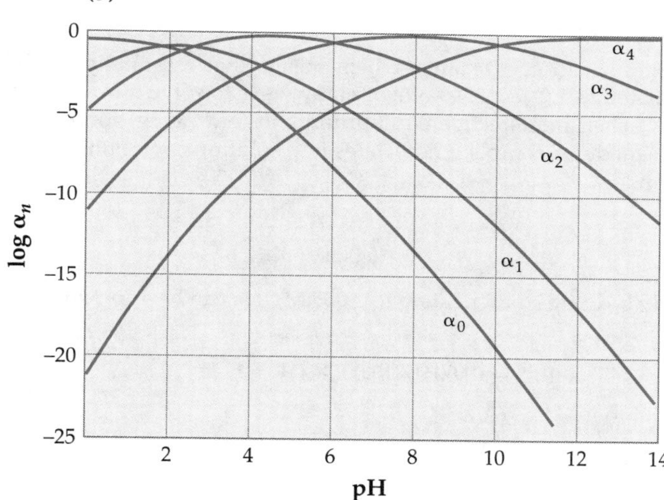

◀ FIGURE 8D.1
Fractional compositions of edta solutions vs. pH.
(a) The values of α_n as a function of pH for $H_n edta^{n-4}$. The curves were calculated as described in the text. The values of the equilibrium constants were determined in 0.1-M ionic strength solution. As a result, the calculated values are for the same conditions. **(b)** Plot of the logarithm of α_n as a function of pH. Each type of plot emphasizes a different portion of the curves. The linear plot is expanded vertically for the major species, whereas the logarithmic plot is spread out vertically for the minor species.

On the other hand, published equilibrium constants involve a specific species of the ligand. For the example of VO_2^+,

$$VO_2^+ + edta^{4-} \rightleftharpoons VO_2\,edta^{3-}; \qquad \log K_f(VO_2^+) \qquad \textbf{(8D-6)}$$

The problem at hand is to relate these two equilibrium constants $K_{eff}(VO_2^+)$ and $K_f(VO_2^+)$. If the pH of the solution is high enough, then the formation constant $K_f(VO_2^+)$ applies since all the edta would be $edta^{4-}$. However, at lower pH, any vanadium present in the solution will not bind as strongly as it would at higher pH values. The protons compete against the metal for the edta ligand. At pH 7.4 only about one part in 800 of the edta added is in the form $edta^{4-}$, the chemical form that appears in Equation 8D-6. The great majority of edta is in other forms: 95% of the edta ligand will be in the form $Hedta^{3-}$ and 5% will be H_2edta^{2-}.

But note that the only difference in the mass-action expressions for Equations 8D-5 and 8D-6 is the quantity [edta(all forms)] and $[edta^{4-}]$. Since

$$[edta^{4-}] = [edta(all\ forms)] \cdot \alpha_4$$

then from substitution, the following important relation is found:

$$K'_{eff}(VO_2^+) = K_f(VO_2^+) \cdot \alpha_4 \qquad \textbf{(8D-7)}$$

A more convenient form of this equation is

$$\log K'_{eff}(VO_2^+) = \log K_f(VO_2^+) + \log \alpha_4 \qquad \textbf{(8D-8)}$$

EXAMPLE 8.27

Taking into account the competition of vanadium with H^+ for the ligand edta, what is the concentration of free VO_2^+ in a solution that has a total of 5 mM VO_2^+ and 10 mM total edta (a large excess) in it?

Solution

We substitute into Equation 8D-8 the α_4 value at pH 7.4 found using the equation in Example 8.20 and the value of log $K_f(VO_2^+)$.

$$\log K'_{eff}(VO_2^+) = 15.55 + \log(1.30 \times 10^{-3})$$

$$= 15.55 + (-2.89) = 12.66$$

$$K'_{eff}(VO_2^+) = 4.57 \times 10^{12}$$

Given the very large formation constant—which means that essentially all the vanadium is coordinated as VO_2edta^{3-}—one of many ways to solve this equation is to let $[VO_2^+] = X$. Then the effective equilibrium constant expression can be written as if 5 mM vanadium complex dissociates in a solution containing 5 mM residual edta(all forms).

$$4.57 \times 10^{12} = K'_{eff} = \frac{(0.005 - X)}{(0.005 + X)(X)}$$

Since X will probably be quite small relative to 0.005 M, this can be approximated quite well by

$$4.57 \times 10^{12} = (0.005)/[(0.005)(X)] = 1/X$$

and

$$X = [VO_2^+] = 2.2 \times 10^{-13}\text{ M}$$

With X this small, the approximation is quite good.

8D.3 Correction for Hydrolysis

In a similar manner, if any of the vanadate is in a form other than VO_2^+, the effective binding constant can be modified to account for simultaneous equilibria involving protons by using α-values. Another chemical form of vanadate arises due to hydrolysis. The reaction is

$$H_2O + VO_2^+ \rightleftharpoons VO_3^- + 2H^+; \qquad \log K_h = -7.0 \qquad \textbf{(8D-9)}$$

which is the predominant reaction of vanadium hydrolysis at pH 7.4. Evaluation of the mass-action expression at pH 7.4 leads to the ratio

$$[VO_2^+]/[VO_3^-] = 10^{-7.8} = 1.58 \times 10^{-8}; \qquad \text{at pH 7.4}$$

This expression says that almost all the vanadium exists as VO_3^- at pH 7.4. The equation expresses a competition for the vanadate, VO_2^+, between the solvent, water/hydroxide, and the edta ligand. This competition results in an increase of the total amount of vanadate in its various chemical forms that is not bound to edta.

To be sure that $K'_{eff}(VO_2^+)$ takes this competition into account, we proceed in a manner similar to accounting for protonation of the ligand. We use a term M_h which is the fraction of vanadium that is VO_2^+.

$$M_h = \frac{[VO_2^+]}{[V^{5+}_{total}]} = \frac{[VO_2^+]}{[VO_2^+] + [VO_3^-]} \approx \frac{[VO_2^+]}{[VO_3^-]} \qquad \textbf{(8D-10)}$$

where $[V^{5+}_{total}]$ represents the concentration of pentavalent vanadium in all forms. The approximation in Equation 8D-10 is excellent in this pH range because of the relatively small amount of VO_2^+. Therefore,

$$[VO_2^+] \approx M_h[VO_3^-] \approx M_h[V^{5+}_{total}]$$

By substitution into the equation for K_{eff} we have

$$M_h \cdot K'_{eff} = \frac{[VO_2edta^{3-}]}{C_L[V^{5+}_{total}]}$$

Equation 8D-8 is then modified to include the M_h term, which accounts for the hydrolysis.

$$\log K_{eff}(VO_2^+) = \log K_f(VO_2^+) + \log \alpha_4 + \log M_h \qquad \textbf{(8D-11)}$$

At pH 7.4, the value of $\log M_h$ is -7.8. Therefore,

$$\log K_{eff} = 15.55 - 2.89 - 7.8 = 4.9$$

EXAMPLE 8.28

What is the concentration of uncomplexed vanadium(V) in a pH 7.4 solution of edta containing a total of 5 mM vanadate and 10 mM total edta in it?

Solution

With compensation for hydrolysis of the metal,

$$\log K_{eff} = 4.9 \quad \text{therefore} \quad K_{eff} = 8 \times 10^4$$

And, as in the previous example,

$$8 \times 10^4 = K_{eff} = \frac{(0.005 - X)}{(0.005 + X)(X)}$$

However, in the equation,

$$X = [V^{5+}_{total}] = [VO_2^+] + [VO_3^-]$$

Again, since X is expected to be small, we try the approximation

$$8 \times 10^4 = 1/X$$

$$X = [V^{5+}_{total}] = 1.3 \times 10^{-5}\,\text{M}$$

Solving with an approximation is still satisfactory since 10^{-5} is only 1/500 of the 5 mM stated concentration.

Note: We calculate that the vanadium is almost eight orders of magnitude less complexed after we made the correction for hydrolysis. Errors due to differences in ionic strength become inconsequential by comparison. For most metals, this correction must be made to have any hope of approximating the chemistry.

The above example shows how the concept of effective equilibrium constants can be used to simplify complicated equilibrium calculations by separately treating the various equilibria that simultaneously affect the components of the solution. You have seen how the formation constant for VO_2^+

with $edta^{4-}$ contains only a small part of the information needed to reproduce the correct concentrations of species in this relatively simple chemical system composed only of vanadate, water, and edta as the reacting chemicals (ignoring spectator ions).

Similar mathematical models can be constructed for other reactions. Zinc with ammonia is shown in Chapter 9.

Multiple-equilibria calculations such as these do not reflect the real solution properties unless at least all the major reactions are included. The possible errors are huge. Consider the vanadium solution above—5 mM total vanadate and 10 mM total edta. With no corrections for "side reactions," the uncomplexed vanadium concentration is calculated to be 2.8×10^{-16} M. Considering the acid-base reactions, this concentration rises to 2.2×10^{-13} M. And finally, including the hydrolysis of the metal ions, the free vanadate rises to 1.3×10^{-5} M.

As a final caution, note that coordinating agents such as edta are not highly specific. This can be seen in Table 8D.2 in which formation constants for some edta-metal complexes are listed. As a result, if edta is used for titration, significant effort must often be expended to eliminate interferences in all but the simplest solutions, such as those containing mostly Ca^{2+} or Mg^{2+} with other metal ions comprising less than ~1% of the total ($[Ca^{2+}] + [Mg^{2+}]$). It is no surprise that metals are determined now more often by spectrometric methods, which tend to be far more specific than titration.

Selected spectroscopic methods and equipment are described in Chapter 11.

Table 8D.2 Formation Constants for $edta^{4-}$ with Various Metals

Temperature at 20°C unless noted. The general reaction is

$$M^{+n} + edta^{4-} \rightleftharpoons M(edta)^{n-4}$$

Metal ion	Log of formation constant	Conditions (M)
Ag^{+}	7.72	0.1 KNO_3
Al^{3+}	16.7	0.1 KNO_3
Bi^{3+}	28.8	0.1 $NaClO_4$
Ca^{2+}	10.7	0.1*
Cd^{2+}	16.9	0.1 $KClO_4$
Ce^{3+}	5.5	0.5 NH_4^+, 22°
Co^{2+}	16.5	0.1 $KClO_4$
Co^{3+}	40.6	0.2 KNO_3, 25°
Cu^{2+}	18.9	0.1 $KClO_4$
Fe^{3+}	25.1	0.1 $NaClO_4$
Hg^{2+}	21.8	0.1 KNO_3
Mg^{2+}	8.7	0.1 KNO_3
Mn^{2+}	14.0	0.1 KNO_3
Ni^{2+}	18.4	0.1 $KClO_4$
Pb^{2+}	18.3	0.1 $KClO_4$
Pu^{3+}	25.8	0.1*
Sn^{2+}	18.3	1 $NaClO_4$
UO_2^{+}	10.4	0.1 NH_4Cl
Zn^{2+}	15.9	0.2 KNO_3
Zr^{4+}	27.7	1 $NaClO_4$

*Unnamed salt used.
Source: Sillen, L. B. Martell, A. E. 1971. "Stability Constants of Metal-Ion Complexes." Supplement 1, Spec. Publ. No. 25. London: The Chemical Society.

Suggestions for Further Reading

LAITINEN, H. A., HARRIS, W. E. 1975. *Chemical Analysis: An Advanced Text and Reference*. 2nd ed. New York: McGraw-Hill.
A rather old, but excellent, description of equilibrium calculations.

RAMETTE, R. W. 1981. *Chemical Equilibrium and Analysis*, Reading, MA: Addison-Wesley.
A carefully written textbook mostly of methodical equilibrium calculations, ignoring, in general, activity effects.

BARD, A.J., PARSONS, R., JORDAN, J., eds. 1985. Standard Potentials in Aqueous Solution. New York: Dekker.
Some of the compendia of data needed for equilibrium calculations of the types presented here.

CONNORS, K. A. 1987. *Binding Constants: The Measurement of Molecular Complex Stability*. New York: Wiley.

MARTELL, A. E., SMITH, R. M. 1974 and later. *Critical Stability Constant*. Vols 1–6. New York: Plenum.

MARTELL, A. E., MOTEKAITIS, R. J. 1992. *The Determination and Use of Stability Constants*. Weinheim: VCH.

Stability Constants of Metal-ion Complexes, (Special Publication #17). 1964 Supplement (Special Publication #25). 1971. London: The Chemical Society.

Concept Review

1. What chemical processes are involved in
(a) redox equilibria?
(b) complexation equilibria?
(c) solubility equilibria?

2. What two rules must be followed when combining E° values for half-reactions to obtain the E° for an overall redox reaction?

3. What part of the Nernst equation takes into account
(a) the number of electrons transferred?
(b) any deviations from standard state conditions?

4. What is the difference between the conditions implied in reporting a standard potential and a formal potential?

5. A metal forms a dichloride and a trichloride, both of which are sparingly soluble. Which salt's solubility will be affected most by changes in ionic strength? (Assume that the ions that determine the ionic strength do not participate in any simultaneous equilibria with the metal salts.)

6. Which simultaneous equilibria must be taken into account in calculations involving complexation by edta?

7. (*Requires use of information in Section 8C*) In the accepted nomenclature for electrochemical cells, how is each of the following indicated?
(a) a boundary between two phases
(b) a salt bridge
(c) two chemical species in the same solution
(d) in which half cell reduction takes place

Exercises

Values of equilibrum constants can be found in Appendices I, III, and IV.

8.1 The following redox reactions occur in acid solution. Balance the equations.
(a) $ClO_3^- + I^- \rightarrow Cl^- + I_2$
(b) $Zn(s) + NO_3^- \rightarrow Zn^{2+} + NH_4^+$

8.2 The following redox reactions occur in basic solution. Balance the equations.
(a) $Al^0 + H_2O \rightarrow Al(OH)_4^- + H_2(g)$
(b) $NiO_2(s) + Fe^0 \rightarrow Ni(OH)_2(s) + Fe(OH)_3(s)$

8.3 Suppose standard reduction potentials were not based on hydrogen but on lead. That is,

$$Pb^{2+} + 2e- \rightleftharpoons Pb^0; \qquad E° = 0.00 \text{ V exactly}$$

(a) What is the reduction potential for $Ni^{2+} + 2e^- \rightleftharpoons Ni^0$ based on this scale?
(b) On the "Standard Lead Electrode" scale, what would be the potential for the following two cells?

$$Ni|Ni^{2+}(a = 1)||H^+(aq, a = 1)|H_2(a = 1)|Pt$$

$$Zn|Zn^{2+}(a = 1)||Ni^{2+}\ (a = 1)|Ni$$

8.4 Consider the abstract electrochemical reaction

$$O + e^- \rightleftharpoons R \qquad \text{at } 25°C$$

where O stands for the oxidized form and R stands for the reduced form. Initially, a solution has equal concentrations of the oxidized and reduced species.
(a) If the solution is changed from its initial conditions until

$$[O]/[R] = 0.10,$$

how much, in mV, will the potential change? Will the change be to more positive or negative voltage?
(b) If the solution is changed from its initial conditions until

$$[O]/[R] = 100.0,$$

how much, in mV, will the potential change? Will the change be to more positive or negative voltage?
(c) If the solution is changed from its initial conditions until

$$[O]/[R] = 0.010$$

how much, in mV, will the potential change? Will the change be to more positive or negative voltage?
(d) If the temperature of the original solution is increased by 10°C, how much, in mV, will the potential change? Will the change be to a more positive or more negative voltage?

8.5 A voltaic cell is constructed with the following overall reaction. Species are all in their standard states.

$$Mg(s) + Ag^{+}(aq) \rightarrow Mg^{2+}(aq) + Ag(s) \quad \text{(unbalanced)}$$

(a) Write the equations for the individual electrode reactions.
(b) Balance the reaction equation.
(c) With the reaction occurring as written above, what is the cell voltage?
(d) Would the reaction proceed spontaneously as written?
(e) What is the value of $\Delta G°$ for the reaction as written?

8.6 The following reaction takes place in a 1 M H_2O_2 solution:

$$H_2O_2 + 2\,H^+ + 2\,e^- \rightleftharpoons 2\,H_2O; \qquad E° = 1.776\ \text{V}$$

(a) What will the half-cell potential be at pH = 4.0?
(b) What will the half-cell potential be at pH = 6.0?
(c) Rewrite the reaction for basic solutions. Does $E°$ change? (yes or no)

8.7 The following cell is set up:

$$Pt|Fe^{3+}(1\ M),\ Fe^{2+}(1\ M)||Fe^{3+}(1\ M),\ Fe^{2+}(1\ M)|Pt$$

(a) What do you expect the cell potential to be?
(b) If, in the left half-cell, the Fe^{3+} concentration is changed to 0.1 M, does the cell potential become more positive or more negative?
(c) What is the numerical value of the voltage change in part (b)?
(d) If, in the right half-cell, the Fe^{3+} concentration is changed to 0.1 M, does the cell potential become more positive or more negative?
(e) If, in the left half-cell, the Fe^{2+} concentration is changed to 0.1 M, does the cell potential become more positive or more negative?
(f) What is the numerical value of the voltage change in part (e)?

8.8 An electrochemical cell is set up between iron (Fe^{3+}/Fe^{2+}) and titanium (TiO^{2+}/Ti^0). The reactions that can occur in the half-cells are, respectively,

$$Fe^{3+} + e^- \rightleftharpoons Fe^{2+} \qquad E° = 0.77\ \text{V}$$

$$TiO^{2+} + 2\,H^+ + 4\,e^- \rightleftharpoons Ti^0 + H_2O \qquad E° = 0.10\ \text{V}$$

(a) Under standard conditions, what is the cell potential when the reaction is written so it will proceed spontaneously?
(b) If the concentration of Fe^{2+} is reduced to 0.15 M, what do you calculate the cell potential will be?
(c) If the concentration of Fe^{3+} is reduced to 0.15 M, what do you calculate the cell potential will be?
(d) If the pH of the original titanium half-cell is raised from 0 to 2, what do you expect the cell potential to be?
(e) If the conditions of part d exist, and then the pH of the iron half-cell is raised from 0 to 2 as well, what do you expect the cell potential to be?

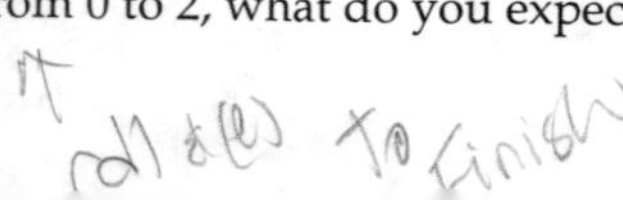

***8.9** The following reactions were used for a voltaic cell.

$$Co^{2+} + 2\,e^- \rightleftharpoons Co^0$$

$$Ni^{2+} + 2\,e^- \rightleftharpoons Ni^0$$

Initially both half-cells were at the standard conditions. Then, the electrodes were connected and the reaction allowed to proceed until the cell's potential was zero. What are the ionic concentrations in each half-cell when $E_{cell} = 0.0$ V at 25°C? Assume activity = concentration.

8.10 Calculate the value of $E°$ (not primed) at 25°C for the following two reactions. Use the values of the formal potentials that are given.
(a) $NO_3^- + 3\,H^+ + 2e^- \rightleftharpoons HNO_2(aq) + H_2O$
$E°' = 0.50$ V vs. SHE, pH 5.0, 25°C
(b) $SeO_4^{2-} + 4\,H^+ + 2\,e^- \rightleftharpoons H_2SeO_3 + H_2O$
$E°' = 0.56$ V vs. SHE, pH 5.0, 25°C

8.11 $Eu(IO_3)_3$ has a K_{sp} of 4.8×10^{-12} at 25°C and zero ionic strength. How many milligrams of europium iodate will dissolve in 100 mL of water at 25°C?

8.12 At 18°C, 1.8×10^{-4} g of Bi_2S_3 dissolves in a liter of water.
(a) What is the molarity of the saturated solution in contact with the solid?
(b) What is the K_{sp} of the compound?

8.13 The following two reactions are related through the K_{sp} for AgCl:

$$Ag^+ + e^- \rightleftharpoons Ag(s) \quad ; \quad E° = 0.799\ \text{V}$$

$$AgCl(s) + e^- \rightleftharpoons Ag(s) + Cl^- \quad ; \quad E° = 0.222\ \text{V}$$

What is the voltage equivalent to the solubility product of AgCl(s)?

8.14 Using your answer from problem 8.13, calculate the solubility product of AgCl.

***8.15** An electrochemical cell is composed of a saturated calomel electrode (E = 0.2412 vs. SHE) and a metallic cadmium electrode in contact with a solution that is 0.100 M in cadmium ion in the presence of 1 M cyanide. The complex $Cd(CN)_4^{2-}$ forms. The dissociation constant for the tetracyano complex is 1.4×10^{-19} under the conditions of the experiment.
(a) What is the value of $[Cd^{2+}]$ (not $Cd(CN)_4^{2-}$) if you assume all activity coefficients are unity?
(b) What is the potential of the full cell written as a cadmium oxidation?

8.16 The K_{eff} for $Ca(edta)^{2-}$ at pH 10 is 1.8×10^{10}. What would be the expected concentration of free Ca^{2+} in a solution of 0.010 M $Ca(edta)^{2-}$?

---

* Indicates more involved exercise.

Additional Exercises

8.17 Balance the following redox reactions, assuming that they take place in an acidic solution.
(a) $ReO_2 + Cl_2 \rightleftharpoons HReO_4 + Cl^-$
(b) $I_2 + Cl_2 \rightleftharpoons H_3IO_6^{2-} + Cl^-$

8.18 Balance the reactions in problem 8.17, assuming that they take place in a basic solution.

8.19 Suppose that the standard potentials were not based on hydrogen but on the Fe^{3+}/Fe^{2+} couple. That is,

$$Fe^{3+} + e^- \rightleftharpoons Fe^{2+}; \quad E° = 0.00 \text{ V exactly}$$

(a) What is the reduction potential for $Ni^{2+} + 2\,e^- \rightleftharpoons Ni^0$ based on this scale?
(b) On the "Standard Iron Electrode" scale, what would be the potential for the following two cells?

$$Ni|Ni^{2+}(a = 1)||H^+(aq, a = 1)|H_2(a = 1)|Pt$$

$$Zn|Zn^{2+}(a = 1)||Ni^{2+}\ (a = 1)|Ni$$

8.20 A voltaic cell is constructed with the following overall reaction. Species are all in their standard states.

$$Li(s) + Ag^+(aq) \rightarrow Li^+(aq) + Ag(s) \quad \text{(unbalanced)}$$

(a) Balance the reaction equation.
(b) Write the equations for the individual electrode reactions.
(c) With the reaction occurring as written above, what is the cell voltage?
(d) Would the reaction proceed spontaneously as written?
(e) What is the value of $\Delta G°$ for the reaction as written?

8.21 The following reaction takes place in solution.

$$H_2O_2 + 2\,H^+ + 2\,e^- \rightleftharpoons 2\,H_2O; \quad E° = 1.776 \text{ V}$$

(a) What will be the half-cell potential at pH = 5.0?
(b) What will be the half-cell potential be at pH = 9.0?

8.22 The following information is known.

$$Ag^+ + e^- \rightleftharpoons Ag(s) \quad E° = 0.799 \text{ V}$$

$$AgBr(s) \rightleftharpoons Ag^+(aq) + Br^-(aq) \quad K_{sp} = 5.0 \times 10^{-13}$$

What would be the $E°$ for the reaction

$$AgBr(s) \rightleftharpoons Ag(s) + Br^-(aq)?$$

8.23 The listed precipitates have the following K_{sp} values.

ppt	K_{sp}
Im(OH)	6.0×10^{-12}
$Un(OH)_2$	6.0×10^{-12}
$Ev(OH)_3$	6.0×10^{-12}
$My(OH)_4$	6.0×10^{-12}

These are, respectively, the hydroxides of improbablium (Im), unlikelium (Un), evanescentium (Ev), and mythium (My).
(a) Assume all four metal ions, Im^+, Un^{2+}, Ev^{3+}, and My^{4+} are together in an acid solution, and each is 1 mM. KOH is then slowly added with vigorous stirring. What is the order of precipitation of these four metal ions?
(b) At what pH will each of the hydroxides just begin to precipitate—assuming equilibrium conditions, no metal hydrolysis, and negligible ionic strength effects?

8.24 These calculations involve the same K_{sp} values given in problem 8.23. Again, the metal ions are each present at 1 mM.
(a) The pH is adjusted to the point where 99.99% of the Im^+ is in the form Im(OH) under equilibrium conditions. At that pH, what percentage of the Un^{2+} present will be precipitated?
(b) The pH is adjusted to the point where 99.99% of the Ev^{3+} is in the form $Ev(OH)_3$ under equilibrium conditions. At that pH, what percentage of the My^{4+} present will be precipitated?

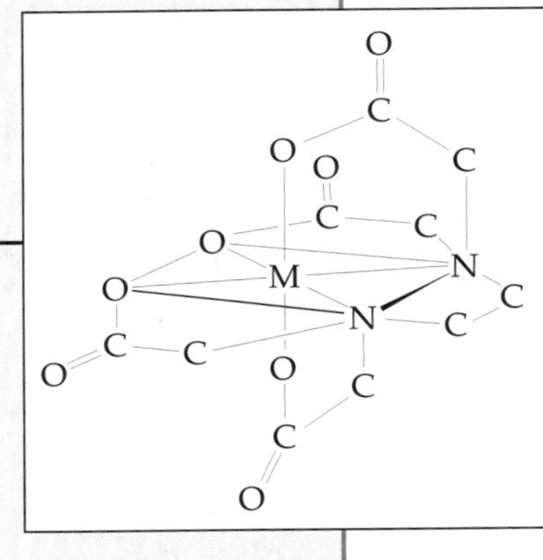

CHAPTER 9

Other Titrimetric Methods

Overview

9.1 Classifying Other Kinds of Titrations

Neutralization titration curves are explained by the law of mass action together with the requirement that a titrant reacts with the analyte to completion. Acid-base equilibria are not unique in these respects. Exactly the same criteria can be fulfilled by the following kinds of reactions:

1. Oxidation-reduction reactions
2. Reactions in which a precipitate is formed
3. Reactions in which a coordination complex is formed
4. Reactions in which tightly bound ion pairs or stable polymer pairs or colloids are formed

Titrimetric assays are based on all these types of reactions.

The principles and practice of all methods of titration are so similar that if you understand the details of acid-base titrations you will be able to comprehend the fundamentals of all other types. The following needs are exactly the same in all:

a. Reproducible stoichiometry
b. Adequate reaction rates (forward and reverse)
c. Standardized titrants
d. Volumetric measurement equipment
e. End-point detection
f. Careful sample preparation

The greatest differences among the types of titrations exist only in the nature of the underlying chemistry. For instance, for all four types noted above, colorimetric indicators have been discovered, but the indicators for each work in different ways. Instrumental detection of end points also may depend on different principles for each type. In this chapter are described some of the similarities and differences in the chemistries between neutralization (acid-base) titrations and oxidation-reduction, complexation, and precipitation titrations. Titrations based on polymer-polymer binding (formation of colloids or ion-pairs) will not be discussed here.

9.2 Oxidation-Reduction Titrations

In oxidation-reduction (redox) titrations, instead of monitoring the pH, we can monitor the electrochemical potential of the solution, *E*. Since measurement of the electrochemical potential is often an integral part of the method, oxidation-reduction titrations are also called **potentiometric titrations**. This name, in fact, applies more broadly. Since electrodes are available to sense many type of solution species, potentiometric monitoring is actually done for all types of underlying chemistries mentioned: acid-base, redox, precipitation, and complex formation.

Similarities between Redox and Acid-Base Titrations

To carry out a redox titration, a measured volume of an oxidizing agent (or reducing agent) of known concentration is added to oxidize (or reduce) the analyte. Compare this to an acid-base titration during which a measured volume of an acid or base titrant of known concentration is added to neutralize the analyte.

The titrant can also be added through an electrochemical reaction in the analyte solution. These are called **coulometric titrations** and are described in Section 15.6.

The electrochemical potential E of a redox titration is analogous to the pH in the acid-base titration. The measured E-value is controlled by an oxidation-reduction equilibrium. This is similar to the way the pH of a solution is controlled by an acid-base equilibrium.

In both types of titrations, at the end point there is a sharp jump in the quantity being monitored. In a redox titration, the jump is in E. This jump occurs where a new redox couple begins to determine the potential of the system. In an acid-base titration, the jump in pH at the end point is where a new acid-base pair begins to determine the pH of the solution.

$E°'$, the formal potential, is defined in Section 8.4.

In a redox titration, the $E°'$ value of the indicator should be near the E-value at the end point. In an acid-base titration, the pK_a value of the indicator should be near the pH of the end point.

The close similarities in the concepts and practice of neutralization and redox titrations are reflected in the forms of two fundamental equations. For neutralization, the pH is described by the Henderson-Hasselbalch equation (Equation 7-5).

$$\mathrm{pH} = \mathrm{p}K_a + \log \frac{[\text{conjugate base}]}{[\text{acid}]}$$

For redox potentials, the half-cell potential is described by the Nernst equation (Equation 8-1)

$$E = E°' - \frac{2.303\,RT}{n\mathcal{F}} \log \frac{[\text{reduced form}]}{[\text{oxidized form}]}$$

The descriptive calculations associated with redox titration curves are analogous to those done for neutralization titrations.

EXAMPLE 9.1

A 50.00-mL sample of 0.1000-M $FeSO_4$ solution is titrated with 0.1000-M $Ce(SO_4)_2$. The stoichiometry of the reaction that occurs is

$$Fe^{2+} + Ce^{4+} \rightleftharpoons Fe^{3+} + Ce^{3+}$$

Calculate the concentrations of Fe^{2+}, Fe^{3+}, Ce^{3+}, and Ce^{4+} at the equivalence point. What is the value of E at the equivalence point? Assume that both activity effects and any metal hydrolysis can be ignored and that no metal-sulfate compounds form. This means that E°′ becomes E°.

Solution

Necessary information:

$$Ce^{4+} + e^- \rightleftharpoons Ce^{3+} \quad E° = 1.61 \text{ V vs. NHE}$$

$$Fe^{3} + e^- \rightleftharpoons Fe^{2+} \quad E° = 0.771 \text{ V vs. NHE}$$

Volume of the solution at the equivalence point = 100.00 mL

At the equivalence point, two conservation equations hold:

$$[Fe^{3+}] + [Fe^{2+}] = (0.1000\text{M} \cdot 50.00 \text{ mL})/100.00 \text{ mL} = 0.0500 \text{ M}$$

and

$$[Ce^{4+}] + [Ce^{3+}] = 0.0500 \text{ M}$$

At *any* point during the titration, at equilibrium

$$E_{Fe^{3+}/Fe^{2+}} = E_{Ce^{4+}/Ce^{3+}}$$

$$E^{\circ}_{Fe^{3+}/Fe^{2+}} - 2.303\frac{RT}{n\mathcal{F}}\log\frac{[Fe^{2+}]}{[Fe^{3+}]} = E^{\circ}_{Ce^{4+}/Ce^{3+}} - 2.303\frac{RT}{n\mathcal{F}}\log\frac{[Ce^{3+}]}{[Ce^{4+}]}$$

At 25°C, the logarithmic multiplier equals 0.0591 V, and

$$0.771 - 0.0591\log\frac{[Fe^{2+}]}{[Fe^{3+}]} = 1.61 - 0.0591\log\frac{[Ce^{3+}]}{[Ce^{4+}]}$$

Carrying out the appropriate algebra, we obtain the general result

$$\log\frac{[Fe^{3+}][Ce^{3+}]}{[Fe^{2+}][Ce^{4+}]} = \frac{1.61 - 0.771}{0.0591}$$

which yields

$$\frac{[Fe^{3+}][Ce^{3+}]}{[Fe^{2+}][Ce^{4+}]} = 1.57 \times 10^{14}$$

The concentrations of the ions are found as for any other competitive equilibrium—by using the mass-action expressions and the conservation relations. From the stoichiometry of the reaction, at the equivalence point

$$[Fe^{2+}] = [Ce^{4+}] = X$$

Then, we can let

$$[Fe^{3+}] = [Ce^{3+}] = 0.0500 - X$$

(Note, since K is large ($\sim 10^{14}$), we expect that both $[Fe^{2+}]$ and $[Ce^{4+}]$ must be relatively small.)

By substituting the algebraic concentrations of the four ions into the equilibrium expression, we find

$$\frac{(0.0500 - X)(0.0500 - X)}{X^2} = 1.57 \times 10^{14}$$

Solving for X (which is made especially easy by taking the square root of both sides of the equation), we find

$$X = [Fe^{2+}] = [Ce^{4+}] = 4.0 \times 10^{-9}\text{ M}$$

Then,

$$[Fe^{3+}] = [Ce^{3+}] = 0.0500\text{ M}$$

The equivalence point is found from the Nernst equation of either metal.

$$E = E^{\circ}_{Fe^{3+}/Fe^{2+}} + 0.059\log\frac{[Fe^{3+}]}{[Fe^{2+}]}$$

$$E = 0.771 + 0.059\log\frac{0.0500}{4.0 \times 10^{-9}} = 1.19\text{ V } vs.\text{ NHE}$$

Note: Three assumptions are mentioned in the question, which, if not true, would change the calculated result.

In this example, the potential at the equivalence point 1.19 V = $\frac{1}{2}$(1.61 + 0.771) V; it is the average of the two E° values (both written as reductions).

This will be true only when the oxidant and reductant react in an equinormal ratio. It is not a general result.

Differences between Redox and Acid-Base Titrations

Acid-base and redox titrations are, indeed, nearly identical in the manner in which competitive equilibria cause the behavior displayed by their titration curves. However, there are also major differences between the two, both in the chemistry and in the titration practice.

The one significant chemical difference arises at the initial point on the titration curve—before any titrant is added. In an acid-base equilibrium system, the solvent will almost always act as an acid or base. For instance, in an acid solution the solvent takes up a proton. Thus, some acid and its base are both present in the solution. These define the value of $[H^+]$, and the origin of the pH is understandable. However, in general, we choose solvents for redox reactions that are not easily oxidized or reduced. Thus, when the species to be titrated is dissolved in the solvent, there is no reaction. But the solution's electrochemical potential can only be established by *some* redox couple.

If the solution contains only one of the **electroactive species**—either the reduced form or the oxidized form alone—the redox equilibrium remains undefined. The potential E will not be determined by the analyte redox couple. In practice, it is nearly impossible to figure out the origin of the measured E before any titrant is added. However, we can predict the potential after a small amount of titrant has been added.

A redox titration sometimes requires that all of the analyte to be titrated be converted to a single oxidation state. The reason for this extra step is straightforward. During the sample pretreatment, it is difficult or impossible to eliminate all undesirable oxidants and reductants from the sample. For instance, oxygen from the air is a powerful oxidant and can only be eliminated with extra effort. Clearly, the presence of any oxidizing and/or reducing impurities can contribute to an error in the titration. For example, one of the common oxidizing titrants is cerium(IV) ion, Ce^{4+}. (In the older literature, this is called ceric ion.) In the redox determination of iron by oxidative titration of its iron(II) form, Fe^{2+}, the quantitative reaction that occurs is

$$Ce^{4+} + Fe^{2+} \rightleftharpoons Ce^{3+} + Fe^{3+}$$

But what happens to the accuracy if, before the titration is begun, some oxygen is present and oxidizes some of the iron(II) into the product, iron(III)? The titration to the end point will not be quantitative, since the iron(II) has already been partially "titrated" by oxygen. A pretreatment is required that will reduce all the iron to the iron(II) form but, in addition, will leave no excess reducing agent in the solution. (Will the assay result be high or low if some excess reducing reagent is present?) Some of the more commonly used reagents and methods for such pretreatments are outlined in Table 9.1.

Redox Titrants

Just as neutralization titrants are strong acids or strong bases, the titrants for redox titrations are strong oxidizing or strong reducing agents. Two other requirements are desired. First, the stoichiometry of the reactions should be well understood. For redox titrants, this means that the number of electrons transferred in a reaction is known. Second, the reaction should be clean. This means that only one reaction occurs, and it goes to completion rapidly. These

Table 9.1 Examples of Reagents Used for Pretreatment of Redox Samples Oxidizing and Reducing Agents*

Agent	Conditions of use	Method used to destroy excess
Preoxidizing agents		
Sodium peroxide	Fusion	Dissolve in water, boil
Hydrogen peroxide†	Alkaline solution	Boiling; $2\ H_2O_2 \rightleftharpoons 2\ H_2O + O_2$
Potassium persulfate [peroxydisulfate]	Acid solution	Boiling; $2\ S_2O_8^{2-} + 2\ H_2O \rightleftharpoons 4\ SO_4^{2-} + 4\ H^+ + O_2$
Ozone, O_3	Acid and alkaline	Boiling or inert gas purge (N_2, Ar)
Perchloric acid [hazardous with organics]	Acid	Dilution with water
Prereducing agents		
Metal amalgams of Zn, Cd, Pb, Ag	Solution	Passage through column of solid amalgam granules
Powdered metals	Solution	Filtration
Sulfur dioxide as gas	Acid	Boil or heat and purge with CO_2
Hydrogen peroxide	Acid	Boiling

*For further information about preoxidation and prereduction agents, see Blaedel, W. J., Meloche, V. W. 1963. *Elementary Quantitative Analysis*. New York: Harper & Row. pp. 818–822 and cited references.

†Hydrogen peroxide is both an oxidizing and reducing agent.

requirements are relatively stringent. As a result, the great majority of redox titrations are done with only a few different titrants. The identities of some redox titrants along with the number of electrons transferred by each are listed in Table 9.2.

Table 9.2 Commonly Used Titrants for Redox Titrations

Name of titrant	Common product of reaction	Electron equivalents per mole reagent
Oxidants		
Permanganate, MnO_4^-	In acid Mn^{2+}	5
	In base MnO_2	3
Iodine, I_2	I^-	2
Cerium (IV), Ce^{4+} (ceric ion)	Ce^{3+}	1
Dichromate, $Cr_2O_7^{2-}$	In acid Cr^{3+}	6
Reductants		
Thiosulfate, $S_2O_3^{2-}$	$S_4O_6^{2-}$	1
Iron(II), Fe^{2+} (ferrous ion)	Fe^{3+}	1

Redox Colorimetric Indicators

Colorimetric indicators can be convenient reagents to show the end points in neutralization titrations. Similarly, indicators have been discovered that can display the end point of redox titrations. These redox indicators change color with a change in the solution's electrochemical potential.

Just as a pH indicator is characterized by a pK_a value or pH range for its color change, redox indicators are characterized by a half-cell electrochemical potential $E^{\circ\prime}$ or an EMF range. Also, in exact analogy to the pH indicators, which change over a range of pH, the transformation of color of redox indicators progresses over a range of potentials. Some of these colorimetric indicators and their E° values (at pH = 0) are listed in Table 9.3. Most of these values change with pH.

There is a special group of redox indicators that are without parallel in aqueous acid-base titrations: the **self-indicators**. They are highly colored reagents that can act both as a titrant and as an indicator of the end point. As titrants, they are deeply colored materials. But after reacting with the analyte, they become pale in color or even colorless. Thus, during a titration, while the titrant is reacting with the analyte (and becoming pale to colorless), the solution appears colorless. However, after the end point, the excess titrant remains and colors the solution. A straightforward way to see how much of the colored, unreacted titrant appears in solution is to look at Figure 7.2a on page 218 and substitute "highly colored material" for the vertical axis label $[OH^-]$. Negligible amounts of the colored form remain until the end point,

Table 9.3 Examples of Color-Changing Redox Indicators*

	Color		
Indicator	Reduced	Oxidized	$E^{\circ}(pH = 0)^{\dagger}$
Tris(5-nitro-1,10-phenanthroline) Iron(II)sulfate [Nitroferroin]	Red	Pale blue	+1.25 V
Tris(1,10-phenanthroline) Iron(II)sulfate [Ferroin]	Red	Pale blue	1.06
Diphenylaminesulfonic acid	Colorless	Red-violet	0.84
Methylene blue	Colorless	Blue	0.34
Self-indicators			
Permanganate $[MnO_4^-/Mn^{2+}]$	Very pale pink	Pink	1.51
Iodine (+ starch) $[I_2/I^-]$	Colorless	Blue	0.536

*All these indicators are used in strongly acid solutions. Further information on redox indicators can be found in *Handbook of Analytical Chemistry*, L. Meites, ed. 1963. New York: McGraw-Hill, Sect. 3, and references therein.

†The potentials are at unit activity of H^+. These potentials will change significantly with pH and less so with other titration conditions such as ionic strength. High-acid concentrations (~1 M) will keep both the pH and ionic strength nearly constant during a titration.

Source: Kolthoff, I. M., Stenger, V. A. 1942. *Volumetric Analysis*, 2nd ed., Vol 1. New York: Interscience. p. 140.

but then its concentration increases linearly with titrant addition. Some self-indicating titrants are listed as a separate set in Table 9.3.

Redox Volumetric Calculations

Calculating analyte content from the data of redox titrations is done in exactly the same way as for neutralization titrations. Similarly, in developing the method, often the most difficult part is to ascertain the stoichiometric relationship between the titrant and analyte. The reactants and products of the reaction must be sufficiently well-known.

A sample calculation of analyte content follows. The method described is a quantitation using an **indirect titration**. In an indirect titration, a quantitative reaction is run to completion between the analyte and an excess of a reagent. The product of the first reaction (the amount is directly proportional to the quantity of analyte) is then titrated to an end point. An indirect titration is less desirable than a direct titration because, due to the extra steps, the precision is inherently lower. Indirect titrations are used when:

1. the analyte cannot be determined by a direct titration, and
2. the reaction of analyte and titrant is so slow that it will not go to completion within a reasonable time.

EXAMPLE 9.2

Chlorine in water can be determined in the following way. An aliquot of the sample is mixed with a solution containing an excess of KI. This reacts quantitatively with the stoichiometry

$$Cl_2 + 2\,I^- \rightleftharpoons 2\,Cl^- + I_2$$

The flask must be closed so that no volatile iodine escapes. The iodine formed is then titrated with $Na_2S_2O_3$ in neutral or slightly alkaline solution.

$$I_2 + 2\,S_2O_3^{2-} \rightleftharpoons 2\,I^- + S_4O_6^{2-}$$

The end point is determined by the disappearance of the iodine color (pale yellow). Starch can be added to the solution as an aid in the colorimetric end-point determination. Starch and iodine form an intensely colored material, so the disappearance of the last color is far easier to detect than for the pale iodine alone.

A 100.0-mL aliquot of chlorinated water was treated with excess KI solution and allowed to react in a closed flask. The pale yellow solution was titrated to an end point with 7.14 mL of 0.0114-N sodium thiosulfate solution. What was the content of Cl_2 in molarity and in ppm(w/v) of the water?

Solution

From the first chemical equation, we see that the normality of the iodine (I_2) titrated equals that of the chlorine (Cl_2) in the solution. From the second equation, we see that it requires two equivalents of $S_2O_3^{2-}$ per *mole* of iodine (and, thus, two equivalents per mole of chlorine). Thus,

$$N_{Cl_2} \cdot 100\text{ mL} = (0.0114\,N)(7.14\text{ mL})$$

and

$$N_{Cl_2} = 8.14 \times 10^{-4}\,N$$

This normality must now be converted to molarity and ppm(w/v) of Cl_2.

$$\text{molarity}_{Cl_2} = \tfrac{1}{2}\,\text{normality}_{Cl_2}$$

Thus

$$M_{Cl_2} = \tfrac{1}{2}(8.14 \times 10^{-4}) = 4.07 \times 10^{-4}\ M$$

The molecular mass of chlorine, Cl_2, is 70.91. The mass of chlorine in a liter of 4.07×10^{-4} M solution is

$$(70.91\ g/mol)(4.07 \times 10^{-4}\ mol/L) = 0.0290\ g/L$$

With a liter taken to be 1000 g, the relative concentration of Cl_2 in a liter is

$$0.0290 g/1000 g = 2.90 \times 10^{-5} \quad \text{or} \quad 29.0\ \text{ppm(w/v)}$$

9.3* Precipitation Titrations

Titrimetric assays can be based on the reaction of a titrant and analyte that form a precipitate. Neutralization and precipitation titrations differ in one interesting aspect. As you have seen, when a neutralization titration is half-way to the end point the solution's pH = pK_a. But consider the precipitation titration for sulfate with lead. The applicable reaction and mass-action expression are

$$Pb^{2+} + SO_4^{2-} \rightleftharpoons PbSO_4(s); \quad K_{sp} = [Pb^{2+}]\,[SO_4^{2-}] = 1.7 \times 10^{-8} \text{ at } 25°C$$

The pPb or pSO_4 equals pK_{sp} only when the other species is at unit activity—around 1-M concentration (the exact concentration depends on the activity coefficient). The K_{sp} and the SO_4^{2-} left in solution control the pPb. The titration curve of the pPb *vs.* volume-of-titrant differs from an acid-base titration curve because the solid does not form a buffer. The solid does not appear in the expression for K_{sp}, and therefore no buffer region is possible. In other words, half-way to the equivalence point pPb *does not* equal pK_{sp}. The pPb will be a large number (that is $[Pb^{2+}]$ is small) until near the end point.

There is one instance where a simple relationship exists between the lead and sulfate concentrations and the K_{sp}. At the equivalence point, the lead concentration in solution equals that of the sulfate.

$$[Pb^{2+}] = [SO_4^{2-}] = \sqrt{K_{sp}}$$

This is a general result only for 1:1 precipitate formation. After the equivalence point, the lead titrant concentration then increases rapidly as more is added.

It is important to recognize that not all precipitation reactions are good candidates for titrimetric analyses. One reason is that many precipitations are quite slow. As a result, the precipitation may require a long time to complete, and the titration's end point can become uncertain.

The mechanism of precipitation as a surface reaction is addressed in Section 10.2.

In a precipitation titration, as the end point is approached, the analyte forming the precipitate becomes more and more dilute. Thus, if the reaction is slow to begin with, the titrant must be added even more slowly. One of the reasons such reactions tend to be slow is that ongoing precipitation occurs at the surfaces of the precipitate particles. Considered as a concentration, these surfaces are relatively dilute.

More details on coprecipitation can be found in Section 10.3.

A second drawback of using precipitation reactions is that they tend *not* to be highly specific. Impurities with similar chemical properties can coprecipitate. As a result, there is a tendency for the results to be high. Sample preparation to remove impurities tends to be difficult, and the extra sample handling may reduce the assay precision.

A third drawback has been that colorimetric indicators for precipitation titrations tend to be tricky to work with. They require carefully controlled conditions of pH and ionic strength to work even moderately well. Another drawback to using such indicators is that the precipitate tends to make the solution opaque and the indicator color difficult to see. Only an especially intense color yields a clearly visible endpoint. However, the problems have been circumvented with the advent of selective sensors. For example, a titration for lead with sulfate titration can be be monitored with a lead-selective electrode to determine a titration curve. Nevertheless, few such titrations can compete with contemporary instrumental techniques with regard to specificity, limits of detection, simplicity of sample preparation, and speed.

Lead-selective electrodes, among others, are described in Section 15.3.

EXAMPLE 9.3

A sample was known to contain only chloride, iodide, and nitrate anions. A titration was carried out to determine the chloride and iodide content. Standardized silver nitrate was used as titrant, and a silver wire immersed in the solution was used to monitor the solution's electrochemical potential as the precipitates formed. The measured potential began with a plateau around −0.25 V, exhibited two end points, and ended with a plateau near +0.43 V as the titrant was added. The standardized silver nitrate solution was 0.1103 M. The burette's initial reading was 1.36 mL, and the volumes at the two end points were 12.77 mL and 42.97 mL. Given that the original sample was 0.4332 g, what are the percentages (w/w) of chloride and iodide in the sample? What is the concentration of silver at each of the equivalence points?

Solution

Necessary information:

$$K_{sp}(AgCl) = 1.78 \times 10^{-10} \text{ at } 25°C$$

$$K_{sp}(AgI) = 9.8 \times 10^{-17} \text{ at } 25°C$$

a.w. iodine = 126.90, and a.w. chlorine = 35.45

Stoichiometries for the halides' reactions with silver are

$$Ag^+ + Cl^- \rightleftharpoons AgCl(s) \quad \text{and} \quad Ag^+ + I^- \rightleftharpoons AgI(s)$$

Of the two halides, the iodide is the less soluble. Therefore, the first end point—when less silver has been added to the solution—is due to the iodide, and the second due to chloride. The volume of reagent for each ion type is

for iodide $12.77 - 1.36 \text{ mL} = 11.41 \text{ mL}$
for chloride $42.97 - 12.77 \text{ mL} = 30.20 \text{ mL}$

With a $AgNO_3$ solution = 0.1103 M, there are

for iodide $0.01141 \text{ L} \cdot 0.1103 \text{ M} = 1.258 \times 10^{-3} \text{ mol}$
for chloride $0.03020 \text{ L} \cdot 0.1103 \text{ M} = 3.331 \times 10^{-3} \text{ mol}$

With the appropriate atomic weights,

for iodide $1.258 \times 10^{-3} \text{ mol} \cdot 126.90 \text{ g mol}^{-1} = 0.1596 \text{ g}$
for chloride $3.331 \times 10^{-3} \text{ mol} \cdot 35.45 \text{ g mol}^{-1} = 0.1181 \text{ g}$

For a sample weighing 0.4332 g, the percent (w/w) composition is

for iodide $(0.1596 \text{ g}/0.4332 \text{ g}) \cdot 100 = 36.84\%$
for chloride $(0.1181 \text{ g}/0.4332 \text{ g}) \cdot 100 = 27.26\%$

At each equivalence point the metal-ion concentration equals the halide concentration. Thus, letting $X = [Ag^+]$,

for iodide $K_{sp} = 9.8 \times 10^{-17} = [Ag^+][I^-] = X^2$ and $[Ag^+] = [I^-] = 9.9 \times 10^{-9}$ M

for chloride $K_{sp} = 1.78 \times 10^{-10} = [Ag^+][Cl^-] = X^2$ and $[Ag^+] = [Cl^-] = 1.3 \times 10^{-5}$ M

assuming that the titration was done at 25°C and ignoring ionic strength effects.

9.4 Complexometric Titrations (Chelometric Titrations)

Coordinating agents and chelates are molecules (such as ammonia) or ions (such as cyanide) that can exist independently in solutions and that are capable of binding to metal ions, such as zinc. Molecules composed of metals and chelates, or metals and coordinating agents, are called coordination complexes.

A titration based on the formation of coordination complexes is called a **complexometric** titration. It is necessary to find some titrant that has a *simple* stoichiometry and also binds strongly to the analyte ion; the coordination reaction must go to completion. The usual procedure employs a chelate as titrant to determine metal-ion analytes. A number of different reagents satisfy these criteria, but only a few are used and most belong to the group of molecules called polyaminocarboxylic acids. Reading this word in parts, it says that there are many (poly-) amino groups along with carboxyl groups (−COOH) on the molecule. The structures of a few of these polyaminocarboxylate reagents are shown in Figure 9.1, edta being the most commonly used. As you saw in Chapter 8, its great advantage is that it reacts with many metals by the simple stoichiometry

Formation constants for several metal-edta complexes are found in Table 8D.2 on page 276.

$$\text{edta}^{4-} + M^{n+} = M(\text{edta})^{n-4}$$

where n is the charge on the metal ion, M. Six sites on the molecule can bind to many metal ions: two nitrogen sites and four carboxylic acid sites, as seen in the structure in Figure 9.1. Competition of the metal with protons dictates that the titrated solution must be well buffered for pH and approximately fixed in ionic strength to obtain reproducible, precise results.

edta As Titrant

One of the more common tests for which edta titration is used is quantitating calcium in water—part of the characteristic called water hardness. Hardness is customarily expressed as ppm of calcium as calcium carbonate. The hardness is expressed this way even though it is caused by calcium and magnesium sulfates and bicarbonates.

The stoichiometry of the reaction of calcium and edta is simply

$$Ca^{2+} + \text{edta}^{4-} \rightleftharpoons Ca(\text{edta})^{2-}$$

with an equilibrium constant for formation expressed as

$$K_{f,\{Ca(\text{edta})\}^{2-}} = \frac{[Ca(\text{edta})^{2-}]}{[Ca^{2+}][\text{edta}^{4-}]}$$

Finding the effective formation constant is presented in detail in Section 8D.

However, as usual, the ability to use edta as a titrant requires that the effective formation constant be large enough. So, once again, we find ourselves

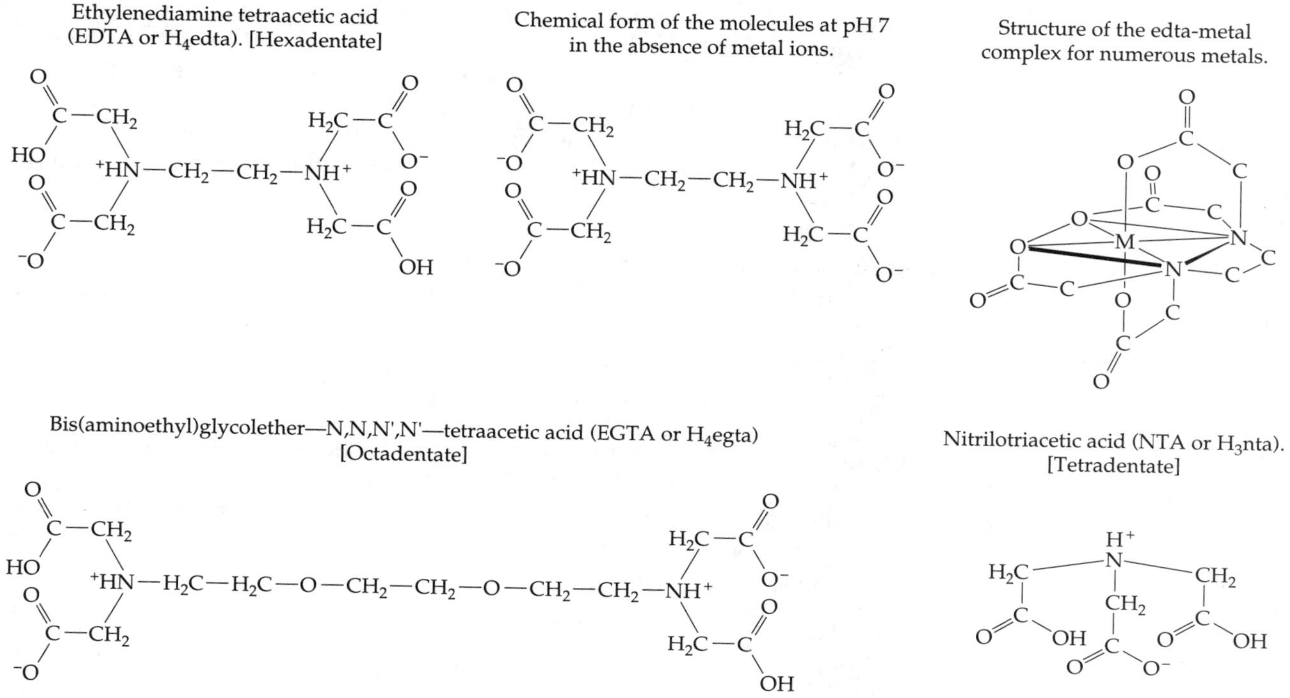

Ethylenediamine tetraacetic acid (EDTA or H_4edta). [Hexadentate]

Chemical form of the molecules at pH 7 in the absence of metal ions.

Structure of the edta-metal complex for numerous metals.

Bis(aminoethyl)glycolether—N,N,N',N'—tetraacetic acid (EGTA or H_4egta) [Octadentate]

Nitrilotriacetic acid (NTA or H_3nta). [Tetradentate]

FIGURE 9.1 s
Three titrants used for complexometric titrations.
The Chemical Abstracts name for H_4edta is N,N′-1,2-ethanediylbis [*N*-(carboxylmethyl)-glycine].

working with K_{eff}. The value of the calcium-edta effective formation constant at pH 9.0 (and only at pH 9.0), at 25°C, and at 0.2-M ionic strength is

$$K_{\text{eff},\{\text{Ca(edta)}\}^{2-}} = 3.8 \times 10^{10}$$

What does this number mean? If we mix 50.00 mL of a 0.100-M edta solution in water strongly buffered to pH 9.0 together with 50.00 mL of 0.100-M calcium chloride solution, the resulting solution at equilibrium has a volume of 100.0 mL and contains the following:

0.05-M Ca(edta)^{2-}

4.4×10^{-6}-M Ca^{2+}

4.4×10^{-6}-M edta not bound to calcium

0.1-M Cl^-

1×10^{-9}-M H^+

Various components of the buffer

Note that of all the calcium and edta only a small part is left unbound—around 1 part per 10,000. As a qualitative description, we can say that the reaction has gone to completion and the ligand is expected to be a satisfactory titrant under these conditions.

However, if any more edta is added above the amount that binds quantitatively with the calcium, no more free calcium remains to bind with it. So the amount of edta added in excess remains free in solution. But this is just the same behavior as when a strong base was added to the solution of acetic acid. After the acetic acid was used up, the excess base remained in the solution unreacted. A plot of the concentration of free edta in the solution during the titration will be similar to Figure 7.4a, with [edta] replacing $[OH^-]$. A number of direct parallels between acid-base titrations and complexometric ones are shown in Table 9.4.

Why Not Use a Simpler Ligand?

Why not use a simpler ligand such as ammonia for a titration? To answer that question, let us calculate the equilibrium properties of Zn with ammonia, which reacts to form the zinc-ammonia complex

$$Zn^{2+} + 4\,NH_3 = [Zn(NH_3)_4]^{2+}$$

In aqueous solution, the zinc ion actually has water molecules bound around it, and the ammonia replaces the water. Numerous studies have shown that the following reactions occur. The equations may look messy at first, but you will see by inspecting them that each step shows the replacement of 1 water by 1 ammonia molecule.

$$[Zn(H_2O)_4]^{2+} + NH_3 \rightleftharpoons [Zn(H_2O)_3NH_3]^{2+} + H_2O \qquad K = 186 \qquad \textbf{(9-1)}$$

$$[Zn(H_2O)_3NH_3]^{2+} + NH_3 \rightleftharpoons [Zn(H_2O)_2(NH_3)_2]^{2+} + H_2O \qquad K = 219 \qquad \textbf{(9-2)}$$

$$[Zn(H_2O)_2(NH_3)_2]^{2+} + NH_3 \rightleftharpoons [Zn(H_2O)(NH_3)_3]^{2+} + H_2O \qquad K = 251 \qquad \textbf{(9-3)}$$

$$[Zn(H_2O)(NH_3)_3]^{2+} + NH_3 \rightleftharpoons [Zn(NH_3)_4]^{2+} + H_2O \qquad K = 11 \qquad \textbf{(9-4)}$$

For a solution consisting of water, ammonia, and zinc ions (and some counter-ions such as Cl^- or NO_3^-), the series of equations above represent a set of simultaneous equilibria of the type with which you are now familiar. However, the ammonia and water also take part in another competitive equilibrium system: the simple acid-base equilibrium with the well-known form

$$NH_3 + H_2O \rightleftharpoons NH_4^+ + OH^- \qquad \textbf{(9-5)}$$

which you can recognize as a hydrolysis reaction.

Table 9.4 Comparison of Neutralization and Complexometric Titrations

Factor	Ca-edta complexometric titration (buffered near neutral pH)	Acetic acid-NaOH neutralization titration
Titrant species	H_2edta^{2-}	OH^-
Titrant reaction (to completion)	$H_2edta^{2-} + Ca^{2+} \rightleftharpoons [Ca(edta)]^{2-} + 2H^+$	$OH^- + HOAc \rightleftharpoons H_2O + OAc^-$
Titrated species	$Ca(H_2O)_6^{2+}$	HOAc
Titration product	$[Ca(edta)]^{2-}$	H_2O
Titrant competing with	H_2O	OAc^-
Titrant binds with	Ca^{2+}	H^+
Species monitored during titration	edta (using indicator) Ca^{2+} (potentiometrically)	OH^- via pH (indicator or potentiometric)
End-point jump with change in	edta	OH^-

These five equations, 9-1 through 9-5, will be tedious to solve as simultaneous equations. It is generally easier to use a computer program to do the calculations to find the species present under a given set of conditions. But here we are interested only in how coordination complex formation might be used as the basis of a titrimetric method to determine metals in solution.

Two factors affect the equilibria 9-1 to 9-5 which, in turn, influence how a complexometric titration is carried out. These two factors are true for all complexometric titrations.

1. The concentration of each of the complexes $[Zn(NH_3)_n]^{2+}$ will depend on the pH of the solution. (This is so because the concentration of NH_3 depends on the pH, by Equation 9-5.) In addition, the protons H^+ compete with the Zn^{2+} for the ammonia. As a result, to have any well-defined equilibria, the pH will have to be buffered.
2. As for any equilibrium involving charged species, the equilibrium constants will depend on the ionic strength. Thus, the ionic strength should be kept approximately constant.

Consider that you have a sample containing an unknown quantity of zinc. The sample is dissolved in water, the solution is strongly buffered for pH, and the ionic strength is brought to a known value. Can you titrate the solution with ammonia and get a good measure of the zinc content? The answer is, "not likely." The difficulty is that the stoichiometry involves so many species simultaneously.

Notice that the equilibria represented by Equations 9-1 through 9-4 have similar equilibrium constants. As the ammonia is added, there will be four different ammonia complexes formed under approximately the same conditions. To observe the end point of the titration, some method would be needed to observe either the disappearance of the last zinc or the appearance of an excess of free ammonia. However, as illustrated by the titration curve in Figure 9.2, no simple jump in the amount of free ammonia or free zinc appears at any point during a titration with ammonia. Thus, ammonia is *not* a good titrant to use for zinc or for metals other than zinc as well.

Colorimetric Indicators Used in Complexometric Reactions

The properties of indicators used in complexometric titrations parallel those of pH-sensitive indicators. The indicator must be intensely colored in at least one form (bound or unbound to the metal). Because the color of these indicators changes upon binding (and dissociation) of metal ions, they have been given the name **metallochromic** indicators. The name simply means *metal-colored*.

From the similarities between the acid-base and complexometric titrations, you might guess what chemical properties are required for a metallochromic indicator. You suspect that the indicator must bind the metal ion less strongly than the titrant, edta. Otherwise the indicator would bind in preference to the titrant. The color would not change if the metal could not be displaced. However, again following the analogy, you might predict that the indicator must bind more strongly to the metal ion than the water does. Otherwise the indicator could not react with the metal ion at all. The water would always win out. Your predictions would be correct on both points.

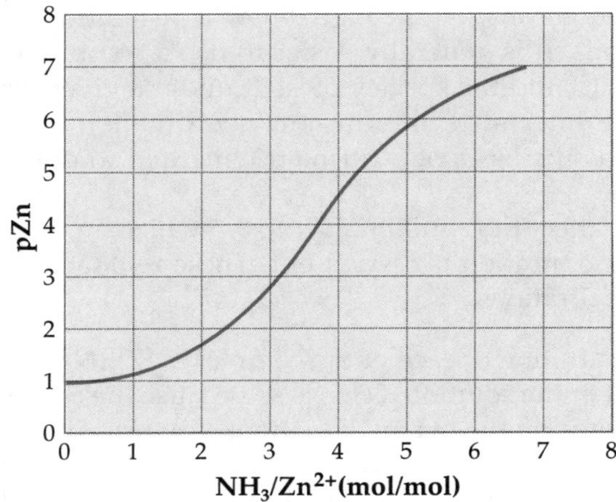

FIGURE 9.2 s
The calculated titration curve of 0.1-M Zn^{2+} in water with ammonia as titrant.

The curve that is plotted is pZn versus the mole ratio ammonia/Zn^{2+}. The pZn is the negative log of the concentration of aquo-zinc ions uncoordinated by ammonia. The pH is assumed to be high enough so that there is negligible NH_4^+. In addition, it is assumed that the zinc remains completely in the form $[Zn(H_2O)n]^{2+}$ and does not

The metallochromic indicator system, however, has one property that differs from those of acid-base systems. The indicator's binding to a metal ion may be dependent on the solution pH. As a result, the complexometric titration must be carried out in pH-buffered solutions. (Certainly not the conditions of an acid-base titration!)

The indicator reaction depends on the ionic strength and perhaps on direct competition of some other metal-ion species in the solution. The metal-ion competition causes one of the major problems with chelometric titrations: the lack of selectivity. Complicated solutions require significant pretreatment to remove possible interferents. The interferents must be removed or **masked** (prevented from reacting). One example of such masking is described in the example that follows in which CN^- is the masking agent.

When the sample has been prepared (interferents removed or masked, and the pH and ionic strength fixed) calculating the content of a sample from the data is as straightforward as for any other titration.

EXAMPLE 9.4

A 100.0-mL sample of tap water was placed in an Erlenmeyer flask. To this was added 1-g ascorbic acid. (The ascorbic acid reduces any oxygen dissolved in the solution and protects the indicator from air oxidation.) To this was added 1 mL of 50%(w/v) NaOH to bring the pH to about 13 and the ionic strength to about 0.2. Then 0.1 g of NaCN was added to reduce interference from any Fe, Cu, Ni, and the like that may be present. (Their metal-cyano complexes are sufficiently stable under the titration conditions so that edta cannot coordinate with the metal ions.) After the solution was warmed slightly with gentle mixing for a few minutes, it was cooled. Eriochrome Black T indicator was then added. (Dilute solutions of the indicator are pink when complexed with metals and blue when uncomplexed in solutions in the range pH 7 to pH 11.) This solution was titrated to the point where the last tinge of purple disappeared, leaving a blue solution. At that point,

14.11 mL of 0.0237-M Na_2H_2edta had been used. What is the hardness of the water as $CaCO_3$ in ppm? Water with a hardness below 100 is generally considered to be soft. Water with a hardness above 300 is considered to be hard.

Solution

Necessary information:

The formula weight for $CaCO_3$ is 100.09.

The stoichiometry for the reaction at this pH is $Ca^{2+} + edta^{4-} \rightleftharpoons Ca(edta)^{2-}$

Magnesium ion also has the same stoichiometry, and the titration includes the sum of Mg^{2+} and Ca^{2+}.

There are

$$0.01411\ L \cdot 0.0237\ M = 3.34 \times 10^{-4}\ \text{mol of } Ca^{2+} \text{ and/or } CaCO_3$$

This is

$$3.34 \times 10^{-4}\ \text{mol} \cdot 100.09\ \text{g mol}^{-1} \cdot 1000\ \text{mg g}^{-1} = 33.5\ \text{mg of } CaCO_3 \text{ in the } 100.0\ \text{mL}$$

$$(33.5\ \text{mg})/(0.1000\ \text{L}) = 335\ \text{mg L}^{-1} = 335\ \text{ppm hardness (Ca as } CaCO_3)$$

Suggestions for Further Reading

SERJEANT, E. P. 1984. *Potentiometry and Potentiometric Titrations*. New York: Wiley.
An advanced textbook on potentiometric titrations. Not much about how to do the work but loaded with cautions about the techniques.

RINGBOM, A. J. 1963. *Complexation in Analytical Chemistry; A Guide for the Critical Selection of Analytical Methods Based on Complexation Reactions*. New York: Interscience.
Best of the breed for theory.

PRIBIL, R. 1982. *Applied Complexometry*. Oxford and New York: Pergamon.
The best place to start if you are actually going to do complexometric titrations.

WILSON, C. L. 1960. In *Comprehensive Analytical Chemistry*. Vol. 1B, Chap. 7. C. L. Wilson and D. W. Wilson, eds. Amsterdam: Elsevier.
Broadly ranging chapter on the theory and practice of titration of inorganic compounds.

FLASCHKA, H. A. 1964. *EDTA Titrations*. Oxford: Pergamon.
A monograph on the theory and practice of edta titrations.

HUBER, W. 1967. *Titrations in Nonaqueous Solvents*. New York: Academic Press.

KUCHARSKY, J., SAFARIK, L. 1965. *Titrations in Non-Aqueous Solvents*. Amsterdam: Elsevier.
Two books with similar, broad coverage and a similar outlook on the principles and practice of non-aqueous titrations. Newer work has not, apparently, been collected in book form.

BARREK, J., BERKA, A. 1984. "Redox Titrants in Nonaqueous Media." *CRC Critical Reviews in Analytical Chemistry* 15:163–221.
An extensive review article describing oxidizing agents and reducing agents that can be used for nonaqueous redox titrations.

SVEHLA, G. 1978. *Automatic Potentiometric Titrations*. Oxford: Pergamon.
Despite its title, mostly titration theory. The chapters on automatic titration (about one-fourth of the book) are quite informative.

Concept Review

1. Define what is meant by an "indirect" titration based on a redox reaction.

2. *Sketch* the shape of the titration curve for each of the following titrations. Indicate where the endpoint occurs on the curve.

(a) redox titration of Fe^{2+} with Ce^{4+} (E vs. mL Ce^{4+})

(b) complexometric titration of Mg^{2+} with edta (pMg^{2+} vs. mL edta)

(c) precipitation titration of Ba^{2+} with SO_4^{2-} (pBa^{2+} vs. mL SO_4^{2-})

Exercises

9.1 A standardized iodine solution is used to titrate hydrazine sulfate in a sodium bicarbonate-buffered solution. The reactions that occur are

$$I_2 + 2\,e^- \rightleftharpoons 2\,I^-$$

$$N_2H_4{\cdot}H_2SO_4 \rightarrow N_2 + SO_4^{2-} + 6\,H^+ + 4\,e^-$$

What is the molarity of a solution of hydrazine sulfate when 27.29 mL of 0.1000 N iodine is required to titrate 25.00 mL of the hydrazine sulfate to the equivalence point?

9.2 $KMnO_4$ is being standardized by titration with As_2O_3 (f.w. 197.82) in acid solution. The reaction is

$$2\ MnO_4^- + 5\ H_3AsO_3 + 6\ H^+ \xrightleftharpoons{ICl(cat)} 2\ Mn^{2+} + 5\ H_3AsO_4 + 3\ H_2O$$

What is the normality of a $KMnO_4$ solution if 45.00 mL are required to titrate a 0.1500-g sample of As_2O_3?

9.3 Standard $KMnO_4$ (f.w. 158.038) was prepared by dissolving about 3 g of reagent grade $KMnO_4$ in 950 mL of distilled water in a beaker. The beaker was covered with a watch glass, heated almost to boiling for 1 hour, and set in the dark to age overnight. The resulting solution was filtered through a glass sinter and stored in the dark. The solution was standardized against primary standard sodium oxalate (f.w. 133.9995) that was dried at 110°C for 1 h. A 1.3653-g portion of the oxalate was weighed out and dissolved in 250.0 mL water. This relatively unstable solution was used immediately. A 50.00-mL aliquot of the $Na_2C_2O_4$ solution was pipetted to a 250-mL beaker, and 50 mL of water containing 5 to 6 mL of conc. H_2SO_4 was added. The solution was heated to 90°C and slowly titrated with the permanganate to the end point of the first perceptible persistent pink color. (Too fast an addition of permanganate results in unwanted side reactions.) 45.43 mL of the $KMnO_4$ solution was required. The reaction for the titration is

$$5\ H_2C_2O_4 + 2\ MnO_4^- + 6\ H^+ = 10\ CO_2(g) + 2\ Mn^{2+} + 8\ H_2O$$

Calculate the normality and molarity of the standard $KMnO_4$ solution.

9.4 You are familiar with the formula for chemical equivalence,

$$\text{volume}_1 \times \text{normality}_1 = \text{volume}_2 \times \text{normality}_2$$

For titration, however, we sometimes like to put the result into a plug-in equation. Consider the following titration reaction.

$$I_2 + S_2O_3^{2-} \rightleftharpoons 2I^- + S_4O_6^{2-}$$

(a) Write an equation that expresses the normality of I_2 (f.w. 253.82) in the original solution in terms of the volume of the sample solution and the volume and normality of the thiosulfate ($S_2O_3^{2-}$) only. All known quantities should be collected into a single algebraic factor:

$$N_{I_2} = \text{factor} \times V_{S_2O_3^{2-}}$$

(b) Do the same as in part (a) but express the concentration of I_2 in mg L^{-1}.

9.5 An aqueous solution was titrated for alcohol in an acidic solution. The procedure was carried out correctly, and the solution was reported as containing 0.550% (w/w) ethanol. The solution contained no ethanol, however, only methanol. The reactions under the experimental conditions were

$$C_2H_5OH \rightarrow H_3CCOOH \quad \text{(unbalanced)}$$

$$CH_3OH \rightarrow CO_2 \quad \text{(unbalanced)}$$

(a) Balance the equations.
(b) What is the w/w content of methanol in the solution?

9.6 50.00 mL of 0.1000 M $FeSO_4$ is titrated with 0.1000 M $Ce(SO_4)_2$ in 1 M H_2SO_4. The reactions occurring in the titration are

$$Ce^{4+} + e^- \rightleftharpoons Ce^{3+};\quad E° = 1.61\ V$$

$$Fe^{3+} + e^- \rightleftharpoons Fe^{2+};\quad E° = 0.771\ V$$

Calculate the expected values, ignoring possible activity effects.
(a) At the equivalence point, what is the total concentration of the two iron species?
(b) At the equivalence point, what is the total concentration of the two cerium species?
(c) At the equivalence point, what is the expected value of E?
(d) At the equivalence point, what is the value of the ratio

$$\frac{[Fe^{3+}][Ce^{3+}]}{[Fe^{2+}][Ce^{4+}]}$$

(e) At the equivalence point, what is the value of $[Fe^{2+}]$ in terms of $[Ce^{4+}]$?
(f) At the equivalence point, what is the value of $[Fe^{3+}]$ in terms of $[Ce^{3+}]$?
(g) Calculate the concentrations of $[Fe^{2+}]$, $[Fe^{3+}]$, $[Ce^{3+}]$, and $[Ce^{4+}]$ at the equivalence point.

9.7 Determination of small amounts of free water in solids or organic solvents can be carried out using the **Karl Fischer** method. The reaction involved is

$$I_2 + SO_2 + CH_3OH + 3\ C_5H_5N + H_2O \rightarrow 2\ C_5H_5NH^+I^- + C_5H_5NH^+ + SO_4CH_3^-$$

A tablet weighing 502.2 mg was crushed and suspended in anhydrous methanol. A standardized solution of I_2 (0.0139 M) was used to titrate the free water in the tablet. The endpoint (I_2 color persisting for >30 s) occurred after 20.16 mL of I_2 solution had been added. What was the w/v percent H_2O in the tablet?

9.8 50.00 mL of 0.1000 M $FeSO_4$ is titrated with 0.1000 M $Ce(SO_4)_2$. What is the expected composition of the solution—$[Fe^{2+}]$, $[Fe^{3+}]$, $[Ce^{3+}]$, and $[Ce^{4+}]$—and the solution's E-value after 40.00 mL of the standard $Ce(SO_4)_2$ is added?

9.9 Total sulfur in nickel can be determined in the following way. A sample of nickel is dissolved in a solution 0.9 M in cupric potassium chloride ($CuCl_2 \cdot 2\ KCl \cdot 2\ H_2O$) and hydrochloric acid. All the sulfur is taken up and precipitated as a copper sulfide residue. This process concentrates the sulfur. The residue is collected on a low-sulfur filter paper, which is dried and then burned in a combustion furnace at above 1850°C for two minutes. The sulfur on the paper produces SO_2 which is passed into acidified water to form sulfurous acid, H_2SO_3. The acid produced is titrated with potassium iodate to a redox end point determined with the aid of thyodene-iodide indicator. The reactants and products in an acid solution are

$$IO_3^- + SO_3^{2-} \rightarrow SO_4^{2-} + I_2 \quad \text{(unbalanced)}$$

A sample of 9.888 g of nickel alloy was dissolved overnight in the dissolution solution. After drying and combustion, a blank required 0.11 mL to titrate. The potassium iodate solution contained 0.0444 g/L of KIO_3 (f.w. 214.00). If the sample required 1.96 mL of titrant, what is the sulfur content of the sample in ppm? [Ref: Burke, K. E. 1974. *Anal. Chem.* 46:882.]

9.10 A sample to be analyzed for bromide and chloride was prepared by the following EPA method.

Summary of the Method The determination of bromide and iodide consists of two separate experiments. The iodide is first determined in the sample, and then a second experiment determines the combined iodide and bromide. The bromide content of the sample is calculated from the difference between the iodide and the combined iodide and bromide determination.

The iodide in the sample is oxidized to iodate, IO_3^-, with saturated bromine water in an acid buffer solution. The excess bromide is destroyed by the addition of sodium formate. Potassium iodide is added to the sample solution with the resulting liberated iodine being equivalent to the iodate initially formed in the oxidation step. The liberated iodine is determined by titration with sodium thiosulfate.

In a second sample, iodide and bromide are oxidized to iodate and bromate with calcium hypochlorite. The iodine liberated by the combined reaction products is measured after destruction of the excess hypochlorite and addition of potassium iodide.

Interferences Iron, manganese, and organic matter interfere with the above methods. Treatment of the initial samples with calcium oxide removes the interferents.

The following procedure was followed in analysis of wastewater. Sodium thiosulfate titrant was prepared by diluting 100.0 mL of a stock 0.0375 *N* (including 5 mL/L of chloroform as preservative) solution to 500.0 mL. The solution was standardized by titration with potassium biiodate, $K_2I_2O_6$, and found to be 0.0076 *N*.

100.0 mL of the sample was placed in a beaker and stirred with addition of H_2SO_4 to make the pH approximately 6.5. The sample was transferred to a 250-mL conical flask, and the beaker was rinsed with distilled water twice, with the rinse being added to the flask. To this solution, 15 mL of sodium acetate solution (275 g/L) and 5 mL acetic acid solution (1:8 acid: water) were added and mixed. Then, 40 mL of bromine water solution (0.2 mL bromine in 800 mL of distilled water) was added. After five minutes, 2 mL of sodium formate solution (500 g/L) was added, and the solution mixed. After five minutes, bromine fumes were removed from over the solution over a period of 30 s using a stream of nitrogen gas. Finally, approximately 1 g of potassium iodide was added to the sample followed by 10 mL of 1:4 H_2SO_4 solution. After a five-minute reaction period, the sample and a blank were titrated with the sodium thiosulfate solution using a 10 mL burette (markings each 0.1 mL). As the end point was approached, amylose indicator was added. (The end point is indicated by a blue/colorless transition.) A similar procedure was followed, with calcium hypochlorite added as oxidant. [Refs: EPA-600/2-79-200, Level 2 Sampling and Analysis of Oxidized Inorganic Compounds; EPA-600/4-79-020, Methods for Chemical Analysis of Water and Wastes.]

The following data were collected.

Sample	Titrant Volume Readings (mL)	
	Beginning	**Ending**
Blank 1	0.03	0.07
Br_2 oxidant	0.07	1.63
Blank 2	1.63	1.82
$Ca(ClO)_2$ oxidant	1.82	8.34

(a) What is the sample content in mg/L of I^-?

(b) What is the sample content in mg/L of Br^-?

9.11 The following half-reactions of nitrate in aqueous acid have been measured.

$$NO_3^- + 4\,H^+ + 3\,e^- \rightleftharpoons NO + 2\,H_2O; \quad E° = 0.96V$$

$$NO_3^- + 3\,H^+ + 2\,e^- \rightleftharpoons HNO_2 + H_2O; \quad E° = 0.94V$$

$$2\,NO_3^- + 4\,H^+ + 2\,e^- \rightleftharpoons N_2O_4 + 2H_2O; \quad E° = 0.80V$$

However, a quantitative analysis of nitrate can be made by reaction with excess $FeCl_2$ in the presence of a catalytic amount of molybdenum and back-titrating the excess Fe^{2+} in acid with dichromate with diphenylamine as indicator. The reactions are

$$2\,NO_3^- + 6\,Fe^{2+} + 8\,H^+ \rightleftharpoons 2\,NO + 6\,Fe^{3+} + 4\,H_2O$$

$$Cr_2O_7^{2-} + 6\,Fe^{2+} + 14\,H^+ \rightleftharpoons 2\,Cr^{3+} + 6\,Fe^{3+} + 7\,H_2O$$

The titrant solutions were 0.0100 M $FeCl_2$ and 0.0100 M $Cr_2O_7^{2-}$. If 25.00 mL of the iron solution was reacted with 100.0 mL of a nitrate-containing solution, and 3.47 mL of the chromate solution was needed to titrate the excess Fe^{2+}, what was the original concentration of nitrate in the sample?

*9.12 The reaction of Cr(VI) (as $Cr_2O_7^{2-}$) with Fe(II) is reasonably rapid. As a result, the titration chemistry can be determined from equilibrium considerations. The following reactions may occur in an $FeCl_2$ solution titrated with $K_2Cr_2O_7$.

$$Cl_2 + 2\,e^- \rightleftharpoons 2\,Cl^- \qquad E° = 1.36\text{ V}$$

$$Cr_2O_7^{2-} + 14\,H^+ + 6\,e^- \rightleftharpoons 2\,Cr^{3+} + 7\,H_2O \qquad E(1\text{-}N\,H_2SO_4) = 1.05\text{ V}$$

$$Fe^{3+} + e^- \rightleftharpoons Fe^{2+} \qquad E(1\text{-}N\,H_2SO_4) = 0.70\text{ V}$$

$$Fe^{3+} + e^- \rightleftharpoons Fe^{2+} \quad E(1\text{-}N\,H_2SO_4;\,0.5\text{ M }H_3PO_4) = 0.61\text{ V}$$

Possible redox indicators:

Ferroin: Transition potential at pH = 0 1.06 V

Sodium diphenylamine sulfate:
Transition potential at pH = 0 0.85 V

(a) Which of these two indicators is the better one to use?
(b) The 0.5 M phosphate is added to decrease the E of the Fe(II)/Fe(III) couple in order to increase the sharpness of the end point. Does it do so by interacting preferentially with the Fe(II) or the Fe(III)?
(c) Will the dichromate oxidize the chloride to any appreciable extent at pH = 0?

9.13 It is sometimes significantly misleading to take $E°$ as a measure of oxidizing power. For example, dichromate in acid reacts as

$$Cr_2O_7^{2-} + 14\,H^+ + 6\,e^- \rightarrow 2\,Cr^{3+} + 7\,H_2O$$

(a) Calculate the potential you expect for this reaction at pH = 0 for a solution containing 1.33 mM $K_2Cr_2O_7$, 1.33 M $CrCl_3$, and 1.00 M HCl. Assume that activity effects can be ignored. T = 25°C.
(b) Chloride should be oxidizable by this system since

$$Cl_2 + 2\,e^- \rightleftharpoons 2\,Cl^-; \quad E° = 1.36\text{V}$$

However, it is not. An experimental measurement of E for the $Cr_2O_7^{2-}/Cr^{3+}$ system shows 1.09 V. Assuming that activity effects are negligible, and assuming that the effect is due entirely to chloride, does the chloride bind more strongly to the $Cr_2O_7^{2-}$ or the Cr^{3+}?

9.14 Water hardness is defined as the combination Ca^{2+} plus Mg^{2+} reported as $CaCO_3$. One method used to determine hardness is to acidify a water sample with HCl, boil the acid solution to remove CO_2, and neutralize with NaOH. The solution is buffered at pH 10 with an ammonia buffer. This solution is titrated with edta to the blue end point of Eriochrome Black T. The stoichiometry of the edta-metal interaction is 1:1. If a 50.00-mL sample of water requires 31.63 mL of 0.0136 M edta solution to titrate it to the end point, what is the water hardness in ppm as $CaCO_3$?

*9.15 Zirconium in the form Zr(IV) can be determined by direct titration with edta in aqueous solution forming a 1:1 complex. However, in order to speed up the equilibration by heating to 70–90°C, significant hydrolysis occurs with a consequent loss of accuracy. Thus, the procedure is to take a sample solution with approximately 3–5 mM in zirconium, add a slight excess of edta solution, adjust the acidity to pH 1.4 with ammonia solution, heat the solution to near boiling, add Eriochrome Cyanine indicator, and back-titrate with zirconium chloride to the first permanent pink color.

To a sample weighing 0.01070 g dissolved in hydrochloric acid was added 5 g of zinc amalgam to reduce any iron present which might otherwise interfere. To this was added 10.00 mL of 0.0477 M standardized edta solution. After adjusting the pH, adding indicator, and heating, the sample required 5.91 mL of 0.0384 M zirconium oxide in 5% HCl solution to titrate to the end point. [Ref: Fritz, J. S., Fulda, M. O. 1954. *Anal. Chem.* 26: 1206–1208.]
(a) What percentage of the sample is zirconium?
(b) Any iron present is reduced from Fe(III) to Fe(II) by pretreatment with zinc amalgam. Given that this eliminates iron interference, how do the Fe(III)-edta and Fe(II)-edta complexes compare in stability with the Zr-edta complex in these conditions?
(c) Is the pink form of the indicator the protonated form or the zirconium complex?
(d) With a direct titration, under the conditions of the experiment, which is more stable: the complex of zirconium with the indicator or the complex with edta?
(e) With back-titration, under the conditions of the experiment, which is more stable: the complex of zirconium with the indicator or the complex with edta?
(f) Write the equations for the zirconium reaction with the edta and the indicator for the process of back-titration. At this pH, the predominant form of edta is H_5edta, and the metal-free indicator is in the form H_2In^-. It has been found that the indicator forms a 2:1 complex with zirconium. The edta as well as the indicator both lose all their protons upon complexation.

Additional Exercises

9.16 A standardized iodine solution is used to titrate a bleach solution that contains hypochlorite.
(a) What is the reaction that occurs?
(b) If we use a 10.00 mL solution of 3.5% sodium hypochlorite, how much of a 0.010 N solution of iodine will be required?

9.17 $KMnO_4$ is being standardized by titration with As_2O_3 (f.w. 197.82) in acid solution. The reaction is

$$2\,MnO_4^- + 5\,H_3AsO_3 + 6\,H^+ \xrightleftharpoons{\text{ICl(cat)}} 2\,Mn^{2+} + 5\,H_3AsO_4 + 3\,H_2O$$

What is the normality of a $KMnO_4$ solution if 25.43 mL is required to titrate a 0.1034 g sample of As_2O_3?

*Indicates more involved exercises

9.18 A 50.00 mL aliquot of a solution containing $Fe(NH_4)_2SO_4$ is titrated with 0.0893 M $Ce(SO_4)_2$. The reactions occurring in the titration are

$$Ce^{4+} + e^- \rightleftharpoons Ce^{3+}; \quad E° = 1.61 \text{ V}$$

$$Fe^{3+} + e^- \rightleftharpoons Fe^{2+}; \quad E° = 0.771 \text{ V}$$

The titration required 27.63 mL of the Ce^{4+} solution. What was the concentration of the Fe^{2+} solution?

9.19 If a 50.00-mL sample of tap water requires 16.74 mL of 0.01073-*M* edta solution to titrate it to the end point, what is the water hardness in ppm as Ca^{2+}? As $CaCO_3$?

9.20 How many mL of a 0.00135 M Cl^- solution is required to titrate the following samples?

(a) 25.00 mL of 0.0123 M Ag^+

(b) 50.00 mL of 0.0257 M Hg_2^{2+}

9.21 What is the concentration of metal ion present at the endpoint of each of the titrations in Problem 9.20?

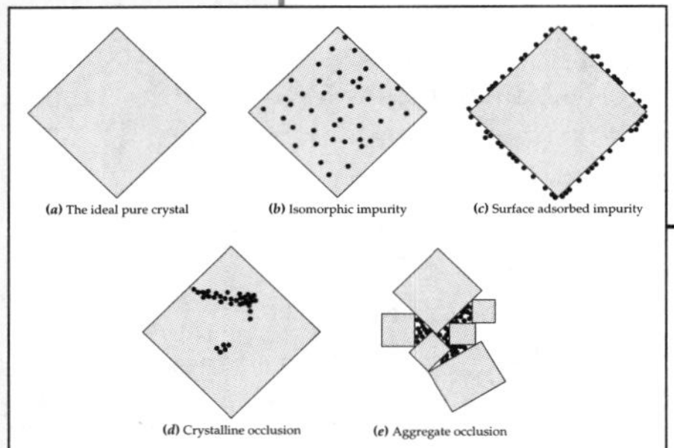

CHAPTER 10

Gravimetric Analysis

Overview

10.1 Summary of Gravimetric Methods

Gravimetric analysis, as its name suggests, depends on the measurement of weight to determine the quantity of analyte in a sample. Because of the capabilities of weighing technology, the precision of gravimetric analysis is potentially in the range of 1 part in 10,000 or even 1 part in 100,000. For instance, it is possible to weigh a one-milligram sample to the nearest microgram and weigh a one-gram sample to within ten micrograms. However, it is necessary to have a highly pure and physically homogeneous material to reach this level of precision in practice. The chemical purity and homogeneity of the weighed material limit the precision of gravimetric analyses.

There are a number of different assay methods that fall under the classification of gravimetric methods. One of these is **electrodeposition** or **electrogravimetry**. This technique involves the electrochemical conversion of metal ions in solution into the metallic form which plates out onto the cathode. The amount of metal present is then found by the difference in weight of the cathode before and after the plating operation. Electrodeposition is generally used to determine the quantity of a specific metal—for instance, cadmium—in a macro sample.

Electrodeposition is a subject of Section 12.6.

A second gravimetric method involves measuring a sample's changes in mass upon heating at a controlled rate in a controlled atmosphere. This is called **thermogravimetry**. Uses for thermogravimetry include the determination of water content of numerous types of samples and the carbon and ash content of coal.

A third gravimetric method involves determining the content of analyte in a sample by measuring the mass of a precipitate that forms from ions in aqueous solution. An example is the determination of sulfate by precipitation of barium sulfate:

$$Ba^{2+} + SO_4^{2-} \rightleftharpoons BaSO_4(s) \qquad \textbf{(10-1)}$$

The precipitate is formed by addition of excess barium to the sample solution. The precipitated barium sulfate is separated from the **mother liquor** (the remaining solution), washed, dried, and weighed. The weight of barium sulfate is then related to the original sulfate concentration of the sample. Such gravimetric precipitation assays are the topic of this chapter.

To a significant degree, **gravimetric precipitation** has been replaced by instrumental methods of analysis. With appropriate internal standards and automated sample manipulation, a number of different instrumental methods can achieve 0.1–0.3% precisions, which equal the common practical limit for the last two types of gravimetric methods mentioned above. In addition, titration methods, especially with automatic titrators, can achieve the same level of precision with less manual work.

Weighing Technique

The major difficulty with gravimetric precipitation assays arises in the production of a chemically pure, homogeneous precipitate with 100% recovery. In addition, the assay method requires reproducible weighings. Precise weighings require, in turn, avoiding or accounting for errors such as the following.

a. The weighed precipitation products must contain reproducible amounts of water. This is often achieved by heating the wet, washed precipitate at an appropriate temperature in an open container in an oven. To ensure that the moisture content remains constant, the sample is covered before cooling. The reason for covering the dried precipitate is that the material may be **hygroscopic** (note: not *hydro*scopic), which means attracting or absorbing moisture from the air. If such a material remains in contact with a humid atmosphere, it continually gains weight as water is taken up. Some compounds take up so much water from the air that the solid dissolves and becomes a concentrated aqueous solution; such materials are called **deliquescent** compounds. Precipitates for gravimetric analyses that are deliquescent or hygroscopic should be avoided if at all possible.

b. In order to obtain precise weighings, containers cannot be touched by bare fingers since oil and dirt (fingerprints) will be deposited. Dirty tongs also must be avoided. Such deposits increase the measured weight. Similar additions to the mass may occur if the container is placed on an unclean surface.

c. A fingerprint or other material inadvertently deposited on a weighing container must be cleaned off. However, rubbing the container with a lintless cloth or paper may also cause the container to become electrostatically charged. Static charges can cause the container to be attracted to or repelled from the surrounding parts of a balance and cause a mistake in weighing.

d. If a sample or precipitate is weighed in a closed container, the pressure inside the container may differ from that in the surrounding room. For instance, if the container was closed when warmer than ambient temperature, the internal pressure will be lower after cooling, and the container becomes buoyant in the air like a ship is buoyant in water. The mass measured will be less than it should be. The internal and external pressures must be equalized to obtain a precise weighing.

In the remainder of this chapter, it is assumed these difficulties have been eliminated.

Desirable Precipitate Properties

For analyses using quantitative precipitation, the following properties of the precipitate are desired:

1. Its solubility should be low enough so that losses are negligible in the concentration range of the analyte.
2. Its physical properties should allow it to be easily filtered from the mother liquor, washed, and transferred quantitatively from container to container. For instance, a gelatinous substance that sticks to glass would be undesirable. Another unwanted precipitate is one in which the particles are too small to be collected quantitatively from the solution either by filtration or centrifugation.
3. The material should be stable and of definite composition in its weighable form. For instance, the unstable $Al(OH)_3$ precipitate can be converted by heating to stable Al_2O_3 to be weighed as was described in Section 2.9.

ANALYTICAL CHEMISTRY IN ACTION
Living Off the Fat of the Land?

It is now routine to find labels on food with percentages of various primary nutrients listed. One of the most interesting to many people is the quantity of fat, which is given in weight percent. A number of different methods are used to extract the fat from the rest of the food matrix, but the final measurement routinely is gravimetric.

The quantity of total fat in food includes fatty acids (triglycerides), cholesterol, and other lipids. The extraction of total fat from the food is carried out with a mixture of chloroform: methanol 1:1 with water present. The extraction inevitably involves mincing the food under test in a powerful blender with the organic solvents present. Freeing the fat from the rest of the matrix may also include a prior step of decomposition with proteases (enzymes that cleave proteins). After the extraction, the mixture is centrifuged. The solids are removed from the liquid, and the liquid allowed to stand. Upon standing, a chloroform layer separates from an aqueous methanol layer. An aliquot of the known volume of chloroform is taken, placed in a predried and weighed container, and the solvent is evaporated. The residue's weight is the aliquot of total fat in the sample.

[Ref: Phillips, K. M., et al. 1997. "Simplified Gravimetric Determination of Total Fat in Food Composites after Chloroform-Methanol Extraction." *Journal of the Association of Official Analytical Chemists (Journal AOCS)*. 74:137–142.] ■

4. The precipitant should have a chemistry that is as selective as possible to minimize interference. The precipitated species may be the analyte itself or a chemical species transformed from the analyte, for instance, SO_4^{2-} from S^{2-}. A large body of descriptive chemistry is available to aid in making this choice.
5. A small mass of the analyte should yield a large mass of precipitate, thus maximizing the assay sensitivity. A good example is when cobalt is reacted with the organic reagent l-nitroso-2-naphthol to form the precipitate $Co(C_{10}H_6O_2)_3{\cdot}2H_2O$, which weighs more than 10 times the cobalt's weight in the original sample.

Of these properties, number five is less important than the first four. To obtain the optimum properties, the conditions of precipitation must be chosen carefully.

10.2 The Mechanism of Precipitation of Ionic Compounds

Since the chemistry of precipitation affects the physical and chemical properties of precipitates, let us look at a representative mechanism of precipitation of ionic compounds from homogeneous solutions. Precipitate formation is a dynamic process—a reaction that is proceeding from an unstable solution toward an equilibrium between solid and solution. The precipitation can be separated into a set of ordered stages as illustrated in Figure 10.1. The stages are as follows.

a. **Supersaturation** is achieved by adding precipitant to the solution. Supersaturation occurs when the solution phase contains *more* of the

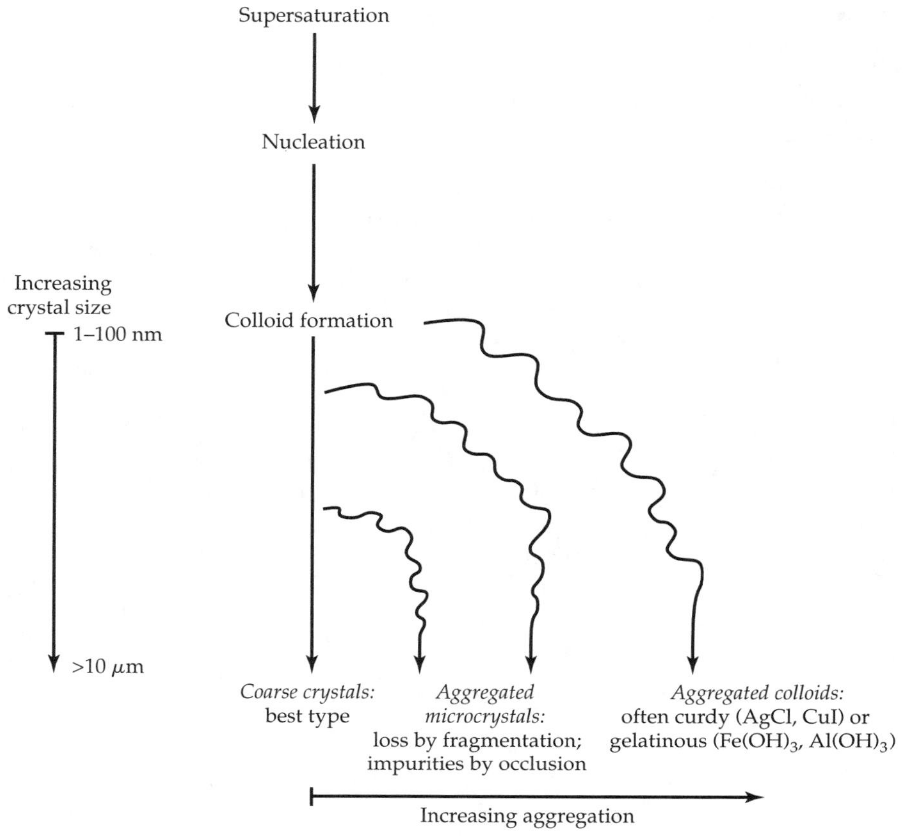

FIGURE 10.1 ▲
Diagram of the stages of crystallization from solution.
After colloid formation, two competing processes occur: crystal growth by deposition of mass at crystal surfaces, and aggregation of already-formed crystals of various sizes. A wavy line indicates that the path of formation of the precipitate particles is a complicated one that depends on conditions near the surfaces of the growing crystals.

dissolved precipitate than can be present at equilibrium. The greater this disparity, the greater the relative supersaturation.

b. **Nucleation** occurs next. The process of nucleation involves formation of the smallest particles of precipitate (the nuclei) that can grow spontaneously from the supersaturated solution.

c. **Crystal growth** follows. This process consists of the deposition of ions from the solution onto the *surfaces* of the solid particles present. The precipitation nuclei grow to form **colloids** or **colloidal particles** which, by definition, are 1–100 nm in diameter.

d. **Aggregation** of the solid particles may begin at this point. Particles aggregate by colliding and sticking together. It is a process that competes with crystal growth to form larger particles from the solution. The structure of an illustrative aggregated particle along with other types of imperfections are shown in Figure 10.2e. Such **aggregates** trap the mother liquor in voids between the composite particles. This trapping is called **occlusion**.

e. Aggregation and crystal growth compete. Depending on the conditions, the final precipitate may be composed of coarse crystals, aggregates of fine crystals, and/or aggregates of colloids. The paths to each are illustrated in Figure 10.1. The lines representing the competitive routes between crystal growth (vertically downward) and aggregation (horizontally to the right) are wavy to show randomness. This reflects the complexity of the precipitation process. The physical and chemical properties of the solution and precipitate are both changing with time and are heterogeneous through the solution. As a result the kinetic processes, which depend on solution composition, are heterogeneous as well. A number of different stages of precipitation most likely occur simultaneously in a single solution.

The following example illustrates this last point. Consider, for instance, adding a drop (about 0.05 mL) of a 0.1-M barium chloride solution to a beaker containing 50 mL of solution 1.0 mM in sulfate. (The solubility product $K_{sp}(BaSO_4) = 1.3 \times 10^{-10}$. A saturated solution of 1 mM sulfate contains approximately 1.3×10^{-7} M Ba^{2+}.) In the region at the edge of the added drop, the solution will be highly supersaturated in $BaSO_4$. When the solution is stirred, the drop disperses, and the level of supersaturation decreases as the local concentrations of Ba^{2+} decrease. Eventually the supersaturation *would* become homogeneous throughout the solution with 1×10^{-4} M Ba^{2+} *except* that both barium and sulfate are continually being removed from the solution to form the solid precipitate. Finally, the solution achieves equilibrium, but only the last-formed precipitate arises under near-equilibrium conditions. The majority of the precipitation occurs under rapidly varying circumstances.

By controlling the conditions under which the nucleation and subsequent crystal growth and aggregation occur—mainly through adjusting the chemical concentrations and temperature—the desired precipitate properties are optimized. The best precipitate form for gravimetric assays are large, well-formed crystals because they tend to have a minimum of contamination and are easiest to filter and transfer quantitatively—actions that are usually necessary at least once during an assay procedure.

10.3 Coprecipitation of Interferences

Crystal growth (step c in the list above) is a two-step reaction. First, the ions must migrate from the solution to the surface of the growing crystal. The migration rate depends on the identities of the ions, the temperature, and the rate of stirring of the solution (whether by manual stirring or simply convection). Following their arrival at the crystal surface, the ions must fit into the lattice. The rate of fitting into the lattice depends on the concentrations and identities of the ions and the chemical characteristics of the surface. These characteristics include such properties as net charge on the surface and whether there are surface imperfections on the atomic level such as ridges or steps. It is a highly complicated process which is still not clear in detail.

Crystal growth is also fast on the atom's scale. Consider a crystal of sodium chloride which grows at a visible rate—say 1 mm an hour. This means that 1 mm^3 of sodium chloride is deposited onto a 1 mm^2 sodium chloride surface each hour. From the density of NaCl (2.165 g/cm^3), its formula weight (58.45), and Avogadro's number (6.022×10^{23}), we can calculate that in one hour, 2.23×10^{19} ions of sodium and an equal number of chloride ions deposit on that mm-square area. This means that about 6.2×10^{15}—six million billion—ions of each element deposit there each second! You might expect that if some other ions, similar in size and charge, were present, they could easily become incorporated into the NaCl lattice. This is the case, and incorporation of such impurities is called **coprecipitation**. The coprecipitants may be distributed in different ways depending on the crystallization conditions. These distributions are illustrated in Figure 10.2 and are described next.

The closer the crystal and ionic structures of the possible impurities match those of the precipitate (the **host**), the easier it is for the impurities to coprecipitate. (The similarity in crystal structures is called **isomorphism**: Greek for *same form*.) When the isomorphism is close, the contaminating species becomes relatively evenly distributed in the host crystal as shown in

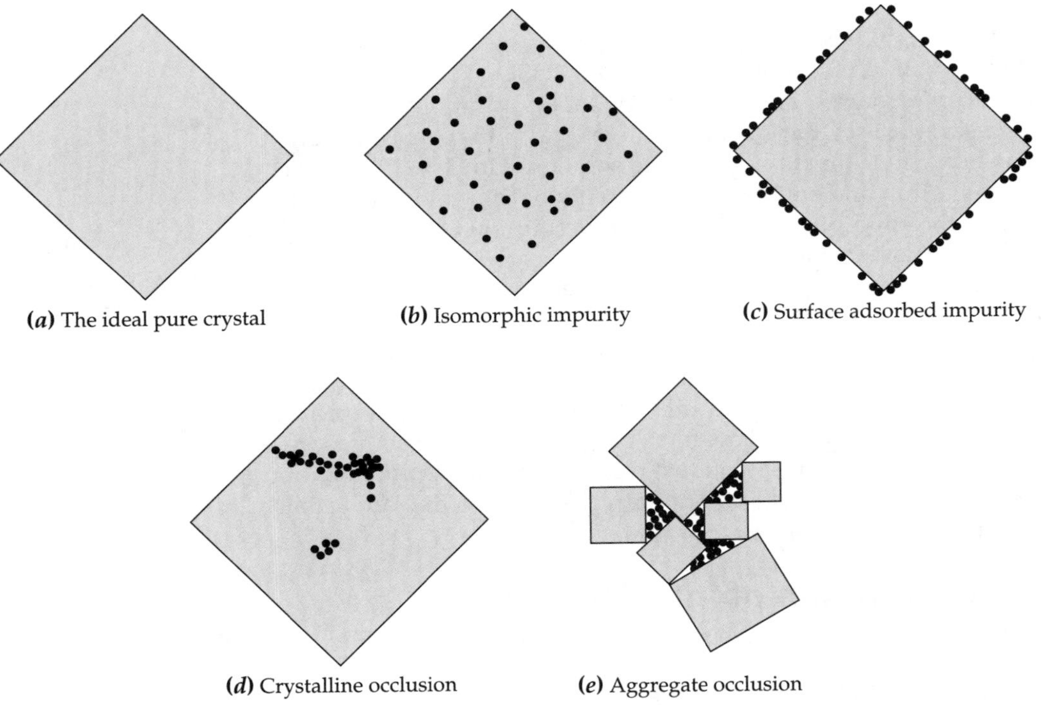

FIGURE 10.2 ▲
Various types of impurity distributions shown as two-dimensional representations.

The impurity locations are indicated by dots. **(a)** An ideal pure crystal has no impurities. **(b)** Isomorphic impurities are distributed throughout the crystal mass. Since the chemistry of the impurities is close to that of the host, it is extremely hard to remove them. **(c)** Surface adsorbed impurities are excluded from the bulk crystal but bind to surface sites. The binding can range from weak (washed off easily) to extremely strong. **(d)** Impurities can be trapped inside faults of a crystal. This happens when surface-adsorbed impurities are subsequently enclosed by further crystal growth. The impurity can be removed only by reforming the crystal. **(e)** Impurities and the mother liquor can be trapped in interstices between crystallites as they aggregate. If an interstitial volume is open to the solution, the impurities in it can be washed out. Otherwise, the impurity can be removed only by reforming the structure.

Figure 10.2b. Isomorphic impurities cannot, in general, be separated from the host by fractional precipitation. Examples of isomorphic precipitates are NaCl, KCl, KBr, and KI. All four form the same type of crystalline lattice (cubic) with atomic sites at the corners of cubes 563, 626, 659, and 710 pm on an edge, respectively. KCl forms high-quality mixed crystals with KBr, its closest isomorph, but only lesser-quality crystals with NaCl and KI, in which the atomic spacing is less similar. Another example is Pb^{2+} in the presence of $BaSO_4$. The lead will be found substituted for barium ions in the lattice. This is interpreted as lead sulfate dissolved in or coprecipitated with the barium sulfate.

Nevertheless, if the ionic electronic structures differ, crystal isomorphism alone is not sufficient to cause coprecipitation. For example, even though KCl and PbS are isomorphic, with atomic spacings 626 and 597 pm, respectively, they do not form mixed crystals due to the differences in ionic charges—a uni-univalent salt vs. a bi-bivalent salt.

A second type of impurity structure is **adsorption** on the surface of the host precipitate. This type of structure is shown figuratively in Figure 10.2c. In adsorption (note: *ad-* and not *ab-*), the impurity, while not fitting into the lattice, can bind relatively strongly to the surface of the precipitate particles. Surface adsorption is significant only when a host precipitate has a large surface area such as is found in the gelatinous hydrated oxides, such as $Al(OH)_3$ and $Fe(OH)_3$. Adsorption seldom needs to be considered for precipitates that are crystalline or are aggregates of crystals.

A third type of impurity distribution, called **occlusion**, is illustrated in Figures 10.2d and e. The impurities are trapped inside of cavities within the precipitate. If the precipitate is *crystalline*, the occluded volumes tend to be along flaws and imperfections of the crystal. Crystalline occlusions occur when the impurities are entrapped during a rapid crystallization. During a slow crystallization, the adsorbed, nonisomorphic impurities tend to be excluded from the crystal matrix, leaving the bulk crystal uncontaminated. However, during a rapid crystallization, the impurities may not **desorb** (leave the surface—the opposite of *adsorb*) rapidly enough and so are covered over and entrapped in localized regions. (See Figure 10.2d.) On the other hand, if the precipitate is an *aggregate*, the impurities are entrapped in the volumes between the particles. Aggregate occlusions not only can trap surface contaminants, but the occlusion volumes can be so large that the mother liquor is trapped as well, as illustrated in Figure 10.2e. Even nonadsorbable contaminants dissolved in the mother liquor can be occluded.

10.4 Minimizing Contamination of Precipitates

Obtaining a precipitate in the form of large, perfect crystals minimizes the problems with impurities. The experimental precautions that tend to enhance such crystal formation are the following.

a. Have a dilute solution. This slows the crystal-growth process, allowing time for the exclusion of impurities from the solid matrix and minimizing chances for occlusion. The concentration level at which the crystallization can occur depends on the solubility of the precipitate under the conditions of the reaction. More soluble precipitates require more concentrated solutions.

b. Add the precipitant slowly. This procedure keeps supersaturation at a minimum and slows crystal growth.
c. Stir the solution well. This keeps the supersaturation constant throughout the solution and keeps the crystal growth as even as possible.
d. Keep the temperature high. A higher temperature speeds the equilibration processes. If the surface equilibrium excludes contaminants, high temperatures speed the rates and reduce both occlusion and surface adsorption.
e. After the precipitate forms, allow the solution to stand for a long time to allow the solid to reach equilibrium with the mother liquor. This period is called **digestion**. A number of different processes occur during the digestion period. The mass of small crystals decreases and larger ones form; the surface area decreases and occlusions are eliminated as the small crystals dissolve. Also, imperfect regions of crystals become more perfect since the imperfect regions are less stable, so this is a spontaneous process. Concomitantly, impurities are released, and the correct ions replace them. This process is called **metathesis**. An example is substitution of Ag^+ for impurity Na^+ in solid AgCl after precipitating Cl^- with $AgNO_3$.

Thus, impurities can be removed from precipitates by reforming them. As was described, reformation does not necessarily mean that the precipitate is completely redissolved and then reprecipitated, although this may sometimes be done.

After obtaining the best crystalline material possible, separating the precipitate from the mother liquor generally involves filtering. The precipitate is held up by the filtering medium, such as filter paper or a glass frit.[1] The residual liquid is then removed by washing. Washing the precipitate is not as simple an operation as you might expect.

Special washing conditions are generally included with the precipitation instructions for specific gravimetric precipitation protocols. The reason that special conditions are needed are threefold:

1. To minimize redissolving any part of the precipitate
2. To minimize breaking up any coagulated precipitates into their composite microcrystals or colloids (that is, to minimize **peptization**), which would then pass through the filter.
3. To minimize any chemical changes, such as those causing further precipitation from the residual mother liquor.

Although peptization can occur from chemical action such as changes in ionic strength between the mother liquor and the wash, breaking up the precipitate into smaller particles can also arise from shaking or stirring. So

---

[1]A glass frit (or fritted glass) is a porous glass plate made by fusing small glass particles together at temperatures below the melting point of the glass. Larger particles produce coarser frits.

minimizing mechanical disruption should be a requirement for all precipitated samples.

The precipitate's chemical integrity should be retained. For example, special washing conditions are needed for $MgNH_4PO_4 \cdot 6H_2O$, which can be used to determine either magnesium or phosphate. If washed with water alone, the material tends to hydrolyze to $MgHPO_4 \cdot nH_2O$. To avoid this problem, the precipitate must be washed with dilute NH_3 solution.

A third example involves washing the solid after precipitating chloride as AgCl with $AgNO_3$. For three reasons, it is necessary to wash with a dilute, cold HNO_3 solution. First, it avoids breaking up the aggregated colloid which forms the precipitate. Water alone would wash off the electrolyte that is essential for coagulation. Second, a cold wash solution is needed because the solubility of AgCl rises rapidly with temperature. And third, any adsorbed Ag, K, and Na nitrates are replaced by the nitric acid, which volatilizes upon drying the precipitate. The adsorbed metal nitrates would not volatilize, and due to the large surface area, they would lead to a high weight.

One other significant source of error has not yet been discussed: the filter paper or fritted glassware that is used to separate the mother liquor from the precipitate. Which of these is selected depends on the precipitation assay method chosen.

If fritted glassware is used to both retain and weigh the precipitate, the glassware's weight must be the same before being used and after the precipitate is washed and dried. Therefore, the glassware must be cleaned, dried, and accurately weighed before using it in the assay. The drying is done at the same temperature at which the precipitate will be dried and is repeated until **constant weight** is achieved—that is, until the glassware weighs the same after each of two sequential drying periods. As usual, "same" means within some specified limit of difference.

On the other hand, filter paper cannot be dried to constant weight since it begins to decompose at temperatures as low as 100°C. Therefore, it must be removed from the precipitate by oxidizing it at high temperatures—as high as 1000°C. The process of oxidizing the paper is called **ashing**, and special low-ash papers are used to reduce the interference from the ash to minimum values. Ashing the filter paper cannot be done if the final form of the precipitate is not stable at such high temperatures. Stability in this case means both unchanging in stoichiometry and being nonvolatile. Most such materials are oxides. Two instances when filter paper ashing may be used are with precipitated $Al(OH)_3$ which is converted to Al_2O_3, and with precipitated $Fe(OH)_3$ which is converted to Fe_2O_3 in the ashing step.

10.5 Stoichiometry and Gravimetric Factors

Consider that the material to be analyzed has been sampled, the sample has been prepared for the precipitation assay, and the analyte has been precipitated. Perhaps the precipitate has been digested. This was followed by separating, drying, and weighing the precipitate. The data that have been obtained for each sample consist of two weights: the weight of the sample and the weight of either the precipitate or a modified form of it. It is now necessary to find how the weight of the product is used to calculate the percentage of the sample that is the analyte. This calculation requires knowledge of the stoichiometry of the reaction(s) involved and straightforward stoichiometric calculations.

A number of examples follow. All the calculations involve finding:

1. The number of grams of analyte, given the number of grams of precipitate
2. The percentage of the sample that is the given analyte

These two steps should be recognizable within the calculations.

EXAMPLE 10.1

Calculate the number of grams of elemental sulfur (a.w. 32.064) expected from reacting 1.3423 g $KMnO_4$ (f.w. 158.04) with an excess of H_2S in a sulfuric acid solution. The reaction is

$$5\,H_2S + 2\,KMnO_4 + 3\,H_2SO_4 \longrightarrow K_2SO_4 + 5\,S(s) + 2\,MnSO_4 + 8\,H_2O$$

Solution

From the stoichiometry,

2 moles of $KMnO_4$ produces 5 moles of precipitated sulfur, S(*s*)

In the general case, where y moles of $KMnO_4$ are used,

y moles of $KMnO_4$ produces $\frac{5}{2} \cdot y$ moles of precipitated sulfur, S(*s*)

The factor of $\frac{5}{2}$ is called the **stoichiometric factor**—just as are the numbers used in chemical equations. Numerically, from the data given,

$$\left(\frac{1.3423}{158.04}\right) \text{ moles of } KMnO_4 \text{ produces } \frac{5}{2} \cdot \left(\frac{1.3423}{158.04}\right) \text{ moles of sulfur}$$

Note that the quantities in parentheses are the same.

Since one mole of sulfur weighs 32.064 g, we expect

$$\frac{5}{2} \cdot \left(\frac{1.3423}{158.04}\right) \cdot 32.064 \text{ g} = 0.6808 \text{ g of sulfur}$$

EXAMPLE 10.2

A metal ore sample is digested with concentrated HNO_3 together with $KClO_3$. This treatment dissolves the ore and simultaneously converts the elemental sulfur present to sulfate. The nitrate and chlorate are removed by repeatedly adding concentrated HCl and heating to dryness. The sulfate is then precipitated as $BaSO_4$, which is separated, dried, and weighed. If the ore sample weighed 1.1809 g, and the dried $BaSO_4$ weighed 0.6068 g, what is the percentage sulfur in the original sample of ore?

Solution

From the stoichiometry,

1 S produces 1 SO_4^{2-} which is precipitated as 1 $BaSO_4$

In other words,

1 mole of S produces 1 mole of precipitated $BaSO_4$

Other information we need is

$$\text{a.w. sulfur} = 32.064$$

$$\text{f.w. } BaSO_4 = 233.40$$

Using reasoning similar to that in the example above, we can write

$$\left(\frac{0.6068}{233.40}\right) \text{ moles of } BaSO_4 \text{ comes from } 1 \cdot \left(\frac{0.6068}{233.40}\right) \text{ moles sulfur}$$

with the stoichiometric factor of one.

Using the weight of a mole of sulfur, we find that

$$1 \cdot \left(\frac{0.6068}{233.40}\right) \cdot 32.064 \text{ g} = 0.0834 \text{ g of sulfur}$$

The percentage of sulfur in the sample is then

$$\% \text{ S} = \frac{\text{weight of sulfur}}{\text{weight of sample}} \cdot 100 = \frac{0.0834 \text{ g}}{1.1809 \text{ g}} \cdot 100 = 7.06\% \text{ S}$$

In the second example, we could have written a single arithmetic expression that takes all the numerical values into account. It is

$$\% \text{ S} = \frac{0.6068 \cdot 32.064 \cdot 100}{233.40 \cdot 1.1809} = 7.06\% \text{ S} \quad \textbf{(10-2)}$$

Another way to write Equation 10-2 would be to separate it into terms, each of which represents one step of the calculation.

$$\%\text{S} = \frac{0.6068}{233.40} \cdot 1 \cdot 32.604 \cdot \frac{1}{1.1809} \cdot 100 = 7.06\% \text{ S} \quad \textbf{(10-3)}$$

- $\frac{0.6068}{233.40}$: moles of $BaSO_4$
- $\cdot 1$: moles of sulfur
- $\cdot 32.604$: mass of sulfur
- $\cdot \frac{1}{1.1809}$: sulfur as a fraction of the sample
- $\cdot 100$: sulfur as a percentage of the sample

Of course, there is no new information in this rearrangement of the equation.

Since the formula weights and stoichiometry remain fixed for each specific assay, it is convenient to calculate a factor that relates directly the mass of the precipitate weighed to the mass of analyte in the sample. This is the **gravimetric factor** (G.F.), and it contains the chemical information. Thus,

$$\% \text{ S} = 0.6068 \cdot 1 \cdot \frac{32.064}{233.40} \cdot \frac{1}{1.1809} \cdot 100 = 7.06\% \text{ S} \quad \textbf{(10-4)}$$

- 1: stoichiometric factor
- $\frac{32.064}{233.40}$: mass ratio of chemical forms
- 0.6068: weight of precipitate obtained
- $1 \cdot \frac{32.064}{233.40}$: gravimetric factor
- $\frac{1}{1.1809}$: (weight of sample)$^{-1}$
- 100: % multiple

For sulfur determined by precipitation as $BaSO_4$, the gravimetric factor is 0.1374. That is

$$1 \cdot \frac{32.064}{233.40} = 0.1374 \quad \textbf{(10-5)}$$

and, using the G.F. shown in Equation 10-5, the mass of sulfur can be calculated simply as

$$\text{grams S} = \text{weight of BaSO}_4 \cdot 0.1374 \qquad \textbf{(10-6)}$$

Once the mass of analyte is obtained, we can report our results using any of the w/w or w/v measures. Tables of gravimetric factors can be found in a number of chemistry handbooks.

EXAMPLE 10.3

Calculate the gravimetric factor for an analysis for chromium as Cr_2O_3 if the chromium is converted to $BaCrO_4$ for analysis.

Solution

2 mole of $BaCrO_4$ is formed from 1 mole of Cr_2O_3
f.w. $BaCrO_4$ = 253.33
f.w. Cr_2O_3 = 151.99
Let y be the weight of $BaCrO_4$ found.

The gravimetric factor as defined above is

$$\text{G.F.} = \text{Stoichiometric Factor} \times \frac{\text{F.W. analyte}}{\text{F.W. weighed species}} \qquad \textbf{(10-7)}$$

The stoichiometric factor is $\frac{1}{2}$ since $(y/253.33)$ moles of $BaCrO_4$ comes from $\frac{1}{2}(y/253.33)$ moles of Cr_2O_3. Thus, we find that the gravimetric factor for this procedure equals

$$\frac{1}{2} \cdot \frac{151.99}{253.33} = 0.3000$$

Equation 10-7 is a general equation.

EXAMPLE 10.4

Oxalate can be determined by precipitation with excess Ca^{2+} to form CaC_2O_4. This is subsequently heated and decomposed to $CaCO_3$ with evolution of gaseous CO. What is the gravimetric factor for this assay method?

Solution

f.w. $C_2O_4^{2-}$ (analyte) = 88.02
f.w. $CaCO_3$ (weighed) = 100.09

From the information above, the stoichiometry is such that

$$z \text{ moles of CaCO}_3 \text{ comes from } 1 \cdot z \text{ moles of C}_2\text{O}_4^{2-}$$

After substitution into Equation 10-7, the gravimetric factor for determination of oxalate by precipitation as CaC_2O_4 and ignition to $CaCO_3$ is found to be

$$\text{G.F.} = 1 \cdot \frac{88.02}{100.09} = 0.8794$$

EXAMPLE 10.5

A 0.8680-g sample of a powdery white material containing oxalate was dissolved, and the oxalate was precipitated as CaC_2O_4, which was subsequently filtered and ignited to $CaCO_3$. The $CaCO_3$ was weighed after cooling in a desiccator to ambient temperature. Its weight was 0.1794 g. What is the percentage oxalate in the original sample?

Solution

From Equation 10-3 and the information in the previous example,

$$\text{g oxalate} = \text{mass } CaCO_3 \cdot \text{G.F.}$$

$$\text{g oxalate} = 0.1784 \text{ g} \cdot 0.8794 = 0.1569$$

$$\% \text{ oxalate} = \frac{0.1569 \text{ g}}{0.8680 \text{ g}} \cdot 100 = 18.07\%$$

The calculation becomes quite simple indeed.

Suggestions for Further Reading

ERDEY, L. *Gravimetric Analysis*. Oxford: Pergamon, 1963, Vol. I; New York: Macmillan, 1963, Vol. II; New York: Macmillan, 1965, Vol. III.
A three-volume treatise on the gravimetric analysis, including illustrations of the experimental equipment. The parts on precipitation analysis encompass the field. Methods for most elements are also included.

Concept Review

1. What process produces the change in mass used as a basis for analysis in the following?
(a) electrodeposition
(b) thermogravimetry

2. Would the following problems cause a high or low analytical result for a gravimetric precipitation analysis?
(a) a deliquescent product
(b) weighing a closed vessel containing the product while it was still warm

3. What are the desirable characteristics of a precipitate used for a gravimetric analysis?

4. How can supersaturation be minimized (and thus lead to more perfect crystals) in a precipitation process?

5. What is the difference between occlusion and inclusion in crystal formation?

6. After precipitation, what is the benefit of heating the precipitate in the mother liquor and allowing it to cool slowly?

Exercises

10.1 Sulfate can be determined by gravimetric precipitation assay weighing it as $BaSO_4$. The precipitate is filtered through filter paper. The filter paper is removed by ashing in a covered crucible. However, under these conditions, carbon from the paper may reduce the $BaSO_4$ through the reaction

$$BaSO_4 + 4\,C \xrightarrow{\text{heat}} BaS + 4\,CO(g)$$

(a) Will this reaction affect the final result of the analysis? If so, will it be too high or too low?
(b) If the above reaction is not too extensive, continuing the ignition in air causes the reaction

$$BaS + 2\,O_2(g) \xrightarrow{\text{heat}} BaSO_4$$

If this reaction occurs for all the BaS, will the final result of the analysis be affected? If so, will it be too high or too low?

10.2 In the determination of Cl as AgCl, the precipitated AgCl can be decomposed partially by light. The reaction is

$$2\,AgCl \xrightarrow{\text{light}} 2\,Ag^0 + Cl_2(g)$$

If this reaction occurs while the precipitate is still in contact with the mother liquor, the chlorine gas reacts with water to form Cl^- again. If the photoreaction occurs after separating the AgCl from the mother liquor, the chlorine gas can escape into the air.
(a) If the photoreaction occurs while the AgCl is in contact with the mother liquor, will the final result be too high, too low, or the same?

(b) If the photoreaction occurs while the AgCl is dry, will the final result be too high, too low, or the same?

10.3 Calculate to four significant figures the gravimetric factors for the following analyses.
(a) Sr precipitated with SO_4^{2-} and weighed as $SrSO_4$
(b) SrO dissolved in acid, precipitated with SO_4^{2-}, and weighed as $SrSO_4$
(c) Cu precipitated with salicylaldoxime and weighed as $Cu(C_7H_6NO_2)_2$

10.4 When electrodeposition was used to determine the copper content of an aqueous sample, a total of 0.0103 g Cu was deposited from 100 mL of sample.
(a) What was the copper content of the solution in ppm?
(b) Calculate the number of moles of electrons transferred if all of the copper was originally present as Cu^{2+}.
(c) At a current level of 100 mA (1 A = 1 C/sec and $\mathcal{F}$ = 96,500 C/mol), how long in minutes would it take to deposit the 0.0103 g copper?

10.5 A batch of tablets was to be assayed for water. Five tablets were ground together, giving 0.5076 g solid. Three portions with masses of 0.1034, 0.1026, and 0.0998 were then analyzed by thermogravimetry. The masses of the samples after drying and cooling were 0.1027, 0.1021, and 0.0992, respectively.
(a) Calculate the water content of each of the samples on a percent (w/w) basis.
(b) Calculate the mean and standard deviation for the determinations.

10.6 $Eu(IO_3)_3$ has a K_{sp} of 4.8×10^{-12} at 25°C and zero ionic strength. Under these conditions, HIO_3 is completely dissociated. Assume no hydrolysis of Eu occurs; that is, no $Eu(OH)^{2+}$ or other hydroxides form. A quantitative precipitation is defined as one that is 99.99% complete. What concentration of HIO_3 is needed to precipitate quantitatively a 10.0-mM europium solution?

10.7 Four hydroxide precipitates have the K_{sp} values tabulated below.

Precipitate	K_{sp}
$Ba(OH)_2$	5.0×10^{-3}
$Ca(OH)_2$	1.3×10^{-6}
$Cu(OH)_2$	1.6×10^{-19}
$Mn(OH)_2$	2×10^{-13}

Assume no metal hydrolysis occurs at any pH (not true in fact). Assume ionic strength effects are equal for all the ions.
(a) Assume all four metal ions, Ba^{2+}, Ca^{2+}, Cu^{2+}, and Mn^{2+}, are together in an acid solution, and each is 1 mM. KOH then is added slowly with vigorous stirring. Which metal hydroxide will precipitate first? What is the order of precipitation after that?
(b) At what pH will the first hydroxide precipitate form, assuming equilibrium conditions? Does the solution need to be basic to precipitate hydroxides?

*10.8 A 100.0 mL solution is 0.0100 M in $Ba(NO_3)_2$ and 0.0100 M in $Pb(NO_3)_2$. To this solution was added a slight excess of SO_4^{2-} over the Ba^{2+} consisting of 101.0 mL of 0.0100 M H_2SO_4. Under the conditions present,

$$K_{sp}(BaSO_4) = 1.00 \times 10^{-10}$$

$$K_{sp}(PbSO_4) = 1.70 \times 10^{-8}$$

Assume that H^+ does not bind to SO_4^{2-}, that ionic strength effects can be ignored, and that the system is at equilibrium.
(a) Calculate the concentrations you expect to find for Pb^{2+}, Ba^{2+}, and SO_4^{2-} in the final solution.
(b) Calculate the expected composition of the precipitate and report it as mole fraction $BaSO_4$ and mole fraction $PbSO_4$ (if any is present).

Additional Exercises

10.9 To monitor the thickness of silver layers deposited during an industrial process, a gravimetric analysis was carried out for silver. A sample with a known surface area was soaked in 20% nitric acid for 24 hours. KCl was then added to the solution, and the AgCl produced was filtered with a preweighed glass-microfiber filter. The precipitate on the filter was then dried at 60° for 2 hours. If the sample has a surface area of 1 cm², and the layer is 2 μm thick, how much AgCl would be expected to precipitate? (ρ_{Ag} = 10.49.) [Ref: Matsuura, K., Matsuura, Y., Harrington, J. A. 1966. "Evaluation of gold, silver, and dielectric-coated hollow glass waveguides." *Opt. Eng.* 35:3418.]

10.10 A new filter cartridge was developed to determine the particulate content of air in a laboratory. The cartridge has a capacity of 50 mg and a limit of quantitation of 0.2 mg. If the pump can pull 50 L/min through the filter, and the normal range of particulates is 3 ppb(w/v) for the filtered air in the laboratory, what are the maximum and minimum times that the pump should be on?

10.11 Calculate to four significant figures the gravimetric factors for the following analyses.
(a) Mn precipitated with KOH and weighed as Mn_3O_4
(b) Lu precipitated with $C_2O_4^{2-}$ and weighed as Lu_2O_3

10.12 When electrodeposition was used to determine the cadmium content of an aqueous sample, a total of 6.12 mg Cd was deposited from 100.0 mL of sample.
(a) What was the cadmium content of the solution?
(b) Calculate the number of moles of electrons transferred if all of the cadmium was present in solution as Cd^{2+}.

*Indicates more involved exercise.

10.13 $La(IO_3)_3$ has a K_{sp} of 1.2×10^{-11} at 25°C and zero ionic strength. Under these conditions, HIO_3 is completely dissociated. Assume no hydrolysis of La^{3+} occurs; that is, no $La(OH)^{2+}$ or other hydroxides form. A quantitative precipitation is defined as one that is 99.99% complete. What concentration of HIO_3 is needed to precipitate quantitatively the lanthanum from a 10.0 mM solution?

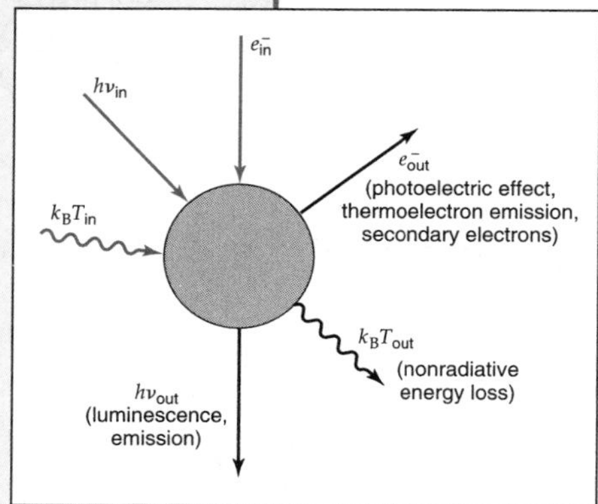

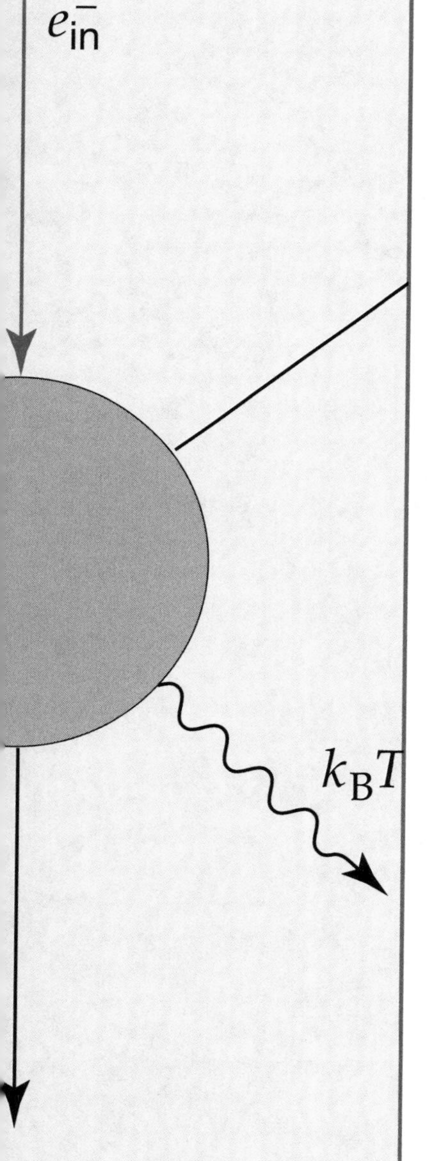

CHAPTER 11

Introduction to Spectrometry

Overview

A Deeper Look

11.1 Spectrometry from Radiofrequency to γ-Rays

The measurement of the **absorption** and **emission** of light by materials is called **spectrophotometry**, often shortened to **spectrometry**, which is the term used here. The terms *absorption* and *emission* have the same meaning that they have in everyday use: Absorption means *to take up*, and emission means *to give off*.

When we use the word *light* colloquially, we usually mean light visible to our eyes. However, visible light is only a small part of the electromagnetic spectrum, which includes radio, microwave, infrared, visible, ultraviolet, X-ray, and gamma-ray radiation. (See Figure 11.1 for a brief overview.) Here, the word *light* will be used as a general term for electromagnetic radiation simply because it is less cumbersome than *electromagnetic radiation from radiowaves to γ-radiation.*

Electromagnetic waves are examined in more detail in Section 11A.

Specific instruments used in spectrometry are referred to as **spectrophotometers, spectroradiometers**, or **spectrometers**, depending on their architectures. The term *spectrometers* will be used here to refer to all types.

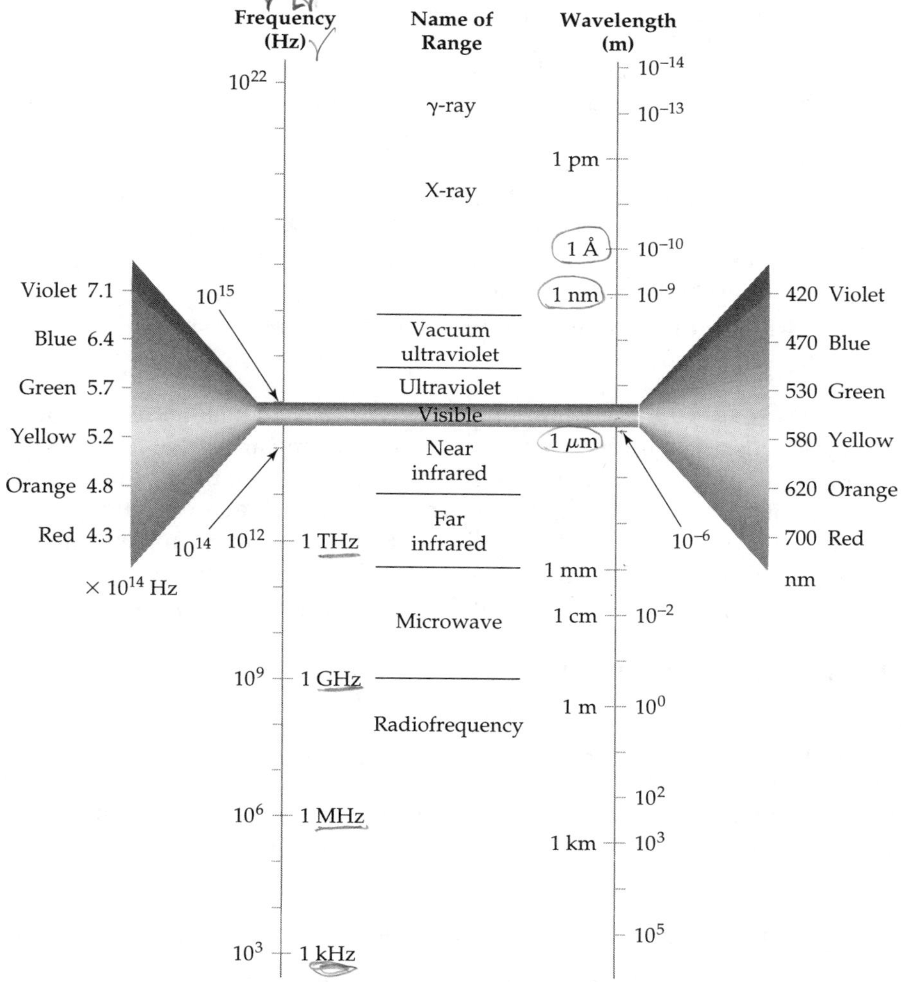

Frequency (Hz) | Name of Range | Wavelength (m)

10^{22}
γ-ray
10^{-14}
10^{-13}
1 pm
X-ray
1 Å — 10^{-10}
1 nm — 10^{-9}
10^{15}
Vacuum ultraviolet
Ultraviolet
Visible
Near infrared
1 μm
10^{14}
10^{12} — 1 THz
Far infrared
10^{-6}
1 mm
Microwave
1 cm — 10^{-2}
10^{9} — 1 GHz
Radiofrequency
1 m — 10^{0}
10^{2}
10^{6} — 1 MHz
1 km — 10^{3}
10^{5}
10^{3} — 1 kHz

Violet 7.1
Blue 6.4
Green 5.7
Yellow 5.2
Orange 4.8
Red 4.3
$\times 10^{14}$ Hz

420 Violet
470 Blue
530 Green
580 Yellow
620 Orange
700 Red
nm

◀ FIGURE 11.1
A diagram of the names, frequencies, and wavelengths (in a vacuum) of electromagnetic radiation. Each region of the spectrum is used in one or more analytical methods. Visible light is centered around 500 nm = 20,000 cm^{-1}.

Production and detection of radiation require different techniques in different regions of the electromagnetic spectrum. However, two fundamental variables are measured in all spectrometric methods:

1. The **wavelength** (or *energy*) of the radiation
2. The *amount* of the radiation at that wavelength

When light energy is absorbed by a specific material, it is absorbed only at certain wavelengths. At other wavelengths, the material may be transparent. Similarly, when light is emitted by an atom or molecule, it does so only at its characteristic emission wavelengths. At other wavelengths, no emission occurs. In sum, the wavelengths that are absorbed and emitted depend on the *identity* of the compound. These wavelengths remain the same regardless of the quantity of the analyte present. However, the amount of light absorbed or emitted does depend on the *amount* of the compound present in the light path (in uncomplicated cases).

Spectrometry can be used to determine both the elemental and the molecular content of materials. However, it is important to realize that each spectrometric method has its own benefits, drawbacks, specificities, and interferences. Expanding on the last of these, possible interferences in spectrometry can be classified into three types.

1. **Spectral interference**: The absorption or emission of other components of the matrix occurs at the same wavelength(s) used for the assay.
2. **Chemical interference**: The material being determined is not in the same chemical form over the determinable concentration range, or else the form varies significantly with changes in the matrix. Thus, part of the spectral power expected may not appear at the wavelength(s) chosen for the analysis.
3. **Instrument interference**: Excess illumination arrives at the transducer due to imperfections in the instrument.

In the first several sections of this chapter, we shall deal with the general fundamentals that apply to all spectrometries, regardless of wavelength region—namely the different general techniques that are used in measuring the intensity and wavelength of light and the results that are obtained. Brief introductions to some of the individual spectrometries appear in Sections 11.10 through 11.15, and an extensive list of references is given at the end of the chapter for those interested in learning more about these.

11.2 Energy, Wavelength, Frequency, and Temperature

As is shown in Section 11A, light can be thought of as wavelike. It is characterized as sinusoidally oscillating electric and magnetic fields. For a light wave moving through a vacuum, its speed c, its wavelength λ, and its frequency, written as the Greek letter ν, are simply related. That is,

$$\text{speed of light} = c = \lambda\nu; \qquad \text{in a vacuum} \tag{11-1a}$$

When passing through any material, the speed is slower. Call that speed v, and

$$\text{speed of light} = v = \lambda\nu; \qquad \text{in general} \tag{11-1b}$$

The speed of light in air is nearly the same as that in a vacuum, so only in calculations involving light passing through solids will we not assume $v = c = 2.9979 \times 10^8$ m sec^{-1}.

On the other hand, when light interacts with atoms and molecules, waves no longer explain its attributes. The light must be considered to have the nature of particles of energy, **photons**. In this situation, we find that the frequency ν of electromagnetic radiation is proportional to the energy E of the photon. The constant of proportionality is called **Planck's constant**, h. Quantitatively the relationship is simply

$$E = h\nu \qquad \textbf{(11-2a)}$$

The photon energy also is related to the wavelength of light. From Equations 11-1b and 11-2a, we see that the energy of the light and the wavelength are inversely related. That is, in a vacuum,

$$E = \frac{hc}{\lambda} \qquad \textbf{(11-2b)}$$

In practice, numerous measures of energy are used. Two measures you have probably seen are the **erg** and the **joule**. For example, for Equation 11-2a: If E is expressed in ergs and the frequency ν in s^{-1}, then Planck's constant,

$$h = 6.62 \times 10^{-27} \text{ erg.}$$

If E is expressed in joules and the frequency ν again in s^{-1}, then

$$h = 6.62 \times 10^{-34} \text{ joule} \cdot \text{s.}$$

Although joules are the recommended units to express energy, in chemical analysis other measures of light and energy are also used. A selection of these are listed in Table 11.1, and factors to interconvert them are shown in Table 11.2.

Sometimes it is convenient to label the radiation with the inverse of the wavelength. An example is cm^{-1}. This inverse of centimeters is called the **wavenumber**, and the unit is used most often in infrared spectrometry. A measurement made in wavenumbers is often denoted by $\bar{\nu}$. Then, the photon energy is the product of hc and the wavenumber. That is,

$$E = hc\bar{\nu} \qquad \textbf{(11-2c)}$$

Table 11.1 Commonly Used Measures of Wavelength and Energy

Name of unit	Symbol	Unit for	Used in range of*
nanometer	nm	λ	UV, visible, near-infrared
Angstrom†	Å	λ	X-ray, UV-visible (in older work)
millimicron†	mμ	λ	Visible (in older work)
micron, micrometer	μm	λ	Infrared (common in physics literature)
wavenumber (reciprocal cm)	cm^{-1}, $\bar{\nu}$	ν/c	Infrared, UV-visible (less common)
electron volt†	eV	E	X-ray, γ-ray
hertz	Hz	ν	Radiofrequency, microwave
= cycles per s	cps	ν	Radiofrequency (less common in current literature)

*See Figure 11.1 for the wavelength ranges.
†1 Å = 0.1 nm, 1 mμ = 1 nm, 1 eV = 1.602×10^{-19} joules

Table 11.2 Conversion Factors for Wavelengths *in vacuo*, Frequency, and Energy

Unshaded regions reflect direct relationships, energy to energy or wavelength to wavelength. For example:

1 nm = 10 Å
5 nm = 5 × 10 = 50 Å

Shaded regions reflect inverse relationships, wavelength to energy or energy to wavelength. For example:

One photon of light with a wavelength of 1 nm has an energy, E = 1.987×10^{-16} J
One photon of light with a wavelength of 500 nm has an energy, E = $(1/500) \times 1.987 \times 10^{-16} = 3.974 \times 10^{-19}$ J.
Algebraically, we write such an inverse relationship as

$$\frac{\lambda_1}{\lambda_2} = \frac{E_2}{E_1}$$

	nm	Å	cm^{-1}	eV	MHz	J*
nm		10	10^7	1.240×10^3	2.9979×10^{11}	1.986×10^{-16}
Å	.1		10^8	1.240×10^4	2.9979×10^{12}	1.986×10^{-15}
cm^{-1}	10^7	10^8		1.240×10^{-4}	2.9979×10^4	1.986×10^{-23}
eV	1.240×10^3	1.240×10^4	8.0655×10^3		2.418×10^8	1.602×10^{-19}
MHz	2.9979×10^{11}	2.9979×10^{12}	3.3356×10^{-5}	4.1355×10^{-9}		6.626×10^{-28}
J	1.986×10^{-16}	1.986×10^{-15}	5.034×10^{22}	6.241×10^{18}	1.509×10^{27}	
Spectrometric region	UV-vis	X-ray UV	Infrared	X-ray γ	Radiofrequency (NMR)	

*4.184 J = 1 cal

It is important to recognize that, regardless of the units of expression, a unique energy can be assigned to each wavelength and frequency of light. The longer the wavelength, the lower the energy and frequency.

EXAMPLE 11.1

Based on the following definition:

$$1 \text{ eV} = (1 \text{ V}) \times (\text{charge on electron in Coulombs})$$

and the equivalency

$$1 \text{ J} = 1 \text{ V} \cdot \text{C}$$

derive the conversion factor in Table 11-2 that converts between eV and J.

Solution

$$1 \text{ eV} = (1 \text{ V})(1.602 \times 10^{-19} \text{ C}) = 1.602 \times 10^{-19} \text{ V} \cdot \text{C}$$

$$1.602 \times 10^{-19} \text{ V} \cdot \text{C} \times \frac{1 \text{ J}}{1 \text{ V} \cdot \text{C}} = 1.602 \times 10^{-19} \text{ J}$$

EXAMPLE 11.2

Using the conversion factors in Table 11-2, calculate the wavelength of a photon with an energy of 4.50×10^{-19} J.

Solution

The table shows that a wavelength of 1 nm corresponds to 1.986×10^{-16} J. Since wavelength-energy conversions involve an inverse relationship, we use the equation

$$\frac{\lambda_1}{\lambda_2} = \frac{E_2}{E_1}$$

Substituting the value in the table for λ_1, the equation becomes

$$\frac{1 \text{ nm}}{\lambda_2} = \frac{4.50 \times 10^{-19} \text{ J}}{1.986 \times 10^{-16} \text{ J}}$$

Then, λ_2 can be found:

$$\lambda_2 = \frac{1.986 \times 10^{-16} \text{ J}}{4.50 \times 10^{-19} \text{ J}} \cdot 1 \text{ nm}$$

$$\lambda_2 = 441 \text{ nm}$$

This wavelength is in the visible region.

EXAMPLE 11.3

Calculate the energy in cm^{-1} of a photon with a wavelength of 4.00 μm (in the infrared region).

Solution

Using the relationships of Table 11-2 and the appropriate conversion factors for lengths,

$$\frac{\lambda_1}{\lambda_2} = \frac{E_2}{E_1}$$

$$\frac{4.00 \ \mu\text{m} \times \dfrac{10^3 \text{ nm}}{1 \ \mu\text{m}}}{1 \text{ nm}} = \frac{10^7 \text{ cm}^{-1}}{E_1}$$

$$E_1 = \frac{10^7}{4.00 \times 10^3 \text{ nm}}$$

$$E_1 = 2.50 \times 10^3 \text{ cm}^{-1}$$

However, in this case, it would be just as easy to use the definition of a wavenumber

$$\bar{\nu} = \frac{1}{\lambda}$$

$$= \frac{1}{4.00 \ \mu\text{m} \times \dfrac{1 \text{ cm}}{1 \times 10^4 \ \mu\text{m}}}$$

$$= 2.50 \times 10^3 \text{ cm}^{-1}$$

Another measure of energy that you know about is the temperature, which is a measure of the average kinetic energy of an object. Energy is directly proportional to the temperature and can be expressed as either the energy per atom,

$$E \propto k_B T \tag{11-3}$$

where k_B = Boltzmann's constant (1.380×10^{-16} erg K^{-1} atom^{-1} *or* 1.380×10^{-23} joule K^{-1} atom^{-1}), or as the energy per mole of material,

$$E \propto RT \tag{11-4}$$

where R = the gas constant (8.3145×10^7 erg K^{-1} mol^{-1} *or* 8.3145 joule K^{-1} mol^{-1}).

Since energy is closely related to wavelength and frequency, they both must be functions of temperature as well. The next example shows the calculations.

EXAMPLE 11.4

Ambient temperature is around 300 K. To what energy, wavelength, and wavenumber is $E = k_B T$ equivalent?

Solution

$$E = (1.380 \times 10^{-23} \text{ joule K}^{-1} \text{ atom}^{-1})(300 \text{ K}) = 4.14 \times 10^{-21} \text{ joule atom}^{-1}$$

$$E = \frac{hc}{\lambda} = \frac{(6.63 \times 10^{-34} \text{ joule s})(3.00 \times 10^{8} \text{ m s}^{-1})}{\lambda}$$

$$\lambda = \frac{1.989 \times 10^{-25}}{4.14 \times 10^{-21}} = 4.80 \times 10^{-5} \text{ m} = 48.0\ \mu\text{m}$$

$$\bar{\nu} = \frac{1}{\lambda(\text{cm})} = 208 \text{ cm}^{-1}$$

Do not confuse the *energy* of the photons with the brightness (more correctly the **irradiance**) of the source. The number of photons and the brightness are directly related, but the photon energy relates to the color of the light.

A short digression on energy and power is worthwhile here. Power (which equals energy per unit time) is measured in quantitative spectrometry, and the power of the light is given by the product of the number of photons and their energy.

$$P = \text{number of photons s}^{-1} \times \text{photon energy}$$

As you can calculate, the above equation's units agree with

$$\text{power} = \text{energy/time}.$$

The light's *power* is the quantity measured in spectrometry.

11.3 The Transformations of Light Energy

As conservation of energy dictates, the sum of all the forms of energy entering the sample must equal the sum of all the forms of energy leaving plus the energy remaining in the material. The results of interactions of an atom (or molecule) with its surroundings and with light can be drawn as shown in Figure 11.2. The energy is carried into and out from the atom as light, as heat, and as the kinetic energy of particles such as electrons. Because light can be

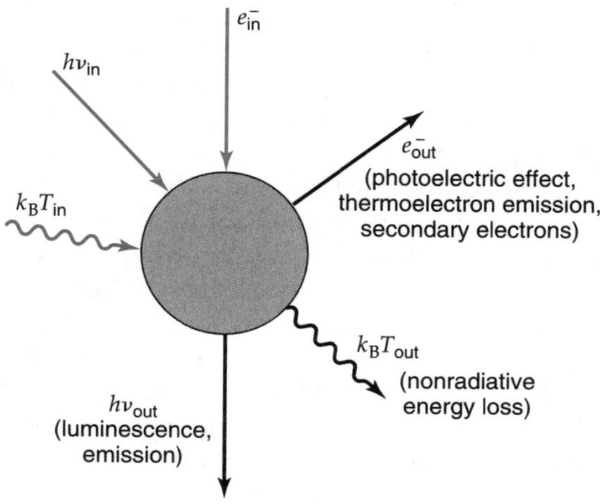

FIGURE 11.2 ▲
Diagram illustrating the transformations of energy by an atom, molecule, or ion.
It is common to illustrate heat generation (nonradiative transitions) as wavy arrows and radiation absorbed or emitted as straight arrows, with the sense of energy flow shown by their directions.

transformed into heat, the wavelength of the light that is emitted can be longer (lower energy) than the wavelength that excites the atom, with the remainder heating the atom and its surroundings.

Here, energy originating or remaining as chemical bonding energy will not concern us in general. Chemical reactions are not desirable in a spectrometric analysis unless the reaction produces the light, a phenomenon called **chemiluminescence**. Chemiluminescence forms the basis for a few standard assays, such as for gaseous NO.

Other reactions that produce energy are the high-energy nuclear decays with γ-emissions. These are the basis for **neutron activation analysis**, a trace analysis method. Slowly moving neutrons react at the atomic nucleus to form excited state nuclei that decay with emission of characteristic wavelengths of γ-radiation. The energies of the radiation identifies the atomic species, and the brightness is proportional to the content—just as for any other spectrometric method.

Spectrometric Names

The names given to the various spectrometries depend on the form of energy that impinges on the atom and the form(s) that leave. These general names are listed in Table 11.3. The first two entries are *emission* and *absorption*.

Table 11.3 The Transformations of Light Energy Interacting with Atoms

Energy in	Energy out	Spectrometry
Heat	Light	Emission (incandescence)
Light	Heat	Absorption
Light	Light	Luminescence (phosphorescence, fluorescence)
Light	Moving electrons	Photoelectron spectroscopies
Moving electrons	Moving electrons	Auger
Bonding energy	Light	Chemiluminescence

The third, luminescence, can be divided into two types: **fluorescence** and **phosphorescence**. These two can be distinguished experimentally by their different light-emission behaviors. That is, after a material is illuminated, the emission may persist for an appreciable time after the illumination is removed. This persistent emission is phosphorescence. Fluorescence emission, on the other hand, ceases virtually immediately. (An accepted, though arbitrary, definition of "immediately" is about 10 ns.) For analytical use, phosphorescence is usually observed in the visible range while analytical fluorescence is used in the visible, ultraviolet, and X-ray regions.

You will notice that we have not used the word *luminescence* in the way you perhaps may have heard it most often—associated with a light bulb. In that case, the term has been shortened from **thermoluminescence**. The classification of the light emitted by the bulb is more correctly characterized as incandescence or **thermal emission**: heat in, light out. We reserve luminescence to refer to fluorescence and phosphorescence: light in, light out.

The next-to-last entry of Table 11.3 is *electron emission*. This is called the *photoelectric effect* and is the basis for a number of **electron spectrometries**, which are most important in the analysis of the material at the surfaces of analytes.

11.4 Spectral Purity and Spectral Resolution

Up to this point in the chapter, the statements you have read indicate that you could look at the interaction of a single wavelength of light with a sample. In reality, no matter what method is used to produce any specific wavelength of light or to select a "single" wavelength from a broad range of wavelengths, the best that can be done is to isolate a *range* of wavelengths. For all of these cases a plot of the power *vs.* the wavelength of the light looks something like that in Figure 11.3.

A quantitative measure of the range of wavelengths of light (the **spectral purity**) is a number called the **spectral bandwidth**, **spectral linewidth**, or, simply, **bandwidth** or **linewidth**. As shown in the figure, the quantitative measurement used is the width of the band at a point halfway between the baseline and the peak. This is called the *full width at half height* (fwhh) or the *full width at half maximum* (fwhm). Incidentally, the band envelope is not in the shape of a Gaussian peak. The lineshape is closer to a shape called *lorentzian*. Compared with a Gaussian, it is narrower in the middle and spreads over a wider range at lower power regions at the edges.

In summary, you should understand that while we might say there is a single wavelength—more correctly, the light is **monochromatic** (from the Greek *one color*)—the truth is that all monochromatic light has a finite spectral bandwidth. For the time being, let us simply continue with the knowledge that

One narrow-bandwidth source is a hollow cathode lamp. See photo EP.6 inside the back cover.

1. Narrow-bandwidth light can be produced at a fixed wavelength, or
2. The light intensity can be measured over a narrow bandwidth. Furthermore,
3. The wavelength can be changed continuously and smoothly over some wavelength range. This last method is called, simply, **scanning** the wavelength.

You should recognize that some electromagnetic sources can be **tuned** to different wavelengths. Numerous radiofrequency sources and some lasers

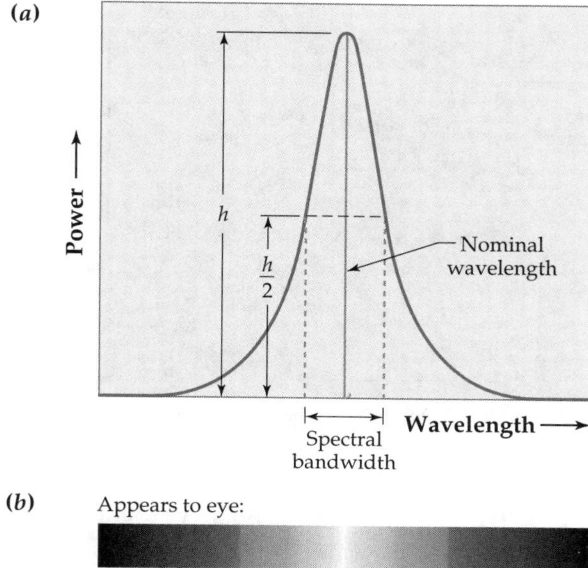

FIGURE 11.3 ▲
Illustration of the definition of spectral bandwidth.
(a) is a plot of radiant power versus wavelength. The nominal wavelength is at the peak of the power spectrum with power decreasing to both sides. **(b)** approximates what your eye would see as the output from a monochromator projected on a screen.

are included in this group. However, the complexity of the tuning process for lasers has, up to now, eliminated them from routine spectrometry, while radiofrequency spectrometries generally use fixed frequencies.

Monochromators and Polychromators in the Optical Range

The wide variety of uses of spectrometry in chemical analysis means that a large variety of spectrometers are made to acquire the data. In this section, we shall discuss briefly the spectrometers that are used in the **optical** range. *Optical* is a somewhat imprecise term used to describe light that can be controlled by lenses and prisms, and is generally considered to include light from the far infrared into the vacuum ultraviolet.

Spectrometers can be separated into two general groups: **dispersive** and **non-dispersive**. *Dispersive* spectrometers depend on the separation of the colors of light over a range of angles. A large number of different designs exist, and the one shown in Figure 11.4a merely indicates their basic parts. Of critical importance is to have a **slit**—or some construction that has the same purpose—to select a narrow angle of the dispersed colors, that is, to pick a narrow range of color. The section of a dispersive scanning instrument that selects a single wavelength at a time is the **monochromator**. (Notice that the name is monochro*mator*, not *-meter*. It passes a single wavelength of light; it does not measure it.) One benefit of a dispersive optical element is that a number of detectors can be distributed across the spread of dispersed light. A dispersive **polychromator** can be fabricated. As illustrated in Figure 11.4b,

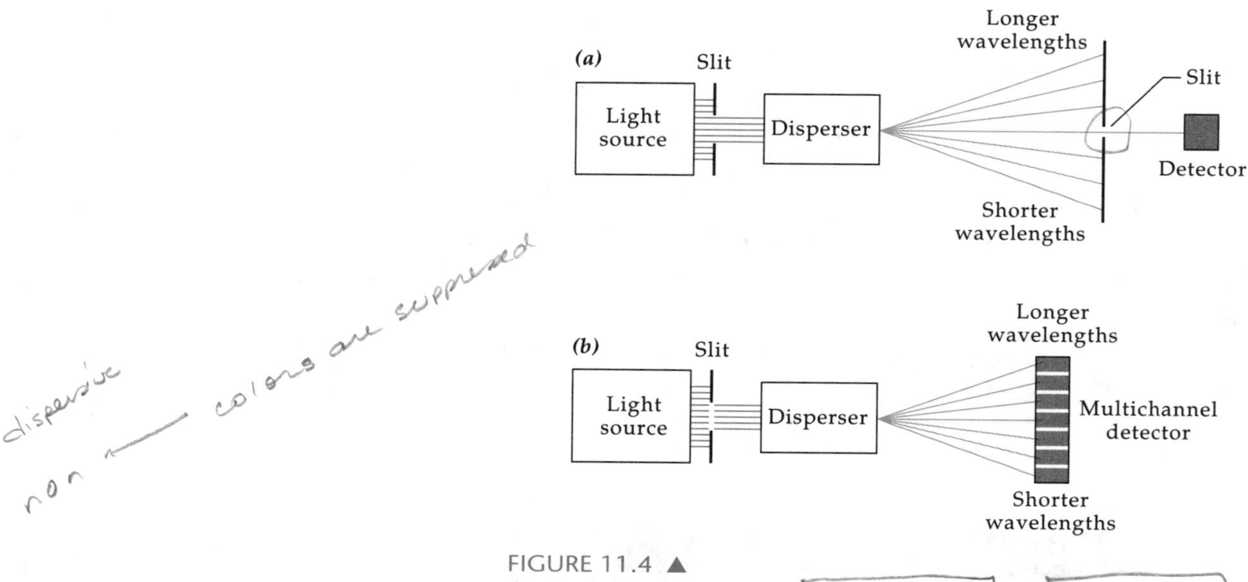

FIGURE 11.4 ▲
General diagram of how a monochromator and polychromator separate light into its colors.

the light at a number of wavelengths is detected simultaneously, and subsequently these intensities can be read out into a computer and the spectrum constructed.

Non-dispersive spectrometers separate light by allowing a specific color to propagate while the other colors are suppressed along the same direction. A simple colored filter can be used, which depends on chemical dyes to block the unwanted colors. A second type of filter, an **interferometer**, does not absorb the unwanted light but instead limits the color by the destructive interference of the unwanted colors and the constructive interference of the desired wavelength. Incidently, notice that the word interfero*meter* ends in *-meter*. The basis for this word is historic; interferometers originally were used to measure properties of light.

Wave interference is discussed in more depth in Section 11B.

Both monochromators and interferometers function because of the constructive and destructive interference of light waves. Only filters based on absorption of light by dyes do not function this way.

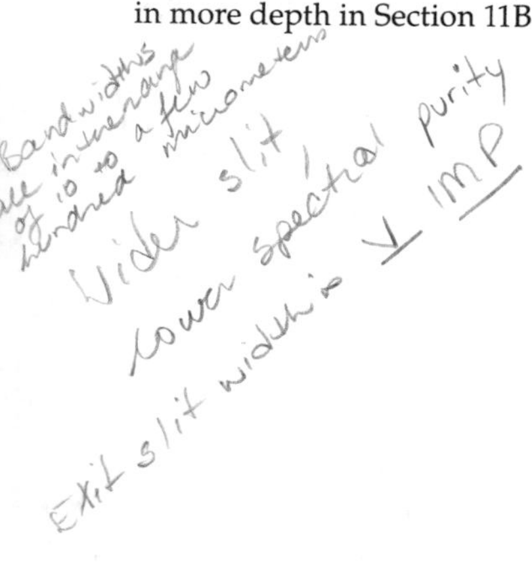

Bandwidth and Spectral Slit Width

Let us consider the operation of a monochromator. Minimally, the monochromator requires a dispersing element and two slits. Slits of a monochromator are exactly that, slits cut out of metal. The **slit widths** are usually in the range of ten to a few hundred micrometers. As pictured in Figure 11.4a, light enters through the **entrance slit**, and is dispersed. The different colors of the light propagate at different angles, and only over a narrow range of these angles can the light pass through a second, **exit slit**. It should be clear from the figure that the resulting spectral bandwidth depends on the width of the exit slit. But the bandwidth also depends on how broadly the colors are spread out and on the distance from the dispersing element to the exit slit. For a *fixed* slit width, the bandwidth that passes through depends on the wavelength of the light. Thus a statement of the spectral slit width will be similar to "0.5 nm at 350 nm." The wider the slit, the greater the spectral

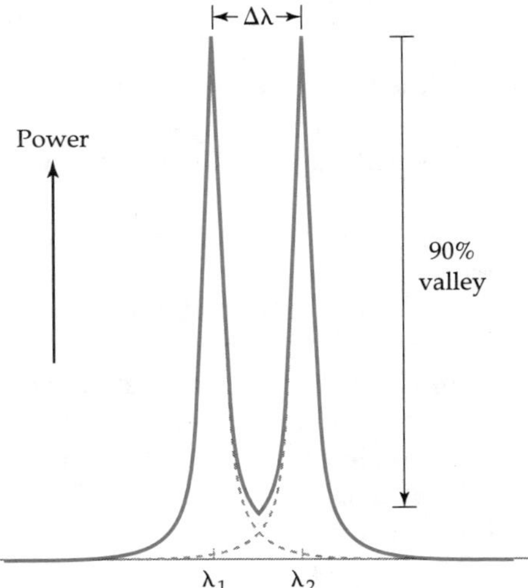

FIGURE 11.5 ▲
Illustration of resolving power and resolution.
Resolving power refers to the capability of an instrument to separate wavelengths. Resolution refers to the separation of wavelengths in a spectrum.

power that passes through it. However, the spectral bandwidth is larger then as well. This tradeoff always exists. When the spectral power is raised by using a wider slit, spectral purity is reduced.

For polychromators with multichannel detectors, the width of the elements of the detectors (the fraction of the total spread) also becomes a factor in the spectral purity. On the other hand, the spectral purity of an interferometric filter does not depend on a slit but on the number of times the light is reflected before passing through. The spectral purity is characterized by a number called the **finesse**.

Resolution

The bandwidth does not tell us when two different colors are "different." Two terms are used to define this difference. One refers to the instrument—the term is the **resolving power**. The other refers to the spectrum—the **resolution** of the spectrum. These are easily defined with reference to Figure 11.5. For both terms, a larger value means that smaller wavelength differences can be ascertained. However, the dip between the two peaks can be any value, so the idea that two wavelengths are "resolved" can have many meanings depending on the depth of the valley. That information must be included. For instance, the resolution of a spectrum can be reported as *resolution xxx with 90% valley*. This means that the peaks are considered resolved when they appear as in Figure 11.5; the valley depth is 90% of the height of *equal* peaks.

11.5 Measurement of Spectra

In spectrometry, the measurement of spectra depends intimately on both the spectral properties of the sample and the instrument used to make the measurement. With respect to the instrument, as much light as possible from a stable source is directed into a scanning monochromator. As light of continuously varying wavelengths passes through the sample and through the exit slit onto a transducer, a graph of the power detected *vs.* time (calibrated as wavelength) results in spectra like those in Figure 11.6.

In some cases, it is useful to consider that the ability to absorb light is present in the molecule even when the absorption is not being measured. The graph of incipient absorbance *vs.* wavelength is called the **absorption envelope** of the molecule.

Let us now depict how the spectrum might change with different spectral slit widths. Note especially the appearance of the narrow feature as the spectral slit width changes. The spectral slit widths are indicated by the distance between pairs of parallel lines on the wavelength axis as shown in Figure 11.6. These lines bracket the range of wavelengths that pass through the monochromator at any **nominal wavelength** setting. The nominal wavelength is the center of the wavelength range. The transducer responds only to the total light power falling on it, with effectively no discrimination of wavelength. As a result, the transducer averages the power arriving at *all* the wavelengths passing through the monochromator. Consequently, as the monochromator scans through wavelength, when the edge of the slit reaches the spectral range of the sharp feature, the *average* power falling on the transducer rises. This is seen as a rise in the power level before the monochromator's nominal wavelength reaches the base of the sharp peak. When the nominal wavelength of the sharp peak reaches the center of the slit, the highest point (the highest power) of the peak appears to be reduced, since the transducer is averaging the peak power with the points of lower power on either side.

If the slit width is optimized, the spectrum will closely resemble the absorption envelope as shown in Figure 11.6a. If a narrower slit width is used, the decreased power arriving at the transducer means that the contribution from noise begins to obscure the spectral features. In this case, the spectrum would resemble that shown in Figure 11.6b.

As the slit is made wider than the optimum, the spectral width becomes greater than that of the sharp feature, and the sharp peak *appears* to be even broader and lower, as illustrated in Figures 11.6c and 11.6d. Notice that the broad peak appears relatively unchanged through all the slit-width changes. The difference in the slit's effect on the broad and narrow features illustrates a general rule. The effect depends on the ratio of the linewidth of each spectral feature compared with the spectral slit width. As a general rule, in order to measure the spectral features precisely,

$$\text{spectral bandwidth} < (1/10)\ \text{linewidth of the feature}$$

Although the power is reduced with a narrower slit, we gain resolution.

To summarize: Three fundamental components of spectrometric instruments are required to carry out measurements in any region of the electromagnetic spectrum.

a. A *source* of electromagnetic radiation must be present.
b. The effects from a *single wavelength* of electromagnetic radiation must be obtained. This single wavelength may arise either from the source or from the emission of the analyte itself.
c. A *transducer* must be present to measure changes in the quantity of electromagnetic radiation falling on it.

The idea of a "single wavelength" was discussed in Section 11.4.

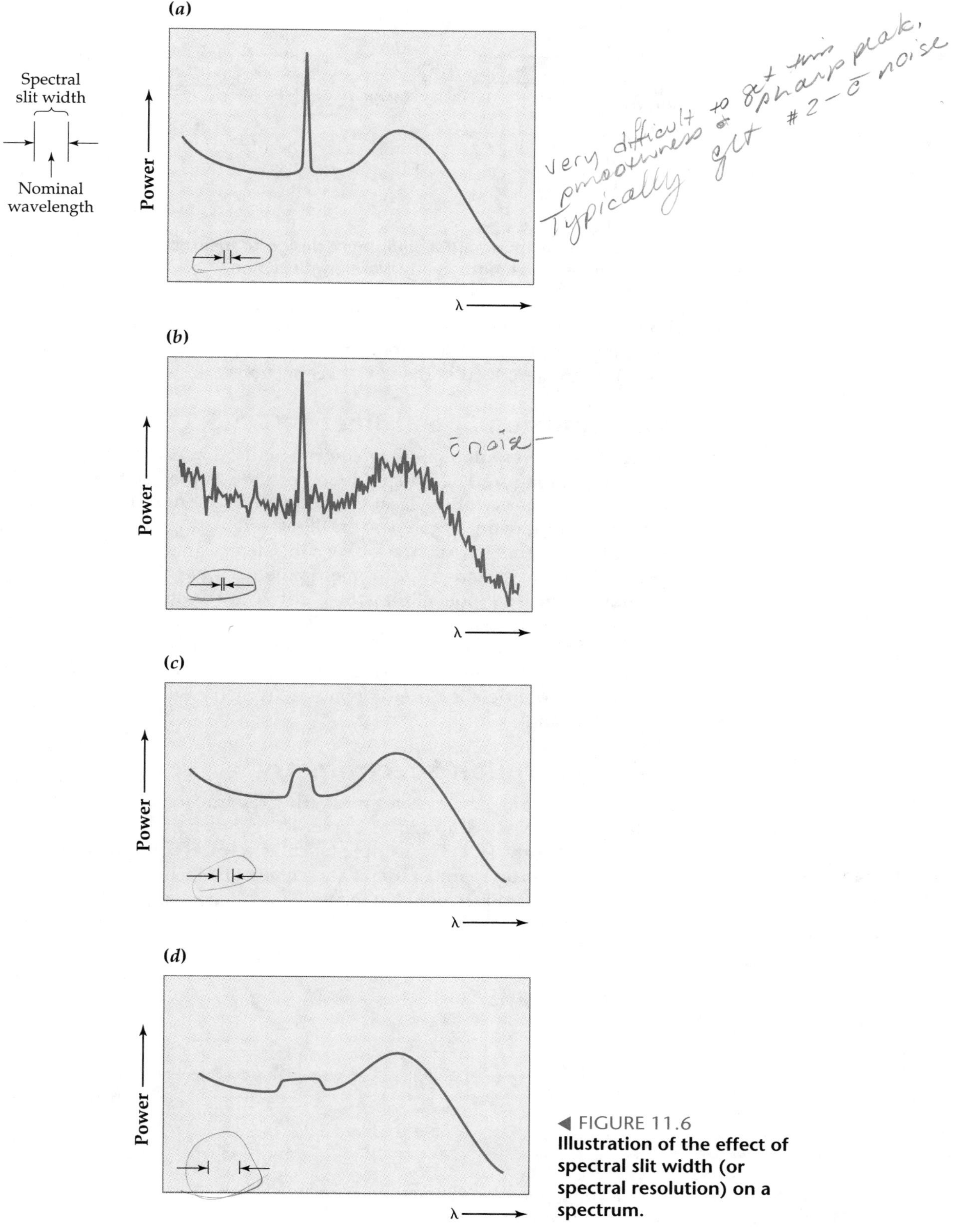

◀ FIGURE 11.6
Illustration of the effect of spectral slit width (or spectral resolution) on a spectrum.

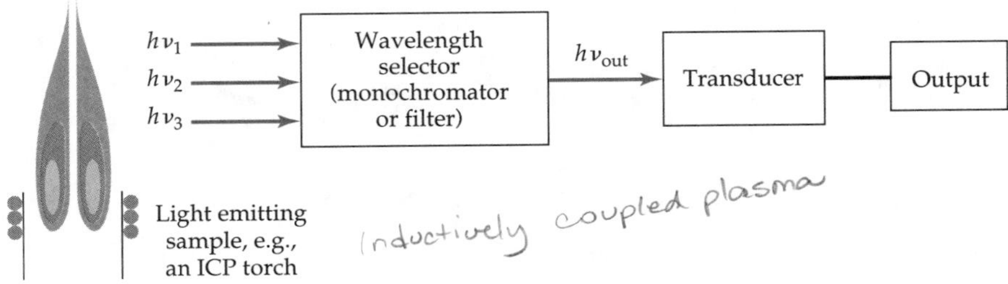

FIGURE 11.7 ▲
Generalized experimental equipment design to measure light emission at a single wavelength in any wavelength region.

With these requirements met, any of the three general types of spectrometric techniques—emission, absorption, and fluorescence—can be carried out. These three are described in the next three sections.

11.6 Emission Spectrometry

A generalized representation of the equipment used in emission spectrometry is shown in Figure 11.7. The emission of the sample itself provides the light, and the intensity of the light emitted is a function of the analyte concentration. The illustration shows the light originating from a luminous flame, but gas discharges, such as are seen in a neon sign, can be used as well.

As with all quantitative spectrometric methods, we require monochromatic light. Here, the monochromator is between the emitting sample and the transducer as shown in Figure 11.7. As shown in Figure 11.4b, this is also the design that is employed when a polychromator is used.

Emission instruments are most commonly used in atomic spectrometry. Accordingly, more details on emission spectrometry will be delayed until Sections 11.10 and 11.11.

11.7 Absorption Spectrometry

The sample need not be in the spectrometer. The light can be taken to the sample through an optical fiber. See EP.3 inside the front cover.

Absorption spectrometry involves measuring the fraction of light of a given wavelength that passes through a sample. The geometry of the equipment is illustrated in Figure 11.8. The sample (such as a colored solution) does not emit light itself, so a separate radiant source must be included.

Most sources produce light having unwanted wavelengths in addition to the ones desired. (The exceptions to this general characteristic are radiofre-

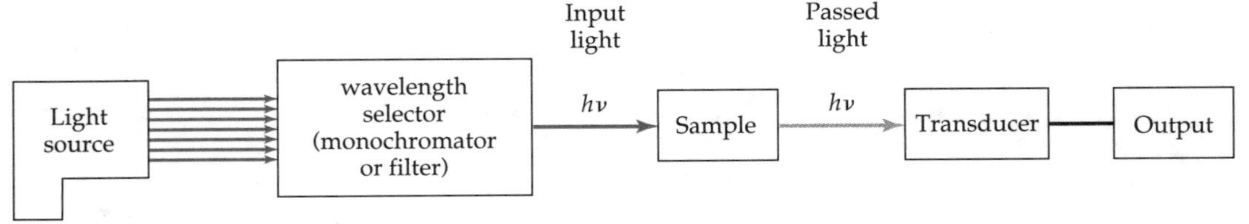

FIGURE 11.8 ▲
Generalized experimental equipment design to measure sample absorption or transmittance at a single wavelength.

quency sources and lasers.) Passing the light through either a monochromator or a filter selects the wavelength for the assay. The figure shows the monochromator between the source and the sample. However, polychromators are placed on the opposite side of the sample, between the sample and the transducer, and in some instruments the monochromator is placed there.

For analyses, two measurements of the amount of light absorbed are made. In the first, we measure the amount of light (at the chosen wavelength) that falls on the transducer when a blank is in place. Call the power P_0 for the blank, that is, when the concentration of the assayed material is zero.

Measurements made when the analyte or calibration (standard) samples are in place are compared with the blank measurement. Let us call the power that is measured with the samples or standards present P. The comparison that we always make involves the ratio P/P_0, with both power measurements made at the same instrument settings—wavelength, geometry, and so forth.

Three different terms are used to express this ratio. The first is simply the ratio P/P_0, which is called the **transmittance**. The usual abbreviation for transmittance is an upper-case T.

$$T = \frac{P}{P_0} \tag{11-5}$$

The second is the percent transmittance,

$$\%T = T \times 100 \tag{11-6}$$

The third is the negative logarithm of T, which is called the **absorbance** and is abbreviated by an upper-case A. The algebraic expression for absorbance is

$$A = -\log \frac{P}{P_0} = -\log T \tag{11-7}$$

Another name, **optical density**, also is used for this quantity, but its use no longer is recommended.

EXAMPLE 11.5

What values of absorbance correspond to 100% T, 10% T, and 1% T?

Solution

100% T, 10% T, and 1% T correspond to transmittances of 1.00, 0.10, and 0.010. From the definition of A,

$$100\%\ T \text{ has } A = -\log 1 = 0$$

$$10\%\ T \text{ has } A = -\log 0.10 = 1.0$$

$$1\%\ T \text{ has } A = -\log 0.010 = 2.0$$

Absorbance and Concentration

The absorbance of a sample is proportional to the total amount of material that absorbs the incident light. Experimentally, it can be shown that

$$A = a\,b\,c \tag{11-8a}$$

In other words, the absorbance is directly proportional to the following:

a, a constant that is a property of the material itself as well as the wavelength of the measurement;
b, the length of the path through which the light travels in the sample; and
c, the concentration of the material that absorbs the light. The difference in use of c as the speed of light and c as the concentration should be obvious from the context.

Equation 11-8 describes the **Beer-Lambert law**. When the concentration c is in (mol L^{-1}) and the path length in cm, then the constant a will have the units (L mol^{-1} cm^{-1}), in which case, it is given the symbol ϵ. ϵ is called the **molar extinction coefficient**, or **molar absorptivity** (the term currently preferred). With labels included, we see

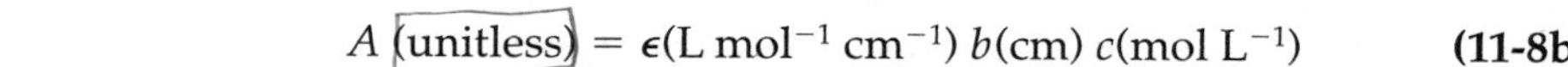

$$A\text{(unitless)} = \epsilon(\text{L mol}^{-1}\text{ cm}^{-1})\, b(\text{cm})\, c(\text{mol L}^{-1}) \tag{11-8b}$$

or, simply,

$$A = \epsilon\, b\, c \tag{11-8b}$$

Quantitation can also be done by comparing the color of a sample to a set of standards by eye. This practice is called colorimetry as used for test strips such as that shown in EP.4 inside the front cover.

It is important to note that ϵ depends on the measurement wavelength. Thus, the value of ϵ is often written

$$\epsilon_{\lambda(\text{nm})}, \quad \text{such as} \quad \epsilon_{530}$$

to indicate that ϵ is measured at $\lambda = 530$ nm.

If no units are noted, then the use of ϵ generally implies molar units. Other units that might be used for the absorptivity are, for instance, (μg L^{-1})$^{-1}$ cm^{-1}. In that case, the accompanying concentration unit c then must be in (μg L^{-1}).

EXAMPLE 11.6

A 5.00×10^{-4} M solution of a nickel macrocycle is put into a sample cuvette with a pathlength of 1.000 cm. The absorbance at 592 nm is found to be 0.446. What is ϵ at that wavelength? If a solution of unknown concentration of the nickel macrocycle has an absorbance of 0.125 at the same λ, what is its concentration?

Solution

To find ϵ_{592}:

$$\begin{aligned} A &= \epsilon_{592} bc \\ 0.446 &= \epsilon_{592}(1.000 \text{ cm})(5.00 \times 10^{-5}\text{M}) \\ \epsilon_{592} &= \frac{0.446}{(1.000)(5.00 \times 10^{-5})} \\ &= 8920 \end{aligned}$$

With the value of ϵ_{592} known, to find the concentration of the unknown

$$\begin{aligned} 0.125 &= (8920 \text{ M}^{-1}\text{ cm}^{-1})(1.000 \text{ cm})\, c \\ c &= \frac{0.125}{(8920 \text{ M}^{-1}\text{ cm}^{-1})(1.000 \text{ cm})} \\ &= 1.40 \times 10^{-5}\text{ M} \end{aligned}$$

A short cut can be formulated since the only difference in the measurements is the change in the concentrations of the measured samples. Since ϵ_{592} and b remain the same,

$$\epsilon_{592}b = \frac{A_{\text{std}}}{c_{\text{std}}} = \frac{A_{\text{unknown}}}{c_{\text{unknown}}}$$

Substitution of the known values into the ratios yields

$$\frac{0.446}{5.00 \times 10^{-5}} = \frac{0.125}{c_{\text{unknown}}}$$

or, $c_{\text{unknown}} = 1.40 \times 10^{-5}$ M directly.

Absorbance, Concentration, and Precision

The random error associated with measurements of light is called the **photometric error**, and the precision of a measurement depends this error. However, because of the logarithmic relationship between T and A, the precision of the measurement of concentration differs over the range of measurable absorbance. Recall that the transmittance P/P_0 is the directly measured quantity. But normally, absorbance is used in analyses because its value is directly proportional to the concentration of the light-absorbing species. To see how the precision changes, let us *assume* that the relative error in measuring T is $\pm 0.5\%$ for all values of T from 0 to 1. This is a relative error of 0.005 for T. As shown in Figure 11.9a, the absolute error in A decreases continuously from high to low values of A.

Next, let us look at the *relative* error in absorbance; it is the measure of our analytical precision. In Figures 11.9b and 11.9c, notice that at the low concentrations, 90% T (A = 0.045), and at relatively high concentrations, $< 2\%$ T (A $>$ 1.70), the relative error is greater than 5%. Clearly, under these conditions the measurement is less precise than in other ranges. The range in which the results produce the smallest relative error occurs between absorbance values 0.4–0.7 or 20–60% T. In other words, to achieve the best precision in an absorption experiment, the concentration of the sample or the pathlength through the sample should be adjusted so that the output of the instrument is in absorbance range 0.4–0.7 if possible. This conclusion is reached by considering only the relationship between the absorbance and transmittance with a fixed, random error in transmittance. In fact, the error in T is not constant over its entire range, and each instrumental design has its own characteristics. However, the general trend described here still is applicable and should be considered when developing an assay.

Other Limits to Photometric Precision

One other often ignored but significant contribution to photometric error in all spectrometries is the lack of reproducibility of the condition and position of samples in the instrument. For instance, for numerous analytical methods the sample is put in a cuvette or other sample cell that is placed in the instrument. If its placement is not reproducible, then some error will be introduced due simply to the slightly different geometry in each measurement.

Differences in sample preparation can, as always, contribute to imprecision. Consider, for instance, a series of samples analyzed with an absorbance technique where each sample has a different amount of particulate matter stirred up in it. The particulate matter blocks some of the light, which results in a changing baseline and inaccurate results. This is just another illustration

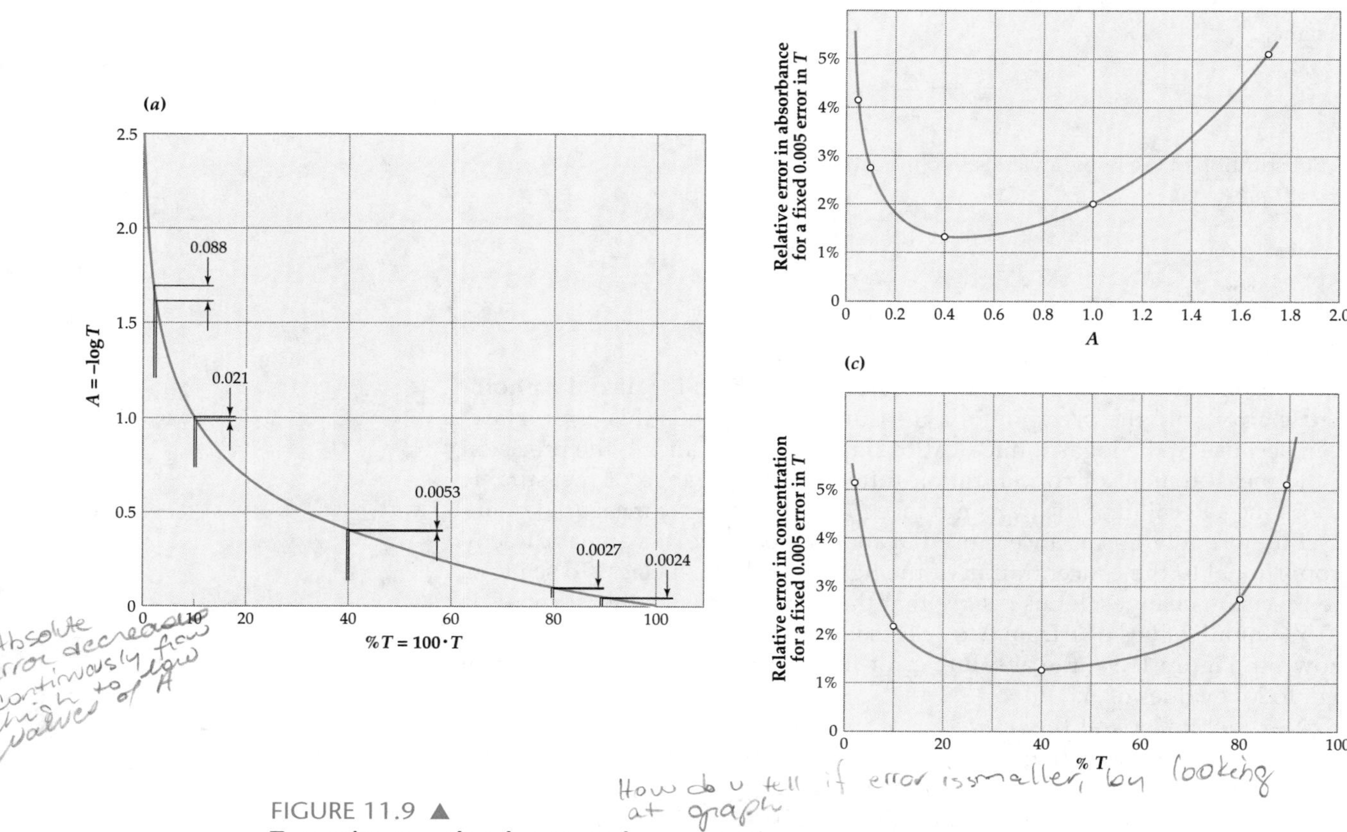

FIGURE 11.9 ▲

Transmittance, absorbance, and concentration error

(a) Graph of the relationship between absorbance and transmittance. The varying errors in *A* (labeled) result from a constant ½% error in *T*. **(b)** Illustration of the changes in *relative* error due to *only* the arithmetic relationship between absorbance and transmittance. For a fixed ½% error in *T*, the relative error in absorbance is shown. The absorbance range from about 0.2 to 0.8 is the optimum range to minimize error in absorbances. **(c)** The relative error in concentration for a fixed ½% error in *T*. This shows that the optimum range for determining the concentrations by transmittance measurements is between 20% and 60% *T*.

of how limitations in precision depend just as much or more on the care and skills of the individual(s) doing the analysis as on the instrument.

11.8 Fluorescence/Phosphorescence Spectrometry

Luminescence (fluorescence and phosphorescence) spectra are obtained with the general instrumental geometry shown in Figure 11.10. A light source for the appropriate energy region is required, and the incident light is monochromatic. In the figure, we indicate that the incident light arrives at the sample after passing through the first monochromator, although monochromatic light sources such as lasers are also used. The light is absorbed by the sample and then reemitted in all directions and at longer wavelengths. The *luminescence* that comes from the sample is measured at an angle that is not in line with the axis connecting the source and sample. This angle is often, but not necessarily, 90°.

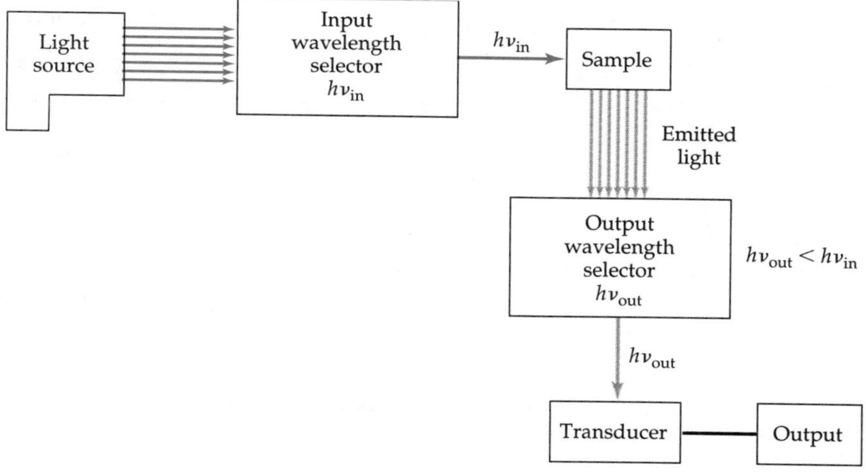

FIGURE 11.10 ▲
Generalized experimental equipment design to measure sample luminescence (fluorescence, phosphorescence) in the wavelength regions from infrared through X-ray.

If the sample luminescence is proportional to the analyte content, it can be used to quantify the analyte. In addition, if a time-dependent spectrometric method is used, the light due to phosphorescence and that due to fluorescence may be separated. On this practical basis, phosphorescent reemission generally can last a much longer time—even after the light source is no longer providing radiation to the phosphorescing sample.

Once again, the sample need not be inside the instrument. See EP.5 inside the front cover for an example of remote sensing. The outside cover photo is the result.

The output measurement can be described simply by

$$\text{Output}_\lambda = \kappa_\lambda c \tag{11-9}$$

Quantitative fluorescence is the basis for the DNA analysis chip shown in photo EP.2 inside the front cover.

where

c is the solution concentration,

κ is a proportionality constant, and

λ is the wavelength at which the luminescence signal is measured. It is *not* the source wavelength.

The factor κ accounts for the power of the source, the efficiency of the luminescence process in converting the source power into emission, the pathlengths of both the source and emitted light, and the geometry of the optics. All these factors must be constant for a series of quantitative measurements.

Notice that the light output should be zero in the absence of the analyte. Equation 11-9 is applicable only when no light is produced or scattered at the analytical wavelength by any other component in the solution.

EXAMPLE 11.7

A fluorescence assay was to be tested to see over what range of concentrations it was linear. The following data were obtained for quinine sulfate standards ranging in concentration from 10^{-3} down to 10^{-9} *M*. Given the wide range to be tested, it was necessary to change the **attenuation** of the signal to keep the readings on scale. Therefore a blank reading was made at each attenuation level to correct for background signal intensity. The intensities are on a linear scale with an arbitrary multiplier.

Log of concentration	Fluorescence intensity of sample	Attenuation	Fluorescence intensity of blank
−3	3,500	10,000	2
−4	5,853	10,000	2
−5	5,928	10,000	2
−6	6,099	1,000	4
−7	6,038	100	42
−8	6,335	10	390
−9	10,005	1	4003

a. Multiply each standard reading by the attenuation and correct for the contribution found for a blank sample run with the same attenuation setting.

b. Plot the corrected intensities *vs.* concentration on a log-log plot.

c. Over what concentration range would the method be useful?

Solution

a.

Log of concentration	Adjusted fluorescence of sample	Adjusted fluorescence of blank	Corrected intensity
−3	3.500×10^7	2×10^4	3.498×10^7
−4	5.853×10^7	2×10^4	5.851×10^7
−5	5.928×10^7	2×10^4	5.926×10^7
−6	6.099×10^6	4×10^3	6.095×10^7
−7	6.038×10^5	4.2×10^3	5.996×10^5
−8	6.335×10^4	3.90×10^3	5.945×10^4
−9	1.0005×10^4	4.003×10^3	6.027×10^3

b. See Figure 11.11.

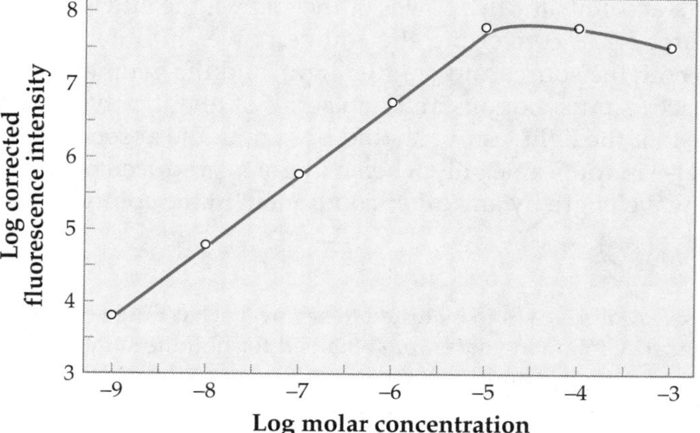

FIGURE 11.11 ▲
Log–log plot of fluorescence intensity vs. concentration.

c. The method could be used at concentrations up to the point where the intensity begins to curve back down to give a **double-valued** plot. *Double-valued* means that one intensity reading can have two different concentrations associated with it—one on each side of the maximum. In order to avoid the ambiguity from a double-valued response, the upper limit for the assay appears to be around 10^{-6} M. On the lower end, the blank has become a significant portion of the intensity reading at the 10^{-9} M level, so it would probably not be a valuable method at concentrations lower than this.

Comparisons of the Methods

When does one consider using luminescence as opposed to absorption methods? Since all compounds that fluoresce/phosphoresce also absorb light, a number of analytical samples can be analyzed either by absorption or luminescence spectrometry. If an analyte does not fluoresce strongly, the choice seems easy: Use absorption. However, the choice of method is often not this straightforward since a number of benefits and drawbacks must be considered when you can choose either.

First, just as your eye finds it easier to see a light spot on a dark background than to see slight changes in brightness, so the limits of detection are inherently lower in luminescence spectrometry than in absorption. But the inherent limits are not necessarily the practical limits. For example, some of the incident light can be **scattered** directly from the source into the transducer from particles in the sample. In order to obtain accurate results, the samples must be carefully filtered to remove particulate matter. This adds an extra step to the sample preparation. In addition, at trace levels, the filter must not interfere by, for instance, adsorbing analyte on its surface. Absorption spectrometry is far less susceptible to errors due to such light scattering.

A second basis for choice is the magnitude of the working range of analyte concentrations. The fluorescence emission intensities can be linear with concentration over three to five orders of magnitude, as you saw in the example above. A nonlinear but usable working curve can add another factor of ten. In contrast, analyses depending on absorbance tend to be linear with concentration over only a factor of 10 to 100 in concentration. In other words, luminescence has a greater linear range than absorbance spectrometry. However, there is a drawback to luminescence spectrometry for more concentrated solutions. Recall that in luminescence spectrometry, the source power is absorbed by the sample and subsequently reemitted. However, the reemitted light can also be absorbed by the sample. This process is called, reasonably, **self-absorption**. As a result of self-absorption, less light will be emitted at higher concentrations of analyte than we would expect by extrapolation of the results from lower concentrations. In fact, if the self-absorption becomes large, the measured intensity of the luminescence can even decrease with increasing concentration. A maximum value is reached and the output decreases with higher concentrations, which produces a double-valued working curve. In general, calibration curves will become nonlinear in fluorescence spectrometry at lower analyte concentrations than for absorption spectrometry. But again, the problem may simply prove to be an inconvenience since the sample can be diluted.

A third basis of choice is that since most molecules do not luminesce, it is sometimes possible to analyze by luminescence in the presence of species that would interfere in an absorbance assay.

So what is the conclusion from these conflicting considerations? In practice, when either method can be used, luminescence is used more for ultratrace analyses, while absorption is applied to problems with more concentrated analytes.

Emission is used when the analyte itself emits light; there are two common cases. The first is when the sample can be heated to high temperatures (500–3000 K). A number of techniques utilizing such hot samples are used for elemental analyses and are described in Section 11.11 below. The second results from an origin of light emission that is quite special, and only a relatively few examples have been found. The energy is introduced from chemical bond rearrangements. This is *chemiluminescence*. Chemiluminescence is not a general spectrometric method but a description of the origin of the light. Where chemiluminescence occurs, it can be the basis of specific and sensitive assays.

11.9 Quantifying Two Species with Spectral Interference Present

If enough spectrometric information is available, all of the components of mixtures can be quantified without separation. Examples of spectra from a two-component mixture are shown in Figure 11.12. You will find the problem is mathematically equivalent to having two simultaneous equations with two unknowns. Similar calculations are done all the time in quantitative spectrometry, both in emission spectrometry as well as in absorption. It is important, therefore, for you to understand the manner in which such calculations are made. The data required for the calculations are as follows:

1. For the unknown sample, the absorbance at a number of wavelengths. The number of wavelengths must be equal to (or greater than) the number of compounds in the sample that absorb a significant fraction of the light.
2. The molar absorptivities *for each component separately* at all the wavelengths chosen for the mixture.

Let us state these requirements algebraically. For a sample containing two components (identified by subscripts 1 and 2) both of which absorb light at the same wavelength λ_i, the total absorbance is

$$A(\lambda_i) = A_1(\lambda_i) + A_2(\lambda_i) \tag{11-10}$$

Equation 11-10 says that at any wavelength the total absorbance measured is the sum of the absorbances of the two components. We use Equation 11-8 with slightly different labels to substitute for each absorbance in Equation 11-10. Then, for component 1 at wavelength i,

$$A(\lambda_i) = \epsilon_{1i} b c_1 \tag{11-11a}$$

A similar expression (only different subscripts) describes the absorbance of the second component:

$$A_2(\lambda_i) = \epsilon_{2i} b c_2 \tag{11-11b}$$

However, by having the same subscript i, we indicate that the wavelengths are the same for both species being measured. The pathlength b also remains the same. Substituting Equations 11-11a,b into Equation 11-10 gives the

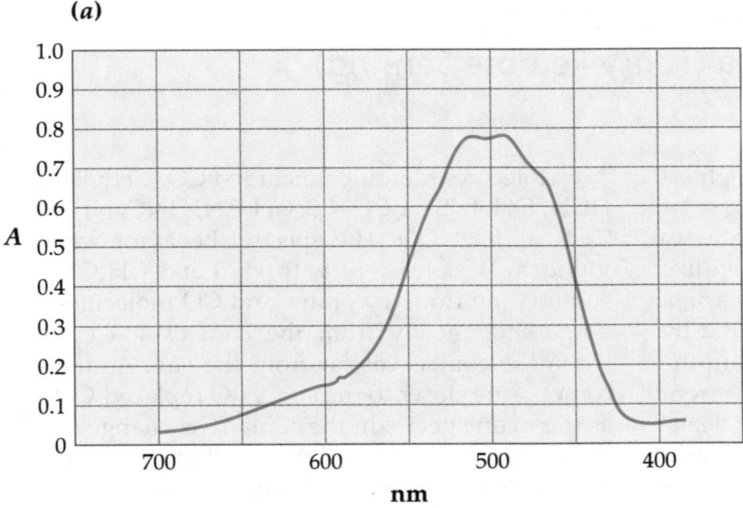

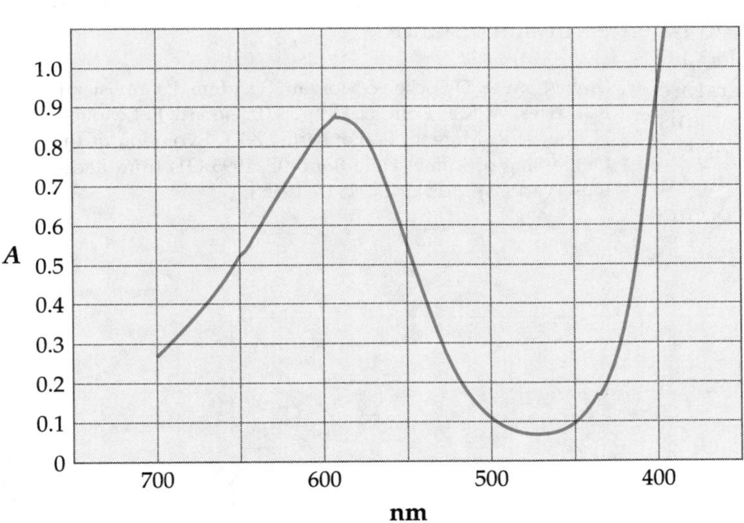

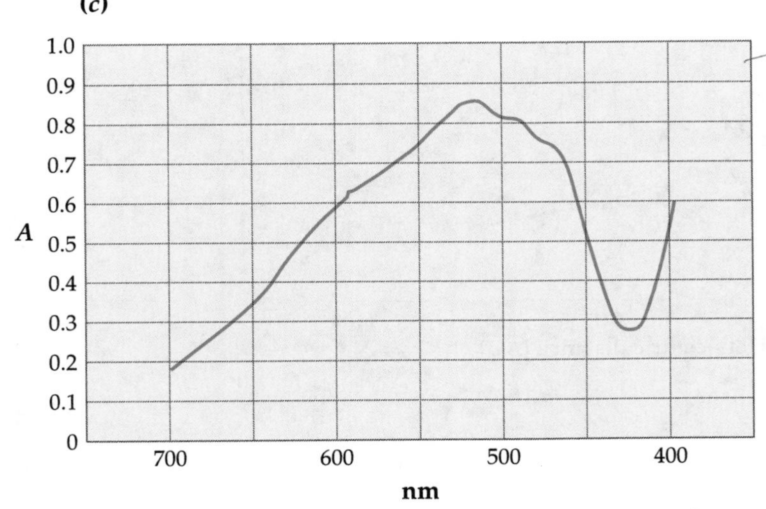

◀ FIGURE 11.12

Spectra of metal-edta solutions in a phosphate buffer.

(a) The absorbance spectrum of a standard Co-edta solution made by diluting 20 mL of 0.150-M $CoCl_2$ to 50 mL with the edta-phosphate solution. **(b)** A similar solution made for nickel. For the Ni standard, 20 mL of 0.250-M $NiCl_2$ solution was diluted to 50 mL. **(c)** A mixture of the two solutions. The spectrum of the mixture is the sum of the spectra of the two components in it.

ANALYTICAL CHEMISTRY IN ACTION

Coming Out of a Coma (*Really* Remote Sensing)

In the spring of 1997, Hale-Bopp, the brightest comet of the century, passed through the solar system and afforded an extraordinary chance to study the chemistry of the *coma*, the cloudlike region around the nucleus of the comet. The data for the chemical characterization was found by spectrometry of the molecular emissions in the high microwave range. The emissions arise from molecular rotations, and the precisely determined frequencies are easily assigned to specific molecules and molecule fragments. By monitoring the brightness of the lines, the relative concentrations of the molecules could be calculated. In addition, the *relative* brightness of each of a series of rotational lines from the same molecular species can be used to calculate the temperature of the molecular species. With the use of this spectral data and the known chemistry of the materials, a model of the chemistry permitted a general picture of the chemical processes occurring as the comet came closer to the sun.

Molecules identified included CO, CH_3OH, HCN, OH, H_2S, H_2CO, CS, CH_3CN, HNC, NH_3, OCS, and HNCO. The spectral behavior was found to be consistent with H_2O and CH_3OH sublimating from icy grains and CO molecules outgassing mostly from the comet's nucleus when the comet was far from the sun. As the comet came closer to the sun, H_2O replaced CO as the main species in the coma. The change in the temperature of the comet as it approached the sun (its heliocentric distance) is illustrated in the figure. The abbreviation AU represents an astronomical unit, which is the mean earth–sun distance.

[Ref: N. Biver, D Bockelée-Morvan, P. Colom, J. Crovisier, J. K. Davies, W. R. F. Dent, D. Despois, E. Gérard, E. Lellouch, H. Rauer, R. Moreno, G. Paubert. 1997. "Evolution of the Outgassing of Comet Hale-Bopp (C/1995 O1) from Radio Observations." *Science* 275:1915–1918.]

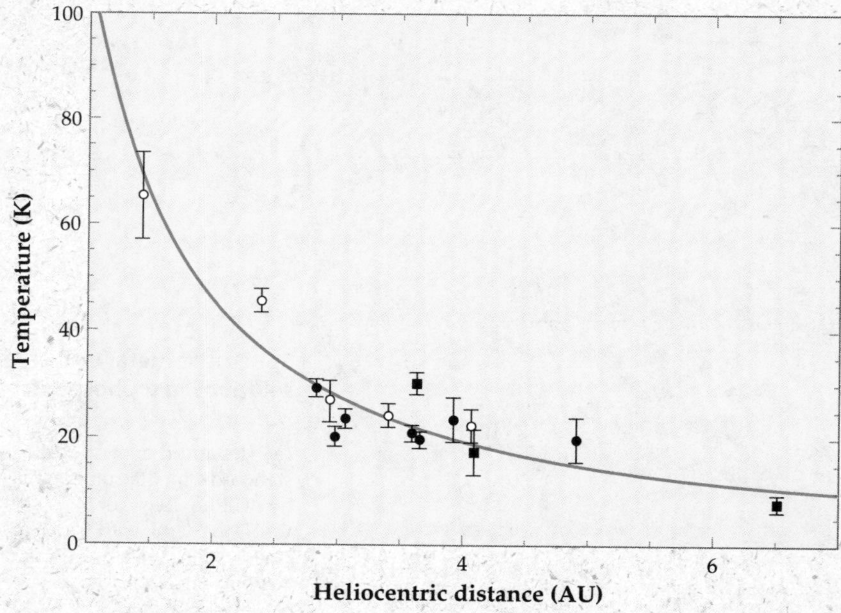

Temperature of Hale-Bopp.

expression for the total absorbance at wavelength λ_i:

$$A(\lambda_i) = \epsilon_{1i} bc_1 + \epsilon_{2i} bc_2 \qquad \textbf{(11-12)}$$

There is nothing different in principle between Equations 11-10 and 11-12.

EXAMPLE 11.8

Two components absorb at the same wavelength λ_i in a solution. Two standards show that at wavelength i, $\epsilon_{1i} = 500$ and $\epsilon_{2i} = 3000$, and the solution has $A(\lambda_i) = 0.450$. Find the concentration of each of the components. The cell path length is 1.000 cm, and the absorptivities are in units of L mol^{-1} cm^{-1}.

Solution

Substituting into Equation 11-12 gives

$$(\text{at } \lambda_i) \quad 0.450 = 500\, c_1 + 3000\, c_2 \qquad \textbf{(11-13)}$$

What values can the concentrations c_1 and c_2 have? It turns out that this equation can be solved with any one of an *infinite* number of sets of c_1 and c_2 values. Thus the information given is *not* sufficient to solve the problem. We must find another equation to limit the possibilities to a unique result.

As you have observed, a measurement at a *single* wavelength will not be sufficient to find the amounts of *two* components. As stated in requirement 1 above, the number of wavelengths must be at least equal to the number of individual components. A second measurement and equation are required to relate c_1 to c_2. Experimentally, the necessary information can be obtained by redoing the same set of experiments at a second wavelength λ_j. Sufficient information is then available to solve unambiguously for the component concentrations.

EXAMPLE 11.9

Two components absorb in the same wavelength range in a solution. At wavelength λ_i,

$$\epsilon_{1i} = 500, \qquad \epsilon_{2i} = 3000, \quad \text{and} \quad A(\lambda_i) = 0.450$$

At a second wavelength λ_j,

$$\epsilon_{1j} = 2100, \qquad \epsilon_{2j} = 160, \quad \text{and} \quad A(\lambda_j) = 0.565$$

Find the concentration of each of the components. The cell path length is 1.000 cm, and the absorptivities are in units of L mol^{-1} cm^{-1}.

Solution

Two components comprise the mixture. Since the information at two wavelengths is now included, the concentration of each component can be found from the data.

$$(\text{at } \lambda_i) \quad 0.450 = 500\, c_1 + 3000\, c_2 \qquad \textbf{(11-13)}$$

$$(\text{at } \lambda_j) \quad 0.565 = 2100\, c_1 + 160\, c_2 \qquad \textbf{(11-14)}$$

So, we have now the two simultaneous equations needed to find a unique solution for the two unknowns. Together, Equations 11-13 and 11-14 can be solved in a straightforward way to find unique values of c_1 and c_2. The result is

$$c_1 = 2.61 \times 10^{-4}\ \text{M}$$

$$c_2 = 1.07 \times 10^{-4}\ \text{M}$$

Problems in emission and luminescence assays on mixtures are solved in a similar way, but using the relative response levels for each species at the chosen wavelengths and the response of the mixture at the same wavelengths.

11.10 Atomic Spectrometry

Up to this point, the subjects covered in this chapter apply to all spectrometries in general. The next four sections (11.10–11.13) include only those spectrometries that measure light from transitions of *atoms* or *ions* and not from molecular species. In other words, the experiments deal either with atoms that have been freed from a condensed-phase matrix or with spectra that are essentially still "atomic" even though the atoms remain bound in a molecule or matrix. Also, from now on, the term *atomic spectrum* will be understood to mean the spectrum of either atoms *or* ions. Either can be used to determine the elemental content of analytes.

When the atoms have been freed from a condensed phase matrix, we measure the absorption or emission of UV or visible radiation by the atoms (or ions) in the gas phase. Their spectroscopic transitions occur at predictable positions with relatively narrow bandwidths. For the bound-atom cases, excitation involves high-energy particles or radiation of energy higher than UV. Either the emitted radiation or the kinetic energy of an outgoing particle may be measured. These bound-atom analyses can be divided into methods that respond to the content of analyte in the bulk and those that only respond to the analyte on a surface in a vacuum. In these surface-sensitive methods the signal arises from the top layer—in the range of 0.1 to 100 nm depth, depending on the method.

Omitted from these sections on spectrometry are methods in which elemental content is determined by measuring spectra of molecular species in solution. Examples of such determinations are those of iron with 1,10-phenanthroline, nickel with dimethylglyoxime, and molybdenum with 4-methyl-1,2-dimercaptobenzene. These have to a large extent been supplanted since these methods require far more complicated sample preparation to avoid interferences.

Spectra and Spectral Notation

The nomenclature of atomic spectra differs from that used to describe the chemistry of solutions and solids. The spectroscopic nomenclature uses Roman numerals that are one unit higher than the oxidation state. In other words, the neutral atom is "I," the +1 oxidation state is "II," the +2 "III," and so forth. Comparisons are shown in the chart below, with copper as an example.

In atomic spectroscopy	Cu(I)	Cu(II)	Cu(III)
In inorganic chemistry	Cu(0) Cu^0	Cu(I) Cu^+	Cu(II) Cu^{2+}

As you are well aware from your previous chemistry studies, the energy levels of atoms and ions are quantized: Only certain, specific energies are observed. The properties of an atomic-emission spectrum, such as that of atomic hydrogen (Figure 11.13a), result from this quantization. In the range from 2000 to 100 nm (near-infrared through ultraviolet), the wavelengths of

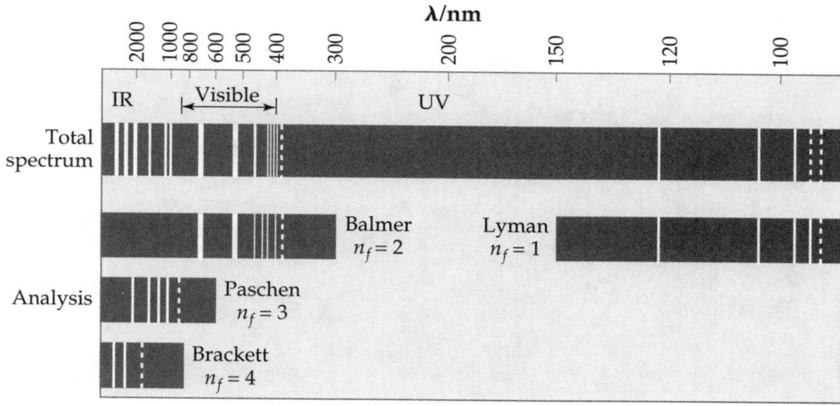

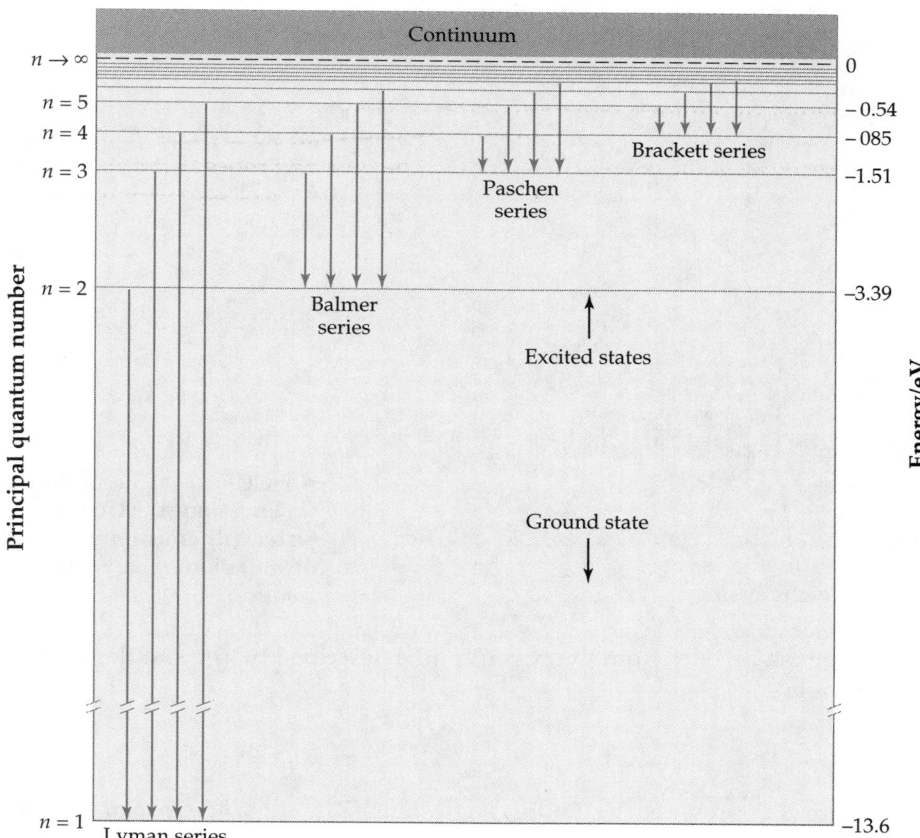

FIGURE 11.13 ▲

(a) The emission spectrum of atomic hydrogen.

(b) The energy-level scheme that describes the spectroscopic lines of atomic hydrogen.

The lines of the spectrum are assigned to sets that have their lower energy levels in common. Each series of lines is named after its discoverer. The arrows indicate the transitions that result in light emission of the Lyman, Balmer, Paschen, and Brackett series. [Ref for (a): Atkins, P. W. 1978. *Physical Chemistry*, San Francisco: W. H. Freeman and Company.]

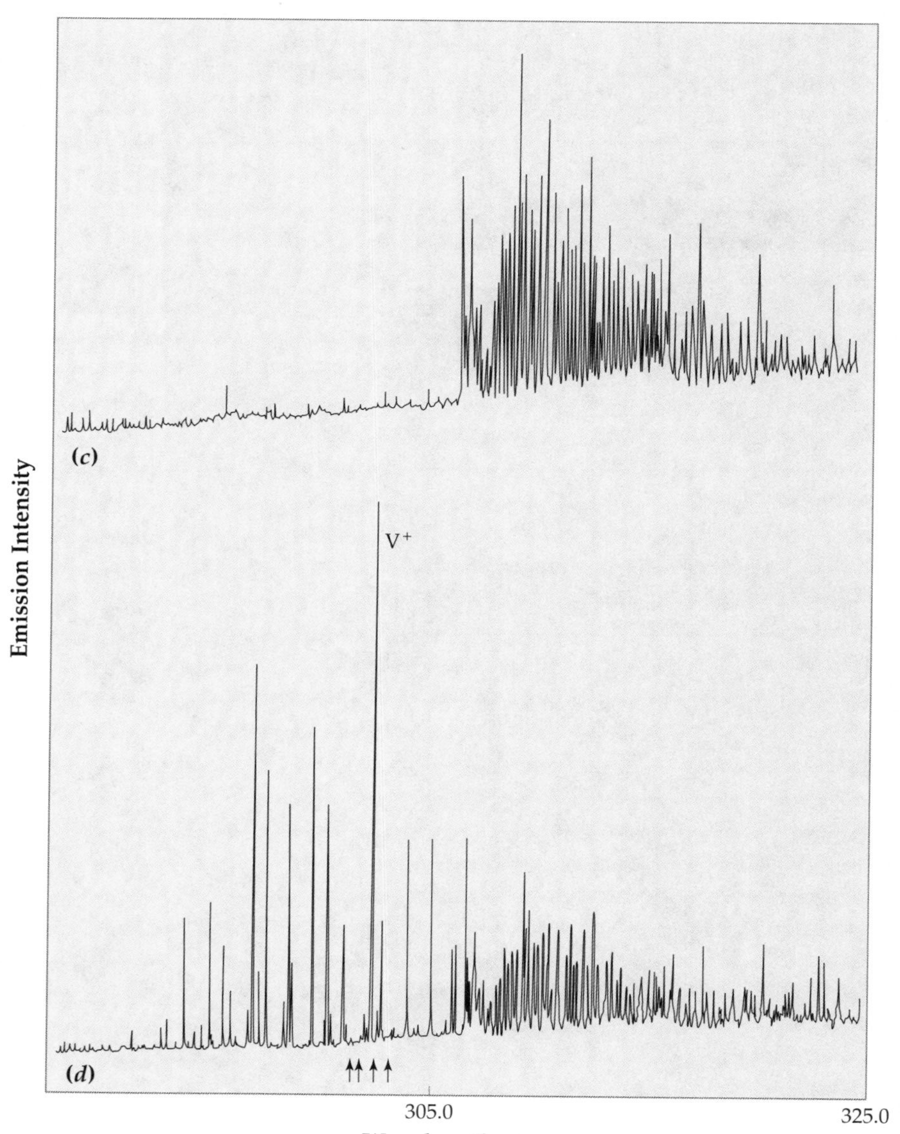

◀ FIGURE 11.13 *(continued)* **(c) Emission spectrum of water. (d) Emission spectrum of vanadium in aqueous solution.**

the emission lines from hydrogen can be described by the simple, algebraic formula

$$\frac{1}{\lambda} = R \cdot \left(\frac{1}{n_f^2} - \frac{1}{n_i^2}\right), \qquad n = 1, 2, 3, \ldots \quad \text{and} \quad n_i > n_f$$

where R represents the Rydberg constant (R = 109,737 cm^{-1} or 3.2898 × 10^{15} Hz). For atomic hydrogen, these lines are usually divided into four sets, depending on the value of n_f (between 1 and 4). Notice in the figure that as the energy levels get higher, the states are closer together.

Each of the n-values is associated with a different quantized energy state of the atom. The n-value is a **quantum number** of hydrogen, and the associated states are shown in Figure 11.13b.

Each spectral line has associated with it a state of lower energy and a state with a higher energy. The energy of the spectral line, and hence its wavelength, equals the difference between the two energies. (Recall $E = h\nu$.) The lowest energy state that an atom or ion can have is called its ground state. For hydrogen, the ground state has the electron in the $n = 1$ quantum level.

All other states—those with higher energies—are called excited states of the atom. Recall that the atomic energy levels with quantum numbers $n = 1, 2, 3, \ldots$ are also called the K, L, M, . . . shells, respectively. The energies associated with the different states are called the electronic energies of the atom.

The process in which an electron jumps between two states of the atom is called an atomic **spectral transition**. When the spectrum is due to changes in electron energies, as it is for hydrogen, the transition is called an **electronic transition**. The point at which the electron ceases being bound to the nucleus is the **ionization potential**. The ionization potential tabulated in beginning chemistry texts is the energy difference between the energy of the electron in the highest occupied orbital in the ground state of the atom and the level where the electron and ion just become completely free of each other.

The spectrum of hydrogen is the simplest of atomic spectra. All the others are much more complicated, and the spectra become even more complicated when molecular species are present. Examples of such spectra are shown in Figures 11.13c and 11.13d, which show a spectrum of water alone and one of an aqueous solution containing vanadium.

11.11 The Practice of Atomic Absorption (AA) and Atomic Emission (AE)

Atomization

Ideally, in analytical atomic spectrometry all atoms are freed of the influence of the surrounding matrix. This means that the analyte atoms are not chemically bound to any other atoms. All particles, compounds, and molecules must be dissociated. The matrix of the sample should be totally destroyed—**atomized** is the correct term. This ideal is not often reached. However, as long as constant conditions are obtainable, precise analyses ($\pm 1\%$) can be carried out.

If the element to be analyzed is a monoatomic gas near ambient temperature, then atomization is quite simple. It is only necessary to hold the element in a closed cell for analysis.

When the sample to be atomized is introduced as a liquid, there are a few additional steps in the atomization process. Liquid samples are commonly blown as a mist into the atomizer. Numerous apparatus designs are in use to convert bulk liquid samples into the mist of small droplets. The process is called *nebulizing* (from the Latin *nebula* meaning *cloud*). As a result, the part of an atomic spectrometer (either emission or absorption) that nebulizes the liquid is called, reasonably, a **nebulizer**. Following nebulization, the solvent must be removed, and the remaining material must be broken down into separate atoms. High temperatures are used to carry out these latter steps. In the next section are described the most commonly used methods of atomization.

Methods of Atomization

Furnaces such as that shown in Figure 11.14a are used to atomize solids, slurries, and solutions, mostly for atomic absorption. It is essential to keep the furnace in an inert atmosphere such as argon during the heating period to

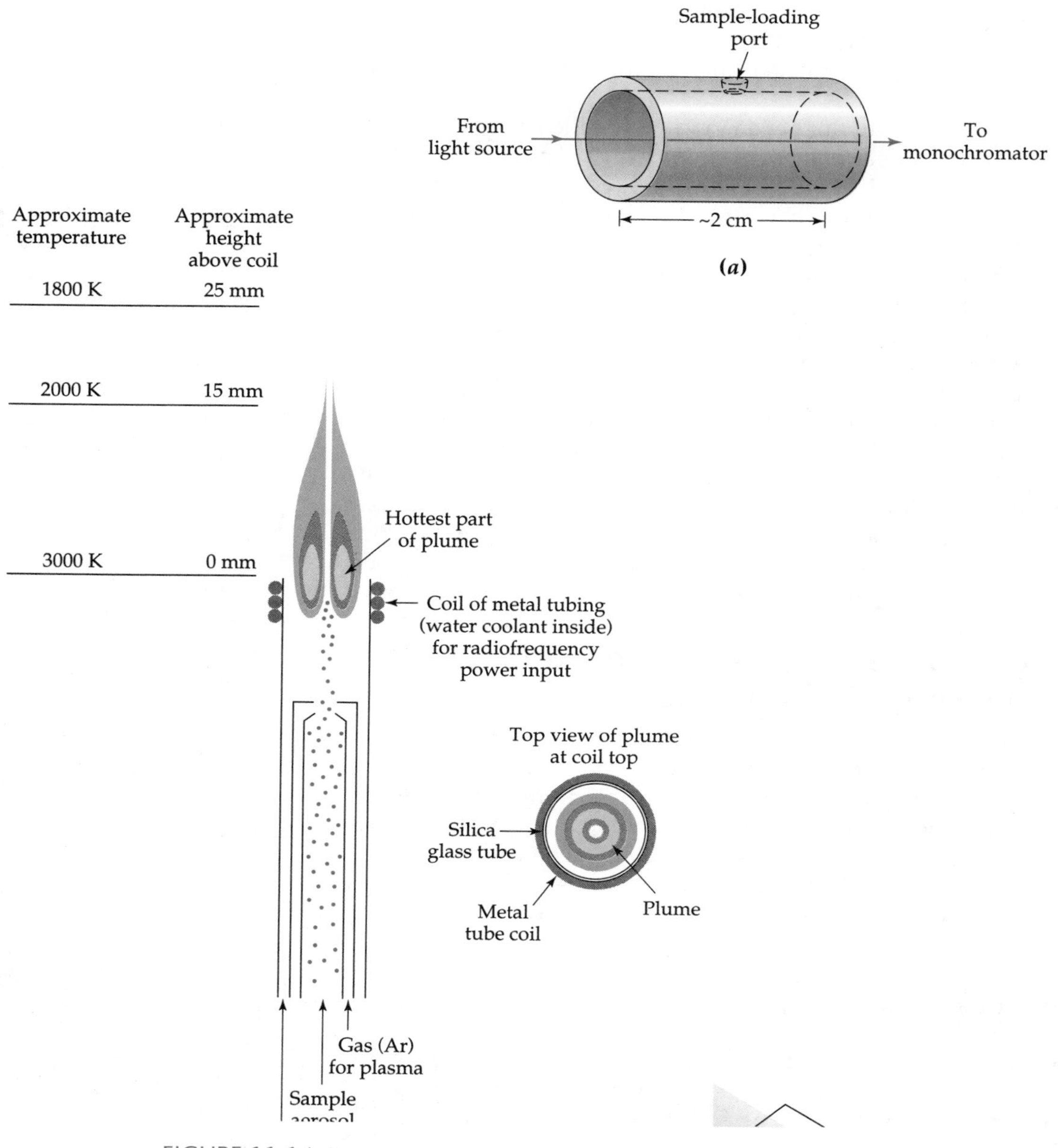

FIGURE 11.14 ▲

Atomization methods for atomic spectrometry.

(a) Illustration of one kind of graphite furnace. The tube is 2–3 cm long and placed between two electrical connections. The cylinder is heated by passing electrical current through it from one end to the other. **(b)** Schematic diagram (not to scale) of the inductively coupled plasma (ICP) torch used as an atomic-emission spectrometry source. A plasma of ionized argon is excited by radiofrequency energy that is "broadcast" from a coil at the base of the plume. The plasma is hollow, and the analyte is blown into the hollow region. The gases carrying the nebulized sample flow up from below together with a separate gas stream for cooling the silica glass. The plasma plume is suspended away from the glass walls through a combination of the helical path of the cooling gas flow and the shape of the radiofrequency electromagnetic field. The spacing prevents the 3,000-K temperature plume from melting the silica. Both the helical gas flow and plasma shape help confine the lighter, hotter gases to the middle of the plume. **(c)** Illustration of a long flame produced by a slot burner. As the light path through the flame is increased, so is the absorbance of the sample.

minimize unwanted chemical reactions with the atomized analytes. Since the furnace is heated electrically, the process is called **electrothermal atomization** and is usually used for AA. Occasionally furnaces are used to vaporize samples to feed into a **plasma torch** for atomic emission.

A photo of a heated graphite furnace can be seen in photo EP.9 inside the back cover.

A plasma is an ionized, high-temperature gas that can function as an outstanding atomizer. The most commonly used type for atomic emission is an *inductively coupled plasma* (ICP) torch. The general structure of an ICP source is shown in Figure 11.14b. The ICP is a continuous (in time) plasma formed in a flowing stream of argon. The energy to keep the torch hot and the gases ionized is supplied by a microwave similar to a microwave oven. (Other methods for generating continuous plasmas also can be used.) Plasmas can atomize gaseous or solution samples, and specially designed torches can be fed slurries of microscopic solid particles without clogging. Water is the preferred solvent, although low-percentage organic-water mixtures can be used. However, organics tend to reduce the performance of the plasma, and, in the extreme case, can extinguish it.

An ICP torch for atomic emission spectrometry is shown in photographs EP.7 and EP.8 inside the back cover.

Flames can also be atomizers for analyses of liquids or solutions by atomic emission and absorption methods. Flame sources of atomization have been brought to a high degree of reliability over their nearly 50 years of development. However, since the gases for the flame dilute the sample, a long flame shape is desired to obtain high sensitivities. A long flame can be obtained easily by using a slot-burner, such as illustrated in Figure 11.14c. A variety of different fuels and oxidants can be used. The choice depends on the desired temperature and the elements that are being determined. However, all the common flame sources remain at relatively low temperatures compared with plasmas. Although still commonly found, instruments that utilize flame atomization are gradually being replaced with those using plasmas. The primary reason is that atomic emission is better suited for simultaneous, multi-element analyses, and plasmas provide, overall, better emission results for a wide range of elements.

Quantitation

Standard addition—Section 4.7.

Just as for other luminescence, absorbance, and emission methods, UV-visible atomic spectrometry exhibits a simple linear dependence between concentration and response. Because of difficulties in maintaining constant experimental conditions, a standard addition method is used in many instances for atomic spectrometry. This approach helps to correct for differences in atomization or nebulization efficiencies, and for any matrix effects on response.

EXAMPLE 11.10

The chromium content of the intermediate treatment steps of effluent from a plating operation was determined using ICP atomic emission spectrometry. A standard addition method was employed, where the volumes of standard added were negligible compared to the volume of the analytical sample. The following data were obtained:

Sample	Emission intensity
Effluent (0 ppm added Cr)	0.223
Effluent + 5.00 ppm added Cr	0.342
Effluent + 10.0 ppm added Cr	0.451

Determine the content of chromium in the effluent.

Solution

A plot of emission intensity *vs.* added concentration yields a line corresponding to the equation

$$\text{Emission intensity} = 0.0228 \cdot (\text{concentration Cr added, ppm}) + 0.2247$$

Recall, for the standard addition method, the sample content corresponds to the negative of the x-intercept for the line. The x-intercept is found algebraically by letting y (that is, the emission intensity) equal zero. Then, you can solve for the chromium concentration, c_{Cr}. That is,

$$I = 0.0000 = 0.0228 \cdot c_{Cr} + 0.2247$$

and

$$c_{Cr} = -0.2247/0.0228 = -9.86 \text{ ppm}$$

Since the negative of the intercept indicates the concentration, $c_{Cr} = 9.86$ ppm.

Instrumentation for Atomic Absorbance and Atomic Emission Spectrometry

The instruments for atomic absorption and emission are similar in concept to those for UV-visible spectrometry. However, instruments for atomic spectrometry have much higher spectral resolution than instruments for solution measurements.

11.12 Bulk Elemental Analysis without Atomization

Bulk elemental techniques include X-ray fluorescence (XRF), neutron activation analysis (NAA), and mass spectrometry with high-energy ionization sources such as a spark or focused laser. X-ray fluorescence and neutron activation are nondestructive methods. The sample remains essentially unchanged by the measurement. They are especially useful for samples that need to be kept intact, such as art objects and forensic samples.

X-ray fluorescence arises from transitions within the quantized energy levels of the innermost electrons of atoms. After excitation, each element emits at characteristic X-ray wavelengths that can be used to identify the elements present in a sample. The intensities of the X-ray fluorescence are used to quantitate the amount of each analyte present. Like the other luminescence methods discussed earlier, a number of elements may be determined during one experiment. The chemical elements that may be quantified by X-ray fluorescence are those with an atomic number ≥ 11 (sodium). On the other hand, since almost all elements produce X-rays under the experimental conditions, there is a fairly high background. Thus, X-ray fluorescence generally is limited to higher concentrations—greater than about 0.1%.

High-energy particles with kinetic energies in the MeV (million electron volt) range penetrate into the bulk of a sample and cause the emission of particles and X-ray fluorescence. The X-rays, since they can escape from the bulk, can be used to analyze the atomic species in the first half millimeter or so of the material. This is **particle induced X-ray emission** (PIXE). If the accelerated particles are protons, the technique is **proton-induced X-ray emission**, also called PIXE. In Chapter 5 you read how PIXE can be used as a technique to do localized analysis. The excitation also can be accomplished by α particles as was done by the Mars lander described later in this chapter.

Figure 5.4 on page 152 shows some PIXE results.

When a sample is immersed in a flux of slowly moving neutrons, many of the nuclei of the elements present capture a neutron and become radioactive. These radioactive nuclei decay with the emission of γ radiation, the wavelengths of which are characteristic of the nuclide. The γ energies are in the keV to the low MeV range—10^3 to 10^6 times the energies of visible light. The γ photons are easily counted individually and summed to obtain the spectrum. The process of neutron capture to form a radioactive nuclide is called **neutron activation**. When the activation is followed by quantitation of the γ-radiation emission, the whole process is called **neutron activation analysis** (NAA) or **instrumental neutron activation analysis** (INAA). Activation analysis is a multielement technique that has an extraordinary linear range of analysis. This range is enhanced by taking larger or smaller subsamples and collecting the spectrometric data for longer or shorter periods of time in order to achieve good precision with the emission counts. The multielement nature of the γ spectrometry means that similar problems exist in NAA as were discussed for emission spectrometry; for example, line overlap. However, a significant difference for activation analysis is that many interferences can usually be eliminated because of the different half-lives. The major drawbacks to the use of NAA are the need for a nuclear reactor and the fact that not all nuclei produced have lifetimes in the range that can be counted precisely. But the advantages are numerous. Most of the elements can be quantitated with NAA. The detection limits are comparable to those of ICP-AES and ICP-MS, as can be seen in Appendix VIII. However, unlike the other atomic methods, NAA can be done on solid samples directly, with no preparation.

11.13 Surface-Sensitive Atomic Spectrometry

The difference between conventional spectrometry and surface spectrometry lies in the thinness of the layer from which the analytical information is obtained. All surface analytical techniques are ultratrace methods since the surface alone is equivalent to concentrations of about 10^{-10} mol cm^{-2}, and their precisions are in the range $\pm 20\%$. To limit the analysis to a surface layer requires that the depth of the excitation for the spectral emission can only have a short **penetration depth** or **mean free path**. This limits us to excitation by high-velocity electrons and ions. By the same token, if radiation excites the sample, then the mean free path for the particles that are emitted must be short enough to allow them to escape only from the surface of the sample.

Electrons can be emitted from atoms and molecules when they are irradiated with light. This is photoemission and the electrons emitted are called **photoelectrons**. Each electron emitted has a kinetic energy, which is related to the wavelength of the exciting light and to the bonding energy of the electron that is emitted. The relationship reflects conservation of energy.

$$h\nu = KE + BE + W_f \quad \textbf{(11-15a)}$$

where

$h\nu$ = the energy of the impinging X-ray

KE = kinetic energy of the photoelectron

BE = bonding energy of the electron

W_f = work function for the sample. This quantity is a constant for a given sample. To account for its effect the carbon 1s electron binding energy is often used to calibrate the spectrum.

The data consist of measurements of the photoelectrons' kinetic energies. That is,

$$KE = h\nu - BE - W_f \qquad \textbf{(11-15b)}$$

It can be seen from Equation 11-15b that weakly held electrons (low *BE*) have higher kinetic energies when emitted; strongly held electrons are emitted with lower kinetic energies. An increase in irradiation intensity only increases the *number* of emitted electrons and not the individual kinetic energies. When ultraviolet radiation is used, the spectrometry is called **ultraviolet photoelectron spectrometry** (UPS or UV-PES). If X-ray radiation is the source of ionizing energy, then the spectrometry is called **X-ray photoelectron spectrometry** (XPES or XPS). Another name for XPS is electron spectroscopy for chemical analysis (or **ESCA**). UPS and XPS spectra are usually obtained on different instruments.

For analysis of chemical content, it is XPS that is more often used. Since the strongly held, inner electrons that are excited by the radiation are only slightly affected by chemical bonding, the values of the electron energies can be used to characterize the composition of the sample surface layer, and, to some extent, the oxidation state present. In contrast, ultraviolet irradiation causes electrons to be ejected only from the outer shells—energy levels that change significantly with changes in molecular bonding. As a result, UPS gives much more information about the chemical environment of the atom, but identification of the elements' identities may not be possible.

Photoelectrons also can be ejected by a more complicated mechanism involving a total of three electrons in the analyte atom. The emitted electrons are called Auger (pronounced *oh-jay'*) electrons, and the analytical method is called **Auger spectrometry**. The mechanism for electron ejection is as follows. The incoming electron or X-ray ejects an electron from an inner orbital of the atom; an outer electron from the atom undergoes a nonradiative transition to fill the hole that is left; and the energy from the non-radiative transition ejects an electron from an outer level. There are a number of different combinations of energy levels that can be involved, so the emitted Auger electrons are classified based on the letter designation for the principle quantum numbers. For example, an *LMM* transition means that the initial ionization came from $n = 2$ (the *L*), then an $n = 3$ electron (the first *M*) underwent a non-radiative transition down to $n = 2$ causing an $n = 3$ electron (the second *M*) to be ejected from the atom. Auger lines can be obtained for all elements except H and He. Auger sensitivity is such that in the most favorable cases, less than 0.01 monolayer can be detected on a surface. Also, Auger spectra are characterized by a relatively high intensity, since the detection of emitted electrons can be carried out with excellent sensitivity. The probability of Auger electron emission is high for elements lighter than zinc, making Auger spectrometry a complementary method to XPS which is most sensitive to elements heavier than sodium. Auger spectrometry, like XPS, is used for finding the surface composition of a sample, but XPS (ESCA) is more useful for characterizing the chemical form of surface species.

A special application of Auger emission involves focussing the excitation beam on a spot and rastering it over the sample surface. (Rastering is the way a television picture is generated by a narrow beam of electrons.) By monitoring the energy of the Auger electrons at each spot, it is possible to find the elements present there. The name given to this type of assay is **electron microprobe analysis**. By looking at a given energy over the whole surface, it

ANALYTICAL CHEMISTRY IN ACTION
I'm a Wanderer, yes, I'm a Wanderer . . . on Mars

In July, 1997, a small robot carrying instruments was landed on the Martian surface to carry out photographic surveys and chemical analyses of the rocks. The chemical analysis was obtained with an alpha proton X-ray spectrometer (APXS). Designed specifically for the Mars Pathfinder mission, it provided a complete and detailed analysis of chemical elements in Martian soil and rocks near the landing site. The APXS instrument was carried on the Pathfinder microrover, which provided transportation to places of interest on the Martian surface. The spectrometer consisted of a sensor head mounted on a movable arm that enabled the head to be placed against soil and rock samples to perform chemical analyses in situ. The general layout of the sensor head is shown in the adjacent figure. You can see the sensor head in the photograph of the microrover, named Sojourner.

The APXS spectrometer works by bombarding the surface with α particles from a Curium source (^{244}Cm). When the particles hit the surface, three different types of interactions occur. The analysis of the spectra of these three provide the quantitative elemental analysis. Two of these interactions—Rutherford backscattering and X-ray fluorescence—are discussed briefly in this chapter. The third interaction, a nuclear reaction, results in the production of protons from the reaction between the α particles and the nuclei of light elements.

Measurement of the intensities and energy distributions of all three spectrometers together yields the information on the abundance of chemical elements in the sample. The three spectrometries are partly redundant and partly complementary: Rutherford backscattering of the α-particles is superior for light elements (C, O), while the proton emission is mainly sensitive to Na, Mg, Al, Si, and S. X-ray emission is more sensitive to heavier elements (Na to Fe and beyond). A combination of all three measurements enables the quantitation of all elements (with the exceptions of H and He) when their concentrations are greater than about 0.2% (w/w).

[Ref: Rieder, R., Wänke, H., Economou, T., Turkevich, A. 1997. "Determination of the chemical composition of Martian soil and rocks: The alpha proton X ray spectrometer." *Journal of Geophysical Research*. Vol. 102, No. E2. pp. 4027–4044.] ■

▲ **Sojourner.**

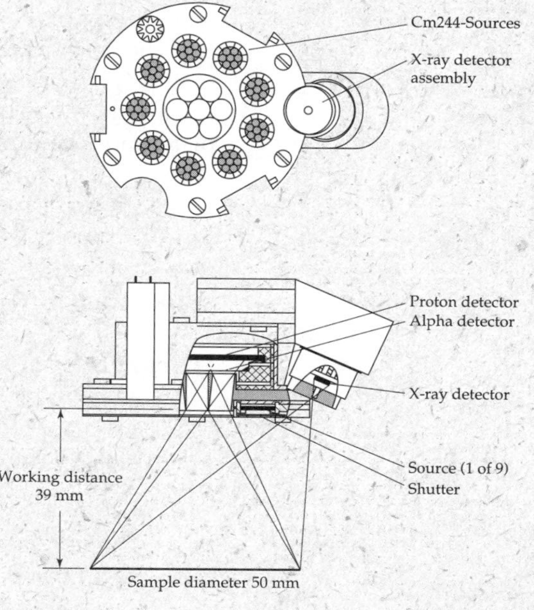

▲ **APXS sensor head.**

is possible to map the areas where a given element lies. With the data storage capabilities available in the computerized instrumentation in use today, it is possible to store multiple spectra for every spot, and a map can be made for every element present, given sufficient time and a sample that can withstand the electron bombardment without degradation.

Rutherford backscattering (RBS) is a surface analytical technique that uses only particles, so RBS is not strictly a spectrometry. The kinetic energies of particles are the quantities measured. Rutherford backscattering involves, in effect, bouncing high-energy particles (usually helium ions with a few million electron volts of kinetic energy) off the nuclei of the surface atoms. Similar to the way the cue ball in a pool game is slowed by hitting the other balls, in RBS some of the energy of the incoming particles is transferred to the stationary atoms. The amount of the energy transferred depends on the target atoms' masses. As a result, the surface composition can be probed by measuring the kinetic energy remaining with the probe particles after the collision. Only the particles that bounce off the surface back in the direction from which they came—the backscattered direction—are measured.

11.14 Vibrational Spectrometry: Infrared and Raman Spectrometry

Infrared spectrometry and Raman spectrometry both are used to probe the characteristic vibrations of molecules, crystals, and glasses, although the sample forms and instrumentation used are significantly different.

The spectra appear with numerous characteristic bands over the infrared spectral range (around 100 cm^{-1} to 4000 cm^{-1}) each with fwhm about 4–7 cm^{-1}. Significantly narrower bands appear in the spectra of gases. Individual bands can often be associated with specific chemical groups such as carbonyl, $>C{=}O$, or organic chlorine, $-C{-}Cl$. As for all spectrometries, the energies at which the bands appear depend on properties of the molecules while the magnitudes of the individual bands can be used to determine concentrations.

The spectra are often complicated, and seldom can every band be assigned to its origin in specific groups in the molecule. Only in the simplest molecules can all the bands be assigned to specific atomic motions. However, an incomplete understanding of the spectra does not detract from their usefulness in both qualitative and quantitative analysis. Infrared spectrometry, in combination with mass spectrometry and nuclear magnetic resonance form the basis for contemporary organic chemical qualitative analysis: identifying the molecular structure of unknown compounds and mixtures. However, after a brief introduction, we will concentrate our attention on their use for quantitation.

Infrared Spectra

The wavelength at which energy is absorbed depends on:

a. the identity of the atoms in a molecule;

b. The molecular structure (for example, CH_3CHO, $H_2C{-}CH_2$ (with O bridging both carbons), and $CH_2{=}CHOH$: all C_2H_4O isomers with significantly different spectra); and

c. the bonding between the atoms (for example, $>C{=}O$ differs from $-CH_2{-}O{-}$).

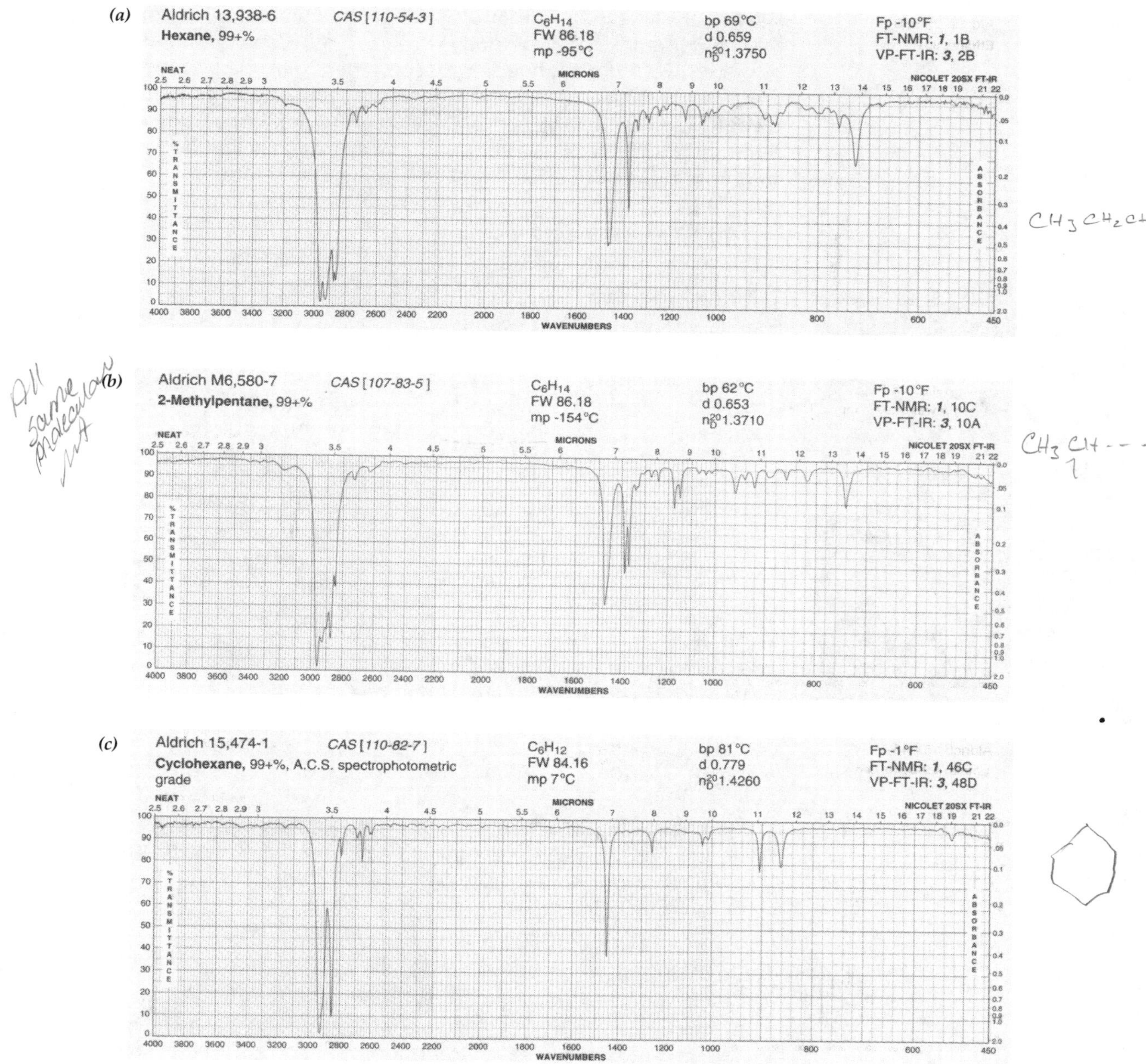

FIGURE 11.15 ▲

Infrared spectra of (a), *n*-hexane; (b), 2-methylpentane; and (c), cyclohexane. Each of the samples is a **neat** (meaning pure) liquid. Spectra I and II are of two hexane isomers; III lacks only two protons. The three spectra together illustrate the sensitivity of infrared spectra to molecular structure as well as the persistence of bands associated with specific chemical groups. [Ref: Spectra from Pouchert, C. J. 1985. "The Aldrich Library of Infrared Spectra." Courtesy Aldrich Chemical Co.]

The spectrum of energy absorption due to excitation of molecular vibrations is quite sensitive to differences in structure. To illustrate this sensitivity, the infrared spectra of three closely related molecules are shown in Figure 11.15. As can be seen, for infrared spectrometry it is customary to present the spectra as a plot of %*T* *vs.* wavenumber or microns (micrometers). Notice that the wavenumber scale is not linear across the entire region. In addition, with contemporary computer-controlled Fourier transform spectral acquisition, it has become a matter of the click of a mouse to convert the output to an absorbance scale on the ordinate. Depending on the instrument used, the *y*-axis will be %*T* or *A* and the *x*-axis will be cm^{-1} or μm: all four possible combinations can be found in the literature.

The ranges of energies for vibrations associated with specific chemical groups are shown in Table 11.4.

Dr. Sir C. V. Raman (1888–1970) demonstrated experimentally the Raman effect in 1928.

Raman Spectra

Raman spectrometers are configured much like fluorescence spectrometers but with higher power illumination. The spectrum is plotted as the intensity of reemitted light *vs.* the energy in cm^{-1}. An example of a typical Raman spectrum is shown in Figure 11.16. Note that although this is a spectrum with which we can probe molecular vibrations of infrared frequencies, the spectrum is obtained with *visible* light. The large peak in the middle, which is in the blue-green region of the visible spectrum, is due to light that is scattered by the molecules in the sample. This light is centered at the same wavelength as the incident light. It is called the **Rayleigh** line.

In addition to the Rayleigh line, there are a number of other spectral lines. They occur in sets, with each member of the set displaced in wavenumber (i.e., energy) equally on either side of the central, Rayleigh, line. These are the **Raman** lines. They arise as a result of the vibrations in the molecular species scattering the light. The displacements of the emissions from the Rayleigh line correspond to the vibrational frequencies of the normal modes of the molecules.

The total amount of light that is emitted as Raman lines is minuscule: about 10^{-5} to 10^{-6} of the light that is scattered by the sample (the Rayleigh line). Because of the small proportion of Raman emission, an intense source of light is needed to obtain good spectra. Therefore, a focused laser illuminates the sample. Also, high-quality optics are required to minimize any instrumental interference from the intense Rayleigh scattering. Otherwise, the small Raman bands would be overwhelmed by a huge background.

Quantitation

The determination of gaseous vinyl chloride by infrared spectrometry is illustrated in Figure 11.17. The whole infrared spectrum and an illustration of instrument response with concentration are shown. The absorbance measured is for a pathlength for transmission of 20.25 m. This long pathlength is achieved by using mirrors to reflect the radiation many times through a shorter cell.

In general, quantitation can be done relatively easily when individual bands are completely separated. It is usually best to leave the instrument settings the same for all samples.

Some complications arise in vibrational spectrometry due to the large number of bands. When bands are close together, deciding how to find the baseline for quantitative determinations can be quite complicated as you can see in Figure 11.17 in the region around 1040 cm^{-1}. It is better to find a well-separated band on a flat baseline that can be used, such as the band near 1600 cm^{-1}.

Table 11.4 Simplified Correlation Chart of Molecular Vibrations by Type*

Type of vibration			Frequency (cm^{-1})	Intensity†
C—H	Alkanes	(stretch)	3000–2850	s
	$—CH_3$	(bend)	1450 and 1375	m
	$—CH_2—$	(bend)	1465	m
	Alkenes	(stretch)	3100–3000	m
		(out-of-plane bend)	1000–650	s
	Aromatics	(stretch)	3150–3050	s
		(out-of-plane bend)	900–690	s
	Alkyne	(stretch)	ca. 3300	s
	Aldehyde		2900–2800	w
			2800–2700	w
C—C	Alkane		not interpretatively useful	
C=C	Alkene		1680–1600	m-w
	Aromatic		1600 and 1475	m-w
C≡C	Alkyne		2250–2100	m-w
C=C	Aldehyde		1740–1720	s
	Ketone		1725–1705	s
	Carboxylic acid		1725–1700	s
	Ester		1750–1730	s
	Amide		1670–1640	s
	Anhydride		1810 and 1760	s
	Acid chloride		1800	s
C—C	Alcohols, ethers, esters, carboxylic acids, anhydrides		1300–1000	s
O—H	Alcohols, phenols			
	Free		3650–3600	m
	H-bonded		3500–3200	m
	Carboxylic acids		3400–2400	m
N—H	Primary and secondary amines and amides			
	(stretch)		3500–3100	m
	(bend)		1640–1550	m-s
C—N	Amines		1350–1000	m-s
C=N	Imines and oximes		1690–1640	w-s
C≡N	Nitriles		2260–2240	m
X=C=Y	Allenes, ketenes, isocyanates, isothiocyanates		2270–1950	m-s
N=O	Nitro ($R–NO_2$)		1550 and 1350	s
S—H	Mercaptans		2550	w
S=O	Sulfoxides		1050	s
	Sulfones, sulfonyl chlorides		1375–1300 and	s
	Sulfates, sulfonamides		1200–1140	s
C—X	Fluoride		1400–1000	s
	Chloride		800–600	s
	Bromide, iodide		<667	s

*Data from Pavia, D. L., Lampman, G. M., Kriz, G. S. "Introduction to Spectroscopy: A Guide for Students of Organic Chemistry."

† s = strong, m = medium, w = weak.

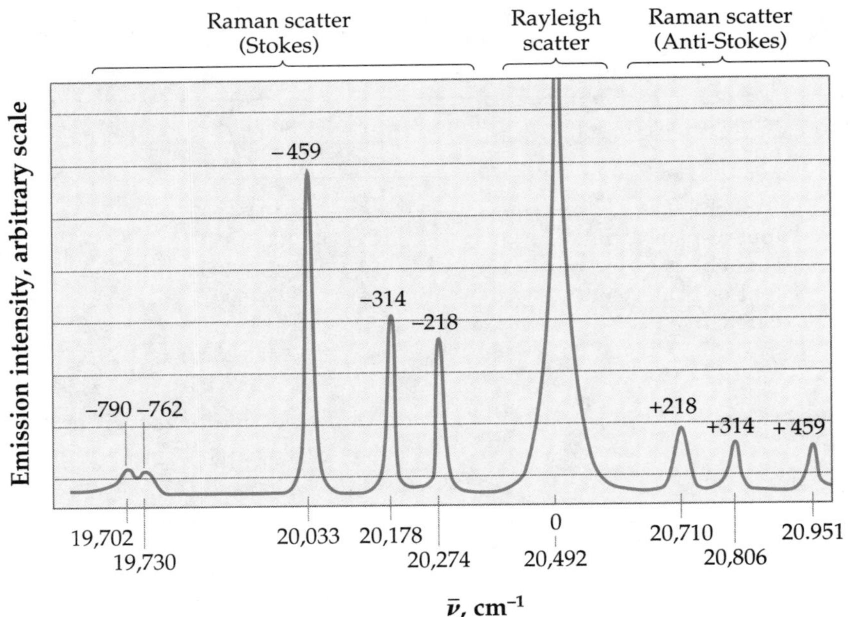

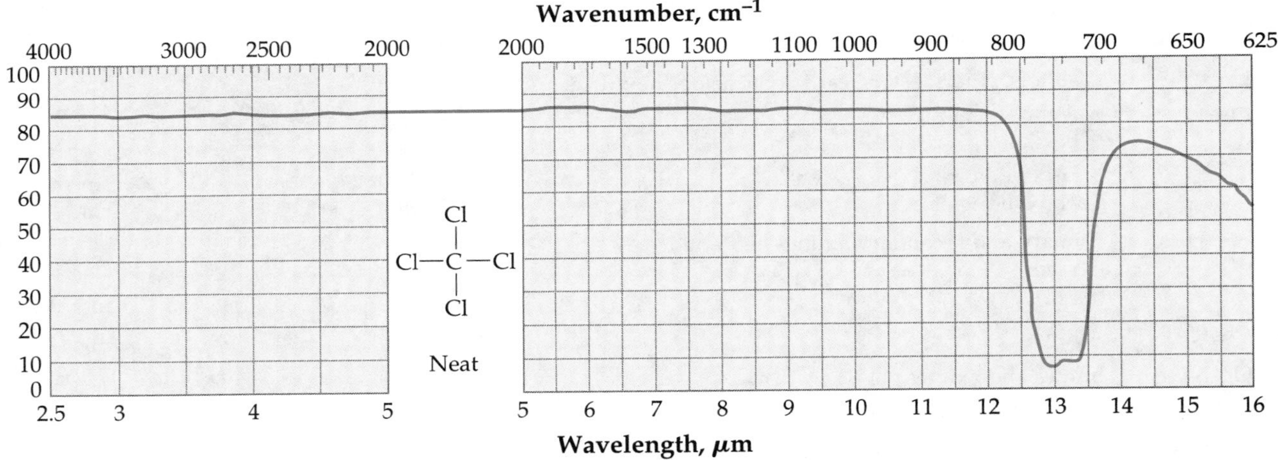

FIGURE 11.16 ▲

Typical (a) Raman and (b) infrared vibrational spectra for carbon tetrachloride, CCl_4.

The Raman spectrum was obtained with laser radiation at 20,492 cm^{-1} (488.0 nm, one line of an argon-ion laser). The *difference* in wavenumber between the central line and a Raman line equals the natural vibrational frequency of the molecules. The infrared spectrum is shown in the conventional form. [Refs: Redrawn from Strommern, D. P.; Nakamoto, K. 1981. *Am. Lab.* Oct. pp. 70–77; Collins, D. W., et al. 1977. *Appl. Optics* 16:252; and Pouchert, C. J. "The Aldrich Library of Infrared Spectra." Courtesy Aldrich Chemical Co.]

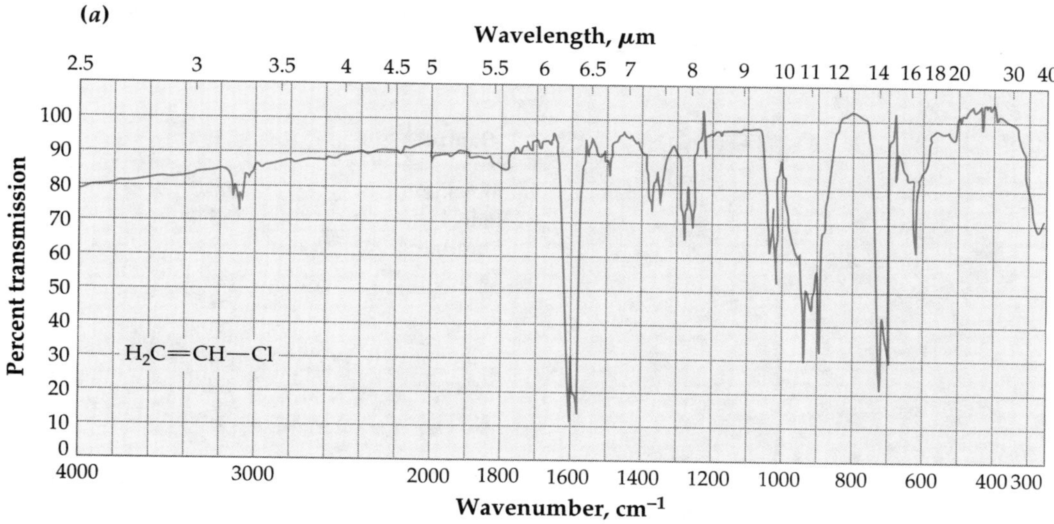

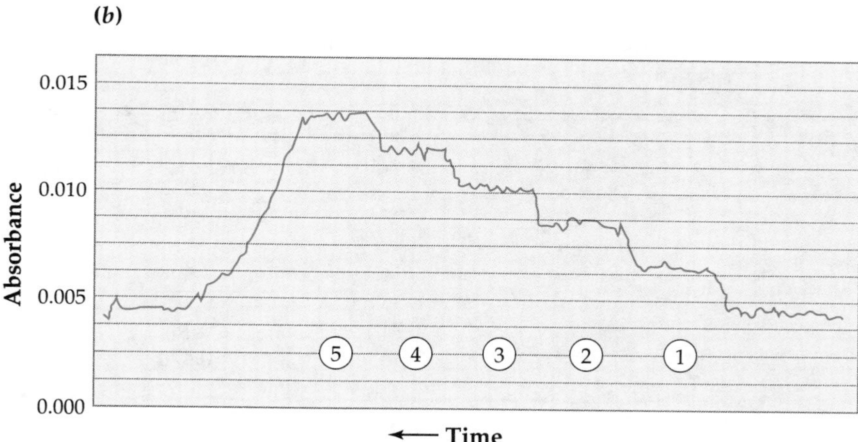

FIGURE 11.17 ▲
IR as a qualitative and quantitative tool.

(a) The infrared spectrum of gaseous vinyl chloride in a multipass cell with a path length of 20.25 m. **(b)** The plot of instrument response over time with changes in concentration of vinyl chloride. Each step (labeled 1-5) is a successive increase of 1 ppm vinyl chloride, beginning at zero on the right. [Ref: Reprinted from Lavery, D. S., Wilks, P. A., Jr. 1977. In *Spectroscopy*, Series II, Vol. II. Fairfield CT: International Scientific Communications, Inc. pp. 127–128.]

EXAMPLE 11.11

Infrared spectrometry was used to monitor the process for producing tablets containing the experimental drug XR375 under study by Miracle Pharmaceuticals, Inc. A placebo was used for the blank sample. (A placebo is identical to the analytical sample except there is no active ingredient present.)

Both the placebo and each tablet were prepared according to a procedure that had been shown to have a (97.0 ± 0.5)% recovery of XR375 from the tablets. The sample was ground to a fine powder and extracted with hexane. The hexane extract was then dried by passing through fresh, dried $MgSO_4$, diluted to 100.0 mL, and 10.00 mL of that solution was then diluted to 100.0 mL with dry hexane. The final solution was then analyzed by IR. The standard was prepared from 25.0 mg XR375 in 100.0 mL hexane, and pure hexane was used to set the 100 %T baseline.

Sample	%T (3255 cm^{-1})
Hexane	100.0
Placebo	88.0
Standard	33.8
Tablets	
#1	34.5
#2	33.8
#3	34.1
#4	33.9

The tablet was supposed to contain 250 (±10.0) mg of XR375. Was the tablet formulated correctly?

Solution

First we must convert to absorbance and then correct for the absorbance of the blank.

Sample	%T (3255 cm^{-1})	Absorbance	Corrected absorbance
Hexane	100.0	0.000	—
Placebo	88.0	0.055	0.000
Standard	33.8	0.471	0.416
Tablets			
#1	34.5	0.462	0.407
#2	33.8	0.471	0.416
#3	34.1	0.467	0.412
#4	33.9	0.470	0.415

The average corrected absorbance for the sample tablets is 0.413. We can now calculate the content of the pellet that was analyzed. Beer's law says that as long as the experimental conditions are the same, the absorbance is directly proportional to the concentrations. Since the final volumes are the same for the weighed, pure standard and the extracted XR375, masses can be used in place of concentrations.

$$\frac{A_{\text{standard}}}{\text{mg standard}} = \frac{A_{\text{sample}}}{\text{mg sample}}$$

$$\frac{0.415}{25\ \text{mg}} = \frac{0.413}{\text{mg sample}}$$

$$\text{mg sample} = 24.9\ \text{mg}$$

This is the amount of active ingredient in the diluted solution, so we must multiply by 10 to "undo" our dilution and then correct for the recovery.

$$24.8\ \text{mg} \times 10 \times \frac{100}{97.0} = 256\ \text{mg XR375}$$

Yes, the tablet was formulated correctly.

Samples for Vibrational Spectrometry

Samples for IR and Raman spectrometries can be mixtures or solutions, and the spectral properties can be measured for solids, liquid films, bulk solutions, or gas samples.

Some significant problems arise when trying to construct sample containers for vibrational spectrometry, because every material has *some* vibrational absorption. What is done is to use material that has a minimum interference in the regions of interest. The material of choice for IR is a solid salt (KBr, NaCl) plate. Such plates are used in a number of ways.

Solids can be probed with infrared spectrometry even if they cannot be dissolved; they can be suspended in the infrared beam as small particles in a viscous liquid such as mineral oil. The mixture of the solid particles suspended in the viscous liquid is called a **mull**. Another popular method to use for solids is to mix and crush the sample intimately with dry KBr and then form it into a disk about 1 cm in diameter and less than 1 mm thick. A solid may be deposited as a film on a single KBr plate by allowing the solvent to evaporate from a solution of the sample. A liquid can be spread as a film on the plate or held in a cell between two salt plates less than a millimeter apart. Water cannot be used as a solvent for two reasons: it is nearly opaque to infrared radiation and it dissolves NaCl and KBr salt plates. Gases are contained in longer cells.

Raman spectrometry differs from infrared in that light in the visible region is used to excite the molecules in the sample. A solution can be held in a glass capillary, a transparent solid can be placed directly in the beam, and a gas can be held in a glass container. Colored samples present a problem, and special techniques are required to prevent the laser beam from boiling liquids and vaporizing solids as well as causing decomposition by photolysis.

11.15 Nuclear Magnetic Resonance Spectroscopy

Nuclear magnetic resonance spectrometry (NMR), together with mass spectrometry and infrared spectrometry, is of major importance in determining the molecular structure of organic compounds as well as macromolecules, both man-made and biochemical.

NMR involves measuring the absorption of radiofrequency radiation by a sample material that is placed in a strong magnetic field. The radiation used is in the 100-MHz range (the frequency of FM radio in the U.S.) up to nearly 1 GHz. The magnetic fields are large; the most sophisticated instruments use some of the largest constant magnetic fields that can be generated.

As in other spectrometries, the energy of the transition is determined by molecular and atomic properties, but in NMR spectroscopy, the energy also depends on the magnitude of the magnetic field. Also, as in other spectrometries, the fraction of the power absorbed is proportional to the concentration of the absorbing species.

An instrument for NMR spectrometry measures the spectrum for one atomic species at a time. The most commonly measured atomic species are ^{1}H, ^{13}C, and ^{32}P. By far, the most commonly measured is ^{1}H NMR, usually called **proton NMR**. Not only is proton NMR the most frequently used, it is also inherently the most sensitive. Also, protons comprise nearly 100% of the natural isotopic abundance of hydrogen, which also enhances the experimental sensitivity. The next most commonly measured NMR nucleus is ^{13}C.

However, ^{13}C-NMR is inherently, experimentally less sensitive, and, further, the natural abundance of ^{13}C—the isotope that is NMR active—comprises only 1.1% of the carbon. For these reasons, either the sample must be enriched in ^{13}C, or highly concentrated samples need to be used.

In ^{1}H-NMR, either solvents containing no hydrogen (for example, CCl_4) or deuterated solvents (for example, heavy water, D_2O) should be used. Otherwise, since the solvents' concentrations are so much greater than those of the analytes in general, the proton NMR absorption from the solvent would overwhelm that of the sample.

Typical Chemical Shifts

One great advantage of NMR spectra is that the chemical shifts for protons in different chemical environments occur in *ranges* that are specific to that environment. That is,

> *The chemical shift of a resonant nucleus depends on the chemical structure near it.*

These relationships are illustrated in Figure 11.18.

As a result of the regularities of the chemical shifts, we can make a **correlation table** of the chemical shifts observed and the identity of the chemical groups in which the protons are bonded. The chemical shift ranges are listed in Figure 11.19 by chemical type. (Similar correlations are seen with ^{13}C-NMR spectra.)

For example, the assignments of the benzyl acetate resonance peaks in Figure 11.20 can be made from the chemical shifts alone. The chemical shifts in Figure 11.19 show that protons of an aromatic ring with a carbon attached should be expected at around 7.1 ppm. The methyl protons of an acetate

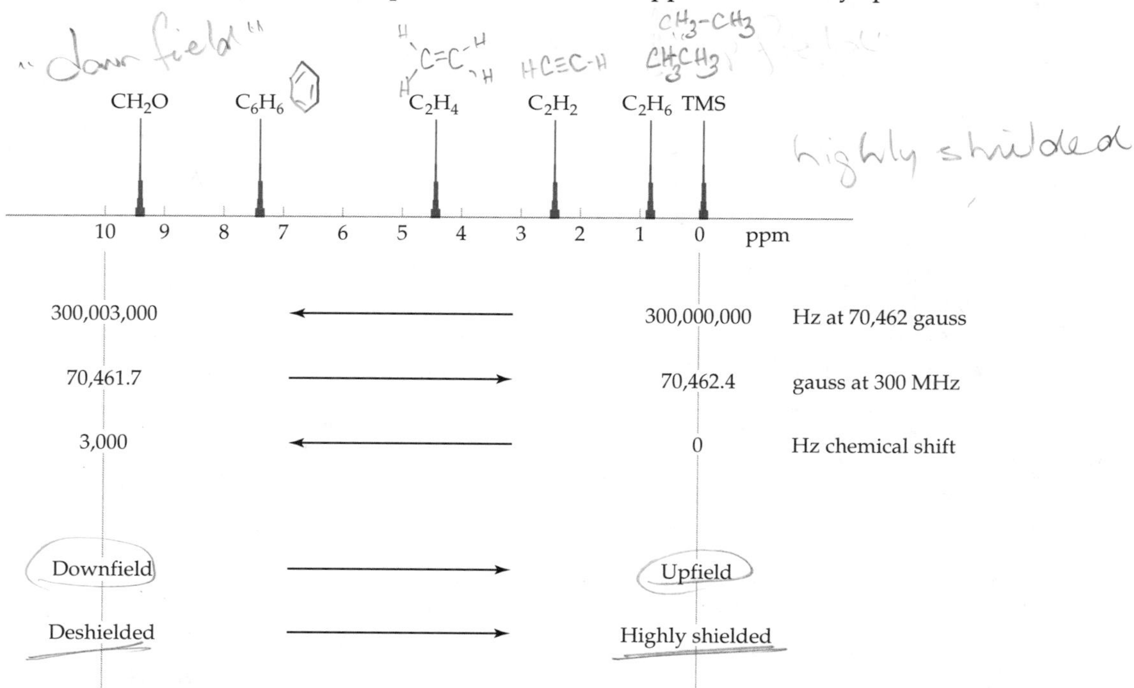

FIGURE 11.18 ▲
Some of the nomenclature and numerical values found for ^{1}H-NMR spectra.
The numbers are typical for the smaller instruments, which operate near 300 MHz.

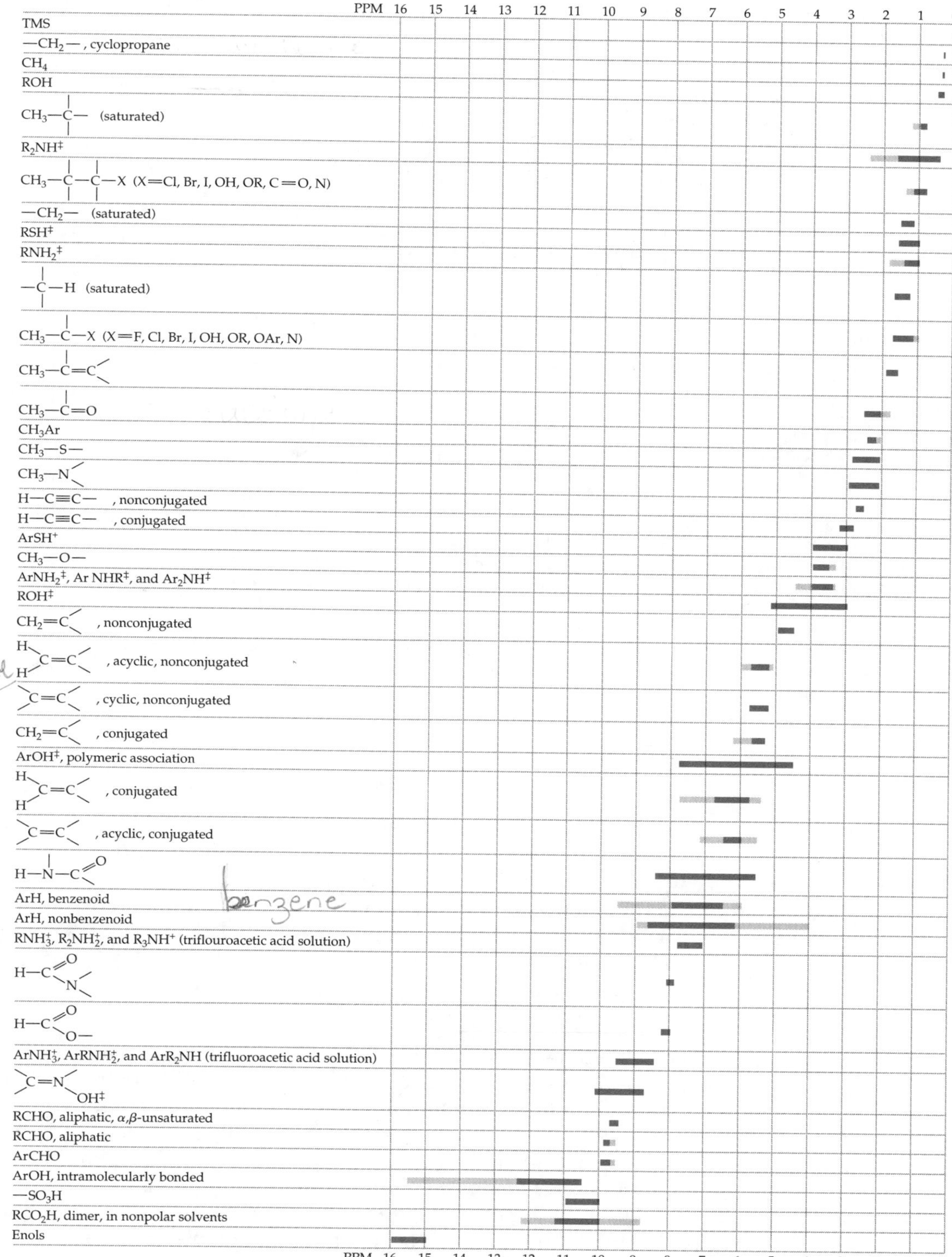

‡ The absorption positions of these groups are concentration dependent; they shift toward zero ppm in more dilute solutions.

FIGURE 11.19 ▲

Typical ranges of NMR absorption chemical shifts for protons in various chemical environments.

The lighter areas are extreme but possible regions. [Based on Dyer, J. R. 1965. *Applications to Absorption Spectroscopy of Organic Compounds*. Englewood Cliffs, NJ: Prentice-Hall. Reproduced with permission.]

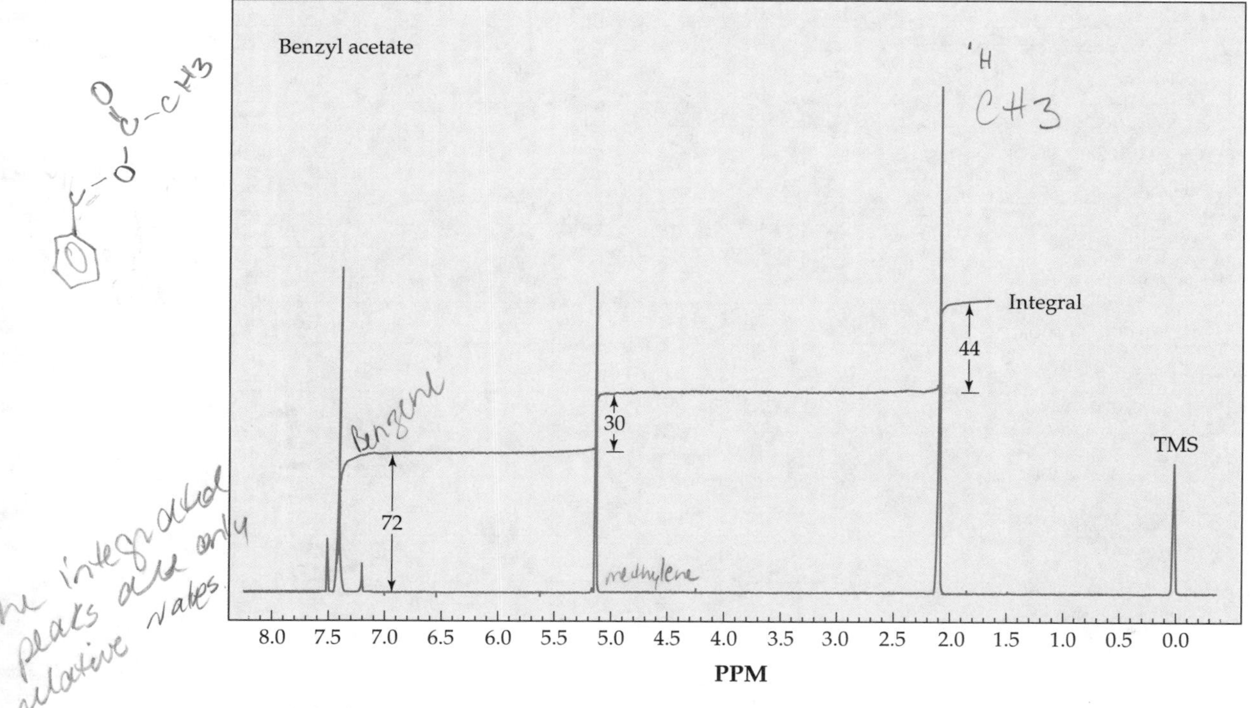

FIGURE 11.20 ▲

A ^{1}H-NMR spectrum with integration of the peaks.

The sample is neat benzyl acetate with an internal standard of TMS.

group are expected at around 2.0 ppm (CH_3–CO–O–Ar). The methylene protons are the only ones left, and they exhibit peaks at 5.0 ppm.

Signal Magnitude and Concentration

In NMR spectrometry, the integrated peak areas are only relative values that depend on the number of nuclei contributing to the peak. This is one of the great strengths of the NMR method:

> *Every proton, when it absorbs energy, contributes equally to a simple NMR absorption spectrum.*

The magnitude of the absorption is independent of the chemical shift when proper care is taken in the experiment. In effect, every proton has the same absorptivity. Obtaining absolute concentrations using NMR spectra is done using concentration standards and the peak heights or integrated peak areas.

The quantitation of other nuclei, such as ^{13}C or ^{32}P, requires more attention to the experimental conditions, since the integral areas per nucleus can depend on the local chemical environment. For example, the areas of ^{13}C peaks tend to vary depending on the number of protons attached.

EXAMPLE 11.12

The NMR spectrum for pure toluene was compared with the spectrum of an unknown mixture of toluene and *p*-xylene. The signals for the methyl protons overlap as do those for the protons in the aromatic region. The following integrated peak areas were found.

	Integrated areas	
	Methyl region	Aromatic region
Toluene	3.7_0	6.1_5
Mixture	4.8_0	5.8_0

(The second decimal is written as a subscript to indicate the measurement's precision of ±5%). What are the mole fractions of xylene and toluene in the mixture?

Solution

This problem can be solved since all protons respond equally. This problem mirrors any mixture problem with absorbance at two wavelengths for two components. First, we should list the number of protons of each type. That is,

	Number of protons	
	Methyl region	Aromatic region
Toluene	3	5
p-Xylene	6	4

Call the integrated signal S, and set up the two equations to be solved for the two frequency regions of the integral just as for any other mixture.

$$S_{methyl} = 6\,R\,X_{xylene} + 3\,R\,X_{toluene}$$

$$S_{aromatic} = 4\,R\,X_{xylene} + 5\,R\,X_{toluene}$$

where R is a factor that scales the signal per proton for the toluene standard. In this case, R can be found by dividing the signal for the toluene's methyl protons by the number of methyl protons it contains (3);

$$R = 1.23 = 3.70/3.$$

The same R value is obtained from the aromatic protons.

$$1.23 = R = 6.15/5.$$

Since there are only two unknowns, the mole fractions of *p*-xylene and toluene, the equation can be solved. But, in fact, the amount of information provided is more than needed to solve the problem, since we also know that

$$\text{Mole fraction } p\text{-xylene} = 1 - \text{Mole fraction toluene}$$

or, using the abbreviation X for mole fraction,

$$X_{xylene} = 1 - X_{toluene}$$

Let us call this the *mole-fraction equation*. Any two of the three equations could be used to obtain an answer. For example, let us use the mole-fraction and one of the integrated intensity equations to solve the problem. Substitute the mole fraction relation into the equation for S_{methyl}.

$$S_{methyl} = 6 \cdot 1.23\,(X_{xylene}) + 3 \cdot 1.23\,(1 - X_{xylene}) = 4.80$$

Now, we can solve for X_{xylene}, and then substitute that value into the mole-fraction equation to find

$$X_{xylene} = 0.30 \qquad X_{toluene} = 0.70$$

The unused integrated-signal equation can be used to check the answer to see whether some third component might be contributing to the spectrum. From the mole-fraction equation and the other integrated-signal equation, you will find

$$X_{xylene} = 0.28 \qquad X_{toluene} = 0.72$$

The agreement between the two calculations lies within the ±5% experimental error that might be expected for integrated peak areas in proton NMR.

11A A Deeper Look: Waves, Wavelengths, and Electromagnetic Radiation

Waves

The description of light as a wave is one of the primary concepts of the action of the physical world. In chemical analysis, you will encounter the concepts of electromagnetic waves—radio, light, and X-rays, for example.

Waves carry energy from one location to another. They have a finite velocity. The "shape" of a wave is usually represented by a graph of height (for a water wave), or electric field strength (for light waves), or pressure (for a sound wave) *vs.* time or *vs.* distance. A graph of a general wave is shown in Figure 11A.1.

Much of the behavior of light can be explained by using the properties of sine waves such as that shown in Figure 11A.2. A sine wave can be described mathematically by

$$A = A_0 \sin \theta \tag{11A-1}$$

where

A is the amplitude at any point,
A_0 is the peak amplitude, and
θ is a continuous variable, the form of which is shown below.

The peak amplitude is usually called simply the **amplitude** of the wave. The distance λ between adjacent peaks (or adjacent valleys) of the wave is the wavelength. In terms of the variable θ, the wavelength is 2π. A simple, static sine wave is completely characterized by the amplitude A_0 and the wavelength λ.

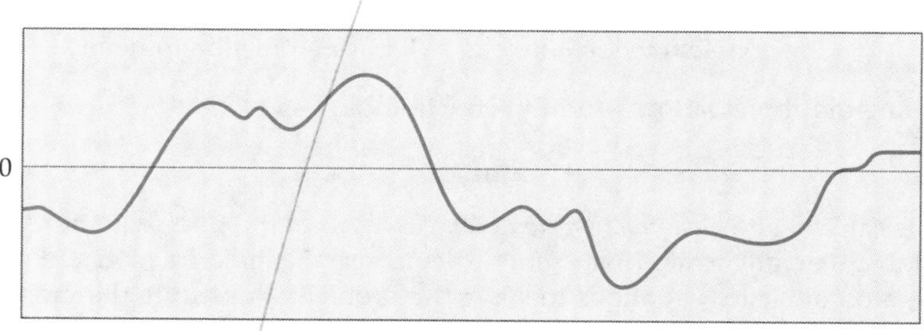

FIGURE 11A.1 ▲
Plot of a general wave.
The vertical direction represents the height or amplitude of the wave. The horizontal direction can be either time or distance. This graph might represent a cross section of an ocean wave or, perhaps, the voltage output from a microphone over a short time duration.

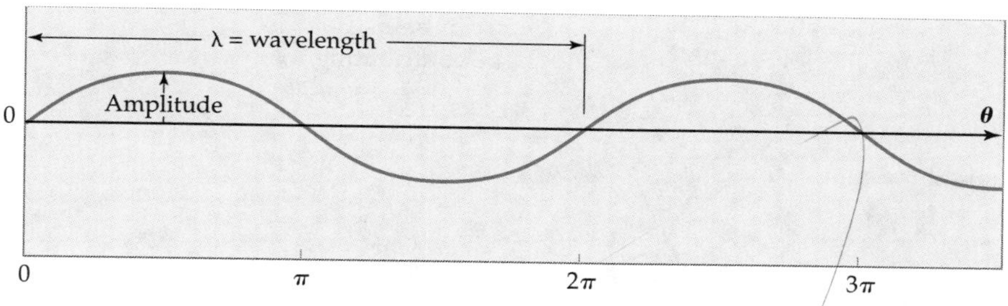

FIGURE 11A.2 ▲
Graph of a sine wave with the amplitude and wavelength annotated. This is a specific example of an harmonic wave. For our interest, it represents an electromagnetic wave. In that case, the amplitude is the magnitude of an electric or magnetic field oscillating in time.

Since the properties of sines and sine waves depend on factors of π, it is convenient to associate a sine wave with the properties of angles. Recall that a change in angle of $360° = 2\pi$ radians. This is equivalent to one complete rotation in a plane.

If the language of angles is used for Equation 11A-1,

θ = an angle (in radians)

2π = a **cycle**

When the angle is a linear function of time, we write

$\theta = \omega t$, where

ω = an angular velocity (radians/unit time), which is a constant

When θ of Equation 11A-1 is written as ωt, the algebraic form for a time-dependent sine wave is obtained,

$$A = A_0 \sin \omega t \tag{11A-2}$$

A complete cycle occurs when ωt has changed in value by 2π. This is also called one complete **oscillation** or one **period**. The time over which one complete cycle occurs is

$$t_{\text{cycle}} = \frac{\omega}{2\pi}$$

The frequency of oscillation ν is, thus,

$$\nu = \frac{1}{t_{\text{cycle}}} = \frac{2\pi}{\omega} \tag{11A-3}$$

Wavelengths

Electromagnetic radiation is describable as electric and magnetic fields that oscillate sinusoidally in amplitude while traveling through space. The direction of travel is perpendicular to the direction of oscillation of both the electric and the magnetic fields. These ideas are illustrated in Figure 11A.3.

What is important to note from the figure is that we can use the properties of simple sine waves to explain the properties of the more complex electromagnetic waves. To do so, we choose a wave that describes either the oscillating electric or magnetic field moving through space.

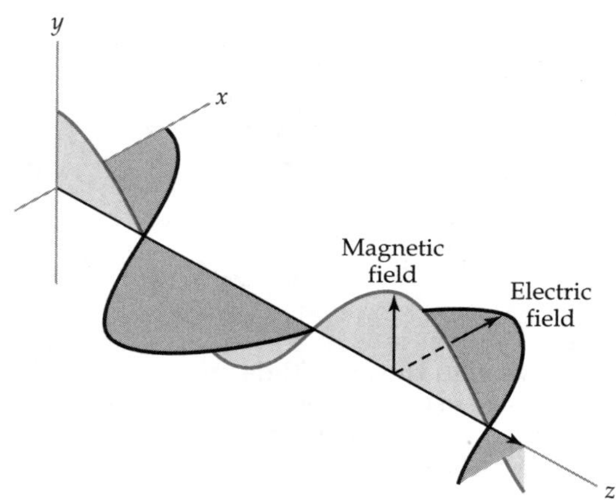

FIGURE 11A.3 ▲
Diagram of the geometric relationship of the oscillating electric and magnetic fields of a polarized electromagnetic wave.
Polarized means that the electric field oscillates in a single plane and not in a random position about the z-axis. Notice that the electric and magnetic fields are perpendicular to each other, and both are perpendicular to the direction the wave is moving, which is along the $+z$-axis.

For any wave moving at a constant velocity, the frequency ν, wavelength λ, and velocity v are related to each other. This relationship is simply

$$v = \nu\lambda \tag{11A-4}$$

For electromagnetic radiation, the units in this equation can be:

v in m s^{-1}

ν in Hertz (cycles per second or s^{-1})

λ in meters

The speed of light in a vacuum has been measured precisely and has been given the label c ($c = 2.9979 \times 10^8$ m s^{-1}). From this value and the frequencies, the wavelengths of electromagnetic radiation *in vacuo* (in a vacuum) can be calculated. (An overview of the ranges of wavelengths in the electromagnetic spectrum is shown in Figure 11.1.)

11B A Deeper Look: Wavelength Separation

Here we describe the basic ideas of wave interference. The aim is to show the fundamentals of how **spectroscopic gratings** and interferometers work when they separate narrow bandwidths from a broad range of light. A spectroscopic grating is a device that can **disperse** electromagnetic radiation. Disperse means to spread the different wavelengths to different angles. A prism also disperses visible light into its component colors. Numerous types of gratings are available for dispersing light in the various ranges from infrared to X-ray wavelengths. Unlike gratings and prisms, interferometers do not disperse the light, but allow only narrow bands to pass through. Interferometers are used for light in the UV-visible-infrared-far infrared region. Narrow bands of radiation at longer wavelengths are selected by radiofrequency filters of various types, which are not discussed here. Wavelengths shorter than X-ray cannot be separated in these ways.

The sine waves of electromagnetic radiation are **linear waves**. Linear is a mathematical term for *being independent of each other*. This means that the presence of one wave does not affect the properties of another wave. As a result, the equations representing waves can be added to represent the physical process of adding the magnitudes of electric or magnetic fields. Let us see how this linear property works and eventually how the mathematics leads to explaining how light of different wavelengths is separated.

Relative Positions of Two Stationary Waves

Assume, as illustrated in Figure 11B.1, that two waves are present in the same space, both with the same amplitude A_0 and frequency ν. However, the peaks (and valleys) do not coincide. As discussed in Section 11A, one wave A_1 can be described by

$$A_1 = A_0 \sin 2\pi\nu t \tag{11B-1}$$

and the other A_2 can be described by

$$A_2 = A_0 \sin\left(2\pi\nu t - \frac{\lambda}{4}\right) \tag{11B-2}$$

where the peak is shifted a quarter wavelength in position. The shift shows up as the second term $\lambda/4$.

A general equation for wave A_2 that describes any relative position is

$$A_2 = A_0 \sin(2\pi\nu t + \phi) \tag{11B-3}$$

The value of ϕ is called the **phase shift**. In the example illustrated in Figure 11B.1, the magnitude of the phase shift ϕ can be expressed in three ways, all of which are equivalent: $\lambda/4$, 90°, or $\pi/2$, which are, respectively, in the units wavelengths, angular degrees, and radians.

FIGURE 11B.1 ▲
Plot of two sine waves of equal amplitude and wavelength (or frequency) in the same region of space.
They are out of phase (or *phase shifted*) by $\lambda/4$ or any value of $(4m + 1)\lambda/4$, where m is an integer.

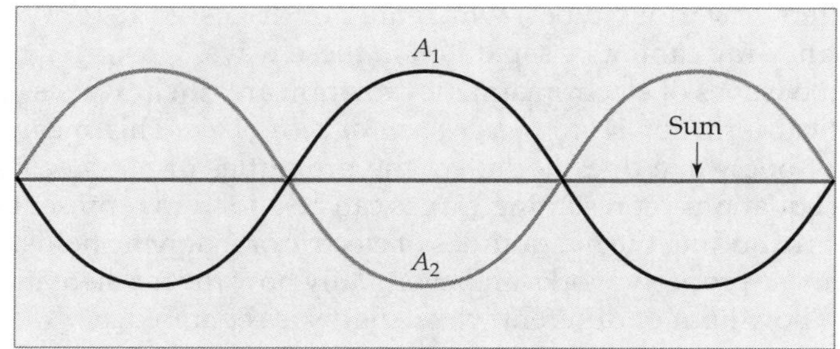

FIGURE 11B.2 ▲
Plot of two sine waves A_1 and A_2 that are in the same region of space and equal in frequency and amplitude.
They are phase shifted by $\lambda/2$ or any value of $(2m + 1)\lambda/2$, where m is an integer. Their sum at every point is zero. This is the explanation for destructive interference.

Interference

Now, look at the waves drawn in Figure 11B.2. Visually, you can add the amplitudes of these waves at each point along the x-axis. Each peak adds to a trough. In fact, at each point along the x-axis, the sum is exactly zero. Mathematically, since the first wave is

$$A_1 = A_0 \sin(2\pi\nu t) \tag{11B-3}$$

and the second wave is[1]

$$A_2 = A_0 \sin\left(2\pi\nu t + \frac{\lambda}{2}\right) = -A_0 \sin(2\pi\nu t) \tag{11B-4}$$

The sum of the two is

$$A_1 + A_2 = 0$$

The resulting wave amplitude is zero. This is called **destructive interference** by two identical waves *out of phase* by exactly $\lambda/2$.

Destructive interference also occurs if the separations of the peaks (or valleys) of the two waves are any *odd* multiple of $\lambda/2$. Examples of the odd multiples are

$$\lambda/2,\quad 3\lambda/2,\quad 5\lambda/2, \ldots, (2m + 1)\lambda/2$$

In all such cases, the waves will exactly destructively interfere. Recall, though, this is only true for two waves of equal amplitude and the same wavelength.

---

[1]Equation 11B-4 can be derived by recognizing that $\lambda/2$ is equivalent to π. Use the equation for the sine of a sum,

$$\sin(a + b) = \sin a \cdot \cos b + \cos a \cdot \sin b$$

to obtain

$$\sin(2\pi\nu t + \pi) = \sin 2\pi\nu t \cdot \cos \pi + \cos 2\pi\nu t \cdot \sin \pi$$

The derivation is completed by recognizing that $\cos \pi = -1$ and $\sin \pi = 0$.

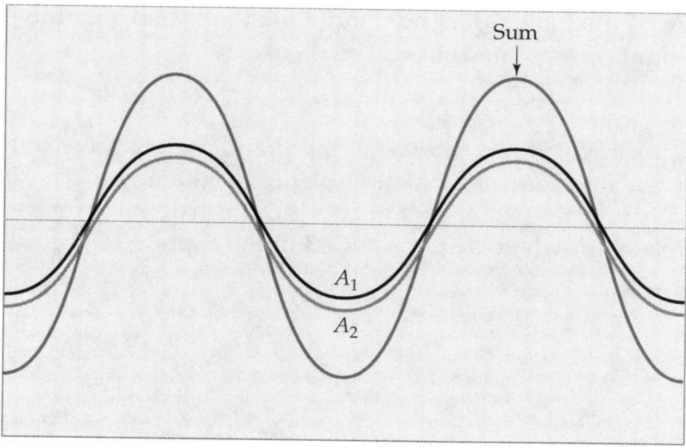

FIGURE 11B.3 ▲
Plot of two sine waves A_1 and A_2 that are in the same region of space. The waves are equal in frequency and amplitude, but are shifted slightly for clarity. They are phase shifted by λ or any value of $m\lambda$. The sum of their amplitude at every point is twice each individual amplitude. This is the explanation for constructive interference.

You can look at Figure 11B.3 and see what happens when the difference between the positions of the waves is *equal* to the wavelength. The peaks add to the peaks and the troughs to the troughs. Then,

$$A_1 + A_2 = 2A_0 \sin(2\pi\nu t)$$

The sum wave has the same wavelength but has twice the amplitude of the original waves. Adding this way is called **constructive interference**. Constructive interference occurs when the difference in ϕ is some multiple of λ,

$$\lambda, 2\lambda, 3\lambda, \ldots, m\lambda.$$

To summarize:

The sinusoidal waves of electromagnetic radiation are linear waves.

Waves of equal amplitude and wavelength that are out of phase by $(2m + 1)\lambda/2$ add to zero.

Waves of equal amplitude and wavelength that are out of phase by $m\lambda$ or exactly in phase add constructively so that the sum amplitude is twice that of the original waves.

Two-Slit Diffraction

Two-slit diffraction and its geometry is illustrated in Figure 11B.4. When light shines onto a distant screen through two identical, parallel slits that are close together in an opaque material, a pattern of light and dark bands is seen on the screen. If the light were simply shining through the slits, we would expect to see a shadow with two bright lines in it, not the pattern we do see. The appearance of these numerous bands of light distributed over a wide angle shows that the light rays have been bent. This is called **diffraction** of the light.

As can be measured, the intensity of the light varies with the angle. (We call the angle θ, which is shown in the figure with its origin between the slits.)

The spacing of the bands depends on the distance between the slits, d. The center of a light (illuminated) band occurs at

$$d \sin \theta = m\lambda; \quad m = 0, 1, 2, 3, \ldots \tag{11B-5}$$

The diffraction and the appearance of bright and dark bands results from differences in the interference, which depends on the angle. The light waves constructively interfere to form light bands or destructively interfere to cause the dark areas. More details are provided in the caption to Figure 11B.4.

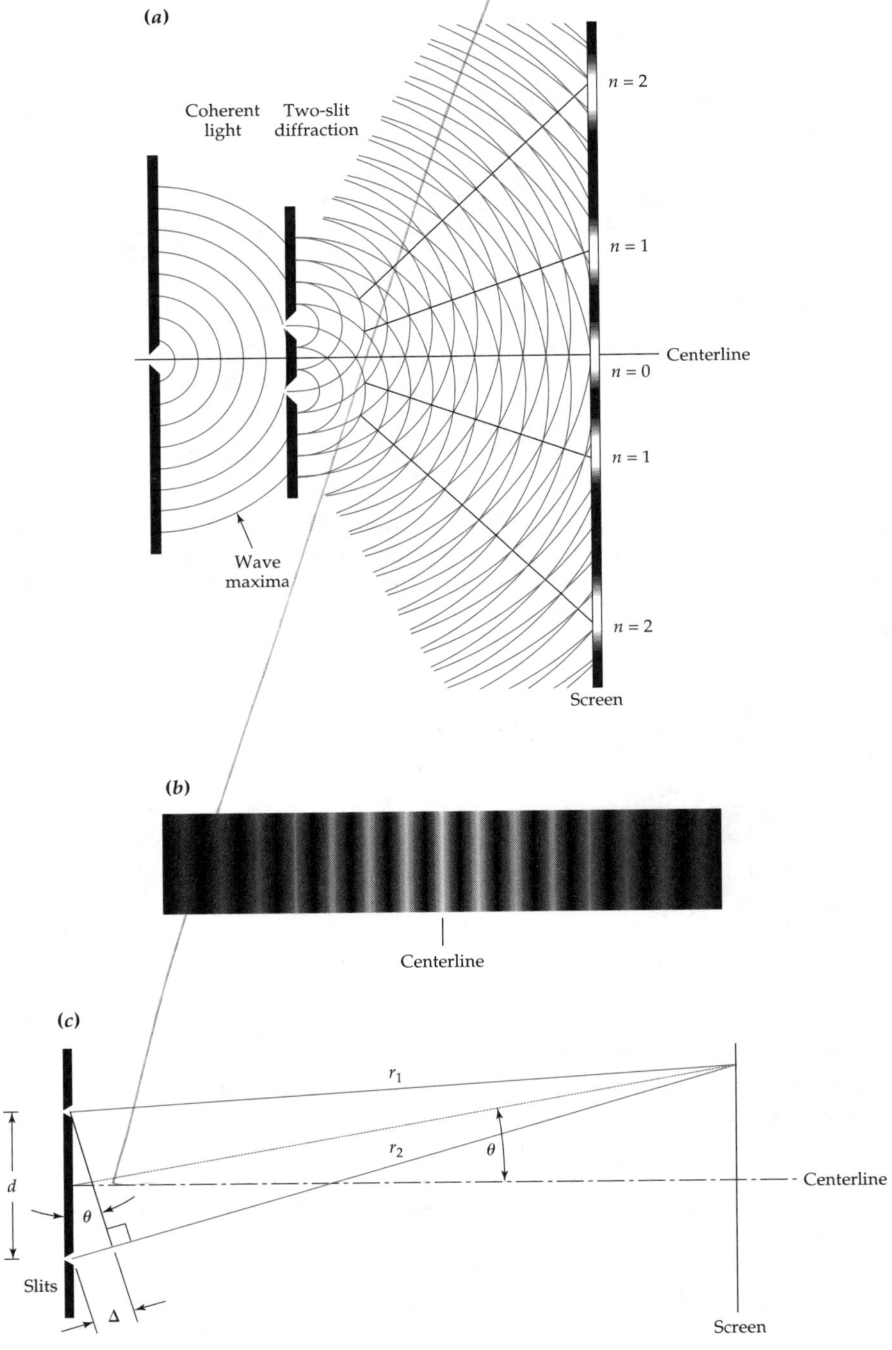

FIGURE 11B.4 ▶ **Illustration of the results of two-slit diffraction of light.** **(a)** Light is passed through a single slit, which causes the light to act as though all the waves are in phase (**coherent**). (A laser could be used instead.) This light then passes through two thin slits perpendicular to the page and falls on a screen represented at the right. Each slit acts like a source of light. The radiation has peaks and troughs of electric field like ripples in water. **(b)** An image appears at the screen: a series of light and dark lines that extend on either side from the center line depending on the angle. These regions of light and dark are explained as regions of alternating constructive and destructive interference. **(c)** The bright regions occur where the path difference from the two slits to the screen ($r_1 - r_2$) is equivalent to a difference of $m\lambda$. The difference in length is shown by Δ in (c). This difference in distance is related to the angle at which the light band appears; algebraically $m\lambda = d \sin \theta$. The central bright line is where $m = 0$ and is called the zero-order line. The two lines on either side of this occur where $m = 1$ and are called first order. This nomenclature continues: second order, third order, and so forth. For more details see any introductory physics textbook.

We can number the bright bands starting from the middle one, which is labeled zero. From there, bands on both sides of the middle are labeled by increasing integer values—1, 2, 3, etc., the further they are from the center. These bands are referred to as **first order**, **second order**, **third order**, and so on. In addition, the numerical values of the orders equal the value of m in Equation 11B-5.

Multislit Diffraction

If, instead of having only two slits in the opaque sheet, there are many hundreds of equally spaced slits, the *centers* of the light and dark areas on the screen will still show the same angular dependence as arose from only two slits. The centers of the illuminated areas will still be at $m\lambda = d \sin \theta$. However, as illustrated in Figure 11B.5, the regions of light will have sharper edges.

Now, assume the light that is shown through the slits is limited to a single wavelength—a single color if the radiation is visible light. Then, each of the narrow bands of light will be the same color.

What, then, is the location of the illuminated region when light of two different wavelengths simultaneously illuminates the slits? Say the colors are blue and red: The blue light wavelength is shorter than the red.

$$\lambda_{\text{red}} > \lambda_{\text{blue}}$$

By inspection of Equation 11B-5 you can see that for *each* m-value, each angle of θ for the blue light will be smaller than the associated angle for the red light ($\theta < 90°$).

$$\theta_{\text{red}} > \theta_{\text{blue}}$$

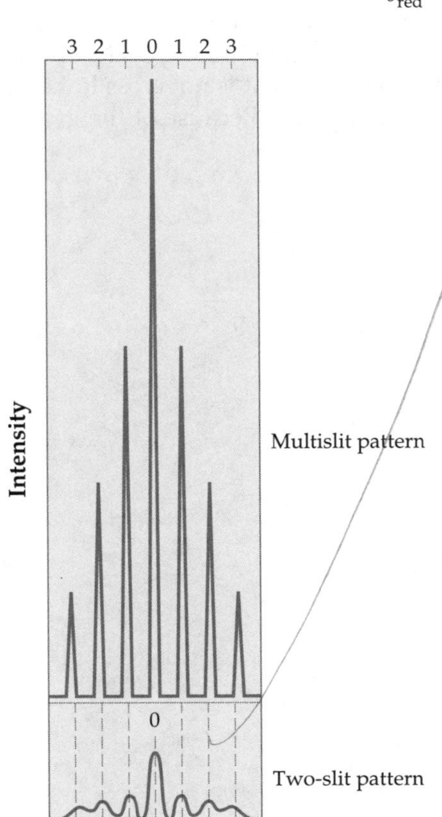

◀ FIGURE 11B.5 **The difference in light-intensity distribution between two-slit diffraction and multislit diffraction.** The peaks are sharper for multslit diffraction. Therefore, if red and blue light are both in the illuminating beam, they are more cleanly separated by the multslit grating than by a two-slit system. In other words, the multislit system provides better resolution of the colors.

The two colors will be separated from each other on the screen. Thus, a multislit diffraction plate can be used to separate colors.

If white light were shining through the slits, a number of small rainbows would show up on the screen. Each of these rainbows can be labeled with an order: first order, second order, etc., just as was done for the individual light and dark bands. The white light has been dispersed into its spectrum.

You may well ask, how can you make a multiple slit diffraction system having slits that close together? The answer is simply that the devices that are made, **diffraction gratings**, only *act* like they have a large number of slits.

Gratings have closely spaced, parallel grooves on their surfaces, and it is these grooves that act like slits. Figure 11B.6 illustrates these surfaces. For gratings used in the visible spectral region, the grooves are spaced about 3μm apart. The number of grooves mm^{-1} is called the **grating constant**, and gratings with 300–1200 grooves mm^{-1} are the most common.

However, no grating is quite perfect. For instance, the groove spacing varies subtly across the grating plate. These imperfections produce spectra with false lines called **ghosts**. These ghosts are dimmer than (and displaced from) the intense lines that were their light source. In addition, slight imperfections in the surface reflect and scatter light into regions where it would not be if the grating were perfect. This is the *stray light*.

Some of these problems of ruled gratings are alleviated by **surface holographic gratings**. Another name for them is **surface-relief** holographic gratings. These gratings have grooves produced in a polymer that is then coated with a thin layer of aluminum. The name holographic comes from the means of producing the parallel, evenly spaced grooves from the interference fringes produced by two beams from the same laser as they intersect: the same type of interference fringes that produce holograms. The alternating light and dark bands cause different amounts of cross linking of the special polymer coating called a **photoresist**. The polymer that is not cross linked is then washed away leaving a thinner area in the film. Because of the greater

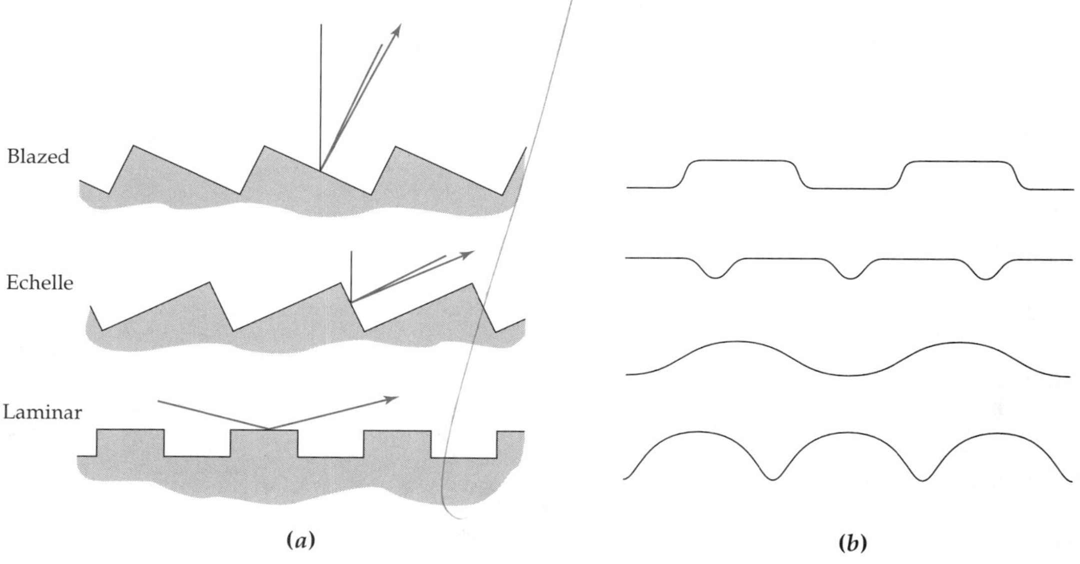

FIGURE 11B.6 ▲
Planar gratings' groove profiles.
(a) Ruled with a diamond tool. **(b)** Surface holographic grooves. [Ref: Horne, D. F. 1983. *Photomasks, Scales, and Gratings*. Bristol: Adam Hilger, Ltd.]

perfection of the line spacing, these planar holographic gratings are virtually free of the ghosts and stray light scatter of ruled gratings.

More recently, both ruled gratings and planar holographic gratings have begun to be replaced by **volume holographic optics** for controlling and dispersing light. Volume holographic gratings received their name because the entire volume of a material acts as a grating, with the grating effect due to planes within a volume having regular variations in the index of refraction.

Interferometers

Interferometers, as their name implies, depend on interference to separate light by wavelength. These devices pass a single band of light and reject the others. An **interference filter** is composed of solid layers of precisely controlled thicknesses. The interfaces between layers are defined by a change in the **index of refraction**. Interference filters are, in essence, fixed interferometers that pass one band of light due to multiple reflections at the surfaces between layers.

The index of refraction, written as n, measures the slowing of the speed of light inside different materials compared to the speed in a vacuum, where the speed is the greatest. Since the frequency is unchanged, the wavelength must be shorter. Algebraically,

$$n = \frac{c}{v} = \frac{\lambda_{\text{vacuum}}}{\lambda_{\text{material}}} \qquad \textbf{(11B-6)}$$

In the equation, v is the speed of light in the material. The index of refraction is a function of both the frequency of the electromagnetic radiation and the identity of the material through which it passes.

Suggestions for Further Reading

ANDREWS, D. L. 1992. *Applied Laser Spectroscopy: Techniques, Instrumentation, and Applications*. New York: VCH Publishers.

Lots of interesting information on how laser spectroscopic experiments are done, and most chapters have sections on applications to chemical analysis. Especially good chapter on Raman techniques.

BARR, T. L. 1994. *Modern ESCA, The Principles and Practice of X-Ray Photoelectron Spectroscopy*. Boca Raton, FL: CRC Press.

An outstanding text on the practice of ESCA.

BRUCH, M. D. 1996. *NMR Spectroscopy Techniques*. New York: Marcel Dekker.

The approach of this book is to give descriptions of what molecular information various types of NMR experiments (solids, 2-D in liquids, etc.) can provide and how the spectra show it. A unique and highly useful approach at a reasonable level of sophistication with excellent figures.

COLEMAN, P. B. 1993. *Practical Sampling Techniques for Infrared Analysis*. Boca Raton, FL: CRC Press.

This book is a mixture of general chapters full of useful tips about IR sample preparation and any needed adjustments that depend on the nature of the sample and instrument. Other chapters are essentially specialist notes about specific types of IR spectrometry. Outstanding if you are planning to do IR.

INGLE, J. D. J., CROUCH, S. R. 1988. *Spectrochemical Analysis*. Upper Saddle River, NJ: Prentice Hall.

A good place to go next for more advanced treatments of the topics mentioned in this chapter. An especially fine treatment of spectroscopic S/N is found in Chapter 5.

KROHN, D. A. 1988. *Fiber Optic Sensors: Fundamentals and Applications*. Research Triangle Park, NC: Instrument Society of America.

This book contains a large amount of information about the title subject presented in a highly readable way with useful, general references up to its publication. The text covers not only chemical sensors, but those for many physical changes as well.

METCALFE, E. 1987. *Atomic Absorption and Emission Spectroscopy*. Chichester: John Wiley & Sons.

A straightforward guide to most aspects of conventional analysis by AA and AE (up to its year of publication).

PARRY, S. J. 1991. *Activation Spectrometry in Chemical Analysis*. New York: John Wiley & Sons.

An outstanding introduction to neutron activation analysis with clear writing and a good balance between basic theory and useful practical information.

RENDELL, D. 1987. *Fluorescence and Phosphorescence*. Chichester: John Wiley & Sons.

An easy-to-read book on most aspects of solution fluorescence and phosphorescence as used for chemical analysis, including applications.

SOMMER, L. 1989. *Analytical Absorption Spectophotometry in the Visible and Ultraviolet: The Principles*. Amsterdam: Elsevier.

A comprehensive and detailed coverage of the many techniques involved in UV-vis spectrometry. A good place to read if you are planning to do careful solution spectrometry.

THOMAS, M. J. K., ANDO, D. J., *Ultraviolet and Visible Spectroscopy*. Chichester: John Wiley & Sons, 1996.

An informal style characterizes this book covering most aspects of absorbance spectrometry applied to analysis. Easy reading and recommended. The book is changed little from the 1st edition by Denney and Sinclair.

VARMA, A. 1991. *Handbook of Inductively Coupled Plasma Atomic Emission Spectroscopy*. Boca Raton, FL: CRC Press.

An outstanding, short, general introduction to ICP atomization with good coverage of factors to be taken into consideration. A handbook with an exhaustive list of applications (both for specific elements and for specific types of samples) and a huge bibliography.

J. N. MILLER, ed. 1981. *Standards in Fluorescence Spectrometry*; London and New York: Chapman and Hall.

This short monograph contains details of procedures to calculate and obtain quantitative fluorescence spectra. Some points apply to absorption as well. Outstanding and concise despite its age.

SILVERSTEIN, R. M., BASSLER, G. C., MORRILL, T. C. 1991. *Spectrometric Identification of Organic Compounds*, 5th ed.; New York: Wiley. Especially Chapters. 3, 4, and 5.

Contains an introductory chapter on assigning bands in infrared spectra of organic molecules. Good references in the area. Also, in another chapter, the authors describe analysis of organic-molecule NMR spectra.

VARMA, A. 1984. *CRC Handbook of Atomic Absorption Analysis*; Boca Raton, FL: CRC Press.

The two volumes list citations of AA analyses along with a separate author and subject index. An unparalleled entrance into the vast older literature of atomic absorption.

VARMA, A. 1990. *Handbook of Furnance Atomic Absorption Spectroscopy*. Boca Raton, FL: CRC Press.

Good general introduction to the technique, good coverage of factors to be taken into consideration, and exhaustive list of applications (both for specific elements and for specific types of samples).

Articles

STEPHENS, R. 1980. "Zeeman Modulated Atomic Absorption Spectroscopy," *Crit. Rev. Anal. Chem*. 9:167–195.

SHARPE, M. R. "Stray Light in UV-vis Spectrometers," *Anal. Chem*. 1984, 56, 339A–356A.

An excellent, brief review of the effects and sources of stray light. Highly recommended.

ALCOCK, N. 1993. "Flame, Flameless, and Plasma Spectroscopy," *Anal. Chem*. 65, 463–469R.

JACKSON, K. W. and CHEN, G. 1996. "Atomic Absorption, Atomic Emission, and Flame Emission Spectrometry," *Anal. Chem*. 68, 231R–256R.

OLESIK, J. W. 1996. "Fundamental Research in ICP-OES and ICPMS," *Anal. Chem*. 68, 469A–474A.

YAN, X.-P., NI, Z.-M. 1994. "Vapour generation atomic absorption spectrometry," *Anal. Chim. Acta*. 291: 89–105.

KRICKA, L. J., THORPE, H. G. 1983. "Chemiluminescent and Bioluminescent Methods in Analytical Chemistry," *Analyst*. 108, 1274–1296.

MILLER, J. N. 1984. "Recent Developments in Fluorescence and Chemiluminescence Analysis," *Analyst*. 109, 191–198.

Two interesting review articles in the area of chemiluminescence.

Concept Review

1. Sketch a box diagram of the components that are required for UV-visible instruments used for
(a) absorption measurements
(b) emission measurements
(c) fluorescence measurements

2. What is the relationship between the $\%T$ for a sample and the absorbance for the sample?

3. Define spectral, chemical and instrument interference.

4. Within what range of absorbances is the error in concentration for a spectrophotometric assay minimized?

5. How are frequency, wavelength, and energy related?

6. Given that IR spectrometry and Raman spectrometry both probe vibrations of molecules, why are the solvent and sample cell requirements so different?

7. What is the relationship between a monochromator's bandwidth and the width of a spectral feature that ensures that it will be accurately recorded?

8. For what elements are Auger, Rutherford backscattering, and X-ray fluorescence most suitable?

Exercises

11.1 What would be the effect on the apparent concentration (increases, decreases, stays the same) of a sample if
(a) A smudge on the cuvette while reading the absorbance of the standard is wiped away before taking readings for the samples in the same cuvette?
(b) The instrument was not set to zero with solvent before absorbance measurements were made for the standard with A = 0.751 and a sample with the same concentration.

11.2 A peak in the UV-vis part of the spectrum has a maximum at 582 nm and a full-width-half-maximum of 100 nm.
(a) What is the energy of the transition based on its λ_{max}?
(b) To what position in wavenumbers do 532 and 632 nm (the fwhm boundaries) correspond?
(c) What is the bandwidth of the peak in cm^{-1}?

(d) What process is most likely to give rise to this transition?
(e) What is the frequency of light which corresponds to the maximum absorbance?
(f) What is the energy of the transition?

11.3 This exercise refers to Figure 11.3.1. Assume that the concentrations are in mM and the absorbance maximum of the least concentrated solution occurs at $A = 0.120$. The compound's molecular weight is 320.4, and the cell holding the sample has a path length of 2.000 cm.
(a) What is the value of the molar absorptivity at the band maximum?
(b) What is the absorptivity in μg^{-1} L cm^{-1} at the band maximum?
(c) What is the sensitivity of the assay in absorbance units mol^{-1} L?

11.4 For a transition with a maximum at 1600 cm^{-1} and a natural linewidth of 20 cm^{-1}
(a) What is the maximum spectral bandwidth that would allow the spectral features of the peak to be measured precisely enough for identification of an unknown?
(b) What would be the maximum spectral bandwidth for quantitation using the peak?
(c) What is the wavelength corresponding to the maximum?
(d) In what region of the electromagnetic spectrum does the peak occur and to what kind of transition does it likely correspond?

11.5 Figure 11.5.1 shows the output from a continuous flow analyzer running at 120 samples h^{-1} and testing for nitrite. The samples are separated by bubbles of air to prevent cross contamination. Each sample is reacted with sulfanilimide, and the nitrite present forms a colored compound which is monitored as it flows through a spectrometric detector. The response for each of the test solutions rises to the measurement level and then falls off as the sample passes through. The protocol of the determination shown here consists of standards of 2-, 6-, 10-, 14-, and 18-μM nitrite (NO_2^-) followed by a test of intersample contamination and then a repeatability test. What is the sensitivity of the instrument in (absorbance unit) (μg $NaNO_2$)$^{-1}$ L? Assume that the samples were 100 μL each. [Figure reprinted with permission from Patton, C. J., et al. 1982. *Anal. Chem.* 54:1113. Copyright 1982, American Chemical Society.]

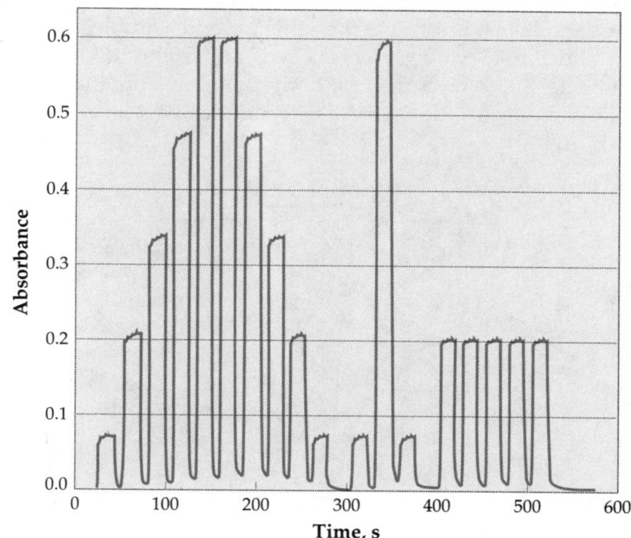

FIGURE 11.5.1 ▲

11.6 With the conditions the same as in problem 11.3, what are the values of $\%T$ at the band maxima for the four concentrations?

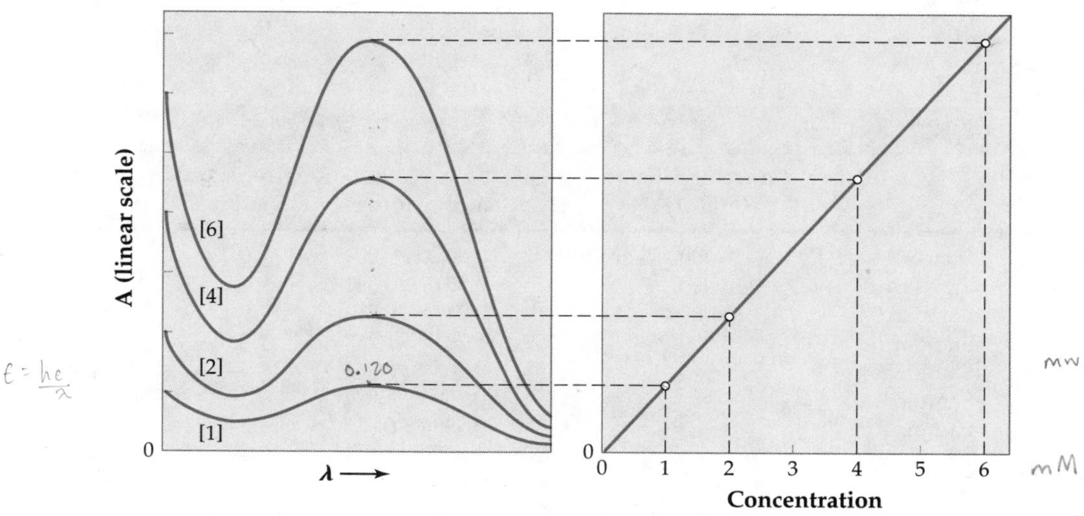

FIGURE 11.3.1 ▲

11.7 Shown in Figure 11.7.1 are two runs of samples on an instrument with %T output. The analyte is the same and the sample conditions are the same for both samples. The 0% and 100% T values were checked and found to coincide and to be correct for both runs. What are the relative concentrations of the two samples, that is, [A]/[B]?

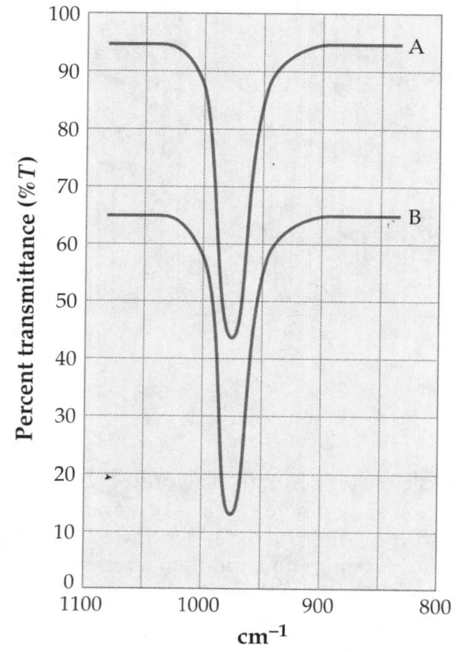

FIGURE 11.7.1 ▲

11.8 An automated flame AA analysis was obtained for a number of samples of river and estuary waters, which were collected and then stored as a solution containing 1% (v/v) HNO_3. The data used in the determination of magnesium are shown in Table 11.8.1. The measurements were done in triplicate and recorded. The operating conditions were wavelength 202.5 nm, lamp current 10 mA, and spectral bandpass 1.0 nm. [Data from Liddell, P. R. 1983. *Am. Lab.* (March), 15:111.]
(a) Complete the table.
(b) It turns out that samples 1, 5, and 6 were from the same river source. They were included to test the sample collection and storage errors. What is the RSD due to sample collection and storage alone?

11.9 Determination of sodium and potassium in blood or serum by low-temperature flame emission on simple instruments is complicated by the instability of the flame and the complexity of the matrix. One way to overcome the problems is to add an excess of lithium to serve as both an ionization suppressor and an internal standard. The reason it can serve as an internal standard is that the lithium emission responds similarly to the emissions of Na and K to changes in the flame conditions. The responses found for calibration solutions are listed in the table below.

Concentration (ppm)			Signal (relative to Na in 1.0-ppm Na sample)		
Na	K	Li	Na	K	Li
0.5	0.5	500	0.43	0.57	31.2
1.0	1.0	500	1.00	1.25	36.3
5.0	5.0	500	4.9	6.4	36.2
10.0	10.0	500	8.8	11.7	31.7

(a) Plot the signal vs. concentration curves for Na and K with the Li internal standard present.
(b) A blood sample of 10 μL is diluted in 1.00 mL of 5000-ppm Li solution. This is further diluted with doubly distilled water to 10.00 mL and fed into the spectrometer.

Table 11.8.1

Solution	Concentration (mg/L)	Absorbance readings			Relative standard deviation (%)
Blank	0.000	0.001	0.001	0.001	0.0
Standard 1	6.000	0.351	0.357	0.361	1.4
Standard 2	15.000	0.714	0.711	0.712	0.1
Standard 3	30.000	1.252	1.264	1.268	0.6
Sample 1	—	0.558	0.566	0.563	—
Sample 2	—	0.364	0.369	0.372	—
Sample 3	—	0.385	0.388	0.385	—
Sample 4	—	1.008	1.020	1.011	—
Sample 5	—	0.555	0.563	0.565	—
Sample 6	—	0.590	0.598	0.595	—

The readings with the blood were Na 2.9, K below detection, and Li 32.5. A second 10-μL sample was taken and brought to 2.00 mL with 500-ppm Li solution. The readings were Na 13.6, K 1.34, and Li 30.4. What are the serum concentrations of Na and K in mM? Assume Na and K do not interfere with each other.

11.10 A 60-MHz ^{1}H-NMR spectrum was run, and six peaks with equal integral areas were found. The peaks were at 0, 83, 98, 112, 228, and 429 Hz downfield relative to TMS. In the sample are tetramethylsilane, acetone, benzene, cyclohexane, *t*-butanol, and dioxane. The hydroxy proton of the alcohol could not be seen.
(a) Assign the six peaks to the six compounds.
(b) If the spectrum were a 100-MHz ^{1}H-NMR spectrum, at what frequencies relative to TMS would the six resonances be?
(c) If the TMS is assigned a concentration of 10.0, what are the concentrations of the other five components of the mixture?

11.11 In an atomic emission assay, lines for two of the analytes occur at 220.7 and 221.2 nm. There is a 90% valley between the peaks when the instrument is working under optimal conditions. What is the resolving power of the instrument (that is, $\lambda/\Delta\lambda$)?

■ *11.12 Infrared absorption spectra can be obtained of small spots on the surface of a sample. If the absorbance at a particular energy is monitored at evenly spaced positions and plotted as a function of position on the surface, a "picture" that shows the presence of certain groups (based on strong absorbances at the monitoring wavelength) can be obtained. Suppose that an FTIR microscope is designed to pass the incident IR radiation out of a slit that is 20.0 microns wide parallel to the sample surface. Since this slit width is very close to the wavelengths for a typical IR spectrum, this light is "spread" as it exits from the slit. The first dark band occurs at an angle θ defined by the equation

$$\sin\theta = \frac{\lambda}{w}$$

where
λ = wavelength
w = slit width
θ = angle from the normal through the middle of the slit.

(a) Calculate the wavelengths corresponding to wavenumbers in the range 4000 cm^{-1} to 2100 cm^{-1}, every 50 cm^{-1}.
(b) Calculate the angle to the first dark band for each of the above wavelengths.
(c) The distance x from the slit to the surface is 10 μm. Let the width of the sampled area be called r, which is defined by the equation

$$r = 2\,x\tan\theta$$

where θ is the angle calculated in part (b). Calculate and plot the value of w as a function of $\bar{\nu}$ for the range 4000 cm^{-1} to 2100 cm^{-1}. The value of w defines the limits of resolution of the microscope.

*11.13 Figure 11.13.1 on page 378 shows the results of the preliminary data taken while developing a fluorometric assay for the amino acid glycine. The glycine was reacted with a reagent, fluorescamine, which forms a fluorescent product with amines. The three scans shown are: **(a)** the relative intensity of emission as the excitation wavelength was scanned—an excitation spectrum at a fixed-emission wavelength; **(b)** the emission spectrum found at a fixed-excitation wavelength; and **(c)** a plot of the relative fluorescence intensity with both excitation and emission wavelengths fixed but varying the pH. To optimize the assay's sensitivity to glycine, at what wavelengths should the excitation and emission monochromators be set, and what should the pH of the solution be? [Ref: Coppola, E. D., Hanna, J. G. 1976. *J. Chem. Educ.* 53:322.]

*11.14 A multiple reflectance cell with an internal length of 0.1 m is used to measure the concentration of a gas sample. The beam traverses the cell 40 times before it exits the cell and passes on to the detector. It hits the reflective ends at an angle of 1° from the normal each time. What is the pathlength of the cell to three decimal places?

Additional Exercises

11.15 An IR spectrum of vinyl acetate shows both a C=C bond vibration and a C=O vibration. Both bands have baselines at 80% *T*. The %*T* at the peak maxima are, respectively, at 23% *T* and 5% *T*. What is the ratio between the absorptivity at the C=O frequency compared with the C=C frequency?

11.16 Figure 11.16.1 on page 379 shows emission spectra (swept from 395.90 to 396.35) from the following sample solutions: (1) pure water, (2) aluminum 250 μg/L, (3) calcium 2500 μg/L, and (4) Al 250 μg/L + Ca 2500 μg/L. [Figure reprinted with permission from Allain, P., Mauras, Y. 1979. *Anal. Chem.* *51*, 2089. Copyright 1979, American Chemical Society.]
(a) If you wish to determine the aluminum content of solutions containing around 2500 μg/L calcium, which peak do you use as a standard, (2) or (4)?
(b) If the calcium solution measured in spectrum (3) was made from high-quality $CaCl_2 \cdot 6H_2O$ (f.w. 208.49), what is the impurity of Al in this calcium reagent in ppm to within the precision possible with the data at hand?

11.17 An NMR spectrum of toluene (that is, methylbenzene) was run. The peaks around 7.2 ppm have

*Denotes more involved exercises.

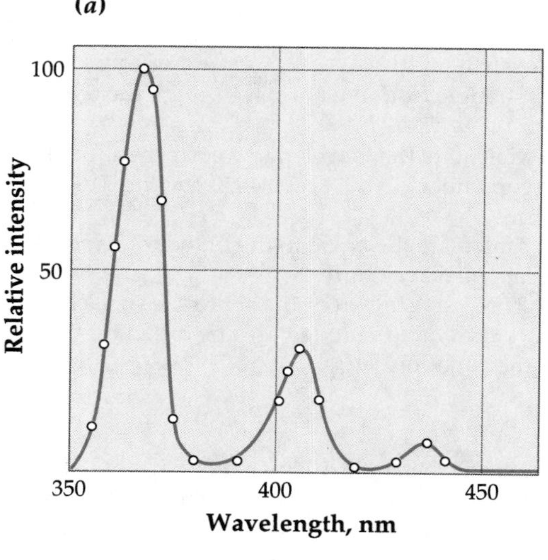

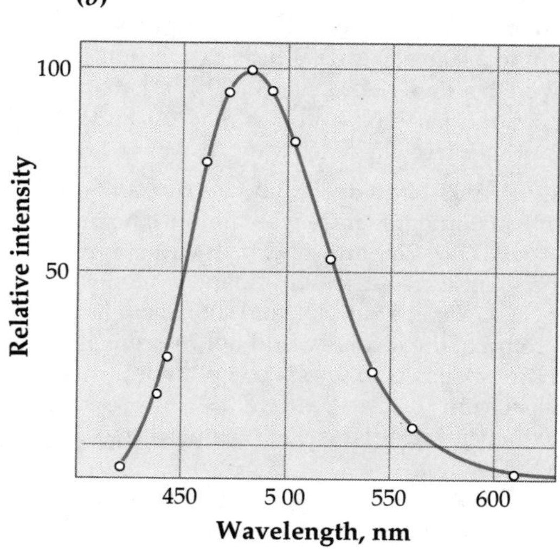

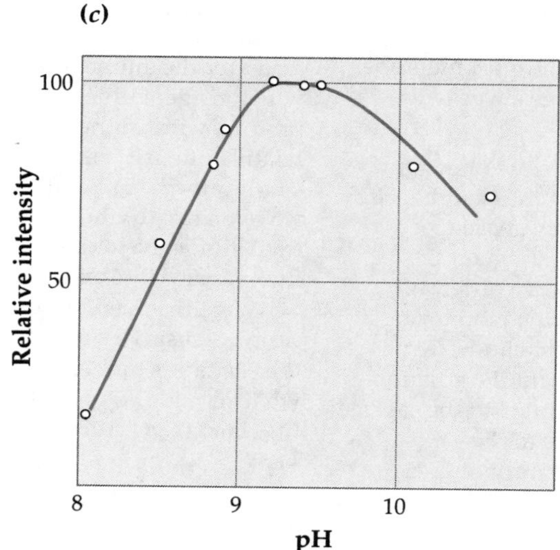

FIGURE 11.13.1 ▲

a total integrated intensity of 72 units. The peak at 2.35 ppm has an integral of 36 units.

(a) What kind of protons appear in an NMR spectrum at the chemical shifts given?

(b) Based on the relative areas of the peaks, is the toluene pure?

11.18 A new sulfonate drug was being developed that contained the following group in its structure:

$$-\overset{O}{\overset{\|}{C}}-C_6H_4-\overset{O}{\overset{\|}{\underset{\underset{O}{\|}}{S}}}-O^-$$

It was decided to find the molecular mass by NMR. To do this, 10.0 mg of the compound was weighed into an NMR tube. To the tube was added 5.0 mg of anhydrous sodium acetate (f.w. 82.04). Both compounds were dissolved in D_2O. An NMR spectrum was run. The sodium acetate peak at 1.90 ppm had an integral area of 82.0 graph spaces (average of 3 runs). A peak at 8.0 ppm (no other peaks in the region) had an integral of 30.5 spaces (average of 3 runs).

(a) To what groups do the peaks correspond?

(b) Based on the structures for acetate and the sulfonate group above, how many protons are in each of the groups responsible for the peaks?

(c) What is the response factor that relates number of protons to integrated area?

(d) Based on the response factor and the ratio between the two peak integrals, how many moles of the drug are present?

(e) What is the formula weight of the analyte?

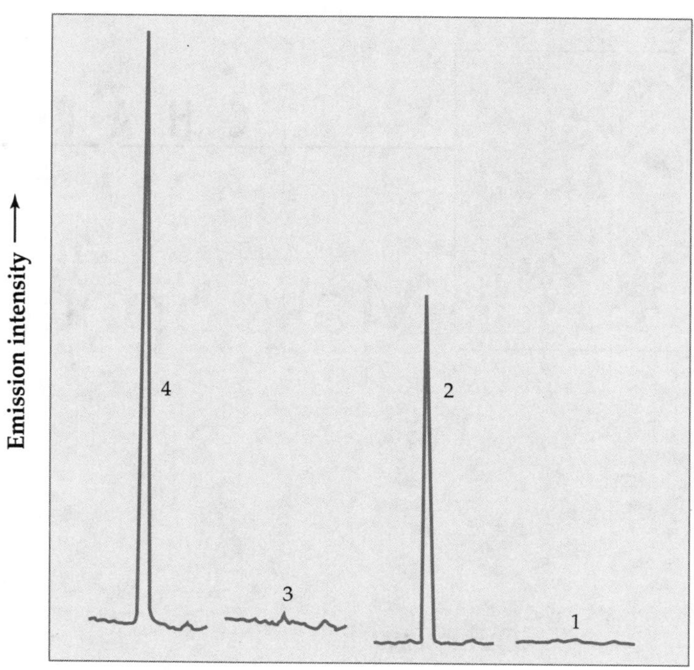

FIGURE 11.16.1 ▲

11.19 A spectrometric assay is designed to work with final concentrations of 0.005 to 0.020 mg L^{-1} nicotine. If a 500 mL sample of urine is used and all of the nicotine present is concentrated during sample preparation into a final solution volume of 10.00 mL, what is the minimum concentration of nicotine (in mg L^{-1}) that can be detected in urine by the assay?

11.20 On an instrument with an absorbance output, a sample has an absorbance of 0.335 at its peak when the baseline is at $A = 0.0$. Somehow, some opaque spot got onto the cuvette in the light path, and the baseline showed up at 0.120. At what absorbance reading will the peak now be for the sample?

11.21 A spectrophotometric assay was to be carried out on a compound with $\epsilon = 5000$. The stock solution of standard had been made with a concentration of 0.0203 M. Given a pathlength of 1.000 cm and the fact that the final absorbance should be around 0.4 to 0.5, by what factor should the stock solution be diluted?

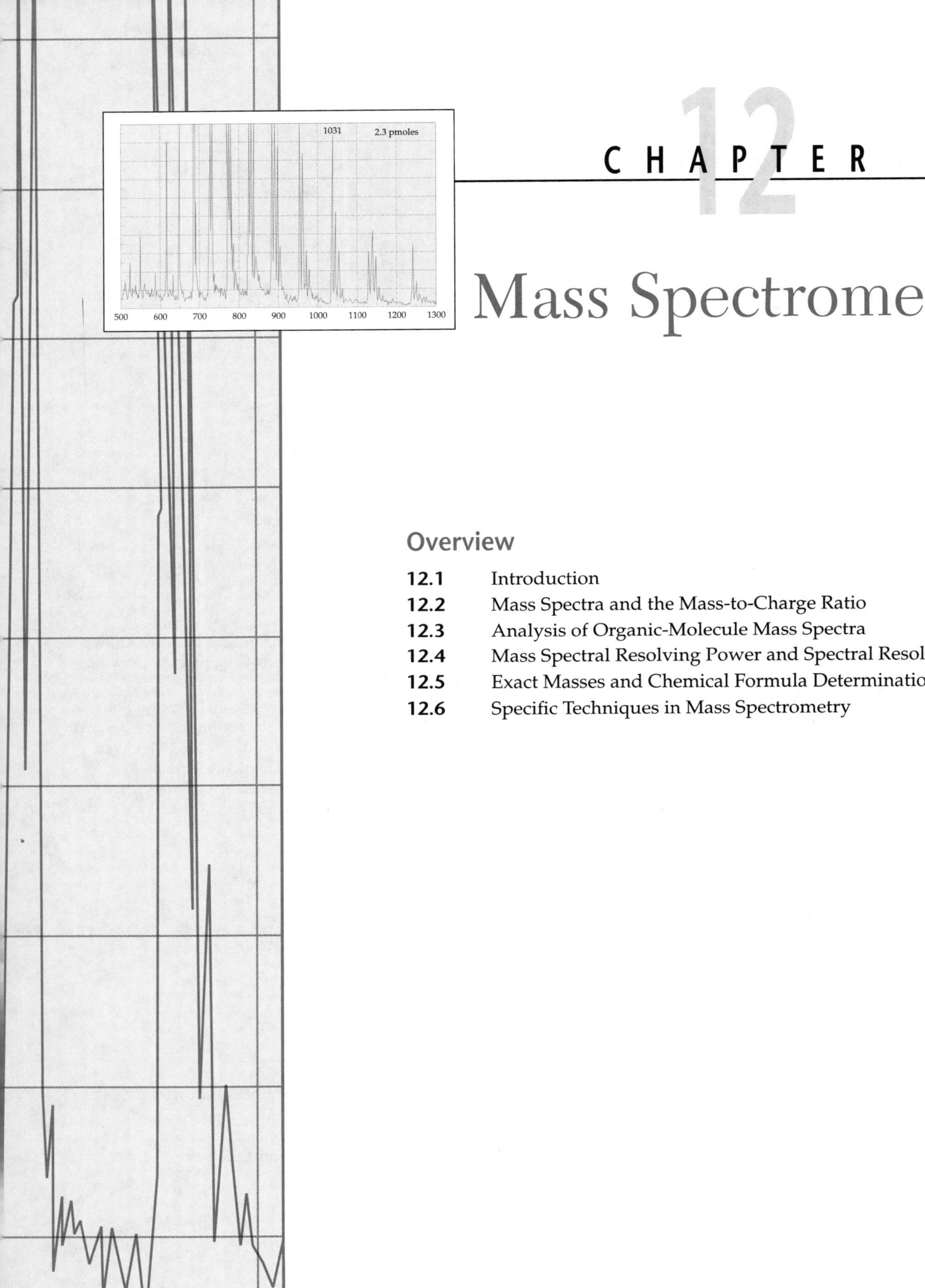

CHAPTER 12

Mass Spectrometry

Overview

12.1 Introduction

Mass spectrometry (MS) is a name for a collection of techniques used to measure the masses of ions and their abundances in the gas phase. Each of the steps of the mass measurement,

a. generation of the gas phase molecules (and molecular fragments and atoms)
b. their ionization, and
c. their separation based on their masses

are now done in a number of different ways. These methods are the subject of this chapter.

The collection of techniques called mass spectrometry is one of the most versatile and powerful tools of chemical analysis. Its broad usefulness is due in part to the wide choice of capabilities possible for each of the three sections of a mass spectrometer. These three basic sections are the mass source, the mass analyzer, and the transducer/detector, as illustrated in Figure 12.1.

Since its inception, mass spectrometry has been used effectively to analyze small molecules and atoms. The more recent expansion of mass spectrometry into the analysis of virtually all types of samples is due in large part to the variety of sources that have been devised to vaporize and ionize the

At the left is an electron multiplier transducer. At the right is shown the same transducer mounted for use.

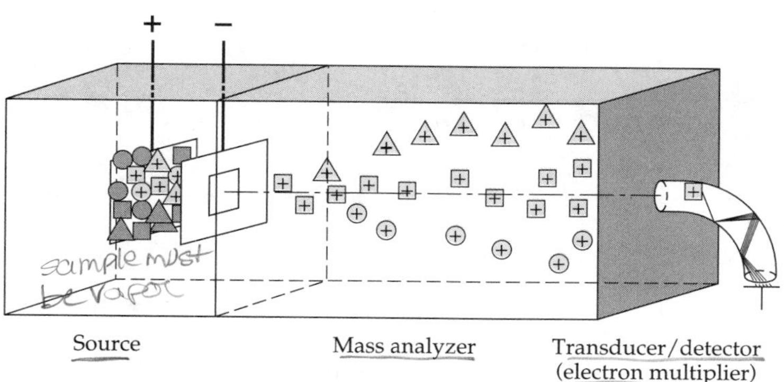

FIGURE 12.1 ▲

Diagram of the major features of a mass spectrometer.
The sample molecules travel from the left, where they are introduced into the instrument, toward the right. The sample is inserted either as a gas, liquid, or solid. (Liquids and solids then must be vaporized by appropriate techniques.) The gaseous molecules are ionized in the space between two charged plates. The ions are accelerated by an electric field and focused using electric fields and/or slits. The acceleration potential is in the range of kilovolts. The region where the molecules are introduced, rendered into a vapor, ionized, and accelerated is called the source. Shown in the source are three types of gaseous, positive ions with different masses, indicated by circles, triangles, and squares. The accelerated ions pass from the source into one of a number of types of mass analyzers. In most mass analyzers the ions of different masses are separated in space, and the ions with different masses impinge on the transducer at different times. This small ion current is amplified by the transducer and its associated electronics to produce the output signal. The output of the ion current vs. time is plotted as ion current vs. mass, giving a mass spectrum similar to that shown in Figure 12.2.

A number of sources different from the one described in the figure are presented in Section 12.6.

Table 12.1 Mass Spectrometric Source Types and Varieties of Samples

Sample	Usual Source Types	Examples
Gaseous species	Electron ionization (EI) Chemical ionization (CI)	He, H_2, D_2, Butane Butane
Vaporizable organics and inorganics	Thermal vaporization-EI Thermal vaporization-CI Thermal ionization (TI)	Derivatized sugars, acetaminophen Isotope ratios for dating of rocks and tracers of drug metabolites
Elements in aqueous solutions	Inductively coupled plamsa (ICP)	Metal ions in water
Molecules/macromolecules soluble in water	Electrospray ionization (ESI)	Polypeptides Surfactants Polyethylene oxide
Molecules/macromolecules soluble in specific liquids (glycerol, thioglycerol, etc.)	Fast atom bombardment (FAB)	Polypeptides Proteins < 5,000 M.W.
Macromolecules and polypeptides dissovled and dried within specific solid matrices (*e.g.*, 3-methoxy-4-hydroxycinnamic acid + trifluoroacetic acid)	Matrix assisted laser desorption/ionization (MALDI)	Polypeptides Proteins up to 10^5 M.W. Nucleic acids < 90-mer
Elements in solid samples	Secondary ion mass spectrometry (SIMS) Laser ablation Glow discharge (GD)	Integrated circuits Window glass Rocks Steel
Non-volatile molecular solids	SIMS, GD, laser ablation pyrolysis	Plastics, elastomers, paint

molecules and atoms of materials that are not initially gases. It is in this area that the development and application of new methods are progressing so quickly. Because of these newer vaporization/ionization methods, mass spectrometry is now firmly established in biochemical, polymer, and inorganic analysis. It is important for you to understand that no single, universal MS source exists. Each type has a range of samples for which it is best. Table 12.1 lists the areas of general application for the most common sources. In practice, the type of sample being analyzed is intimately tied to the operational characteristics of the appropriate source. However, the details are beyond the scope of this chapter, and we will only mention the names of sources in discussing the mass spectral data.

12.2 Mass Spectra and the Mass-to-Charge Ratio

The presentation of data for mass spectrometry usually is done in one of the three ways illustrated in Figure 12.2. All show the masses of the ionized sample molecules (or atoms) and any ionic fragments that arise from decomposi-

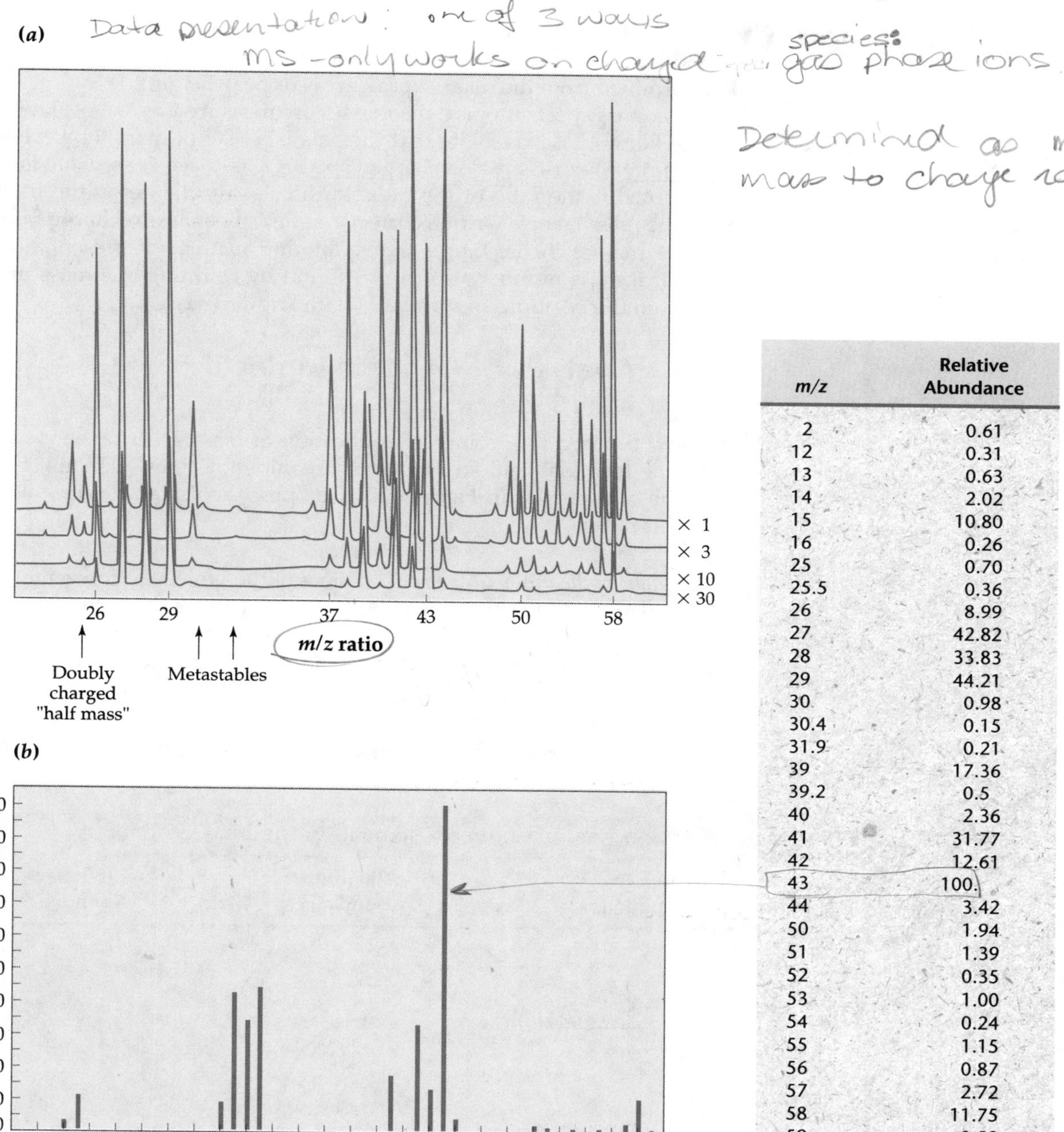

m/z	Relative Abundance
2	0.61
12	0.31
13	0.63
14	2.02
15	10.80
16	0.26
25	0.70
25.5	0.36
26	8.99
27	42.82
28	33.83
29	44.21
30	0.98
30.4	0.15
31.9	0.21
39	17.36
39.2	0.5
40	2.36
41	31.77
42	12.61
43	100.
44	3.42
50	1.94
51	1.39
52	0.35
53	1.00
54	0.24
55	1.15
56	0.87
57	2.72
58	11.75
59	0.52

FIGURE 12.2 ▲

Three different presentations of the mass spectrum of a compound.
(a) A plot of the ion current with four different scales plotted at the same time. Note two special features: There is a peak at "mass" 25½. This is due to a mass-51 particle with two charges: a doubly charged ion with m/z 25½. Also, there are broad, low peaks between m/z 30 and 40. These are **metastable ions**—ions that have decomposed while in flight after they have left the source region. **(b)** A computerized histogram output of unit-mass peaks. The largest peak is scaled to be 100. **(c)** A list of the peak m/z values and relative abundances. Computerized systems generally list both (b) and (c). [Ref: (a) Courtesy of DuPont, Wilmington, DE. (b), (c) Data from NIST library.]

tion as well as the **relative abundances** of each. In mass spectrometry, relative abundance is another name for relative concentration. The relative abundance is determined from the relative peak areas or peak heights.

In a mass spectrum plot, the masses are measured by the peak positions on the horizontal axis. Note that the axis label is not mass, however. Mass spectrometry works only on charged species: gas-phase ions. What is always determined is the ratio of the mass to the electric charge on the fragment. Thus the mass label is written correctly as m/z, the **mass-to-charge ratio**. The table of masses in the figure shows another manner of presentation. The numerical mass number also can be found by counting one mass unit at a time from the positions of standards with known masses.

12.3 Analysis of Organic-Molecule Mass Spectra

We now proceed to a common application of mass spectrometry: to determine the formulas and structures of organic molecules. Our first example uses the data shown in Figure 12.2. Our process of identification proceeds with this order.

1. *Identify the* ***molecular ion***. The mass of the molecular ion is that calculated from the molecular formula by using the atomic weights of the isotopes with the highest natural abundance (that is, ^{12}C, ^{14}N, ^{35}Cl; see Table 12.2). For example, in Figure 12.2, the peak at m/z 58 is the molecular ion of n-butane.
2. *Study the* ***isotope distribution pattern***. This is the pattern of peaks that are due to combinations of the different atomic isotopes in a

Table 12.2 Natural Isotopic Abundances of Common Elements in Organic Compounds*

Element	Mass	Relative Abundance	Mass + 1	Relative Abundance	Mass + 2	Relative Abundance
Elements with one isotope						
F	19	100				
P	31	100				
I	127	100				
Elements with two major isotopes						
H	1	100	2	0.016		
C	12	100	13	1.08		
N	14	100	15	0.36		
Cl	35	100			37	32.5
Br	79	100			81	98.0
Elements with three major isotopes						
O	16	100	17	0.04	18	0.20
Si	28	100	29	5.1	30	3.4
S	32	100	33	0.80	34	4.4

*Radioactive isotopic abundances are negligible.

specific molecule or fragment. In Figure 12.2, peaks with m/z of 58 and 59 form an isotope distribution pattern, as do 43 and 44. This point is explained in more detail later.

3. *Explain the **fragmentation pattern**.* When the sample molecules are ionized in the source, some of the energy introduced to cause ionization serves to break up the molecule. The fragments from the breakup help confirm the identification of the molecule. In Figure 12.2, all the peaks below m/z 58 are part of the fragmentation pattern. However, masses 43 and 29 are the most useful.

You should understand that this sequence of steps is not the only way to approach the problem of identification. Also, computer-aided search routines do not work this way; they use methods of pattern identification which, in essence, look at the major peaks of a mass spectrum as a characteristic group.

Determining the identity of even a small, unknown molecule from its mass spectrum is a skill that can be obtained only with practice. (Mass spectral databases make the identification of the molecules included in the database *much* easier.) To identify or analyze unknown and more complex molecules, often other spectral information is exceedingly helpful, such as that from ultraviolet, infrared, or nuclear magnetic resonance spectra. Also, stoichiometric data might be useful, such as the wt% C,H,S, and Cl. Let us now look in more detail at the three steps of assigning a mass spectrum to the molecule from which it comes.

UV, IR, and NMR spectrometries are briefly discussed in Chapter 11.

Identify the Molecular Ion

When you assign the molecular ion to a peak, in effect you have determined the molecular mass of a molecule. This is the most important part of identification in mass spectrometry. However, the gas-phase chemistry of the ions occasionally makes the task less than straightforward. The properties of some specific molecular ions are discussed below, and a few problems are briefly noted.

Notice that for each of the organic-compound elements listed in Table 12.2, the most abundant stable atomic isotope is that with the lowest mass. As a result, the molecular ion peak is often the largest (tallest) peak among the highest m/z group. This largest peak is called the **base peak**. However, the peak corresponding to the molecular ion is not always the largest, as you will see from the next few examples.

In Figure 12.2a, the peak corresponding to the heaviest of the ions observed appears at m/z 59. Of the group around m/z in this region, that at mass 58 is the base peak. The base peak here also is the molecular ion of *n*-butane.

Figure 12.3 shows a less straightforward example: the mass spectrum of methanol, CH_3OH. We expect the molecular ion to be at m/z 32 from $^{12}C^{1}H_3{}^{16}O^{1}H$. Indeed, there is a large peak at m/z 32. But the peak at 31 is larger. The reason that m/z 31 is more abundant than m/z 32 is because this molecule loses one hydrogen quite easily. The gas-phase ion $CH_3O^{\bullet+}$ forms the base peak and is more abundant than the molecular ion.

The method of sample ionization similarly can cause a change in the mass that is the base peak. For instance, perhaps the majority of the mass spectra you will see will be run with the ionization done by knocking an

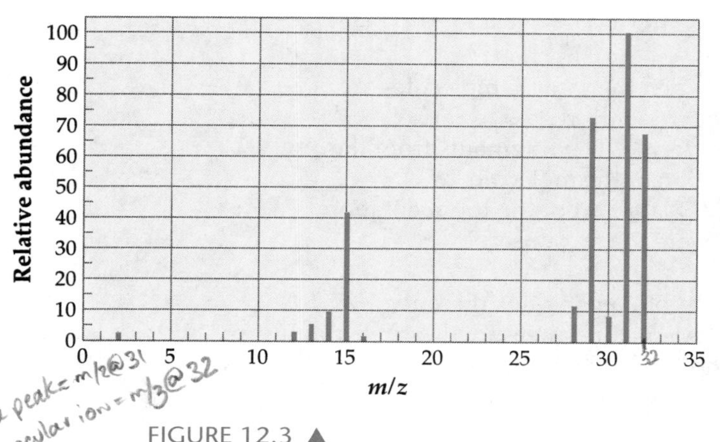

m/z	Relative abundance
12	0.33
13	0.72
14	2.4
15	13.
16	0.21
17	1.0
28	6.3
29	64.
30	3.8
31	100.
32	66.
33	0.98
34	0.14

FIGURE 12.3 ▲
The mass spectrum of methanol.

electron off the molecule using a beam of accelerated electrons. This is called **electron ionization** (EI). This gas-phase interaction yields a positive radical ion, $M^{\bullet +}$. (In mass spectral nomenclature M is used to represent *any* **parent** molecule. Italic *M* is often used to represent the mass of the molecular ion.) However, mass spectra can also be obtained using **chemical ionization** (CI). The ionic species that is formed in a CI source is MH^+; a proton is added to the gas-phase molecule. The protonated species is called a **quasimolecular ion**, MH^+, and has a mass of $M + 1$. This unit-mass difference between EI and CI can be seen in the diagrams of the mass spectra in Figure 12.4b and c. However, the isotope distribution patterns of the shifted set of m/z lines are the same in both. It is this unchanging isotope pattern that we shall discuss next.

Study the Isotope Distribution Patterns

The phrase isotope distribution pattern means the appearance of a characteristic set of peaks with fixed *relative* intensities. For example, isotope patterns of -Cl, $-Cl_2$, $-Cl_3$, and $-Cl_4$ are shown in Figure 12.5. The isotope distribution pattern is determined by two properties of the molecules (or fragments of molecules) producing them:

a. the identities of the atoms in the molecule (C, H, N, and so forth), and
b. the number of each type of atom.

In the example of Figure 12.5 there are two isotopes (Table 12.2) ^{35}Cl and ^{37}Cl, for which the $^{35}Cl:^{37}Cl$ relative abundances are in the ratio of 100 : 32.5.

If we obtained a mass spectrum of (ionized) atomic chlorine alone, it would appear as shown in Figure 12.5: a peak at m/z 35 rises to about three times the height of m/z 37.

When two atoms of chlorine are in one molecule, even though the molecular structure is fixed, either isotope can lie at each chlorine position of the molecule. For example, two general molecules that differ in *isotopic* structure can be written as $^{37}Cl\text{-R-C-}^{35}Cl$ and $^{35}Cl\text{-R-C-}^{37}Cl$. The results for three and four chlorine atoms in the same molecule are less straightforward to explain. However, the reasoning is based on simple statistics and the 100 : 32.5 ratio of the two chlorine isotopes.

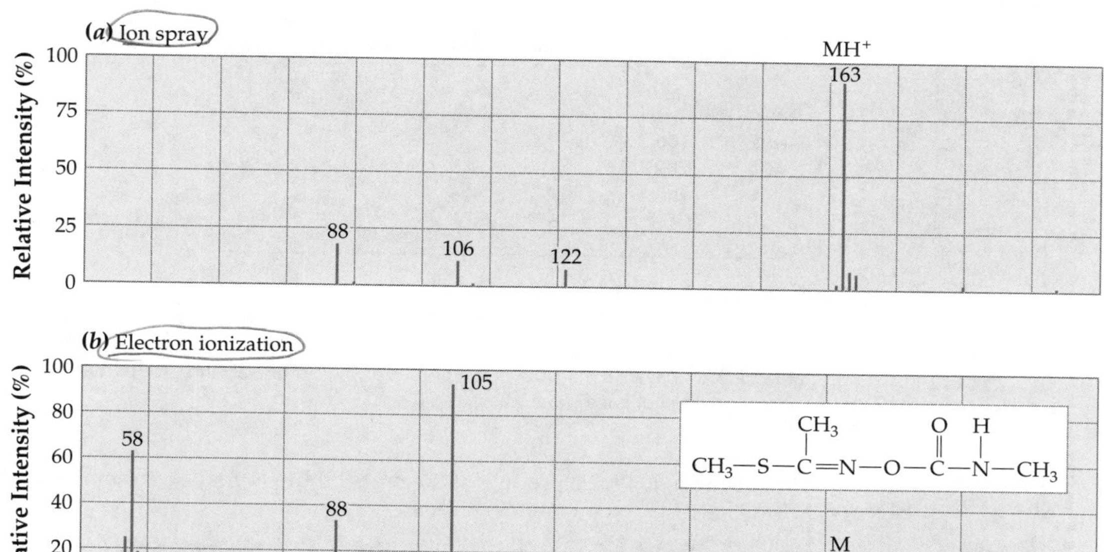

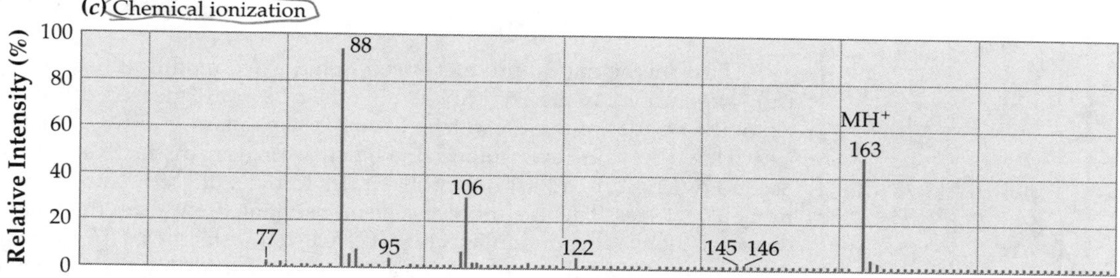

FIGURE 12.4 ▲

Mass spectra of the same compound with three different types of sample vaporization and ionization.

The sample is the carbamate pesticide Methomyl. Its structure is indicated in (b). Spectrum **(a)** results from an ion spray (electrospray). Spectrum **(b)** arises from electron ionization and **(c)** from chemical ionization. Notice the significantly different fragmentation patterns. Ion spray and chemical ionization produce many ions with an added proton, so you see m/z 163 in (a) and (c) but m/z 162 in (b). All have m/z 88. [Ref: Pleasance, S., Anacleto, J. F., Bailey, M. R., North, D. H. 1992. *J. Am. Soc. Mass. Spectrom.* 3:378–397.]

EXAMPLE 12.1

The question to be answered is: If ^{35}Cl:^{37}Cl occur in the ratio 1.00:0.32, with what probability will each of the combinations occur when there are two chlorines in a molecule?

Solution

This is a calculation of probabilities. The probability for each isotope at a specific position on the molecule is different since ^{35}Cl is about three times more likely to be present than ^{37}Cl when each isotope is *randomly* present in each chlorine position.

To be more exact: Of the chlorine atoms that are found, 75.77% are ^{35}Cl and the rest, 24.23%, are ^{37}Cl. We ask the question, If we have a combination of two chlorine atoms, what are the probabilities that each possible combination of the two isotopes will be present? As you know, the probability for finding each set of

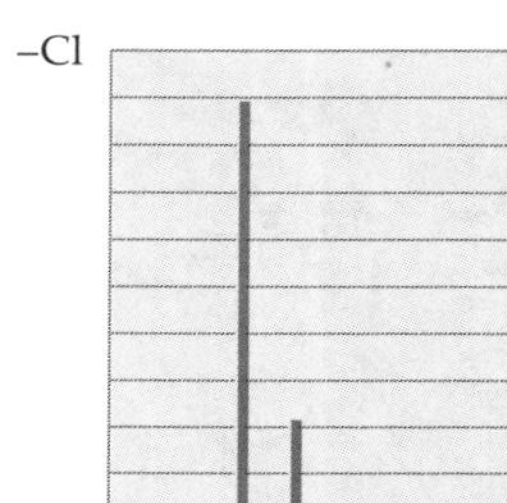

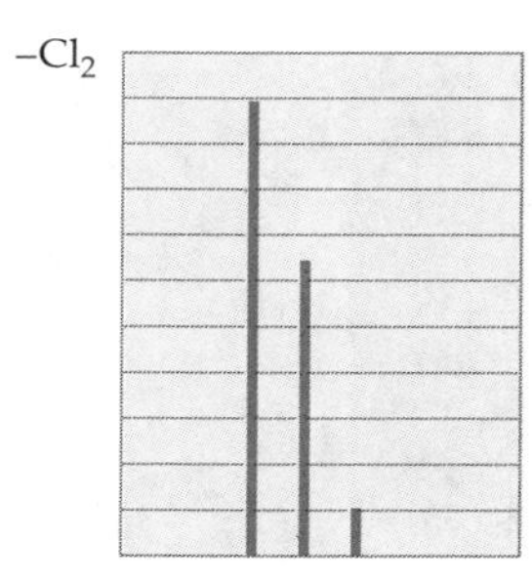

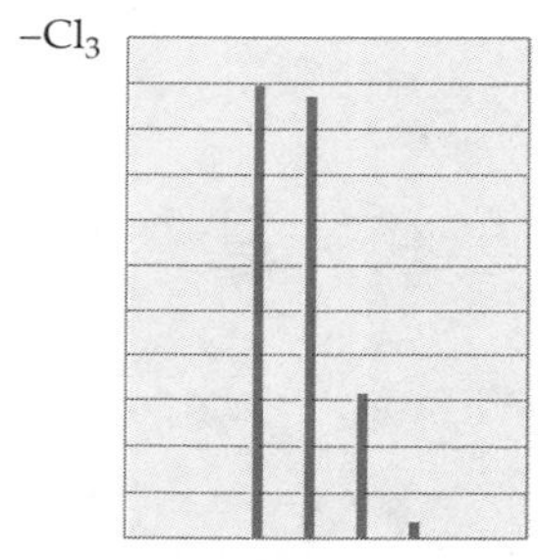

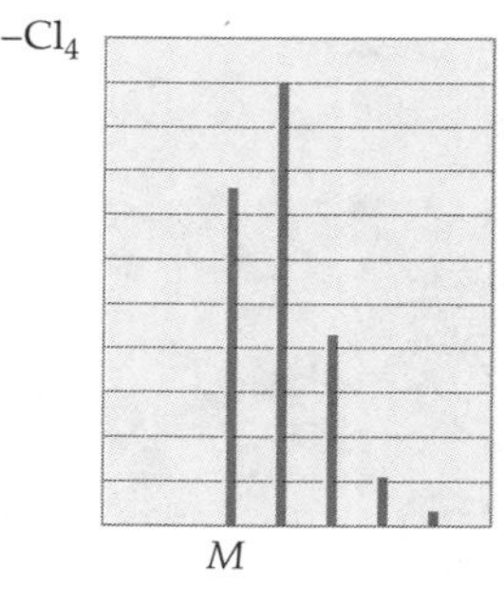

	Relative abundances				
	M	$M + 2$	$M + 4$	$M + 6$	$M + 8$
–Cl	100	32.5			
$–Cl_2$	100	65	10.5		
$–Cl_3$	100	98	32	3.4	
$–Cl_4$	77	100.0	44	10.5	0.9

◀ FIGURE 12.5 ▲

A histogram and list of the isotope patterns for molecules with one, two, three, or four chlorines.

Table 12.3 on page 389 illustrates the origins of the isotope abundance patterns. The largest peak of each set is scaled to 100.

chlorines equals the product of the probabilities to find each member present. We calculate,[1]

$^{35}Cl^{35}Cl$	$^{35}Cl^{37}Cl$
$0.758 \times 0.758 = 0.57$	$0.758 \times 0.242 = 0.18$
$^{37}Cl^{35}Cl$	$^{37}Cl^{37}Cl$
$0.242 \times 0.758 = 0.18$	$0.242 \times 0.242 = 0.059$

These mathematical products need to have two modifications to relate them to an experiment. In the experiment, only the masses are measured, and two of the isotopic formulas of the above four have the same mass. ^{37}Cl-R-C-^{35}Cl and ^{35}Cl-R-C-^{37}Cl have the same mass although their structures are distinctive, as you can see. Second, we need to adjust our probabilities to the scale of 100 units for the most probable species. This is simply a scale adjustment. So, we call the relative abundance of the most common species, $^{35}Cl^{35}Cl$, 100 instead of 0.57. Combining the different structures that have identical masses and adjusting the probabilities to a scale of 100 produce the relative-abundance numbers for the following table.

Mass of peak	M	$M + 2$	$M + 4$
Number of distinct isotopes × relative peak height	1 × 100	2 × 32	1 × 10
Relative abundance (base peak 100)	100	64	10

The three peaks will be in the ratio 100:64:10 for M, $M + 2$, and $M + 4$, respectively.

[1]For those of you familiar with probability theory, you will recognize that the total probability for all possible results should be unity. Indeed, if we use enough significant figures, 0.5741 + 0.1836 + 0.1836 + 0.0587 = 1.0000.

The table in Example 12.1 illustrates the general rule for such calculations:

> *The relative abundance of each peak is found by multiplying the relative probability of each isotopic isomer's mass times the number of distinct isomers having that mass.*

Table 12.3 shows the possible chlorine isotope patterns for up to four chlorines per molecule. Calculations of isotope patterns quickly get more com-

Table 12.3 Possible Chlorine Isotope Compositions in a Molecule[a]

$-Cl_2$	$^{35}Cl^{35}Cl$	$^{35}Cl^{37}Cl$	$^{37}Cl^{37}Cl$		
Mass	M	$M+2$	$M+4$		
No. Isomers	1	2	1		
$-Cl_3$	$^{35}Cl^{35}Cl^{35}Cl$	$^{35}Cl^{35}Cl^{37}Cl$	$^{35}Cl^{37}Cl^{37}Cl$	$^{37}Cl^{37}Cl^{37}Cl$	
Mass	M	$M+2$	$M+4$	$M+6$	
No. Isomers	1	3	3	1	
$-Cl_4$	$^{35}Cl^{35}Cl^{35}Cl^{35}Cl$	$^{35}Cl^{35}Cl^{35}Cl^{37}Cl$	$^{35}Cl^{35}Cl^{37}Cl^{37}Cl$	$^{35}Cl^{37}Cl^{37}Cl^{37}Cl$	$^{37}Cl^{37}Cl^{37}Cl^{37}Cl$
Mass	M	$M+2$	$M+4$	$M+6$	$M+8$
No. Isomers	1	4	6	4	1

[a]You might notice that the multiplicities (the number of isomers) for each composition are the coefficients of the terms for $(x + y)^n$ where n equals the number of chlorines. That is, they correspond to the coefficients in a Pascal's Triangle.

plicated as a molecule or fragment has increasing numbers of atoms with two or more isotopes.

It is interesting to note a peculiar effect in Figure 12.5. Look at the pattern for molecules that contain four chlorines. Unlike all the other patterns, the molecular ion, which is the leftmost (lowest mass) peak does not have the highest relative abundance. This result arises directly from the probabilities noted above.

Explain the Fragmentation Patterns

As you have observed in the illustrations of mass spectra, numerous peaks lie at m/z values less than that of the molecular ion. These peaks result from some of the molecules breaking up into fragments. The sets of fragments that appear and their relative abundances are called *fragmentation patterns*. The patterns depend on the identity (and chemistry) of the parent molecule and the techniques used to vaporize and ionize the sample. Such differences in fragmentation patterns can be seen in Figure 12.4. Notice especially the small amount of fragmentation in the mass spectrum from the ion-spray source. Also note that whatever type of source is used, many of the peaks in the three mass spectra are the same.

In fact, there is a well-studied descriptive chemistry of organic molecules in the gas phase, and, with practice, specialists can identify significant parts of the structures of unknown molecules from the mass spectrum alone. This skill is far beyond the level of expertise that we shall need here. However, even at an introductory level, structural information can be obtained from the *separations* (in mass units) of the peaks in the fragmentation pattern. We can identify the masses of some common **neutral fragments** that split off from the molecular ion. The name *neutral fragments* simply describes their electric charge. Neutral fragments break off from a **parent ion** that may or may not be the molecular ion. *Parent ion* is a generic term for an ion that fragments. For a singly charged parent, the fragmentation generates a neutral fragment and an ion with a lower molecular mass. Two types of simple fragmentation reactions are

$$M^{\bullet+} \rightarrow \{M \text{ minus fragment}\}^{+} + \text{fragment}^{\bullet}$$

and

$$M^{\bullet+} \rightarrow \{M \text{ minus fragment}\}^{\bullet+} + \text{fragment}$$

Table 12.4 Common Neutral Fragments Lost by Gas-Phase Ions

m/z Lost	Formula	Particular Sources of Loss
1	H	Labile H, aldehydes
15	CH_3	Favored from branched site
16	O	Sulfoxides, nitro compounds
16	NH_2	Amides, aromatic amines
17	OH	Acids, oximes
18	H_2O	Alcohols (primary), aldehydes, ketones, ethers
19	F	Fluoroalkanes
26	CN	Alkylcyanides
27	HCN	Cyanides, aryl-NH_2
28	C_2H_4	Alkanes, ethyl esters
28	CO	Aromatic carbonyls, phenols
29	CHO	Alcohols
29	C_2H_5	Favored from branched site, ethyl esters
30	NO	Aromatic nitro compounds
30	CH_2O	Aromatic methoxy compounds
31	CH_3O	Methyl esters
32	S	Sulfides, aromatic thiols
34	H_2S	Thiols
35, 37	Cl	Labile chloride
41	C_3H_5	Propylesters, propylamides
42	CH_2CO	Methylacetates
43	C_3H_7	Propyl groups
45	OC_2H_5	Ethyl esters
46	NO_2	Aromatic nitro groups
79, 81	Br	Alkyl bromides
127	I	Alkyl iodides

Mass losses of 4 to 14, 21 to 23, 33, 37, and 38 are unlikely. In addition to the above, the homologous series of most of the fragments also exist. These fragments differ by an additional C_nH_{2n}: masses 14, 28, 42, and so forth.

The phrases in brackets indicate that the ion has a mass less than the parent's mass; the difference is the mass of the lost fragment. Table 12.4 lists some neutral fragments that are commonly lost from organic molecules. Let us identify some of them in the examples of *n*-butane (Figure 12.2) and methanol (Figure 12.3).

EXAMPLE 12.2

For the example of *n*-butane, we have already identified the molecular ion as *m/z* 58. What fragments have been lost to produce the two other important peaks at 43 and 29?

Solution

The fragments lost have masses of 15 and 29, respectively. We recognize these losses as likely to be $CH_3^\bullet$ and $C_2H_5^\bullet$. The remaining masses, 43 and 29, are consistent with $C_3H_7^+$ and $C_2H_5^+$. Masses 28 and 27 correspond to $C_2H_4^{\bullet+}$ and $C_2H_3^+$.

EXAMPLE 12.3

As noted above, for methanol the neutral loss of an extremely labile proton from the alcohol causes the peak at m/z 31 to be larger than that of the molecular ion. To what fragment does the peak at mass 15 correspond?

Solution

Mass 15 is CH_3^+ and results from the loss of $O^\bullet$. If we thought that m/z 31 were the molecular ion, then the formula of the compound would be CH_3O, which is not a stable compound. The fragmentation pattern helps us make the correct assignment of the molecular ion.

12.4 Mass Spectral Resolving Power and Spectral Resolution

A figure of merit to describe the experimental capabilities of mass spectrometers is the **resolving power**. For mass spectrometry,

$$\text{resolving power} = \frac{M}{\Delta M} \tag{12-1}$$

In the equation, ΔM is the difference between the masses of two species, and M is the mass of the lighter one. With this definition of resolving power, a larger value means that smaller mass differences can be resolved. But this definition is not sufficient to denote the quality of the separation. Why not? The answer is we have not specified the experimental widths of the individual mass peaks. That information must be included. For instance the resolving power for an instrument can be reported as *resolving power xxx with 90% valley*. This means that the peaks are considered resolved when they appear as in Figure 12.6; the valley depth is 90% of the height of *equal peaks*. However, it is common to leave this information out, which results in some ambiguity. The figure of merit for the *instrument* is the unitless number resolving power.

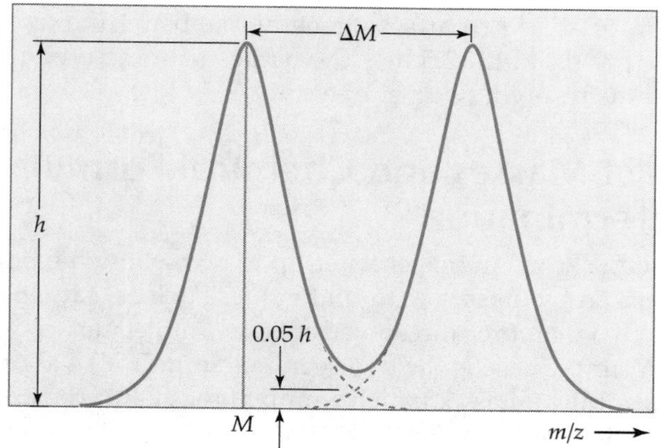

FIGURE 12.6 ▲
Illustration of a definition of resolution in mass spectrometry.
The resolution is defined experimentally as $M/\Delta M$ when the two identical experimental peaks overlap at 5% of their height producing a 90% valley between the peaks. Values you see reported may have the resolution defined with different valley depths. See Example 12.4 delineating resolution and resolving power.

Table 12.5 Commercial Spectrometer Figures of Merit

Type	Extremes of m/z	Resolving power
Double focusing	2–5000	20,000–100,000
Single focusing	1–1400	1500
Time of flight (Reflectron)	1–10,000	3500
Quadrupole	1–1000	1000
Ion trap	10–650*	~500
Fourier Transform	18–100,000¶	up to $10^{6\dagger}$

*Research instruments may be 10–5000.
¶Masses to 1 can be measured, but the frequencies are high and it is not commonly done.
†Depends on the time spent accumulating data and the mass range scanned.

When applied to the mass spectrum itself, the correct term is **resolution**. The resolution has units such as the **unified atomic mass unit** that is abbreviated *u*. These two figures of merit are distinguished in the following example.

EXAMPLE 12.4

Find the resolving power needed to distinguish $(^{14}N)_2$ from $^{12}C^{16}O$ in a gas mixture.

Solution

The resolution required distinguish the peaks of $(^{14}N)_2$ and $^{12}C^{16}O$ is 0.0057 *u*. Since the lighter species has mass 27.99, you would need an instrument with

$$\text{resolving power} = \frac{27.99}{0.0057} = 4900$$

The number calculated here says nothing about how deep the valley is between two experimental peaks of equal height in a mass spectrum.

The resolving power of a mass spectrometer is a function of its design and the quality of all of its parts. In general, the better the resolving power, the higher the price. Table 12.5 lists the names of some types of mass spectrometers and the resolving power of each.

12.5 Exact Masses and Chemical Formula Determination

The mass spectra shown in this chapter up to now show the peaks as if each were characterized by masses with unit values such as, say, 48 or 70. In fact, there may be two or more unresolved, distinct molecules contributing to a single peak. A simple case is the one seen in Example 12.4 above, $(^{14}N)_2$ and $^{12}C^{16}O$; both appear at mass 28, their nominal mass. But the exact masses of atoms are not integer values. The exact masses of some isotopes are listed in Table 12.6. From these isotope masses, more exact molecular masses can be calculated for $(^{14}N)_2$ and $^{12}C^{16}O$, namely, 28.00061 and 27.99491 *u*.

Therefore, if a measurement of the molecular mass is made accurately enough, we can determine the molecular formula from the mass measurement alone. This is the basis of high-resolution mass spectrometry. Experimentally, a high-quality instrument must be used.

Table 12.6 Exact Nuclidic Masses*

Isotope	Atomic Mass	Isotope	Atomic Mass
^{1}H	1.00782522	^{19}F	18.9984046
^{2}H	2.01410222	^{28}Si	27.9769286
^{12}C	12.00000000	^{31}P	30.9737633
^{13}C	13.00335508	^{32}S	31.9720728
^{14}N	14.00307440	^{35}Cl	34.96885359
^{16}O	15.99491502	^{79}Br	78.9183320
^{18}O	17.99915996	^{127}I	126.9044755

*Superscripts are nominal masses.

EXAMPLE 12.5

A molecule is found to have an exact molecular mass of 203.1522. If the molecule can contain only C, H, N, and/or O, what is its formula?

Solution

By looking in tables such as Table 12.7, we find that the formula is $C_{10}H_{21}NO_3$. The molecule is most likely *n*-decyl nitrate. Incidently, the molecular weight listed in a handbook is 203.28. The difference is due to the natural isotopic distribution: The exact molecular mass applies to the isotopically pure species $[^{12}C]_{10}[^{1}H]_{21}{}^{14}N[^{16}O]_3$.

12.6 Specific Techniques in Mass Spectrometry

Inductively Coupled Plasma Source Mass Spectrometry (ICP/MS)

Inductively coupled plasma torches have become the sources of choice for quantitative trace and ultratrace elemental analysis by mass spectrometry. The torch operates at temperatures around 3,000 K, and the molecules in the solution samples are broken down into their elemental forms. Some fraction of each elemental species is simultaneously ionized. The monocationic (1+) form is preferred for the assay, and conditions are set to enhance that form. Because the transducer responds similarly to any ion striking it, most ions are relatively equally detectable. The combination of an inductively coupled plasma and mass spectrometer is abbreviated ICP/MS. ICP/MS has become one of the two most broadly used methods for elemental analysis, with the other being ICP atomic emission spectrometry. An illustration of an ICP/MS output plot is shown in Figure 12.7.

The ICP torch was introduced in Section 11.11 and Figure 11.14b.

The limits of detection of ICP/MS are listed in Appendix VIII.

High-Mass Mass Spectrometry

Vaporization/ionization methods now exist for molecules greater than 10,000 daltons. Only a few research instruments in the past could scan up to 10,000 daltons for singly charged ions. However, **electrospray** sources have expanded the capabilities of mass spectrometry up to the 100,000 *u* range on commonly available mass spectrometers. This ability arises since electrospray ionization (ESI) introduces large molecules into the gas phase with

The spray of an electrospray can be seen in Photo EP.11 inside the back cover.

Table 12.7 Molecular Ions with Exact Masses between 203.0000 and 203.2000 (Limited to molecules containing C, H, and less than 4 N and 4 O only.)

(Mass − 203.0000)	Empirical Formula	(Mass − 203.0000)	Empirical Formula
0.0007	$C_{13}HNO_2$	0.1032	$C_8H_{15}N_2O_4$
0.0093	$C_9H_3N_2O_4$	0.1060	$C_{11}H_{13}N_3O$
0.0120	$C_{12}HN_3O$	0.1072	$C_{13}H_{15}O_2$
0.0133	$C_{14}H_3O_2$	0.1158	$C_9H_{17}NO_4$
0.0218	$C_{10}H_5NO_4$	0.1185	$C_{12}H_{15}N_2O$
0.0246	$C_{13}H_3N_2O$	0.1271	$C_8H_{17}N_3O_2$
0.0331	$C_9H_5N_3O_3$	0.1284	$C_{10}H_{19}O_4$
0.0344	$C_{11}H_7O_4$	0.1298	$C_{11}H_{15}N_4$
0.0359	$C_{12}H_3N_4$	0.1311	$C_{13}H_{17}NO$
0.0371	$C_{14}H_5NO$	0.1396	$C_9H_{19}N_2O_3$
0.0457	$C_{10}H_7N_2O_3$	0.1424	$C_{12}H_{17}N_3$
0.0484	$C_{13}H_5N_3$	0.1436	$C_{14}H_{19}O$
0.0497	$C_{15}H_7O$	0.1509	$C_8H_{19}N_4O_2$
0.0570	$C_9H_7N_4O_2$	0.1522	$C_{10}H_{21}NO_3$
0.0583	$C_{11}H_9NO_3$	0.1549	$C_{13}H_{19}N_2$
0.0610	$C_{14}H_7N_2$	0.1635	$C_9H_{21}N_3O_2$
0.0695	$C_{10}H_9N_3O_2$	0.1648	$C_{11}H_{23}O_3$
0.0708	$C_{12}H_{11}O_3$	0.1675	$C_{14}H_{21}N$
0.0736	$C_{15}H_9N$	0.1761	$C_{10}H_{23}N_2O_2$
0.0821	$C_{11}H_{11}N_2O_2$	0.1801	$C_{15}H_{23}$
0.0861	$C_{16}H_{11}$	0.1873	$C_9H_{23}N_4O$
0.0934	$C_{10}H_{11}N_4O$	0.1886	$C_{11}H_{25}NO_2$
0.0947	$C_{12}H_{13}NO_2$	0.1999	$C_{10}H_{25}N_3O$

Data source: Benyon, J. H. 1960. *Mass Spectrometry and Its Applications to Organic Chemistry*. Amsterdam: Elsevier.

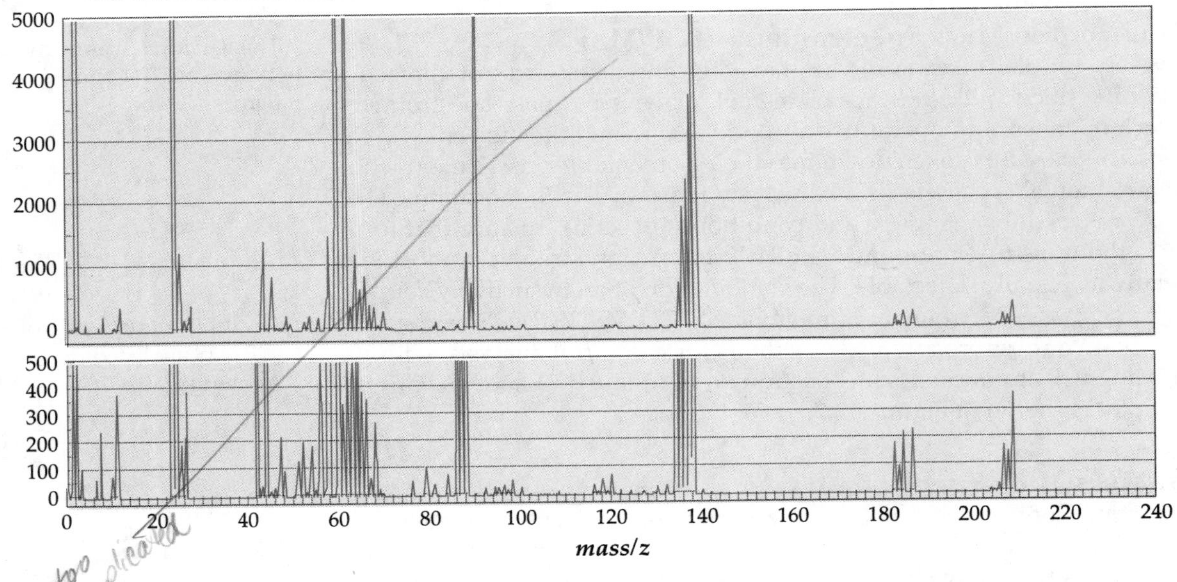

FIGURE 12.7 ▲
An ICP-MS spectrum of a mixture of ions.
Two different scales are shown to encompass the wide response of the technique. On the scales shown, 3000 counts is approximately 10 ppb of each isotope.

ANALYTICAL CHEMISTRY IN ACTION
Heavy Metal (in Clay)

The ability to carry out high-speed, multielement analysis by inductively coupled plasma mass spectrometry (ICP/MS) provides the possibility of answering essential, basic questions about metal ions (some toxic) in soils. For example, What is their mobility and availability to be taken up by crops, and how do these properties change with soil conditions? Without the possibility to do large numbers of analyses inexpensively, it would not be feasible to separate the natural background levels of the metal ions from heavy metal contamination. The reason is, a great number of samples are needed to map the concentrations over a wide enough area and at enough different depths. Metals of significance include Cu, Zn, Se, Pb, Cd, V, Cr, Co, Ni, Se, Sn, Sb, and Tl.

The reference cited here shows that significant correlations exist between the clay content of a soil and levels of trace elements in the soil. The silicates in the clay apparently adsorb the ions, and the highest concentrations of the metals are found in the heavy clay soils. In agreement with that point, when the surface was depleted of clay, the heavy metal levels were lower on the surface than below. On the other hand, because of adsorption by the clay minerals, only 5% to 50% of the trace elements included in fertilizers were available to the plants after the majority is adsorbed by the soil minerals. The study further showed that Zn and Cd were higher on the surface than at lower depths, a result likely to be due to 30 to 40 years of phosphate fertilizer application.

Studies such as this one provide a basis for the descriptive chemistry of soils, both for environmental assessment and for farming.

[Ref: Mermut, A. R., Jain, J. C., Song, L. Kerrich, R., Kozak, L. and Jana, S. 1996. "Trace Element Concentrations of Selected Soils and Fertilizers in Saskatchewan, Canada." *Journal of Environmental Quality.* 25:845–853.] ■

multiple charges on them. Since m/z determines what ions are detected, when z is large, m/z is brought into a "lower mass" range, which can be detected. The limitation in obtaining mass spectra, then, lies only in whether such large molecules can be brought intact into the gas phase. Figure 12.8 shows an ESI-MS of a biological polymer, the protein cytochrome *c*.

Figure 12.9 shows an example of a **Matrix-assisted laser desorption/ionization (MALDI)** spectrum. The analyte molecules are dissolved in a special matrix, and a pulsed laser vaporizes the matrix and simultaneously ejects the analyte molecules into the gas phase. MALDI mass spectra are sensitive to both the method of forming the matrix and its composition. Such differences in the mass spectra are obvious from the figure. The differing conditions are listed in the figure caption.

Not only can molecules as large as proteins be the subject of mass spectral studies, but short polypeptides can be sequenced. Sequenced simply means that the order of the amino acids of the segment can be determined.

Tandem Mass Spectrometry: MS/MS

With the right equipment, one species of ion that is separated by mass spectrometry from all the others produced can be fragmented and the fragments transmitted to another mass spectrometer for further analysis. Probably the

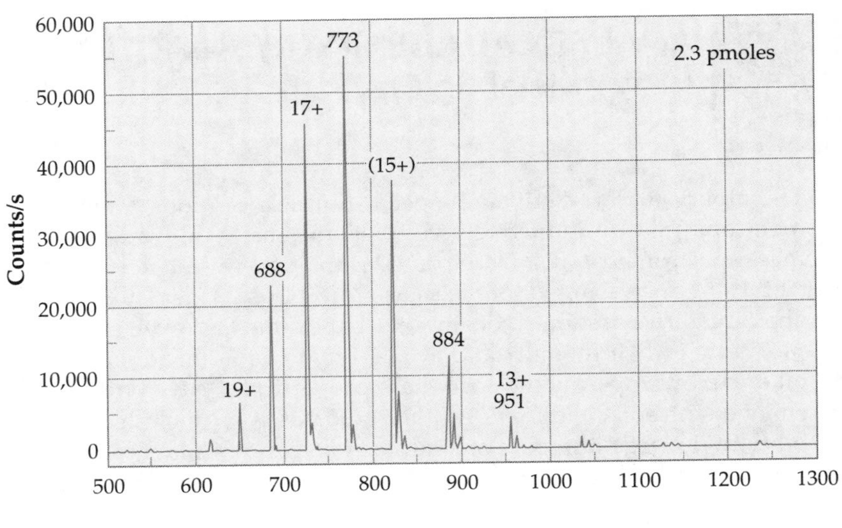

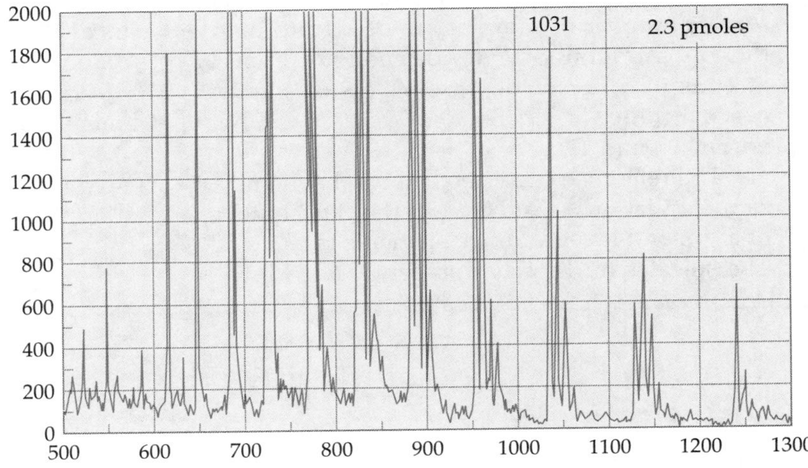

FIGURE 12.8 ▲
Example of an electrospray ionization mass spectrum.
This is of cytochrome *c*, a protein. The spectrum was obtained in 90 s on a flowing solution that had a concentration of 1.5×10^{-6} M in the protein. The two spectra shown are at two different scale expansions in counts s^{-1}. The number of charges on the ions is indicated over the respective peaks. [Ref: Smith, R. D., Loo, J. A., Edmonds, C. G., Barinaga, C. J., Udseth, H. R. 1990. *Anal. Chem.* 62:882–899.]

most clearcut way to visualize the instruments for MS/MS is shown in Figure 12.10, where the sample is introduced through the source, separated, passed through an aperture into a chamber where a given fragment is further fragmented, and then analyzed in the second mass spectrometer. This sequential mass spectrometry is an exquisitely powerful analytical method, called **tandem mass spectrometry**. *Any* initial ion is called the **parent ion**, which is somewhat confusing since for conventional mass spectrometry only one ion is labeled that way. The fragment ions that pass to the second mass spectrometer are called **product ions**. (Until recently, these usually were

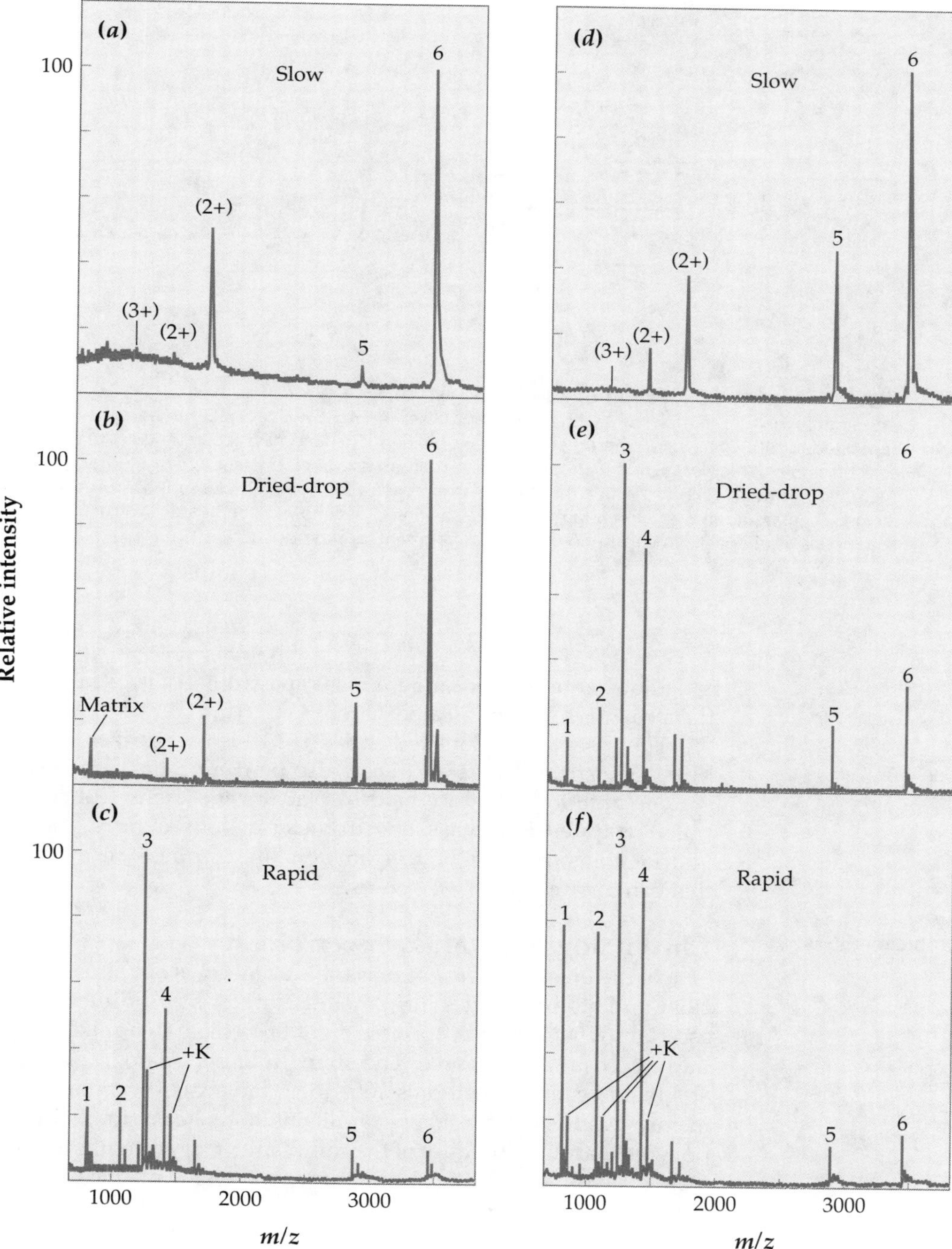

FIGURE 12.9 ▲

Illustration of the influence of the matrix composition and growth conditions on the MALDI mass spectrum of a protein.

The recombinant protein has a known molecular weight of 10,826. Part of the sample was cleaved by proteolysis using an enzyme that cleaves the protein at glutamate residues. The digest was crystallized in two different matrices, each under three different sets of conditions. The peaks labelled 1–6 are the six different singly protonated fragments, the (2+) and (3+) are multiply protonated fragments, and the +K indicates fragments with associated potassium, AW 39. [Ref: Cohen, S. L., Chait, B. T. 1996. *Anal. Chem.* 68:31–37.]

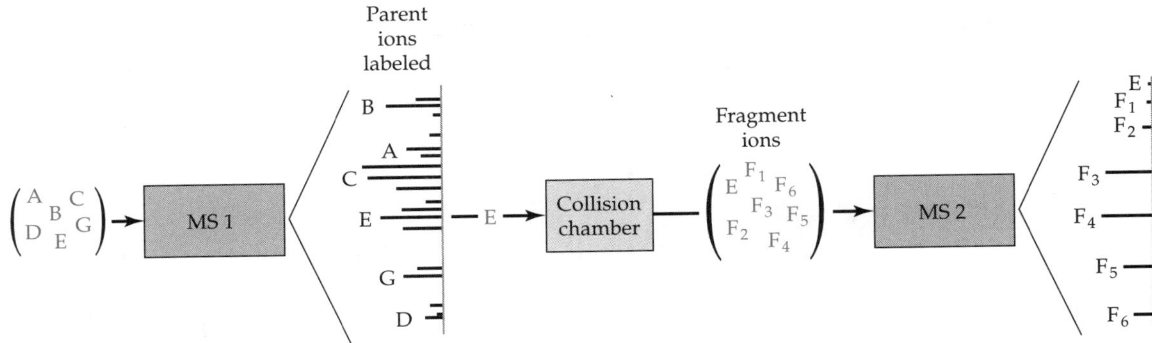

FIGURE 12.10 ▲
Illustration of the operation of a MS/MS system.
The sample mixture of molecules with molecular masses B > A > C > E > G > D can be separated, but the parent ions and their fragments overlap. The first mass separator is set to select a single parent ion, which is passed into the collision chamber where it undergoes fragmentation through collisions with a gas such as argon or helium. For example, suppose E is selected. The parent ion E and ion fragments F_1–F_6 pass into a second mass separator, which produces a mass spectrum consisting of E and its fragmentation products.

called **daughter ions**.) Tandem mass spectrometry is used mainly to unravel the structures of complicated parent ions.

The instrument depicted in Fig. 12.10 is categorized as a **transmission** MS/MS instrument. MS/MS can also be run on instruments that can measure m/z on ions that are held in place inside them. Collectively, these are called **trapped-ion** mass spectrometers. Both quadrupole ion trap and Fourier-transform mass spectrometers are trapped-ion mass spectrometers that can be run in MS/MS mode.

Chromatographic methods; Chapter 13.

Chromatography/Mass Spectrometry

One of the great strengths of mass spectrometry is our ability to identify molecular and atomic species from the MS data. A major difficulty, however, is to ascertain what different species might be present in a mixture of compounds introduced into the source. The strength of chromatographic methods is the ability to separate materials into their molecular components. However, a major weakness of chromatographic methods lies in identifying all the components that elute. Because of the complementary strengths and weaknesses of mass spectrometry and chromatography, the use of a mass spectrometer as a detector to identify chromatographically separated components of unknown mixtures has become probably the most powerful class of analytical methods.

Mass spectrometry is used as a detector for gas, liquid, and supercritical fluid chromatographic columns. Unfortunately, the effluent cannot be fed directly to the inlet of a mass spectrometer. **Interfaces** between the chromatographic columns and the mass spectrometer are needed. The purpose of the interfaces is to get rid of as much of the matrix (in other words, the mobile phase) as possible while keeping as much of the analytes as possible. Such a task is not particularly easy, and, in general, only one part in a thousand to one part in a million of the analyte eluting from chromatographic columns gets into the mass spectrometer.

ANALYTICAL CHEMISTRY IN ACTION
Silicon Valleys

When manufacturing microchips, the surfaces of silicon wafers are carefully cleaned to ensure reproducible production. For example, the cleaning procedure has an influence on process steps like thermal oxidation to form SiO_2 insulation layers and molecular beam epitaxy to modify the chemistry of the surface to form the transistors. The two most widely used cleaning processes are an "HF dip," which forms a hydrophobic surface containing SiH groups, and the "RCA cleaning process," which results in a hydrophilic surface of SiOH groups. To enable better control of the manufacturing process, these two surfaces were characterized by mass spectrometry of the surface elements. By alternately making the MS measurements and eroding the surface, a depth profile of the elements can be made. The depth profiles for a number of important elements after both types of cleaning are shown in the adjacent figures.

Of special interest are the organic contents of the hydrophobic surface. The fragments $C_2H_3^+$ and SiC^+ are found, both of which result from hydrocarbons under the conditions of the experiment. Also, K^+ is enriched at the hydrophobic surface. On the other hand, Mg^{2+} and Na^+ are enriched on the hydrophilic surface, perhaps incorporated into the silica network. Under the manufacturing conditions, these components might react and migrate deeper into the surface and, perhaps, change the semiconductor or insulator properties.

[Ref: Stingeder, G., Grundner, M., and Grasserbauer, M. 1988. "Investigation of Surface Contamination on Silicon Wafers with SIMS." *Surface and Interface Analysis.* 11: 407–413.]

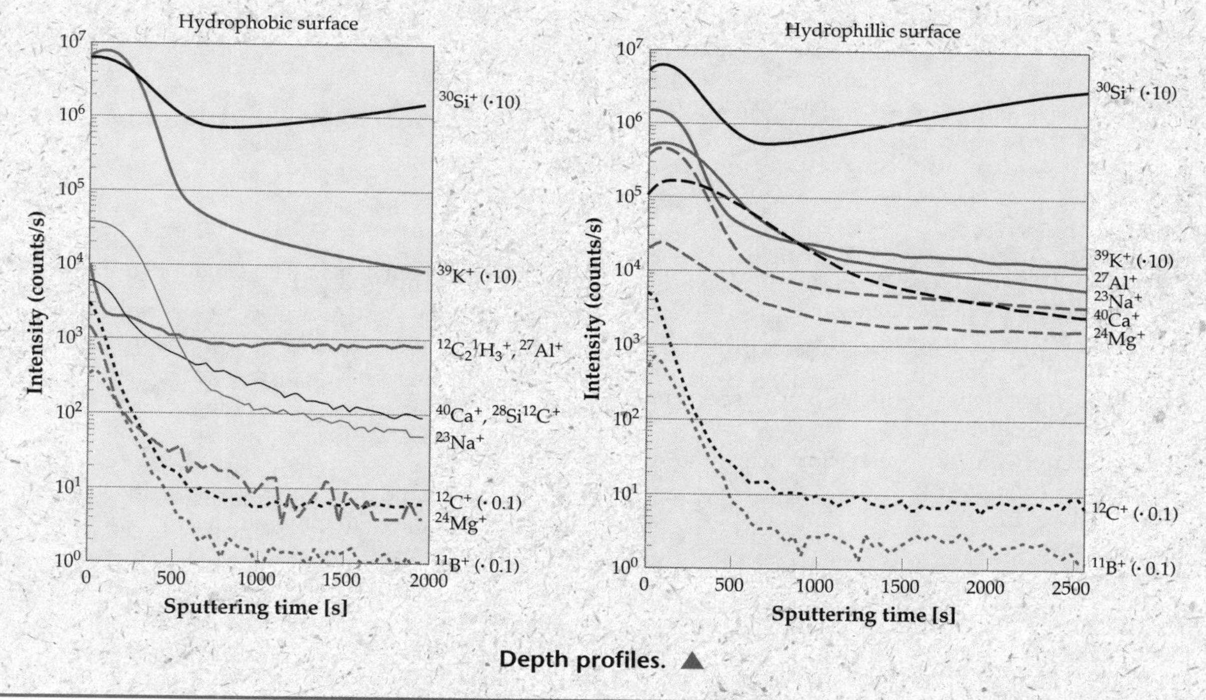

Depth profiles. ▲

Spatial Distribution by Mass Spectrometry: SIMS

Energetic (keV range) atomic or molecular particles impinging on a solid or liquid target cause ejection of atoms, clusters of atoms, and ions from the target's surface. This eroding of material from the surface leaves a pit, and the ejection/erosion process is called **sputtering**. The sputtered material, once in the gas phase, can be mass analyzed just as from any MS source.

Examples of the bombarding particles are ions such as Cs^+ or O_2^+ in the energy range 0.5–100 keV. These energetic ions are called **primary ions**, and any ions that are ejected from the surface are called **secondary ions**. Mass spectrometry of the products resulting from the ion bombardment is called **secondary-ion mass spectrometry**, abbreviated SIMS. The samples on which the technique is used are generally solids. However, since the material that desorbs comes only from the top few molecular layers, the technique is a method of surface analysis. Qualitative elemental analysis can be done with the mass spectra alone. However, the fraction of each species that is ejected as ions depends greatly on the matrix.

Suggestions for Further Reading

SILVERSTEIN, R. M., BASSLER, G. C., MORRILL, T. C. 1991. *Spectrometric Identification of Organic Compounds*. 5th ed. New York: John Wiley & Sons.
Clearly written introductions to organic-molecule mass spectrometry with electron ionization.

MCLAFFERTY, F. W. 1980. Interpretation of Mass Spectra, 3rd ed. Mill Valley, CA: University Science Books.
The classic introduction to identification of small organic molecules by mass spectrometry.

HARRISON, A. G. 1983. *Chemical Ionization Mass Spectrometry*. Boca Raton, FL: CRC Press.
Monograph describing the gas-phase chemistry and equipment of mass spectrometry with chemical ionization. Numerous examples of the mass spectra.

JARVIS, K. E., GRAY, A. L., HOUK, R. S. 1992. *Handbook of Inductively Coupled Plasma Mass Spectrometry*. Glasgow: Blackie.
An outstanding book on the practice of ICP-MS for element analysis. Good chapter on sample preparation.

BRIGGS, D., BROWN, A., VICKERMAN, J. C. 1989. *Handbook of Static Secondary Ion Mass Spectrometry*. Chichester: John Wiley & Sons.
A useful, brief introduction with lots of examples and numerous, interesting case studies of surface organic-molecule SIMS.

Concept Review

1. What are the axes on a typical mass spectrum?
2. What is meant by the following terms?
(a) base peak
(b) molecular ion
3. What are the three steps in utilizing a mass spectrum to help identify a molecule?
4. What MS source would be suitable for each of the following types of analytes?
(a) protein (MW = 10,000)
(b) glycylglycine
(c) an aqueous sample containing a mixture of heavy metal ions
(d) an experimental drug with a molecular mass of 237
5. What do you do to molecules to ionize them in each of the following sources?
(a) MALDI
(b) EI
(c) ESI
(d) ICP/MS
6. How can a tandem mass spectrum help to verify the identity of a molecular ion produced in the first stage of a MS/MS instrument?

Exercises

12.1 The mass spectra in Figure 12.1.1 (I, II, III, and IV) are of important industrial chemicals that are regulated in the workplace. (For spectral data, see Table 12.1.1.) Identify the molecular formula of each and the structure if possible. [Ref: Middleditch, B. S., et al. 1981. *Mass Spectrometry of Priority Pollutants*. New York: Plenum.]

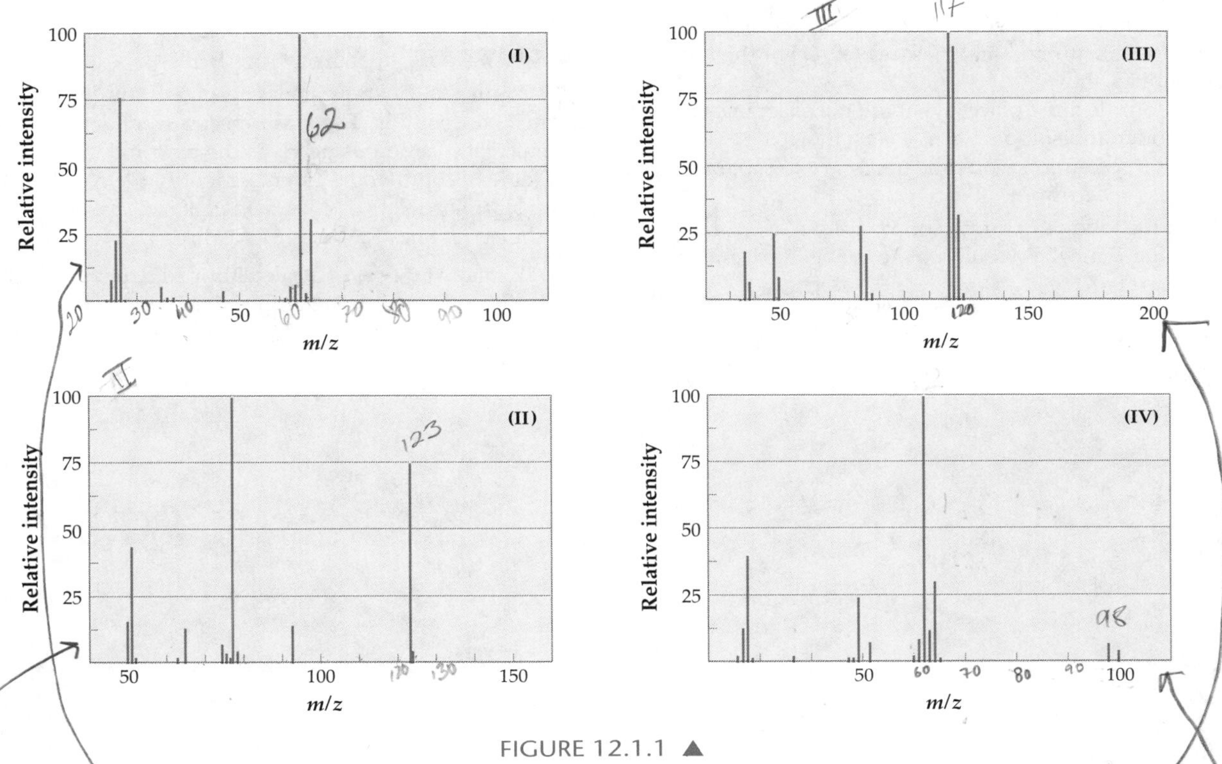

FIGURE 12.1.1 ▲

Table 12.1.1 Spectral Data

I

Mass	Abundance	Mass	Abundance
24	2.4	48	2.0
25	10.3	49	1.7
26	24.2	59	2.0
27	76.4	60	6.5
28	2.3	61	8.4
35	7.3	62	100.0
36	2.7	63	5.0
37	2.6	64	30.8
47	4.8		

III

Mass	Abundance	Mass	Abundance
35	18.7	121	30.7
37	6.6	123	3.2
47	25.6		
49	8.9		
82	28.3		
84	18.3		
86	2.9		
117	100.0		
119	93.7		

II

Mass	Abundance	Mass	Abundance
50	16.0	78	6.7
51	43.8	93	14.8
52	2.5	123	75.6
63	2.1	124	5.4
65	14.0		
74	7.9		
75	4.6		
76	3.7		
77	100.0		

IV

Mass	Abundance	Mass	Abundance
25	3.0	51	7.8
26	13.1	60	3.2
27	39.8	61	8.8
28	3.3	62	100.0
35	4.2	63	12.5
36	2.6	64	31.1
47	2.1	65	3.2
48	2.2	98	7.3
49	24.9	100	4.5

12.2 Figure 12.2.1 illustrates the structures of four compounds with nominal masses of 194. For the four compounds shown

(a) To four decimal places what are the exact masses of the most abundant isotopic form of each molecule?

(b) What resolution would be necessary to separate them in a mass spectrum?

12.3 If the following groups or atoms are lost from a molecular ion, what would be the change in m/z measured (in integral u)?

(a) CH_3

(b) phenyl

(c) *t*-butyl

12.4 Figure 12.4.1 shows mass spectra of four closely related compounds. The mass spectra were obtained by chemical ionization using H_2 as the proton donor. Using the table of common fragments, identify the compounds. [Ref: Harrison, A. G. 1983. *Chemical Ionization Mass Spectrometry*. Boca Raton, FL: CRC Press.]

12.5 Explain, by way of a calculation like that in the example on page 388, the origin of the relative intensities of the peaks for $-Cl_3$ and $-Cl_4$ as shown in Figure 12.5.

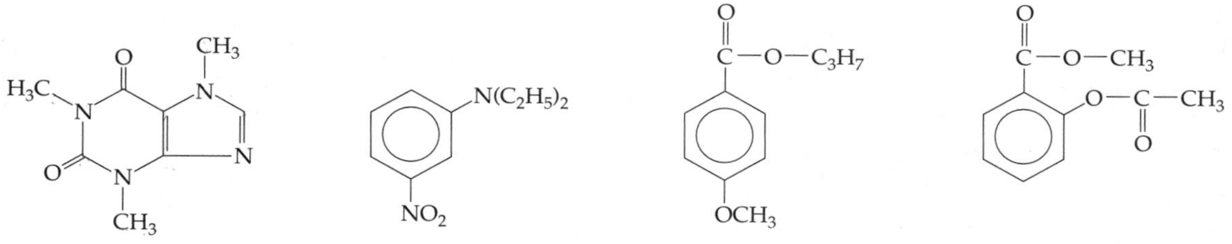

(a) *(b)* *(c)* *(d)*

FIGURE 12.2.1 ▲

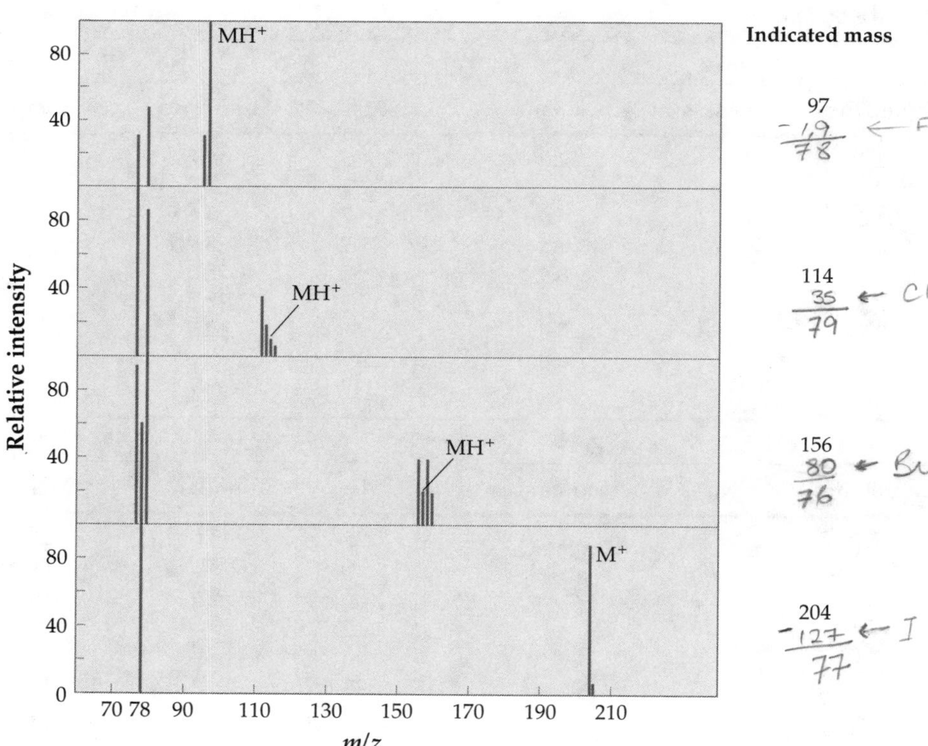

FIGURE 12.4.1 ▲

Additional Exercises

12.6 Using the data in Table 12.2, calculate the relative abundances you would expect to see due to the bromine isotopes for the peaks corresponding to the molecular ions for a $-Br_3$ compound.

12.7 For the compound C_6H_5Cl,

(a) What are the exact masses of $^{12}C_6{}^{1}H_5{}^{35}Cl$ and $^{12}C_6{}^{1}H_5{}^{37}Cl$?

(b) What resolution would be necessary to separate them in a mass spectrum?

12.8 A mass spectrometer has a resolving power of 20,000. To what value of ΔM is this equivalent if m/z is

(a) 300?

(b) 1000?

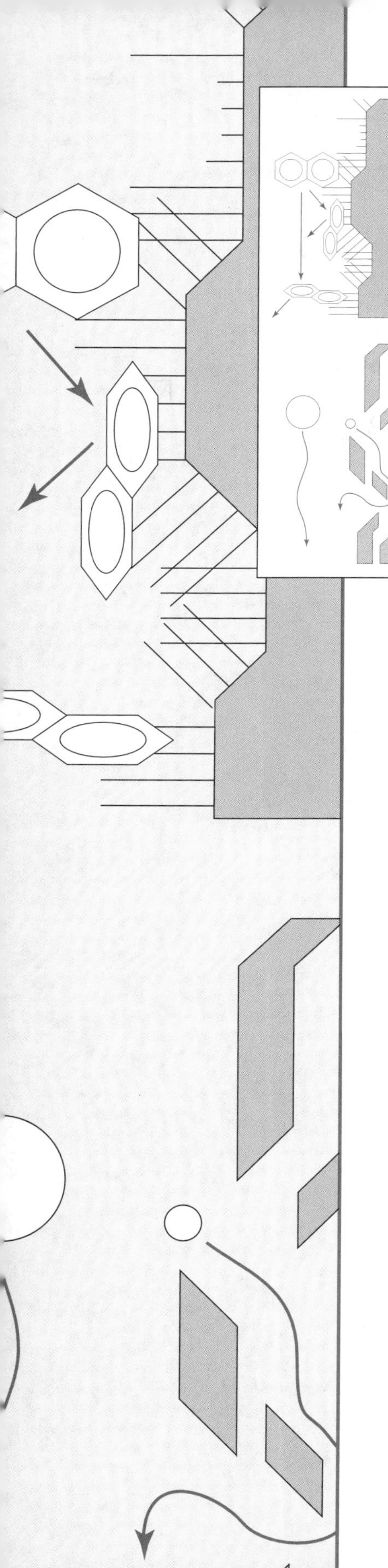

CHAPTER 13

Separations and Chromatography

Overview

A Deeper Look

13.1 The Chromatography Experiment

Chromatography is the science and art of separating the components of materials from each other. Such separations are achieved using a wide variety of techniques, and the molecular differences upon which the separations are based are quite diverse. For example, molecules can be separated by their differences in molecular charge, molecular size, molecular mass, bond polarities, redox potential, ionization constants, and arrangement of bonds such as isomer structure or chirality.

Not generally included in chromatography are separations that use electric fields to drive charged molecules so that they separate. These other methods fall under the class called **electroseparation** or **electromigration** and **electrophoresis**. You will find them described in Chapter 14.

A quote from Michael Tswett, the originator of contemporary chromatography, will be used to introduce the concept of chromatography. In this quote, he uses the word **adsorbent**. This is a solid material to which molecules bind on the surface. A **chromatographic column** is illustrated in Figure 13.1.

Michael Tswett

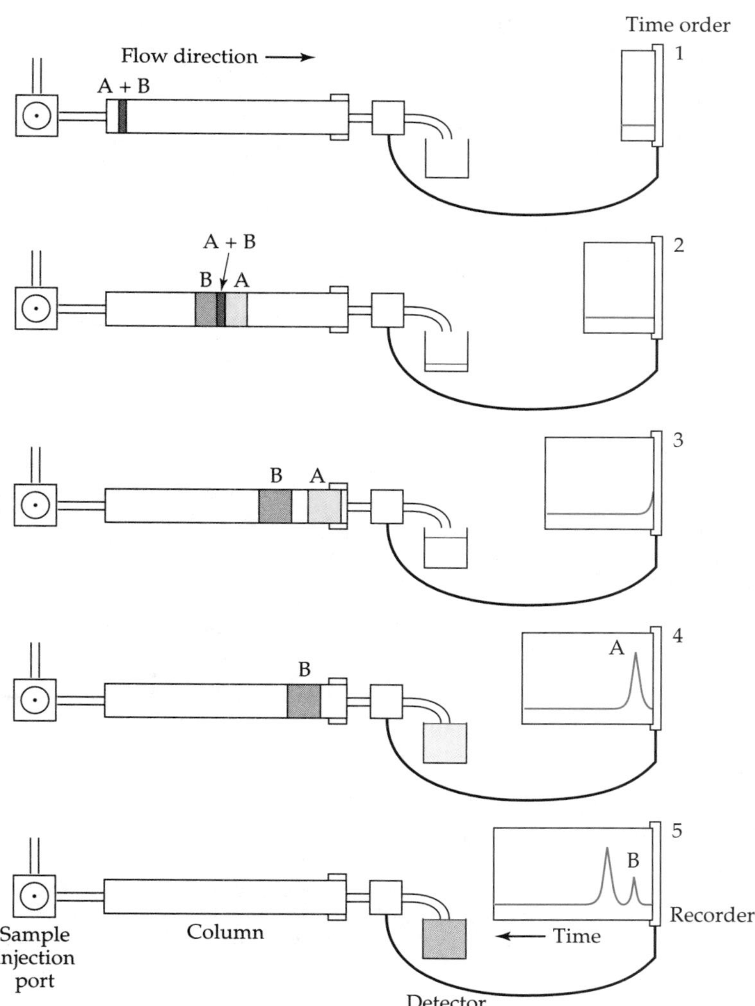

◀ FIGURE 13.1
Illustration of a column liquid chromatographic experiment.
The system is shown at different times (not at regular intervals) after the sample enters the column. The equipment shown includes a column packed with a solid adsorbent. The mobile phase passes through this column—here from left to right. Connected at the head end of the column is an injection port, which allows the sample to be introduced into the flowing stream without leakage and without stopping the flow. The sample, composed of mixed components A and B, is carried through the column. The materials that elute from the column pass through a detector, the response of which is printed on a strip-chart recorder with the paper moving at a constant speed. This is the **chromatogram**. Note that the concentrations of A and B in the eluted volume are always less than in the injected volume. Chromatographic separations always cause dilution of samples.

In this case, it is a tube packed with adsorbent through which flows a liquid which is called the **eluent**. The analyte is placed at the top of the column and washed through with the liquid eluent. The components are said to be **eluted**: They are carried through the packed column by the flowing liquid. In this example, the eluent is petroleum ether, which is a low-boiling-point, organic solvent. Tswett's column was a vertical tube open at the top with the eluent driven under its own weight through the adsorbent.

> If a petroleum ether solution of chlorophyll is filtered through a column of an adsorbent (I use mainly calcium carbonate which is stamped firmly into a narrow glass tube), then the pigments, according to the adsorption sequence, are resolved from top to bottom into various colored zones. . . . Like light rays in the spectrum, so the different components of a pigment mixture are resolved on the calcium carbonate column according to a law and can be estimated on it qualitatively and quantitatively. Such a preparation I term a chromatogram, and the corresponding method, the chromatographic method.
>
> It is self-evident that the adsorption phenomena described are not restricted to the chlorophyll pigments, and one must assume that all kinds of colored and colorless chemical compounds are subject to the same laws. [*Ref*: Tswett, Michael, 1906. As translated and quoted in *J. Chem. Ed.* 1959. 36:144 and *Ibid*. 1967. 44:235.]

Tswett used his eyes as a detector since the compounds he separated were colored, and he could detect the **bands** or **zones** of different materials by their colors. As the compounds separated, there were bands of color on his column from the numerous components in the sample. This experiment showed that chlorophyll is only one of many pigments found in plant leaves.

Tswett's simple column has evolved into the contemporary equipment illustrated in Figure 13.2.

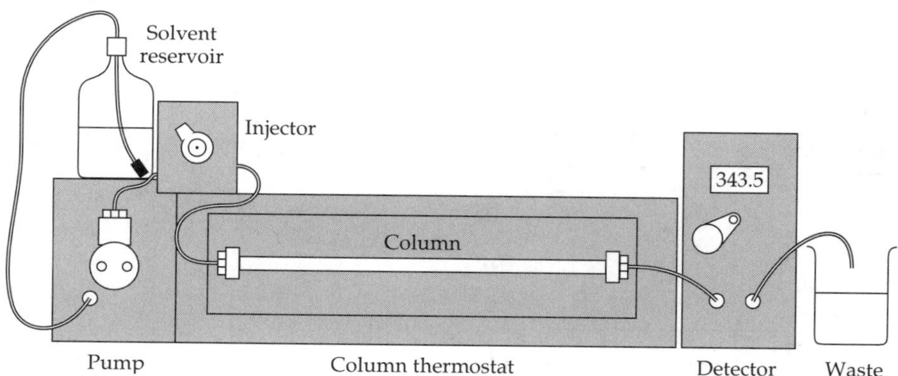

FIGURE 13.2 ▲
Diagram of the fluid-handling components for high performance liquid chromatography (HPLC).

The pressures needed to obtain an adequate flow through a bed of particles in the 3-μm to 10-μm range require that the pump piston press directly on the solvent. To keep bubbles from forming, the solvent should be degassed. For reproducibility, the column should be thermostatted. More details of the construction and operation of a loop injector are shown in Figure 4E.2.

13.2 Nomenclature of Chromatographic Separations

Chromatographic separations can be carried out in the liquid and gas phases. In both **liquid chromatography (LC)** and **gas chromatography (GC)**, the sample is introduced rapidly into a moving fluid phase—a liquid or gas, respectively—more generally called the **mobile phase**. In common usage, the term *eluent* refers to liquids. However, the mobile phase may also be a fluid that is not classifiable as a gas or a liquid: a supercritical fluid. That is the basis for the name **supercritical fluid chromatography (SFC)**.

Supercritical fluids are introduced in Section 4C.

During all separations, the components of the sample are carried by the mobile phase through a volume or layer of particles of solid material. The solid material has a number of different common names. Among these are **solid support, sorbent, static phase,** and **packing material**. Probably the most widely used term is **stationary phase**.

In contemporary analysis, instead of using our eyes, we can detect the components that have been separated by measuring changes in a number of different chemical or physical properties. Three examples are conduction of electrical current, light absorption, and ability to conduct heat. These are only a few. Essentially, measurement of *every* chemical and physical property of the analytes has been exploited to detect the components separated chromatographically.

An eluent that has passed through a column is often called the **effluent**. The results of chromatographic separations can be obtained by collecting the effluent in a series of **fractions** and carrying out further tests on them. For instance, we might let the effluent drip into a test tube and change the tube each time 1 mL was collected. Then, a chemical test could be run on each tube to see when the desired component eluted from the column.

On the other hand, we can make the measurements continuously. Using the example of chlorophyll, instead of simply ascertaining the color of the effluent, we might pass the liquid through an instrumental detector. The transducers of the detectors are placed at the end of the column, as seen in Figures 13.1 and 13.2.

The detector registers the changes in some property of the eluent passing through or by it. An example is the changes with time in the proportions of 610-nm-wavelength light absorbed by material in the effluent as the liquid passes through. You will observe a rise and fall (or fall and rise) from the baseline level in the level of light absorbed. (The change occurs as each component arrives, flows through, and leaves the detector.) As plotted on a moving chart or stored in a file on your computer and plotted later, it will be evident that a series of peaks appears over time. It is this chart of detector response vs. time that is now called a chromatogram (see Figure 13.1) in contrast to Tswett's nomenclature.

The wide applicability of chromatographic methods results from the variety of conditions that can be used to separate components of materials. For instance, we can use different mobile phases (gases, liquids, supercritical fluids) and different stationary phases (minerals, organic or inorganic polymers, or liquid-coated solids). In addition, we can change the geometry of the stationary phase (inside columns or on plates) or the amount of it (large for preparative and smaller for analytical). The techniques and specific names of these chromatographies will be considered in detail later. But remember, all of them have a stationary phase with a mobile phase flowing past it as well as some way to introduce the sample and to detect the presence of separated components.

13.3 Descriptions of Experimental Chromatograms

Figure 13.3 shows chromatograms obtained using liquid chromatography. The first two show individual solution samples. These two samples were mixed, and the third chromatogram was run on the mixture. The vertical axis is proportional to the response from a detector as the liquid effluent from the chromatographic support passed by.

In the chromatogram, the horizontal axis is volume or, equivalently, time. However, in much of modern chromatography the fluid **flow rate** (that is, volume/time) is kept constant, and time is used as a label for the horizontal axis. The relationship is simply

$$\text{flow rate} \times \text{time} = \text{volume} \tag{13-1}$$

The plot of time was made by using a strip chart recorder which passes paper at a constant rate under a pen that follows the response of the detector. The horizontal plot of Figure 13.3 can be calibrated for volume simply by measuring the flow rate of the effluent and knowing the speed of the plotter.

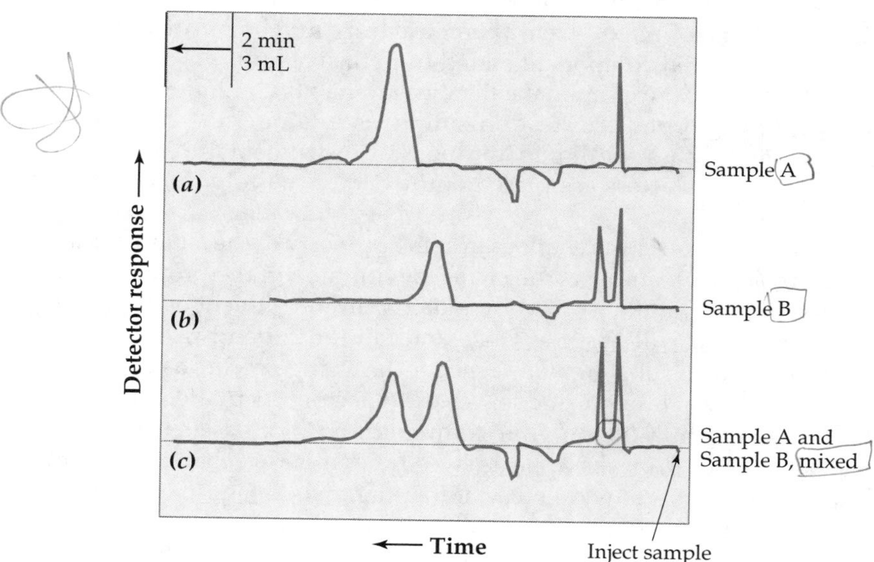

FIGURE 13.3 ▲
Chromatograms showing the independence of components in chromatography.

Chromatograms **(a)** and **(b)** are from two different samples. When these two samples are mixed and run **(c)**, all the peaks appear and remain in the same positions in time as they did in the two original samples. The sample A is from a nerve toxin that sometimes occurs in shellfish. Sample B is a chemically modified form of A. The components were separated using liquid chromatography. [Ref: Rubinson, K. A. 1982. *Biochim. Biophys. Acta* 687:315–320.]

The chromatogram that results from a blank sample establishes the baseline. Here, it is the straight line drawn in each graph. Notice that some of the chromatographic peaks fall below the baseline. This is merely a characteristic of the type of detector being used.

What do the peaks in the chromatogram mean? Since, in this case, the detector response is linearly proportional to the concentration of materials in the mobile phase, the shape of each peak shows the concentration distribution of the component associated with the peak. Often we simply say *peak* or *band* as shorthand for the material that is associated with the output response that forms a peak in the chromatogram. Let us be consistent here and use *peak* to refer to the chromatogram and *band* to refer to the material itself.

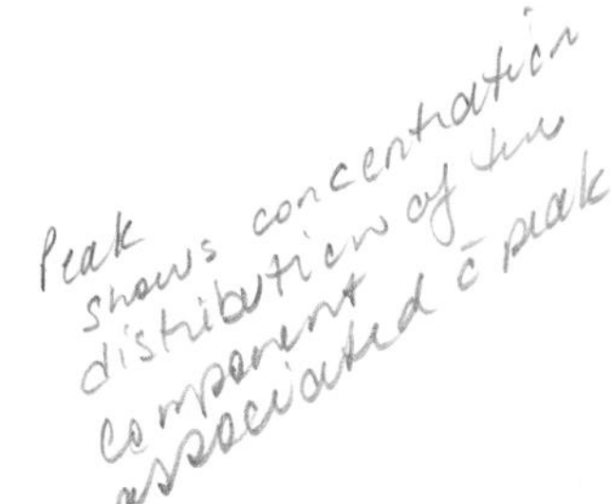

From inspection of the chromatogram in Figure 13.3, we can discern some of the characteristics of a chromatographic separation. Notice that the third chromatogram, that of the mixture of the two samples, is a sum of the two separate chromatograms. It appears that whatever is the cause of the separation, we shall be able to treat each component independently.

Notice that some of the peaks reach the baseline before the next peak begins. However, in the chromatogram of the mixtures, the graph does not reach the baseline between the two largest peaks. These two components are *not* **baseline-resolved**. Having every component baseline resolved is desirable. On the other hand, we say that the components are **unresolved** only if there are two or more components that elute together so that only one peak appears. For the intermediate case, we say the components are **partially resolved**.

ANALYTICAL CHEMISTRY IN ACTION
Not Cabbage Again, Mom!

What's for dinner? Some idea of the diets of ancient peoples can be obtained from the tools they used, and also from seeds, bark, and pollen. However, evidence for leafy vegetables has been difficult to find. Nevertheless, some archaeologists wanted to determine the diet of people that lived during a period that is called the dark ages in Europe, and found a solution in analysis of some chemical compounds and unique decomposition products recovered from the pieces of pots that were used for storage and cooking.

The samples are extracted from the inorganic matrix of ceramic pots into an organic solvent, and the dissolved substances subjected to analysis by gas chromatography and GC/MS. The pots from a Late Saxon (9th–13th centuries) settlement yielded a relatively simple set of compounds: nonacosane, nonacosan-15-one, and nonacosan-15-ol, which together comprised more than 70% of the volatile constituents of the organic extracts. This set of compounds and their relative levels are consistent with *Brassica oleracea*, which is more commonly called cabbage. Other closely related varieties of vegetables include cauliflower, Brussels sprouts, kale, turnips, and broccoli. But Brussels sprouts and cauliflower were not introduced into the area until the 15th and 17th centuries, respectively, so they can be eliminated as candidates. The presence of other alkanes and oxygenated aliphatic components eliminates all but cabbage and turnips, with the leaf wax of cabbage the closest match. Given the close association with the organic components, the pot was clearly used for cooking vegetables—most likely cabbage.

[Ref. Evershed, R. P., Heron, C., and Goad, L. J. 1991. *Antiquity* 65:540–544.] ■

The samples were injected at a time when the pen was on the right-hand side of the graph. The paper moved to the right, and the chromatogram was plotted. Thus, elapsed time is plotted increasing to the left. Note that the peaks are broader the later they pass through the detector. This is a general property of chromatograms. In addition, notice that the first, narrow peak that appears is not at the time of injection. There is a delay. This delay is also a general property of chromatographic separations. The first peak here corresponds to some components that simply pass through the column as fast as the mobile phase does. The volume of mobile phase that passes through the system between the time the sample was injected and the top of the first peak is called the **hold-up volume** V_M. It is composed of two pools: the liquid between the particles of the stationary phase, called the **void volume** V_o, and the volume of the tubing, injector, and detector, which are collectively called the **extra-column volume**. (A term you may see used for the extra-column volume is the **dead volume**. Its use is now discouraged.) More will be said of these terms later.

Now, notice the direction in which time is plotted in the chromatograms shown in Figures 13.2 and 13.3. They are opposite. If we wish to describe the earlier eluting portion of a band, obviously, referring to the directions right or left can thus be confusing if both types appear together. Instead, we refer to the **leading edge** and **trailing edge** of the band. The leading edge refers to the material of a band that elutes at an earlier time (or volume). The trailing edge results from the material eluting at a later time (or volume).

13.4 Parameters of Chromatography

As chemists use the term, a **parameter** is a quantity that can take on different values and characterizes some process, operation, or result. A well-known example in mathematics is the radius of a circle. The radius can change in

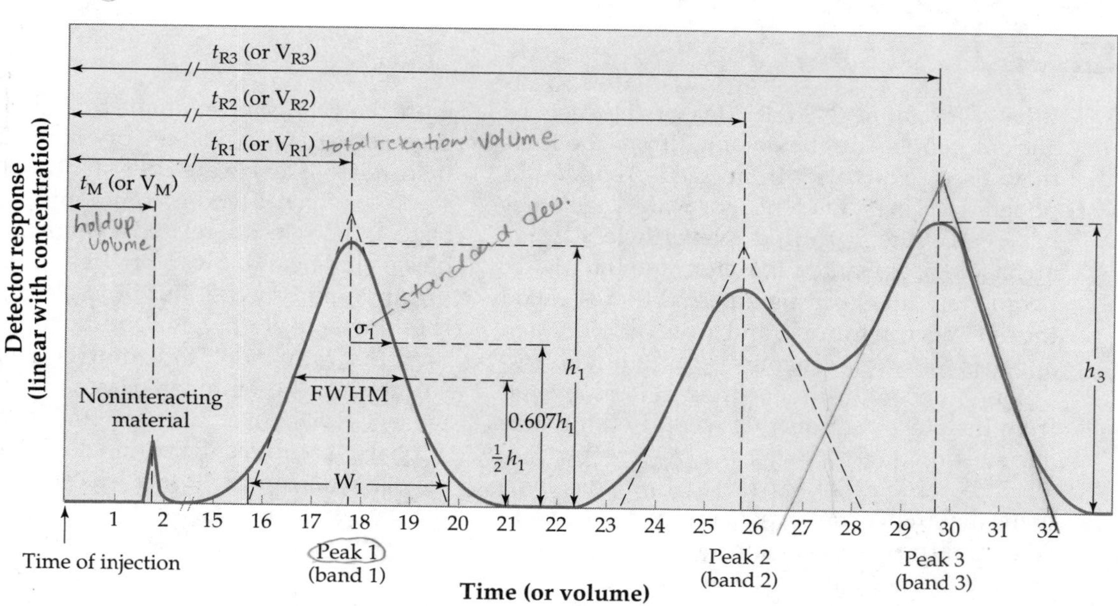

FIGURE 13.4 ▲

Hypothetical chromatogram showing the parameters that are used to characterize a chromatogram.

Each band can be described by a peak position and a peak width. Pairs of bands are characterized by a separation factor or by resolution of the corresponding peaks.

value from circle to circle, but it always characterizes a circle. Parameterizing data in chromatography, as in other methods, allows for ease of tabulation and communication of that data.

The shapes, positions, and resolution of characteristic bands can be parameterized. The parameters and their labels are illustrated in Figure 13.4. These parameters may be correlated successfully with chemical factors. In the case of chromatography, parameters describing the chromatograms can be successfully correlated with descriptions of the molecular processes underlying the separations. However, the data is primary, and the descriptions are models that can explain the data. The models are mathematical analogs that follow from the data and descriptive chemistry.

Parameters for Individual Bands

The first parameter that is used to characterize a chromatogram is the volume that elutes from the column between the injection of the sample and the maximum of the first peak that elutes. As noted above, this volume is called the hold-up volume. Its value, written V_M, corresponds to the liquid (or gas) volume surrounding the stationary phase. This is the minimum volume of eluent that can carry any component of the sample from the point of injection to the detector.

However, occasionally the first peak of a chromatogram may not appear at the hold-up volume of the column. There are two reasons for this.

1. The detector is not sensitive to the material in the band and, thus, will not respond.
2. All the components in the mixture interact with the stationary phase and, thus, none elute in the hold-up volume.

Each peak that elutes after the hold-up volume requires a number of other parameters to describe it. One is the volume at which the maximum of the peak appears. This is called the **total retention volume** (preferred), **retention volume**, or **elution volume**. Time can be used in place of volume. (See the illustration in Figure 13.4.) The IUPAC recommends the abbreviations V_R or t_R. The total retention time alone can be used to confirm the identities of the components in a sample.

The width of the peak—its **peakwidth** or **bandwidth**—is also parameterized. (The same term, bandwidth, is used in spectrometry for a different quantity. Do not confuse them.) One parameter is the *full width at half maximum*, abbreviated FWHM or $W_{1/2}$. For a band having height h_i, the FWHM is the volume eluted between the edges of the band at a position one-half of the total height. This volume is labeled on band 1 in Figure 13.4. Another parameter that can be used to characterize the width derives from assuming that the band is Gaussian in shape—the standard deviation, abbreviated σ_1. A third possible parameter (also shown in Figure 13.4) conveys the same information: it is the peak width at the base, W_i. This is found by drawing tangent lines at the inflection points on both sides of the band outline and extending them down to the baseline. The time between these points on the baseline is a useful measure of the width. A Gaussian band shape is often a good approximation to the shape observed experimentally. In that case, for any band i,

$$W_i = 1.698 \text{ FWHM} = 4\,\sigma_i \qquad \text{(Gaussian band shape)} \tag{13-2}$$

In this text, we will use the abbreviation FWHM here since the use of $W_{1/2}$ can be confused with the baseline width, written W_i for the ith component. Further, the use of FWHM allows us to describe peaks that are not baseline-resolved and are not perfectly Gaussian.

You saw that the peaks in the chromatograms of Figure 13.3 are independent of each other. As a result, descriptive parameters of each band must be independent as well. The primary parameters are:

a. The hold-up volume for the run
b. For each analyte peak, the effluent volume when the peak occurs
c. For each analyte peak, the peak width

These quantities are sufficient to characterize the quality of an experimental separation (baseline or partial) and the possibility that a good separation can be done. However, it is more common to characterize the peaks using a few other quantities derived from these measured parameters. These resultant parameters are defined next.

The **net elution volume** or **net retention volume** is defined as the difference in the eluent volume of a peak and the hold-up volume.

$$\text{net elution volume (peak } i) = V_{Ri} - V_M \tag{13-3}$$

A parameter, the **capacity factor** k_i can be defined for each band. It is

$$k_i = \frac{V_{Ri} - V_M}{V_M} \tag{13-4a}$$

For a constant eluent flow rate, the respective times may be used instead of volumes in the equation for the capacity factor.

$$k_i = \frac{t_{Ri} - t_M}{t_M} \quad ; \qquad \text{constant flow mobile phase} \tag{13-4b}$$

The variable k_i is especially useful since it is independent of the system geometry, unlike the total retention times.

Efficiency

The major goal of chromatography is to separate the components of a sample. The separation of components depends on the widths of the peaks and the spacings in time (or volume) between the peaks. As a result, an effective separation of pairs of components can be achieved with either narrow or broad bands. However, if the bands are broad, the separation may take more time, as illustrated in Figure 13.5. All other effects being equal, we prefer narrow bands of material.

We quantify this preference for narrow bands with the parameter called the chromatographic **efficiency** (N). This property can be found from the characteristics of any individual band, but each individual band of a chromatogram may not result in the same value of the efficiency. The defining equation is

$$N = \left(\frac{V_{Ri}}{\sigma_i}\right)^2 = \left(\frac{t_{Ri}}{\sigma_i}\right)^2 = 16\left(\frac{t_{Ri}}{W_i}\right)^2 \tag{13-5a}$$

The values of σ_i and W_i are measured in the same units as the numerators, so N is a dimensionless ratio. This ratio has the width of the band to the time between injection and detection. With this insight, you can see that N is a measure of the *band broadening per unit time.*

Equation 13-5 is also written

$$N = \left(\frac{2.355\, t_{Ri}}{\text{FWHM}}\right)^2 = 5.54\left(\frac{t_{Ri}}{\text{FWHM}}\right)^2 \quad \textbf{(13-5b)}$$

More about the importance of Equation 13-5 will be presented in Section 13.6.

Parameters Describing Pairs of Bands

The parameters describing each band of a chromatogram do not express information about the relationships between the bands. For example, in the fictitious but representative chromatogram in Figure 13.4, at around 21 time units no material of any band is eluting; the response follows the baseline. Any fraction collected between 15 and 21 time units would contain none of the material that makes up bands 2 and 3. (We hope by now you have become suspicious of the term *none*. Let us define it here as being less than one ppt—0.1 mol %—of the component.)

If a fraction containing the eluent from time 22 to about $27\frac{1}{2}$ were collected, you might guess from the appearance of the two bands that the fraction would have some of the material comprising peak 3 along with the material comprising peak 2. Peaks 2 and 3 are not completely separated.

Two parameters are used to quantitate the amount of mixing of the materials contained in two eluted bands: the **separation factor** and the **resolution**.

(*a*)

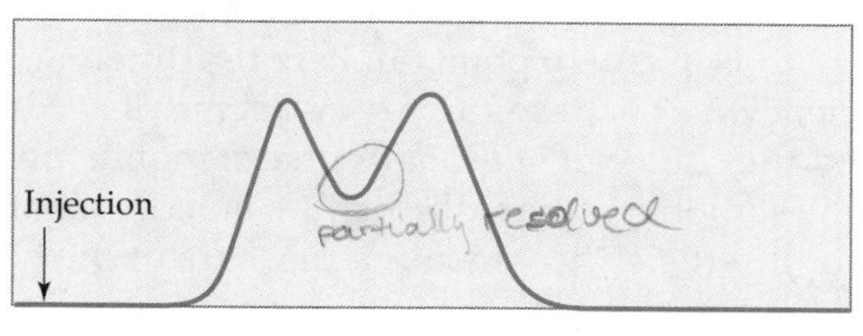

(*b*)

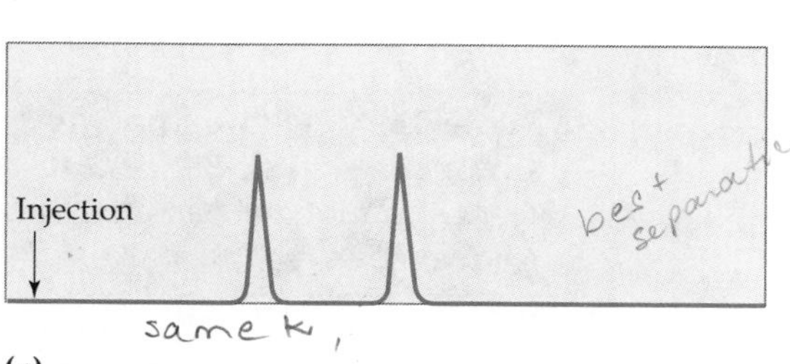

(*c*)

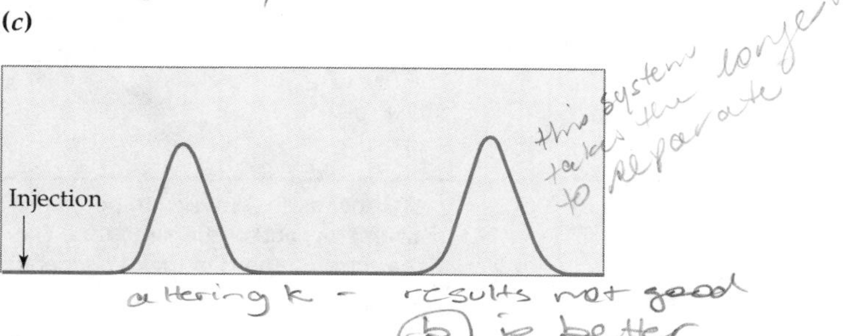

FIGURE 13.5 ▶ **Illustration of two ways to improve resolution.**

Assume the flow rate is the same for all three chromatograms and the samples are identical: **(a)** shows two partially resolved components; **(b)** illustrates the chromatogram after improving the efficiency; the components have their same *k* values as in (a); **(c)** illustrates the effect of improving the separation by altering the *k* values. The conditions of (c) are not as good as those in (b); it takes longer to obtain a separation of the same resolution.

Note that the sensitivity of the vertical scale in the middle chromatogram is decreased. If the vertical scales were all the same, the peak heights of (b) would be significantly larger than those of (a) or (c). However, the peak areas in all three would be equal since the quantities of analytes are equal.

The separation factor for two *adjacent* bands (call them 1 and 2) is defined as

$$\alpha_{1,2} = \frac{V_{R2} - V_M}{V_{R1} - V_M} = \frac{k_2}{k_1} \qquad \textbf{(13-6a)}$$

The calculation is made with the larger volume (more slowly eluted band) as numerator. Under the condition of constant eluent flow, we may also write

$$\alpha_{1,2} = \frac{t_{R2} - t_M}{t_{R1} - t_M}; \quad \text{constant flow mobile phase} \qquad \textbf{(13-6b)}$$

However, α compares only the k_i values of the bands. It does not account for the widths of the peaks, which also determine how well the components are separated.

For any pair of bands the resolution is defined by

$$\text{resolution} = R_s = \frac{V_{R2} - V_{R1}}{\frac{1}{2}(W_2 + W_1)} = \frac{\text{peak separation}}{\text{average peak width}} \qquad \textbf{(13-7)}$$

The denominator in this equation is the average of the two baseline widths, and the numerator is the separation of the peaks. The parameter R_s provides a quantitative measure of how much mixing of materials there is between two adjacent bands. Some representative values are shown in Table 13.1. It is helpful to remember that when $R_s = 0.5$, the bands are quite mixed. At $R_s = 1$, the points of two triangles that approximate the peaks just touch where they meet at the baseline. At $R_s = 1.5$, the overlap between the actual peaks is about 0.1%.

Note that resolution is a dimensionless quantity, so for Equation 13-7, the units of the peak widths must be the same (volume or time) as the numerator units. All calculations of R_s are made so that its value is positive.

It is useful to point out here the difference between resolution R_s and efficiency N. The resolution—or degree of separation achieved—is determined by the choice of stationary phase, mobile phase, temperature, and length of the stationary phase through which the separation occurs. This is in contrast to the efficiency, which is a measure of the band spreading. The efficiency is determined by factors of column construction and packing and the velocity of the mobile phase.

Table 13.1 Relative Impurity of Two Adjacent Chromatographic Bands with Equal Concentrations of Material in Each if Cut Midway between Peaks*

R_s	Relative Impurity Content
1.5	0.001
1.0	0.023
0.8	0.045
0.5	0.16

*The bands are assumed to be Gaussian in shape, symmetric, and of the same area. The results are different if the concentrations of the two separated species are unequal.

EXAMPLE 13.1

Calculate the values of V_R, k, and W for bands 1 and 2 of Figure 13.4. Calculate $R_{1,2}$ and $\alpha_{1,2}$. Finally, calculate the efficiency for component 1. Assume a constant fluid flow rate of 0.80 mL time^{-1}.

Solution

The elution volumes are

$$V_{R1} = 17.8 \times 0.80 = 14.24 \text{ mL}$$

$$V_{R2} = 25.7 \times 0.80 = 20.6 \text{ mL}$$

The capacity factors are

$$k_1 = \frac{17.8 - 1.7}{1.7} = 9.5$$

$$k_2 = \frac{25.7 - 1.7}{1.7} = 14.1$$

The widths (in time units) are

$$W_1 = 4.0 \text{ units}$$

$$W_2 = 5.0 \text{ units (uncertain due to overlap)}$$

The resolution of 1 and 2 is

$$R_{1,2} = \frac{t_{R2} - t_{R1}}{\frac{1}{2}(W_2 + W_1)} = \frac{25.7 - 17.8}{\frac{1}{2}(5.0 + 4.0)} = 1.76$$

The separation factor is

$$\alpha_{1,2} = \frac{k_2}{k_1} = 1.49$$

The column efficiency is

$$N = 16\left(\frac{17.8}{4}\right)^2 = 317 \quad \text{for component 1.}$$

It is important to realize that for any two specific substances these parameters can change. They all depend on the exact natures of the stationary and mobile phases as well as on temperature. The ways some factors affect chromatograms are shown in Figure 13.6. The term **isocratic** there literally means equal power. In chromatography, it means that the composition of the mobile phase is constant throughout the time of the separation.

Comparing Column Efficiencies

Recall that the chromatographic efficiency N is a measure of the relative retention of the solute compared with the width of its peak. As you read above,

$$N = 16\left(\frac{t_{Ri}}{W_i}\right)^2 \qquad \textbf{(13-5)}$$

FIGURE 13.6 ▶
Effects of mobile and stationary phases.

(a) Chromatographic separations can be changed by changes in mobile phase composition. In these column LC chromatograms of some antidepressant drugs, all conditions are the same except the solvent composition. The conditions are noted in the figure. Not only are the elution times different, but the order of elution is changed. Note the loss in resolution in the chromatogram on the right. The arrow indicates the injection time. *Isocratic* means the mobile phase has a constant composition. Peak identification:

1. Imipramine; 2. Desipramine; 3. Amitriptyline; and 4. Nortriptyline. [Ref: Reprinted from *Am. Lab.* 1979, 11 (August):9. Copyright 1979 by International Scientific Communications, Inc.]

(a)

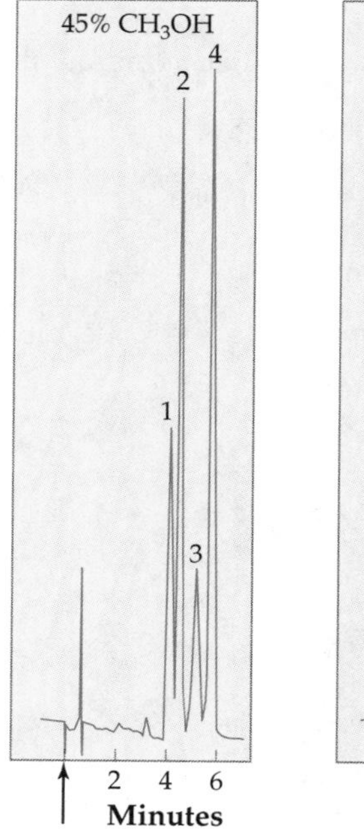

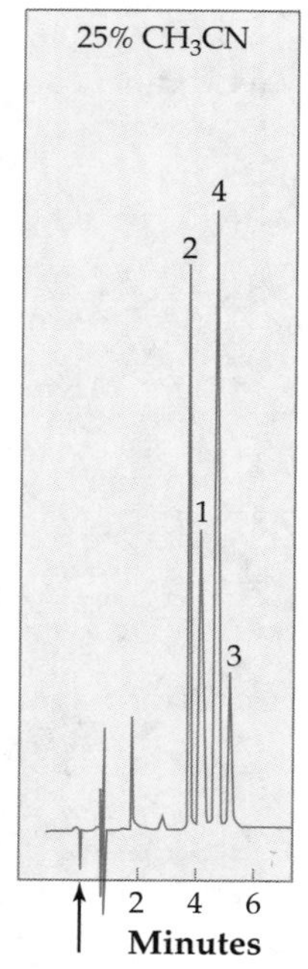

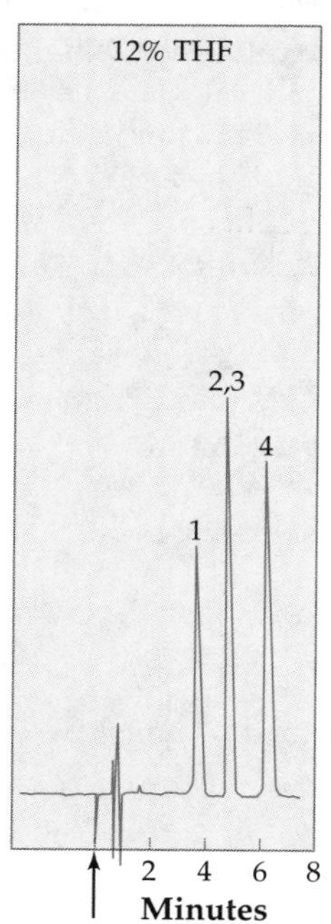

Instrument: Altex Model 322 MP with Model 155 Detector
Column: Ultrasphere-IP, 150 × 4.6 mm
Mobile phase: (Isocratic)
A = 0.01 M KH_2PO_4
0.01 M Nonyl Amine
pH = 3.0
B = Organic modifier as noted (v/v)
Flow rate: 2.0 mL/min

Peak identification
1. Imipramine
2. Desipramine
3. Amitriptyline
4. Nortriptyline

(b)

(b) Even small changes in the stationary phase cause changes in the separation. The stationary phases of these materials are the same in name, but the packings are from two different manufacturers. Their slight differences cause the shifts in peak positions shown (most easily seen in the second and sixth major peaks) as well as relative peak heights (especially clear in the fourth through sixth peaks). Left graph $N = 15{,}500$, right graph, $N = 14{,}000$. [Ref.: Courtesy B. G. Archer, Altex]

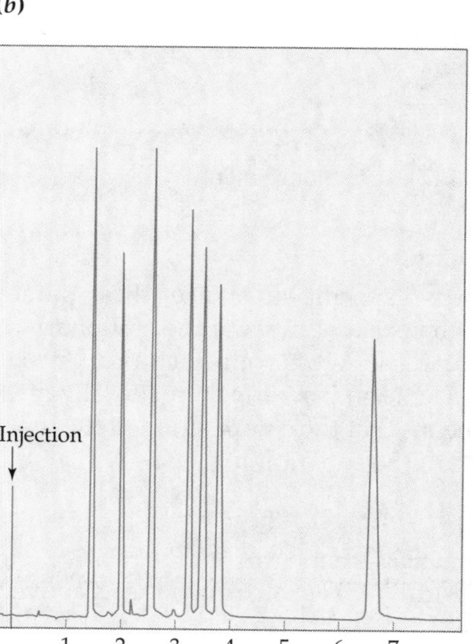

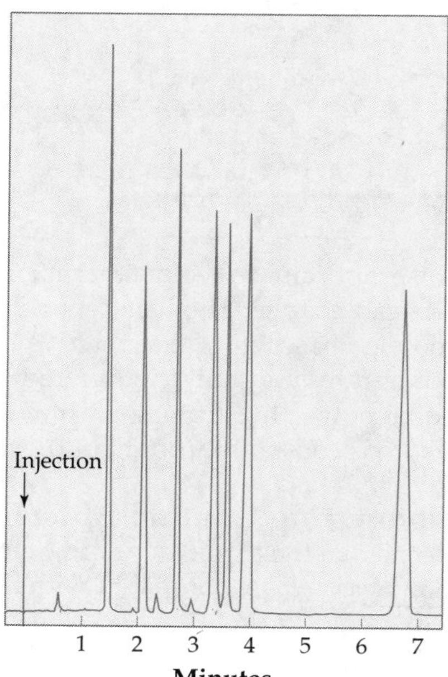

The efficiency parameter N is useful when comparing chromatographic separations under different conditions.

The efficiency parameter is usually called the **number of theoretical plates** hence the use of N. (This name comes from early distillation separations of volatile organics in fractionation towers which contained "plates" on which liquids with a given range of boiling points condensed and then drained into a collection vat.) Another useful measure is the number of theoretical plates *per unit length* of stationary phase. Assume that the packing has an externally measured length L and that the entire column length can be used to separate components. If the column is found to have N theoretical plate equivalents (using Equation 13-5 and data from a chromatogram), then the **height equivalent to a theoretical plate,** H, is

$$H = \frac{L}{N} = \frac{L}{16}\left(\frac{W_i}{t_{Ri}}\right)^2 \qquad \textbf{(13-8)}$$

Height equivalent to a theoretical plate is usually abbreviated as the capitalized initials HETP. A chromatographic system is better when the HETP is smaller. In other words the system becomes more "perfect" as the HETP approaches zero.

EXAMPLE 13.2

If the chromatogram of Figure 13.4 was obtained on a column 10 cm long, what is the HETP of the system?

Solution

In Example 13.1, the efficiency of the separation was calculated; $N = 317$. Therefore,

$$H = \frac{L}{N} = \frac{10.0 \text{ cm}}{317} = 0.032 \text{ cm}$$

13.5 Quantitation in Chromatography

You have seen how to characterize each chromatographic band and pairs of these bands. With fixed conditions (type of stationary phase, identity of mobile phase, flow rate, temperature, and detector type), these parameters alone often can be used to identify the presence of specific components in a material. As a result, chromatograms can be used to provide *qualitative* analyses. However, we have not yet considered how to determine the *quantity* of each component. Quantitation can be done quite precisely: With internal standards, precisions can approach ±0.1%. Thus, contemporary chromatographic equipment is arranged both to separate and allow a quantitative analysis of the components nearly simultaneously.

Loss of Material on the Column

Regardless of how the materials composing the bands are detected, there can be a more fundamental problem. In an extreme case, it is possible that some materials in a mixture will not elute at all. In other words, the material is effectively lost in the analysis; its recovery in the analysis is 0%. In this extreme case, even the qualitative analysis will be incorrect since that component would be completely missed. With a few exceptions, the quantity of material lost *cannot* be detected experimentally. (A possible exception: a radioactively labeled compound could be measured while still on the column.)

A practical way to detect an irreversible adsorption is to see a progressive change in retention characteristics of the eluting components of the sample as multiple samples are run. The irreversibly binding components occupy the surface sites at the head of the column and, as a result, change the characteristics of the column. Each sample injection changes the column characteristics, and the retention times of the eluted compounds change with each progressive injection.

Another way the qualitative and quantitative analysis can be misleading is for the components to change chemically during the separation. Examples are catalyses of decomposition, isomerization, or racemization. For the remainder of this chapter, let us assume that all components manifest 100% recoveries and remain chemically unchanged.

Detector Response

The precision and detection limits of a chromatographic method depend intimately on the characteristics of the detector used. The choice of detector depends not only on the chromatographic conditions but also on the chemical and physical characteristics of the analytes present, their concentrations, and the sample size that can be used. Precise quantitation can be accomplished most easily if the output from the detector is on a scale that is exactly linear with the concentration of each component. Algebraically, this situation can be described by a simple equation.

Sensitivity is introduced in Section 5.2.

$$\text{Detector response} = S_{\text{component}}\,[\text{component}] \tag{13-9}$$

S represents the sensitivity. However, as inferred by Equation 13-9, the detector response for each component depends not only on the characteristics of the detector and the analytes, but also on dilution, which is inherent in the separation process. For example, if two analytes have the same sensitivity and occur in the injected sample at the same concentration, the band eluting later becomes broader (more dilute), and the response at the peak of the later band will be less than the earlier-eluting one.

Quantitation Techniques

If all components of a material are baseline-separated, you can use a number of different techniques to quantitate them. (See Figure 13.7a.) For example, each component could be collected as a separate fraction, and each of these fractions quantitated individually. The quantitation could by done by titration or some instrumental detection method such as one of the electrochemical methods described in Chapter 15. No matter what analytical method is used, the total quantity of a component in the fraction equals the mathematical product of concentration times the volume of the fraction.

However, this average concentration is usually measured more conveniently without collecting fractions. Recall how the plot of the detector response vs. time was obtained. The measurement is made on a small volume of the effluent as it passes through the detector. When the output response is *linear*, the concentration of a species is proportional to the *area* under the graph of response vs. time. (Those of you who have studied calculus will recognize that the area is the integral of the graphed output curve.)

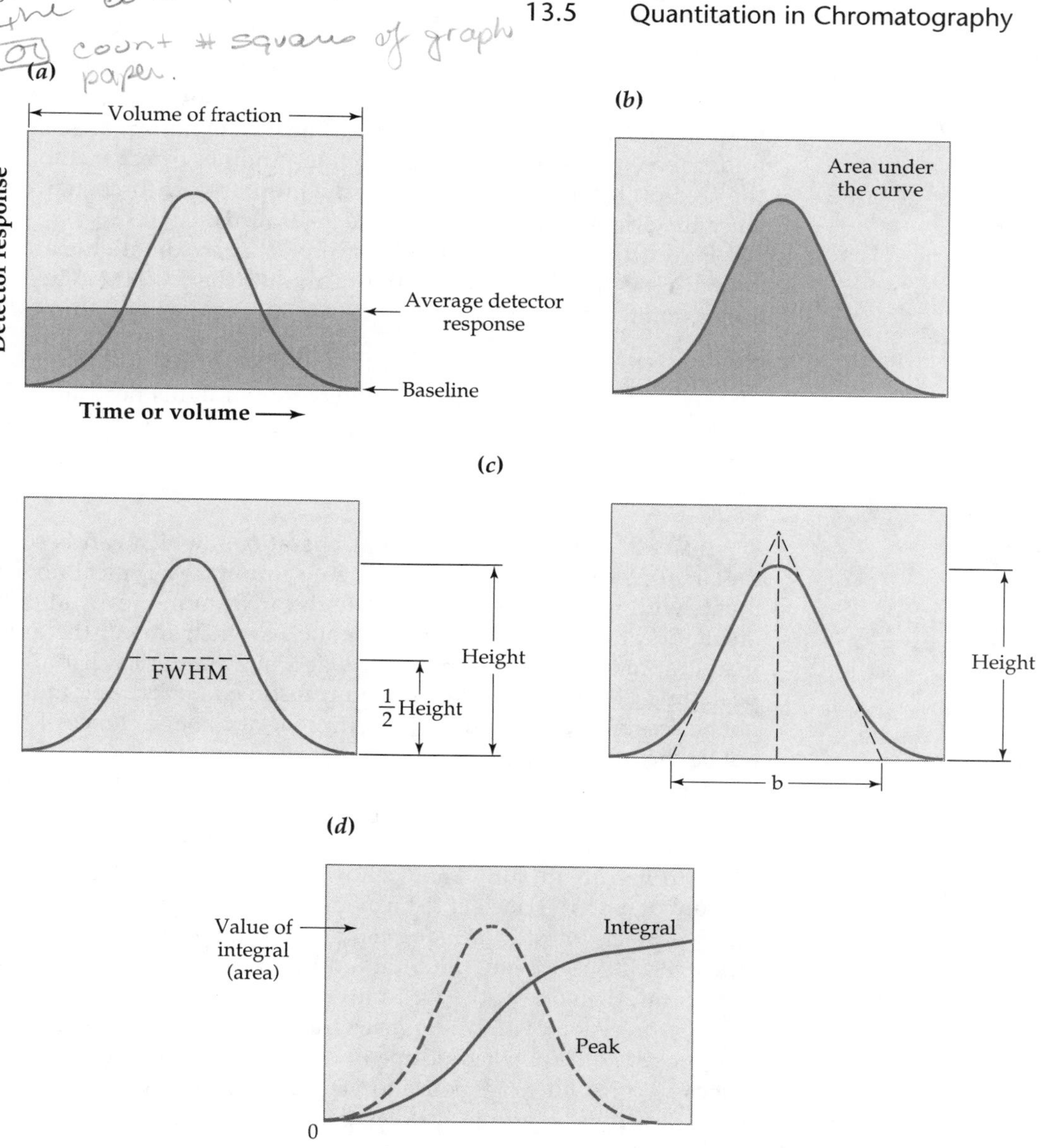

FIGURE 13.7 ▲
Illustration of a number of ways to quantify chromatographic bands.
Each peak area (or height) must be calibrated using standards if it is to be related to a specific concentration of the species it represents. A constant mobile-phase flow rate is assumed. **(a)** The content of a zone is the product of the concentration and volume. A fraction may be collected that contains all the material of a single band, and its content can be measured in a separate determination. **(b)** The area under the peak's curve can be measured and calibrated with standards. **(c)** The area can be approximated as a triangle, and is calculated as FWHM × height or as ½ base × height. **(d)** The peak is integrated electronically—the magnitude of the integral is proportional to content. Usually the integral curve is not drawn but the integral value is printed near the peak on the display or in a summary at the end of the chromatogram. The integral curve is drawn on a different scale than the plot of the output.

A number of techniques are used to measure the areas under chromatogram peaks. Some are quite exact, and some are approximations which are simpler to measure while still allowing satisfactory precision.

When the plot is on graph paper as for older instruments that are not interfaced to a computer, there are two ways to determine the area (Figure 13.6b). One is simply to count the number of squares of graph paper under the curve and above the baseline. Another direct method is to weigh the area under the curve cut out of the graph paper (or, better, a photocopy since the original data should not be destroyed).

Figure 13.7c illustrates two methods to approximate the area under the curve. Measure the height of the peak and the FWHM. Then the area is approximated by

$$\text{area} \approx \text{FWHM} \cdot \text{height}$$

The alternate method requires drawing draw straight lines tangent to the sides of the output plot. The area is then approximated by the area of the triangle.

$$\text{area} \approx \frac{1}{2}\ \text{base} \cdot \text{height}$$

Instead of doing these manual operations, which can become tedious, the area measurements can be done electronically. Typical now are instruments with **integrators** that calculate the areas numerically after "deciding" on the location of the baseline. On such an instrument, the areas of chromatographic peaks are usually supplied in a numerical list along with the peak times. However, for incompletely resolved peaks, their numbers should not be accepted without question, since the automatic choice of baseline may not be the best choice.

The above methods all account for the areas under chromatographic peaks even if the widths of the bands change. However, when the experimental conditions can be adequately controlled, the widths will remain the same from run to run. That control allows an excellent measure of the amount of material present to be made from the height of the peak alone. The relative precision of a measurement made this way can be better than 1%. However, if there is any doubt about the constancy of the chromatographic conditions, then one of the area-measurement methods should be used.

The precision of all of the measurements made from a chromatogram—as opposed to being made directly on the fractions—depends on having a detector with a linear response to the concentration. In addition, the mobile-phase flow must continue at a constant rate for the method to work. If a chart recorder is used, as opposed to a computerized unit, the chart movement also must be constant if area measurements are used to quantitate analytes. If any of these three conditions do not exist, then the precision of the analysis is certain to decrease.

13.6 Explaining Chromatographic Separations

So far in this chapter, only descriptions of chromatograms and chromatographic efficiencies have been presented. Little has been said about why the components separate and elute in a certain order. Nor have the bands' widths been explained. Let us do so now.

Extractions

Extractions are also used for sample preparation. See Section 4.4.

As we shall see in this section, chromatographic separations can be thought of as similar to carrying out a large number of **extractions**, one after the other. It is worth some time to understand extractions in order to understand one

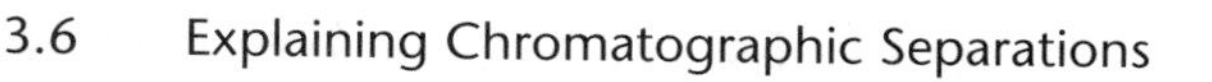

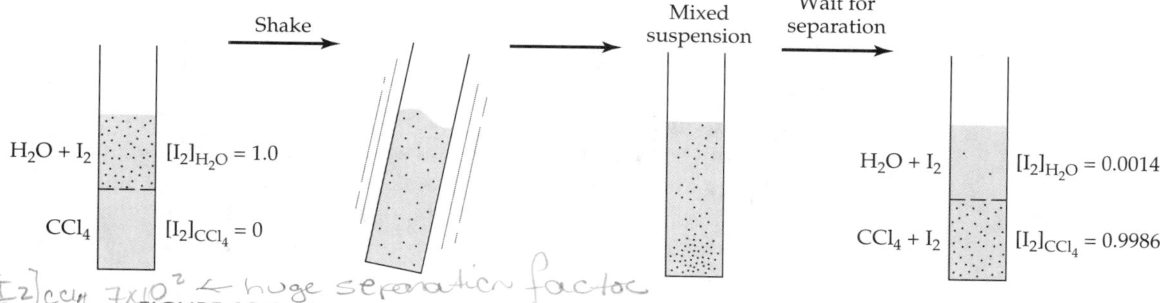

FIGURE 13.8 ▲

A simple extraction experiment.

Two immiscible liquids are shaken and allowed to separate. At equilibrium, some components in the solution might be soluble in both phases. The relative concentrations of the solute in the two phases are described by a distribution coefficient K_D. The value of K_D depends on the identities of the solute and the two solvents, the temperature, and possibly on the other components present as well.

view of the underlying chemistry of chromatography as well as the origin of much of the vocabulary of chromatography.

Illustrated in Figure 13.8 is a simple **liquid-liquid extraction**. The technique entails using two liquids that are not too soluble in each other—they are immiscible. Let us say the two liquids in the illustration are water and carbon tetrachloride, CCl_4. Since the CCl_4 is more dense, it composes the bottom layer. Dissolved in the water is a solute: In this example, it is iodine. The mixture is shaken vigorously and then allowed to sit while the two liquids again form separate layers. If the iodine concentration is high enough, we can see its purple color in the organic phase.

After the extraction, we determine the concentrations of iodine in both the aqueous and organic phases. (For instance, the iodine that is in each phase can be determined by a redox titration.) If we did this experiment a number of times using a number of different initial concentrations of iodine but at a fixed temperature, we find that the results are described by the following expression.

$$K_D = \frac{[I_2]_{CCl_4}}{[I_2]_{H_2O}} = 7 \times 10^2 \tag{13-10a}$$

K_D is an equilibrium constant called the **partition coefficient**, or **distribution coefficient**. We can also write

$$K_D = \frac{C_{I_2,CCl_4}}{C_{I_2,H_2O}} = \frac{C_1}{C_2} \tag{13-10b}$$

where the C_i are the concentrations (mol/unit volume) in the respective phases. (By convention, the organic phase is "1.")

The partition coefficient is a ratio of concentrations. Finding the *amounts* of material in each phase requires knowing the volumes of both phases. Let us call the volume of phase i, V_i. For the example of iodine partitioning between water and carbon tetrachloride,

$$\text{total moles of molecules} = \text{moles in } H_2O + \text{moles in } CCl_4$$

$$= [I_2]_{H_2O}V_{H_2O} + [I_2]_{CCl_4}V_{CCl_4} \tag{13-11a}$$

And, for the general case of partitioning between two phases,

$$\text{total moles of molecules} = C_1V_1 + C_2V_2 \tag{13-11b}$$

To here: 3/11/09

Sample Loading

Inherent in the equations above is the assumption that the volumes of the two phases are large enough to be able to dissolve all the iodine that was added to them. If too much iodine were added, some solid would remain. In that case, the total number of moles of iodine would no longer be described by $C_1V_1 + C_2V_2$. When a similar situation occurs in chromatography, the system is said to be **overloaded**. For now, it is only necessary for you to understand that the correct **sample loading** is important in achieving good chromatographic separations. The total mass of a sample that is run should be limited so that the stationary phase does not become overloaded; that is, its **capacity** is not exceeded.

K_D and Elution Times

Now let us see how the elution time of a chromatographic band is related to the partition coefficient of the compound in the band. The important ideas are illustrated in Figure 13.9.

We begin by asking the question, What fraction of the time is a solute molecule in the mobile phase? When in the mobile phase, the solute molecules are being carried along at the velocity of the mobile phase. When

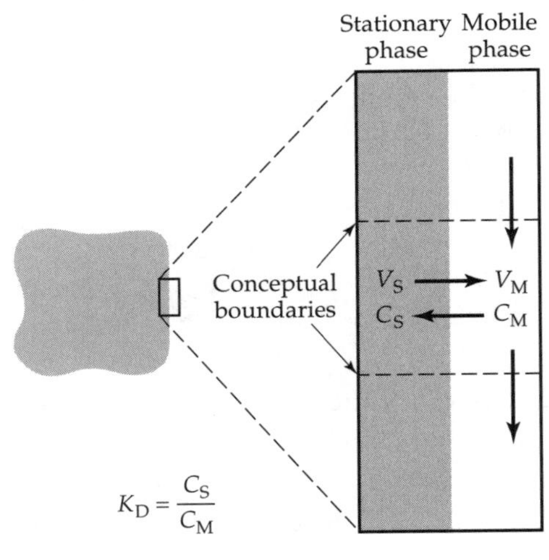

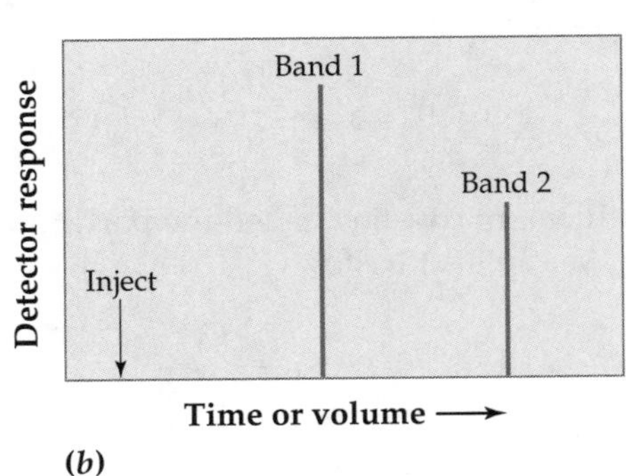

FIGURE 13.9 ▲
Illustration of the idea of chromatography as an extraction.

(a) Here, an extraction occurs as the mobile phase moves past the stationary phase. Included in the model is the idea that a separation, which is *equivalent* to a single *equilibrium* extraction, occurs over a specific length between two "conceptual boundaries." The length (or height) between these conceptual boundaries is a theoretical plate. **(b)** If we considered only the average time spent in the mobile phase relative to the average time anchored on the solid phase, a chromatogram would appear as shown here. The separation of the components is explained, but the finite widths of the bands are not.

a molecule is associated with the stationary phase, it does not move. Thus, we can state the law underlying all chromatographic separations:

> *The average velocity of a single molecule depends on the average time it associates with the stationary phase compared to the time it moves with the mobile phase.*

An equation can be written expressing this law. The equation contains the quantity **linear velocity**. The linear velocity represents the *average velocity* with which the mobile phase flows along the long direction of the column.

average rate of travel=
  linear velocity of mobile phase
    × fraction of time molecule is in mobile phase
  + linear velocity of stationary phase
    × fraction of time molecule is in stationary phase

Because the linear velocity of the stationary phase is zero, the second term contributes zero to the average. If we call the average linear velocity of the mobile phase $\overline{u}$, then

$$\text{average rate of travel} = \overline{u} \times \text{fraction of time molecule is in mobile phase} \qquad \textbf{(13-12)}$$

After studying this section, you may be interested to go back and read the quote from Tswett at the beginning of the chapter. You will see that he already understood the basis for the relative retention times and the elution order of the components being separated.

A More Complete Model

In the previous pages, we have considered a single extraction as a model to explain chromatographic retention times and, hence, the underlying cause of separations. However, this simple model does not tell the whole story. If it did, each peak would come out at a single time that is characteristic for the eluted material. However, in chromatography, each band elutes over a span of time.

The idea of a sequential series of extractions is quite useful and can be used to explain the widths of chromatographic peaks. However, by consensus, the model presented in Section 13A is a better model, although slight disagreements about various points still exist. The model is described by the **van Deemeter equation**, which explains the underlying causes of the retention and shapes of chromatographic bands.

13.7 Liquid Chromatography

Types of Liquid Chromatography

As its name implies, liquid chromatography (LC) has a liquid mobile phase. The great advantage of liquid chromatography resides in the combination of a wide range of possible properties for the mobile phase together with the choice of numerous, significantly different kinds of stationary phases and detectors. (See Table 13.2 for examples of detectors.)

Within the broad definition of liquid chromatography are a myriad of techniques, and many of them are classified with more than one name. The names focus on many different aspects of LC methods. For instance, one of the primary classification schemes of LC is by the overall physical shape of

Table 13.2 Characteristics of Selected Liquid Chromatography Detectors

Type	Approximate Limit of Detection*	Approximate Linear Range†	Comments
Ultraviolet and visible absorption	10^{-11} g	10^{4}	Specific for light-absorbing compounds (*Cf.* Chapter 11)
Differential refractive index	10^{-9}–10^{-10} g	10^{3}	Universal detector. Measures changes in refractive index. Cannot be used with gradients.
Electrochemical: amperometric	10^{-10}–10^{-11} g	10^{5}	Specific detector. Compound must be electroactive. (*Cf.* Chapter 15)
Electrochemical: conductometric	10^{-8} g/mL‡	10^{5}	Specific detector but for all ions (*Cf.* Chapter 15)
Fluorescence	10^{-14} g§	10^{5}	Specific detector. Compound must be fluorescent. (*Cf.* Chapter 11)
Mass spectrometry	10^{-7}–10^{-9} g	10^{5}¶	Universal detector. Also can be used to identify analytes with great certainty. (*Cf.* Chapter 12)
Solution light scattering	10^{-6} g/mL	10^{5}	See footnote *a*.
Evaporative light scattering	10^{-9} g	10^{6}ϕ	Universal except for volatile analytes. Not a linear response.

*Depends on the peak width. Value for a measurement without preconcentration of the sample. The lower limit of their usefulness in every case is in the trace to ultratrace range.
†Of a detector (transducer + electronics) without chemical interferences present.
‡Limited by the inherent conductance of the eluent.
§This value is for strongly fluorescing compounds without interferents.
¶For a sector instrument. Quadrupole, ion-trap, and Fourier-transform instruments are narrower.
a. Light scattering is used to determine the molecular weights of polymers (range 500–10^{8} daltons) as they elute. The concentrations are usually far above the limit of detection.
ϕThe response is of the form (constant · [analyte]^{m}), so a log-log calibration plot is linear.

the stationary phase such as *column* chromatography (like that done by Tswett and described at the beginning of the chapter and shown in Figure 13.1.), *thin-layer* chromatography (a layer on a plate), and *capillary* liquid chromatography (a layer on the inner surface of the capillary). Other names arise based on the direction of flow of the mobile phase: *ascending* chromatography, *descending* chromatography, and *flat-bed* chromatography. Classification is also based on the efficiency of the separations, such as *high-performance* liquid chromatography or *high-performance* thin-layer chromatography. Sometimes names of LC methods identify the solutes that are separated and detected, such as *ion* chromatography and *amino acid* analysis (both usually done on columns).

The types of liquid chromatography also are named after the general type of interaction that occurs between the stationary phase and the solutes in the eluent. The four basic types of interactions are shown in Figure 13.10. The classifications are denoted **normal-phase**, **reversed-phase**, **ion-exchange**, and **gel-filtration** (or *exclusion* or *gel-permeation*) chromatographies. Some stationary phases are designed to interact with specific chemical groups. Chromatography with such site-specific groups on the stationary phase is called **affinity chromatography**. The stationary phase also may be made with **chiral** groups linked to it. You might recall from organic chemistry that chiral molecules are those that contain at least one atom—usually carbon—that has four different groups bonded to it. Such a chiral center can have two different structures, which are mirror images of each other. These chiral phases are used to separate chiral molecules.

It is imperative to understand that no LC separation occurs by means of only one type of interaction between the analytes and the stationary phase. There are always interactions besides the named, predominant, type. So if you read "the separation was done by normal-phase chromatography," it means that polar adsorption was the *predominant* mechanism of interaction with the solid. However, there also was some contribution to the separation from at least one of the other mechanisms discussed here, such as ion exchange or gel filtration.

It is useful to be aware of all these special names. However, the underlying principles of all these methods are those you have read about in the earlier sections of this chapter. That is, doing effective liquid chromatography is an exercise in manipulating all the equilibria that affect the reaction

$$\text{analyte (mobile phase)} \rightleftharpoons \text{analyte (stationary phase)}$$

Any change in the solvent results in a shift of the equilibrium between the mobile and stationary phases. Chromatographic separations can be improved by switching from an **isocratic** elution (same eluent throughout) to a **gradient** elution. This is done by mixing two or more different eluents such that the mobile phase composition changes over time. The gradient always changes the mobile phase toward better desorbing eluents.

An example of an isocratic separation was shown in Figure 13.6.

Let us now look in some more detail at each of the four general types of liquid chromatography interactions—normal-phase, reversed-phase, ion-exchange, and gel-filtration—and the practice of liquid chromatogarphy.

Normal-Phase Liquid Chromatography

In normal-phase chromatography, the solid phase is more polar than the mobile phase. Polar, in this context, is used the same way that we speak of polar bonds in molecules: the bonds of the stationary phase have larger dipole moments than the bonds in the solvent molecules. The stationary phases in normal-phase chromatography are mostly inorganic polymers that have a large number of pores of molecular size (several nm to tens of nm) so that their surface areas are large. The two most common materials are hydrated silicon-oxygen (**silica** or silica gel) and hydrated aluminum-oxygen polymers (**alumina**). When strong bonds result between these supports and the sample components, the support is said to be **active**. Their activity is dependent on the density of hydroxy groups on the surface and their chemical form—that is, whether they are deprotonated or not. You might expect that the Si—O bonds might hydrolyze in base, and, in fact, silica slowly dissolves in mobile phases where the pH is high. But at all pH's, it is usually worthwhile to slow the dissolution of the analytical column by attaching a

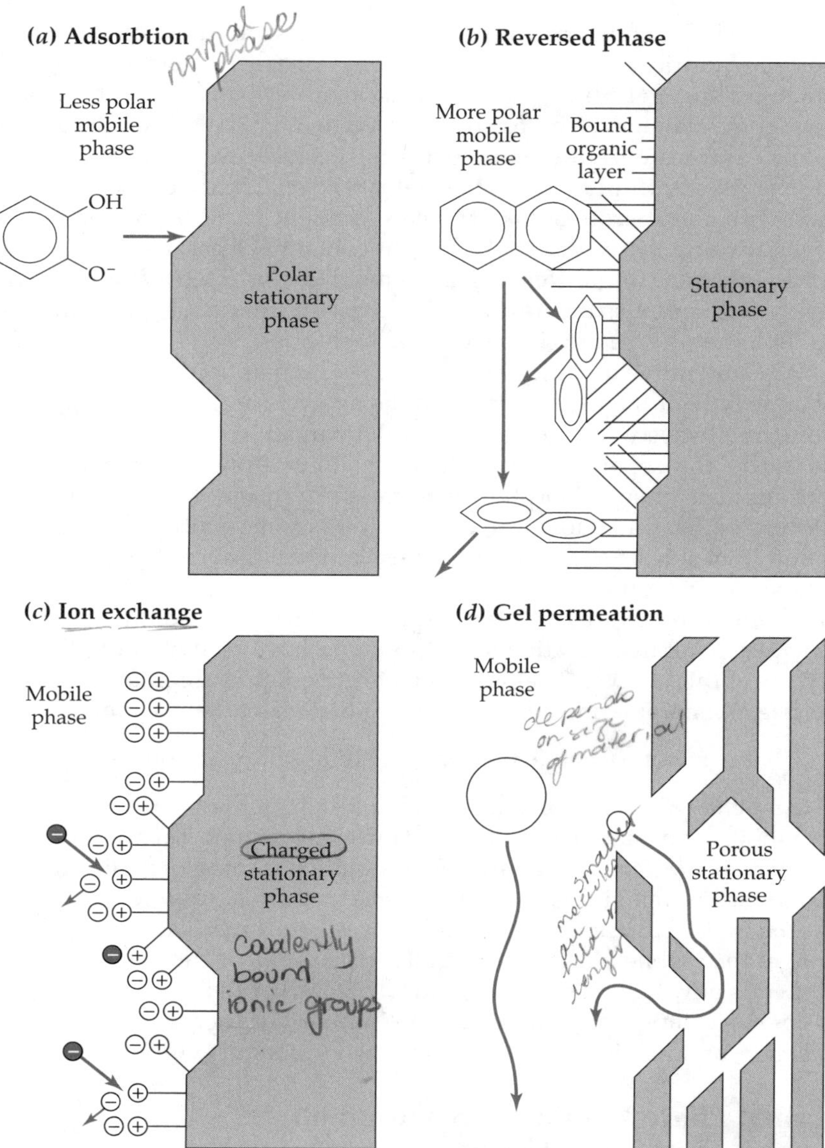

FIGURE 13.10 ▲

Illustration of the way that four different mechanisms of LC retardation are explained.

Although one of these four mechanisms may be predominant, there is always some contribution from at least one other. **(a)** Adsorption, in which a polar stationary phase is in equilibrium with a less polar mobile phase. **(b)** Reversed phase, in which the stationary phase is less polar than the mobile phase (60–70% of HPLC assays are done with reversed-phase techniques). **(c)** Ion exchange, in which the stationary phase has covalently bound ionic groups. **(d)** Gel permeation, in which the smaller molecules, which are able to enter pores in the stationary phase, are held up longer than larger molecules, which are unable to penetrate the gel matrix. (Chiral chromatography and affinity chromatography may be thought of as adsorption chromatographies, but the binding in them may have elements of adsorption, reversed phase, and ion exchange.)

short **guard column** at the front end. The guard column contains the same material as the analytical column, and its packing is sacrificed; its dissolved products help preserve the analytical column packing.

For normal-phase chromatography the solvent polarity is the primary factor that determines the elution volumes. The solutes move more rapidly and elute earlier with more polar solvents. The explanation is simple: more polar solvents compete for the solutes better than do less polar solvents.

It follows that predicting the elution properties of solvents would help in developing assays involving normal-phase separations. For instance, if the elution is too slow, you could look on a list and find a more polar solvent than the one used. One such list is called the **eluotropic series**. Eluotropic series classify solvents into a semiquantitative order based on their abilities to elute solutes from a *specific* stationary phase. Table 13.3 lists the eluotropic series for alumina. The order is mostly in the order of the polarity (dipole moment) of the solvent molecules. Hexane is low, ether in the middle range, and water on the high end. However, the polarity of the solvent is not the only factor that will cause better solvation and faster elution. Another could

Table 13.3 Solvent Strength Parameter $e°$ for Alumina Supports: The Eluotropic Series

In Alphabetical Order		In Numerical Order Low to High	
Solvent	***e°***	**Solvent**	***e°***
Acetic acid	1.0	Pentane	0.00
Acetone	0.56	Petroleum ether	0.01
Acetonitrile	0.65	Hexane	0.01
Benzene	0.32	Cyclohexane	0.04
Carbon tetrachloride	0.18	Carbon tetrachloride	0.18
Chlorobenzene	0.30	Xylene	0.26
Chloroform	0.40	Toluene	0.29
Cyclohexane	0.04	Chlorobenzene	0.30
Dimethylsulfoxide	0.62	Benzene	0.32
Dioxane	0.56	Ethyl ether	0.38
Ethyl acetate	0.58	Chloroform	0.40
Ethylene dichloride	0.49	Methylene chloride	0.42
Ethyl ether	0.38	Tetrahydrofuran	0.45
Hexane	0.01	Ethylene dichloride	0.49
iso-Propanol	0.82	Methylethylketone	0.51
Methanol	0.95	Dioxane	0.56
Methylene chloride	0.42	Acetone	0.56
Methylethylketone	0.51	Ethyl acetate	0.58
Pentane	0.00	Dimethylsulfoxide	0.62
Petroleum ether	0.01	Acetonitrile	0.65
n-Propanol	0.82	Pyridine	0.71
Pyridine	0.71	*iso*-Propanol	0.82
Tetrahydrofuran	0.45	*n*-Propanol	0.82
Toluene	0.29	Methanol	0.95
Water	Large	Acetic acid	1.0
Xylene	0.26	Water	Large

be the hydrogen bonding ability of the solvent or other, perhaps more specific, factors that affect the equilibrium of the adsorption.

We are not limited to separations based only on interactions with silica and alumina. Virtually any compound can be linked covalently to the solid particles. With silica supports, for instance, silicon-oxygen-carbon bonds on the surface can anchor carbon compounds onto the surface. Some of the groups that have usefully been bonded to silica surfaces for normal-phase separations are listed in Table 13.4. These packings with covalently linked surface groups are called **bonded phases**.

Reversed-Phase Liquid Chromatography

In reversed-phase chromatography, the stationary phase is less polar than the mobile phase. Two fundamental types of stationary phases are used, the most common being nonpolar groups bonded onto silica. Of these, the most often used are the organic groups —CH_3, —C_8H_{17}, and —$C_{18}H_{37}$. Of these, the 18-carbon chain (the octadecyl group) is the most common. The abbreviations ODS and C18 are used for this type of stationary phase. These bonded organic groups have an effect similar to that which would be produced by an extremely thin organic solvent layer on the surface of the silica particles. Thus, the solutes partition between the surface coating and the mobile phase much like a liquid-liquid extraction. In addition, the longer the carbon chain length, the more "organic" these bonded layers become. As a result, the longer chains interact more strongly with solutes that prefer to dissolve in an organic phase.

The second type of stationary phase used for reversed-phase chromatography is composed of organic polymer beads. A typical polymer is a resin composed of polystyrene and divinylbenzene. The divinylbenzene component forms bonds between the polystyrene polymer chains—**crosslinks**. The crosslinks cause a physical stiffening of the polymer; a greater density of crosslinks is associated with a stiffer material. Stiffness is necessary to resist deforming under high pressure.

Reversed-phase chromatography is quite popular since the peaks in a reversed-phase separation tend to be narrow and symmetrical and the adsorption/desorption equilibrium reactions tend to be fast.

In contrast with the eluotropic series for stationary phases such as silica and alumina, less polar solvents are more powerful eluents in reversed-phase separations. In general, the reversed-phase eluting power is the opposite the order of Table 13.1 for normal-phase elution. This reversed-phase eluting power is put on a numerical scale that is called the **polarity index**.

Table 13.4 Representative Groups Bound to Silica for Normal-Phase HPLC

Type	Application
Iminodiacetic acid–Ni^{2+} complex	Proteins, polypeptides
Nitrilotriacetate–Cu^{2+} complex	Proteins, polypeptides
Cyclodextrins (cyclic oligo D-glucopyranose)	Racemate separations
Bovine serum albumin (a protein)	Chiral separations
Amino acids (Cu^{2+} complex in mobile phase)	Chiral separations
Ferrocenylpropyl amine	Chiral separations

A comparatively new stationary phase is a form of graphite. The material is called **porous graphitic carbon** (PGC), which is essentially pure carbon, and the surfaces are molecularly flat. There are few non-carbon atoms, and those present are located at the edges of the large (on the molecular level) plates. PGC is less polar than the solvents, in general, and so falls into the class of reversed-phase supports. However, the most intriguing property of PGC is that the strength of the interaction between the PGC and solute molecules depends almost entirely on the molecular area that is in contact with the plate. As a result, large molecules tend to bind more tightly than small ones and flexible molecules more than rigid ones. Ionic charge does not affect the binding. And, finally enantiomers are separated when the structures allow a larger number of atoms to get closer to the surface in one enantiomeric form than another. This mixed mechanism of interaction offers an alternative to normal and bonded phases for certain classes of compounds.

Separation of enantiomers is discussed further below in "Chiral Separations."

Reversed-phase chromatography evolved from an older type called **partition chromatography**. For partition chromatography, a solid support was coated with an organic solvent such as *n*-butanol, benzene, or chloroform, and the mobile phase was water and/or a polar organic solvent. Partition chromatography is, thus, truly a liquid-liquid extraction chromatography, but the unbonded, organic coating tends to wash off over time. You will see the name partition chromatography referring to the original form in the older literature, but now the term sometimes is used to refer to chromatography on C18 and similar phases.

Ion-Exchange Liquid Chromatography (Ion Chromatography)

In solution, one of the causes for two ions to bind is the attraction of unlike charges. As an example, we explain the strengths of acids by the magnitude of the negative charge at the site of proton binding. Highly-charged sites tend to hold the protons more strongly. In the same way, if, for example, positively charged groups are anchored to a stationary phase, then ions of the opposite charge will be attracted to them. These ionic interactions can be used to separate eluting species by differences in their average charge.

Stationary-phase materials with bound charges are called **ion-exchange resins**. A few are listed in Table 13.5. The reason they are called ion-*exchange* resins is because ionic solutions always contain equal numbers of positive

Table 13.5 Ion Exchange Groups

Type	Active Group	pH Range of Operation	Application Example
Strongly acidic cation exchanger	$—SO_3^-$	1–14	Amino acids, inorganic separations
Weakly acidic cation exchanger	$—COO^-$	5–14	Transition elements, organic bases
Strongly basic anion exchanger	For example, $—N^+(CH_3)_3$	1–12	Alkaloids, fatty acids
Weakly basic anion exchanger	For example, DEAE, $—C_2H_4N(C_2H_5)_2$	1–9	Organic acids, amino acids

and negative charges; that is, they are electrically neutral. As a result, the ions of the resin support are always associated with some ion of the opposite charge from the mobile phase. Thus the ion initially associated with the site can leave only when it is exchanged with a new one, as illustrated in Figure 13.10.

Positive ions (cations) exchange with stationary phases composed of cross-linked polymers with fixed negative charges. These are called **cation-exchange resins**. In analogy, polymers with covalently linked positive charges on them exchange negative ions (anions). These resins are called **anion-exchange resins**.

However, the ion exchange is not quite as straightforward as it may sound since competitive chemical reactions occur simultaneously involving *all* the ionic species in the solution and on the resin. These ions include the bases and conjugate acids of the analytes as well as the bases and conjugate acids of the groups of the resin, and any buffers that may be present. Fortunately, we do not need to consider all the details of this complicated, simultaneous-equilibrium system, since the strengths of the various analytes' binding interactions with the stationary phase are proportional to the *average* charge of each acid-base pair involved. To the extent that we can control the fraction of each solute/analyte that is neutral, we can control the elution time for each. Because of the crucial dependence on pH, the mobile phase is almost inevitably buffered to a fixed and carefully chosen pH.

At present, the primary analytical use of ion exchangers is in **ion chromatography**. With ion chromatography, a single separation experiment can allow us to quantify numerous anions or cations down to the ppb level.

Gel-Filtration Liquid Chromatography

Gel-filtration separations differ from the other three types of interactions since the molecules are not separated through differences in chemical interactions. The separation is made on the basis of the effective sizes of molecules. (Again, though, let us remind you that no separation is accomplished due to a single, pure mechanism.)

The mechanism of gel filtration, illustrated in Figure 13.10, is perhaps best described using the alternate name for the process, **exclusion chromatography**. The stationary phase used in exclusion chromatography is a solid that has pores with controlled (as carefully and precisely as possible) cross sections. These pores permeate the volumes of the particles packed in the columns. The separation occurs because only smaller molecules can enter the pores, where they are removed from the flowing mobile phase. Thus, the larger molecules, which do not enter the pores of the gel, flow along with the eluent and elute first followed by the smaller molecules.

The retention times of molecules follow an interesting behavior. As illustrated in Figure 13.10, if a molecule is larger than the largest pore size (above the **exclusion limit**), then it will not enter into the pore throughout its passage through the column. On the other hand, all molecules *below* a certain size will spend an equal time tortuously making their way through the pores of the stationary phase. Thus there is a lower limit to the molecular size resolution. These low-molecular-mass molecules elute from the column in a volume called the **total permeation volume**, V_t. Between the upper and lower limits of size, the retention volume is a linear function of the *logarithm* of the molecular weight. That is,

$$\text{retention volume} = a_0 + a_1 \cdot \log \text{(molecular weight)}$$

The relatively linear part of the calibration curve corresponds roughly to the molecular weight region between about $\frac{1}{3}$ of the exclusion limit up to the exclusion limit.

Chiral Separations

As you know, isomeric molecules have the same atom content with different structures. **Chiral** isomers are **stereoisomers**, molecules that have the same atomic connectivities but differ in how the atoms are oriented in space. In addition, chiral isomers cannot be superimposed on their mirror images, just as your left and right hand cannot be superimposed on each other even though they are "mirror images." The molecules in a set of chiral stereoisomers that are mirror images of each other are also called **enantiomers** or an **enantiomeric pair**. An important point to understand about the chemical properties of stereoisomers is that *enantiomers have identical physical properties* (for instance, melting and boiling points) other than their interaction with polarized light.

Separations of enantiomers (chiral separations) are usually carried out by utilizing their differential interaction with chiral reagents in one of two general ways.

1. A chiral reagent is added to the eluent—a **chiral mobile phase**. The reagent is also called a **chiral additive** or a **chiral selector**. The reagent reacts differentially with the stereoisomers, which allows a separation to be achieved on an **achiral** support such as an ODS column. The explanation for the separation is that transient diastereomers are formed that are separable by the chiral reagent.
2. A chiral reagent is bound onto a solid support, and the differential interaction of the stereoisomers with this chiral phase causes a separation of the stereoisomers.

The choice of conditions and reagents is not uncomplicated, so details are not presented here. See the bibliography of this chapter for references.

Planar Stationary Phases

So far, liquid chromatography has been presented in the context of a column packed with a stationary phase. However, this is not the only shape that is used for chromatographic analysis. In another geometry, the stationary phase is composed of a layer a fraction of a millimeter thick of the particles attached to a solid backing plate of aluminum, plastic, or glass. A separation carried out on these planar, stationary phases is called **thin-layer chromatography** (TLC). (Unfortunately, sometimes *thick-layer chromatography*, with layers 0.5–2 mm thick and primarily used for preparations, is also abbreviated TLC.)

Figure 13.11 illustrates how TLC separations are done in its simplest form. A few microliters of the sample dissolved in a minimum volume of solvent are placed onto the *dry*, stationary phase a few cm from one edge of the plate. The solvent from the sample is then allowed to dry. To **develop** the chromatogram, the edge near the samples is placed into a shallow pool of development solvent. The solvent, drawn upward by capillary action, is allowed to rise to a predetermined height on the plate. The plate is then removed from the solvent tank and dried. Figure 13.11 shows a closed container, which keeps the atmosphere saturated with the vapor of the solvent. This saturated vapor prevents solvent from evaporating from the face of the

Solvent front on plate

Filter paper wick for solvent (helps vapor saturation)

Developer (mobile phase)

(*a*)

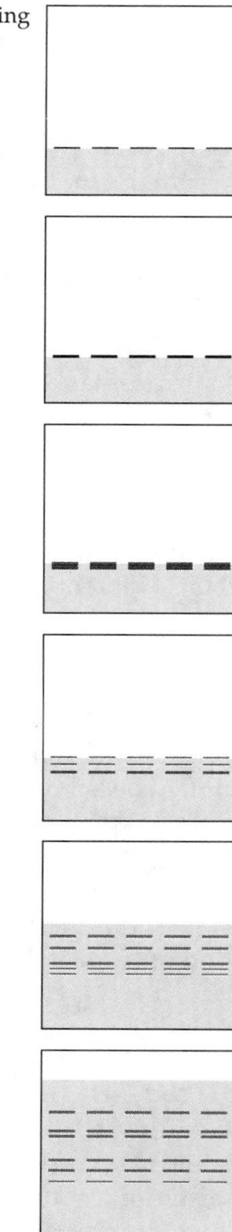

FIGURE 13.11 ▲

Illustration of the development of a TLC plate in a closed tank.

(a) The mobile-phase liquid is in the bottom of the development tank and rises by capillary action through the solid support, carrying the solutes with it. Before the plate was developed, it was spotted with three samples on a marked line. The different components are moved as spots traveling different distances relative to the solvent front. **(b)** A plate at various times during its development is shown; the five samples are the same material. [Chromatograms courtesy E. Merck, Darmstadt, Fed. Rep. of Germany.]

plate. Alternatively, depending on which conditions work best, the plate may be developed in an open tank where the atmosphere is not saturated with solvent.

The primary advantage of TLC is that a number of samples and standards can be analyzed simultaneously; with column chromatography, individual samples must be analyzed sequentially. Another advantage arises

1 Progesterone

2 11-Deoxycorticosterone

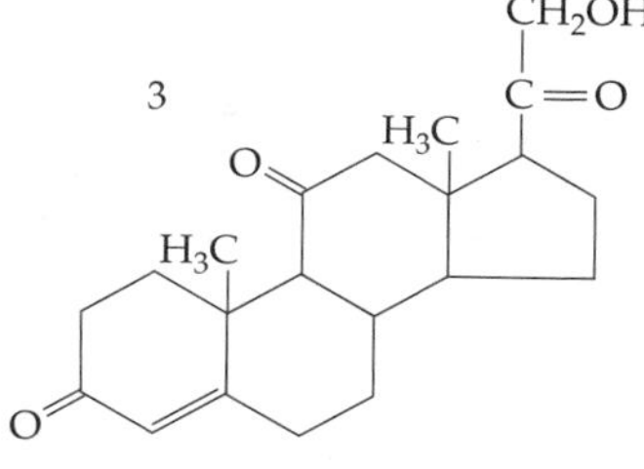

11-Dehydrocorticosterone

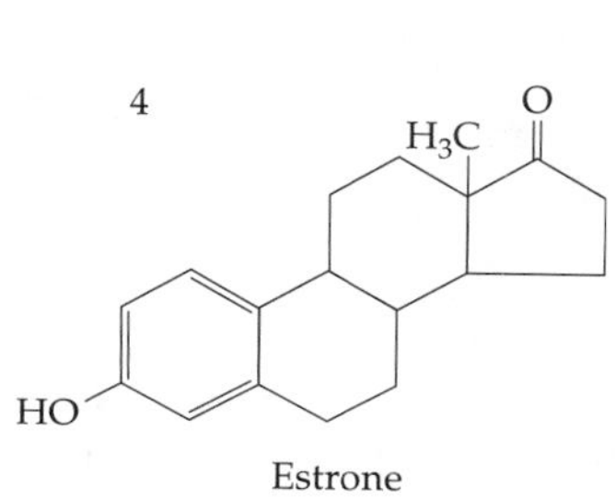

Estrone

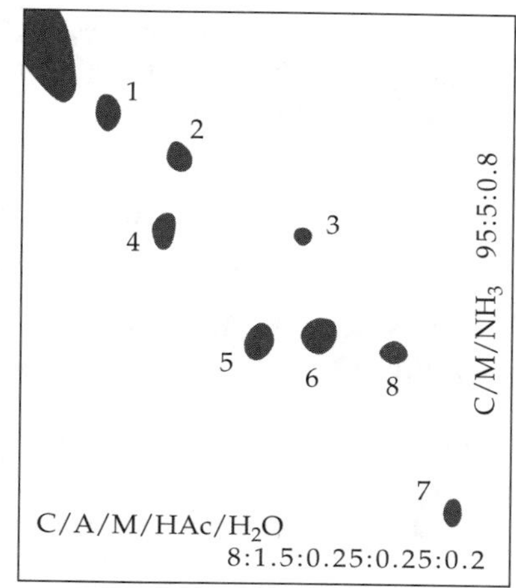

8 Corticosterone

5 Estradiol

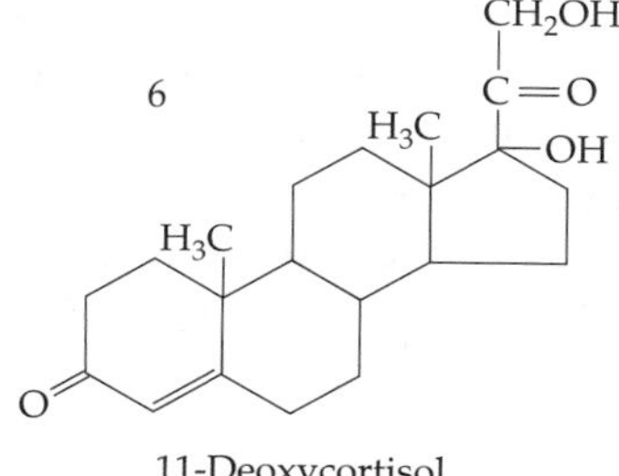

11-Deoxycortisol

7 Hydrocortisone

FIGURE 13.12 ▲

Results of a TLC plate developed in two dimensions in order to achieve separations that could not be done with a single development.

The spots are numbered with the various steroids that were separated. The development was first done in the bottom-to-top direction. Compounds 3 and 4 travel together, as do compounds 5, 6, and 8. After the plate was dried, these compounds were then separated by development from right to left with a different developing solvent system (see below). The sample was originally spotted at the point of compound 7, which was not moved by either solvent. [Ref: Rouser, G. J. 1973. *Chromatog. Sci.* 11:63. Reproduced from the *Journal of Chromatographic Science* by permission of Preston Publications, Inc.]

when samples that are difficult to resolve may be developed in two different solvents run in perpendicular directions. An example of the results of such a two-dimensional development scheme is shown in Figure 13.12. In addition, as long as no components are lost to the vapor surrounding the plate, all components are somewhere on the plate. This is in contrast to the case in column chromatography, in which some components may never elute and are thus lost. A further advantage is that since the TLC plate is used only once, harsh separation conditions that could degrade the support and would rapidly destroy a column can be used for TLC.

As can be seen in Figure 13.12, each of the individual components has traveled a different distance through the stationary phase. Thus the values of k (which depend on the time spent passing through the entire length of the stationary phase) cannot be determined in exactly the same way for TLC as for column chromatography.

For *qual*itative analysis, spots on a TLC plate are characterized by the distance they travel relative to the distance that the mobile phase traverses. The parameter used to characterize this ratio is called the **retention factor** or **retention index**, abbreviated R_F. (Do not confuse R_F with the resolution R_s.) For TLC,

$$R_F = \frac{\text{distance from origin to spot}}{\text{distance from origin to solvent front}} \qquad \textbf{(13-13)}$$

The R_F values depend on the same experimental conditions as do the k values of column chromatography: the composition of the mobile phase, the identity of the stationary phase, the temperature, and the identities of the compounds separated. Usually, standards in TLC are run simultaneously with a number of samples on the same plate.

Another type of planar configuration is **paper chromatography**. Historically, a wide range of materials has been separated by elution along the length of a paper sheet. The mobile phase is usually a water-organic mixture. The stationary phase is the main component of paper—cellulose (or chemically modified forms of it). It is generally accepted that the cellulose adsorbs water onto its surface. The interaction of the stationary phase with the solutes is, then, a combination of two types: a liquid-liquid extraction between the mobile phase and the water adsorbed on the cellulose surface as well as some contribution from direct adsorption on the underlying cellulose. Paper chromatography is relatively slow and has been replaced almost completely by reversed-phase HPLC and reversed-phase TLC.

13.8 Gas Chromatography

Gas chromatography (GC) is an analytical method that has been studied and used intensively for more than 40 years in solving a wide variety of analytical problems. As a result, conditions have been found for the separation of virtually all classes of compounds for which GC is an appropriate separation method. With this knowledge base as a good starting point, development efforts continue in order to optimize specific analyses.

The parameters describing the chromatograms for both gas and column liquid chromatography are calculated in the same manner. With consistent conditions, the retention times or volumes are used in the same way to aid identifying the chemical species eluting. The techniques used to measure the areas under the chromatographic peaks are also the same.

A block diagram for a typical GC instrument is illustrated in Figure 13.13. Gas chromatography equipment differs significantly from HPLC. GC requires precise control of the flows of gases instead of liquids, and the columns are longer and usually narrower. Also, the stationary phases differ.

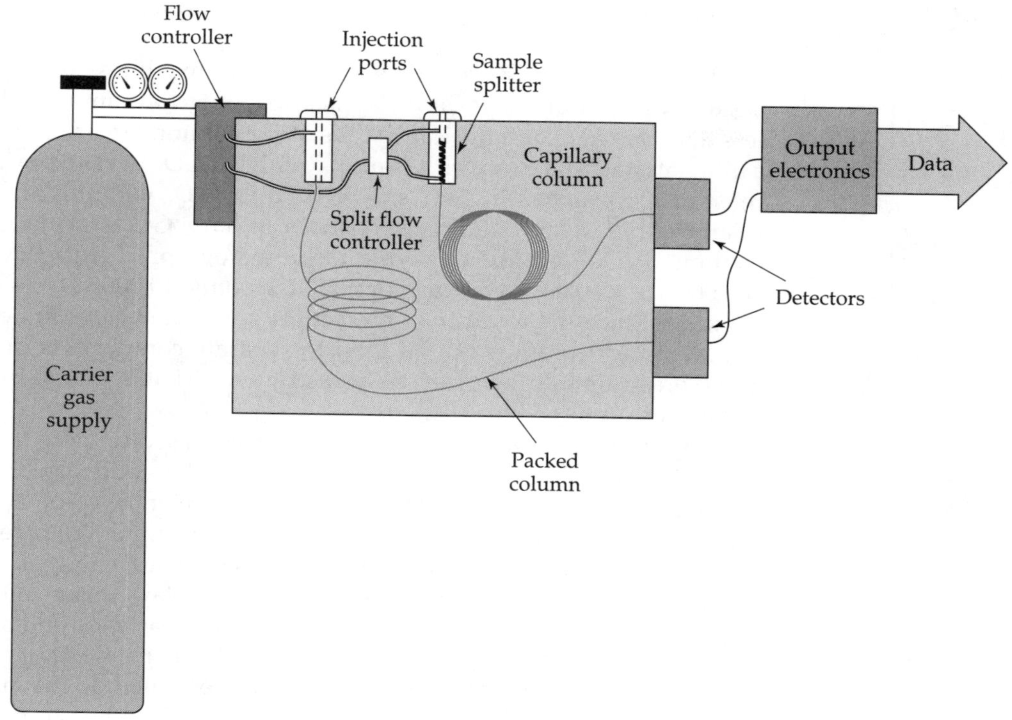

FIGURE 13.13 ▲

Diagram (not to scale) of the basic parts of a GC/GLC chromatograph.

The carrier gas stored in a tank under high pressure is connected to the system through an appropriate gas regulator. This provides gas at a fixed pressure to the instrument's gas-flow controller, which is another regulator. The sample is introduced into an injection port (usually heated) and passes into the chromatographic column. If the capillary column is chosen, the sample is usually diluted with carrier gas, and only a portion continues through the column. That is the function of the splitter. At the end of the column is the detector, the transducer that responds to the presence of analytes. Signals from the detector are amplified and recorded.

Many, but not all, detectors differ both in construction and in operation from those used for HPLC. Finally, since the carrier gas does not act as a solvent, the temperature range of the column is a critical variable in optimizing separations and must be strictly controlled.

The Nomenclature of Gas Chromatography

Gas chromatography may be classified first by the type of columns used and then by the categories of stationary phases.

In **packed-column** GC, the stationary phase consists of particles packed inside a glass or stainless steel column with an inside diameter (I.D.) that is typically 2–4 mm. The stationary phase can consist of various porous solid materials alone—**gas-solid chromatography**, GSC—where the retardation of analytes is due to an equilibrium involving adsorption onto and desorption from the surface. Alternatively, the stationary-phase particles may be coated with one of a number of high-boiling liquids—**gas-liquid chromatography**, GLC. The separation of the analytes depends on the different fractions of time they are dissolved in the stationary, liquid phase. The more soluble

Capillary column etched onto a silicon wafer. The wafer is about 5 cm in diameter. The column is 100μm diameter and 5 m long. [Courtesy of Conrad Yu, LLNL]

components are retained longer since they spend less time traveling in the carrier gas.

Packed-column GC methods are currently much less widely used than **capillary** gas chromatography. Capillary columns have internal diameters of less than one mm, and the inner walls of the column are usually coated with a film of stationary phase. Columns with an I.D. of 530 μm are usually referred to as **megabore**, and smaller, 100 μm I.D., columns are designated **microbore**. The most obvious reason capillary GC is replacing packed-column GC is because the typical chromatographic efficiency can reach 200,000 theoretical plates, as compared to only 10,000 or less for packed columns. Since the number of theoretical plates translates to better resolution, capillary columns can be used to separate components more quickly, and more samples can be run on the instrument. That is, the **sample throughput** is increased.

Capillary columns are further categorized based on how the stationary phase is immobilized inside the column. When the stationary phase consists of a film of material coated onto the inside surface of the column, the column is said to be a **WCOT** (*wall-coated open tubular*) column. To increase the column capacity the surface area can be increased by attaching packing material to the wall. There are currently two different ways this is done. A **SCOT** (*support-coated open tubular*) column has support particles on its wall; these are, in turn, coated with the liquid phase. A **PLOT** (*porous-layer open tubular*) column has a thin layer of a porous polymer deposited on the inner surface. The WCOT and SCOT columns are gas-liquid chromatography columns. SCOT columns are used far more often that WCOT columns because of their greater capacity. PLOT columns are used for gas-solid chromatography.

A photograph of a capillary GC column can be seen in EP.12 inside the back cover.

Capillary GC is one of the most powerful separatory techniques that now exists to resolve the components of complicated mixtures. The chromatogram in Figure 13.14 illustrates this point. Each peak represents at least one compound in an extract of oak leaves, and these are only the volatile components!

Samples Analyzed

Gas chromatography is the separation method of choice to analyze both organic compounds that have boiling points below about 250°C and the fixed gases; both groups have vapor pressures that are high enough so that they will be carried in the gaseous mobile phase. It is not low vapor pressure that usually limits the use of gas chromatography, however. Decomposition does. When higher-boiling materials are being separated, the compounds may decompose on the column at the temperatures required to obtain a chromatogram in a reasonable time.

On the other hand, thermal decomposition done before the chromatography can be useful. In **pyrolysis GC**, the analyte is deliberately decomposed at high temperatures, and the decomposition products are chromatographed. The peaks appearing in a pyrolysis-gas chromatogram provide an analytical "fingerprint" of the analyte. It is difficult to do quantitative analysis with pyrolysis GC, but it is routinely used to compare batches of complex materials, such as paints, heavy fractions of crude oil, or plastics.

At the other extreme of temperature, the GC column may be cooled to effect a separation of low-boiling materials such as methane or H_2. For such materials, gas chromatography can be run conveniently under conditions ranging from ambient down to dry-ice temperature, −80°C.

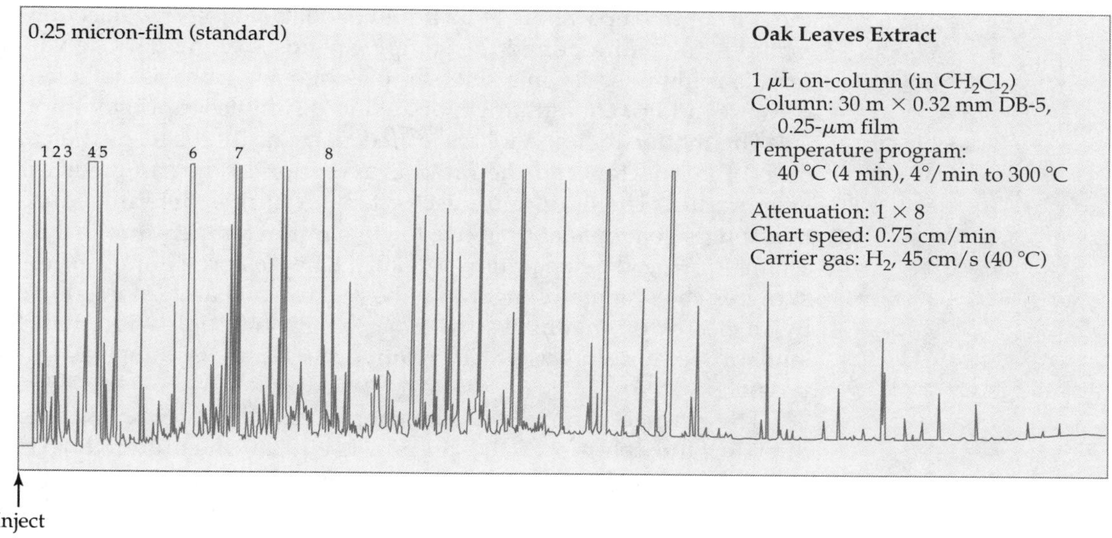

FIGURE 13.14 ▲
Chromatogram of the volatile components of oak leaves.
The conditions of the separation are listed on the figure. This separation was carried out on a capillary column with hydrogen as the carrier gas. The column temperature was gradually increased over time as the sample proceeded through the column. Total time for the chromatogram was about one hour. Even though a large number of components have obviously been separated, a statistical calculation suggests that a significant fraction of the peaks that appear actually are due to two or more coeluting species.

For some groups of compounds, chromatographic analyses can be done equally well by either GC or LC. Examples are sugars, fatty acids and amino acids. Since GC requires volatile samples, such compounds must be **derivatized**. Derivatization in GC means carrying out a reaction on the chemical groups that cause low vapor pressure (high boiling points) in order to produce a compound that is more easily vaporized. A simple example of derivatization is the conversion of a carboxylic acid to its methyl ester: for example, hexanoic acid, $C_5H_{11}COOH$ (b.p. 187°C), to its methyl ester, $C_5H_{11}COOCH_3$ (b.p. 127.3°C). Derivatization can also be useful when components are reacted to yield chemical groups that provide strong responses to the detector used with the instrument. An example is to add chlorines (such as a $—CCl_3$ group) when using an **electron capture detector** that is highly sensitive to halogens.

Both LC and GC also can be used to separate moderately high-boiling organics (<200°C). For those that boil above 200°C, liquid chromatography is more likely to be chosen.

Samples and Sample Introduction

Samples for GC can be gases, liquids, or solids. Solid and liquid samples must be vaporized in some manner. If the solid samples are easily vaporized, then the molecular components can be chromatographed. However, if the solids are decomposed, only the decomposition products can be analyzed to characterize the sample.

To obtain the best separations, the samples should be injected over the shortest possible time. The longer the injection time, the broader the bands at the start of the separation. For a liquid or solid sample, the injection time also includes the entire period the sample spends vaporizing. The vaporization time is minimized by injecting samples into a region that is hotter than the column and hot enough to vaporize liquid samples rapidly without pyrolysis. This hotter region and associated parts at the head of the column are called the **injector**, and the rapid evaporation of the sample is called **flash evaporation**. The optimum injector temperature is determined experimentally, but is generally at or above the maximum temperature that the column attains during the separation. Usually, samples are simply injected with a syringe. The syringe plunger may be pushed by hand (manual injection) or by an electric or pneumatic driver (automatic injection). Automatic injection tends to be more precise, with relative standard deviations as low as 0.1% possible.

The syringe can deliver the sample into the injector or, alternatively, directly onto the top of the column itself (an **on-column injection**). The choice between these two methods is often a consequence of the column diameter. Either method can be effective for packed columns but sample introduction is a rather tricky process for capillary columns less than 1 mm diameter for two reasons: first, the open cross section of the capillary is very small, and second, the sample must be small in order not to overload the stationary phase and cause distortion of the band. Enough stationary phase must be present to interact with *all of the components of the analyte mixture at one time*.

To keep the quantity of sample in the correct range, a special injector reduces the quantity of sample that reaches the head of a capillary column: a **split/splitless injector**. Samples with adequately low analyte concentrations are injected in the *splitless mode*, and carried onto the column by carrier gas just as with a conventional injection block. When the *split mode* is used, carrier gas flows continuously past the inlet at a rate 50 to 1000 times the rate the carrier flows through the column. The excess flow is vented, and only a fraction of the sample is carried onto the column.

Columns

The choice of diameter, length, and stationary phase depend first on whether column or capillary chromatography is being done. Some figures of merit for packed columns and capillary columns are given in Table 13.6. For most samples the choice of stationary phase can be made by relying on specific information (for example, polarity) about each type of stationary phase and on the specific conditions for various types of samples, which can be found in vendors' catalogs/handbooks.

Another choice to be made in GLC is the amount of liquid phase present. The effect of using more liquid phase is twofold. First, the capacity is higher, so larger amounts of sample can be used. More sample leads, generally, to a better quantitation, a better limit of detection, and a greater measurable range in the concentrations of analytes in the sample. However, as more stationary phase is available, analytes spend more time dissolved in the stationary phase and the retention time is increased. So capacity must be balanced with analysis time. For capillary columns, the column diameter and the thickness of the stationary phase layer are critical in determining how well the mobile phase-stationary phase equilibrium is established, and thus the quality of the separation. The liquid phase in packed columns is measured by the **percent loading**, which is the weight/weight measure of

Table 13.6 Comparison of Typical Packed Column and Capillary Column Characteristics

	Packed	WCOT and SCOT	PLOT
Column I.D.	3–6 mm	0.10–.53 mm	0.32–0.53 mm
Column length	1–3 m	30–50 m	30–50 m
Particle diameter (d_p)	120–185 μm	—	—
Coating thickness	—	0.1–0.5 μm	12–25 μm
Theoretical plates/m	1000–5000	1000–8000	1000–8000
Typical number of total theoretical plates (N)	1000–5000	25,000–200,000	25,000–200,000
Total capacity (μg/component)	100	0.05–5	0.05–5
Surface area of stationary phase (m^2)	0.2	10^{-6}–10^{-7}	10^{-6}–10^{-7}
Volume flow rate (mL/min)	20–100	0.5–5	5–10
Linear flow rate (cm/s)	10–50	20–50	50–100

stationary phase compared to the total weight of the packing. The amount of stationary phase available is also a function of the particle diameter. The particle diameters are expressed in terms of the **mesh size** of the packing. These numbers define the hole size in standard screens for sifting powders. The higher the number, the more lines in the mesh and the smaller the particles in the packing. For example "80/100 mesh" means that the particles can pass through the "80" screen, but are trapped by the "100" screen. These particles are approximately 150–175 μm in diameter.

Temperature

The temperature is crucial to the separations in gas chromatography. Control of the temperature within ±0.5°C is worthwhile. Temperature affects the positions of the distribution equilibria of analytes between the mobile carrier and the stationary phase as well as how quickly the equilibria are established. Raising the column temperature speeds both the elution and the approach to equilibrium between mobile and stationary phase. An optimum temperature is found between the constraints of needed resolution and the desire to minimize the time for the separation. The separation can be conducted at a constant temperature (an **isothermal** separation) or with increasing temperatures programmed for optimum results. The increase in temperature as the separation proceeds is called a **temperature gradient**. A *gradient elution* in GC means a *temperature* gradient. The gradient can take many forms. *Ramps* in temperature can be linear, or one can proceed through a number of ramps and plateaus as illustrated in Figure 13.15. This increase in temperature with time allows the analysis of compounds with widely varying boiling points and retention characteristics to be accomplished in shorter periods of time. Note that the temperature gradient is a change in temperature with time, dT/dt. It is *not* a gradient of temperature over the length of the column. The highest temperature during a separation is not only limited by the sample decomposition mentioned earlier, but the stability and vapor pressure of liquid stationary phases also must be considered. The stationary phase itself can be vaporized and/or decomposed and the material then passed along the column and eluted. This is called **column bleeding**—a graphic description. For a specific separation, the low end of the

(a)

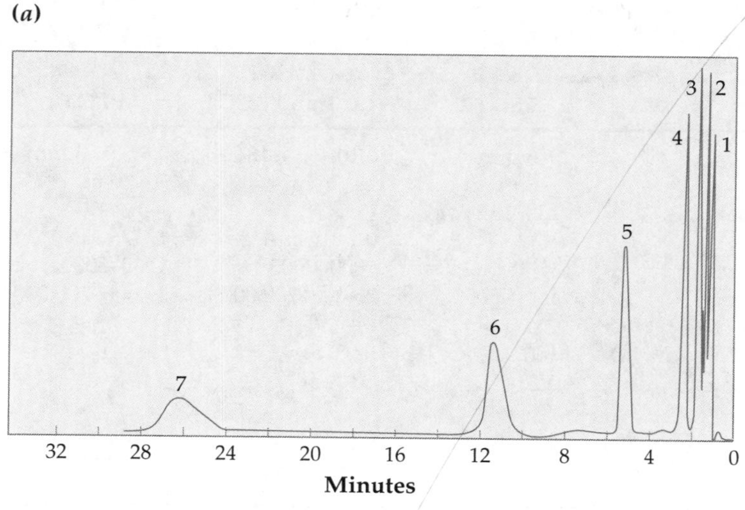

(b)

◀ FIGURE 13.15 **Chromatogram comparing a separation obtained under isothermal and gradient conditions.**

(a) $T = 168°$c. **(b)** T increasing at 6°C min^{-1} from 50°C. Use of the gradient causes the early eluting components to have longer retention times and be better resolved, while the peaks corresponding to components with very long retention times are sharpened. *Sample*: 1, pentane; 2, hexane; 3, heptane; 4, 1-octene; 5, decane; 6, 1-dodecene; 7, tetradecane. [Ref: Nogare, S. D., Bennett, C. E. 1958. *Anal. Chem.* 30:1157. Copyright 1958 American Chemical Society.]

temperature range is determined by the properties both of the stationary phase and analyte components. In the extreme case, at excessively low temperatures, all the compounds injected will merely stay at the entrance to the column or the top of the stationary phase, never to move again. At higher, but still too low, temperatures, the adsorption/desorption process is sluggish, and the peaks are broadened.

Flow Rate of the Mobile Phase

Typical flow rates for a variety of GC columns are listed in Table 13.6. While packed columns are generally operated at flow rates of tens of milliliters per minute (in some cases up to 200 mL/min), capillary columns are normally operated at flow rates ranging from well below 1 mL/min up to a maximum of about 5 mL/min. An increase in flow rate can often speed up elution without having a great effect on resolution.

Detectors

A number of different detectors are available for indicating the changes in eluent composition in gas chromatography. The most widely used of these, along with brief synopses of their **figures of merit**, are listed in Table 13.7. Figures of merit for detectors include their limits of detection and linear ranges. Even with capillary GLC and the low-capacity columns that require small samples, it is possible to do ppm determinations, which require quantifying 10^{-10} g of components of real samples. This 10^{-10} g is the mass integrated over the entire band. At any time, less is present in the transducer, and the limit of detection is usefully given in g s^{-1} or smaller units such as pg s^{-1}. As you may now have come to expect of all transducers, the detectors are not equally sensitive to all compounds. For example, two of the most common detectors, **flame ionization** and **electron capture**, have significantly different selectivities. Also, the least complicated detector—the **thermal conductivity** detector—responds to the widest range of compounds but is the *least* sensitive of those in common use. (Curses, foiled again!)

Combining the separating power of chromatography with the identification capabilities of infrared spectrometry or mass spectrometry provides

Table 13.7 Properties of Selected Gas Chromatography Detectors

Type	Approximate Limit of Detection (g s^{-1})	Approximate Linear Range	Comments
Thermal conductivity (TCD)	10^{-5}–10^{-6}	10^{3}–10^{4}	Universal detector. Measures changes in heat conduction.
Flame ionization (FID)	10^{-12}	10^{6}–10^{7}	Universal detector. Measures ion currents from pyrolysis.
Electron capture (EC or ECD)	10^{-14}	10^{2}–10^{3}	Selective detector for compounds containing atoms with high electron affinities.
Flame photometric (FPD)	10^{-13}	10^{2}	Selective detector for compounds containing S, P.
Nitrogen-phosphorus	10^{-8}–10^{-14}	10^{5}–10^{7}	Selective for N, P containing compounds.
Photoionization (PID)	10^{-8}–10^{-12}	10^{5}	Universal (some selectivity due to identity of gas in lamp).
Hall Detector	10^{-11}	10^{5}	Specific detector for compounds which contain halogen, S, or N.
Mass spectrometer (MS)	10^{-12}	*a*	Universal detector.
Fourier-transform infrared (FTIR)	10^{-10}	10^{2}	Polar molecules.

a. Variable and dependent on the type of mass spectrometer as well as the kinds of compounds being analyzed.

powerful tools for screening complicated samples. Both the infrared spectrometer and mass spectrometer responses are specific to each component even if both elute in the same band. If a component is a known compound, it is highly likely that pattern-searching algorithms can help identify the component, and quantitative analysis can be done as well. These are **combination** or **hyphenated techniques**. These combinations are called, respectively, gas chromatography with Fourier-transform infrared detection, abbreviated **GC/FTIR**, and gas chromatography/mass spectrometry, abbreviated **GC/MS**.

13.9 Supercritical Fluid Chromatography

Supercritical fluids combine characteristics of both gases and liquids. For example, supercritical fluids' diffusion coefficients and viscosities approach those of the carrier gases of gas chromatography, while their densities approach those of liquids.

We can vary three conditions of supercritical fluid chromatography that parallel changes for HPLC: temperature, eluent composition, and mobile-phase velocity. One condition can be varied like in GC: the pressure, which influences both the flow rate and the density of the supercritical fluid. However, unlike GC or HPLC, in supercritical fluids an increase in pressure at a given temperature increases the density, which increases the solubility of most analytes. Also unique to SFC, an increase in temperature at a given pressure leads to a *lower* fluid density, which leads to a *lower* solubility. As a result, even at constant temperature and constant eluent composition, the pressure drop from the inlet to the outlet of the column means that the solubilities of analytes decrease as they proceed through the column. In short, method development is less straightforward than for GC or HPLC.

Analytes

At present, SFC has not found the widespread popularity that exists for GC and HPLC. However, the technique is nearly ideal for separations of some surfactants and polymers. (An example is shown in Figure 13.16.) SFC is preferred for these compounds since volatilization into a gas is not necessary and the mass transfer is an order of magnitude higher than in HPLC mobile phases. This means that the analysis time can be decreased significantly due to the faster equilibration time between mobile and stationary phases. SFC also has been shown to be useful for separating some chiral compounds. The retention order tends to be the same as for chiral HPLC, but, again, separations may be faster.

The Mobile Phase

The SFC mobile phase begins, just as with GC, with a cylinder of gas under high pressure. However, in this case the fluid remains compressed and is heated above T_c (its critical temperature) to obtain supercritical conditions. Carbon dioxide is by far the most often used mobile phase because of its low cost, non-flammability, and non-toxic nature. Since it is non-polar, polar modifiers are added to allow analysis of polar compounds. **Modifiers** that commonly are added include methanol and ethanol and, for basic substances, amines. Gradients in SFC can be imposed by changing any conditions that modify the analyte solubilities in the supercritical mobile phase.

ANALYTICAL CHEMISTRY IN ACTION
Detecting Steroid Abuse in Athletes

After many international sport events, especially the Olympics, the winners are tested for misuse of anabolic androgenic steroids. The method used to determine the steroid abuse is gas chromatography/mass spectrometry, and, more specifically, high-resolution mass spectrometry with specific-ion monitoring. The original drugs are modified by the body's metabolism, so the urine is tested for the metabolites. The most effective tests measure the species that are retained the longest time, since their presence probes behavior for the longest period prior to the competition. For example, stanozolol is shown here. The analytes are stripped from the urine by *t*-butylmethyl ether extraction and further cleanup can be done with an immunoabsorbant chromatography column made with antibodies that react with methyltestosterone. After derivatization with trimethyliodosilane, the analytes are screened by capillary gas chromatography with a high resolution ($\Delta M/M = 3{,}000$) mass spectrometer with selected ions monitored sequentially. An example of the results is shown in the Figure for stanozolol metabolites labeled 1 and 2 in the scheme. The peak numbered 3 is an internal standard of 4α-hydroxystanozolol added to the derivatization mixture at a level of 100 fg $(\mu L)^{-1}$. The high-resolution peak's positions are noted on the chromatograms with the elution times on the ordinate.

Anabolic steroid metabolism. ▲

◀ **Monitoring of GC of steroids by high resolution MS.**

[Refs: Schänzer, W., Delahaut, P., Geyer, H., Machnik, M., Horning, S. 1996. *Journal of Chromatography B*. 687:93–108. Schänzer, W., Donike, M. 1992. *Anal. Chim. Acta*. 27:23.] ■

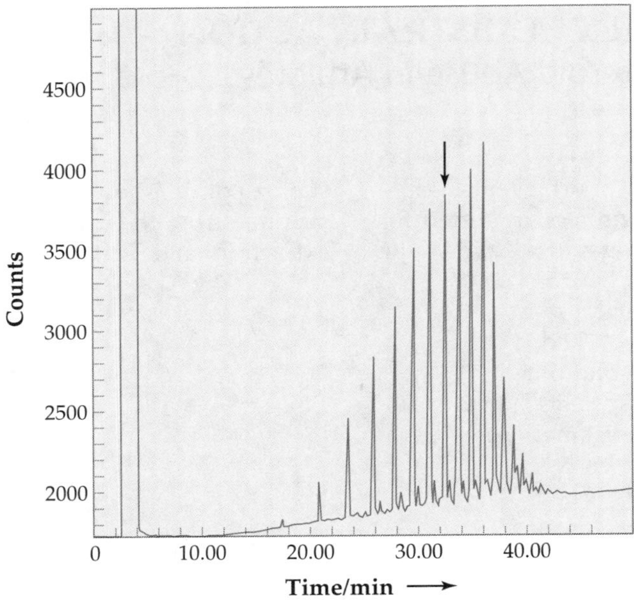

◀ FIGURE 13.16 **Separation of oligomers in the surfactant Triton X-100 by SFC.**

[Ref: Chromatogram courtesy of Hewlett-Packard.]

Sample Form and Sample Injection

Gas phase samples are not suitable for SFC. Solids are generally extracted by on-line supercritical fluid extraction and the extract concentrated. By far the most common form for an SFC sample is a solution. Liquid sample introduction in SFC is done with an injector valve equipped with a sample loop similar to that used for HPLC.

Columns and Packings

Columns used for SFC are essentially those used for HPLC and some GC column types. Columns made of stainless steel are packed with 5 μm particles; small bores of 250 μm and 530 μm are the most common. These packings can be any of the types used in normal- or reversed-phase HPLC: porous silica, alumina, and silica with bonded organic groups.

Detectors

Detectors that are commonly used for GC (such as flame-ionization detectors) as well as those used in HPLC (such as measuring the absorption of light) have been adapted for use with SFC. Detectors adapted from GC must be constructed to function with the relatively large volume of gas that is produced upon depressurizing the supercritical fluid, while detectors such as are used for HPLC must be modified to withstand the high pressures required to sustain supercritical fluids.

13A A Deeper Look: The van Deemter Equation

Thus far in this chapter, the widths of peaks in chromatograms have been understood as reflecting the distribution of analyte in the eluent. The chemical factors involved in this spreading are understood by the model that will be described in this section. It is important to realize that the details of the model are not generally used on a daily basis. However, the concepts in it are used when thinking about how to run better separations and what conditions or materials should be varied to do so.

The model explains variation in H, the HETP, as it depends on the average linear velocity of the mobile phase, $\overline{u}$. Recall that H, defined in Equation 13-8, is a measure of the ratio between a band's width and its retention time. (This result arises because H depends on N, the efficiency. A change in efficiency is illustrated in Figure 13.5.)

In a simplified form, the model equation can be written as

$$H = A + \frac{B}{\overline{u}} + C\overline{u} \qquad \textbf{(13A-1a)}$$

You can read this equation as: The HETP (or, equivalently, efficiency) is determined by some constant value, a factor that is inversely proportional to the mobile-phase velocity, and some factor that is directly proportional to it. Equation 13A-1 is called the **van Deemter equation**. Each of the parameters A, B, and C are constants determined from certain physical properties of the stationary and mobile phases. All are positive numbers.

As we shall discuss, the A-term is not needed for well-packed GC columns and for GC capillary columns. In other words,

$$H = \frac{B'}{\overline{u}} + C'\overline{u} \qquad \text{for GC}$$

The superscript primes indicate that the B and C values differ from those used when an A-term is included.

A slightly more complex form of the equation is more helpful. The second term, $C\overline{u}$, can be divided into two parts, ascribing part of the effect to the stationary phase and part to the mobile phase. Thus, C is describable as a sum of C_M and C_S from the mobile phase and stationary phase, respectively. (These are *not* the concentrations in each phase. The use of C for two different quantities in the same field of study is an historical accident. However, in any reading in this area the correct meanings should be clear to you from the context.) With this separation of the C-term, the model equation becomes

$$H = A + \frac{B}{\overline{u}} + (C_M + C_S)\overline{u} \qquad \textbf{(13A-1b)}$$

In the following sections let us investigate the general algebraic properties of this equation and more details of the origins of the parameters A, B, C_S, and C_M. The description of these origins will be mostly pictorial.

Recall that $\overline{u}$ is the actual velocity (for instance, in cm s^{-1}) of the mobile phase inside the column. It is *not* the volume flow rate of solvent (in, for instance, mL min^{-1}) through the column.

Equations 13A-1 describe the *relative broadening* (not the retention times) of chromatographic bands. If you were given a set of $C = (C_S + C_M)$, B, and A values, then using Equation 13A-1 alone, you could recreate the experimentally found HETP as it depends on $\overline{u}$. A plot of such values for both HPLC and GC is illustrated in Figure 13A.1.

At $\overline{u}$ near zero, the $(B/\overline{u})$-term is the largest. It decreases in value with increasing $\overline{u}$ and tends toward zero. In an opposite trend, the $C\overline{u}$-term is zero at $\overline{u} = 0$, and it increases with increasing $\overline{u}$. The $C\overline{u}$-term increases without limit as $\overline{u}$ increases. The sum of these terms, the HETP, has a value that is large and decreasing at very low flow velocities and large and increasing at very high mobile-phase velocities. As a result, there must be a minimum

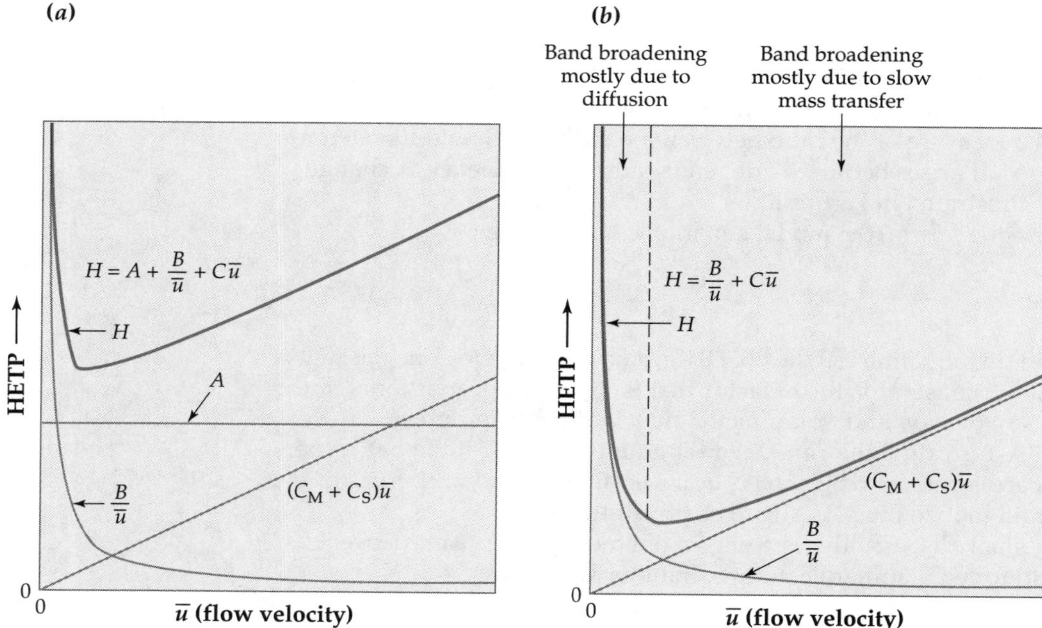

FIGURE 13A.1 ▲
Plots of the van Deemter equation's factors.
The graphs show van Deemter plots of the behavior of the height equivalent to a theoretical plate (HETP) as the mobile-phase flow velocity changes for **(a)** HPLC; **(b)** for capillary GC and GC with well-packed columns. The curves labeled *H* show the experimental behavior. We can attribute the flow-dependent behavior to two independent causes. The first is longitudinal diffusion (along the column length) characterized by *B*. The second is the broadening effect of transverse diffusion (perpendicular to the long axis) and rates of reaction at the support, characterized by *C*. A third factor, characterized by *A*, is needed for LC and for less perfect GC systems. The contributions from each are calculated from the terms in Equation 13A-1. The highest efficiency occurs at the minimum. Graphs (a) and (b) are plotted on different scales, as can be seen from the quantitative van Deemter plots for specific gas, liquid, and supercritical fluid chromatography experiments shown in Figure 13A.2.

value of H at some value of $\overline{u}$ in between, and the $\overline{u}$-value at which the minimum occurs depends on the values of $C = (C_S + C_M)$ and B. Contributions from processes described by A raise and lower the entire curve, so the minimum value of H depends on all three parameters.

Figure 13A.2 illustrates representative experimentally determined plots of the HETP as it depends on $\overline{u}$ for gas chromatography, supercritical fluid chromatography, and liquid chromatography. As you see, there is a significant difference between gas chromatography and liquid chromatography, both in the values of H and in the optimum $\overline{u}$-range. GC separations are often done near the optimum H vs. $\overline{u}$ point. However, the flow rates at which the optimum occurs for LC are so low that LC separations are usually done somewhat above the optimum value. Also, as can be seen in Figure 13A.2, the HETP for SFC separations remains almost constant at flows much above the optimum. As a result, SFC can be run at higher flow rates and gains even more advantage over HPLC. On the other hand, the technical difficulties of SFC have prevented its widespread adoption despite this advantage in speed.

The values of the chromatographic efficiency parameters (HETP, N) for a specific set of conditions are calculated from the properties of a single band.

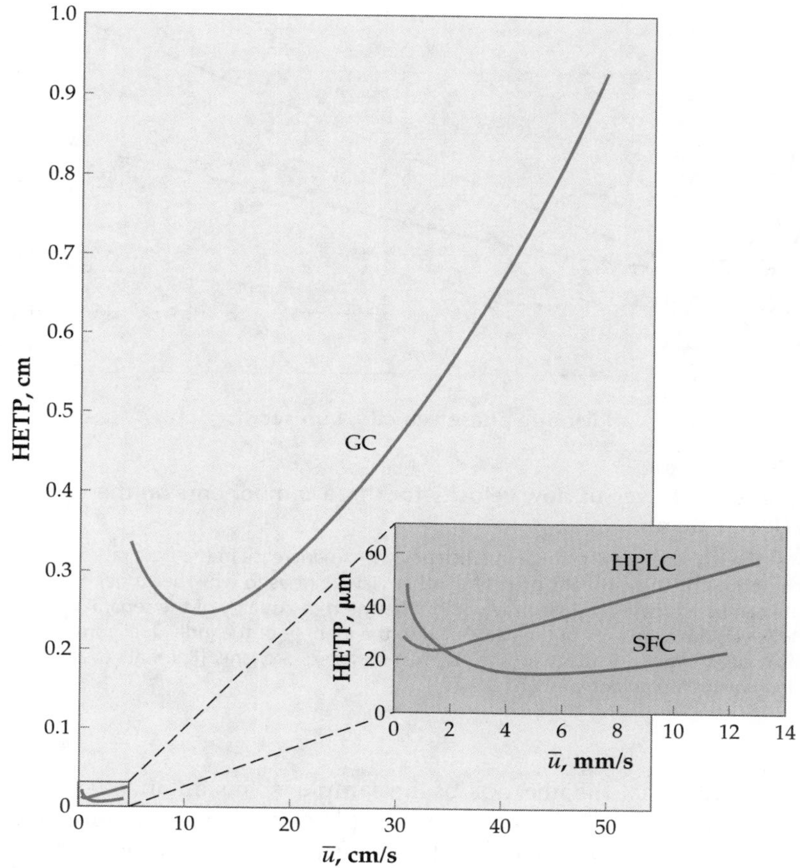

◀ FIGURE 13A.2 **Plots of HETP versus flow velocity for specific gas, liquid, and supercritical fluid chromatography systems.** The gas chromatography experimental data are calculated from a peak with $k = 7.90$ on a chromatogram run with nitrogen as the mobile phase carrier gas. (The temperature was 175°C.) The liquid chromatography plot is from data using a 4.6-mm inside diameter column packed with particles 10 μm in diameter.The curve for supercritical fluid chromatography is for a column 4.6-mm inside diameter packed with particles 5 μm in diameter with a fluid of supercritical CO_2 at 50°C and an average column pressure of 240 bar. All three chromatographies are significantly different in practice as indicated perhaps by the significant differences in these three plots.

The conditions that affect them include the identity and physical characteristics (shape, physical dimensions, surface characteristics, and so forth) of the stationary support, the identity and flow rate of the mobile phase, and the temperature. We want to emphasize here that the efficiency is a function of all these characteristics, and intimately depends on the chemical properties of the analytes as well. The sensitivity and complicated nature of the measurement of N is illustrated in Figure 13A.3. The curves shown are calculated from three different components of a multicomponent mixture run at different flow rates over the range plotted. Clearly, the value of N is highly sensitive to operating conditions, and comparing the efficiency of one type of chromatography to another or even one column to another must be done with extreme care. In addition, the values of the variables A, B, and C in Equation 13A-1 will be different *for each component eluted*. With that precaution in mind, let us probe more deeply into these variables' origins. The behavior of each of the terms in Equation 13A-1 is not fortuitous, as you will see next. Each term corresponds to one or more physical effects that affect chromatographic separations.

The Zone-Broadening (van Deemter) Equation in More Detail

Next, the origins of A, B, C_M, and C_S are explained in more detail. This description is presented primarily in a pictorial way. The purpose is to help you understand how and why chromatographic conditions can be varied to obtain better separations.

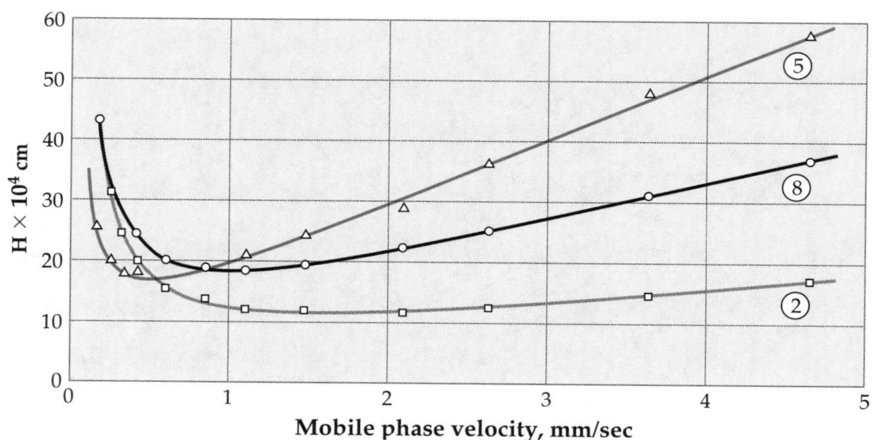

FIGURE 13A.3 ▲

Examples of HETP versus flow velocity for three components on the same HPLC column.

Note first the relatively slow change in efficiency with an increase in the flow rate past the optimum. These graphs present another viewpoint as to why the carrier flow rate can be increased significantly without harming the quality of the separation. However, the sensitivity of the value of *H* to the identity of the individual components is clear. *Sample*: 5, dioctylphthalate; 8, *o*-cresol; 2, *p*-xylene. [Ref: Katz et al. 1984. *Journal of Chromatography* 270:51–75.]

Let us consider that the injection of the sample is "instantaneous." This means that the time for the injection to take place is very short compared with the width (in time) of any effluent band. Thus, if no band broadening occurred at all in the elution, each band in the chromatogram would appear to be instantaneously narrow. However, obviously they are not. Each term of Equation 13A-1 represents a different process contributing to the broadening.

The Broadening Process Accounted for by $B/\overline{u}$

Diffusion is the random motion that tends to spread molecules uniformly throughout space. In the mobile phase, diffusion causes the molecules to migrate from the higher concentration region both forward and backward along the flow direction. This process is illustrated in Figure 13A.4. The broadening caused by diffusion in these directions is called **longitudinal diffusion** and is characterized by $B/\overline{u}$.

Why is the term inversely proportional to $\overline{u}$? The reason is that when the flow is faster, the time between injection and elution/detection is shorter. So there is less time for the material to spread out by longitudinal diffusion.

The Broadening Process Accounted for by $C\overline{u}$

Not only does diffusional spreading of the materials occur along the flow direction, as illustrated in Figure 13A.4, the molecules in the mobile phase diffuse *across* the flow direction as well. In this way, the solute molecules reach the surface of the stationary phase and also reenter the mobile phase. In addition, once the analytes are near the stationary surface, some time is required for the chemical interaction to occur. Adsorption and desorption have kinetic rates just as any other chemical reaction.

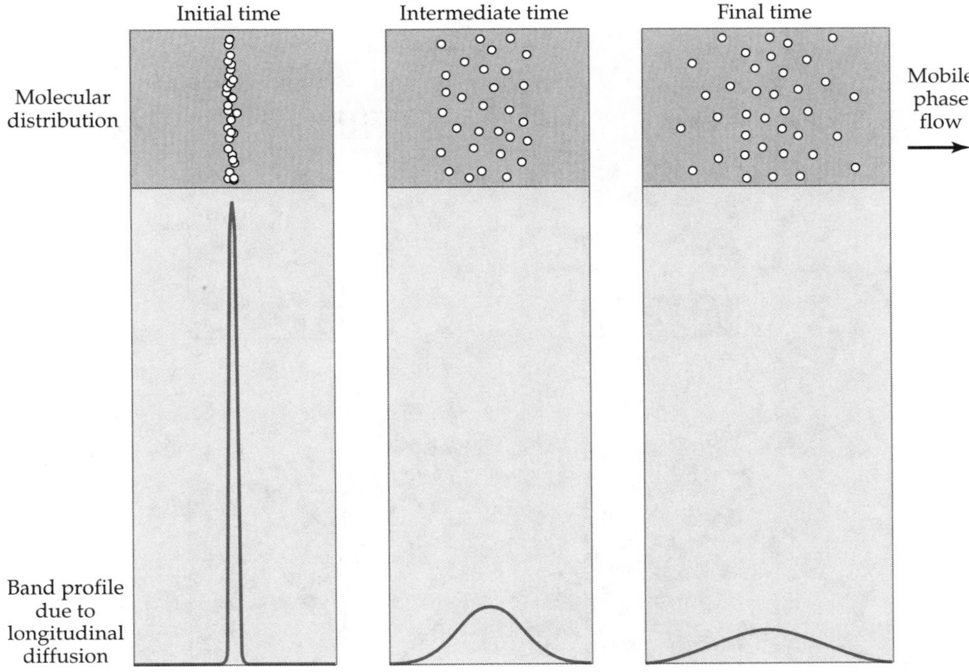

FIGURE 13A.4 ▲
Illustration of the mechanism of longitudinal diffusion in chromatography.
This is the *B*-term in Equation 13A–1. The molecules of the sample initially are together in a plane. As they pass through the particle bed, the molecules tend to diffuse randomly. The diffusion takes place both along and against the direction of mobile phase flow, as indicated in the figure. The longer time the sample is in the column, the wider will be the distribution of the component molecules. Thus, the contribution to the band's width increases with a slower flow rate. This process contributes little to the width of the bands in liquid chromatography.

We describe the two processes separately.

1. A diffusion across the flow direction—the C_M part
2. A reaction rate of the analyte and the stationary phase—the C_S part

Either or both of these processes can be "fast" or "slow" relative to the mobile phase linear velocity $\overline{u}$. The line between fast and slow, as you may guess, is not sharp. Fast and slow are extremes.

As illustrated in Figure 13A.5, if the mobile phase flow rate is slow relative to the diffusion across the flow direction, each solute remains within the same small volume as it is carried along. However, if the mobile phase flows faster than the diffusion, then the molecules that must diffuse over a range of distances to reach the support will be carried different lengths before adsorbing. The band thus spreads out. This spreading will be directly proportional to the velocity of the mobile phase. Since the entire broadening process occurs in the mobile phase, we express the effect as $C_M\overline{u}$.

The same sort of effect will occur if the actual adsorbing or desorbing process is relatively slow. The mobile phase carries the unbound material beyond the point where the same solute remains bound because of slow adsorption/desorption kinetics. The band then broadens, with the spread directly proportional to the flow velocity. This effect is expressed as $C_S\overline{u}$.

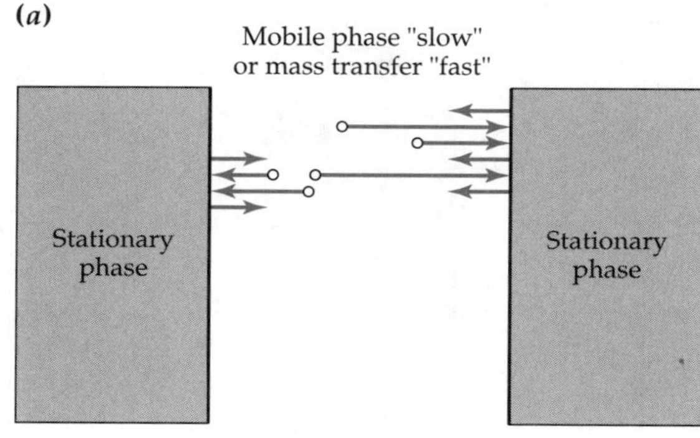

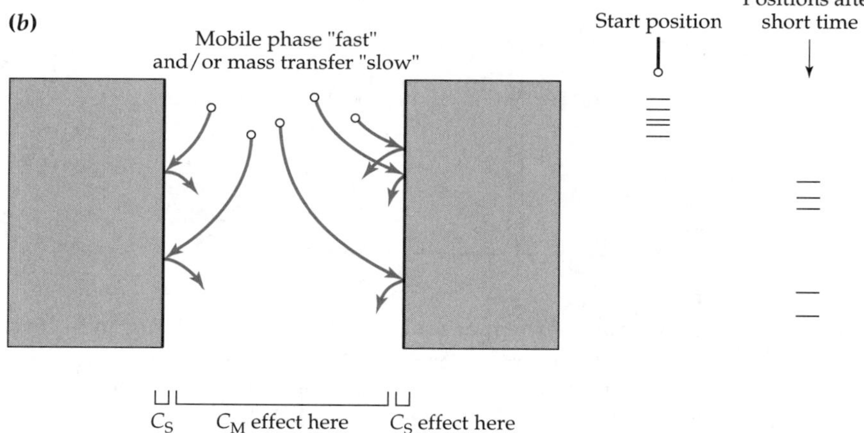

FIGURE 13A.5 ▲

Illustration of the mechanisms of mass transfer resistance broadening in chromatography.

The *C*-term in Equation 13A–1 depends on these effects. The molecules of the sample initially are together. As they pass through the particle bed, the molecules tend to diffuse randomly. The diffusion is perpendicular to the solvent flow. **(a)** shows this perpendicular diffusion at equilibrium and *ignores* diffusion in other directions. However, when the mobile phase is started, the molecules follow the paths shown in **(b)**. The positions of the particles are spread out as shown on the scale to the right. This spreading can be considered due to the system not being at equilibrium. The amount of spreading is proportional to the rate of flow and contributes the term $C_M\bar{u}$ in Equation 13A–1b. In addition, slow interactions at the surface will hold back the molecules more than we would expect from K_D alone. The effect is to cause more spreading in the zone as the solute in the mobile phase continues moving past. This latter process contributes to the HETP as $C_S\bar{u}$.

Another process causes the same effect as slow adsorption/desorption kinetics and is included in the $C_S\bar{u}$ term. The mechanism is illustrated in Figure 13A.6. The particles usually are not "solid" on the molecular level. There are uneven pores and channels, smaller and larger, filled with the mobile phase into which the molecules that are being separated can enter. The presence of these pores also causes band broadening by changing the path length a molecule travels. The molecules can be trapped inside a pore only to emerge later into the mobile phase. The amount of band spreading depends on the depth of these pores, since the deeper the pore, the longer the unwanted excursion can take. Therefore, one more benefit of having smaller particles is

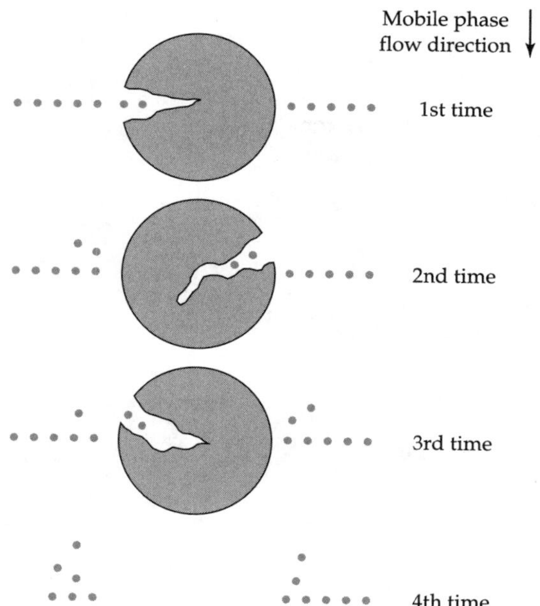

FIGURE 13A.6 ▲

Illustration of how the inner structure of packing materials can cause band broadening.

A number of molecules all start at the same position in the mobile phase. These molecules interact with a particle of the solid support. Some of the molecules migrate into a crevice and then migrate out again. However, they are now behind the original line. This process continues throughout the particle bed. The deeper the pores, the greater the spreading will be. This is because a deeper pore will allow a greater migration length and, thus, a greater delay. Also, the faster the mobile phase moves, the greater the spreading will be. The process also contributes to the term $C_S\bar{u}$ in Equation 13A–1b. One way to reduce this broadening is to make the particles smaller. That way, the longest possible pore length is reduced.

that the maximum length that these pores and channels can have is limited by the size of the particles. The contribution to H by the particle size is found to be approximately proportional to the square of the diameter. That is,

$$H \propto d_p^z \ ; \ z = 1.4 \text{ to } 2.0$$

These two effects—slow exchange kinetics and diffusion into heterogeneous pores—both cause spreading of the band proportional to the flow velocity. The faster mobile phase simply moves the unassociated molecules further before those associated with the solid reemerge into the mobile phase. These two effects cannot be separated and collectively contribute the term $C_S\bar{u}$ to Equation 13A-1b.

Packing of the Stationary Phase and the *A*-term

All the molecular processes described above assume a homogeneously packed stationary phase. One of the arts of chromatography is creating such a well-packed column from a small pile of particles; if there are any significant inhomogeneities, broadened bands result and, consequently, the chromatographic efficiency decreases. Inhomogeneities consist of larger channels through which the mobile phase passes more rapidly than through the rest of the bed. This spreads the analyte zones, as illustrated in Figure 13A.7. The process is called **eddy diffusion**.

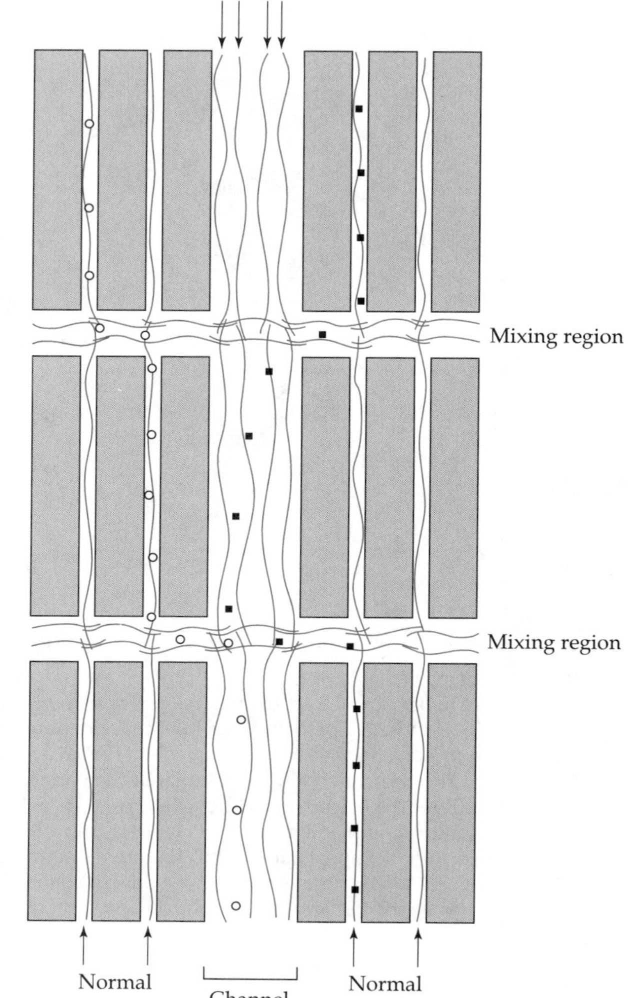

FIGURE 13A.7 ▶
Diagrammatic representation of how a larger channel affects the quality of a chromatogram.

The solutes travel faster in the channel. However, if there is easy transfer between the channel and normal packing regions, as shown here, all the particles will experience the same average rate of travel. Only if the channel is isolated will there be zone spreading due to its presence. *Isolated* in this situation means that not all the solute molecules will pass through the channel for some period of time as they pass through the region. Isolation is reduced with smaller particles and smaller column cross sections. If some isolation remains, it is generally accounted for by the *A* in the van Deemter equation. The process is called **eddy diffusion**.

The problems with such channels can be diminished in four ways.

1. Having highly regular particles. Spheres of equal size seem to be particularly good. Irregular channels have less tendency to form.
2. Having smaller particles. The larger channels that do form tend to be closer to the size of the average. Also, there are more openings for analyte to diffuse back into the main bed.
3. Having a thinner (smaller diameter) bed. Any channels formed will be less isolated. The term *isolated* is described below.
4. Having only one channel. That is, use a capillary column.

Since the analyte molecules diffuse across the flow direction, in effect they can jump randomly back and forth between channels that may be of different size and have different longitudinal transport rates. (See Figure 13A-7.) If the jumping between the channels is rapid, all the molecules tend to experience the same *average* mobile-phase velocity. The rapid jumping means the channels are not isolated, and small variations in channel size do not create much eddy diffusion broadening. Thus, the effect of minor channels becomes simply one of speeding up the average flow rate somewhat.

The *A*-term of the van Deemter equation reflects the interaction between the quality of the packed bed and the isolation of the irregularities. Regardless of the quality of the packing of the stationary phase, it appears that an *A*-term is necessary to describe the HETP of the columns in LC. However, for capillary GC columns and highly homogeneous packed GC columns, the *A*-term apparently is not needed. The larger channels are small enough and the transfer among the regions so rapid that the effect described by the constant *A* value disappears.

Suggestions for Further Reading

BRAITHWAITE, A., SMITH, F. J. 1996. *Chromatographic Methods*, 5th ed. London: Blackie.

This text is a good reference for your next level of study of chromatography.

HAWKES, S. J. 1983. *J. Chem. Ed*. 60:393.

A discussion of the theory of zone dispersion. Highly recommended.

JANCA, J. ed. 1984. *Steric Exclusion Liquid Chromatography of Polymers*. New York: Marcel Dekker.

A collection of chapters by different authors on gel-permeation chromatography. They cover the details and problems of GPC. Especially recommended is Chapter 7 on precision and accuracy.

GRINBERG, N. 1990. *Modern Thin-Layer Chromatography*. New York: Marcel Dekker.

A fine book on the supports, solvents, detection, theory, and techniques of TLC. (Much is applicable to all liquid chromatography.) The chapter on mobile phases is excellent for all of LC. The chapter on quantitation for TLC is also particularly well done.

WEISS, J. 1995. *Ion Chromatography*. Weinheim: VCH Verlag.

This single-author book covers the field of ion chromatography thoroughly and well: instrumentation, reagents, theory, and applications. The next place to go if you want to do ion chromatography.

SUBRAMANIAN, G. ed. 1994. *A Practical Approach to Chiral Separations by Liquid Chromatography*. Weinheim: VCH.

Some theory and a lot of practical information about all the methods used to separate chiral molecules.

FOWLIS, I. A. 1995. *Gas Chromatography*. 2nd ed. Chichester: John Wiley & Sons.

Easy to read but thorough summary of the basic principles and an introduction to the most widely used types of GC techniques and detectors.

GROB, R. L. ed. 1995. *Modern Practice of Gas Chromatography*. New York: Wiley.

Somewhere in this book is nearly everything you will want to know about gas chromatography (although it might be difficult to find).

CORTES, H. J. 1990. *Multidimensional Chromatography: Techniques and Applications*. New York: Marcel Dekker.

An advanced but clear mathematical introductory chapter shows the potential for multidimensional chromatography, and the subsequent chapters describe how it is done for LC, GC, and SFC.

For GC, HPLC, and SFC: In these mature and widely used techniques, it is worthwhile to peruse the catalogs of suppliers such as (alphabetically) Alltech, Bio-rad, Hewlett-Packard, Phenomenex, J&W, SGE, Supelco.

Concept Review

1. How are the following calculated?
(a) net elution volume
(b) capacity factor
(c) separation factor
(d) number of theoretical plates
(e) height equivalent of a theoretical plate

2. What are the advantages of using integrated peak areas instead of peak height for quantitation?

3. If component A has a much longer retention time than component B for a separation using the following liquid chromatographic methods, what does this indicate about the analytes?
(a) ion-exchange chromatography
(b) gel-permeation chromatography
(c) reversed phase chromatography

4. Sketch a block diagram and identify the basic components of a
(a) liquid chromatograph
(b) gas chromatograph

5. Under what circumstances might you want to derivatize an analyte before carrying out a gas-chromatographic analysis?

6. What do the following packed GC column specifications mean? "1/8 in × 8 ft stainless steel containing 5% SE-30 on Chromosorb 100/120 mesh"

7. What process(es) do each of the terms in the van Deemter equation take into account?

8. What does the term "gradient" mean with respect to gas chromatography and to liquid chromatography?

Exercises

HW 1-4 all

13.1 Assume that the solvent flow accidently stopped for 20 s just as the peak of zone 1 of Figure 13.4 reached the detector.
(a) Assuming the horizontal axis is volume in mL, draw a diagram of the appearance of peak 1 as it would appear.
(b) Will the peak height appear to increase, decrease, or remain the same?
(c) Will the peak area appear to increase, decrease, or remain the same?

13.2 Assume that by accident the sample that was used for the chromatogram shown in Figure 13.4 was injected over a period of 10 s. The solvent flow is precisely controlled by a pump.
(a) Will the capacity factor k_1 increase, decrease, or remain the same?
(b) Will the peak heights appear to increase, decrease, or remain the same?
(c) Will the peak areas appear to increase, decrease, or remain the same?

13.3 Use Figure 13.3.1 to do the following:
(a) Calculate the elution volumes of the two peaks.
(b) Calculate the capacity factors of the two peaks.
(c) Calculate the W_i (in time units) of the two peaks.
(d) Calculate the resolution of the two peaks.
(e) Calculate the separation factor.
(f) What is the column efficiency?
(g) What is the HETP of the column?
(h) If the column were doubled in length, what would the capacity factors then be?

13.4 An experiment was run to determine percent recoveries. The peak heights were used for quantitation. After the standards were run and then the samples, all the recoveries were found to be significantly greater than 100%. The operator decided that there was some change in the experimental conditions between the time the standards were run and the time the samples were run.
(a) Could the results occur if all the k_i values remained constant?
(b) Could the results occur if all the t_{Ri} values remained constant?
(c) Is there a simple way to change the assay to avoid the problem in the future?

13.5 Shown in Figure 13.5.1 is a chromatogram of normal fatty acids in CCl_4. This was done with a packed column with I.D. 0.4 cm, length 150 cm, filled with Apiezon L on Celite (50 to 100 μm) in a 1:4 ratio. Assume that the response of the detector is the same per mole for all the acids, and the sample contained 240 μg of the C_{18} acid. Using the triangle approximation, what masses of C_{12}, C_{14}, and C_{16} acids are present? [Ref: Beerthuis, R. K., et al. 1959. *Ann. N. Y. Acad. Sci.* 72:616.]

13.6 From each of the bands of the fatty acid chromatogram in Figure 13.5.1, calculate the efficiency of the column.

13.7 Shown in Figure 13.7.1 is a separation obtained on a microbore HPLC column (column I.D. = 0.05 mm, flow rate = 0.05 mL/min) compared to that on a standard HPLC column (column I.D. = 4.6 mm, flow rate = 1.0 mL/min). Peak 1: 0.5 μg benzene; peak 2: 0.05 μg naphthalene; peak 3: 0.013 μg biphenyl. [Ref: Figure courtesy of Phenomenex.]
(a) What is the area and the FWHM (in mL) for each peak in each chromatogram?

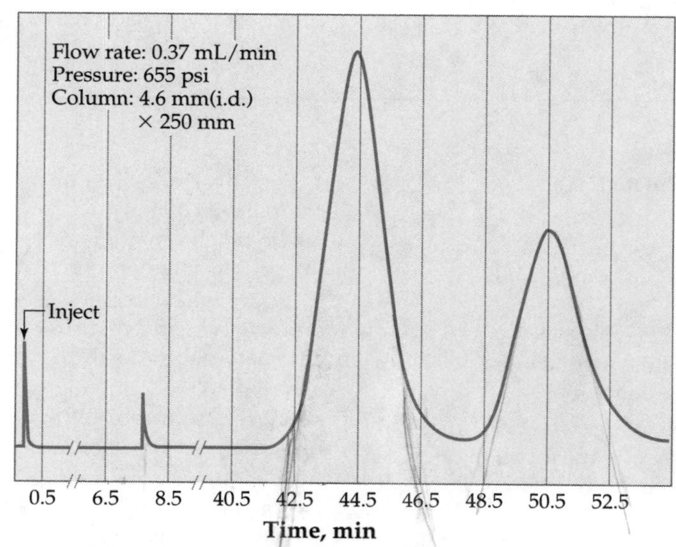

FIGURE 13.3.1 ▲

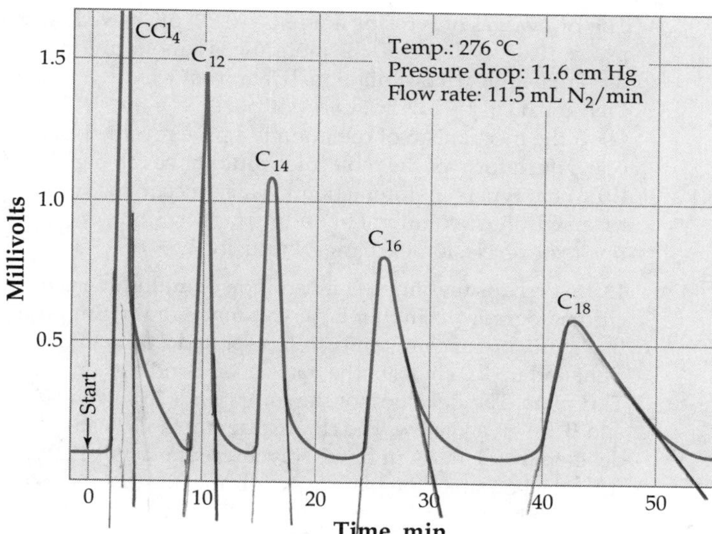

FIGURE 13.5.1 ▲

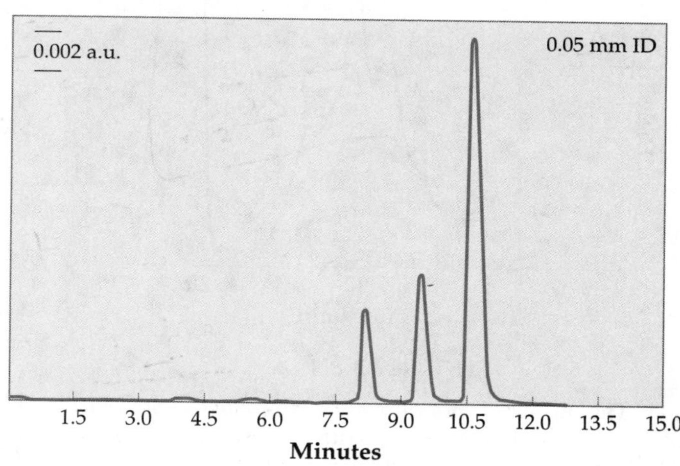

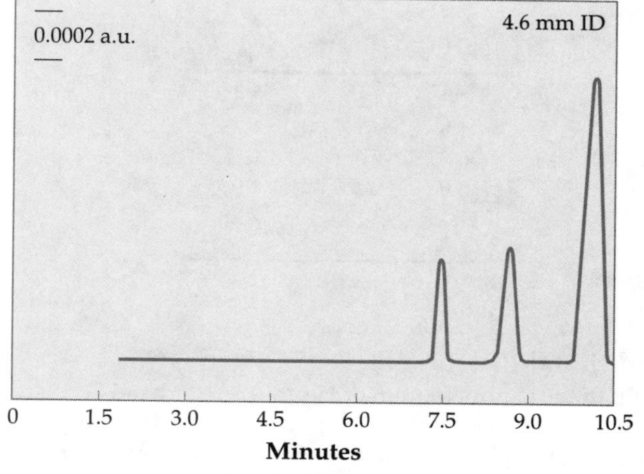

FIGURE 13.7.1 ▲

(b) Calculate the relative response (compared to benzene) for naphthalene and biphenyl for each of the chromatograms.
(c) After correcting for the difference in the flow rates, do your results in (a) explain the difference in detector response for the two chromatograms?

13.8 On the gel-filtration chromatogram in Figure 13.8.1 four of the five peaks are labeled with the molecular weights of the fractions (278,000, etc.). What is the average molecular weight of the middle fraction? [Ref: Fuller, E. N. et al. 1982. *J. Chromatog. Sci.* 20:120. Reproduced from *Journal of Chromatographic Science* by permission of Preston Publications, Inc.]

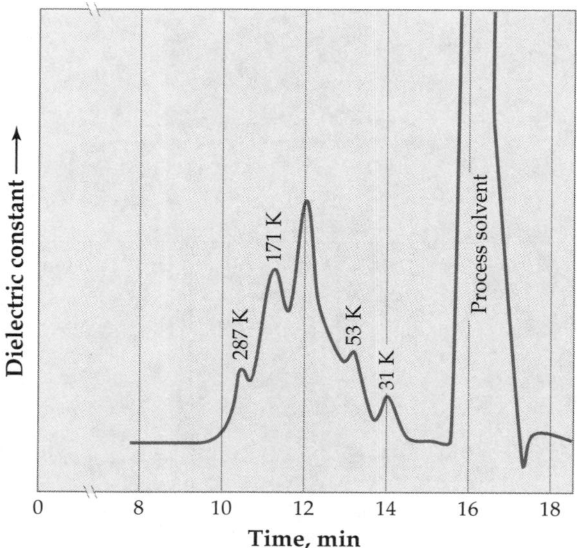

FIGURE 13.8.1 ▲

13.9 The capacity factor of DL-tyrosine changes with pH on a reversed-phase support. With a phosphate buffer 0.02 M and solvent of ethanol:water 5:95, the pH values and the associated k values are:

pH	k
2.3	3.33
4.8	1.75
6.5	1.56
8.9	1.94
11.4	7.28

The pK_a values of tyrosine are pK_1 = 2.20, pK_2 = 9.11, and pK_3 = 10.07. At low pH, the molecule is a monocation. At high pH values, it is a dianion. [Data from Kroeff, E. P. and Pietrzyk, D. J. 1978. *Anal. Chem.* 50:502.]
(a) If the mechanism of retention is a pure reversed-phase one, which form of the molecule would have the highest k?
(b) What type(s) of interaction(s) is/are occurring on this reversed-phase column: H-bonding (adsorption), reversed-phase, ion-exchange, exclusion?

13.10 Perhaps the simplest illustration of multidimensional chromatography is in thin-layer chromatography. Suppose that a mixture of compounds 1–6 is spotted 2.0 cm to the right and up 2.0 cm from the bottom left hand corner of a TLC plate. The R_f values for the compounds in solvents A and B are given below. The chromatogram is allowed to develop in Solvent A until the solvent front reaches 12.0 cm from the bottom of the plate. After drying, the plate is then rotated 90° counterclockwise, and the chromatogram is allowed to develop in solvent B until the solvent front is 12.0 cm from the immersed edge of the plate. Sketch (approximately to scale) what the plate would look like after the first stage and then after the second stage of the separation.

Compound	R_f in Solvent A	R_f in Solvent B
1	0.2	0.2
2	0.2	0.7
3	0.5	0.6
4	0.7	0.2
5	0.8	0.5
6	0.8	0.6

13.11 One μL of hexane (density = 0.66 g/mL) is injected into a GC. Assume that the density of hexane vapor is 2.5 g/L at 69°C and 1 atm. The temperature of the column and injector are 120°C. The gas pressure at the column head is 2.2 atm.
(a) If we assume that the hexane vapor behaves as an ideal gas, what is the volume of hexane vapor that enters the column?
(b) If the flow inside the column is 2 mL/min, for how long a period (in seconds) is the sample entering the column? (This is the minimum width of the peak; that is, if no broadening occurs on the column before detection.)

Additional Exercises

13.12 In the chromatogram of Figure 13.4, for band 3, calculate the elution volume, capacity factor, bandwidth (volume), and FWHM (volume). Assume that the flow rate is 3 mL/time unit. Assume that the band is exactly symmetrical so that the triangle approximation may be made by reflecting the tangent from the trailing edge to give the tangent for the leading edge of band 3.

13.13 Calculate for Figure 13.4 the separation factor and resolution for band 3 compared with band 2.

13.14 Shown in Figure 13.16 is a chromatogram of Triton X-100. Assume that the response of the detector is the same for a given mole fraction for all of the components, and that the peak heights can be used for quantitation. The mole fraction of the component marked with an arrow was 0.09.

(a) What are the mole fractions of the seven highest concentration components?

(b) What is the mole fraction made up by the rest of the components?

13.15 You have probably used plates for thin-layer chromatography in your organic or biochemistry laboratory. Just as there is high performance liquid chromatography which is analogous to column chromatography, there is also a "high performance" analog of thin layer chromatography. Four radio-labelled standards (all 50 ng/mL) and two samples of a mixture of radioactively-labelled proteins were spotted on an HPTLC plate. The mobile phase was allowed to travel for 15.00 cm and then each sample's elution profile was scanned with a counter. The following results were found for the samples.

Sample	Distance traveled (integrated intensity of spot)
Standard	
Protein A	2.67 (27,502)
	2.63 (28,001)
Protein B	1.45 (14,666)
	1.46 (14,299)
Protein C	3.89 (38,622)
	3.75 (39,001)
Protein D	6.99 (17,257)
	7.04 (17,237)
Protein E	8.82 (23,456)
	8.90 (23,676)
Sample	2.67 (13,076)
	6.95 (22,445)
Sample	2.70 (12,989)
	6.99 (22,223)

(a) What proteins seem to be present in the sample?

(b) Calculate the content in ng/mL of each of the proteins in the sample?

13.16 What are the R_f values of spots 1, 2, and 3 in Figure 13.16.1?

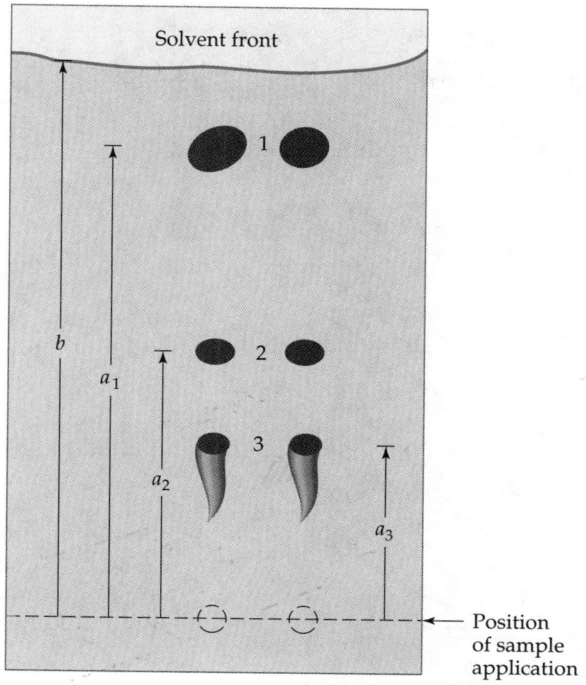

FIGURE 13.16.1 ▲

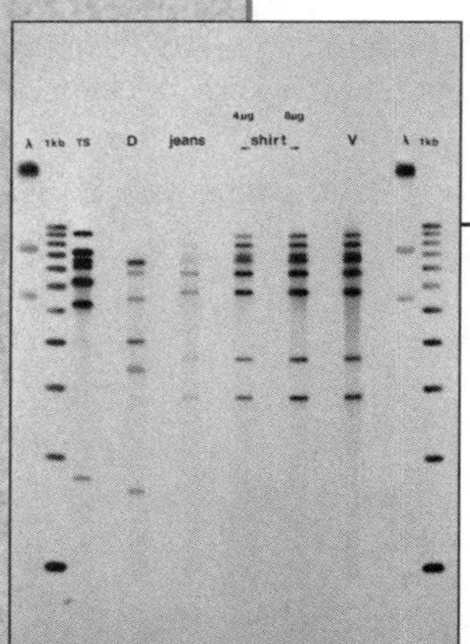

CHAPTER 14

Separations by Applied Voltage: Electroseparations

Overview

14.1 The Basis of Electroseparations

The analytical separation methods that are the subject of this chapter all have one characteristic in common: Separations of different species are based on their dissimilar behavior under the influence of an applied electric field. These are classified generally as **electroseparations**.

There are six phenomenological rules that describe the factors that underlie the various types of electroseparations. The first rule is one you have probably known for a long time.

> *Negative ions tend to migrate toward the higher (positive) voltage and positive ions tend to migrate toward the lower (negative) voltage whether the ions are in solution or in a gas.*

The next two phenomenological rules of the six are straightforward. When a charged body is placed in an electric field, it has a force on it that is proportional to the charge on the moving particle and on the voltage applied to it. In electroseparations,

> *The average* rate *of migration of an analyte species is proportional to the average* charge *on the ion.*

> *The average* rate *of migration of an analyte species is proportional to the average* voltage *applied.*

The fourth phenomenological rule can be visualized easily since even ions of atomic size appear to behave in solutions and gases just as do any macroscopic body. Assume you have a pea and a small feather that both weigh the same and you hold them at the same height above the ground and drop them both at the same time. Gravity acts on both to pull them toward the ground. But you know that the feather will hit the ground well after the pea does, and that they reach a fixed velocity called the terminal velocity. The terminal velocity in electroseparations is called the ion velocity, and the quantitative relationship between the ion velocity and the electric field (electric field = voltage/distance) is given by a proportionality factor called the **ion mobility**. Ion mobility in solution is written as u or μ, and

$$v_{\text{ion}} = \mu V/L$$

where V is the applied voltage, v is the velocity, and L is the distance over which the voltage is applied. The direction of motion depends on the voltage polarity.

Differences in terminal velocity are due to the drag exerted by the fluid: either a liquid or a gas. For our example in air, the pea has little drag compared to the feather's, so the pea has a higher terminal velocity. The effect of drag is similar in an electric field. The reason for these different terminal velocities are described by the fourth phenomenological rule for electroseparations.

> *The average rate of motion of an ion is inversely proportional to its average cross section.*

The **cross section** is the area of the ion that is perpendicular to the direction of the motion. This simply means, e.g., that a larger monovalent ion moves more slowly than a smaller monovalent ion. Also, a larger divalent ion moves more slowly than a smaller divalent ion, and so forth.

To make a quantitative estimate of the ions' cross sections, we usually assume that a molecule is a sphere, which is an excellent approximation since most ions tumble so quickly their shapes average to a sphere. For long-chain polymers, such as **denatured** (unfolded) proteins and DNA, the molecules can be approximated as a chain of spheres connected by weak springs. However, each sphere acts nearly independently of the others, so the rate of motion is inversely proportional to the number of spheres. That is, the average rate of motion is still inversely proportional to the average cross section even though it is not a single sphere.

With this fourth phenomenological rule we can now understand the basis for most of the methods that fall under the label of **electrophoresis** (from the Greek *phoresis* meaning *being carried*). Electrophoresis separates ions because of differing **charge-to-size ratios.** Electrophoresis can be done both in solution and in the gas phase. The gas phase separation is called **ion mobility spectrometry** (IMS). Both electrophoresis and ion mobility spectrometry are based on the first four phenomenological rules together. However, we will limit this chapter to electrophoresis techniques in solution, which have two more phenomenological rules that should be understood.

Just as in chromatography, diffusion tends to cause longitudinal spreading of the bands. But in electroseparations in liquids, a process that tends to be more important in decreasing the quality of separations is **convection.** Convection is the macroscopic mixing of liquids due to differences in density. The density of a homogeneous liquid decreases as it is heated, and so any liquid that is heated unevenly (almost impossible to avoid) tends to mix by convection. This brings us to the statement of the fifth phenomenological rule.

The quality of electroseparations in liquids improves as the convection is suppressed.

There are two ways to suppress convection in a liquid:

a. Performing the electroseparation in a polymer network, where the molecular strands of the network stop macroscopic convection by trapping the liquid between the strands, and
b. Performing the electroseparation inside a small capillary, where the inner wall of the capillary has the same beneficial effect as the polymer molecules and, in addition, the small size of the capillary minimizes the temperature differences that can arise.

Heating of a liquid in electroseparations *cannot* be avoided since the electric current passing through the solution heats it. The solution is heated by an amount equal to the power. ($P = VI = I^2R$, where V is the voltage, I the current in the ionic solution, and R the solution's resistance.) The heat produced by the electric current is called the **Joule heat.**

In addition to the problems caused by convection, the Joule heat that is generated must be dissipated to hold the temperature constant overall for three more reasons. First, the pK_a of each ionic group and of the surrounding buffer ions varies with temperature, so to keep the bands homogeneous and the pH fixed, the temperature must be constant throughout. Second, macromolecular structures usually depend on temperature, so constant temperature keeps their average size—and average migration rate—the same. And third, the viscosity of the solvent decreases with increasing temperature. A lower viscosity allows an increase in the speed of migration of all the ions. As a result, different temperatures at different positions produce a range of migration rates that spread the band. In sum, the quality of the calibration and the sharpness of the homogeneous analyte bands is increased if the temperature is constant overall. This brings us to the sixth and final phenomenological rule.

The quality of electroseparations in liquids improves when the temperature can be held constant over the time of the experiment.

The analytical methods that fall under the classification of electroseparations in solution are used to separate and quantitate analytes ranging from small molecules (formula weight ~100) to molecules that can be seen under a microscope (formula weight up to 10^8) such as DNA chains or the largest proteins.

The quality and ease of the separation of species in electroseparations also depend on how narrow the distribution of each species can be made at injection—just as in chromatography. That is, if all the ions of the same type can be kept spatially closer together in the direction of migration, it is easier to separate one species from another. It is also necessary to keep the charges on each species of ion constant so the charge-to-size ratio (and the migration rate) remains in the narrowest possible range. To ensure that the charge remains the same on any acidic or basic sites means that electrophoretic separations require that the solutions be buffered.

Electrophoresis is *the* primary separation method in molecular biology, and, together with various preparation and detection methods, is the analytical basis of nearly all molecular biochemistry. Here, we will focus almost exclusively on the separation methodology, and not on the sample preparation and detection techniques that give such power to the biochemical methods.

14.2 Negative, Neutral, or Positive?

The first piece of information you need before running an electrophoretic separation is an estimate of the charges on the various analytes under the proposed experimental conditions. When you understand the concepts of chemical equilibria this is easy to do, especially when a molecule has only one charged group. Let us take as an example benzoic acid with its pK_a of 4.20 (at 25°C). Essentially all the acid molecules will carry a negative charge when in a (buffered) solution at pH 6.2. Half of the molecules will carry a negative charge and half will be neutral at pH 4.2, where the solution pH = pK_a. At pH 2.2, most of the molecules are uncharged, but not all. However, if you wanted to use electrophoresis to analyze the benzoic acid, you would never see two separate species—the protonated and deprotonated forms. The reason for this behavior is that each benzoic acid molecule is exchanging protons at a rate that is very fast compared to the time over which you ran the experiment. So, while an individual molecule is negatively charged, it moves quickly, and while the molecule is neutral, it does not migrate, and the overall migration rate is characteristic of the *average* charge of the benzoic acid molecules. As a result, you would see at pH 6.2 and above the benzoic acid moves at its fastest speed toward the positive pole, and at pH 2.2 it would move quite slowly, as if it has only about 1% of the full charge (if it does not precipitate out of solution). At pH = pK_a the rate is half the maximum since the molecules spend half the time protonated (neutral) and half the time as an anion.

Although it is unlikely that benzoic acid itself would be determined by electrophoresis when so many other techniques could be used, its properties do illustrate why you can adjust an electrophoretic separation by changing

Table 14.1 Isoelectric Points of Some Proteins

	Isoelectric pH		Isoelectric pH
Pepsin	1.1	γ_1-Globulin	6.6
Casein	4.6	Hemoglobin	6.8
Egg albumin	4.7	Myoglobin	7.0
Serum albumin	4.9	Ribonuclease	9.5
Urease	5.0	Chymotrypsin	9.5
β-Lactoglobulin	5.2	Cytochrome *c*	10.65
Insulin	5.3	Lysozyme	11.0

the pH of the buffer. If a separation is poor at one pH, then a different pH may cause the ion mobilities to change to allow a better separation.

When a molecule has both positive and negative groups on it, it is called a **zwitterion** (from the German *Zwitter*, meaning hybrid or hermaphrodite). Simple examples are the amino acids, and of these the glycine zwitterion is the simplest: $^+H_3N—CH_2—COO^-$. Glycine is *not* a zwitterion at high pH, where it remains in its anionic form $H_2N—CH_2—COO^-$, nor at low pH, where it remains in its cationic form $^+H_3N—CH_2—COOH$. At one pH in between, the average charge on the glycine is zero. This is its **isoelectric point**.

The situation gets slightly more complicated when the analytes have many charged groups, both positive and negative—a characteristic of proteins. Just as for glycine above, proteins too have an isoelectric point, where the number of remaining negatively charged groups, such as $—COO^-$, just equals the number of remaining positively charged groups, such as $—NH_3^+$, even though the multiple equilibrium is complicated. Isoelectric points for some selected proteins are listed in Table 14.1.

The isoelectric point provides a useful classification for electroseparations. At pH values lower than the isoelectric point, there is a *net* positive charge on the protein from the protons taken up. This is not a jump by integral numbers of charge, but, like the benzoic acid example, the charge is incrementally more positive. Similarly, at pH values above the isoelectric point, the protein takes on a *net* negative charge. Clearly, for electroseparations, changing the buffer pH from one side of the isoelectric point to the other will reverse the direction of migration of a molecule, as can be seen in Figure 14.1. In theory, at the isoelectric point the protein will not migrate. Indeed, they do not, but not simply from a lack of charge, but because proteins tend to precipitate at their isoelectric point. They do not move in the field because they sit in a lump somewhere in the buffer.

14.3 Electrophoretic Separations within a Gel Matrix

The classic methods of electrophoresis are done by carrying out the separation in a polymeric **gel** (as in *gelatin*). The gels that are used are three-dimensional webs of polymer strands with the spaces between the strands filled with liquid. Not only do the polymer networks suppress convection, but they also act like sieves that can retard and even block the migration of larger, polymeric analytes. However, small ions can freely move through the porous structure of the gel. Thus, **gel electrophoresis** has two different separating

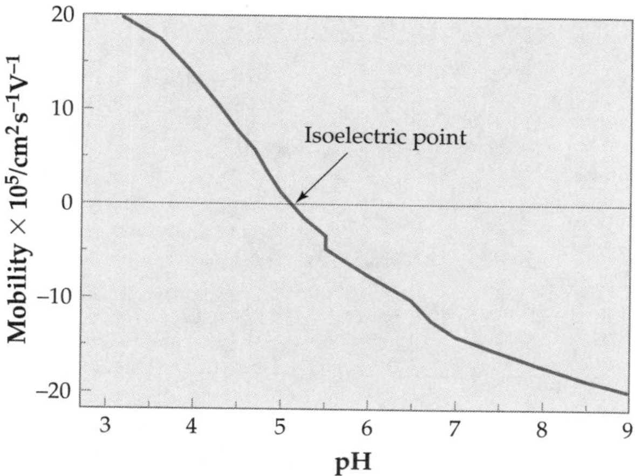

FIGURE 14.1 ▲
A graph of mobility *versus* pH for β-lactoglobulin, a protein from cow's milk.

The mobility is proportional to the charge on the protein and follows its changes with pH. Experiments suggest that the overall charge of the protein results from 57–60 carboxyl groups, 5–7 guanidinium groups, 33–35 amino groups, and 6 imidazole groups. The isoelectric point is 5.19 in the presence of acetate buffers. [Data from R. K. Cannan, A. H. Palmer, and A. C. Kibrick. 1942. *Journal of Biological Chemistry* 142:803–822.]

mechanisms: electrophoresis, which separates by the charge-to-size ratio, and sieving, which separates mostly by size. The sieving effect is proportional to the size of the macromolecules and the packing density of the gel molecules. The reason for this is that the macromolecules' migration is slowed by colliding with gel molecules. So the longer the macromolecules, the more often they collide with the gel, and the slower the migration. Similarly, if the gel molecules are more concentrated, more collisions occur and the migration slows. We say that the **pore size** is smaller for more concentrated gels. This sounds as if the gel were like a filter with small pores in it. However, gels are three-dimensional nets of molecules, and instead there are channels in the gel that are smaller on average and more tortuous when the matrix is more concentrated.

The Gels

Two different materials are used to make the great majority of electrophoresis gels: **agarose** and **polyacrylamide**. Agarose is a highly purified polysaccharide—that is, a polymer of sugar units—which is refined from agar. To make the gel, agarose is dissolved in boiling buffer and poured into a mold where it gels at about 40°C. It will remain as a gel until raised again to about 100°. The range of concentrations used are usually around 0.4–2% w/v, where the lower concentrations have larger channels. These gels are quite fragile.

Polyacrylamide is sturdier than agarose, and it is irreversibly formed by a free-radical polymerization process: the acrylamide monomer is covalently linked to form chains. The reaction is shown in Scheme 1. Simultaneously, the chains are **crosslinked** by a **crosslinker**, most commonly N,N′-methylene-*bis*-acrylamide, that ties two of the polymer chains together. The sizes of the pores in the structure are determined by both the amount of acrylamide

%T, due to its context, should not to be confused with either temperature or the percent transmittance (Chapter 11).

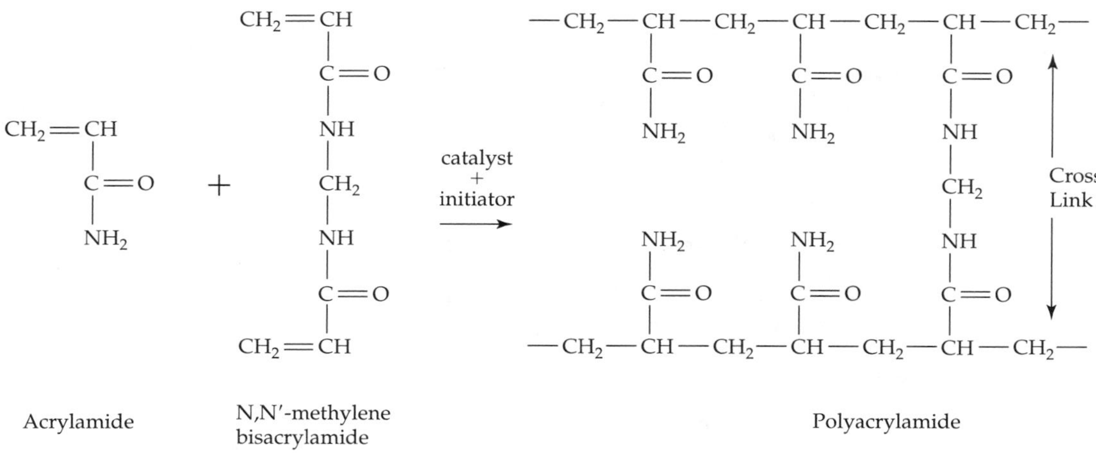

catalyst + initiator

Cross Link

Acrylamide

N,N′-methylene bisacrylamide

Polyacrylamide

Scheme 1

monomer and the amount of crosslinker used. There is a shorthand to express this formula. One term is %T. %T indicates the w/v percentage of the *combined* acrylamide and crosslinker. The second term is %C, which represents the crosslinker as a *percentage of the total acrylamide weight*. For example, a gel written as 20%T 5%C_{bis} has 20% w/v of acrylamide and crosslinker, and of this 20% weight, 5% is the crosslinker. In other words, 1% of the total buffer and acrylamide species is crosslinker. The subscript *bis* is the standard shorthand for N,N′-methylene-*bis*-acrylamide. Finally, for any %T of acrylamide and crosslinker, the 5% crosslinker concentration produces a gel with the smallest channels in the porous structure.

The agarose and polyacrylamide gels have two characteristics in common. First, the shapes into which they are cast are the same, such as the **slab gel** illustrated in Figure 14.2. They are also cast inside tubes with inner diameters of about 5 mm. Such gels are called **tube gels** or **disc gels**. Gels are also used in capillaries less than 0.5 mm I.D.

Second, both of these types of gels are nearly free of any ionic charges on their structures. This is important to prevent buffer from being driven through the gel when the electric field is turned on. The cause of the buffer's flow is called **electroosmosis**. We can ignore electroosmosis when the gels are non-ionic. But we will need to examine electroosmosis later in this chapter. (See Section 14.6.)

One final point: Making excellent gels is an art. That is, there are so many details to control in casting a gel that a great deal of practice is required. Also, this fact should immediately indicate one requirement for an analysis using gel electrophoresis: the need for internal standards.

Some Modes of Operation

Gel electrophoresis can be used alone to analyze the presence or absence of molecules of specific charge-to-size ratios. However, when joined with special detection methods, such as those discussed in Section 14.4, absolute identification of specific macromolecules can be made. Here, macromolecules include the DNA and RNA fragments that are separated in the methods of gene sequencing, as well as proteins and polypeptides.

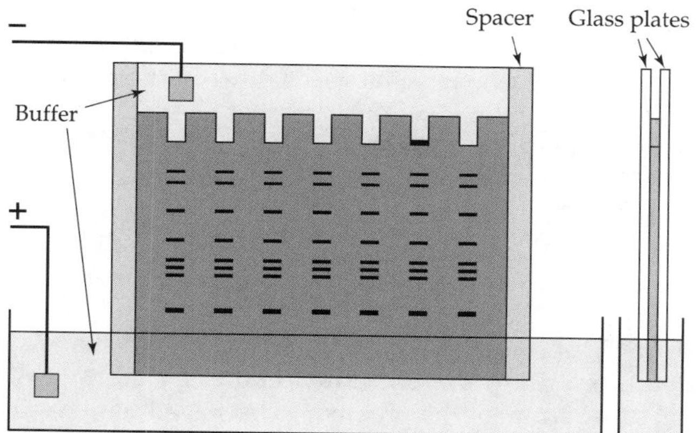

FIGURE 14.2 ▲
Slab gel electrophoresis.

The gel (shaded) has notches in which the samples are placed. A **tracking dye** is included in each. This dye runs through the gel without being retarded and indicates the leading edge of the distance moved so the experiment can be halted at the correct time. Gels are generally between 0.25 and 1 mm thick although in specific instances they may be as much as 5 mm thick. The thickness is set by the spacers that hold the glass apart. The gel is poured in between them. Slabs are made in a wide range of sizes, from 5 × 5 cm to 33 × 60 cm. The size depends on the number of samples and the desired separation. Not shown in the figure are the clips that hold the plates onto the spacers. The bands in the gel are only indicative of the kinds of differences that might be seen in a separation if the bands were colored while the gel was running. In one **track** (2nd from the right), some material is too large to enter the gel. Most often, the bands are stained after the gel is run.

Conditions for gel electrophoresis can be divided into two classes—those of **denaturing** and **nondenaturing**. To *denature* means to make the molecules biologically inactive. In denatured biopolymers, the monomers are covalently linked in their original order, but the polymer no longer is folded into its active form. The order in which this set of monomers is covalently linked is called the **primary structure**. The formation of helices and sheets is called the **secondary structure**, and then the folding of these parts into the functional structure is called the **tertiary structure**. The **quaternary structure** is the association of specific, individual proteins to form distinct, larger assemblies. To run a gel electrophoresis on the denatured form indicates that only the primary structures of the analytes are kept intact.

Detergents such as sodium dodecyl sulfate (SDS) and other non-detergent agents are known to cause such denaturing. Sodium dodecyl sulfate has a twelve-carbon chain attached to a sulfate group—its formula is $CH_3(CH_2)_{10}CH_2OSO_3^{-}Na^{+}$. Urea is one example of non-detergent denaturing reagents. Both the detergents and the other denaturing agents help break up the structures that form by non-covalent binding: structures such as the helices and sheet structures of proteins. The methods have names such as *SDS polyacrylamide gel electrophoresis;* this name tells you the type of gel and also that it is run under denaturing conditions. Proteins in the presence of SDS migrate a distance that depends on the molecular weight and on the average size of the channels in the gel: the larger the protein, the slower the migration. So SDS electrophoresis is similar in many ways to gel exclusion chromatography in its effect. However, there is an assumption involved in the calibration of the SDS: we assume the detergent binds to all the denatured proteins with the same number of SDS molecules per unit length of the chain. This SDS binding overwhelms the charges on the protein amino acids, and essentially every protein has the same charge-to-size ratio. As a result the sieving effect of the gel becomes the predominant separation mechanism. SDS polyacrylamide gel electrophoresis (SDS-PAGE) is used to determine molecular weights. The experimental parameter used is R_f, which is the ratio of the distance traveled by each polymer relative to the distance moved by a unretained compound. The logarithm of the molecular weight plotted vs. the R_f of the band is linear. In order to measure R_f, however, the proteins must be detected. Most of the time, the proteins are dyed and made visible. More about this detection will be presented in Section 14.4 below.

Sometimes the nature of the problem to be solved requires that the separation not destroy the proteins. The separation must be done under nondenaturing conditions. For instance, a solution of numerous components may be separated in order to find how many different acid phosphatases are present. (Acid phosphatases catalyze the hydrolysis of phosphate esters to produce inorganic phosphate.) The whole solution mixture can be loaded onto a gel, separated, and the phosphatases may then identified by their reaction with one of a number of standard assays for acid phosphatases.

Finally, two-dimensional gel electrophoresis is a powerful method to fingerprint a mixture of proteins. Hundreds of proteins in a mixture can be resolved. However, the gel cannot be run and then the slab turned 90° and run again. For 2D electrophoresis, the first dimension usually is run in a tube gel. Then, the tube is grafted onto the top of a slab gel, and the second dimension run across the tube gel and then in the slab. Generally, the gel conditions (ionic strength, pH, and so forth) differ between the tube and slab gels.

Improving the Resolution

Improvement of the resolution in gel electrophoresis can be done with a process called **stacking**. Stacking is a phenomenon unique to electroseparations in which the width of the band of the analytes can actually be made narrower than the width over which they initially were applied to the gel. Stacking is also called **electrofocusing** and is done both in gel and capillary electrophoresis. The general way it works is illustrated in Figure 14.3. The analytes are loaded as a segment of buffer solution that differs in composition, pH, and ionic strength from the buffers on either side of it. When the voltage is turned on across all the solutions, a higher resistance (lower conductivity) develops across the sample region. As a result, the electric field is higher in the sample region and the analyte ions migrate more rapidly in this region. When the analyte ions arrive at the interface with the lower resistance solution, they slow down. The analyte ions still in the low conductivity region keep moving fast and keep piling up at the interface, where they are said to be focused or stacked. (The various buffer and analyte ions are moving in such a way as to keep the solutions electrically neutral overall and to allow the analyte to move faster than the buffer ions of the same charge.) As the electrophoresis proceeds, the solutions mix and the interface dissipates leaving a homogenous buffer in which the separation proceeds.

Another way to improve the resolution of gel electrophoresis is to construct gels in which the channel sizes (gel concentrations) change along the path of the migration. These changes in the gel can be in steps, in which case they are called **discontinuous** gels. Also, the changes can be continuous,

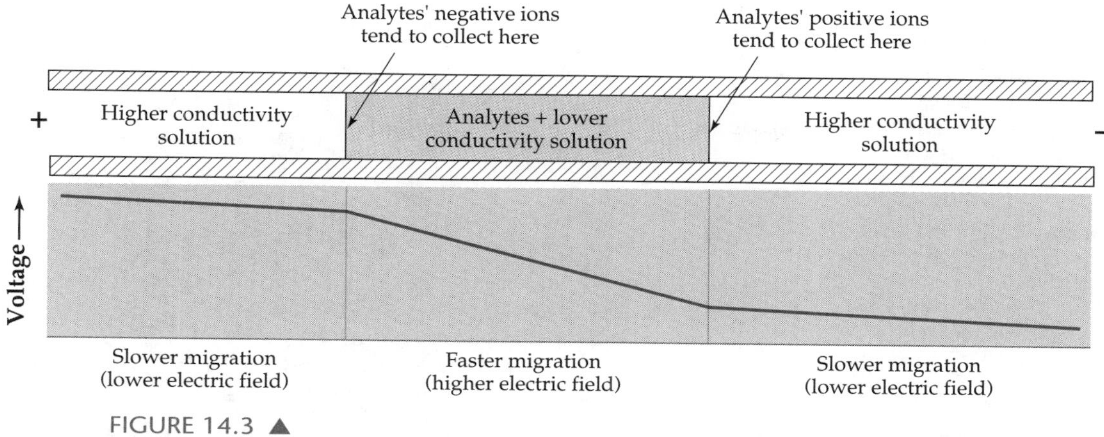

FIGURE 14.3 ▲
Electrofocusing.

Electrofocusing, illustrated here, is done in both gel and capillary electrophoresis. The ions are focused because the high field in the sample region forces them to accumulate at the interface with a low-field region. The difference in voltage drop can be understood in analogy with a simple voltage divider, made with a number of resistors in series. The largest voltage drop occurs across the resistor with the highest resistance.

such as having the channel sizes continuously decrease from the top to the bottom. This may be called a **pore-gradient gel** or simply **gradient gel**. The gradient's effect is that the sieving becomes more restrictive as the sample migrates. Such a gradient gel allows a larger range of molecular weights to be separated in a single run and also has the effect of narrowing the individual bands somewhat.

The resolution of an electrophoresis run can also be improved by changing the applied voltage in a controlled way. The changes can be pulses or gradual changes in the voltage over longer times. This slower change may be called **ramping** the voltage. Such protocols require a more sophisticated power supply, one that can be programmed to produce the desired voltages and timing. The language used to describe these voltage protocols varies. For example, the voltages might be given directly as volts or indirectly as the power in watts. With the information given in most publications, it is usually not possible to translate quantitatively between these different descriptions.

14.4 Some Detection Methods for Gel Electrophoresis

When the analytes are separated in a slab gel electrophoresis and then made visible, you can see a number of separated, dark or fluorescent bands such as are shown in Figure 14.4. However, few biopolymers are themselves intensely colored enough and can be run at high enough concentration in the gel to be seen by eye or even detected by sensitive spectrometric methods. As a result, after a separation is performed, further treatments must be done to make the bands detectable.

The spectrometric methods are those described in Chapter 8.

The bands of protein can be made visible using various intensely colored reagents (dyes) that bind with the analyte molecules and do not associate with the gel molecules. (Those that associate with the gel matrix are interferents that raise the detection limit.) The gels are filled with dye and then the excess washed out; the gels are **stained** and then **destained**. The stain that

FIGURE 14.4 ▶
DNA sequencing gel image.
The DNA is digested with four different restriction enzymes. Each cleaves the DNA at only one type of the four base pairs. When these digests are separated by electrophoresis, the four adjacent tracks contain, then, segments of every length. The smaller fragments run faster, and the position along the gel, then, indicates the fragment mass. At every length, if the separation is done well, only one track of the four will have a band. The base at that length is then known from the track. The next band toward the origin is one base longer, and it too is found on only one track. In this way, the base sequence is read off the gel image. Here, approximately 300 bases can be read from each set of four tracks on the image on the membrane. [Source: Copyrighted material used with permission of Hoefer Pharmacia Biotech, Inc.]

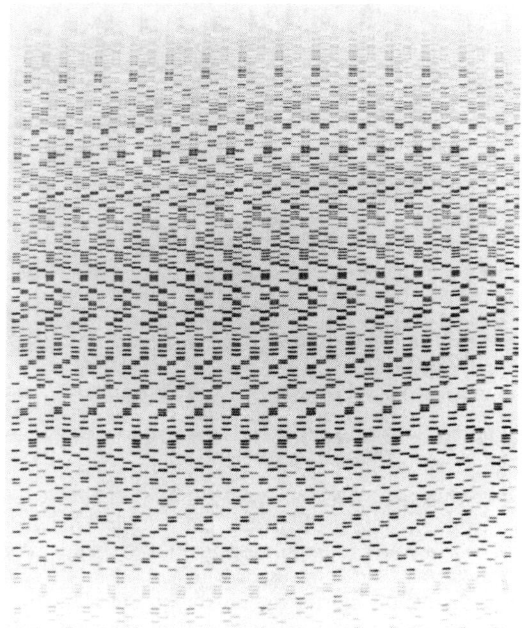

associates with the analytes remains, and bands of color are observable in the gel. Proteins are usually stained with the dye Coomassie Blue or made visible with silver in a process similar to production of an image on a photographic negative. Coomassie Blue gives usable images with about 1 μg of protein while the silver method can make visible about 10 ng of protein. (Note that these effective limits of detection are not as precise as can be stated for solutions of small molecules since the concentrations of proteins in the gel are not homogeneous, and different proteins react differently with the reagents.)

Fluorescence—see Chapter 11.

Staining nucleic acids is done usually with ethidium bromide, which is a reagent that becomes strongly fluorescing with an orange color under ultraviolet light only when it becomes bound within the nucleic acids. (Specifically, ethidium bromide becomes **intercalated**; it slides in between parallel rings like the filling in a sandwich.) The detection limit is about 10–50 ng of DNA.

Radiochemical methods have lower limits of detection than does staining. Radioactive isotopes can be incorporated into the analyte bands to produce images such as that shown in Figure 14.4. These images are obtained by clamping a photographic film onto the gel and exposing it for a sufficient time to produce the darkened bands. When the analytes themselves have radioactive isotopes incorporated in them, the image of the gel is called an **autoradiogram**. When a radioactive isotope is bound to a reagent that reacts with the analytes, it is called **radiographic** detection.

A wide variety of specific radiographic detection methods are done by using reagents that specifically bind to only one analyte or maybe a few of many that may be present. A common way to do this specific binding is first by removing the analytes through the thin direction of the gel onto an adjacent, tougher and dimensionally more stable plastic sheet. The locations of the analytes are preserved; an image of the gel is produced. A diagram of the

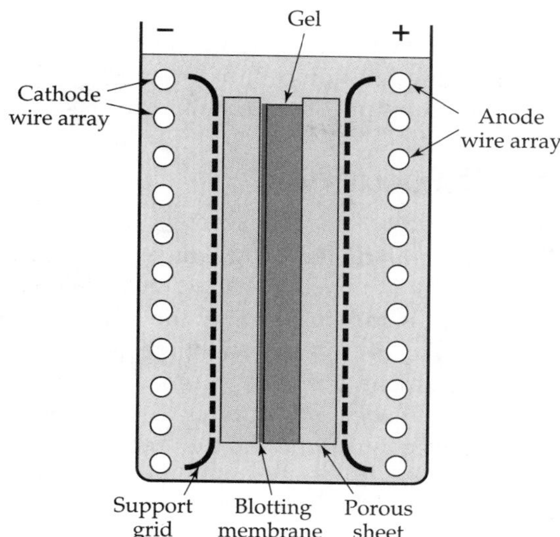

FIGURE 14.5 ▲

Cross section of the geometry for electrophoretic blotting.

From left to right, the parts are cathode wire array, rigid support grid, porous support sheet, blotting membrane, gel, porous support sheet, rigid support grid, and anode wire array. The support grid and sheets are present so that there is a firm and even contact overall between the gel and the membrane. The entire apparatus may be immersed in a tank of buffer or, alternately, in the **semi-dry method**, the materials are all wetted and pressed together. The buffer and pH are chosen so that the material being transferred has the correct charge sign to be driven onto the blotting membrane. Approximate conditions: current, 1 mA cm^{-2}; time, in the range of tens of minutes.

geometry is shown in Figure 14.5. One reason this transfer is done is to keep the image preserved since the gel is relatively fragile and changes size depending on how much water is in it. In addition, the transfer can concentrate the analytes; the analytes can be drawn out from a gel that may be as much as 5 mm thick onto a thin layer at the surface of the plastic sheet. Transferring the reagents out of the gel is called **blotting**, and it allows many powerful techniques of immunochemistry to be used subsequently for identification of proteins and peptides. Similarly, the equally powerful techniques of nucleic acid recognition with complementary strands or reactive enzymes are more effective on blotted samples.

For all these blotting methods, the various components of the buffers can interfere with the blotting or detection. As a result, analyses are usually begun using the carefully optimized protocols that are available in the literature for RNA and DNA. However, there are large numbers of different protocols for specific applications for the nucleic acids and an even more diverse range for the proteins. Included in the protocols are instructions on the arts of sample preparation, gel casting and loading, and then stopping further migration (**fixing**), blotting, and detection. Nevertheless, quantitation of the amounts of materials can be effective only with the utmost care and where the conditions can be tightly controlled. Quantitation of the molecular weights is more reliable; differences of one base in polynucleotide chain lengths are resolvable. (See Figure 14.4.) Protein molecular weights may be

A set of fluorescently labeled bands can be seen in EP.13 inside the back cover.

less certain since the ratio of SDS to chain length may vary between the standards and the analytes. Nevertheless, it is clear from the progress contributed by gel electrophoresis to biochemistry and molecular biology that the ability to quantitate the results of the separations has been adequate.

14.5 Isoelectric Focusing

The differences in isoelectric points for proteins can be exploited to separate them by the method of **isoelectric focusing**. The method is illustrated in Figure 14.6. Isoelectric focusing is based on setting up a stable pH gradient in a solution or gel. A gradient of pH means that the pH changes continuously over a distance. For example, in the x direction, the gradient is $d\text{pH}/dx$. Optimally the change in pH is linear: $d\text{pH}/dx = \text{constant}$. The pH gradient is set up by a set of special buffers called **ampholytes** that themselves migrate until they reach their own isoelectric points. At the steady state, the most acidic one will be closest to the anode and the most basic one will be at the end toward the cathode.

The analytes migrate through the medium until they become neutral, at which point they stop. The position of each analyte lies where the pH of the buffer equals its isoelectric point. (See the first and second rules in Section 14.1.) Each analyte becomes **focused** at a specific position in the pH gradient. By *focused*, we mean that it is more concentrated there than in the original solution. Each protein is driven into a narrow band, and the electromigration even overwhelms the random diffusion that tends to disperse the molecules composing the band.

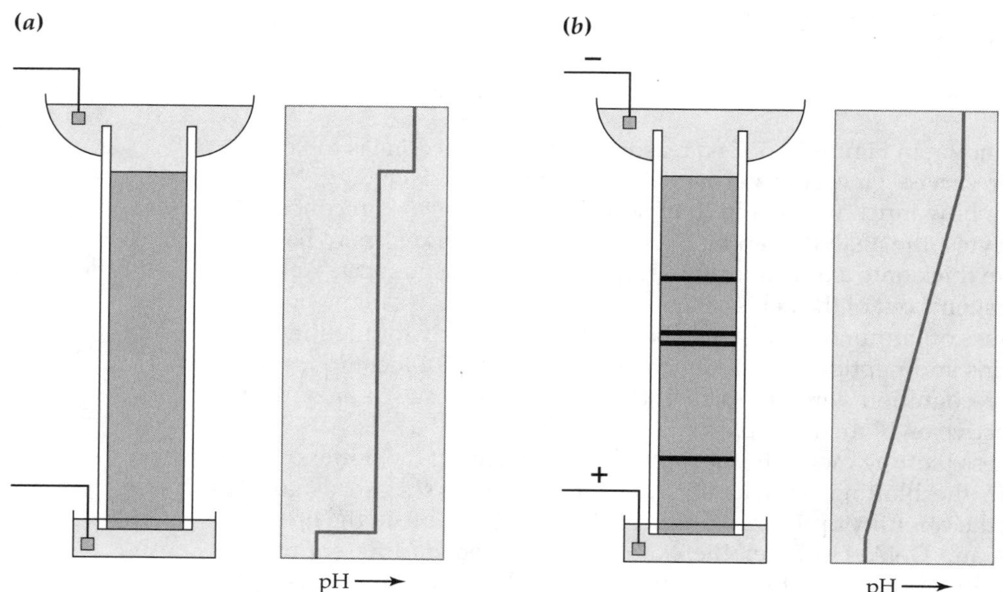

FIGURE 14.6 ▲
Diagram of an isoelectric focusing experiment.

(a) One type of experiment has the analytes dispersed throughout the gel with the ampholytes. The pH in the system is indicated in the graph to the right. **(b)** The voltage is applied and the ampholyte buffers migrate to establish a pH gradient along the gel as shown in the pH plot. The proteins move until they become neutral at their respective isoelectric points in the pH gradient.

Isoelectric focusing is done in slab gels, in tube gels, and in free solution inside a capillary tube. The gels for isoelectric focusing are made with channels so large that sieving does not influence the separation. Isoelectric focusing can separate proteins differing by only 0.01 pH units in their isoelectric points by using shallow pH gradients. Steeper pH gradients cause bands to focus closer to one another and give less spatial resolution.

Finally, you should recognize that isoelectric focusing is used only for proteins and peptides, which are amphoteric. When describing proteins, *amphoteric* is an alternate way to say they are zwitterions able to both donate and accept protons. On the other hand, nucleic acids are all negatively charged, and they are negatively charged over the entire pH range that the gels can tolerate. As a result, the nucleic acids are not candidates for separation by isoelectric focusing.

14.6 Capillary Electrophoresis

In an electrophoretic separation, both the speed of the separation and the resolution of the components improve as the applied electric field is increased. Since the electric field is proportional to the applied voltage, why not just increase the voltage? The answer is that the Joule heating increases to the point where the heating ruins the quality of the separation. Is there some way that this problem can be overcome? The answer is yes. The cross section of the solution in the electrophoresis apparatus can be reduced drastically by running the electrophoresis inside a capillary tube with an inside diameter less than 0.1 mm and a length about 50 cm to 1 meter. The technique is called **capillary electrophoresis** (CE) or **capillary zone electrophoresis** (CZE). The apparatus used is illustrated in Figure 14.7. Running in a capillary is effective because the solution's electrical resistance is so high that the current remains low (less than about 10 μA). As a result, higher voltages can be sustained

Heating effects in electrophoresis are introduced in Section 14.1.

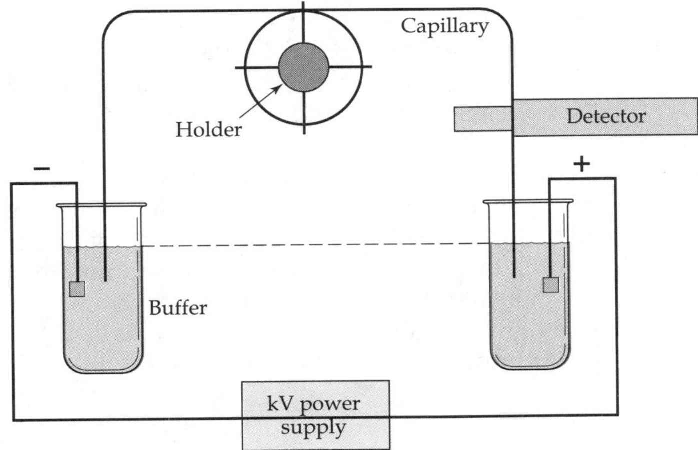

FIGURE 14.7 ▲
Diagram of the basic parts of a capillary electrophoresis apparatus with spectroscopic detection.

The sample is introduced at the left side by placing the end of the capillary in the sample solution and loading it either hydrodynamically or electrokinetically. The entire unit shown is held at a constant temperature. The capillary is wound around a holder. During the analysis, the levels of the two buffer solutions must be the same (as indicated by the dotted line) to avoid bulk flow due to a pressure difference.

ANALYTICAL CHEMISTRY IN ACTION

There May Be Only One

DNA profiling (or DNA "fingerprinting") has become admissible evidence in courts, and may become a general method of identification of, for example, members of the armed forces. The theory behind classifying an individual's DNA is straightforward. At certain, highly variable locations on human chromosomes, molecular biologists have discovered that specific sequences are repeated different numbers of times for different individuals. For example, a sequence of DNA components such as cytosine (C), guanine (G), adenine (A), CGA, may be repeated five times in one individual, that is, CGACGACGACGACGA, while in another person, the CGA sequence may occur, say, 66 times.

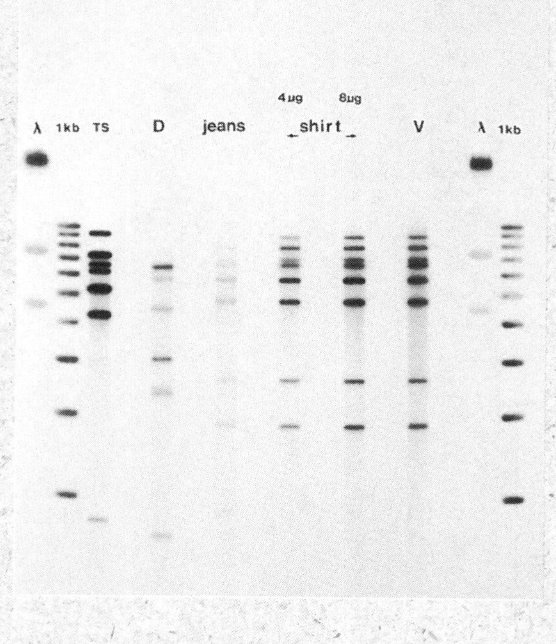

To measure these differences requires chopping a person's DNA with one of many specific restriction enzymes. Each restriction enzyme cuts the DNA only at specific sequences of base pairs. The products of the reaction are millions of DNA fragments. However, because of the differences in the DNA repeats, each individual's DNA produces fragments of different lengths; a 5-repeat chain is shorter than a 66-repeat chain. These fragments are sorted by size employing electrophoresis. Then, a blot is done on the DNA pattern, and the chains are labeled with radioactive probes that bind only to specific sites on the fragments. The resulting radiograph of the labeled fragments provides a pattern for that specific restriction enzyme and specific radioactive probe. Of the entire human population, the specific pattern seen may be found in, say, 1 of 20 individuals. A second restriction enzyme produces a different pattern, with, again, say, a 1 in 30 chance. However, the probability of both patterns occurring is $1{:}20 \times 1{:}30 = 1{:}600$. With five different enzymes, each having similar selectivity, the probability of two individuals having the same pattern reaches around 1 in 6×10^7. Profiles using 6 or 8 well-chosen restriction enzymes are expected to enable the identification of everyone in the world.

Shown in the accompanying figure is evidence from a trial where the defendant (column D) claimed that the blood stains on his jeans and shirt were his own blood and not from the fatally stabbed victim (V). From the set of tests, the probability that the blood was not the victim's was 1 in 50 million. He was lying.

[Ref: Courtesy of Cellmark Diagnostics, Germantown, Maryland.] ■

with low Joule heating. The technique has great resolving power, and methods have been developed so that electrophoresis can even be done for neutral, non-ionic molecules.

In addition, the analyte zones separated by capillary electrophoresis can be detected as in column chromatography. This contrasts with gel electrophoresis where separate steps are required to make the bands visible after the separation.

But (by now you are probably realizing there is *always* a "but") there is a tradeoff to using a narrow capillary; the samples must be about a nanoliter or less, and, after separation, the analytes are each also in sub-nanoliter volumes. With such small liquid samples, only a few methods of detection can be used effectively. For example, light absorption is not routinely used to detect concentrations below 1 μM in such small volumes. As a result, to detect analytes from real samples (without preconcentration) requires solution concentrations around 10 μM to 0.1 mM. These limitations are found for both spectrometric and electrochemical detection and are described in the following way: Separations by capillary electrophoresis allow excellent *mass* detection (ng to pg quantities of analytes are detectable on the column), but the limited volume allows only poor *concentration* detection. The numbers associated with *excellent* and *poor* should be considered temporary; what is considered "poor" now was "excellent" a few decades ago. It is this progress in detection that has led to the revival of electrophoresis in this newer, capillary, form.

Capillary electrophoresis is used to separate macromolecules (usually of biological origin) and also smaller, simple ions such as Cl^- and SO_4^{2-}. In addition, with various additives in the mobile phase, chiral separations can be done.

Next, let us look more closely at why reducing the cross section of the electrolysis solution is so effective at improving the quality of electrophoretic separations.

Temperature Differences and Convection

Recall that the primary problem with Joule heating is that it causes convection in the electrophoresis buffers, and the convective mixing degrades the quality of the separation. You might expect that the convection would be reduced if the temperature differences in the buffer can be reduced. By placing the buffer inside a capillary, the temperature differences are reduced in two ways.

First, a decrease in the cross section reduces the current that flows at a given voltage. For a fixed voltage and buffer, the current that flows decreases as the cross-sectional area of the liquid is reduced. Since the cross section is πr^2, the Joule heating (power) decreases essentially as the *square* of the radius. So upon decreasing the diameter from 100 μm to 50 μm we expect the current at a fixed voltage to decrease by a factor of four. The Joule heating is, then, also reduced by a factor of four since $P = I \cdot V$.

Second, the smaller cross section decreases the distance the heat has to flow to equalize the temperature within the buffer. Leveling the temperature differences within the buffer reduces the convection. Also, the total heat rise is made smaller by the short distance to the surroundings. Figure 14.8 shows how the temperature changes from its highest point at the center of the buffer in the capillary to the lowest temperature, that of the surroundings. During a capillary electrophoresis separation, the highest temperature is in the range of 10° to 40°C above the surroundings.

Small size not only reduces the Joule heating, but the magnitude of the convection itself is decreased. In effect, over these smaller distances to the surface, the water solution is more resistant to convection than for larger distances. The capillary wall acts like the polymers within a gel to reduce the tendency for convection. Both the reduction in temperature rise and the

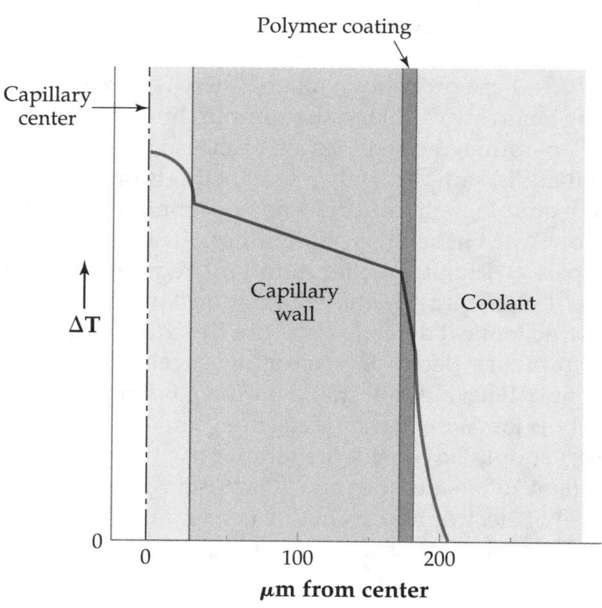

FIGURE 14.8 ▲
A plot of the approximate relative temperatures across the capillary during an electrophoretic separation.

The details depend on the capillary internal diameter, the voltage, the conductivity of the buffer, and the coolant (gas or liquid). The center of the capillary is in the range of 10° to 40° above the temperature of the bulk of the coolant during a regular run. The temperature falls off faster in the polymer coating than in the wall because the polymer is a better conductor of heat. The rapid falloff at the outside surface occurs because the coolant is circulating. If the coolant were still, the temperature would drop more slowly in it, and the buffer temperature would be higher.

effect of the small distances to the capillary wall allow us to put a much larger voltage on the capillary (5 to 40 kV) compared to a larger system, and, thus, improve the quality of the separation. The change in radius also reduces other effects of temperature on capillary electrophoresis separations, which are large and complex, as indicated in Figure 14.9.

Electroosmotic Flow in Capillaries

When a voltage is applied between the ends of an insulating tube that contains a liquid, the liquid is found to move along the entire length of the tube as a single long plug. This motion is called electroosmosis (EO), **electroendoosmosis**, or **electroosmotic flow** (EOF). The flow rate is proportional to the voltage applied and the buffer viscosity and depends on the charge at the inner surface of the capillary. Control of the EOF is essential for reproducible results in capillary electrophoresis.

One great benefit of having electroosmotic flow present is that electrophoretic separations and detection can be run for anions and cations at the same time. This outcome is illustrated in Figure 14.10 for a sample containing a single type of anion, a single type of cation, and a single type of non-ionic component. Without electroosmotic flow, either only anions or only cations would migrate toward the detector. The oppositely charged ion would migrate back out the injection end, and the neutrals would just sit at the end and spread by diffusion. With electroosmotic flow, all species are swept past

(a)

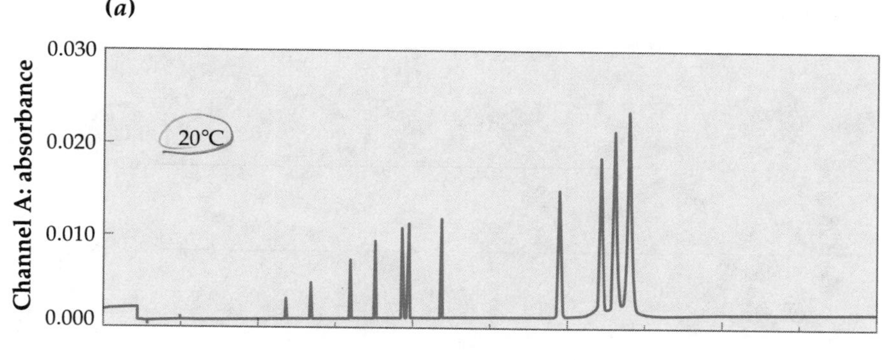

(b)

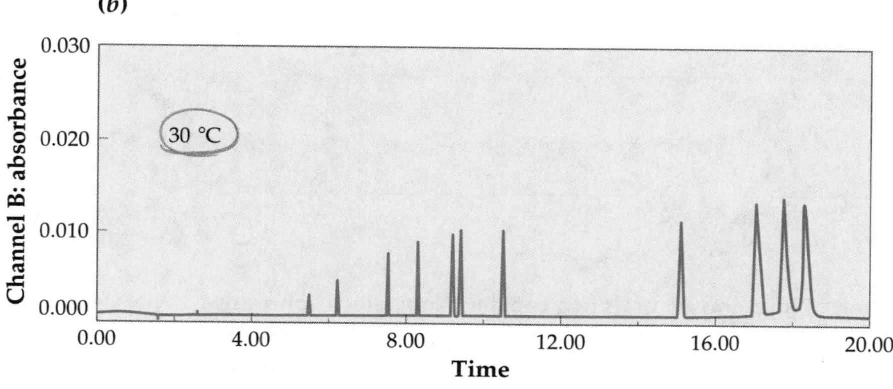

(c)

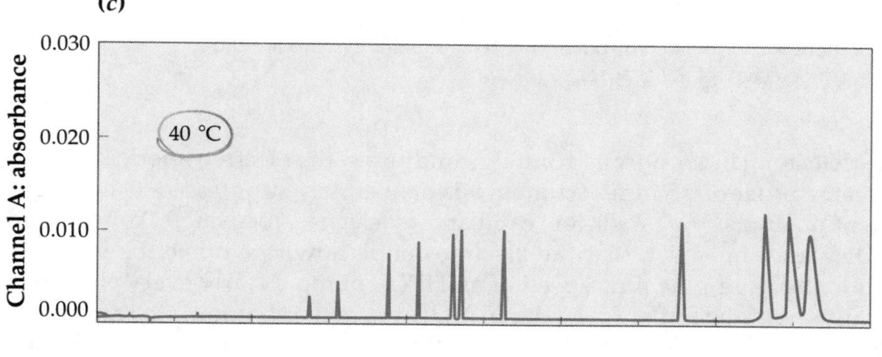

(d)

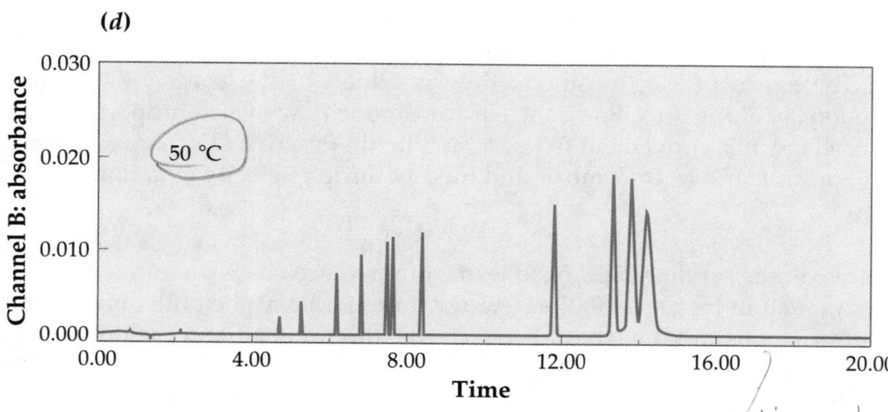

◀ FIGURE 14.9

The effects of temperature on a capillary gel electrophoresis separation.

The sample consists of fragments of DNA cleaved by restriction enzymes. The temperatures of the individual runs are labeled. The separation was carried out at constant current. As the temperature is increased, the migration time first increases and then decreases. These results reflect the interplay of the various influences on the gel electrophoretic separations. In addition, notice how the changes in the widths of the rightmost three bands reverse the relative heights as the temperature changes and the peaks are more poorly resolved. [Ref: Guttman, A., and Cooke, N. 1991. *J. Chromatography* 559:285–294.]

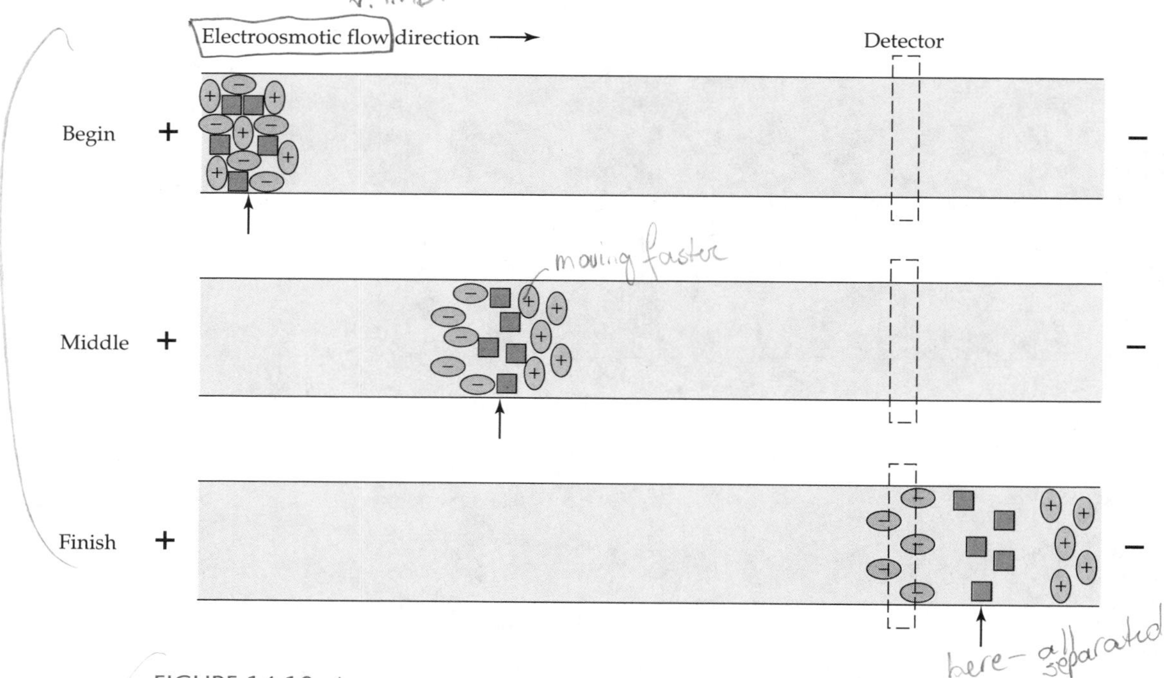

FIGURE 14.10 ▲

Diagram of the motion of ions and neutrals in a capillary zone electrophoresis separation.

The small vertical arrows indicate the position to which the neutrals have been moved by the electroosmotic flow. Relative to the detector, the positive ions move faster and the negative ions move slower. But relative to the electroosmotic flow, the positive ions have moved ahead and the negative ions have moved backward. Note that the solution remains electrically neutral since the anions and cations are surrounded by buffer counterions. Buffer systems are chosen that do not interfere with detection of the analytes.

the detector. Under normal running conditions, the electroosmotic flow rate is on the order of 0.5 mm s^{-1} unless it is deliberately suppressed. (The EOF is intentionally suppressed for capillary isoelectric focusing.) You should understand, however, that the electroosmotic flow rate *cannot* be set independently as can the flow speed of an HPLC pump. Nearly every change in conditions of the buffer (pH, special additives in the solution, etc.) affects the electroosmotic flow. That is, any alteration in the solution that modifies the charge on the surface of the capillary, modifies the viscosity, or requires a change in the voltage alters the rate of EOF.

The speed of electroosmotic flow is added to the speed of the electrophoresis of the ions. Since the electroosmotic flow is also proportional to the voltage, it is convenient to use a mobility to describe EOF as well. So we write μ_{eo} for the electroosmosis and modify the ion velocity equation to be

$$v_{ion} = (\mu + \mu_{eo})V/L$$

relative to the capillary as a fixed frame of reference.

As seen in Figure 14.7, the detector does not lie at the outlet end of the column. Let us call the length between the injection end and the detector ℓ. Then the time for a zone to migrate is less than the time to traverse the entire capillary length. So the time to arrive at the detector is given by

$$t_{zone} = \ell L/(\mu_{zone} + \mu_{eo})V$$

which follows from the definition of the ion mobility. Both ℓ and L appear in the equation because the detector resides at ℓ, while the voltage is applied across L.

Efficiency

Since the analytes separated by capillary electrophoresis can be detected as they pass a detector, electropherograms have the same general appearance as chromatograms. The nomenclature used to describe the separation of the bands in chromatography has come to be used for capillary electrophoresis results as well. The major difference here is that the positions of the peaks are determined by the electrophoretic mobilities of the constituents and *not* by differences in their interaction with the stationary phase.

We use the relationships given for chromatography in Equation 13-5 to define a value for the efficiency from the characteristics of a single band. The defining equation is

$$N = \left(\frac{t_{Ri}}{\sigma_i}\right)^2 = 16\left(\frac{t_{Ri}}{W_i}\right)^2 \qquad \textbf{(13-5)}$$

To understand the qualities of capillary electrophoresis and the approaches to increase N, we need to appreciate, at least qualitatively, what chemical processes contribute to σ_i. There are four:

1. Diffusion of the analyte along the direction of migration,
2. Interactions with the wall,
3. The temperature dependence of differential migration due to the effect of temperature on the viscosity and on the pH of the buffer, and
4. The length of the sample when it is loaded.

Collectively these are the causes of **dispersion**. Let us look at each.

Without proof, we state that diffusion along the direction of migration produces a Gaussian bandwidth $\sigma_i = (2D_it)^{1/2}$. In this equation t is the time and D_i is the diffusion coefficient of the ith analyte; for small ions in water $D \approx 1 \times 10^{-5}$ cm^2 s^{-1} and for proteins $D \approx 1 \times 10^{-6}$ cm^2 s^{-1}. σ_i is in cm.

The interactions with the wall are analogous to retardation by the stationary phase in chromatography. However, any retardation by the wall significantly degrades the electrophoretic separation. Figure 14.11 shows how rapidly N is reduced by *any* interactions with an immobile surface. This degradation of the efficiency is so great that capillary electrophoresis has been called *anti-chromatography* to emphasize that it is best to have no chromatographic interactions and that chromatographic interactions are not the cause of the separations.

The temperature effect enters because the solution is warmer in the center of the capillary, and the migration there is faster. The zone spreads somewhat since the center tends to get ahead of the periphery.

An excessively long sample zone sets the limit for the maximum value of N. However, a rule of thumb says that if the zone's length is 3 mm or less in a 100 cm capillary, you can generally ignore this contribution to the band

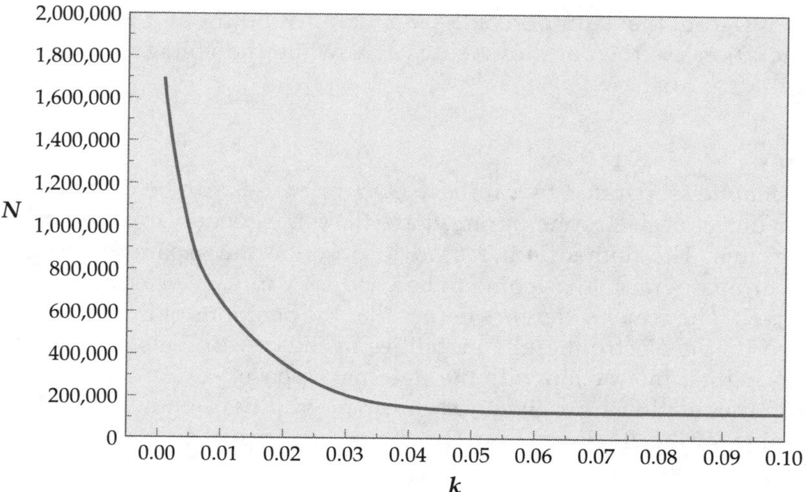

FIGURE 14.11 ▶
Effect on the efficiency of reversible analyte binding to the capillary in CE.
The analyte binding is accounted for as a capacity factor *k*, such as used in chromatography. The efficiency *N* rapidly decreases with binding.

broadening. Longer sample zones of charged analytes can be shortened by electrofocusing with a concomitant increase in the efficiency.

It is interesting to note that *N* is used to describe both chromatography and electroseparations even though it derives from the *number* of theoretical plates, as if these techniques depended on a number of sequential distillations. Chromatography was found not to be interpreted correctly that way or even as a series of sequential extractions, but *N* was retained nevertheless. Electrophoresis even more clearly does not resemble such chemical processes. However, again we ignore the origin of *N* and retain this useful concept to describe the quality of electrophoretic separations.

Resolution

Electrophoretic resolution is defined in exactly the same way as for chromatography.

$$\text{resolution} = \mathrm{R_s} = \frac{t_{\mathrm{R2}} - t_{\mathrm{R1}}}{\frac{1}{2}(W_2 + W_1)} = \frac{\text{peak separation}}{\text{average peak width}} \qquad \textbf{(13-7)}$$

The W_i are the baseline widths of the peaks, and the numerator is their separation, which results from differences in electrophoretic mobilities.

Sample Injection

Water does not flow easily through capillaries with 50–100 μm I.D.s, the size range most commonly used in CE. As a result, sample injection is done in two ways unique to capillary electrophoresis. One involves placing a pressure difference between the capillary ends. For example, one end of the capillary is placed into the sample, and for a fixed time the surface of the liquid in the container is raised above the level of the buffer at the detector end. Afterwards, the injection end is placed into the running buffer and the electrophoresis begun. This method is called **hydrostatic** or **hydrodynamic** sample injection. Alternate ways to get the same result are to put a vacuum on the detector-end buffer compartment or a positive air pressure on the sample. In any case, an unchanging protocol should inject a fixed zone length of sample into the capillary. The amount actually injected depends on the viscosity as well as on the pressure difference across the capillary and on the capillary length; that is, all the factors that make it easier or harder to suck a liquid through a straw. With care, realistic relative standard deviations of sample injection can be 3% in volume, although claims to better precisions are made.

The other general injection method is to place the capillary's input end into the analyte solution and pass a current through the solution and capillary. The components then enter the capillary by electrophoresis. This is called **electrophoretic** or **electrokinetic** injection. At first, it might appear that this method would be more accurate since it is so easy to measure precisely the total charge that is passed into the capillary. However, electrokinetic injection proves less accurate for three reasons. First, ionic species other than the analyte(s) can carry the charge, and we cannot easily discover what fraction of the total charge is carried by the analyte as it enters the capillary. Second, the electroosmotic flow also contributes to the volume injected. And third, discrimination against low mobility (and opposite mobility) ions causes their undersampling.

Modes of Operation

When the anode (+) is on the injection side of the capillary, this is called running in the **normal mode**. When the cathode (−) end is on the injection side, it is the **reversed polarity mode**. Voltages up to 1000 V cm^{-1} may be applied.

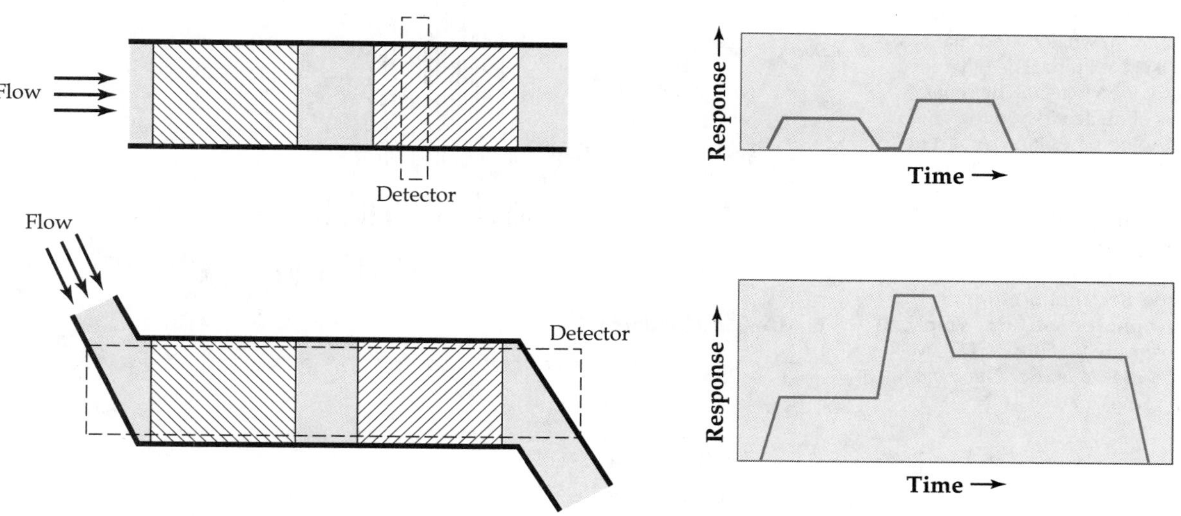

FIGURE 14.12 ▲

An illustration of the tradeoff between sensitivity/limit-of-detection and efficiency/resolution for a small volume.

We assume that the zones are boxes of concentrations rather than Gaussians, and the detector's response to the second band is 1½ times the response to the first. The two zones are moving along the capillary as indicated by the flow arrows, and the dashed-line boxes represent the volumes to which the detectors respond. The top figure shows a measurement across the capillary and the response to the bands, which is baseline resolved. The edges of the output are linear slopes, since that is the expected response as the zones enter, fill, and then leave the detected region at a fixed rate. The lower figure shows a detector that is along the capillary. This provides a greater response since the path through the zone is longer than the width of the capillary. However, two bands can be in the path at the same time and they cannot be resolved by the detector. A less drastic effect on a real sample is shown in Figure 14.13.

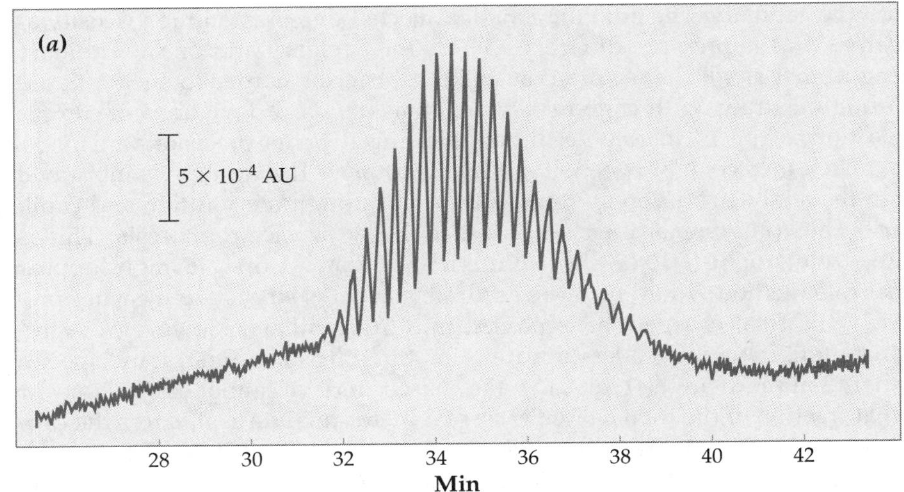

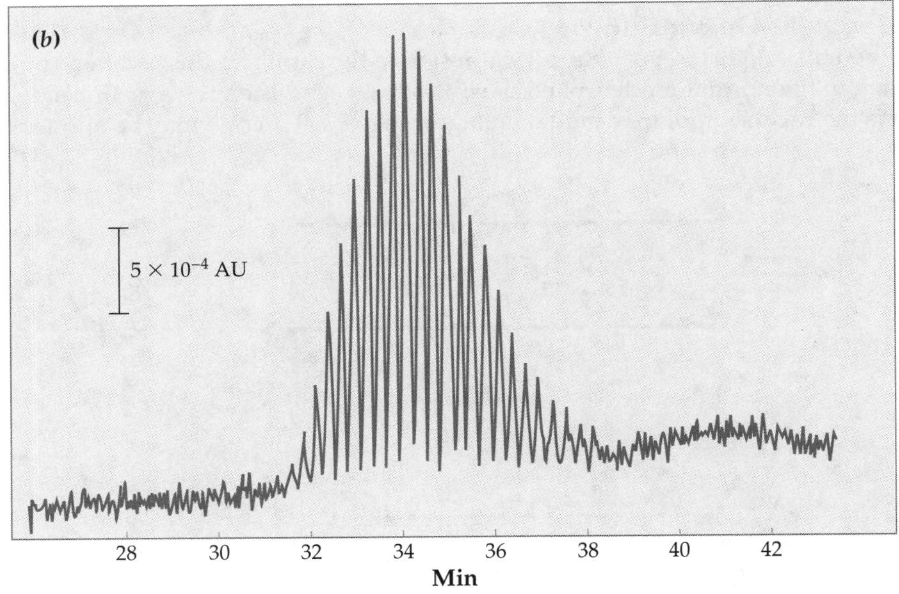

FIGURE 14.13 ▶
Detector response and resolution in capillary gel electrophoresis.
The detector cell length along the capillary of **(a)** is estimated by the authors of the reference article to be 1.0 mm, while the estimate for cell **(b)** is 0.3 mm. The difference in resolution is clear. The separation is of nucleotides in the range of 40 to 60 bases long. Adjacent peaks represent oligonucleotides differing by one nucleotide base. Output such as this is called an **electropherogram**. Other terms that are used for response vs. time include **electrophoretogram**, **electrophorogram**, and, simply, **pherogram**. [Ref. Bruin, G. J. M., et al. 1991. *Journal of Chromatography* 559:163–181.]

The power supplies in commercial instruments usually can be run in a number of different ways: at constant current, constant voltage, or constant power. In addition, it is useful to be able to run gradients in all three. The gradients can be used to improve quantitation by ensuring the entire sample is launched into the capillary, and once it is in, does not get expelled during the initial expansion of the electrolyte from Joule heating.

Different behaviors arise in constant-current and constant-voltage protocols. In the constant-current mode, the electric field changes (the voltage decreases) internally as the temperature changes. In the constant-voltage mode, the electric field stays the same, but the current varies as the temperature (and thus internal resistance) changes. These properties all are easily understood from the simple description given by Ohm's law, $V = IR$. Analyte quantitation may be better in the constant-voltage mode than in the constant-current mode.

Detection

The small volume of an analyte zone in capillary electrophoresis requires a trade-off between the efficiency and sensitivity of the detection. The basis of this trade-off is illustrated in Figure 14.12. The effect of the detector volume on resolution is shown in the electropherograms of Figure 14.13. Similar reductions in efficiency will result from any extra-column volume, and the demands on construction techniques are great indeed. Many detectors are similar to those used for HPLC but with optics designed for much smaller capillary.

In addition, the precision of quantitation in CE depends critically on controlling the electroosmotic flow. Any change in flow rate changes the calibration of both the heights and the integrals of the peaks. One other effect should be kept in mind. Unlike chromatography, where the analyte bands are carried past the detector at a fixed flow rate, the capillary zones still are migrating as they pass the detector, and the faster migrating bands pass through the detector faster than do the slower migrating ones. This effect can be a source of error even though the detector response is calibrated. One situation where this effect is particularly evident is for chiral separations. A photodetector's response is the same, but the faster isomer passes through the detector more rapidly then the slower one.

Suggestions for Further Reading

Darling, D. C., Brickell, P. M. 1994. *Nucleic Acid Blotting: The Basics*. Oxford: IRL Press.

It would be wonderful if all laboratory technique books were as well-written and as well-integrated as this one.

Dunbar, B. S. 1994. *Protein Blotting: A Practical Approach*. Oxford: IRL Press (at OUP).

A book full of useful details if you know what specific compounds and the blotting and detection techniques you will use. For a general guide, see the book by Darling and Brickell.

McLaughlin, G. M., Nolan, J. A., Lindahl, J. L., Palmieri, R. H., Anderson, K. W., Morris, S. C., Morrison, J. A., Bronzert, T. J. 1992. "Pharmaceutical Drug Separations by HPCE: Practical Guidelines." *J. Liquid Chromatog.* 15:961–1021.

An excellent introduction to the practice of capillary electrophoresis with illustrations of the effects for each factor contributing to the separations. Probably the best next place to read.

Landers, J. P., ed. 1994. *Handbook of Capillary Electrophoresis*. Boca Raton, FL: CRC Press.

A useful book (although not a handbook in the usual sense) of chapters on aspects of CE such as UV detection, thermal effects, and applications such as for proteins. Uneven in scope and level, but loaded with the authors' experience and most worthwhile to read carefully. Finding specific information is difficult since the index is poor.

Walker, J. M., ed. 1994. Totowa, NJ: Humana Press. "Basic Protein and Peptide Protocols" in *Methods in Molecular Biology*.

A useful book that has the details and references to a number of protein electrophoretic and blotting methods.

The guides to CZE by Hewlett-Packard and Beckman provide guidance in the subject.

Concept Review

1. What six phenomenological rules underlie electroseparations?
2. What is meant by the term "nondenaturing gel?"
3. What is the difference between a pore-gradient gel and a discontinuous gel?
4. Why is temperature control so important in electroseparations?
5. What is the advantage of hydrostatic injection over electrophoretic injection?
6. What is meant by "normal" vs. "reversed polarity" mode in capillary electrophoresis?

Exercises

14.1 The fastest traveling analyte of DNA gel electrophoresis travels 5.0 cm hr^{-1}. How fast does analyte ending up halfway between origin and front analyte travel?

14.2 Table 14.1 shows the isoelectric points of 14 proteins. What are the signs of the charges on the proteins in the list when in buffers with

(a) pH 3.0?
(b) pH 7.2?
(c) pH 10?

■ **14.3** An SDS gel electrophoresis was run with the following proteins as calibrants. An unknown protein was found to migrate 2.90 cm in the same gel. What is the molecular weight of the unknown protein? Recall that the migration distance follows the logarithm of the molecular weight. Either a regression calculation or graphical interpolation can be used to find the answer.

Protein Name	Molecular Weight/kDa	Position on Gel/cm
Ovalbumin	45.0	1.50
Chymotrypsinogen	25.7	3.70
Trypsin	20.0	4.90
Myoglobin	17.7	5.50
Lysozyme	14.5	6.30
Cytochrome c	12.4	7.00

14.4 Figure 14.4.1 shows a peak of a capillary electrophoresis separation for the same analyte run under non-stacking and on-column stacking conditions. From the data calculate the efficiency, N, for both runs by determining the baseline width for each. The retention time is shown in the plot. [Ref: 1991. *J. Chromatog.* 559:3–26.]

14.5 The table below gives the name, pK_a values, and retention times for a capillary electrophoretic separation of a series of organic acids commonly found in wines. The separation was carried out at a pH of 5.6. [Data from Levi, V., Wehr, T., Talmadge, K., and Zhu, M. 1993. *Am. Lab.* 25:29.]

Acid	$pK_{a,1}$	$pK_{a,2}$	$pK_{a,3}$	t_R
Oxalic	1.23	4.19		1.75
Citric	3.14	4.77	6.39	2.15
Tartaric	2.98	4.34		2.30
Malic	3.40	5.11		2.65

(a) Calculate the fractional compositions α_i for each of the four acids at pH 5.6.
(b) Calculate the average charge expected at pH 5.6 for each acid.
(c) If we assume that the cross section of a molecule is directly proportional to (molecular weight)$^{2/3}$, calculate the charge/cross-section ratio for each molecule. Do the components elute in the expected order?

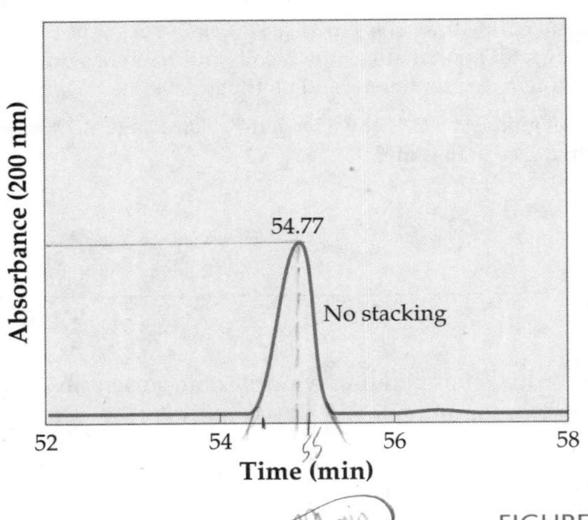

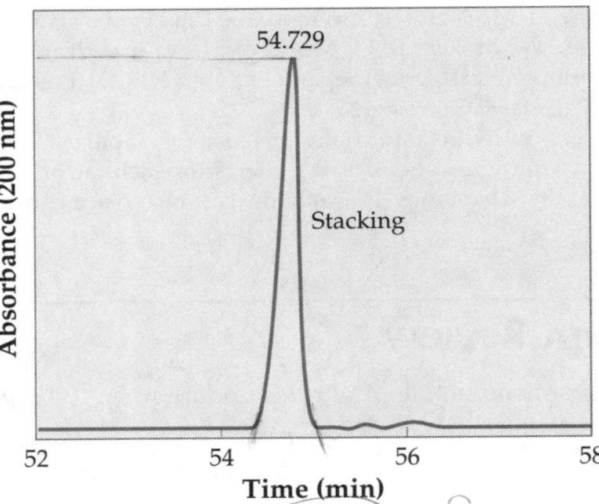

FIGURE 14.4.1 ▲

Additional Exercises

14.6 Table 14.1 shows the isoelectric points of 14 proteins. What are the signs of the charges on the proteins in the p*I* list when in buffers with

(a) pH 5.0?
(b) pH 8.0?
(c) pH 1.0?

14.7 The table below gives the name, pK_a values, and retention times for a capillary electrophoretic separation of a series of organic acids commonly found in wines. The separation was carried out at a pH of 5.6. [Data from Levi, V., Wehr, T., Talmadge, K., and Zhu, M. 1993. *Am. Lab.* 25:29.]

Acid	p$K_{a,1}$	p$K_{a,2}$	p$K_{a,3}$	t_R
Fumaric	3.83	4.44		3.70
Formic	3.75			3.85
Acetic	4.75			4.22

(a) Calculate the fractional compositions α_i for each of the four acids at pH 5.6.
(b) Calculate the average charge expected at pH 5.6 for each acid.
(c) If we assume that the cross section of a molecule is directly proportional to (molecular weight)$^{2/3}$, calculate the charge/cross-section ratio for each molecule. Do the components elute in the expected order?

14.8 Using the data for the protein standards from Exercise 14.3, find the molecular weight of a protein that migrates 3.95 cm during the same gel separation.

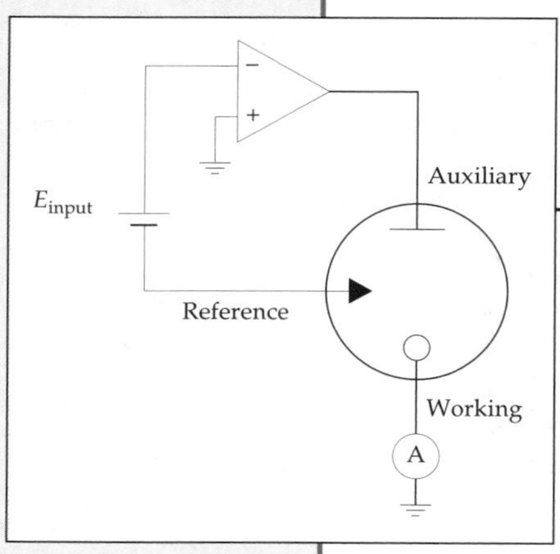

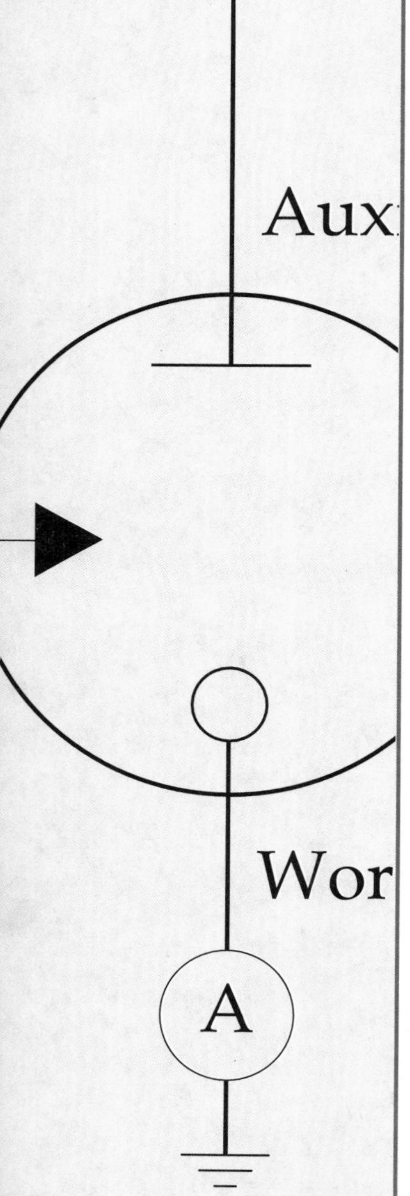

CHAPTER 15

Electrochemical Methods

Overview

A Deeper Look

15.1 The Variety of Electrochemical Methods

In this chapter are presented some of the instrumental methods used to measure electrical currents, charge, and/or potentials for analytical uses. In many cases, the details of the chemistry that occurs are not well understood. Yet, the methods can still be exceedingly useful. Given the sophistication of contemporary electronics, most of the problems with the electrochemical methods lie in controlling the chemistries of the samples and the transducers. We shall, therefore, concentrate on this aspect of the measurement. The processes that permit electroanalysis to be used are those in which oxidation and/or reduction occurs and those in which dilution or diffusion of charged (and related uncharged) species occurs.

The relationships of the concentrations, charges, and energies in chemical reactions involving charged species were discussed in Chapter 8.

The chapter is organized as follows. Since you are already familiar with the theory of electrochemical cells, as described by the Nernst equation, experimental methods to which the Nernst equation is applicable are presented first. These are the techniques of **potentiometry**, the measuring of potentials.

Following these, the techniques for measuring the changes in **resistance** of electrolyte solutions are described. Ohm's law relates potential to current and resistance, and is stated

$$V = IR$$

where

I = current in amperes (A ≡ coulombs s^{-1}),

R = resistance in ohms (as the name implies, the "opposition" to the flow of electrons)

V = the potential difference in volts. (In electrochemistry, the potential difference is usually denoted by E.)

The inverse of resistance, $1/R$, is called **conductance**, which is directly relatable to ionic concentrations. The measurement of conductance is called, reasonably, **conductimetry**.

The next set of techniques relates the number of moles of a substance oxidized or reduced to the total charge. These **coulometric** techniques are presented in the third portion of this chapter.

The last group of techniques covered combines the measurement of potential, current, and time. In these methods, we look at the changes in the current (over time) while the voltage applied to the electrochemical cell is changed in a precisely controlled manner. Because the voltage and current are measured simultaneously, these are called the **voltammetric** methods. They are highly sophisticated in concept and are, perhaps, experimentally the most elegant methods for trace analysis.

As was noted in Section 8.3, the Faraday constant $\mathcal{F}$ is the proportionality constant that relates equivalents of electrons to coulombs.

15.2 Potentiometry

Reference Half-Cells

In the potentiometric methods, we measure the potential of an electrochemical cell to find the concentration of some analyte. The concentration is related to the potential through the Nernst equation or a modification of it. The unknown solution is in one a half-cell, and to measure a voltage another half-cell is needed as a reference. Further, a precise measurement of the unknown quantity requires a reproducible and stable reference half-cell. Then, if we want a measurement that can be compared to those made in other laboratories or on other days, the half-cell must possess an accurately known potential.

The standard hydrogen electrode reference cell was introduced in Section 8.3 (page 239).

One reference half-cell that could be used is the normal (or standard) hydrogen electrode (NHE), but it is rather hard to work with experimentally. Although the NHE is used as the fundamental reference cell, two other reference half-cells are used most commonly on a daily basis: the **calomel** and the **silver-silver chloride** half-cells.

Calomel is an old name for mercurous chloride, Hg_2Cl_2. The redox reaction in the half-cell is

$$Hg_2Cl_2(s) + 2\,e^- \rightleftharpoons 2\,Hg^0 + 2\,Cl^-$$

When the chloride solution present is aqueous KCl that is saturated and has excess KCl crystals present, the electrode is called a **saturated calomel electrode** (SCE). The SCE half-cell contains mercury, and the practical half-cell has the configuration

The notation representing the half-cell shown here is described in Section 8C.

$$\|KCl(saturated), Hg_2Cl_2(s)|Hg$$

$$E = +0.241\text{ V } vs.\text{ the normal hydrogen electrode (NHE) at 25°C}$$

SCE electrodes can have any one of literally scores of different shapes without affecting the potential of the half-cell as long as the chemical configuration is correct. However, as for all electrochemical cells, the SCE potential varies with temperature, as can be seen in Table 15.1. Incidently, the normal hydrogen electrode's potential is *defined* to be zero exactly *at all temperatures.*

Sometimes, instead of using a saturated solution of KCl, a 1-M solution is used. The cell is then referred to as a **normal calomel electrode**. The normal calomel electrode is

$$\|KCl(1\text{ M}), Hg_2Cl_2(s)|Hg$$

$$E = +0.280\text{ V } vs.\text{ NHE at 25°C}$$

The silver-silver chloride half-cell utilizes the redox couple

$$AgCl(s) + e^- \rightleftharpoons Cl^- + Ag^0$$

Table 15.1 Temperature Dependence of a Saturated Calomel Electrode (SCE)*

Temperature °C	Potential E *vs.* NHE=0 V (experimental)
10	0.2507
25	0.2412
40	0.2308
70	0.2078

*The potentials can be calculated at other temperatures using the following empirical formula, with E in volts *vs.* NHE and T in °C.

$$E_{SCE} = 0.2412 - 6.61 \times 10^{-4}(T - 25) - 1.75 \times 10^{-6}(T - 25)^2 - 9.0 \times 10^{-10}(T - 25)^3$$

The cell can be as simple as a silver wire coated with AgCl and immersed in a KCl solution. The resulting half-cell is represented by

$$\|\text{KCl(saturated)}, \text{AgCl}(s)|\text{Ag}$$

$$E = 0.197 \text{ V } vs. \text{ NHE}; E = -0.045 \text{ V } vs. \text{ SCE both at } 25°\text{C}$$

Like the calomel reference, the Ag/AgCl cell potential is strongly temperature dependent. The Ag/AgCl cell has the advantage of greater stability at higher temperatures but has the drawback that the cell potential is sensitive to light. Incidentally, this cell may also be constructed with nonaqueous solvents that dissolve, even sparingly, a silver salt or a chloride salt.

15.3 Ion-Selective Interfaces and Ion-Selective Electrodes

Whenever an interface forms between two phases (that is, the two do not mix freely) an electrochemical potential is generated, which is called the **interface potential**. Interface potentials create difficulties in making accurate measurements of the voltages of electrochemical cells. However, they are the basis of the measurements made with **ion selective electrodes**. (An example is the ion-selective pH electrode with which we can measure the activity of H^+ in the presence of any number of other species.)

The interface potential of ion-selective electrodes arises in the following way. See Figure 15.1 and think of an imaginary plane separating the two half-cells. This plane is indicated in the figure by a dashed line in 15.1b. The chloride ions can move back and forth across the position of the plane. However, because the concentrations of chloride differ in the half cells, a significant imbalance arises in the number of chloride ions that transfer across the imaginary plane to the left: five times more. Since the ions are negatively charged, and since more of them will move to the left than to the right, a net current of negative charge moves to the left, and an excess of negative charge builds up

(a)

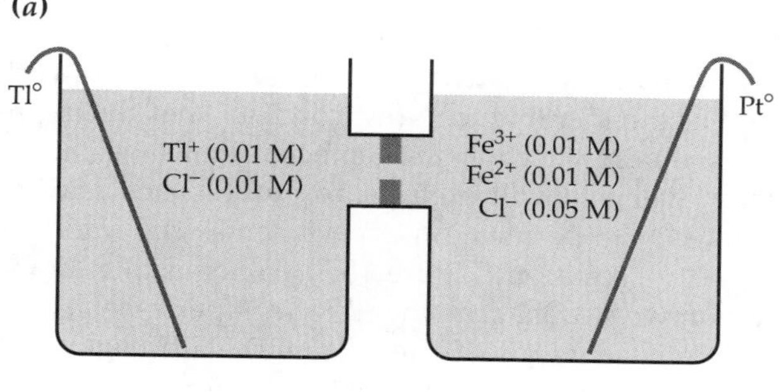

Cl will transfer to LHS cell 5x faster than LHS cell → RHS cell.

(b)

Cl^- (0.01 M)
Tl^+ (0.01 M)

Cl^- (0.05 M)
Fe^{2+} (0.01 M)
Fe^{3+} (0.01 M)

Transfer rate = k_{Cl} [0.01 M] →

← Transfer rate = k_{Cl} [0.05 M]

− +

Liquid junction potential due to chloride alone

◀ FIGURE 15.1
The Origin of Interface Potentials.

(a) Example of a cell that will have an interface potential between the two half-cells. **(b)** Schematic diagram illustrating the imaginary boundary somewhere in the neck between the two reservoirs.

on the left. This excess of charge causes the left side to be more negative than the right: The interface electrochemical potential forms in this **concentration cell**. In addition, as long as electrical current can be conducted across the solutions and interface, the potential can be measured even if the interface is between a liquid and a solid.

An ion-selective pH electrode is composed of a soft glass membrane placed between two aqueous solutions of different pH. For the measurement of other ions, membranes are constructed from thin liquid layers or crystals of various kinds or from chemically modified polymers (special plastics). These materials will be described briefly after presenting the general principles of ion selective electrodes (ISEs) as well as some of the difficulties associated with them. A list of various selective electrodes is presented in Table 15.2. The majority of electrodes that are currently available commercially are specific for ionic species. The nonionic species for which ISEs are used are generally gases. However, these electrodes make a measurement only after the gas separates from the matrix by passing through a selective membrane and then reacts to form ionic species.

We present here a simple model to explain why an ISE (specifically, a pH electrode) works. The reasons that such a simple model is presented are the following:

a. The analytical determinations using ion-selective electrodes are carried out under reproducible conditions of ionic strength, temperature, and sample pretreatment. Thus, knowledge of the exact nature of the error, if conditions change, is not necessary in order to use ISEs in analysis.
b. The interferences that result from these factors can be approached empirically by the use of appropriate standards.
c. Not all the points of detail are generally agreed upon.

Consider a thin membrane separating two liquid volumes (Figure 15.2). For simplicity, we can think of the membrane as having two surfaces and a region of bulk material between them. In many cases, these surfaces are not well understood in their chemical details. Accordingly, in the figure they are shaded gray without specific features. The two surfaces are in contact with the solutions on either side of the membrane. Now, consider that the membrane surface is special; it will allow only certain specific ions to enter (or permeate) into this surface layer and will exclude any others that may be present. More properly stated, the permeation occurs across the interface between the membrane and the solution (Figure 15.3). When a membrane exhibits this type of behavior, we say that the membrane is **perm-selective**. The complexities of ionic charge, activity coefficients, and the different diffusion rates of each species of ion all affect the potential at the surface.

Suppose that the solution in contact with the surface is hydrochloric acid in water. The protons, with their positive charges, penetrate into the glass leaving the chloride with its negative charges excluded outside. This creates a voltage across the interface. How does this voltage change with changes in the concentration of protons in the solution? In the presence of a higher proton concentration, a greater number of protons diffuse into the interface. (This is exactly equivalent to the chloride interface potential described above, but here, the interface consists of one solution and one solid phase.) The higher the concentration of protons diffused into the solid, the greater the

Table 15.2 Ion-Selective Electrodes and Their Uses

Ion for Which the Electrode Is Selective	Useful Concentration Range (M and ppm)	Uses
Ammonia, ammonium NH_3, NH_4^+	10^{-6}–0.1 M 0.02–2000 ppm	The electrode measures dissolved ammonia. Other substances can be measured by chemical pretreatment including ammonium nitrate and organic nitrogen. Ammonia is commonly measured in foods, effluents, soil, and biological systems.
Bromide Br^-	10^{-6}–0.1 M 0.08–8000 ppm	Water pollution analyses, photographic processing
Cadmium Cd^{2+}	10^{-7}–0.1 M 0.01–11,000 ppm	Cd^{2+} activity can be measured directly or through EDTA titrations. Used in plating bath analysis and as end-point detector in titration of sulfide in wood-pulping liquors.
Calcium Ca^{2+}	10^{-5}–1 M 0.4–40,000 ppm	Direct measurement of Ca^{2+} in biological fluids and total calcium by EDTA titrations. Also soil, food, and water.
Carbon dioxide CO_2	5×10^{-5}–0.01 M 2–400 ppm	Dissolved CO_2 and carbonate can be measured directly. Samples analyzed include serum, wine, water, and soft drinks.
Chloride Cl^-	3×10^{-5}–0.1 M 1.0–3500 ppm	Chloride activity is critical in processing of foods, resins, and insecticides. Also used for biological material and in medical research of cystic fibrosis.
Copper Cu^{2+}	10^{-6}–0.1 M 0.06–6000 ppm	Plating and etching baths, soil, sewage, pharmaceuticals, food, and beverage products
Cyanide CN^-	10^{-6}–10^{-3} M 0.03–250 ppm	Analysis of plating baths, mineral extractions, and waste water.
Fluoride F^-	10^{-6}–0.1 M 0.02–2000 ppm	Widely used in analysis of fluoridated water supplies, toothpaste, bone, plating baths, water discharge, and fluoride metabolism.
Fluoroborate BF_4^-	3×10^{-6}–0.1 M 0.3–9000 ppm	Quantitating fluoroboric acid in plating baths and analysis of boron in soil, plant tissue, and glass upon conversion to fluoroborate
Hydrogen H^+	0–14 pH	Measures a_{H^+} of solutions
Iodide I^-	10^{-7}–0.1 0.01–10,000 ppm	Commonly used in analysis of salt brine, industrial waste and in pharmaceutical research
Lead Pb^{2+}	10^{-6}–0.1 M 0.2–20,000 ppm	Major uses in analyses of electroplating baths and to sense lead titrant in the indirect determination of sulfate
Nitrate NO_3^-	6×10^{-6}–1 M 0.4–60,000 ppm	Primarily in analyses of soil and fertilizer, food, water treatment biologicals, and in pharmaceutical and photographic manufacturing
Nitrogen oxide NO_X	4×10^{-6} 5×10^{-3} M 0.2–200 ppm	Gas-sensing electrode used primarily for determining nitrite, NO_2^-, in foods, ground water, plating baths and atmospheric NO_X
Perchlorate ClO_4^-	3×10^{-6}–0.1 M 0.3–10,000 ppm	Primary use has been in explosives and solid-propellant industry.
Potassium K^+	10^{-5}–1 M 0.4–40,000 ppm	Analyses of soil, fertilizers, biological fluids, and wines
Silver/sulfide Ag^+, S^{2-}	10^{-7}–0.1 M Ag^+ 0.01–10,000 ppm	Applications in the pulp and paper industry and in silver recovery from photographic solutions
Sodium Na^+	10^{-5}–0.1 M 0.2–2000 ppm	Sodium analyses of dietetic foods, high-purity water sources, tissues, glass, pulping liquor, and biologicals
Thiocyanate SCN^-	5×10^{-6}–1 M 0.3–60,000 ppm	Most commonly used in the plating industry

Primary reference: The *Beckman Handbook of Applied Electrochemistry*. Fullerton, CA: Beckman Instruments, Inc., 1980.

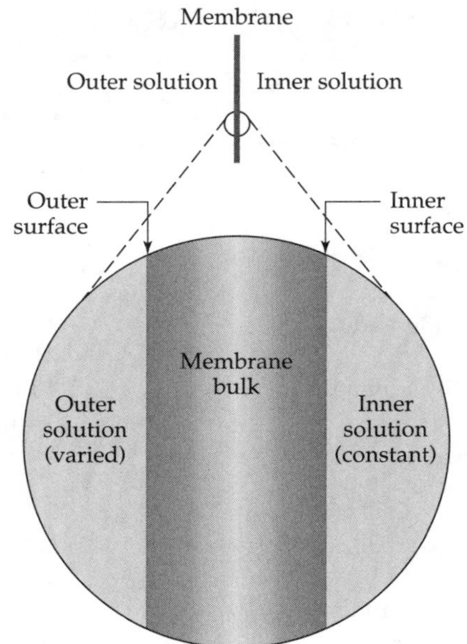

FIGURE 15.2 ▶
Ion Selective Membrane Structure.
The structure of a membrane is conceptualized as consisting of separate regions having surface and bulk properties. The experimentally measured properties are attributable to all the regions—the surfaces, the bulk, and the solutions. However, the details of the explanation are still debated.

excess of anions left outside, and the larger the potential that develops across the interface.

There are two more points to notice in this figure. First, there are two surfaces. There are potentials across the inner surface and the outer surface. The potential that is measured from one side of the membrane to the other is the sum of these, as illustrated in Figure 15.4. Because of this structure, we must keep a standard solution in contact with the inside surface of the electrode. The membrane is then exactly analogous to the two half-cells in a concentra-

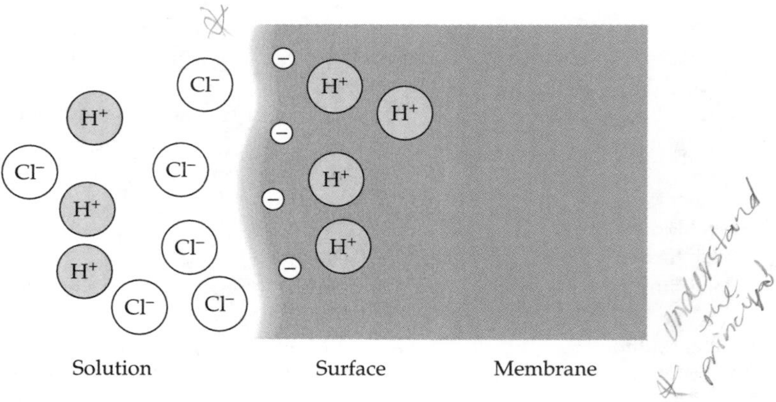

FIGURE 15.3 ▲
Ion-selective surface of a proton ion-selective electrode (pH electrode).
The protons are able to enter the surface region, but the chlorides are mostly excluded. The excess positive ions in the surface region and the excess negative ions in the solution region produce a voltage at the interface. At least part of the selectivity can be attributed to negative charges fixed at the surface. They produce a barrier to the passage of negative charges such as the chlorides.

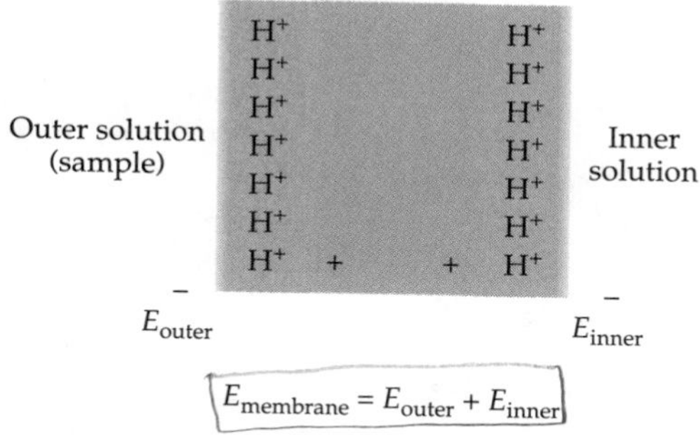

FIGURE 15.4 ▲
Operation and structure of a H^+ ion-selective electrode.
The total potential is the sum of the two interface potentials. They are opposed and so have opposite signs: the sum is smaller than the larger potential. The resistance between the surfaces is much larger than the resistances of the ionic solutions.

tion electrochemical cell with a salt bridge, and the potential is described by the Nernst equation.

$$E = E^\circ + \frac{RT}{\mathcal{F}} \ln a_{H^+} = E^\circ - 0.059 \text{ pH} \quad \text{at } 25^\circ\text{C} \tag{15-1}$$

So the potential that develops across the glass membrane can be used to quantify the hydrogen ion activity of the solution.

Second, the circuit of the electrochemical cell must be complete. Two connections must be made: one of them through the membrane and the other from the inside of the electrode back to the solution. In a glass pH electrode, the connection through the membrane is made by ions such as sodium and lithium moving through the solid glass. However, they are not highly mobile, so to get a good connection—a lower resistance—the glass membrane must be as thin as possible. (Almost every practicing analytical chemist has been reminded of this after inadvertently breaking one.)

The connection needed to complete the circuit from the electrode through the electrolyte solution is made through a salt bridge from the reference electrode. An illustration of a glass pH and reference electrode pair is shown in Figure 15.5a. Combination electrodes that have an internal Ag/AgCl reference (Figure 15.5b) are more common than separate pH and reference electrodes.

Since the two surfaces of the membrane act together like a concentration cell, you might reasonably ask, Why do you need a reference electrode? The answer is that the reference electrode eliminates problems caused by changes in junction potentials when making measurements on solutions with different compositions. With the junction potentials throughout the system at steady values and the internal pH of the electrode fixed, the response of the glass electrode to protons becomes simply

$$E_{\text{electrode}} = \text{constant} + \frac{RT}{\mathcal{F}} \ln a_{H^+}(\text{sample})$$

where the constant is the sum of the junction and reference potentials. The value of the constant is unimportant for practical use since the electrode potential corresponding to a specific pH is found by calibrating with standard

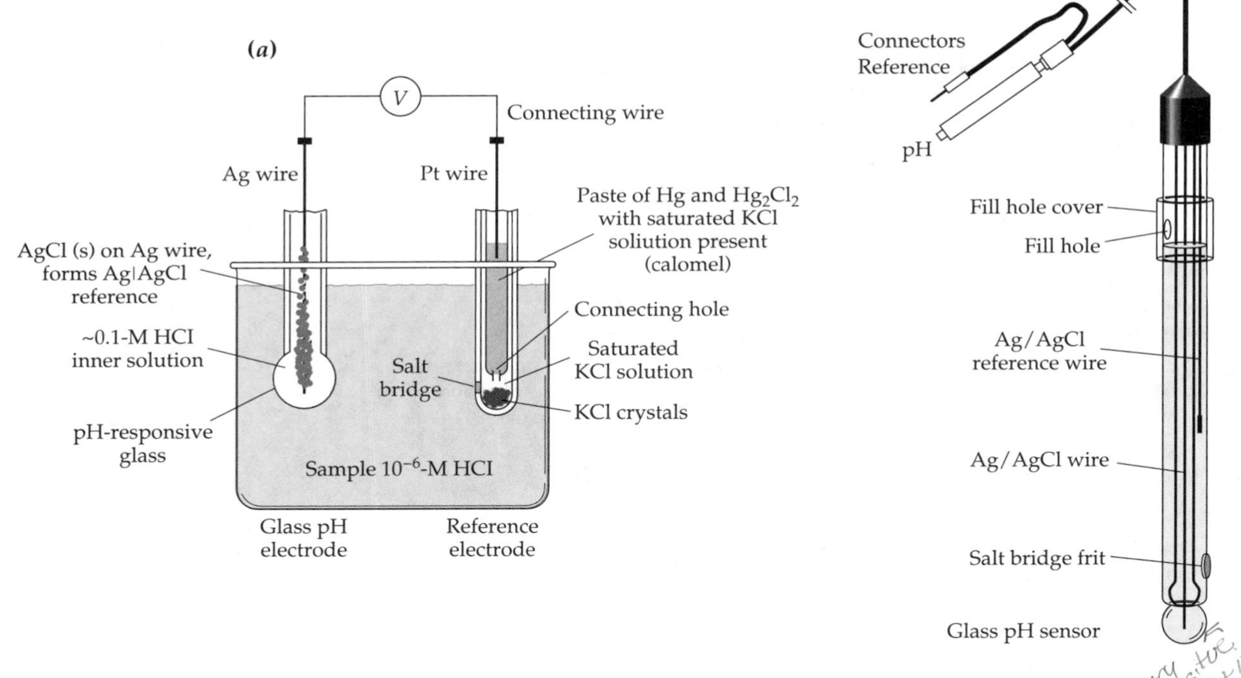

FIGURE 15.5 ▲
Measurement of pH.
(a) Illustration of the most important parts of a pH electrode and a calomel reference electrode, as used to determine $[H^+]$. The KCl salt bridge minimizes differences in the interface potential between the reference electrode and different test solutions. The design provides two reference potentials in series with the pH-sensitive glass membrane. **(b)** A combination pH electrode. The reference electrode resides within the single electrode body.

buffers. Then, changes in pH follow the regular Nernstian form of the second term of the right-hand side. Notice that the response of the electrode depends on the temperature.

Range of Response of Ion-Selective Electrodes

Ion-selective electrodes, like all transducers, have a limited range of response. These ranges are listed for some representative electrodes in Table 15.2. Note that although the electrodes respond to ion activity levels, the limits are stated in molar concentrations. For most ISEs, the concentration range extends down to about 10^{-6} M. However, in practice, the usable range also depends on the other components in the solutions. As an example, see Figure 15.6 in which the response of an ISE to lithium is shown at different values of sodium activity. From a study of the figure, you should see that the linear range for lithium decreases as the sodium concentration is raised. At the extreme, the potential being measured does not respond to changes of lithium concentration, and the electrode becomes useless for analytical purposes. This lower limit for the working range for lithium occurs at higher concentrations as the sodium concentration is raised. The effects of interference, such as sodium on a lithium ISE, can be treated quantitatively in a manner described next.

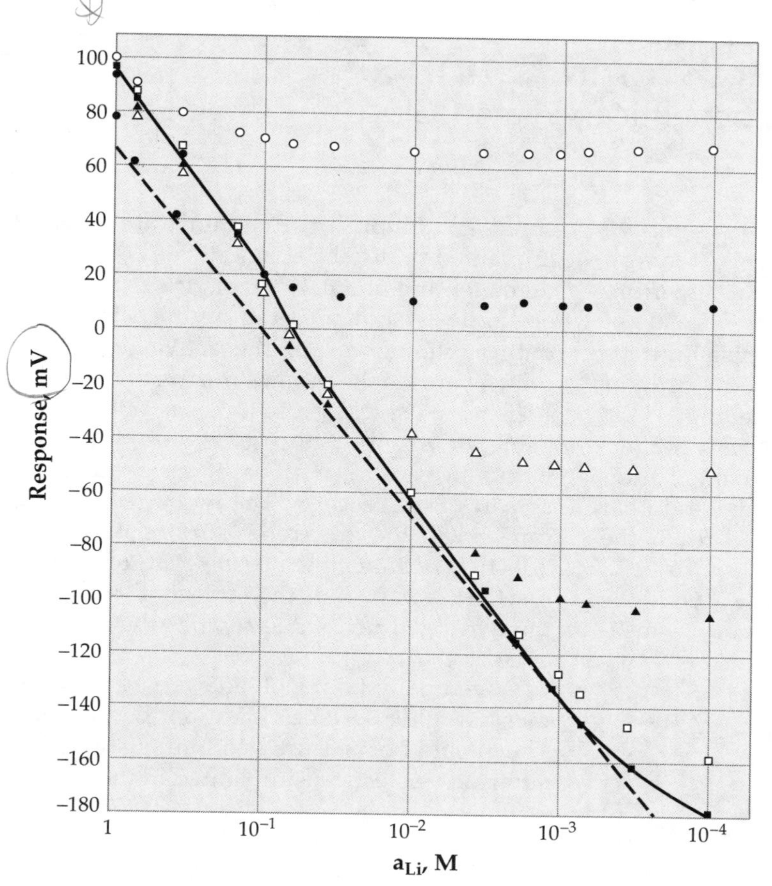

FIGURE 15.6 ▲

Plot of the lithium response of an ISE as it varies with the concentration of sodium.

The plot shows the measured potential in mV *vs.* activity of lithium. The dashed line is the Nernstian response: 59 mV per decade at 25°C. The electrode was a Beckman sodium ion electrode (Serial No. 39278). The temperature was 25°C, and the pH was buffered by keeping the solution saturated with $Ca(OH)_2$. Sodium concentrations are ○, 1 M; ●, 0.1 M; Δ, 0.01 M; ▲, 0.001 M; □, 0.0001 M; ■, pure LiCl response. Note that the electrode could be used for either sodium or lithium ion-selective detection. [Ref: Reprinted with permission from Buck, R. P. 1974. *Anal. Chem.* 46:261. Copyright 1974 American Chemical Society.]

Interferences and Ion-Selective Electrodes

Look back at Figure 15.3, which illustrates how the charge separation of chloride (and other anions) from hydrogen occurs across the interface of the glass electrode. The electrode is selective for hydrogen because of the special chemistry of the protons with the glass surface. At least part of this special interaction is due to the negative charges at the surface, which attract the adjacent hydrogen cations and repel the adjacent chloride anions.

Now consider what will happen at the surface of the glass membrane in a sodium hydroxide solution with pH = 13. The activities of the ions in the solution are

$$a_{H^+} = 10^{-13} \quad \text{and} \quad a_{Na^+} \approx 0.1$$

ANALYTICAL CHEMISTRY IN ACTION
Keeping an (Electrochemical) Eye on Glucose

Amperometric methods are not highly selective chemically: at a given potential the current can arise not only from the analyte but also from any other compound that reacts more easily at the electrode. However, the inherent ultratrace detection limits and the relatively low price of electrochemical detection is compelling. Thus, electrochemical methods and especially amperometric methods are excellent in combination with some chemical separation method that allows only the selected species to reach the electrode and react there.

Glucose can be detected in this way, and the blood glucose levels of diabetics can be measured continuously by a miniature electrode implanted under the skin. It is only 0.25 mm in diameter and about 1 cm long. A sketch of the electrode is shown in the figure below along with a photograph of the active region. The sensing area is the thinnest region at the left.

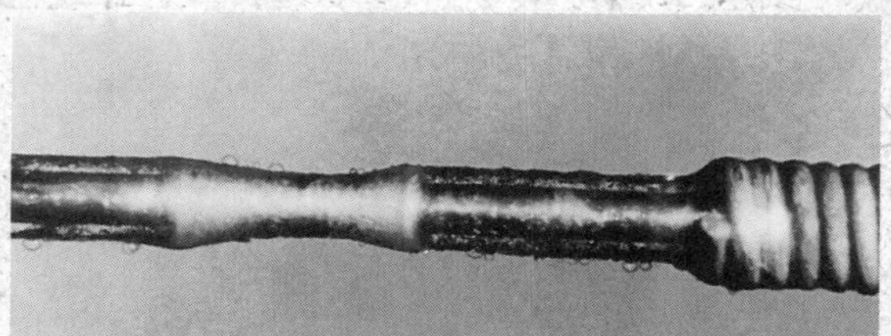

The chemical selectivity is obtained by using an enzyme (glucose oxidase) that reacts with glucose and produces hydrogen peroxide as part of the reaction. The hydrogen peroxide can reach the electrode surface through a selective polymer membrane that keeps other reactants away from the surface. Oxidation of the hydrogen peroxide provides the measured current.

Eventually, taking blood every few hours to measure glucose levels may not be necessary. They could be measured continuously by such a miniature electrode implanted in the skin.

[Ref: 1994. *Diabetologia* 37;610–616.] [Photo courtesy of George Wilson.] ■

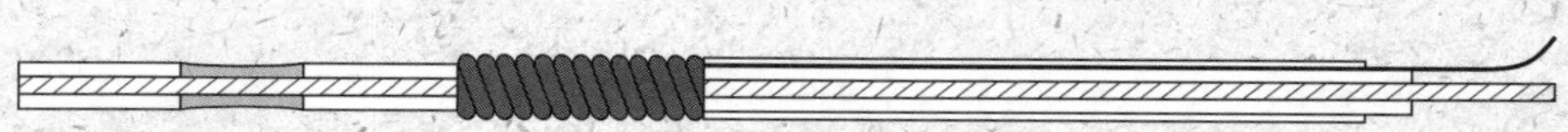

The sodium is about 10^{12} times more concentrated than the protons. Such an overwhelming excess of sodium ions might be expected to interfere to some degree with the protons since both are cations. Some of the sodium ions pass across the interface and contribute to the potential measured by the pH electrode.

Interference such as by Na^+ on the pH measurement is expressed quantitatively using the **selectivity coefficient**. The selectivity coefficient is usually written as k_{ij} or K_{ij}. The equations in which it appears, Equations 15-3 and 15-4, will be used without proof of their validity.

For the case of sodium interfering with protons, the potential determined experimentally with the electrode is expressed algebraically,

$$E_{\text{ISE}} = \text{constant} + \frac{RT}{\mathcal{F}} \ln (a_{\text{H}^+} + k_{\text{H,Na}}\, a_{\text{Na}^+}) \qquad \textbf{(15-3)}$$

Equation 15-3 says that the sodium activity contributes to the measured potential, and the contribution is proportional to the selectivity coefficient: A smaller selectivity coefficient indicates less interference. As an example, values for $k_{H,Na}$ can be as low as 10^{-15} but vary depending on the type of glass composing the electrode. If $k_{H,Na} = 10^{-15}$, then Equation 15-3 indicates that the Na^+ contribution will equal that from H^+ when $a_{Na^+} = 10^{15}\, a_{H^+}$.

A more general form of Equation 15-3 is slightly more complicated. If the ionic charge of the analyte z_i and the interferent z_j are not the same, then the equation for that interferent becomes

$$E_{ISE} = \text{constant} + \frac{RT}{\mathcal{F} z_i} \ln \left(a_i + k_{ij} a_j^{z_i/z_j}\right) \tag{15-4}$$

This equation reduces to the simpler form of Equation 15-3 when $z_i = z_j$.

Potentiometric Precision

It is difficult to measure a cell potential more accurately than about ± 1 mV = ± 0.001 V. (Greater precision is possible; greater accuracy is difficult to obtain because of the junction potentials.) This error of ± 1 mV in potential leads to an error in analyte concentration, and the concentration error equivalent to 1 mV increases with the number of electrons transferred. An uncertainty of 1 mV produces an uncertainty of 4%, 8%, or 11% in a measurement involving, respectively, a mono-, di-, or trivalent charge transfer. These percentages represent the approximate limits of accuracy for the potentiometric measurement of concentration.

15.4 Electrochemical Methods Using Current Flow

The response of ion-selective electrodes depends only on the motions of charged species and not on the occurrence of oxidation-reduction reactions. The remainder of the electrochemical methods discussed in this chapter depend on oxidation and/or reduction of analytes.

Quantitative analyses using oxidation–reduction chemistry can be done in a number of different ways. The most direct of them, perhaps, is simply to reduce all of an ionic metal species to its metallic form, which removes the ions from solution. This is **electrodeposition.** It is then straightforward to dry the metal-coated electrode and weigh (by difference) the metal that is plated out. Overall, analyses done this way are called **electrogravimetry.** Electrogravimetry can be highly precise (better than $\pm 0.02\%$) and accurate if done under optimum conditions.

Electrodeposition was introduced in Section 10.1.

Alternatively, we could measure for a relatively short time (seconds) the electric current that oxidizes or reduces the species of interest: the **amperometric methods.** A third possibility is to run the oxidation or reduction for a time long enough to oxidize or reduce all the analyte and measure the total charge: the *coulometric methods*. All three of these methods require a moderate (μA and greater) current during at least some part of the analysis. In contrast, the currents allowed to flow when measuring potentials in potentiometry tend to be less than picoamps—a millionfold less.

Let us now look at these methods that require current flow. The initial question is, What happens in an electrochemical cell that has an applied voltage high enough to cause a continuous current to flow?

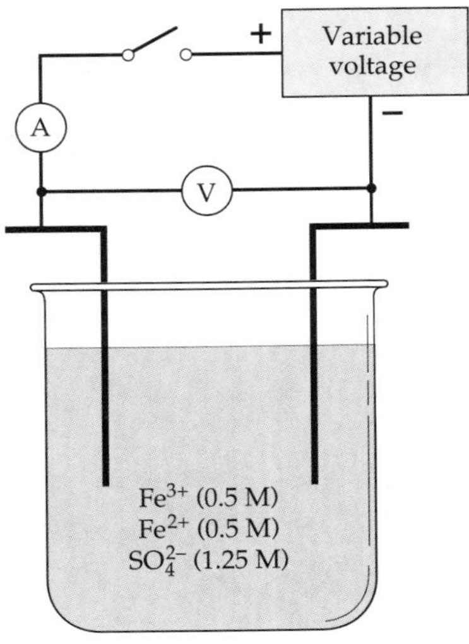

FIGURE 15.7 ▲
Diagram of an electrochemical cell with $E_{cell} = 0$.

Such a system might be used to measure the current when the cell has an external potential applied across nominally inert electrodes. A labels the ammeter and V labels the voltmeter.

Behavior of an Electrochemical Cell with an Applied Potential

Consider the apparatus illustrated in Figure 15.7. It has two electrodes which are initially at the same potential; if the voltmeter is connected by closing the appropriate switch, the measured potential will be zero. After making this reading, the voltmeter is disconnected.

Any voltage between zero and the source's maximum can be applied to the electrodes. In an experiment, the external voltage is applied to the cell for a few seconds by closing the switch. The ammeter in the circuit is read to find the steady current flowing through the whole circuit (including the cell) before disconnecting the source. After disconnecting the source and a pause of 15–20 seconds, the source again is connected and another measurement made at a different applied voltage. After a number of such measurements and pauses, the results from a series of known voltages can be plotted as a graph of current *vs.* applied potential as represented in Figure 15.8.

Let us look at the behavior of the current that is plotted there. No current flows when the applied potential is zero. The current then exhibits a region in which its increase is a linear function of the applied potential. At the end of this range there is a departure from linearity; less current flows than we would have expected had the linear behavior continued. This departure becomes quite large, and, finally, a plateau is reached where there is no further increase in current as the potential is increased. The value of the current at this highest level is called the **limiting current** and is written in equations as i_L.

We have been quite vague about the numerical values of the current and voltage in Figure 15.8, because the values can depend on the surface areas

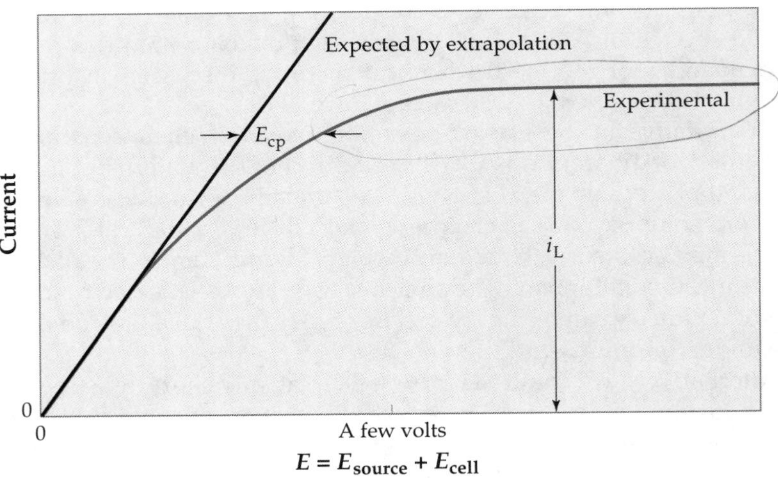

FIGURE 15.8 ▲
Current *vs.* applied potential.

Graph of the current that flows (vertical axis) when an external potential (horizontal axis) is applied to the cell illustrated in Figure 15.7. Initially the current rises linearly with potential—the cell exhibits ohmic behavior. Then, the current decreases relative to that expected if the ohmic behavior were to continue (dashed line). Finally, the current levels off at the limiting current value. The limiting current depends on the identity of the electrolyzed species and its concentration. The leveling off is due to concentration polarization, which is characterized by the potential E_{CP}.

and surface qualities of the electrodes, both of which may be highly variable. The surface area is determined by the interface structures and compositions at the atomic level, both those of the electrode surface and of the adjacent solution. Let us ignore the effects of surface structure and function and seek only to explain the behavior illustrated in the figure.

The Linear Region

When the difference between the applied potential E_{app} and the electrochemical cell potential E_{cell} is zero, no current flows. We can write this fact algebraically as

$$E_{app} - E_{cell} = 0 \quad ; \quad \text{no current flow}$$

Given that the cell is based on two half cells with potentials E_1 and E_2, we can write

$$E_{cell} = E_1 - E_2$$

Therefore, when no current flows we can also write

$$E_{app} = E_1 - E_2; \qquad \text{no current flow} \tag{15-5}$$

However, once the current begins to flow, this simple equation, 15-5, no longer describes the relationship correctly. The reason is that part of the circuit through which the current flows is the electrolyte solution, and, like all electrolyte solutions, it has a finite resistance. (As a simple number to remember, a 1 M KCl solution has a resistance of about 1 ohm if the areas of the electrodes are 1 cm^2 and the electrodes are about 1 cm apart. All the numbers are ones, and the electrodes form opposite sides of a cube 1 cm on each edge.)

When E_{app} differs from E_{cell}, a current flows in the circuit, and the voltage drops across the electrolyte solution in the cell. The current that flows is described by Ohm's law, $I = V/R$. For the cell with electrolyte,

$$I = \frac{E_{app} - E_{cell}}{R_{electrolyte}} \tag{15-6}$$

If Equation 15-5 is substituted into Equation 15-6, then

$$E_{app} = E_1 - E_2 + IR_{electrolyte} \tag{15-7}$$

Equation 15-7 describes the linear portion of the curve. It is a region of Ohmic behavior—that is, $E_{app} - E_{cell} = IR$.

The Nonlinear Region

As the voltage applied to the cell is increased further, the current does not increase as Ohm's law suggests. The reason is illustrated in Figure 15.9. At the electrode surface, the relative concentrations of Fe^{2+} and Fe^{3+} can be described by the Nernst equation, where E is the potential at the electrode's surface. Meanwhile, the potential away from the electrode (the bulk solution) corresponds to a different value of $[Fe^{3+}]/[Fe^{2+}]$, namely, the ratio in the bulk solution. This concentration ratio in the bulk is tending to change (by diffusion) from the bulk value to the ratio defined at the electrode surface by the electrode's potential. That is, if the surface has $[Fe^{2+}]$ greater than the bulk, that form diffuses into the bulk and tends to increase $[Fe^{2+}]$ there.

The relationship between the concentrations at the surface and in the bulk are described in more detail in Section 15A.

As the potential at the cathode becomes more negative, a point is reached where essentially all the iron ions near the surface are going to be Fe^{2+} ions.

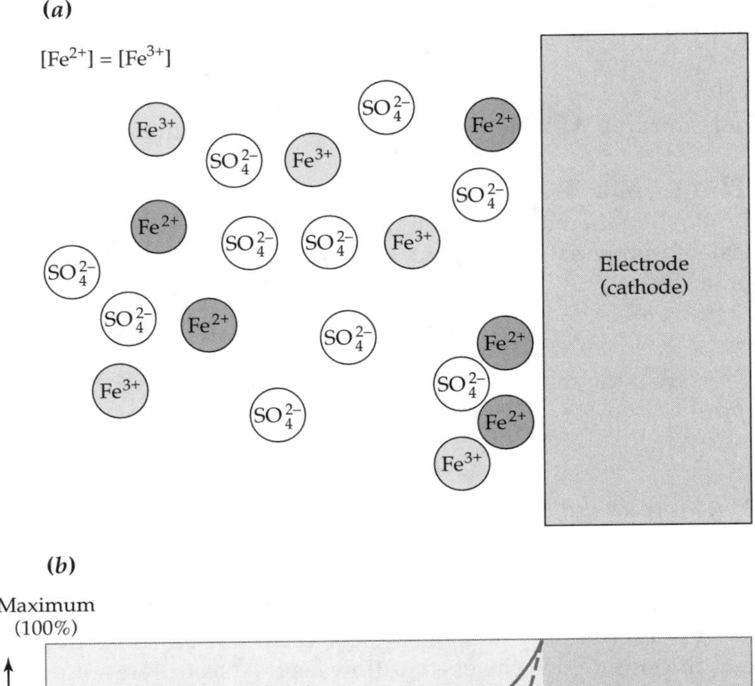

FIGURE 15.9 ▶
Concentration Polarization.
(a) Illustration of the conceptualization of the events occurring at a cathode/solution interface. The solution is a mixture of equal concentrations of Fe^{3+} and Fe^{2+} sulfates. As the Fe^{3+} ions arrive in proximity to the electrode surface, an electron may be transferred to form an Fe^{2+} ion. The ratio $[Fe^{3+}]/[Fe^{2+}]$ at the surface of the electrode is described by the Nernst equation with E being the electrode potential. **(b)** The Fe^{2+} ions formed at the interface are in excess at the electrode in comparison with the bulk. There is thus a net diffusion of Fe^{2+} away from the interface. However, since the Fe^{2+} ions were formed from Fe^{3+}, there is a dearth of Fe^{3+} at the electrode surface. Thus, there is a net diffusion of Fe^{3+} toward the electrode. After a time, the concentration distribution in space of Fe^{3+} and Fe^{2+} occurs as is plotted in the graph. The bulk solution in this case begins where both Fe^{3+} and Fe^{2+} are equal. The distance at which the concentrations are equal to the bulk-solution values moves with time:
---- = Early time;
—— = Later time.

Then, the only way to get more electrons into the solution is to wait for more Fe^{3+} ions to get close enough to the electrode surface to pick up electrons. At this point, the flow of current does not depend on the potential at the electrode but only on the rate that Fe^{3+} ions diffuse to and contact the cathode surface. The current reaches its limiting value, the limiting current, i_L. (See Figure 15.8.)

Focus now on the change in concentration that occurs between the surface of the electrode and the bulk solution. As illustrated in Figure 15.9b, at the surface of the electrode in the presence of a sufficiently negative potential, all the ions are Fe^{2+}. At the same time, if the current has flowed for only a short period, in the bulk solution the ratio $[Fe^{3+}]/[Fe^{2+}]$ is still near unity, and $E_{\text{bulk soln}} = E^\circ_{Fe^{3+}/Fe^{2+}}$. The difference in ionic concentrations between the surface and the bulk is equivalent to an electrochemical concentration cell. Thus, there is a voltage that is equivalent to this concentration change. This voltage is called the **concentration polarization potential** since the concentration difference between the surface and bulk is called **concentration polarization**. Concentration polarization occurs at both electrodes and tends to limit the current that can flow. However, concentration polarization is not the only phenomenon that leads to lower-than-expected currents.

Kinetic Overpotential

Using an apparatus similar to that drawn in Figure 15.7, we can **electrolyze** (from *lysis*, to cut) water, with the reactions yielding hydrogen gas at the cathode and oxygen gas at the anode. It is likely that you have seen a demonstration of this experiment in a previous chemistry course. The chemical free energy needed to convert water to H_2 and O_2 has an equivalent electrochemical potential. It is 1.23 V and is independent of pH. In other words, only after a potential of 1.23 V is applied do we expect to see the decomposition of the water begin; at that voltage the electrochemical energy just equals the decomposition energy. However, when an experiment is run, the **electrolysis** of water does not begin to any extent until about 1.7 V is applied. These facts are illustrated in Figure 15.10. The difference in these two voltages, 1.23 V and 1.7 V, is called an **overpotential**.

The cause of the overpotential is not the concentration polarization, as can be shown from two pieces of evidence.

a. The overpotential does not change with the rate of diffusion to the surface of the electrode.
b. The overpotential even shows up in the limit of zero current. Compare Figures 15.8 and 15.10 to see the difference graphically.

This overpotential is attributed to the process of electron transfer from the electrode to the adjacent ions in the solution, but the detailed mechanism

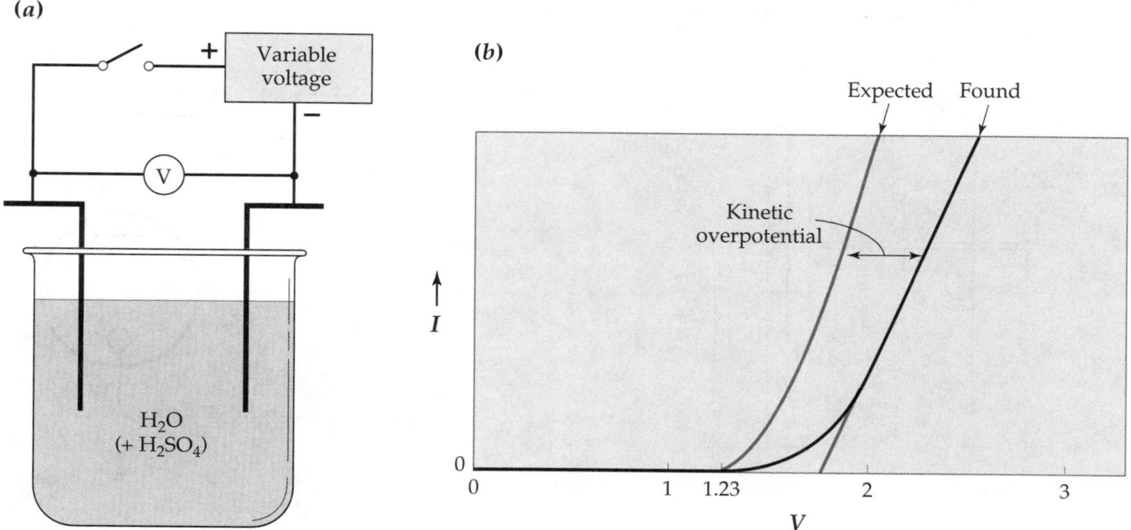

FIGURE 15.10 ▲
Kinetic overpotential.
(a) Line drawing of the experimental setup that might be used to measure the effects of the kinetic overpotential on the electrolysis of water to hydrogen and oxygen. **(b)** The graph shows the current expected if no effects of overpotential occurred (colored line). The equilibrium potential is 1.23 V. The current measured as the applied potential changes is shown by the black line. The electrolysis does not begin until a potential greater than 1.23 V is applied across the cell. The difference between the lines is the kinetic overpotential; it is slightly current-dependent. As indicated by the total current I, the electrolysis occurs at a faster rate as the potential is raised.

is not fully understood. In addition, it depends significantly on the material composing the electrode and the specific nature of the electrode surface. This is one reason that clean liquid mercury is used as an electrode in many electrochemical amperometric analyses. Compared with other electrode materials, its surface is significantly more reproducible.

In review: When a current is allowed to flow through an electrochemical cell, electrochemical reactions will not take place at the potentials described by the simple Nernst equation. There are three causes for this behavior. The first is the resistance of the electrolyte solution in the cell. The second is the concentration polarization that develops at the electrodes. And, in some cases, there is a third, the kinetic overpotential. These fundamental effects must be considered when using any of the following electrochemical methods.

Three-Electrode Potentiostat

A three electrode cell can be seen in EP.14 inside the back cover.

We now take a brief digression to describe a key tool in electrochemical measurements, the **three-electrode potentiostat**. As its name implies, a three-electrode potentiostat consists of three electrodes and holds a specific potential constant. The electrodes include a **working electrode** at which a given potential (compared to the **reference electrode**) is applied, as well as an **auxiliary electrode**. Such a cell is illustrated in Figure 15.11. The circuit and electrodes operate so as to define the potential at the working electrode at a preset value regardless of the changing resistance or concentration polar-

(*a*)

Potentiostat
Reference electrode
Working electrode
Auxiliary or counter electrode
Salt bridge
Frit
Fine glass frit (sintered glass)

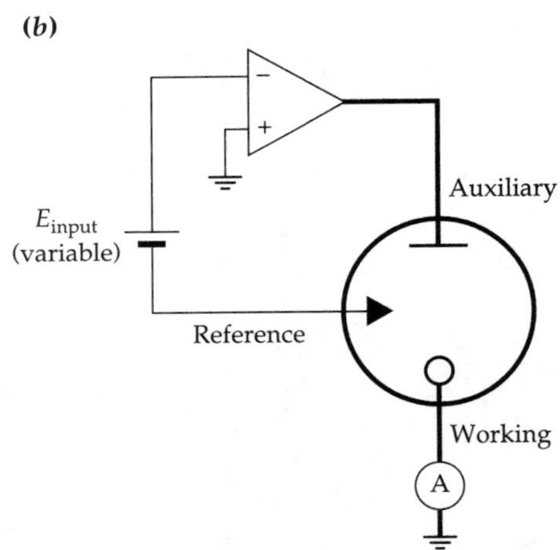

FIGURE 15.11 ▲
Three electrode electrochemical cell.

(a) Line drawing of the components of a three-electrode potentiostat and **(b)** the schematic diagram for the system. The commonly used symbols for the types of electrodes and the cell are shown. The current at the working electrode is measured at potentials maintained automatically by the potentiostat. In order for the system to operate properly, the reference electrode should be closer to the working electrode than is the auxiliary electrode. The electronic circuits of a contemporary three-electrode potentiostat are considerably more sophisticated than the one shown here for illustration where the triangle represents an amplifier. In (b), the heavier lines indicate the higher-current path.

ization of the solution. The figure shows the three different shapes that are used to represent the three types of electrodes schematically.

We do not want to pursue the details of its operation here, but only emphasize that the variable voltage source E_{input} sets the voltage desired at the working electrode *relative to the reference electrode potential*. The current that is measured flows between the auxiliary and working electrodes.

15.5 Conductimetry

An ionic solution conducts electrical current by means of the motions of ions in the solution. One way of measuring how well the solution conducts would be to place two electrodes into the solution and measure its resistance. The resistance measured depends on the following factors:

a. the surface areas of the electrodes,
b. the shapes of the electrodes,
c. the positions of the electrodes relative to each other in solution,
d. the identities of the species in the solution,
e. the concentrations of the species, and
f. the temperature.

As noted in this list, the resistance measured depends not only on the solution properties but also on the instrument geometry—that is, a through c of the list. Within a single laboratory and with the same electrodes, this measurement could be useful for analysis. But the information would not be directly comparable among laboratories. As a result, the *apparent* resistance—or its reciprocal, the conductance—is not used to describe the properties of ionic solutions.

The units that we use for analytical measurements account for the differences in electrode area and geometry. A resistance is converted to a unit *as if* the measurement had been carried out in a cell of volume 1 cm^3 between two electrode plates, each 1 cm^2, that are placed parallel to each other and 1 cm apart. This measure of resistance in a specific geometry is called **resistivity**. The units for this quantity are, for instance, ohm-cm. [(1 ohm) · (1 cm length)$^{-1}$ · (1 cm^2 area) = 1 ohm-cm.] Other units are shown in Table 15.3. The reciprocal of resistivity is called the **conductivity** or **specific conductance**, with units as shown in Table 15.3. Typical values of ionic specific conductance approximate those for KCl, which are listed in Table 15.4.

Table 15.3 Units Used in Conductimetry

Quantity	Name of Unit	Unit Symbol
Resistance	ohm	Ω
Conductance	ohm^{-1} = mho = siemens*	Ω^{-1}=Ω^{-1} = S
Resistivity (ρ)	ohm cm	Ω cm
	ohm meter	Ω m
Conductivity (κ)	mho cm^{-1} = siemens cm^{-1}	Ω^{-1} cm^{-1} = S cm^{-1}
	siemens m^{-1}	S m^{-1}

*Siemens is the SI unit.

Table 15.4 Specific Conductances of KCl Solutions

Concentration (M)	Conductivity ($S\ cm^{-1}$)		
	19°C	20°C	21°C
1	0.1001	0.1020	0.1040
0.1	0.01143	0.01167	0.01191
0.01	0.00125	0.00128	0.00130

Conductivity and Ionic Concentration

As noted in the preceding list, the solution conductivity depends on the identities of the ions present and their concentrations. All the ions present in the solution participate in the conduction process. To divide up these contributions, it is helpful to consider the conductivity per equivalent of each ion (per ionic charge). This is called the **equivalent conductivity** and is designated by a Greek Λ. The equivalent conductivity attributable to a specific ion is labeled λ_i. It is always true that

$$\Lambda = \sum_{\text{all ionic species}} \lambda_i \qquad \textbf{(15-8)}$$

Equation 15-8 states quantitatively that the total conduction consists of contributions from all the ions present. Representative values of equivalent conductivities for solutions with *only one salt* present are listed in Table 15.5. You will note from inspection of the table that the equivalent conductivities are dependent on concentration. This is due to ion-ion interactions.

The greater conductivities of KOH, NaOH, and HCl can be explained by the important observation that H^+ is about five times more efficient in conducting charge in water than are the other cations, and OH^- is about three times as efficient as other anions. This means that a conductimetry experiment in water will be about five times as sensitive to changes in proton concentration as to other cations. Similarly, a threefold increase in sensitivity is expected for hydroxide.

Table 15.5 Equivalent Conductivity in Water at 25°C ($S\ cm^2\ equiv^{-1}$)*

Electrolyte	Limit $\tilde{c} \to 0$ $\Lambda°$	Concentration ($equiv\ L^{-1}$)		
		0.001	0.01	0.1
KCl	149.86	146.95	141.27	128.96
NaCl	126.45	123.74	118.51	106.74
HCl	426.16	421.36	412.00	391.32
$AgNO_3$	133.36	130.51	124.76	109.14
KNO_3	144.96	141.84	132.82	120.40
NH_4Cl	149.7	—	141.28	128.75
LiCl	115.03	112.40	107.32	95.86
KOH	271.5	234	228	213
NaOH	247.7	244.6	239.9	—

* The units are ($S\ cm^2\ equiv^{-1}$) since $\Lambda = \kappa/\tilde{c}$ where $\tilde{c}$ is the equivalent concentration in $equiv\ cm^{-3}$.

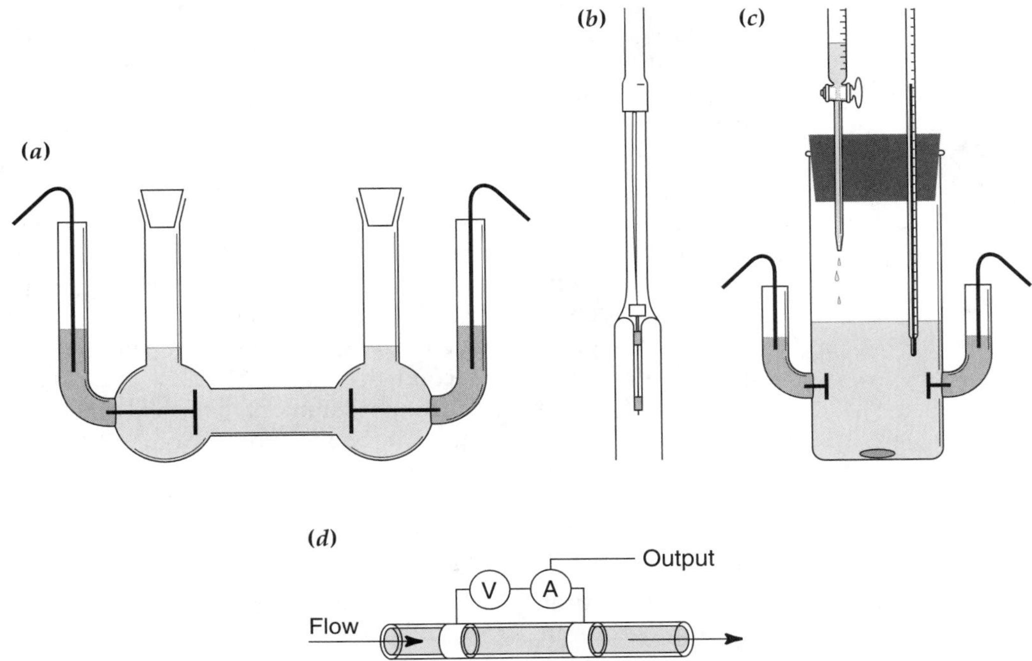

FIGURE 15.12 ▲
Four different types of conductivity cells.
In all four, the electrodes are constructed from platinum foil covered with platinum black. Samples fill the cells so that the geometry is the same for all runs. The temperature should be held precisely constant. Control of the temperature to 0.1°C allows precision in the measurements to be in the range of ±1 to 2%. **(a)** The sample cell is filled with the solution. **(b)** A dipping cell with an open bottom that is immersed in the sample. **(c)** A cell used for titration with conductrimetric monitoring. **(d)** One of many designs for an HPLC conductivity detector used for ion chromatography. The two hollow conducting electrodes are separated by an insulating tube.

Conductimetry in Practice

Experimental values of conductivities are almost always found by using a cell with rigidly fixed electrodes. Four types of such conductivity cells are illustrated in Figure 15.12. With such cells, quantitation using conductance measurements can be done simply by calibrating the instrument with known concentrations of the ion-forming analyte and then measuring the unknowns. The calibration curves are slightly nonlinear, as can be seen from the values in Table 15.4 for different concentrations of KCl. Alternatively, the cell can be calibrated with a material of known specific conductance. Sets of solutions having a range of concentrations of KCl in highly purified water are common calibrants. Then the specific conductance of the unknown material can be determined. Not only can analyses be done in this way, but fundamental properties of ionic materials can be determined, such as the ionization constants of acids.

Solid-state conductimetry is the basis for the electronic nose pictured in EP.10 inside the back cover.

Conductance measurements adequate for numerous applications can be carried out with a battery, an ammeter to measure the current, and two electrodes immersed in the solution to be tested. However, we encounter some major problems with such a simple system if more precise measurements are needed. The difficulty lies in the influence of overpotentials and concentration polarization on the currents that pass through an electrolytic cell when an external voltage is connected to the electrodes.

The problems due to the kinetic overpotential can be overcome by choosing the right electrodes. These are usually made of platinum foil that has more platinum plated onto the surface, but the plated metal is not shiny. The surface appears to be black, and the electrodes are said to be coated with *platinum black*. The possible problems due to concentration polarization are overcome by doing the measurement so quickly that no appreciable concentration polarization can build up near the surfaces of the electrodes. In practice, two identical electrodes with opposite polarities (one positive, one negative) are rapidly switched between positive and negative at a rate in the range 100–1000 times per second. See Figure 15.13.

The utility of conductance measurements is broad since concentration changes in any ionic solution can be monitored. Some of the uses for conductimetric techniques are for monitoring:

contamination in streams and rivers,

salt content in boiler systems,

the ion concentration in the effluent from an analytical liquid chromatography column,

acid concentrations of solutions used in industrial processes,

the concentration of liquid fertilizer as the fertilizer is applied, and

the end points in titrations where an ion concentration of the analyte or titrant changes (for example, neutralization titrations by $[H^+]$ or $[OH^-]$).

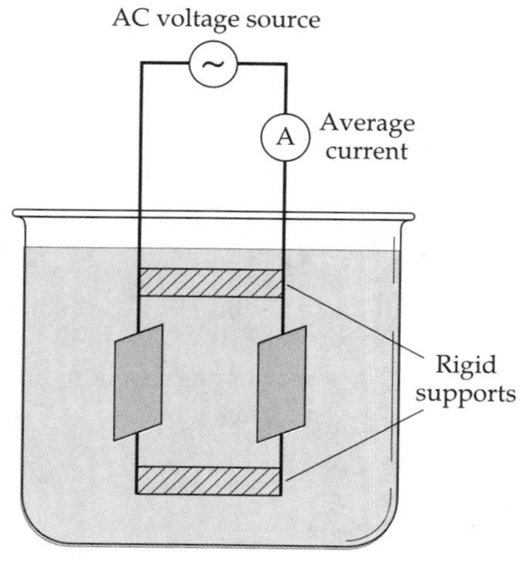

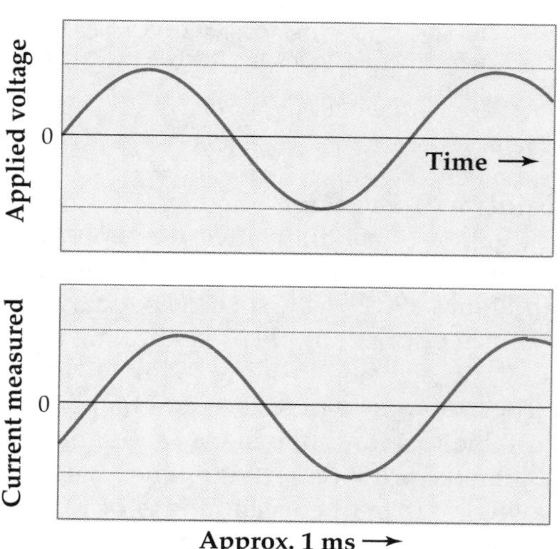

FIGURE 15.13 ▲
AC conductivity measurement.

An alternating voltage is used in these measurements to minimize effects of concentration polarization. The applied potential is graphed on the top right as a function of time. It is a sine wave. The current that flows through the cell is plotted below. The current will follow the voltage, reversing direction twice each cycle. The output of the instrument is the average current as related to the average applied voltage.

With proper standards and temperature control, high-precision conductimetry can be done with accuracies and precisions in the range of 0.1%. However, these are not routine, primarily because of the strong temperature dependence of conductivity. As seen from Table 15.4, ionic conductance typically changes by about 2% per °C. Quantitative conductivities can be determined to about ±1-2% if the resistance between the electrodes is in the range of 1000 ohm, and the temperature is thermostated to within ±0.1°.

15.6 Coulometry

Electrogravimetry can be used to analyze for metals by plating the metal(s) onto a preweighed cathode. However, numerous materials can be reduced or oxidized that do not plate out onto an electrode. In addition, there is a lower limit to electrogravimetry: when the amount of material becomes too small to weigh. The method of coulometry can be used to overcome both of these limitations.

To carry out a coulometric assay, we measure the total charge that is needed to complete an electrochemical oxidation or reduction. This charge is directly proportional to the total amount of material electrolyzed. Recall the simple relationship

$$96{,}485 \text{ coulombs} = 1 \text{ mole of charge} = 1 \text{ Faraday}$$

Further, recall that

$$1\,\text{C}\,\text{s}^{-1} \equiv 1\,\text{A}$$

Thus, the current in amps passing for a given time can be related to the number of moles. The fundamental equation for coulometric analysis is

$$\text{number of moles} = \frac{\text{coulombs for total electrolysis}}{96{,}485 \cdot n} \qquad \textbf{(15-9)}$$

or

$$\text{number of moles} = \frac{\text{current (A)} \cdot \text{time (s)}}{96{,}485 \cdot n} \qquad \textbf{(15.10)}$$

The letter n represents the number of electrons transferred per mole, as in the Nernst equation. Note, of all the electrochemical techniques, coulometry is the only one for which temperature control is not crucial for quantitation.

EXAMPLE 15.1

It is relatively easy to measure a total charge of millicoulombs (mC) over periods of minutes. Calculate the number of moles of electrons to which this corresponds and the mass of Cu(II) that would be reduced to metallic copper with 1.00 mC of electrons.

Solution

$$1.00 \text{ coulomb} = (1/96{,}500) \text{ mol of } e^- = 1.036 \times 10^{-5} \text{ mol of } e^-$$

Thus, 1.00 mC is equivalent to 1.036×10^{-8} mole of electrons. For the reduction of Cu^{2+} to copper metal, 2 moles of electrons are required per mole of Cu^{2+}, so 5.15×10^{-9} mol copper metal could be produced. Copper has a molar mass of 63.4 g, or

$$(5.18 \times 10^{-9} \text{ mol}) \times (63.4 \text{ g/mol}) = 3.28 \times 10^{-7} \text{ g copper}$$

could be produced—truly a mass difficult to weigh with any reasonable degree of precision or accuracy.

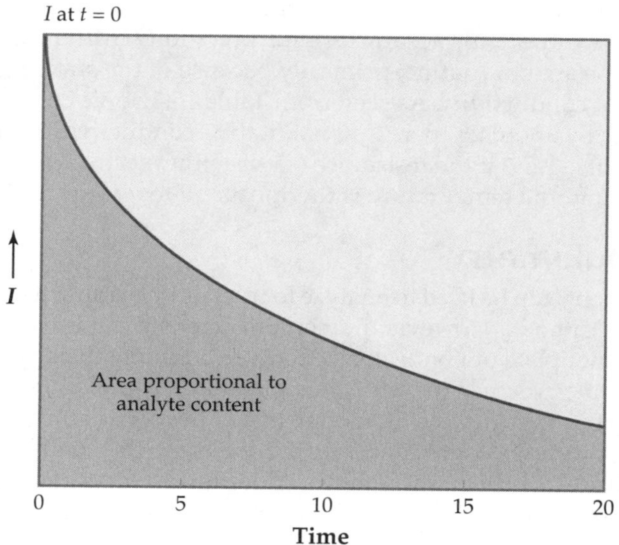

FIGURE 15.14 ▲
Constant-potential electrolysis in a stirred solution.

If an electrode at a fixed potential were electrolyzing some species in a vigorously stirred solution, the electrolysis current would vary with time as shown in this graph. The current is proportional to the amount of electrolyzed species left in the solution. After all of the material is electrolyzed, the current will stop flowing, assuming no other species could be electrolyzed. The curve has an exponential form since the rate of electrolysis is proportional to the amount of the material that is left in solution—analogous to a reaction with first-order kinetics. The electrolysis rate depends on the stirring rate, electrode area, electrode material, electrode potential, initial concentration of the electrolyzed species, temperature and, perhaps, concentrations of other species in the solution.

In order to do a coulometric determination, an analyte is exhaustively oxidized or reduced, usually in a stirred solution. The redox reaction is driven by an external voltage source. The working electrode—where the analyte reaction occurs—can be composed of any of a number of electrically conducting materials. Electrolysis of the analyte eventually depletes it from the bulk solution, as is illustrated in Figure 15.14. At a constant potential the electrolysis current decreases with time as the **electroactive species** is depleted, since a steadily lower concentration is available at the electrode surface to react.

EXAMPLE 15.2

A sample of L-dopa, a drug used for treatment of the symptoms of Parkinson's disease, is electrolyzed exhaustively with 42.0 μC. The process requires two electrons per molecule. The formula weight of L-dopa is 197.2. What is the mass of L-dopa in the sample?

HO—C_6H_3(OH)—CH_2—CH(NH_2)—COOH

L-dopa

Solution

The conversion from coulombs to moles is

$$\frac{42.0 \times 10^{-6}}{96{,}485 \cdot 2} = 2.17_6 \times 10^{-10} \text{ mol}$$

The total mass of L-dopa is then

$$2.17_6 \times 10^{-10} \text{ mol} \cdot 197.2 \text{ g mol}^{-1} = 4.29 \times 10^{-8} \text{ g or } 42.9 \text{ ng}$$

Background Currents: Competitive Electrolysis

In the example above, it was assumed that all of the current is used in the reaction of interest. In an ideal assay, only the analyte would be electroactive. However, materials other than the analyte species can *always* be oxidized or reduced. These materials include the electrodes themselves, the ions of the **background electrolyte**, any gases that might be dissolved in the solution, as well as reagent impurities and matrix components of the sample. At some potential, even the solvent itself will electrolyze.

Some examples of interfering electrolysis are: water converted to gaseous oxygen and hydrogen at the positive and negative electrodes, respectively; chloride oxidized to chlorine or higher oxidation states such as hypochlorite and chlorite; oxygen reduced to peroxide and hydroxide/water; and platinum or gold electrodes oxidized to their respective ions. The currents from these interfering reactions are called **background currents**. Examples of such currents are shown in Figure 15.15. The solutions have been purged with an inert gas to remove most of the oxygen. Note where the background currents change rapidly. For the mercury, the change starts over a narrow range of voltage and clearly defines the electrode's **potential window**, which is where analyses can be carried out. Outside the window, the background currents overwhelm any analyte's. These large background currents are either from electrolysis of the background electrolyte or from the electrodes themselves. For the mercury, the sharp change in current at the extreme negative potential results from breakdown in the electrolyte. On the other end of the range, at the left side of the figure, the mercury electrode surface is being oxidized. This fact means that when potentials more positive than about zero volts on this scale are needed, of the two metals, platinum would be the better electrode material under the conditions here.

In practice coulometry requires that the analyte being measured is completely electrolyzed by the time the experiment is finished. Otherwise, the relation between the charge transferred and the quantity of analyte will be uncertain or even meaningless.

Coulometric Titrations

If we can control the experimental conditions so that virtually all the electrons are transferred to the analyte, then a second, simpler, coulometric method can be used in analyses. This simpler method involves use of a constant-current source and an accurate timer. The constant-current source delivers, as its name implies, a constant current regardless of the potentials or resistances of the electrolysis cell. In order to measure the total number of coulombs, all that has to be done is to measure the time from the beginning of the experiment to the time when the electrolysis is finished. This method

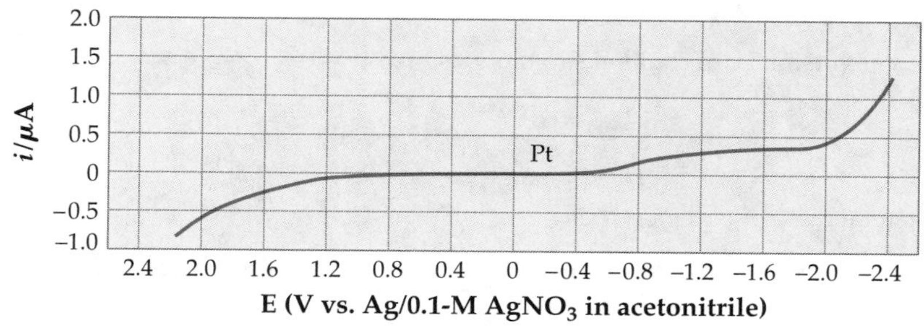

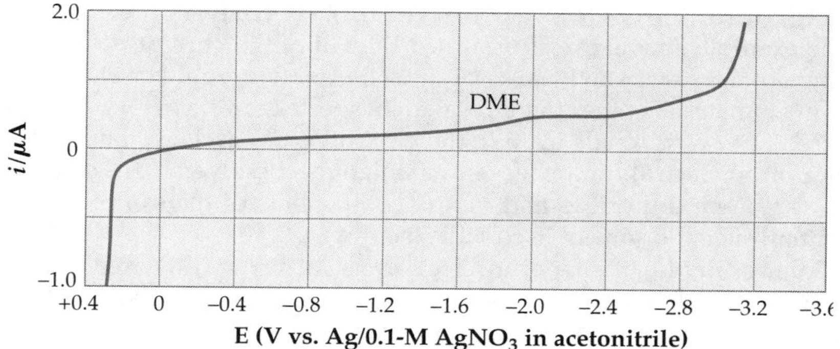

FIGURE 15.15 ▲
Potential windows.

Background currents are shown for two different electrode materials as they vary over a potential range. The solvent for both is purified acetonitrile with background electrolyte of 0.1-M tetraethylammonium perchlorate, $Et_4N^+ClO_4^-$. The upper curve was obtained with a platinum electrode, the lower curve with a dropping mercury electrode. The horizontal scale is in volts relative to a Ag/AgCl reference electrode. The reference electrode was made with a AgCl-coated silver wire immersed in a 0.1-M $AgNO_3$ solution in acetonitrile, a nonaqueous electrode. As a result, the potential of the voltage scale cannot be compared directly with an experiment that employs an aqueous Ag/AgCl reference. Note the relatively flat baseline until the ends, where solvent breakdown occurs or the electrode itself electrolyzes. These two steep portions of the curve limit the usable voltage range. The usable voltage range is also called the **window** or the **useful range**. This range varies depending on the specific conditions. [Ref: Reprinted with permission from Sherman, E. O., Olson, D. C. 1968. *Anal. Chem.* 40:1174–1175. Copyright 1968 American Chemical Society.]

is unlike coulometry done at a constant potential, where the current inherently falls to the level of the background (recall Figure 15.14).

As a consequence, we are required to find some other method to discover the point at which the analyte's redox reaction is finished and some other redox reaction (perhaps the background electrolysis) begins. This description sounds like a titration. In fact, the constant-current coulometric method is called **coulometric titration** because of its similarity to regular titration. Professor G. W. Ewing has called it "titrating with electrons,"[1] a most concise descriptive phrase. In general, the analyte itself is not directly electrolyzed. Instead, a titrant is generated quantitatively by electrolysis at an electrode rather than by adding a reagent solution.[1]

---

[1]Ewing, G. W. 1981. *American Laboratory* 13:16–22.

Table 15.6 Examples of Coulometric Titrations

Assayed Substance	Reagent Generated	Precursor	Titration Type
Br^-	Ag^+	Ag° anode	Precipitation
Fe^{2+}	Cl_2	HCl	Redox
H_2O	I_2, I_3^-	KI (pH $<$ 9)	Karl Fisher reagent
Organic acids	OH^-	H_2O	Neutralization
Bases	H^+	H_2O	Neutralization
Ca^{2+}, Zn^{2+}	$Hedta^{3-}$*	$HgNH_3edta^{2-}$	Complexometric
Olefins	Br_2	KBr (pH $<$ 5)	Olefin addition (redox)

*edta = ethylenediaminetetraacetate.
Further refs: Stock, J. T. Annual fundamental reviews in *Anal. Chem.*; Farrington, P. S. 1963. In *Handbook of Analytical Chemistry*. L. Meites, ed. New York: McGraw-Hill. Table 5.55.

The reagent used is generated electrolytically from a precursor that is in excess compared to the analyte. In Table 15.6 are listed a number of reagents that can be generated coulometrically. Note that not all of these titrations are redox titrations. Also included are H^+ and OH^- generated for neutralization titrations. In addition, chelates can be generated by reducing a strongly binding metal ion (usually mercury) which releases the chelate to bind with another species. Precipitation reactions can be done by generation of precipitant metal ions such as silver.

One of the primary advantages of coulometric titrations over volumetric ones is that standardized solutions need not be prepared. Thus, there are neither problems with primary standards nor with storing unstable reagent solutions.

Another benefit of coulometric titrations is the relative ease of handling smaller quantities of titrant and sample. As you saw in the first example of this section, 0.3 μg of copper corresponds to 1 mC of electrons, which can easily be measured accurately. In fact, the generation of enough reagent for large or concentrated samples can take so long that coulometric titrations become inconvenient.

As for any titration, some method is needed to determine the end point. You are aware of a number of such methods. We could use redox indicators or a potentiometric method. For the latter, a voltmeter is used to sense the potential jump of an indicator electrode, in the same way that a pH electrode is used to detect the pH jump at the end points in acid-base titration.

Coulometric titrations are especially amenable to automation, since not only can the end point be determined directly from the signal of a transducer, but the quantity of titrant generated can be controlled precisely, and the electrochemical reagent addition can be stopped at the end point—all by the same electronic circuit. A few specific examples of coulometric titrations will be described in the following section on amperometry. Among its other uses, amperometry can be used to detect the end points of coulometric titrations.

15.7 Amperometry

Amperometry, as the name implies, uses measures of current: specifically, the current that passes through a solution to cause oxidation or reduction of the analyte. Amperometric assays depend on quantitative measurements of

The phenomenon of the limiting current is illustrated in Figure 15.8.

the limiting current. The reason that quantitative analysis can be done by amperometry is that the limiting current is directly proportional to the concentration of the species being oxidized or reduced at the working electrode. The relationship for i_L is, simply,

$$i_L = \text{constant} \times \text{concentration of electroactive analyte} \qquad \textbf{(15-10)}$$

The constant is determined through the use of standards.

Two general types of amperometric methods are commonly used: those with a *changing potential* and those with a *fixed potential*. With a changing potential, the amperometric measurement involves measuring the current as a function of potential at the working electrode of a three-electrode cell. These **voltammetric** methods—which include **polarography**, **cyclic voltammetry**, and **anodic stripping voltammetry**—are used to measure the concentrations of electroactive species in solutions. The types of materials that can be quantitated include metals and organic species. Usually, for the organics, the determination is done by reducing a bond or bonds of the molecule. Some of the organic-bond types that can be reduced in this way are listed in Table 15.7.

The second group of amperometric methods involves measuring the changes in the current at a fixed potential at the working electrode. Abrupt changes in current can be used to monitor the end point during a titration. These titrations are often called **amperometric titrations**, a name that is deceptive: the reagent is being added either coulometrically or volumetrically. Amperometry can also be used to monitor chromatographic separations. Let us look in slightly more detail at some fixed-potential amperometric methods.

Fixed-Potential Amperometry: Amperometric Titrations

In order to determine the end points of titrations, two different fixed-potential amperometric measurement techniques are common. In one, the current is monitored as it flows between two identical electrodes immersed in the titrated solution. In the other, only one electrode is in the titrated solution.

The Two-Electrode System

With *two* identical, inert electrodes (such as platinum wires) in the titrated solution, the technique is called **biamperometric detection** of an end point. Diagrams of the experimental setups for volumetric and coulometric additions of titrant are shown in Figures 15.16a and 15.16b, respectively.

Table 15.7 Bond Types That Can Be Reduced for an Amperometric Determination

C—C	C—N	C—O	C—S	C—X[a]
C=C	C=N	C=O	C=S	
C≡C	C≡N			
	N—N	N—O	N—S	
	N=N	N=O		
		O—O	O—S	O—X
			S—S	S—X
Condensed benzenoid rings				
Some heterocyclic rings				

[a]X = halogen

Source: *Chemical and Engineering News*, March 18, 1968, p. 96.

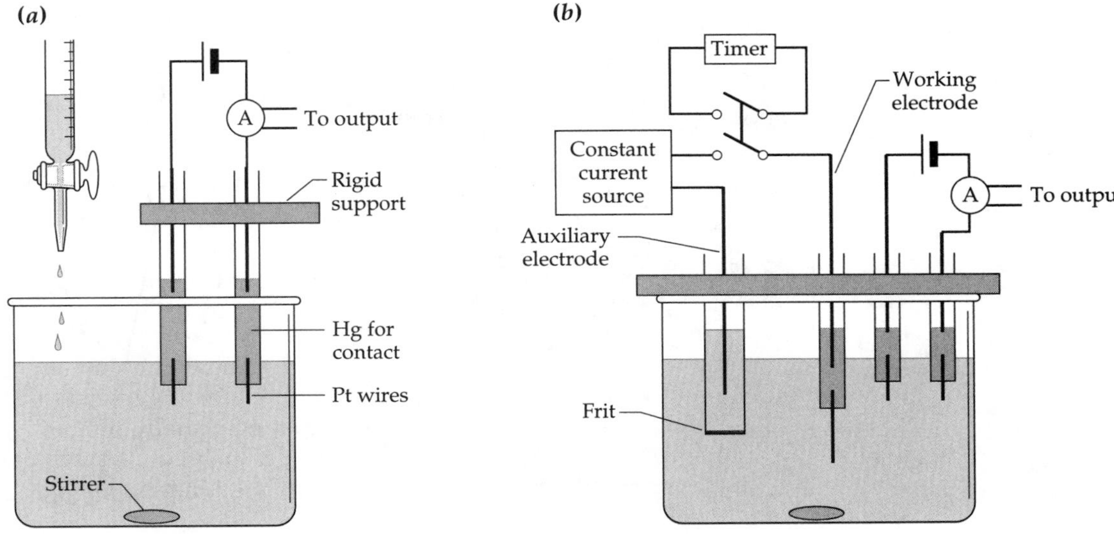

FIGURE 15.16 ▲
Titration with biamperometric detection.
(a) shows the equipment for use when the titrant is added volumetrically; **(b)** shows the equipment for use with coulometric generation of titrant. The biamperometric detection electrodes and electronics are the same for both. For the coulometric generation of titrant, the exact time needed to generate the reagent is recorded. With a highly stable current source, the number of coulombs required to titrate the sample is calculated from the mathematical product (current × time). If the electrolysis efficiency is 100%, then no calibration of the titrant is needed.

A potential at some value between a few millivolts to a volt is placed across the two elecrodes. During the titration, the current is monitored. From the biamperometric data shown in Figure 15.17, you can see that the current is kept in the range of microamps. No potentiostat, reference electrode, or salt bridge is needed. Thus, the technique is experimentally quite simple and inexpensive. It is also quite effective.

Let us consider a specific example of biamperometric detection in more detail: the aqueous titration of a 0.1 *N* thiosulfate solution with 0.1 *N* potassium iodide/iodine. The descriptive chemistry of the reaction of iodine and thiosulfate is shown by the equation

$$2\,S_2O_3^{2-} + I_2 \rightleftharpoons S_4O_6^{2-} + 2\,I^-$$

The sulfur-oxygen species on the right side, called tetrathionate, forms specifically when iodine is the oxidizing agent. The results of this titration, as monitored by biamperometric apparatus, are illustrated in Figure 15.17a. At the end point there is a sharp break in the curve of the measured current. This is just what we want in order to determine a titration end point.

What is the chemistry that underlies this simple and useful method? The electronic flow of current through the wires is converted at the electrode surfaces into an ionic flow of current through the electrolyte. The ionic current flows *only* when some species becomes oxidized at the more positive electrode while *simultaneously* some species becomes reduced at the more negative electrode. This idea is extremely important to understand. Electrons must be transferred between the electrodes and the solution for the circuit to be complete. Thus, with the use of inert electrodes, the potential applied to

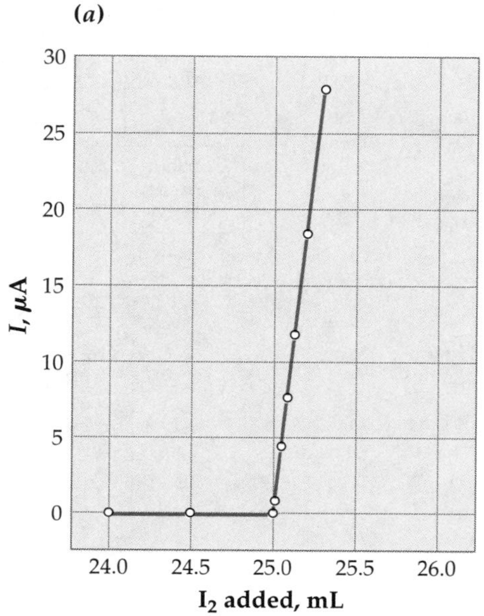

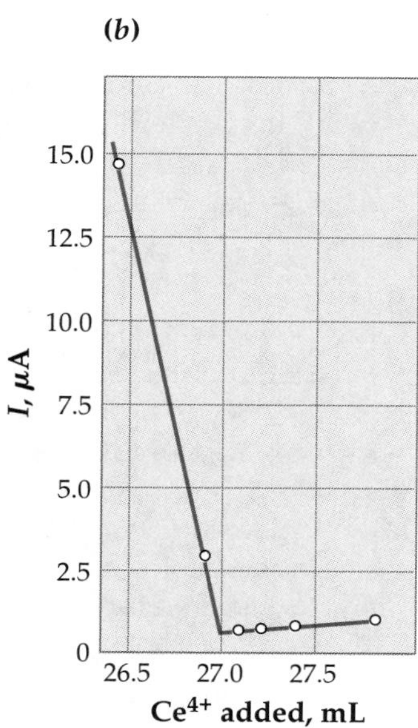

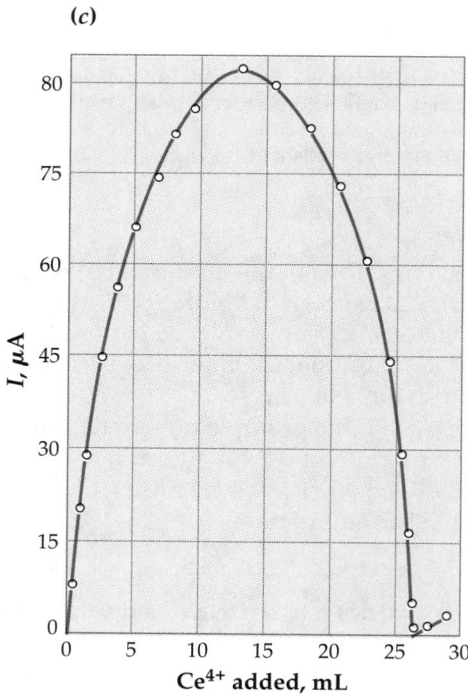

FIGURE 15.17 ▶ **Graphs of the data from volumetric titrations with biamperometric detection.**

The graphs are plotted as μA *vs.* volume of reagent added. **(a)** The region near the end point for a titration of 0.1 *N* thiosulfate with a solution of 0.1 *N* I_2 in aqueous KI. The electrodes were two platinum wires with a 50-mV potential between them. This is a "dead-stop" end point. **(b)** The region near the end point for a titration of $K_4Fe(CN)_6$ with Ce^{4+}. The products are $Fe(CN)_6^{3-}$ and Ce^{3+}. The applied potential was 50 mV. Current flows after the end point due to electrolysis of the titrant. **(c)** The same as Figure (b) except the entire titration curve is shown, not just the region of the end point. Note that no current flows initially. [Redrawn with permission from Stone, K. G., Scholten, H. G. 1952. *Anal. Chem.* 24:671–674. Copyright 1952 American Chemical Society.]

the electrodes must be large enough to produce a redox process for *some* species in solution at *each* electrode. If there is no reaction at *either one* of the electrodes, no current will flow. This straightforward idea is the basis for biamperometric amperometry.

In our example, the two platinum electrodes were placed into the thiosulfate solution, and a fixed voltage of approximately 50 mV was placed across them. Prior to the endpoint, any iodine added is converted to iodide.

Therefore the species present in solution are iodide, thiosulfate, and tetrathionate. The tetrathionate cannot be reduced at a platinum electrode under these conditions; nor can the thiosulfate be oxidized. So, before the endpoint, no current flows. Once the endpoint is reached, there both iodide and iodine are present in solution, and current flows. At the cathode, where reduction is occurring, the reaction is

$$I_2 + 2\,e^- \rightarrow 2\,I^-$$

while at the anode, where oxidation is occurring, the reaction is the reverse

$$2\,I^- \rightarrow I_2 + 2\,e^-$$

The current measured by the detecting electrodes increases as more iodine is added after the endpoint.

Consider what would happen if the "opposite" titration were carried out—titration of iodine/iodide with thiosulfate. In this case, the current would decrease as the concentration of iodine decreased, finally ceasing at the endpoint. This behavior—having a current flow that "stops dead" at the endpoint—led to the naming of this latter titration behavior. It is called a **dead-stop end point**.

Different behavior is observed when both anodic and cathodic reactions can occur with *both* the reactant and the titrant. This situation is illustrated in Figure 15.17b for the titration of $Fe(CN)_6^{4-}$. The current has a finite value on both sides of the end point. However, the end point is still easy to determine since it corresponds to the intersection of two straight lines.

"One-Electrode" System

The second amperometric titration monitoring technique utilizes a three-electrode potentiostat as illustrated in Figure 15.18a. Only the working electrode is in direct contact with the titrant solution. The auxiliary and reference electrodes are separated from the test solution by salt bridges. The electrode potential is set in the limiting-current range (due to either the titrant or the analyte). The current measured is then directly proportional to the concentration of the electrolyzed species. This idea is expressed in Equation 15-11.

$$i_L = \text{constant} \times \text{concentration of electroactive species} \qquad \textbf{(15-11)}$$

As illustrated in Figure 15.18b, the species concentration can be monitored and graphed as a function of added titrant. The end point of the titration appears as an intersection between two straight lines.

Note the resemblance of the curve in Fig. 15.18b to that in Fig. 7.2c.

All that you have learned about titrations holds for amperometric titrations. The techniques of amperometric detection are simply alternate methods to determine the progress of a titration and its end point.

Chromatography/Amperometry

Amperometry at a fixed potential tends to be a nonspecific assay method. As long as the electrode potential is adequate, all species coming into contact with the electrode will be electrolyzed. However, with adequate separations prior to electrolysis, highly specific amperometric detectors for chromatography have been developed. A working electrode is placed so that electroactive analytes pass over it after they are separated. The electrolysis currents can be used to quantify the amounts of a number of analytes sequentially. Subnanogram amounts of electroactive compounds can be determined directly.

More about liquid chromatography can be found in Chapter 13.

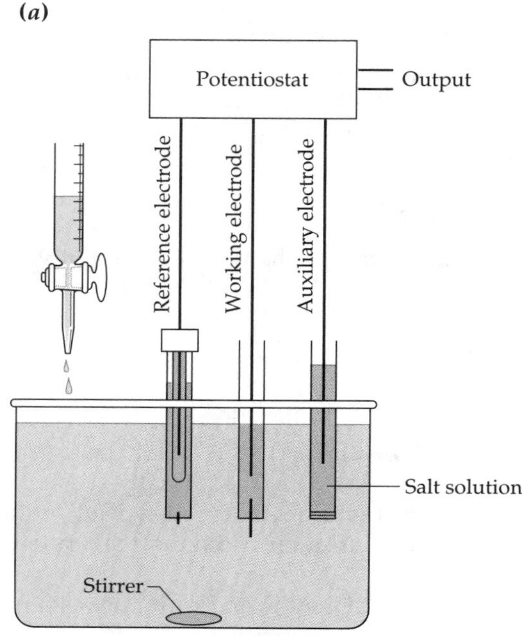

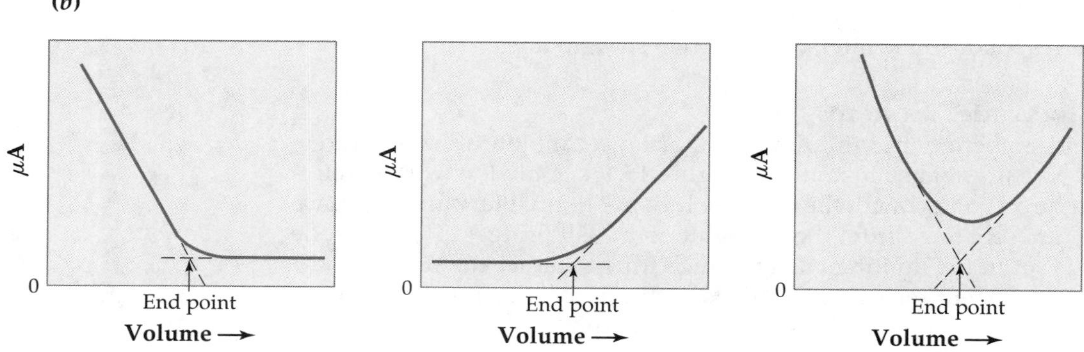

FIGURE 15.18 ▲
Amperometric titration.

(a) Diagram of the experimental apparatus required to do a volumetric titration with "single-electrode" amperometric monitoring. The reaction is followed by measuring the current at the working electrode of a three-electrode potentiostat with the auxiliary electrode isolated.
(b) Three types of behavior that might be seen plotted as the amperometric output *vs.* the volume of titrant added. (Compare these with Figure 7.4.) The end points are the point of the breaks in the output response. The only difference between them is the slopes of the lines on either side. Three cases are illustrated. (Left) Analyte electrolyzed at the working electrode, titrant is not. (Middle) Titrant is electrolyzed at the working electrode, analyte is not. (Right) Both analyte and titrant are electrolyzed at the working electrode. The break is usually not as sharp as desired for a precise measure of the end point. The intersection of line segments (the dashed lines here) can give a more precise answer.

15.8 Voltammetry

One problem with amperometric measurements is that the measured current is a sum of the current from background interferents and from the electroactive analytes. The background currents could be determined by running a blank, *if* every possible electroactive species that might contribute to the background is known. The blank then consists of all the interferents in the right proportion; such a blank can get exceedingly complicated. However, there is an easier way.

In a single solution, both the background current and the limiting currents from the analytes can be measured and separated by scanning the potential over a range of voltages. Refer to Figure 15.19. Shown is a graph of the current for reduction at a working electrode *vs.* the slowly scanned potential. As shown in the lower graph, the potential is changed at a constant rate from zero volts to more negative potentials over minutes. This constantly changing voltage is called a **linear voltage scan**.

Initially, the electrode potential is set so no electrolysis of the analyte species occurs, although there may be some background currents. The voltage scan is then begun. As the potential moves toward the formal potential of the species being investigated $E^{\circ\prime}$, some additional electrolysis occurs. The current continues to increase as the potential is scanned further, and, eventually, the region of the analytically useful limiting current is reached.

As shown in Figure 15.19, the current that is due to the background electrolysis can be extrapolated into the region where the analyte's limiting current is present. The difference between the baseline extrapolation and the limiting current is taken to be due to the analyte alone.

In order to obtain reproducible limiting currents, voltammetric assays are inevitably done with a high concentration of background electrolyte present. Under these conditions, electrostatic effects that might perturb the current are minimized; the magnitude of the limiting current remains, then, proportional to the bulk solution concentration of the electroactive species.

While the limiting current can be used for quantitation of the electrolyzed species, the position along the voltage axis where the rise of the current occurs can aid in identifying that species. The horizontal location of the

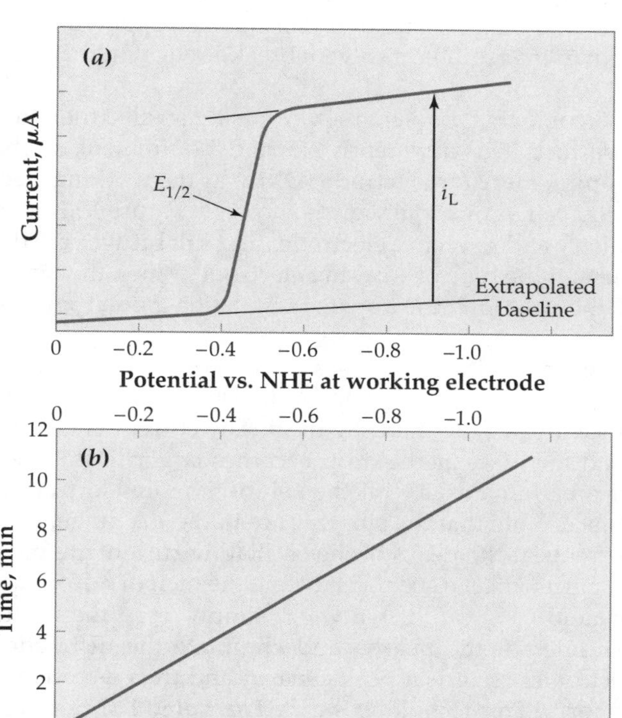

◀ FIGURE 15.19
Linear-scan voltammetry.
Illustration of the output of a voltammetric experiment using a working electrode consisting of a few-mm diameter flat disk, one side of which contacts the solution. It is rotated about its axis at a moderate rate to keep bulk solution passing across its surface. (This is called either a **rotating disk** or **rotating button** electrode.) **(a)** The current flow is plotted versus potential. At the left is the background current. As the potential is scanned, the electrolysis of analyte begins, and the current rises rapidly until it reaches the limiting current. The limiting current is defined as the difference between the extension of the background current and the plateau region. The midpoint of the rise (and point of maximum slope) is $E_{1/2}$. The potential is scanned at a fixed rate of 100 mV min^{-1} for these experiments. **(b)** Graph of the scan voltage *vs.* time. Note that (a) and (b) share the same voltage axis.

voltammetric curve is characterized by $E_{1/2}$, which is the midpoint of the rise in current. $E_{1/2}$ is equal to $E^{\circ\prime}$ if the redox reaction of the species being measured is **reversible**.

Reversible in the electrochemical sense is slightly different from a *reversible reaction*. In most of chemistry, reversible means that a reaction *can* proceed in both forward and reverse directions. In electrochemistry, it means that the redox reaction *does* go easily both forward and backward. Following from this, you should understand that electrochemical reversibility requires that both the oxidized and reduced species be chemically stable. (If either reacted further, it would not be present to be reduced or oxidized repeatedly.) Another requirement for electrochemical reversibility is that the rates of electron transfer for *both* oxidation and reduction must be extremely fast.

In this sense, relatively few electrochemical reactions are truly reversible. In aqueous solutions, most reversible electrochemical reactions involve metal ions. Organic aromatic groups such as those containing benzene or naphthalene rings also can be electrolyzed reversibly at some electrodes, but only in certain solvents. The solvents must be nonaqueous and cannot donate a proton to the reactant(s). Organic aprotic solvents such as acetonitrile (CH_3CN) and methylene chloride (CH_2Cl_2) are representative of such solvents. Only analyses involving reversible reactions will be considered here.

DC Polarography and the Dropping Mercury Electrode

The scanned potentials in voltammetry are usually produced using a three-electrode potentiostat so that the working electrode's potential can be known accurately. However, the experiments require some care since the currents can become unreproducible due to "gunk" deposited on the electrode surface during electrolysis. Examples of deposits are salt precipitates and organic "tar" or polymers that coat the electrode when some organic species is electrolyzed. If the voltammetry is done in organic solvents, polymerization is quite likely to occur at the regions of solvent breakdown, when the solvent itself electrolyzes.

The region of solvent breakdown was discussed in association with Figure 15.15.

One approach to overcome this problem requires using an electrode that continually renews its surface. This apparently strange requirement can be satisfied by using a **dropping mercury electrode (DME)** as the working electrode in an electrochemical cell. A pictorial view is shown in Figure 15.20a. In the illustration, the auxiliary and reference electrodes and salt bridges are the same as usual—regardless of the type of working electrode. When the working electrode is a DME, the voltammetric apparatus is called a **polarograph**, for historical reasons.

Representative data from a voltammetric scan with a dropping mercury electrode is shown in Figure 15.20b. The "hash" of the DC scan is due to the changing size and surface area of the mercury drop. The current is smaller when a drop is small and increases as the drop becomes larger. The larger-sized drop has a larger area in contact with the solution; more electrolysis occurs on the larger surface. Note that the currents are in the μA range.

For an assay, the currents at the peaks of the oscillating current are most useful. (The peaks of the current curve occur just before the mercury drops are knocked off by the solenoid.) We can, if we want, simply keep the useful information—the magnitudes of the peaks—and eliminate the distracting "hash" current. The electronics records a peak current, and then records the next peak value, and so forth. This technique is called **sampled DC polarography**. An example of a sampled DC polarogram is shown in Figure 15.20b.

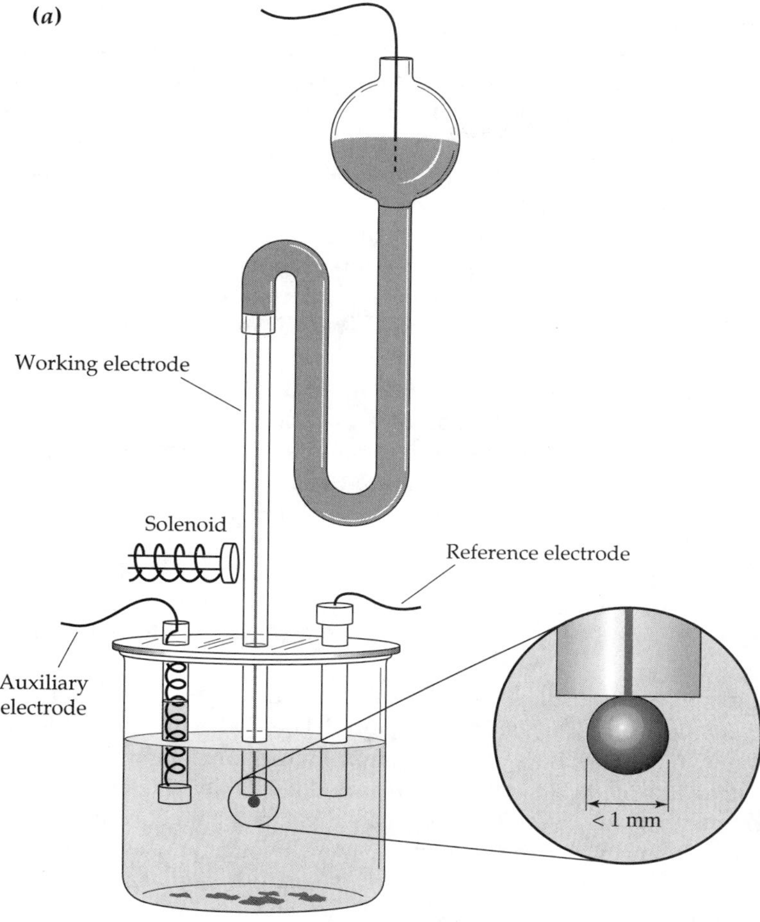

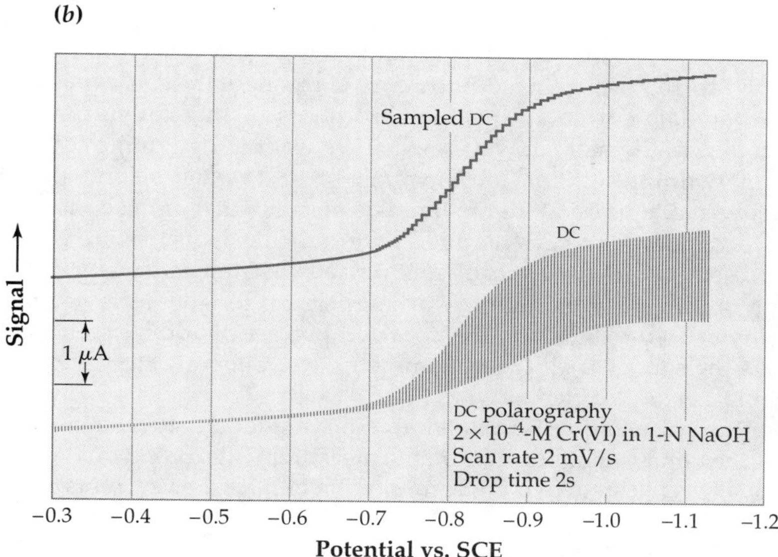

◀ FIGURE 15.20
A polarograph and its output.

At the top is a drawing of the apparatus used for forming a working electrode from a constantly flowing mercury stream. This is a dropping mercury electrode, DME. The reference and auxiliary electrodes are set up in chemical isolation and connected either with a frit (common for the auxiliary) or KCl salt bridge (common for the reference). The working electrode itself is a small drop of mercury at the tip of a thin-bore capillary tube. The drop falls off when it gets larger than about one millimeter in diameter, or it can be knocked off at a constant size by activating the solenoid at regular intervals. This apparatus can be used for all the forms of voltammetry. At the bottom is a DC polarogram with the conditions noted on the plot. Quantitation for the DC is done using the limiting current on the top of the "hash." A sampled DC polarogram is shown offset above the DC polarogram.

Professor Jaroslav Heyrovský (1890–1967) and his polarograph, which he first described in 1922. The output was recorded on photographic film.

Pulsed Voltammetry

Electrochemists call a current that is due to electrolysis a **faradaic current**. However, whenever an electrode has a changing potential applied to it, there is another contribution to the observed current besides electrolysis. This other contribution, called a **nonfaradaic current**, is a result of having to charge the electrode as the voltage is changed to a new value. The total current measured is the sum of the two types.

An illustration of how these currents arise is shown in Figure 15.21. The system pictured is a **hanging mercury drop electrode** (HMDE) in contact with an aqueous ionic solution. Immediately adjacent to the mercury surface is a layer of adsorbed water molecules. If there are any ions in this first layer, they tend to be anions, such as I^-, rather than cations, such as Na^+. In either case their numbers are relatively small compared with the water. When the metal electrode is uncharged, the anions and cations outside the first water layer are randomly mixed. However, when the electrode becomes charged, the ions of the opposite charge are attracted to it, and those with the same charge as the surface are repelled. For example, the electrode is negatively charged. As a result, an excess of positive charges will locate nearer to the surface, outside of the surface water layer. The negative charges that are expelled from this region cause an excess of negative charge to be located further away. This region of separated charges is called the electrical **double layer**.

What do we see in our experiment as a result of the formation of the double layer from randomly distributed ions? The motion of the anions and cations as they rearrange is an electric current in the solution at the interface. This current of double-layer formation is called the **capacitive current**. Since no electrons are transferred in the process, the capacitive current is a nonfaradaic current, and it is transient. In a 1-M salt solution, the double layer forms in less than 100 microseconds after the electrode is charged. When the electrode surface is discharged, the double layer disappears within the same length of time. A transient capacitive current in the opposite direction accompanies the disappearance. After a rapid change (a step) in potential, the nonfaradaic currents rise or decay with a simple, exponential time dependence.

In order to assay many materials voltammetrically at concentrations in the range 10^{-4} to 10^{-5} M, we need not worry about nonfaradaic, capacitive current; the faradaic currents from the electrolysis of the analyte will be much larger and last longer. However, when we want to determine concentrations of more dilute solutions, in the range of 10^{-6} M and below, special efforts must be made to separate the faradaic currents that are used for the assay from the nonfaradaic currents and background currents that are simultaneously present. To do this, two different **pulsed voltammetry** methods are used. (With a dropping mercury working electrode, the term **pulsed polarography** is used.) Understand before you proceed that the experimental cell looks the same for all the voltammetric experiments. However, the potentiostat is used in a different way.

The purpose of using pulsed voltammetric methods is to decrease the influence of the nonfaradaic currents on the measurements. Based on the discussion above, we can state two properties of the nonfaradaic currents:

a. The faster the change in potential, the larger the nonfaradaic currents.
b. After a voltage change, the nonfaradaic currents decrease very rapidly relative to the faradaic currents.

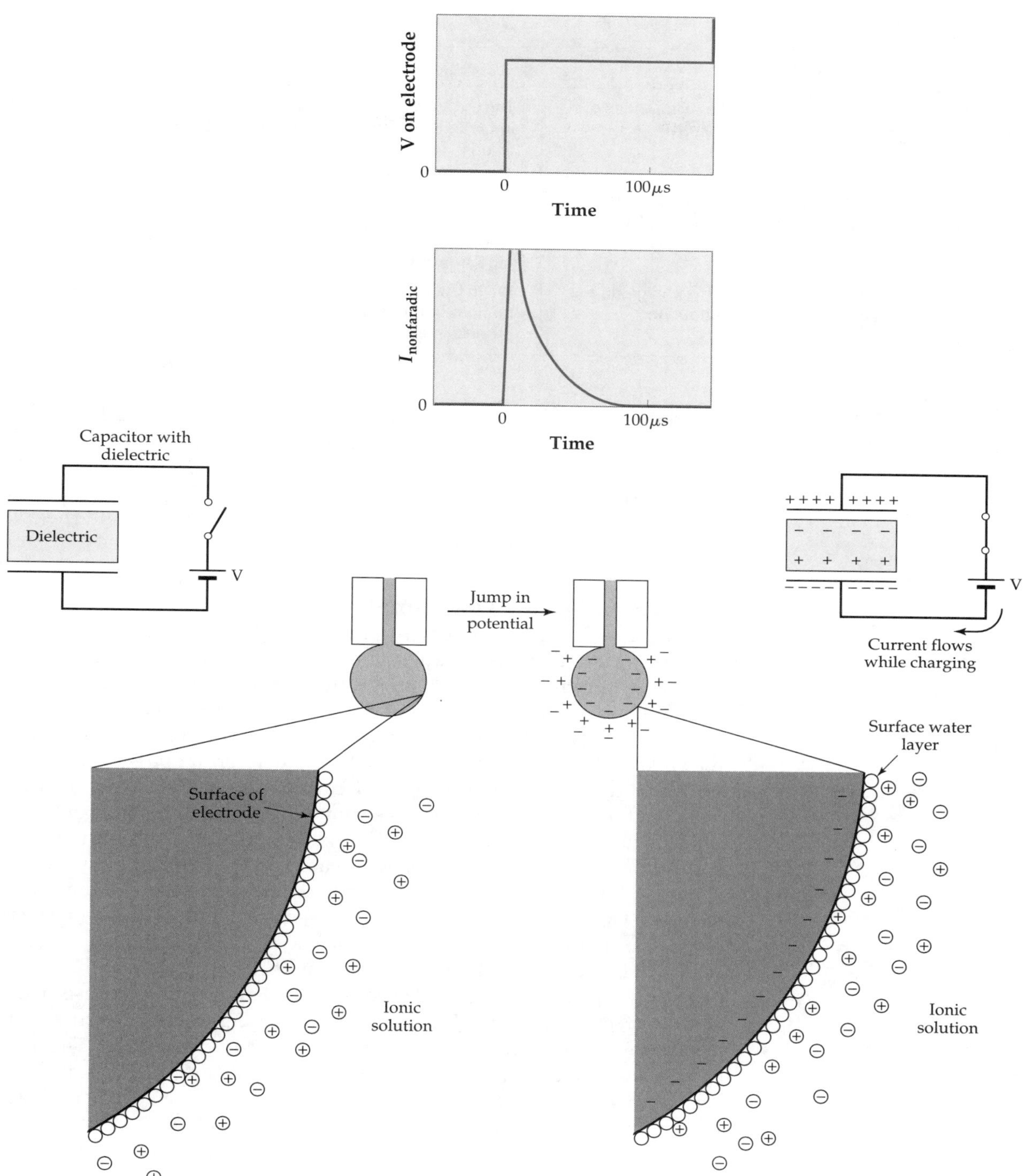

FIGURE 15.21 ▲

Illustration of the similarity of behavior for the surface of a metallic electrode in an ionic solution and a capacitor with a dielectric between its plates.

We consider a hanging mercury drop. The mercury is rapidly charged with a step potential jump. The jump and the non-Faradaic currents are shown at the top. The capacitive current rises and rapidly decays (in about 100 μs in a 1-M solution). The analogy with a capacitor is shown in the middle and bottom. The table on the next page compares electrodes in ionic solution with capacitors.

FIGURE 15.21 *(continued)*

Capacitor	*Electrode*
The capacitor is uncharged, and the charges within the dielectric are random.	The electrode is uncharged, and the ions are randomly distributed in the region of the interface. If ions are adsorbed, they tend to be anions.
The voltage is applied to capacitor plates, causing an electric field to appear between the plates. Charges appear at the surface of the dielectric. For the charges to appear, electric currents must flow within the dielectric.	A negative potential is applied to the electrode, causing the nearby positive ions to tend to move toward the surface and the accompanying negative ions to tend to move away. This charge rearrangement is a current and produces a less random distribution of charge at the interface—the double layer.
When the capacitor plates are discharged, the charges in the dielectric revert to the original random distribution.	When the electrode discharges, the ions return to the more random distribution, and the double layer "collapses," again producing a current.

To minimize the interference of the capacitive currents, the total current is sampled just prior to the pulse and again near the end of the pulse (after hundreds of microseconds). The early measurement provides a baseline, and the latter one occurs when the nonfaradaic currents have decreased significantly. This manipulation of the potential on the working electrode and the collection of data is done electronically.

Differential Pulse Voltammetry

You have read that improvement in the sensitivity of voltammetry can be obtained with careful sampling of the currents to separate the interfering capacitive currents from the analytically useful faradic currents. However, the faradaic background currents from the solution itself still are not removed in this way. An experimental method has been devised to minimize the effects of background faradaic currents, **differential pulse voltammetry**. With this technique, the background currents are sampled and subtracted from the total faradaic currents (the sum of the faradaic currents of the sample plus those of the background). The experimental apparatus is the same as before. However, the pulsing voltages placed on the working electrode are as shown in Figure 15.22, and the resulting output is a measure of the *slope* of a sampled DC voltammogram, as can be seen in Figure 15.23. The steepest slope of a standard voltammogram is at the $E_{1/2}$ of the material being electrolyzed. In a differential pulse experiment, the *peak* appears at the $E_{1/2}$. You can also see that differential pulse voltammetry allows more precise analyses at lower concentrations. The data shown in Figure 15.23 resulted from 0.36 ppm of tetracycline—equivalent to a 8×10^{-7}-M solution. The area of the peak above the baseline is directly proportional to the concentration of the electroactive species being assayed. Alternatively, the peak height can be used to measure concentration if voltammograms of the standards and sample are the same shape. With effective baseline extrapolations under the analytes' peaks, concentrations down to 10^{-6} to 10^{-7} M can be analyzed with precisions around $\pm 3\%$.

Cyclic Voltammetry

All the voltammetric methods described thus far employ voltage scan rates around 100 mV per minute. This allows time for the limiting currents to be established at reproducible values, thereby optimizing precision and accuracy.

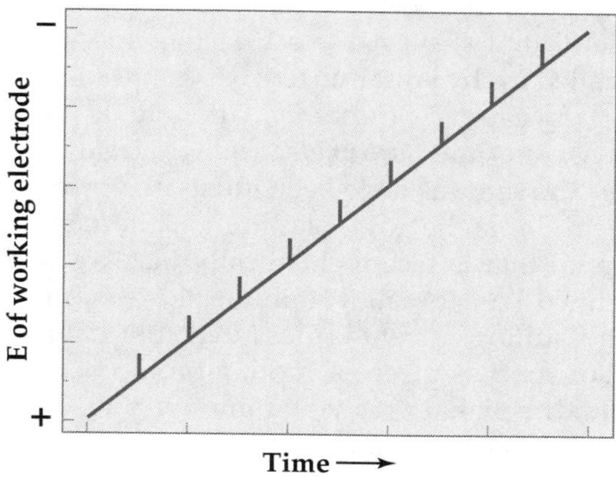

FIGURE 15.22 ▲

Graph of the voltage on the working electrode during differential pulse polarography.

Pulses are applied after the drop has had time to grow. Immediately after the pulse, the drops are mechanically knocked off the capillary so that the drop size is reproducible for each pulse. The pulses are all the same size (the **modulation amplitude**; 12–100 mV), and the base potential is changed at a fixed rate as is done in DC polarography. The instrument output has the shape of a derivative of a sampled-DC polarogram as seen in Figure 15.23.

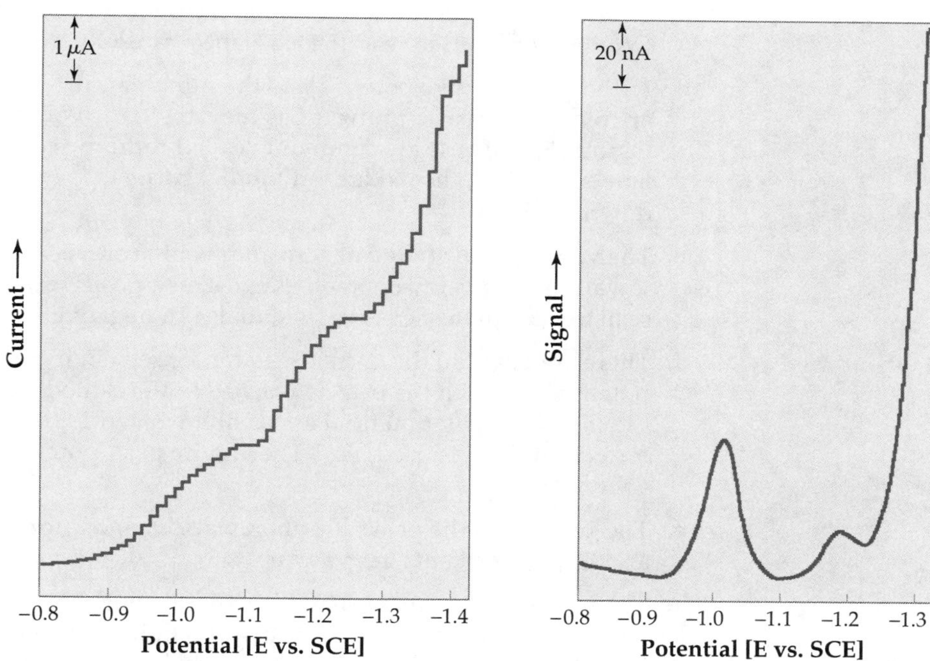

FIGURE 15.23 ▲

Sampled DC polarography compared with differential pulse polarography.

The polarograms are of tetracycline hydrochloride (an antibiotic). Both polarograms were obtained from samples in a 0.l-M acetate buffer, pH 4, with a dropping mercury electrode as the working electrode. A drop was generated each second. Notice that the current scales of the polarograms are different. Also, the concentration for the sampled DC polarogram (left) is 180 ppm while the concentration for the differential pulse determination (right) was 0.36 ppm—a factor of 500 more dilute.

However, the potential does not have to be scanned so slowly, and a faster scan is useful in a survey of electroactive species in a solution. The method used is called **cyclic voltammetry** (CV), and the voltage scan may be up to 100 V s^{-1}. However, in normal use, the rate of voltage scan is nearer 0.01–1 V s^{-1}. In CV experiments, the electrodes and other components of the sample cell can be the same as have been illustrated throughout this chapter. The working electrode can be gold, platinum, carbon, mercury, or any other conducting material that is electrochemically inactive within the potential range to be scanned. As for all voltammetric techniques, the voltage is scanned, and the resulting current is measured simultaneously. However, no pulsed-potential or special current-sampling protocols are used. The voltage scan also is not unidirectional as in voltammetry.

An example of a cyclic voltammogram and a graph of the values of time and voltage associated with the potential scan are shown in Figure 15.24. In this voltammogram, the linear scan of the potential starts at 0.0 V. It is taken to slightly over +0.9 V, then reversed to about −0.1 V, and finally brought back to about +0.2 V. This is slightly more than one **cycle**, which was completed upon returning to 0.0 V.

Before the origins of this strangely shaped curve are described, we need some definitions and an explanation of the conventions of the plot.

a. The horizontal axis, the voltage axis, is written with more negative potentials to the right and more positive to the left. This convention arose from polarography. However, the opposite convention is just as frequently found, so read the axis labels carefully.

b. With more negative potentials to the right, the currents for reduction are plotted increasing upward as they are in the voltammograms in Figure 15.17. These are commonly called **cathodic currents** and are labeled i_c. Thus, the oxidative (**anodic**) currents i_a are plotted downward.

c. The voltage scan started at zero and went positive. A voltage scan toward more-positive potentials is called a **positive scan**. A voltage scan toward more-negative potentials is a **negative scan**.[2]

d. There are peaks in the currents, as can be seen in Figure 15.24. The potential at which the peak is reached in the cathodic current is called the **cathodic peak potential** and is abbreviated E_{pc} or $(E_p)_c$. The former notation is used here. The **anodic peak potential** is, as shown, abbreviated E_{pa}.

e. The currents at the peaks are abbreviated i_c and i_a for cathodic and anodic peak currents, respectively.

f. The peak currents are measured as the current above background levels. The background extrapolation is indicated by the dashed lines in the figure. Notice how similar this measurement is to finding the limiting current above background for a voltammogram (illustrated

[2]In place of positive scan and negative scan, you may find the terms *anodic scan* and *cathodic scan*. Their use is discouraged.

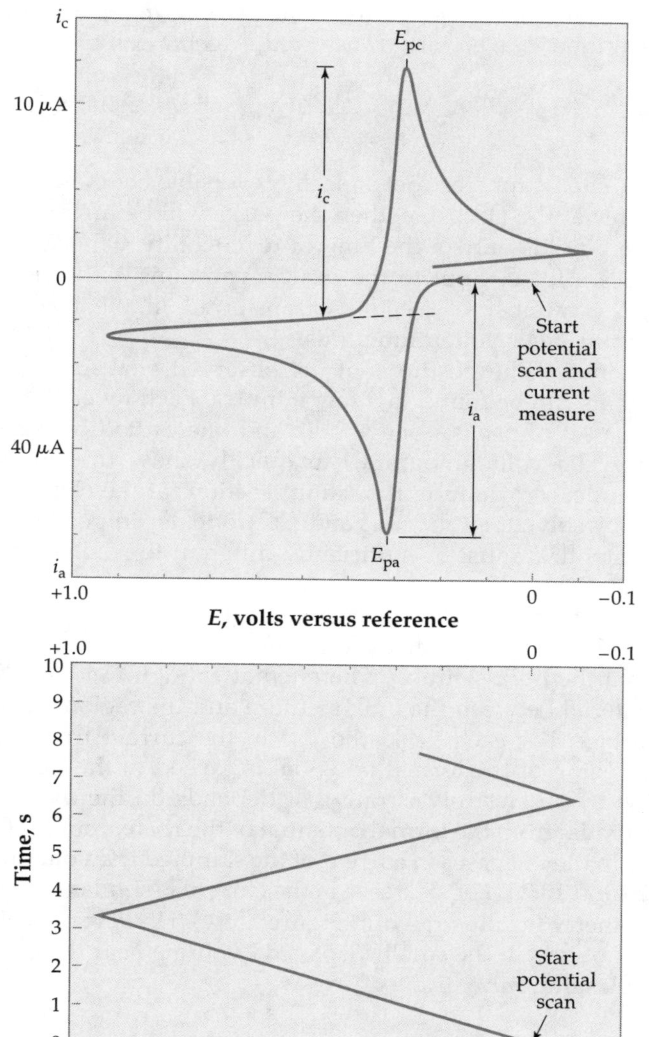

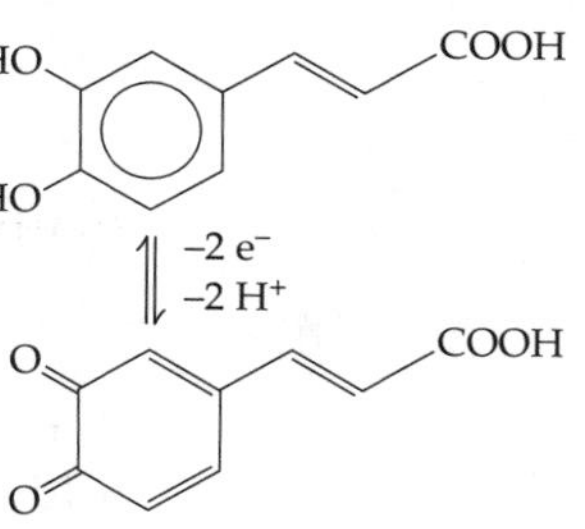

FIGURE 15.24 ▲

A cyclic voltammogram of the compound shown.

The anodic and cathodic peak potentials and peak currents are defined. The peak currents are measured from the baseline extended from a region before the electrolysis begins. The potential is scanned as shown in the lower graph. The voltage axes of both graphs are the same. The "cyclic" shown here begins at a potential of zero volts relative to the reference potential. The direction of the first scan is toward more positive potentials and results in oxidation of the compound. Then the scan is reversed. The cathodic peak is due to reduction of the material oxidized earlier in the scan. The original oxidation produces the anodic peak.

in Figure 15.19). This should be no surprise since much is the same except the rate of the voltage change on the working electrode. The magnitude (above background) of each peak currents (all other conditions remaining the same) are proportional to the concentrations of the electroactive species involved.

g. The difference in voltage between the peaks is inversely proportional to the number of electrons transferred. Algebraically,

$$E_{pa} - E_{pc} = \frac{0.059}{n} \quad \text{at } 25^{\circ}\text{C} \tag{15-13}$$

Equation 15-13 holds for electrochemically reversible couples. If the reaction is not reversible at the electrode, then the peaks will be further apart. Thus, the relationship holds only in the limit of reversibility. Otherwise, Equation 15-13 describes a lower limit to the peak separation. With this caveat, the value of n, the number of electrons transferred in the reaction, can be determined from cyclic voltammograms.

As you can see, the positions of the observed peaks are quite useful. Redox potentials can be found in a few minutes. Each reversible redox couple produces a set of peaks—one anodic and one cathodic—centered at its $E^{\circ\prime}$. Thus a cyclic voltammogram can quickly show the presence of all species that undergo oxidation-reduction reactions at the electrode—within the limits set by solvent electrolysis and electrode decomposition. However, the formal potentials must be sufficiently different to resolve peaks due to each redox couple.

But why are there peaks in the current at all? A simplified answer is that the voltage scan is so fast that an experiment "isolates" the region around the electrode from the bulk solution. There is not enough time for diffusion to transport material between the bulk solution and the region adjacent to the electrode surface. For each peak, the rise in the current from the baseline through the region of the formal potential is caused in the same way as in voltammetry: by an electrolysis rate that depends on the difference of the electrode potential from the formal potential of the redox couple. The rise can be compared to the increases in current of the sampled DC voltammogram in Figures 15.20 and 15.23. The decrease in the current is similar to the decreases seen in coulometry (as illustrated in Figure 15.14). However, the decrease in concentration occurs in the small, "isolated" volume near the electrode and not in the bulk solution.

15.9 Stripping Voltammetry

The electroanalytical technique that has the lowest limits of detection for metals is **anodic stripping voltammetry** (ASV). Its limits of detection range to ppb and below: 10^{-6} M to 10^{-9} M solutions. ASV combines coulometric and voltammetric techniques, and the steps of a stripping analysis are illustrated in Figure 15.25 along with an output scan.

The lower limit of concentration for which differential pulse polarography can yield good results is approximately 10^{-6} to 10^{-7} M in the electroactive species. However, if the material being assayed can be concentrated, the limit of the working range *as a fraction of the sample* can be lowered significantly. In other words, if the analyte is first concentrated by a factor of 100, then the assay method's limit of detection improves by that factor of 100. This is the principle behind stripping voltammetry. A hanging mercury drop electrode is commonly used as the working electrode, although a thin film of mercury deposited on a platinum or carbon (graphite) surface can be used as well. We shall describe only a mercury drop electrode.

Initially, a solution containing the analyte is electrolyzed for 0.5 to 5 min with stirring. In the example shown in Figure 15.25, the electrolysis was done

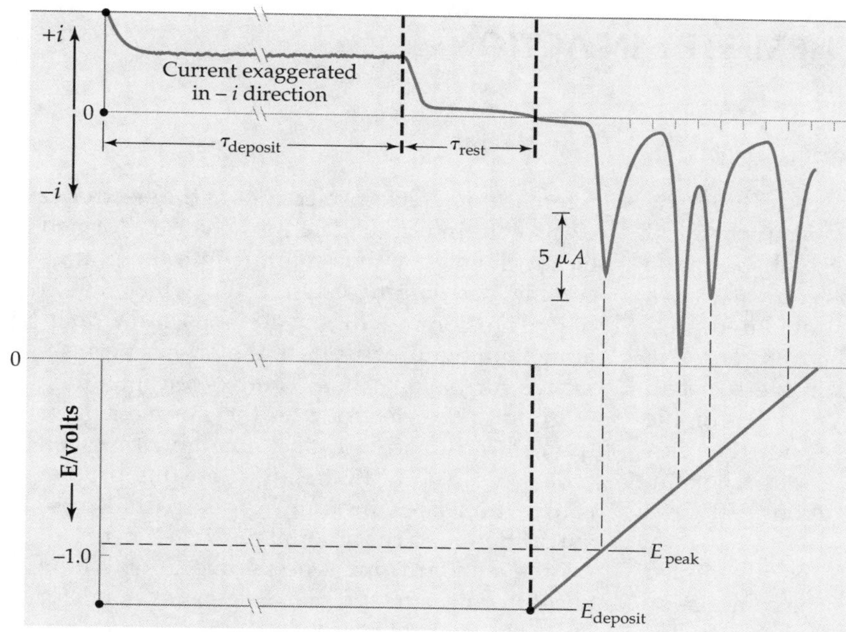

FIGURE 15.25 ▲
Anodic stripping voltammetry.

The solution is stirred while the metals are reduced and amalgamated in the mercury-drop working electrode. Then there is a rest period, while the solution is allowed to become still and the amalgam is allowed to become homogeneous. Following this, the potential is scanned in the positive direction. The metals reoxidize at their characteristic potentials, producing current peaks that rise when each metal begins to oxidize and fall off due to depletion of the metal in the mercury drop. The area under each peak above its baseline is proportional to the concentration of one metal. The stripping peaks are near the $E_{1/2}$ of each metal: 2.5 ppm in Cu and Zn and 5 ppm in Pb and Cd. The solution was 0.1 M sodium acetate with 0.01 M HNO_3 added, and the pH was 5.5.

in a stirred solution for 90 s at −1.3 V vs. SCE. At this negative potential, ions of metals that form amalgams with mercury are reduced and dissolve in the mercury drop. (These metals are listed in Table 15.8.) In this way, the elements become concentrated from the large solution volume into the small—less than 2 mm diameter—mercury drop.

Electrodeposition as a preconcentration method was introduced in Section 4.5.

Table 15.8 Elements Determinable by Anodic and Cathodic Stripping

By anodic stripping
Ag*, As, Au*, Ba, Bi, Cd, Cu, Ga, Ge, Hg*, In, K, Mn, Ni, Pb, Pt, Sb, Sn, Tl, Zn
By cathodic stripping†
Br^-, Cl^-, I^-, S^-, thio compounds

*Determined on solid electrodes, such as carbon or gold.
†These form mercury precipitates at the electrode, which are subsequently stripped off in a negative scan.

ANALYTICAL CHEMISTRY IN ACTION

Some Like It Hot!

Electrochemical cells do not have to be made with liquid electrolytes. An electrochemical sensor for oxygen in hot gases can be made with a solid electrolyte, and it is equivalent to a concentration cell. The cell illustrated in the figure can be used to measure the differences in the partial pressure of oxygen in the two half cells. The selectivity for oxygen depends on the ion-selective solid membrane and the gas-porous (or gas-permeable) electrodes. The temperature of the cell is held at about 850°C—a dull red heat.

The gas stream being analyzed is sampled continuously by passing part of the combustion gases through the chamber shown on the left side of the figure. The partial pressure of oxygen in the sample is less than the partial pressure of oxygen in the reference side, which has air passing over it. Oxygen molecules on the reference side pick up electrons at the electrode-electrolyte interface. The ions that are formed electrochemically then migrate into the porous ceramic zirconium oxide, ZrO_2, which has a high diffusion rate for oxygen ions at the operating temperature. In effect, the ZrO_2 is permselective for oxygen ions. The oxygen ions then migrate to the electrode on the sample side of the cell, are oxidized, and emerge as oxygen molecules. The EMF arising from this process can be translated into an oxygen partial pressure difference. The amount of oxygen in the air (as the standard) must be known to obtain an absolute measure of the oxygen level in, for example, exhaust gases.

[*Ref*: U.S. EPA Continuous Air Pollution Source Monitoring Systems, EPA 625/679-005, 1979, pp. S-27, S-28.] ■

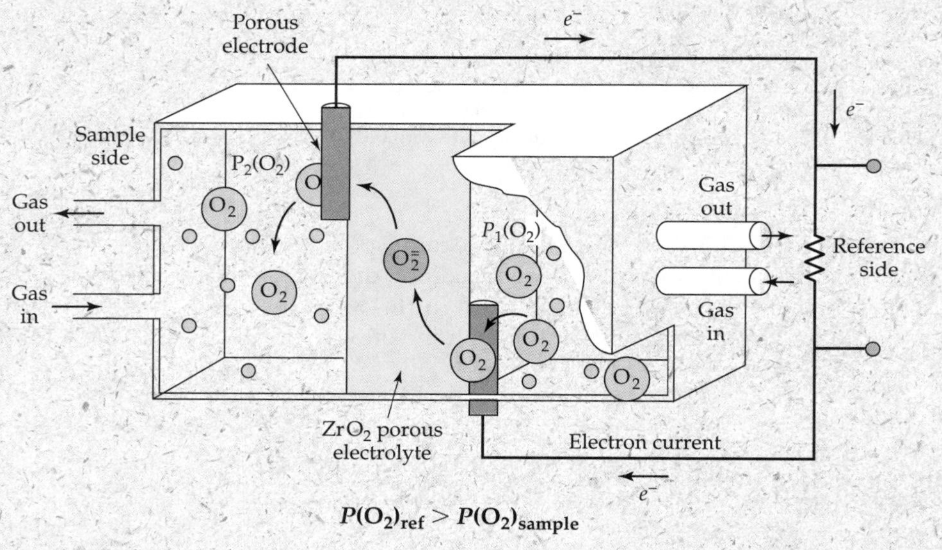

$P(O_2)_{ref} > P(O_2)_{sample}$

Next, the drop is isolated electrically and given time to equilibrate. This equilibration time and isolation are necessary for three reasons. If the circuit to the working electrode remained connected, more ions would be reduced, or some of the amalgamated (dissolved) metals might reoxidize. Second, the metals that have been reduced have entered the drop through its outer surface. It takes some time for the metal atoms to diffuse and become equally distributed inside. And third, the external solution, which was stirred during the reduction-concentration step, needs time to become still. In other words, after the concentration step, some time is needed to obtain an equilibrated, unstirred solution both inside the mercury drop and in the aqueous solution surrounding it.

Following the rest period, the potential is swept linearly in the positive direction starting from the deposition voltage. In turn, the metals are reoxidized. A peak is produced for each species at a characteristic potential. In addition, the height of the peak is proportional to the concentration in the sample solution. The peak heights are also proportional to the deposition (preconcentration) time; more ions deposit during longer deposition times. As a result, standards must be used with the technique.

Why are there peaks in the voltammogram as opposed to having a series of plateaus, as with polarography? The answer to that question is similar to the answer given for cyclic voltammetry. The rise is a result of the formal potential being reached for the reaction

$$M^0(\text{amalgam}) \rightleftharpoons M^{n+}(\text{aqueous}) + n\,e^-$$

The falloff is due to the exhaustion of the metal atoms in, or on, the mercury drop. The falloff continues until the material has been completely electrolyzed. To see this similarity more clearly, look at the shapes of the anodic currents in Figure 15.25 and compare them with the anodic current in Figure 15.24. Both have a sharp rise on the leading (time) edge with a slower falloff on the trailing side.

Even greater sensitivity can be obtained if the scan toward more positive potentials is done with differential-pulse voltammetry. Concentrations in the range of 10^{-9} M can be determined with precisions of 3–5%. However, before that level of metal can be determined, all the problems of sample preparation for ultratrace analysis will be encountered. Anodic stripping is one of the most effective trace analytical methods for the amalgamating metals, and the equipment needed is relatively inexpensive compared with the spectroscopic instruments that can be used at the same concentration levels.

15A A Deeper Look: How an Inert Electrode Senses the Potential of a Redox Couple

In constructing half-cells, the element composing the electrode need not be the same as the ions in solution—for instance, a platinum wire can be used with a Fe^{3+}/Fe^{2+} couple. Other materials could be employed in addition to platinum—among them carbon, in the form of graphite or glassy carbon, and gold. Why can these so-called inert electrodes be used to measure the potential of the redox reaction in solution? This question is answered in this section.

In the solution, the electrochemical equilibrium, like any equilibrium, is a continuously active process. For example, in a solution that contains a mixture of Fe^{3+} and Fe^{2+}, electrons are constantly transferred from the iron(II)—Fe^{2+}—ions to the iron(III)—Fe^{3+}—ions, causing the pairs to exchange oxidation states. The reaction can be written

$$Fe^{2+} + {}^*Fe^{3+} \rightleftharpoons Fe^{3+} + {}^*Fe^{2+}$$

The asterisk is used here to label a specific iron atom regardless of its ionic form. The equilibrium concentrations of both Fe(II) and Fe(III) ions are maintained, but electron transfers of this type are occurring between them all the time. Interpretation of experimental data has produced a picture of the **electron transfer reaction**, in which two separate processes are occurring:

a. The two ions must move sufficiently close to one another. This happens by random collisions. The name given to this motion is **mass transfer**.
b. During a collision between the components of a redox pair, such as Fe^{2+}/Fe^{3+}, the electron is transferred from the reduced species to the oxidized species: the **electron transfer** step. It is characterized by a rate constant, just like any other reaction.

In all the examples that we shall consider, the electron transfer reaction is very fast once the two ions are sufficiently close together. As a result, the rate of the oxidation-reduction reaction will be controlled by the rate of mass transfer—by how often the ions collide.

The potential of the redox couple, here Fe^{3+}/Fe^{2+}, is determined by the ratio of the relative concentrations of the Fe^{3+} and Fe^{2+} ions. As you know, the potential's dependence on the concentrations can be described by the Nernst equation. But what happens when the wire is placed into the solution? How does the wire measure the solution potential?

The answer involves finding how the potential of the wire itself is set. From electrostatic theory, we explain that the potential of a metal is determined by the excess or defect of electron charges. Thus, to change the potential of a wire, some electrons must be added to or removed from it: Some current must flow. Let us assume for now that the wire is not connected to any measuring device. Then, the current to change the wire's potential must result from electrons being transferred between the ions in the solution and the wire. The molecular events leading to this electron transfer can be depicted as shown in Figure 15A.1.

Let us assume that when the wire is placed into the solution, its potential is more negative than the redox couple in solution: The wire has an excess of electrons. As a result, electrons will be transferred from the wire to any Fe^{3+} ions that are in the *immediate vicinity* of the wire surface. (The fast electron transfer, indicated in the figure by the arrow from the wire to an Fe^{3+} ion next to the surface, will not occur unless the wire's surface and the ion are sufficiently close together.) As a result of this electron transfer, the potential on the wire becomes less negative, because it has lost a negative charge to the solution. Such electron transfers occur until the potential of the wire is the same as that of the solution.

What happens if the wire initially is more positive than the solution potential? Then electrons will be transferred to the wire from any Fe^{2+} ions that are sufficiently close. This is indicated by the arrow from an adjacent Fe^{2+} to the wire. Such an addition of negative charge to the wire causes the wire's potential to become more negative. Electrons flow into the wire until its potential is the same as that of the solution. Thus, at equilibrium, a wire that can exchange electrons with the ions in solution must have the same potential as the solution. If we sample the potential carefully so that very little current flows, the inert electrode's measured potential still equals that defined by the solution. With contemporary instruments, this is quite easy to do.

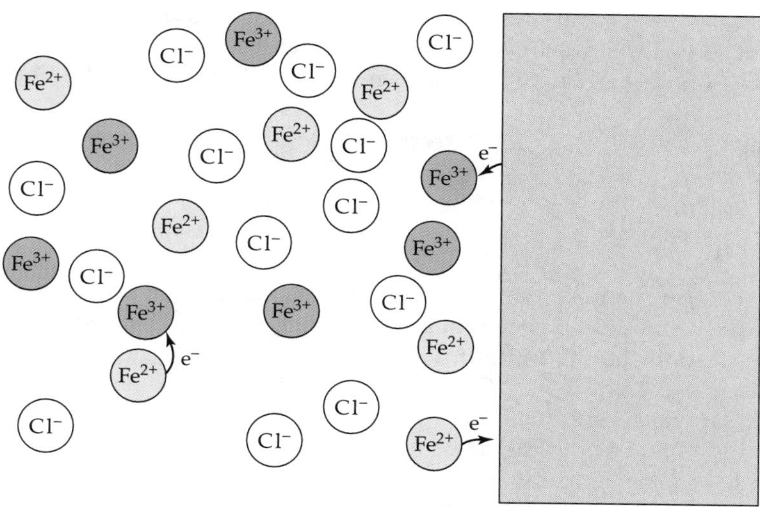

FIGURE 15A.1 ▲
Sensing a solution's electrochemical potential.
The inert electrode represented by the block at the right responds to the electrochemical potential of a redox couple in solution in the following way. When the electrode and ions are sufficiently close, electron transfer can occur: either transfer from Fe(II) ions to the electrode or from the electrode to Fe(III) ions. Thus, the inert electrode can act as an intermediary in the electron exchange between Fe(III) and Fe(II).

The Experimental Measurement

As illustrated in Figure 15A.2, an electrochemical cell can be compared with an electrical circuit. The circuit's voltage source is equivalent to the potential difference of the two electrochemical half-cells. The circuit's resistor represents the sum of the resistances of the cell's ionic solutions, the salt bridge, and the wires connecting the electrodes to the voltmeter. In most real cells, the resistance of the wires is negligible compared with those of the solution and salt bridge.

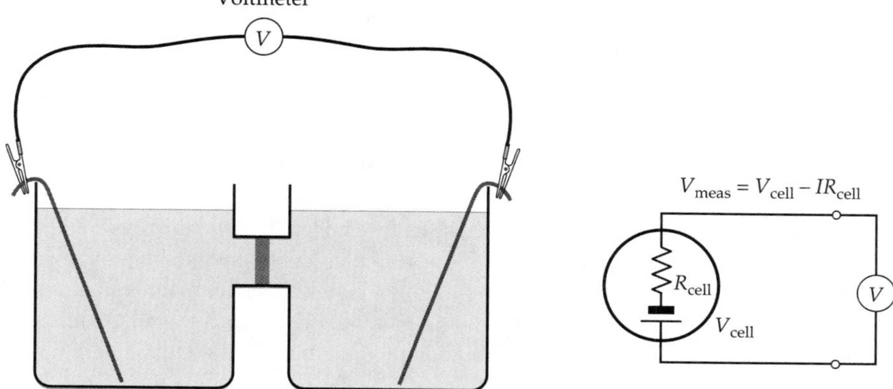

◀ FIGURE 15A.2
Diagram of an electrochemical cell shown with an electrical circuit equivalent.
The cell acts like a battery in series with a resistance. The resistance arises from the limited ability of the ionic solution to conduct charge.

Ideally, the measurement of electrochemical potentials should be done without allowing any current to flow. The reason is clarified by using the equivalent electrical circuit. For the simple circuit shown in the figure, the measured potential, V_{meas}, is

$$V_{meas} = V_{cell} - V_{cell\ resistance}$$

However, from Ohm's law, we can write

$$V_{meas} = V_{cell} - IR_{cell}$$

The IR_{cell} term becomes zero only when there is no current flow ($I = 0$). Only then will the measured potential be the potential of the electrochemical cell.

A question that naturally follows from this is, Isn't there some current flowing into the voltmeter? Otherwise, how could you get a response? It is, in fact, true that a voltmeter draws some current. However, if the current is small, the potential can be measured as accurately as needed. This point is illustrated in Figure 15A.2.

From inspection of the circuit, we note that the resistances of the meter and the cell are in series. What will the measured voltage be in this case? The meter measures the potential drop across its resistor, R_{meter}. Thus,

$$V_{meas} = IR_{meter}$$

According to Ohm's law, the current that flows through the circuit depends on the total resistance of the circuit: the resistances of both the cell and the meter. This is described algebraically by

$$V_{cell} = I(R_{cell} + R_{meter})$$

The ratio of the measured potential to the true potential is

$$\frac{V_{meas}}{V_{cell}} = \frac{R_{meter}}{R_{cell} + R_{meter}} \tag{15A-1}$$

The behavior of this equation shows that the larger the meter resistance, the closer the measurement will be to the true value of the cell's potential.

EXAMPLE 15.3

Find the error in a typical potentiometric experiment, given that typical values for electrochemical cells and simple, contemporary voltmeters are

$$V_{cell} = 0.5\ \text{V}$$

$$R_{cell} = 10^3\Omega = 1\ \text{k}\Omega$$

$$R_{meter} = 10^{12}\Omega = 1\ \text{T}\Omega$$

Solution

From Equation 15A-1, the ratio of the measured potential to that of the cell is[3]

$$\frac{V_{meas}}{V_{cell}} = \frac{10^{12}}{10^3 + 10^{12}} = 1 - (1 \times 10^{-9})$$

---

[3]The solution to the equation uses the relation

$$\frac{1}{1 + x} = 1 - x + x^2 - \cdots$$

and the absolute error in the measured potential is expected to be

$$V_{meas} - V_{cell} = (1 \times 10^{-9}) \times 0.5 = 5 \times 10^{-10}\ \text{V}$$

This is only 0.5 nV, not enough to worry about since the errors due to interface potentials have values above a millivolt or so.

15B A Deeper Look: How Selectivity Coefficients Are Determined

Equations such as 15-4 and tables of measured selectivity coefficients comprise a technique for bookkeeping—for storing the information derived from an experiment. Selectivity coefficients depend on the temperature, the type of electrode and the manufacturer, and even the production batch if production of the surface material is difficult to control. From such selectivity coefficients we can reconstruct the changes in potential due to different interferents. However, the calculation matches the measured behavior only if the conditions are the same.

In Equations 15-3 and 15-4, the value of the constant includes a number of unknown junction potentials. An experiment to determine the selectivity coefficients must be organized so that the junction potentials are as constant as possible. Then, the selectivity coefficients can be determined by measuring the cell potential with a number of different concentrations of each interferent present. First, the potential is measured without the interferent. The potential can be expressed as

$$E_{ISE}^{(1)} - \text{constant} = \frac{0.059}{z_i} \log a_i \qquad \text{at } 25°\text{C} \qquad \textbf{(15B-1)}$$

The potential is then measured again after some interferent is added. The potential will be characterizable by

$$E_{ISE}^{(2)} - \text{constant} = \frac{0.059}{z_i} \log (a_i + k_{ij}\, a_j^{z_i/z_j}) \qquad \text{at } 25°\text{C} \qquad \textbf{(15B-2)}$$

EXAMPLE 15.4

Data to Determine $k_{Ca,Mg}$ for a Ca^{2+} ISE

Concentrations Ca^{2+}, Mg^{2+} (mM)	Potential vs. Reference (mV)
0.10, 0	−353.0
0.10, 10	−341.8
0.10, 50	−326.4
0.10, 100	−318.3

Given the data in the table above, determine $k_{Ca,Mg}$ for a Ca^{2+} ISE. The experiment was carried out by measuring the potential of the electrode *vs.* a standard half-cell and then adding aliquots of concentrated magnesium chloride to the solution to change the magnesium concentration while leaving the volume nearly constant. The solution was stirred, and the potential at the new equilibrium measured. The solution used was at 25°C and contained 0.25-M KCl as background electrolyte.

Solution

To carry out the calculation, we first find the difference between expressions 15B-1 and 15B-2. For clarity, the subscript ISE will be dropped.

$$E^{(2)} - E^{(1)} = 0.0295 \log \left[\frac{a_{Ca}{}^{(2)}}{a_{Ca}{}^{(1)}} + k_{Ca,Mg} \left(\frac{a_{Mg}}{a_{Ca}} \right) \right]$$

The ISE measurement is sensitive to the activities of the ions. However, since in day-to-day use concentrations are measured, we do the calculations with the ion concentrations. At constant ion strength (here 0.25 M), the ratios of ion activities that are measured will be close to the ratios of the respective ion concentrations. For the experiment here, $a_{Ca}{}^{(1)} = a_{Ca}{}^{(2)}$, and

$$\Delta E = E^{(2)} - E^{(1)} = 0.0295 \log \left\{ 1 + k_{Ca,Mg} \cdot \frac{[Mg^{2+}]}{[Ca^{2+}]} \right\}$$

Rearranging and taking the antilog of each side, we obtain

$$10^{\Delta E/0.0295} = 1 + k_{Ca,Mg} \cdot \frac{[Mg^{2+}]}{[Ca^{2+}]}$$

Using the data in the table, we plot the left hand side *vs.* the magnesium-calcium ratio in Figure 15B.1. The slope of the graph is $k_{Ca,Mg}$ and equals 0.014 in 0.25 M KCl at 25°C.

You might wonder about our determining the selectivity coefficient in 0.25 M KCl. After all, might not the K^+ ion interfere too? Yes, it might; but, from appropriate tables, the interference by potassium is found to be small. More importantly, however, is that any K^+ interference will be constant with a constant K^+ concentration. Notice that the answer to the question was that the selectivity coefficient is 0.014 in 0.25 M KCl. If more than one interfering ion is present, a sum of correction term is needed.

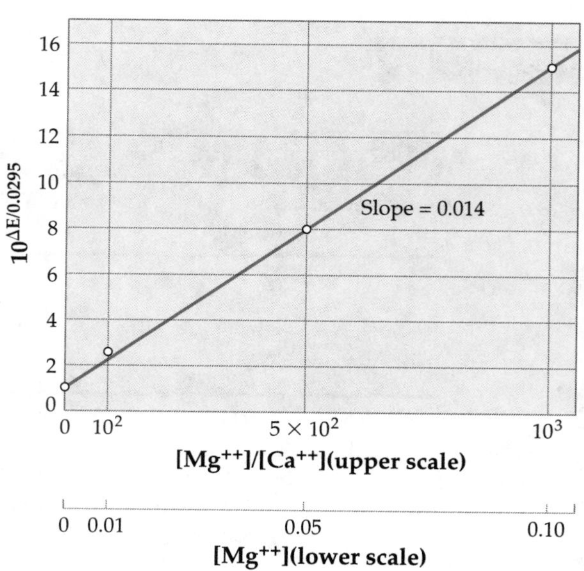

FIGURE 15B.1 ▶ **A graph to find the selectivity coefficient $k_{Ca,Mg}$.** This is a plot of

$$10^{\Delta E/0.0295} \textit{ vs. } [Mg^{2+}]/[Ca^{2+}]$$

The data are in Table 15B.1. The slope, equal to 0.014, is the selectivity coefficient.

Suggestions for Further Reading

SAWYER, D. T., ROBERTS, J. L., Jr. 1974. *Experimental Electrochemistry for Chemists*. New York: Wiley.
Despite its age, the best next place to go for more electrochemistry if you are going to do experiments.

COVINGTON, A. K. 1979. *Ion-Selective Electrode Methodology*. Boca Raton, FL: CRC Press. Vols. I and II.
A practical approach to ion-selective electrodes with manufacturers' data and numerous examples of applications.

KORYTA, J., STULIK, K. 1983. *Ion-Selective Electrodes*. 2nd ed. Cambridge and New York: Cambridge Univ. Press.
A physical-chemistry and descriptive-chemistry approach to ion-selective electrodes. The text covers the factors responsible for ISE response. There is not a list of uses.

LATIMER, W. M. 1952. *The Oxidation States of the Elements and Their Potentials in Aqueous Solutions*. New York: Prentice-Hall.
Still the classic work on the subject of oxidation potentials. Note: these are *oxidation* potentials.

KISSINGER, P. T., HEINEMAN, W. R., eds. 1984. *Laboratory Techniques in Electroanalytical Chemistry*. New York: Dekker.
A textbook balanced between theory and practical information with mostly pictorial explanations of the details of electroanalytical chemistry. It is encyclopedic, so know what you are looking for first.

BARD, A. J., FAULKNER, L. R. 1980. *Electrochemical Methods*. New York: Wiley.
An advanced textbook treatment of the topics in analytical electrochemistry. Significant mathematical and thermochemical sophistication required.

BOCKRIS, J. O'M., REDDY, A. K. N. 1970. *Modern Electrochemistry*. New York: Plenum. Two volumes.
A complete coverage of the physical chemistry underlying electroanalytical chemistry. Extremely well-written and highly recommended.

RILEY, T., WATSON, A. 1987. *Polarography and other voltammetric methods*. Chichester: Wiley.
An elementary book of voltammetric methods.

STOCK, J. T. 1965. *Amperometric Titrations*. New York: Interscience.
A complete book of amperometric titrations.

HEINZE, J. 1984. "Cyclic Voltammetry." *Angewandte Chemie*, International Edn. 23:831–847.
A brief, clear description at an intermediate level of the capabilities of cyclic voltammetry. A good entry into the primary literature. Emphasis on the electrochemical mechanisms rather than the analytical capabilities. (In English.)

SERJEANT, E. P. 1984. *Potentiometry and Potentiometric Titrations*. New York: Wiley.
An advanced textbook on potentiometric titrations. Not much about how to do the work but loaded with cautions about the techniques.

WANG, J. 1985. *Stripping Analysis, Principles, Instrumentation, and Applications*. Deerfield Beach, FL: VCH Publishers.
A nonmathematical, practical treatment of stripping analysis approximately equally divided between the principles, descriptions of the instrumentation, and applications. A useful next place to read if you plan to do stripping analysis.

MEITES, L., et al. 1977. *CRC Handbook Series in Organic Electrochemistry*. Boca Raton, FL: CRC Press.
MEITES, L., et al. 1980. *CRC Handbook Series in Inorganic Electrochemistry*. Boca Raton, FL: CRC Press.
Sets of multivolume compendia of electrochemistry data for various systems listed by element and/or compound.

BARD, A. J., ed. 1973. *Encyclopedia of Electrochemistry of the Elements*. New York: Marcel Dekker.
A multivolume compendium of the electrochemical information, both on EMF and kinetic data, of all the elements. Later volumes contain similar information for organic compounds.

BUCK, R. P. 1978. *Anal. Chem.* 50:April 17R–29R.
An entry to the literature of ion-selective electrodes.

Concept Review

1. What measured quantity provides information about analyte concentration in each of the following methods?
(a) coulometry
(b) pontentiometry
(c) amperometry
(d) conductimetry

2. What is the reaction that is the basis for each of the following reference half-cells?
(a) saturated calomel
(b) standard calomel
(c) silver-silver chloride
(d) normal hydrogen electrode

3. What kind of plot gives a linear relationship between concentration of analyte and potential for an ion-selective electrode?

4. What three factors can cause the current in an electrolytic cell to be lower than expected for a given E_{app}?

5. In a system which uses a three-electrode potentiostat, between which two electrodes is the potential measured? Between which two is the current measured?

6. In an aqueous solution of 5% dextrose (a monosaccharide), what species would contribute to the conductivity of the solution?

7. What is the difference (if any) between the specific conductance of a solution and its conductivity?

8. Sketch the current *vs.* time curve that would result from an exhaustive coulometric experiment carried out under constant potential conditions with stirring. What quantity

from the graph would be proportional to the amount of analyte in solution?

9. If we say an electrochemical reaction is reversible, what is implied?

10. Sketch how the applied potential varies with time for:
(a) differential pulse voltammetry
(b) cyclic voltammetry

Exercises

15.1 A fluoride ion-selective electrode was used to analyze for fluoride in drinking water. The range for which the water treatment plant aims is 1.00 ($\pm$0.1) ppm. Each day three standards and six periodic samples were analyzed. The pH and ionic strength were buffered for all solutions before analysis. One day in the spring, the following values were obtained:

Solution	*E* (mV)
1.00×10^{-4} M F^-	40.5
5.00×10^{-5} M F^-	60.2
1.00×10^{-5} M F^-	100.0
Sample 1	57.6
Sample 2	61.0
Sample 3	60.7
Sample 4	59.3
Sample 5	59.6
Sample 6	59.4

(a) What kind of algebraic relationship do you expect between E and $[F^-]$?
(b) Plot the data for the standards and find the best straight line through your points. Based on your calibration standards, what is the level of F^- in ppm for the water samples?
(c) Is the average for the day within the range set by the water treatment plant?

15.2 A fluoride ion-selective electrode responds to both F^- and OH^- but not to HF. A measurement is made and interpreted assuming that all the fluoride ion is unassociated and no other species interferes with the ISE. Data: $K_a(\text{HF}) = 7.2 \times 10^{-4}$ at 25°C; $k_{\text{F,OH}} = 0.06$ under the conditions of the experiments. Answer the following questions.
(a) Will the formation of HF tend to make the measured result low or high?
(b) Will the presence of OH^- tend to make the measured result low or high?
(c) Given a true fluoride concentration of 1 ppm, at what pH will the presence of HF create an error of 3% in the measured fluoride?
(d) Given a fluoride concentration of 1 ppm, at what pH will the presence of OH^- create an error of 3% in the measured fluoride?

15.3 A measurement of the potassium concentration difference between the interior and exterior of kidney tubules (microscopic tubular structures involved with excretion) were made using a double microelectrode. Two microelectrodes are fused together side-by-side. One electrode is a reference electrode. The other is a K^+ ion-selective electrode with a liquid membrane. The response of the ISE to changes in potassium activity is + 50 mV per decade increase in a_{K^+}. The activity coefficient for potassium at the solution ionic strength is 0.77. With the double electrode, the measurement inside the kidney tubule produces a potassium ISE potential of +68 mV relative to the potential measured with the double electrode in the solution surrounding the outside of the tubule. A spectrometric method was used to show the external potassium concentration to be 2.5 mM.
(a) What is the external potassium activity?
(b) What is the internal potassium activity?
(c) What is the internal potassium concentration?

15.4 For the method of standard additions in potentiometry, the potential developed in a known volume of sample V(s) is recorded. An aliquot of volume of standard V(a) is added, and the new potential recorded. Show that the concentration of the sample C(s) is related to the concentration of the aliquot C(a) by

$$C(\text{s}) = \frac{C(\text{a})\,V(\text{a})}{\{V(\text{s}) + V(\text{a})\}\,10^{\Delta E/S} - V(\text{s})}$$

ΔE is the difference in potential between the (sample + aliquot) and the sample alone. The letter S represents the slope of the electrode response in mV per decade (ten-fold) change in activity. The slope of the electrode response must be known under the conditions of the experiment to obtain accurate values with the method.

15.5 One way to enable accurate determinations of ion concentrations is to keep the ionic strength of the test solution high and relatively constant. In this way, the activity coefficients do not change appreciably, and concentrations of standards and samples can be related directly to each other. To this end, it is useful to add a total ionic strength adjustment buffer (TISAB) to a solution for an ISE analysis. For fluoride, the TISAB has the following composition: NaCl 1.0 M, acetic acid 0.25 M, sodium acetate 0.75 M, sodium citrate 0.001 at pH 5.0. The total ionic strength is 1.75. Equal volumes of sample and TISAB are added in one fluoride ISE assay. The purpose of the citrate is to coordinate metal ions such as Fe^{3+} and Al^{3+}

which might bind F^- and lower its concentration. Other chelating agents may be used as well.

A sample of toothpaste weighing 188.0 mg was placed in a 250-mL beaker containing 50 mL of TISAB. The mixture is boiled for 2 min and cooled. The solution is transferred with two washes into a 100-mL volumetric flask and diluted to the mark with distilled water. This solution had a reading of +175 mV with a F^- ion-selective electrode which had a sensitivity of −60.0 mV (decade increase)$^{-1}$. Then, separately, two 0.010-mL standard additions 10 mg/mL F^- were added to the sample, mixed, and measured. The readings were 73.6 and 55.3 mV, respectively. What was the concentration in ppm of F^- in the original toothpaste? (You will need to use the equation given in Exercise 15.4.) [Ref: Light, T. S., Capuccino, C. C. 1975. *J. Chem. Ed.* 52:247.]

15.6 Using a calcium ion-selective electrode in a determination, what is the maximum concentration level of magnesium that can be tolerated in a 10^{-4} M Ca^{2+} solution and still produce less than a 10% error due to magnesium interference? Under the conditions, $k_{Ca,Mg} = 0.014$. Assume that the activity coefficients of Ca^{2+} and Mg^{2+} are equal.

15.7 For its reduction reaction, hydrogen has a kinetic overpotential η (see Figure 15.10) of −0.09 V on platinum and −1.04 V on mercury at low current densities. For its reduction, cobalt has no kinetic overpotential on either material. If a solution contains 0.1 M Co^{2+} at pH 1, will cobalt metal or hydrogen gas be the primary product (more easily reduced) when the cathode is

(a) mercury?

(b) platinum?

15.8 The following questions refer to Table 15.6 on page 509.

(a) Write the half-reactions for the generation of reagents (2nd column) from their precursors (3rd column) for the examples of coulometric titrations shown in the table.

15.9 A Karl Fischer reaction with coulometric generation of reagent was run to determine the water content of a sample. A sample of 1.000 mL was run to determine its water content. The reaction end point was reached after 158 s with a current of 0.1000 mA. What was the water content on a ppm weight/volume basis if each water requires $2e^-$? (Use correct significant figures.)

15.10 Figure 15.10.1 shows the currents due to material arriving in a flowing stream at the working electrode after the three components present have been chromatographically separated. The sample consists of 90 ng of 6-hydroxydopa, 87 ng of L-dopa, and 100 ng of tyrosine. This is the order of arrival in time from first to last. The three different response curves are determined with the working electrode held at three different potentials: 1.0 V (curve A), 0.6 V (curve B), and 0.3 V (curve C) *vs.* Ag/AgCl. [Ref: Last, T. A. 1983. *Anal. Chim. Acta.* 155:287.]

(a) List the three different substances in order of ease of oxidation.

(b) If the experiment were run with a reference electrode of SCE, would the order of oxidation change?

15.11 Shown in Figure 15.11.1 is a set of cyclic voltammograms of a compound in aqueous solution at various pH values.

(a) What are the values of $E°'$ at the pH values shown?

(b) What is the standard potential $E°$ of the compound *vs.* SCE?

15.12 An assumption made in coulometry is that the analyte being measured is completely electrolyzed by the

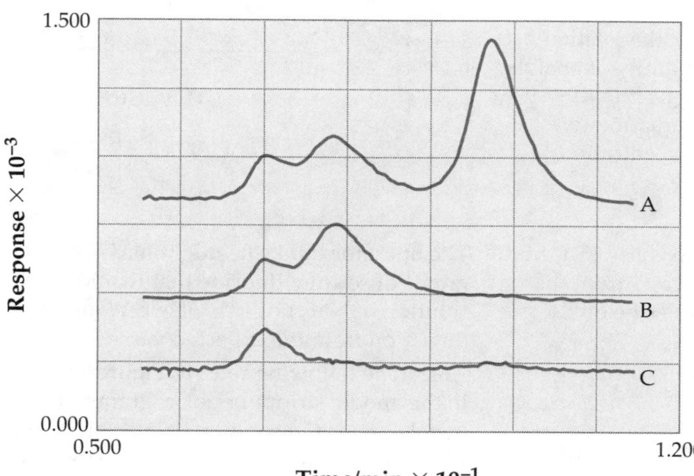

FIGURE 15.10.1

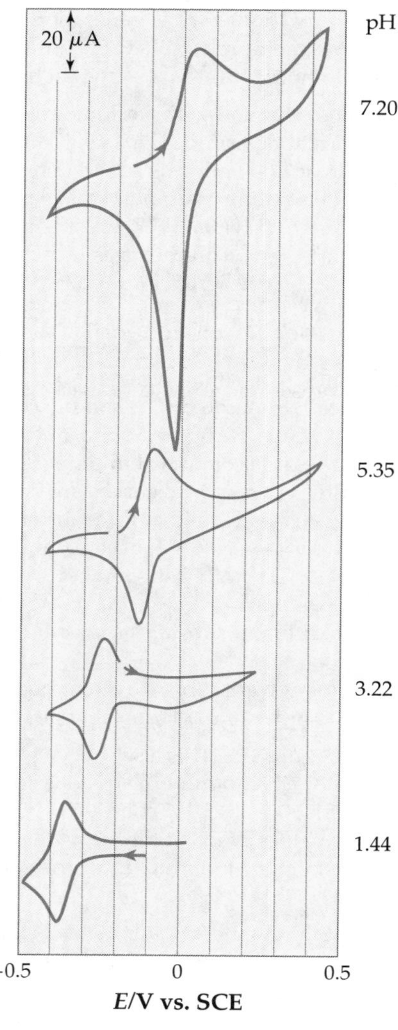

FIGURE 15.11.1 ▲

time the experiment is finished. Otherwise, the relation between the charge transferred and the quantity of analyte will be uncertain or even meaningless. Let us *define* 99.99% reduction of the material to be considered quantitative. Assume that neither overpotentials nor concentration polarization need be considered. You measure the potential of the solution before the deposition and it is E_{meas}. What potential relative to E_{meas} must be used to obtain quantitative results for reduction of an M^{2+} ion to the metal itself?

■ ***15.13** Figure 15.13.1 shows the differential pulse stripping voltammogram for the simultaneous determination of copper, lead, and cadmium in a sample of shark meat. Deposition time was 90 s at −0.9 V *vs.* SCE. The unlabeled scan is the sample. The scans labeled A, B, and C are with standard additions as follows (all concentrations are in μg/L): (A) +0.05 Cd, 0.5 Pb, 0.5 Cu; (B) +0.10, 1.0, 1.0, respectively; (C) +0.20, 1.5, 1.5, respectively. The baselines of the scans are displaced to separate them. In other words, the left-hand sides of the scans would coincide if they were not shifted. Draw in the appropriate baselines, assign the peaks of the voltammogram, and determine the concentrations of each of the three metals. [Ref: Adeloju, S. B., et al. 1983. *Anal. Chim. Acta*. 148:59.]

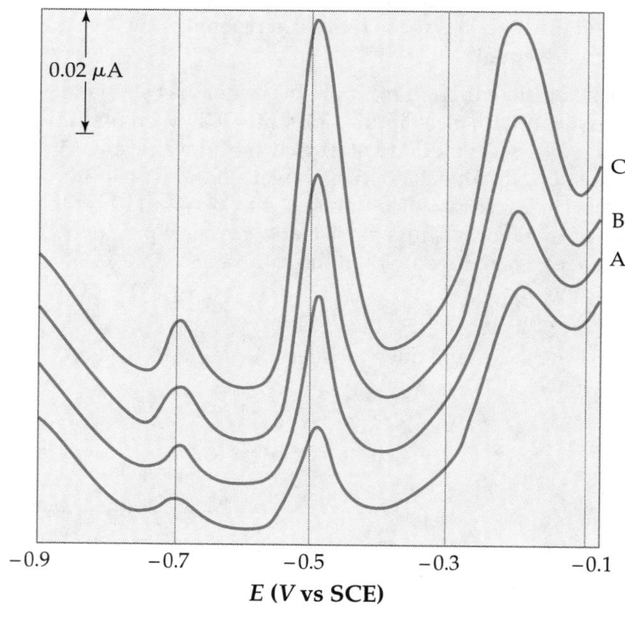

FIGURE 15.13.1 ▲

Additional Exercises

■ **15.14** Using the calibration standards in Exercise 15.1, find the concentration of fluoride in drinking water from the three different water treatment plant samples below:

Solution	E (mV)
Sample 1	76.5
Sample 2	76.8
Sample 3	77.3

The operator thinks he might have used the number of grams of sodium fluoride that would give a 1 ppm solution of NaF, not of F^- itself. Would this explain the deviation from the expected value?

15.15 The following refer to Figure 15.25. Assign the peaks in the anodic stripping polarogram to each of the four metals present in solution (from right to left).

15.16 A coulometric titration of Fe(II) by electrochemically generated Ce(IV) was carried out in the following manner.

A stock solution of Ce(III) sulfate was prepared. This was composed of 100 g of $Ce_2(SO_4)_3 \cdot 8\ H_2O$ and 100 mL of concentrated sulfuric acid brought to 1.000-L volume. Of the solution, 30 mL was placed in a 100-mL beaker, and a platinum electrode was placed in the solution as the generating anode. A second platinum wire, isolated by a fritted separator, was used as the cathode. Two additional platinum wires were used in a biamperometric end-point circuit. The constant current source supplied 0.1000 mA. A 5.00-μL aliquot of the Fe^{2+} sample was added, and the titration completed in 132 s. How many micrograms of iron were in the aliquot?

15.17 A coulometric titration was run to determine the Ca^{2+} content of a 0.750 g sample. The reaction end point was reached after 48 s with a current of 0.1000 mA. What was the calcium content on a ppm weight/weight basis if release of each edta requires two e^-?

15.18 In Exercise 15.11 you found the standard potential of a compound in aqueous solution at various pH values *vs.* SCE. What is the standard potential $E°$ of the compound versus NHE?

15.19 The data of Figure 15.19.1 are from a coulometric determination of lead in a complex biochemical mixture. The lower trace shows the charge transferred due to the background current alone run on a blank sample. The top trace is from the lead-containing species. The difference between the two traces results from the lead-containing sample. The potential used was −0.700 V *vs.* SCE, low enough to reduce the lead ions to metallic lead while keeping the background current small. Calculate the quantity of lead present from the measured charge.

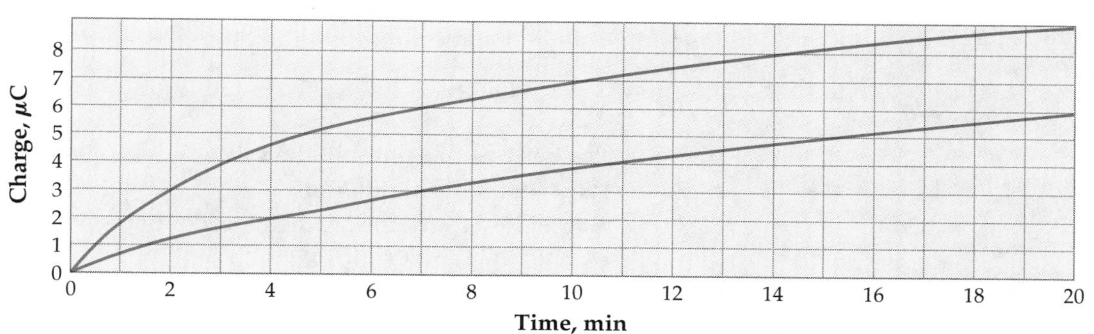

FIGURE 15.19.1 ▲

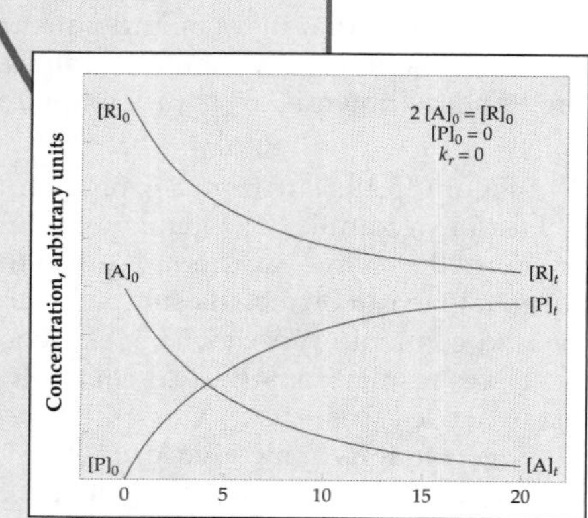

CHAPTER 16

Kinetic Methods

Overview

16.1 The Chemistries of Kinetic Analyses

16.2 Why Kinetic Methods?

16.3 A Brief Review of the Mathematics of Rates of Reactions, Decay Times, and Half-Lives

16.4 Assay Kinetics and Assay Types

16.5 Methods for Determining $[A]_0$ or $k_{catalytic}$ from Rates

A Deeper Look

16A Obtaining Simple Kinetic Behavior

16.1 The Chemistries of Kinetic Analyses

Kinetic methods of analysis are those in which we determine analyte content by measuring its effect on the rate of a chemical reaction. The analyte can have one of two different chemical actions:

1. The analyte reacts with the reagent and is transformed in the reaction. This is called the direct method.
2. The analyte acts as a catalyst for a reaction. This is called the catalytic method.

Experimentally, there is little difference between these methods. The rates of reaction must be determined under a set of standard, reproducible conditions. The rates are determined by monitoring the change in the reagent or product concentration. The results are then related to the initial concentration of the assayed material.

The great majority of kinetic assays are run in solution. If the samples are solids, the concentration of the analyte in solution must be relatable to its mass in the solid.

16.2 Why Kinetic Methods?

Precise timing is important in kinetic assays. So why do we bother doing them? Why not use an equilibrium assay method which can be left for a few extra seconds or minutes while we answer the phone?

There are four possible reasons for developing kinetic assays.

1. Kinetic assays can be extraordinarily specific. As a result, most of the other components that remain after digestion of a sample may not interfere enough to require extra steps in the sample preparation. For example, enzymes accelerate reactions for a narrow range of compounds, which means that one (or at most, a few) of possibly thousands of components can be determined selectively. Enzymatic assays often can be run without sample pretreatment, especially for blood serum and urine: Just add reagents, stir, and measure, or, perhaps, dip in a test strip and check the intensity of the color.
2. Kinetic assays also can be quite sensitive, as seen in the ACA on the next page. Each enzyme molecule catalyzes the production of a large number of colored product molecules. Many other catalytic kinetic assays can amplify the effect of the analyte and be useful for trace analysis.
3. When the rate of a reaction is so slow that waiting for equilibrium to be reached would be impractical, the rate of reaction itself can be observed relatively quickly and used for a kinetic assay.
4. A useful assay reaction might be irreversible; equilibrium cannot be attained. Nevertheless, the rate of reaction can be observed to quantify the analyte.

Kinetic assays comprise many of the colorimetric reactions of test strips such as pictured in EP.5 inside the front cover.

Incidentally, all chemical and biochemical methods that use enzymes as reagents are kinetic methods. As a result, in numbers of individual determinations, the majority of all chemical and instrumental analyses are carried out using kinetic methods. This is due, in large part, to the millions of clinical analyses done by kinetic means every day throughout the world.

ANALYTICAL CHEMISTRY IN ACTION
To Be More Specific. . .

Enzyme-Linked Immunosorbent Assays (ELISAs) are vitally important in analyzing a large number of trace and ultra-trace biological compounds in clinical medicine. As you can tell from the presence of "enzyme" in the name, ELISAs are kinetic methods using the chemical specificity of enzymes to catalyze one (or a very few) reactions in a complex matrix. The product of the enzyme reaction is a colored material.

The quantity of enzyme present is determined by the competitive binding with a specific antibody to the material being assayed. That is the *enzyme-linked* part of the name. At least one of the reagents of the assay is bound to a fixed matrix. That is the *immunosorbent* part of the name. The fixed matrix is now likely to be the surface of one of a set of small wells, each a few mm across and a few mm deep, in a clear plastic plate: the **microtiter plate**. When the color is developed, its intensity can be measured by the absorption of light passed through the plate at each well. A series of standards is usually run with each set, and a set commonly consists of 96 samples and standards on a plate—one in each well. The reason an immunosorbent surface is used is to allow easy washing of the adsorbed material to separate reagents after a step of the assay is done. The steps of an ELISA assay are illustrated in the figure on the opposite page and outlined in the caption.

While the great majority of ELISAs still are used in biomedical work, an increasing area of use is for general trace analysis. For instance, an ELISA assay has been developed for Clomazone, the active ingredient in a herbicide used to control grass and broadleaf weeds among soybeans. If Clomazone persists in the soil longer than expected because of differences in soil type, pH, rainfall levels, temperature, or misapplication, then corn planted later in the same field could be damaged. So an assay that could be used at the field would be desirable. The specificity of an immunoassay means that extensive sample preparation could be reduced significantly. Such an assay was produced, and the results were reported in the reference noted below. However, the assay's utility was limited for two reasons: the different varieties of corn differ greatly in their sensitivities to the herbicide, and both the bioavailability and recovery of the analyte vary greatly depending on the type of soil.

Clomazone

[Ref: Dargar, R. V., Tymonko, J. M., Van Der Werf, P. 1990. "An Enzyme-Linked Immunosorbent Assay for Clomazone Herbicide." In *Immunochemical Methods for Environmental Analysis*. J. M. Van Emon and R. O. Mumma, eds. Washington, D.C.: American Chemical Society. pp. 170–179.]

16.3 A Brief Review of the Mathematics of Rates of Reactions, Decay Times, and Half-Lives

A number of simple chemical processes can be described with the following algebraic equation:

$$\frac{\Delta[X]}{\Delta t} = -k[X] \qquad \textbf{(16-1)}$$

where $\Delta[X]$ is the change in [X] that occurs in the time interval Δt, and k is a proportionality constant. Here k has the units time^{-1}. The equation states that the change in concentration $\Delta[X]$ of some substance X over the time interval Δt is proportional to the amount of X present. The minus sign in the equation

To Be More Specific. . . (*continued*)

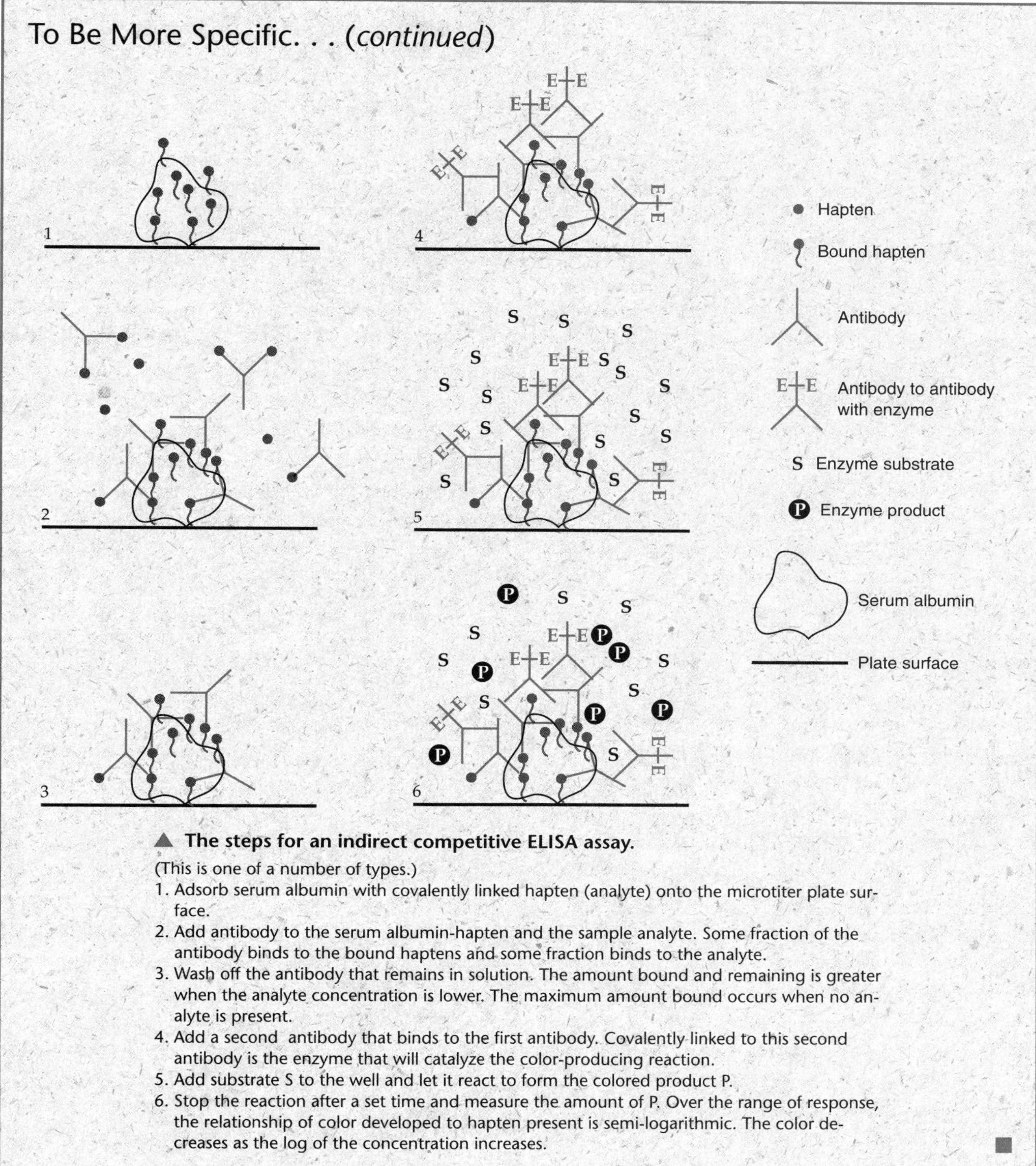

▲ **The steps for an indirect competitive ELISA assay.**

(This is one of a number of types.)

1. Adsorb serum albumin with covalently linked hapten (analyte) onto the microtiter plate surface.
2. Add antibody to the serum albumin-hapten and the sample analyte. Some fraction of the antibody binds to the bound haptens and some fraction binds to the analyte.
3. Wash off the antibody that remains in solution. The amount bound and remaining is greater when the analyte concentration is lower. The maximum amount bound occurs when no analyte is present.
4. Add a second antibody that binds to the first antibody. Covalently linked to this second antibody is the enzyme that will catalyze the color-producing reaction.
5. Add substrate S to the well and let it react to form the colored product P.
6. Stop the reaction after a set time and measure the amount of P. Over the range of response, the relationship of color developed to hapten present is semi-logarithmic. The color decreases as the log of the concentration increases. ■

indicates that [X] decreases with time. This type of behavior is called **exponential decay**. By inspecting Equation 16-1, we can infer that if k is large, then [X] changes relatively rapidly. If k is small, then [X] changes relatively slowly.

It is easier mathematically to manipulate Equation 16-1 with time intervals that are infinitesimally small. This is signified by putting dt in place of Δt and $d[X]$ in place of $\Delta[X]$. With this substitution, the equation becomes

$$\frac{d[X]}{dt} = -k[X] \tag{16-2a}$$

Equation 16-2a describes experimentally measurable behavior. The data consist of measurements of [X] at various times, and from the data, the characteristic value k can be calculated in two different ways.

The first procedure is illustrated in Figure 16.1. It requires measuring the change in [X] as a function of time—that is, $d[X]/dt$—together with the value of [X] at that time, $[X]_t$. From now on, when the concentration is written as [X], it should be understood to be $[X]_t$. Occasionally, the subscript will be used when it adds clarity. Algebraically, we can rearrange Equation 16-2a to obtain

$$\frac{d[X]/dt}{[X]} = -k \tag{16-2b}$$

The value of $d[X]/dt$ in the equation is the slope of a tangent at the point $[X]_t$ on the graph. However, even though $[X]_t$ can be found relatively accurately, a major problem follows because it is often difficult to measure the slope precisely.

An alternate, and inherently far more accurate, method requires measuring $[X]_t$ at a series of times and calculating k from all the data together. To do so we need an equation that relates $[X]_t$ directly to k. Equation 16-3 satisfies these criteria and is written without proof:

$$[X]_t = [X]_0 e^{-kt} \tag{16-3}$$

Equation 16-3 describes the results of an exponential decay in which the *first* measurement of $[X]_t$ is made at a time that we shall call $t = 0$. Call the value of $[X]_t$ at this first measurement $[X]_0$. The t indicates a time t after $t = 0$. (As you know, e is the natural logarithm base.) A graph of the function [X] *vs.* t is plotted in Figure 16.2.

However, the value of k determined from the graph in Figure 16.2 is no more accurate than the slope determined under these conditions; an alter-

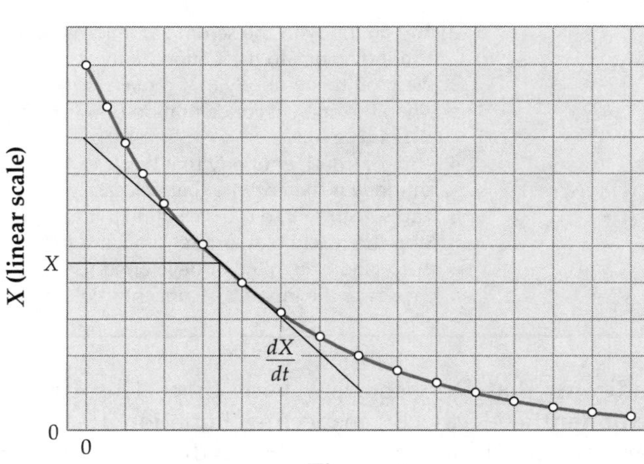

FIGURE 16.1 ▶
Exponential decay.
One way to determine k is to measure $d[X]/dt$ together with [X] at a point on the data curve. This technique tends to be less accurate since little of the data is used.

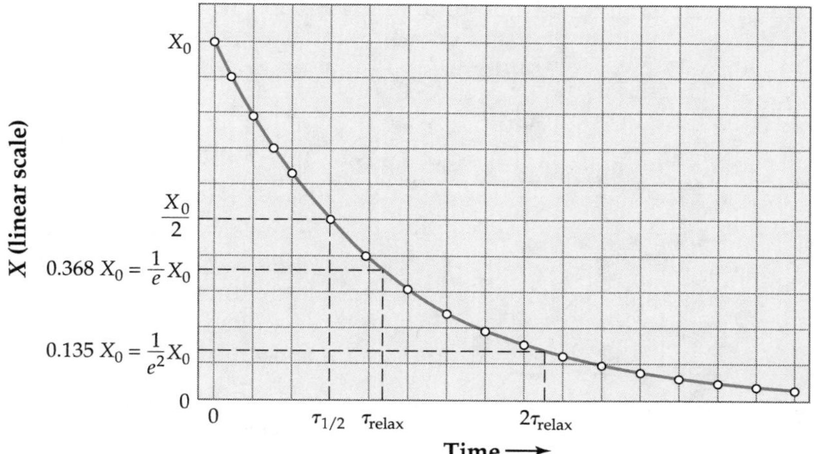

◀ FIGURE 16.2
[X] in units of $[X]_0$ *versus* time.
$[X]_0$ is the value of [X] at $t = 0$. The half-time for the reaction is also shown.

nate form of Equation 16-3 is far superior. This alternate form is obtained by rearranging the equation to give

$$\frac{[X]}{[X]_0} = e^{-kt}$$

and then taking the logarithm of both sides.

$$\ln \frac{[X]}{[X]_0} = -kt \tag{16-4}$$

Equation 16-4 says that if the natural logarithm of the ratio $([X]/[X]_0)$ is plotted vs. time, the slope will be $-k$. A simple way to obtain such a plot is to create a semilogarithmic graph of $([X]/[X]_0)$ vs. t. Such a plot appears in Figure 16.3.

If the base-10 logarithm is used, the slope is less than the base-e logarithm by a factor of 2.303. In other words, Equation 16-4 becomes

$$\log \frac{[X]}{[X]_0} = \frac{-kt}{2.303} = -k't$$

EXAMPLE 16.1

The value of k' from a plot of log [X] vs. time is $0.283\ s^{-1}$. What is the value of k that produces the same plot for ln [X] vs. time?

Solution

$$k = 2.303\ k'$$

and

$$k = 2.303 \times 0.283 = 0.652\ s^{-1}.$$

As the size of [X] decreases from $[X]_0$, there will be a time when

$$[X] = \frac{1}{2}[X]_0 \tag{16-5}$$

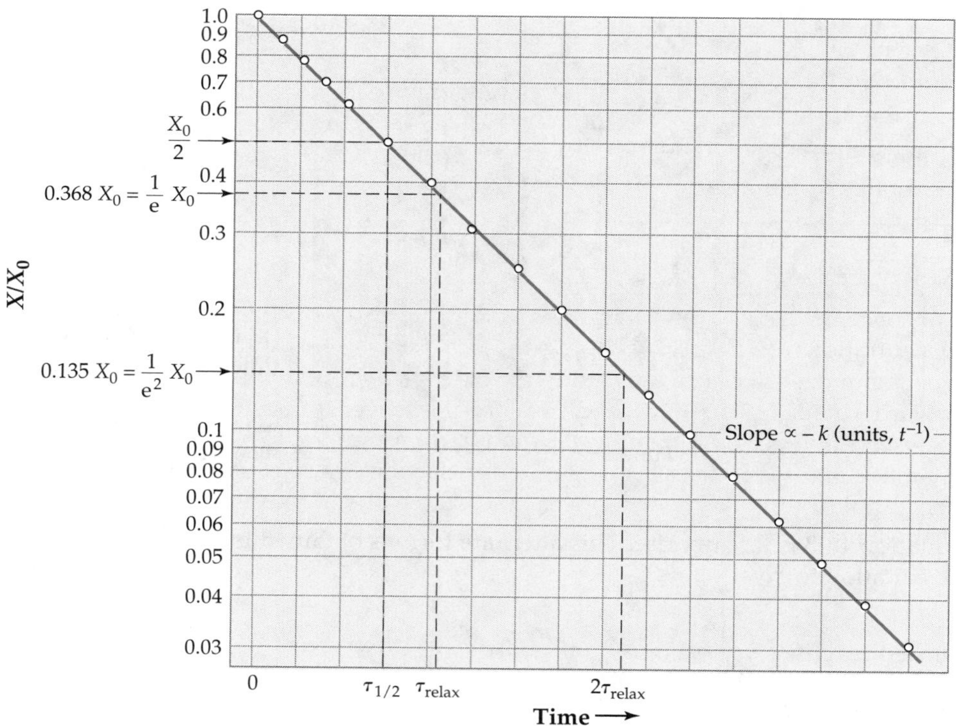

FIGURE 16.3 ▲
A semilog plot of $[X]/[X]_0$ *versus* time.

[X] has the value $[X]_0$ at $t = 0$. The value of $[X]/[X]_0$ is on the left scale. The slope of the plot is proportional to $-k$. The value of k is the characteristic rate constant for the process of exponential decay. The same value of k would be obtained from the same type of plot of [X] alone. The locations of the half-time, the relaxation time, and twice the relaxation time are shown.

This special time is called the **half-time** or **half-life**, abbreviated $t_{1/2}$ or $\tau_{1/2}$. Using $t_{1/2}$, we write

$$[X]_{t_{1/2}} = \frac{1}{2}[X]_0 \quad \textbf{(16-6)}$$

by definition. The value of $t_{1/2}$ is related to k in the following way. Let us substitute Equation 16-5 into 16-4. Then

$$\ln \frac{[X]_{t_{1/2}}}{[X]_0} = \ln \frac{1}{2} = -kt_{1/2}$$

Since $\ln \frac{1}{2} = -\ln 2$, k can be found by rearrangement.

$$k = \frac{\ln 2}{t_{1/2}} = \frac{0.693}{t_{1/2}} \quad \textbf{(16-7)}$$

16.4 Assay Kinetics and Assay Types

For many kinetic methods, we utilize reactions that can be described by

$$A + R \underset{k_r}{\overset{k_f}{\rightleftharpoons}} P \quad \textbf{(16-8)}$$

where

A is a reactant (it may be the material to be assayed),
R is an added reagent,
P is the reaction product,
k_f is the forward rate constant, and
k_r is the reverse rate constant.

All the reactions are started at zero time ($t = 0$). The symbols $[A]_0$, $[R]_0$, and $[P]_0$ are the initial concentrations of A, R, and P. The symbols $[A]_t$ or [A], etc. represent the concentrations of the species at any time t that comes after $t = 0$.

The kinetic behavior of such a system is more complicated than desirable for a straightforward kinetic assay. But, we can choose the reagent concentrations and the observation period that make the quantitation simpler. How these choices are made and their effects are described in Section 16A. Let us assume that such simplification has been made by the choice of conditions. The reactions that we monitor then *appear* to be of the type

$$A \xrightarrow{k} P \tag{16-9}$$

This is a **first-order reaction**[1] which is not reversible. The experimentally observed kinetics for this reaction can be described by

$$\text{rate} = \frac{d[A]}{dt} = -k[A] \tag{16-10}$$

The negative sign indicates that A is disappearing.

When the reaction's stoichiometry is that shown in Equation 16-9, the rate described in Equation 16-10 can also be written as a function of the reaction's product:

$$\text{rate} = \frac{d[P]_t}{dt} = k[A]_t \tag{16-11}$$

As Equations 16-10 and 16-11 show, we can measure either the rate of disappearance of the reactant A or the rate of appearance of the product P since

---

[1]For example, suppose the rate of product production is given by

$$\frac{d[P]}{dt} = k[A]^n[B]^m$$

The exponents n and m are generally integers or half-integers. Here, n is the order of the reaction with respect to A, and m is the order with respect to B. If $n = 1$, the reaction is said to be first order with respect to A; if $n = 2$, it is second order with respect to A, and so forth. The sum $(n + m)$ is the overall **order of the reaction.** These are empirical values and often cannot be predicted from the stoichiometric equation for the reaction. For example, the apparently analogous gas-phase reactions

$$H_2 + I_2 \rightleftharpoons 2\,HI \quad \text{and} \quad H_2 + Br_2 \rightleftharpoons 2\,HBr$$

have the respective rates of formation given by

$$\frac{-d[H_2]}{dt} = k[H_2][I_2] \quad \text{and} \quad \frac{-d[H_2]}{dt} = k'[H_2][Br_2]^{1/2}$$

Rosalyn Yalow, inventor of the radioimmunoassay. The method evolved into ELISA.

both measures contain the same information. For the experiment we simply choose the one that is easier to measure. In the following discussion, let us assume that the concentration of the product of the reaction is monitored.

The Catalytic Method

When the analyte acts as a catalyst, its effect is to change the rate constant k of the reaction. Therefore, for the catalytic method the rate measure can be used directly, as indicated in Equations 16-3 and 16-4. The analyte concentration can be related to the rate through the use of standards. From the working curve (rate vs. concentration) the unknown catalyst concentrations can be determined. If the assay is well-designed (and the chemistry cooperative), the rate $d[P]_t/dt$ and rate constant k are linearly related to the catalyst concentration. The changes in rates are ascertained by measuring the changes in the amount of product after a fixed time interval. The analyte's concentration is directly proportional to the amount of product.

The Direct Method

The direct method is not quite as straightforward. For the direct method, k is a constant, and the rate varies with [A]. Therefore, as the reaction progresses and the reactant concentration decreases, the rate changes. So we must find the amount of analyte present before any reaction occurred; we seek $[A]_0$. Somehow, the measurement(s) must allow us to determine that original concentration.

Because of this behavior, Equations 16-3 and 16-4 are not adequate for the direct method. In order to find an equation for $[A]_0$, one approach is to solve the differential equation 16-11. The solution is

$$[P]_t = [A]_0(1 - e^{-kt}) \tag{16-12}$$

As you see, the equation relates the amount of product to $[A]_0$.

Alternately, $[A]_0$ can be found from the *rate* of the reaction. To find the applicable equation, we substitute $[A]_0\,e^{-kt}$ for $[A]_t$ in Equation 16-11 to get

$$\frac{d[P]_t}{dt} = k[A]_0 e^{-kt} \tag{16-13}$$

Using either Equation 16-12 or 16-13, we can calculate $[A]_0$ from $[P]_t$ and t, because k is known from calibration runs.

16.5 Methods for Determining $[A]_0$ or $k_{catalytic}$ from Rates

Assume that you have made a choice of reagent concentrations so that the kinetics of the reaction act as described by Equations 16-3 and 16-4. Then, the complete set of data from an assay might appear as shown in Figure 16.4. These data might be recorded directly on a chart as the instrument response *vs.* time. Alternatively, the data might be plotted with a graphics program from kinetic data measured individually at a number of sequential times. As will be shown below, only a few data points are needed to determine $[A]_0$ from the assay.

The data that are collected can be related to the concentration of the material to be assayed in three different ways. Each of them has advantages in different circumstances. These are the **derivative method**, the **fixed-time method**, and the **variable-time method**, which are illustrated in Figures 16.4 through 16.6, respectively.

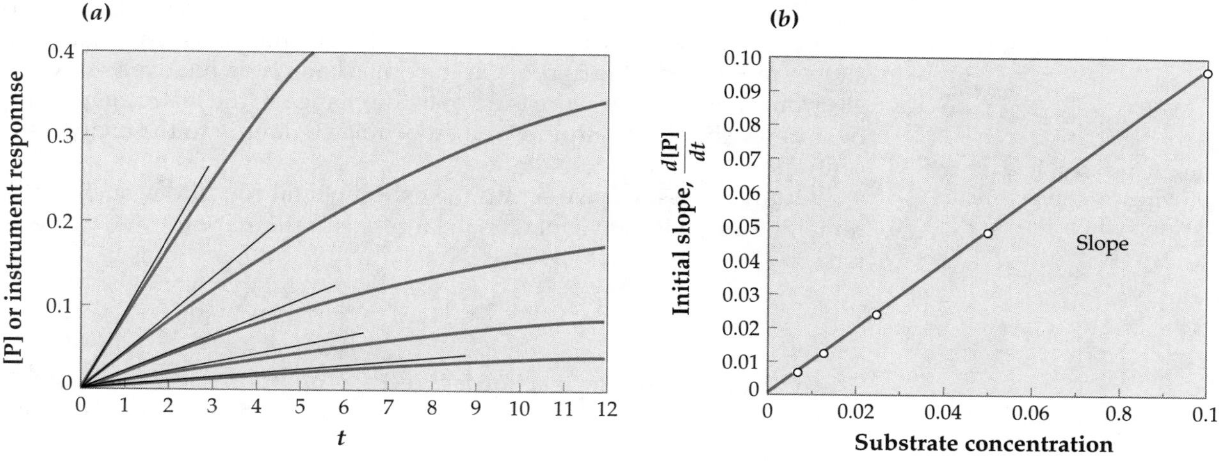

FIGURE 16.4 ▲
The derivative method for a direct kinetic assay.
(a) A series of kinetic runs were carried out with initial analyte concentrations $[A]_0$ of 0.100, 0.050, 0.0250, 0.0125, and 0.0063 in arbitrary units. The time units are arbitrary as well. The tangents to these curves at $t = 0$ are drawn. In **(b)**, the values of the tangent slopes are plotted *vs.* the substrate concentration. The working curve is a straight line. Different experimental methods would be used to obtain the data if the time units were milliseconds to seconds, as opposed to minutes or hours.

When using these methods, the experimental measurement for a direct assay and a catalytic assay are the same. However, the experimental conditions are somewhat different. A direct kinetic assay requires conditions such that the rate constant k remains the same for the standards and unknown.

To obtain simple kinetic behavior in a catalytic assay, it is best to have the reactant concentration(s) remain constant, which can be done by having $[A]_0$ in large excess (as well as by keeping all other conditions fixed). More details about the experimental conditions that simplify the kinetic behavior are to be found in Section 16A.

The Derivative Method

Direct Assay The derivative method involves measuring the analyte concentration from the slope of the concentration-versus-time curve, as illustrated in Figure 16.4. As expressed by Equation 16-13, the slope of the curve is proportional to the value of $[A]_0$. This link between the slope and $[A]_0$ can, perhaps, be seen more clearly if we rearrange Equation 16-13 to

$$[A]_0 = \frac{d[P]_t}{dt} \cdot \frac{1}{k} e^{kt} \tag{16-14}$$

Notice that all tangents to the curves in Figure 16.4 are drawn to reflect the slopes of the curves at the beginning of the run. According to Equation 16-13 or 16-14, a tangent drawn at any other single time works as well. How can this be? The reasons can be seen by studying Equation 16-14. With the sample and standards run under exactly the same conditions, the value of k is fixed for all runs. As a result, the factor $(1/k)$ is a constant. Again, since k is constant, at any single test time, the factor e^{kt} will be the same for all the samples. Thus, the slopes $(d[P]_t/dt)$ are still directly proportional to $[A]_0$, and the assay method still works.

The derivative method for determining analyte concentrations is commonly used. The experiment can be run, plotted, and analyzed in the manner shown in Figures 16.4a and b. On the other hand, it is relatively easy to use electronic means to measure the slope: the change in the instrument's output over time. The slope output can then be related simply to the analyte's concentration.

Having A in excess and other simplifying conditions are described in Section 16A.

Catalytic Assay If we set up the experimental conditions such that $[A]_0$ remains essentially constant (A is in large excess), Equation 16-11 can be written

$$\frac{d[P]_t}{dt} = k[A]_0$$

Therefore, the rate of the reaction is directly proportional to the rate constant. Assuming a linear relationship between k and the analyte (catalyst) concentration, we can write

$$\frac{d[P]_t}{dt} = [\text{catalyst}] \times \text{constant concentration}$$

The initial slopes of the kinetic plots are proportional to the analyte concentrations.

Fixed-Time Method

Direct Assay Equation 16-12 shows the relationship of P_t, the concentration of product P at a time t, to $[A]_0$, the initial concentration of A. It is useful to ask, What sort of relationship between $[P]_t$ and $[A]_0$ can we find if we consider the concentrations at two different times? Let us call the times t_1 and t_2. The concentration of P at time t_1 is

$$[P]_{t_1} = [A]_0(1 - e^{-kt_1}) \quad \textbf{(16-15)}$$

whereas at t_2 this is

$$[P]_{t_2} = [A]_0(1 - e^{-kt_2}) \quad \textbf{(16-16)}$$

Now, subtract Equation 16-15, which describes the concentration of product at t_1, from Equation 16-16, which describes the concentration at t_2. Then,

$$\Delta P = [P]_{t_2} - [P]_{t_1} = [A]_0(e^{-kt_1} - e^{-kt_2}) \quad \textbf{(16-17)}$$

But what happens if the times t_1 and t_2 are kept fixed throughout the series of experiments in the analysis? (Hence the name *fixed-time method*.) In that case, the quantity in parentheses in Equation 16-17—the difference between the two exponential functions—remains at a constant value. So with a fixed-time interval, Equation 16-17 becomes quite simple.

$$\Delta P = [A]_0 \times \text{constant} \quad ; \quad (\text{fixed time interval}) \quad \textbf{(16-18)}$$

The constant of Equation 16-18 can be determined by running standards, and the unknowns are then run with the same fixed-time range. The change in the concentration of P over that time range is directly proportional to $[A]_0$, the initial amount of A, which is what we set out to find. This whole process is described pictorially in Figure 16.5.

Catalytic Assay When experimental conditions are established such that $[A]_0$ remains essentially constant (in large excess), the product produced at a fixed time t for two different rates k_1 and k_2 is (from Equation 16-11)

$$[P]_{k_1} = k_1[A]_0 t \quad \text{and} \quad [P]_{k_2} = k_2[A]_0 t$$

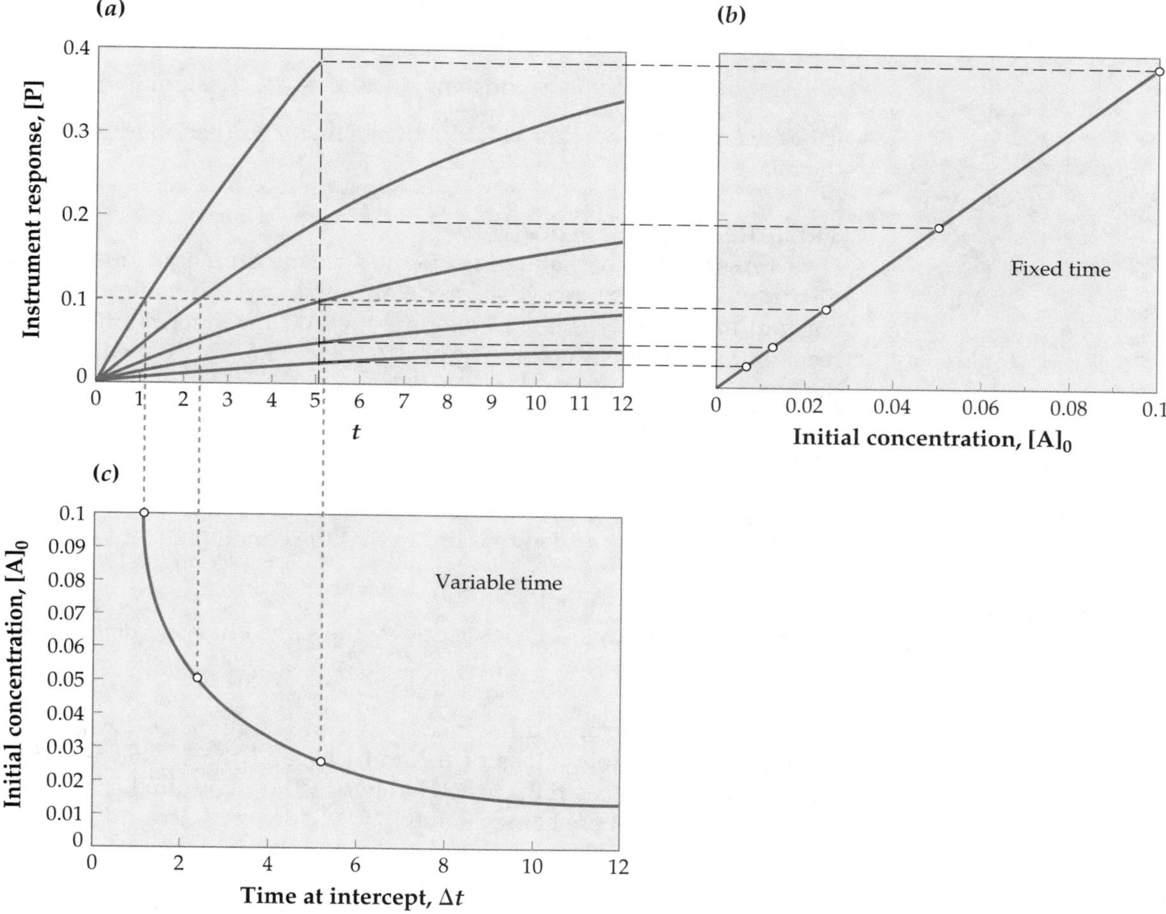

FIGURE 16.5 ▲

Fixed-time and variable-time method.

The fixed-time method (a and b) and variable-time method (a and c and Figure 16.6) for a direct kinetic assay are analyzed pictorially here. The kinetic runs are the same as in Figure 16.4a. For the fixed-time method, the concentrations are measured at a single time. Only one concentration measurement is needed for each run. Here, the concentrations are measured at five time units (the vertical line). The points are projected to **(b)** where the values proportional to [P] are plotted *vs.* the value of $[A]_0$. This working curve is a straight line through zero. For the variable-time method, the time required to reach a fixed concentration is measured. The horizontal line at 0.1 indicates the chosen, fixed concentration. However, measurement at only one time point is necessary for each run. These times are projected to **(c)** a plot of $[A]_0$ *vs.* the times. The working curve is not linear since linearity is expected only for a plot of $1/\Delta t$. A $1/\Delta t$ plot is shown in Figure 16.6.

These mathematical expressions are connected to the catalytic assay since the rates are proportional to the amount of catalyst present, and the analyte is the catalyst, not A.

In the assay, we take the amount of product formed and relate it to the analyte concentration. The difference of the amount of product is

$$\Delta P = [P]_{k_1} - [P]_{k_2} = (k_1 - k_2)[A]_0 t \quad \text{or} \quad \Delta P = (k_1 - k_2) \times \text{constant}$$

One of these two runs can be a blank. An ideal blank will have no catalytic activity, and, algebraically, having such a blank means that $k_2 = 0$. For a zero blank sample, the catalytic assay is, then, described by an equation similar to Equation 16-18. The appropriate relationship is

$$\Delta \mathrm{P} = k_1 \times \text{constant} = [\text{catalyst}] \times \text{constant}'$$

If $k_2 \neq 0$ (such as a non-zero blank), then a linear correction must be made to the result.

Variable-Time Method

Direct Assay In the variable-time method we measure how long the reaction mixture requires to reach a predetermined, *fixed concentration* level of reactant or product. The variable-time method for determining changes in reaction rates is illustrated in Figures 16.5 and 16.6. To see the limitations on the technique, let us look at its algebra.

Assume that we want to analyze two samples that contain different concentrations of analyte. The reactions for both are started at the same time t_0. To describe the changes in $[\mathrm{P}]_t$ for the two samples, we can write two equations. The reaction of sample 1 is described by Equation 16-17. However, an extra subscript is added to indicate that the equation is for sample 1.

$$\Delta[\mathrm{P}]_1 = [\mathrm{A}]_{01}\,(e^{-kt_0} - e^{-kt_1}) \qquad \textbf{(16-19)}$$

The reaction of sample 2 is described by the same kind of equation.

$$\Delta[\mathrm{P}]_2 = [\mathrm{A}]_{02}\,(e^{-kt_0} - e^{-kt_2}) \qquad \textbf{(16-20)}$$

As you can see, the only differences between Equations 16-19 and 16-20 are the subscripts on the values of $\Delta\mathrm{P}$ and $[\mathrm{A}]_0$ and the second exponentials' times. The differences reflect the changes in the two initial concentrations, $[\mathrm{A}]_{01}$ and $[\mathrm{A}]_{02}$, that are being assayed.

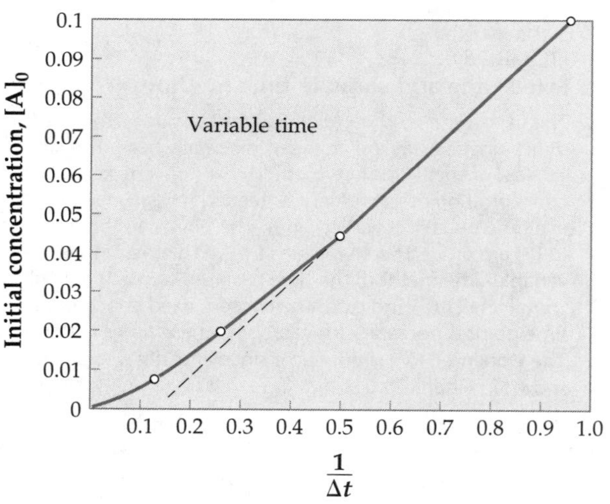

FIGURE 16.6 ▲

The variable-time method (continued from Figure 16.5).

The graph is of $[\mathrm{A}]_0$ (the initial concentration of analyte) *vs.* $1/\Delta t$ (the inverse of the time required to reach a fixed concentration of P). Notice that the working curve is not linear nearer zero because the approximations made for the method no longer hold. Linearity is not expected if the reaction proceeds for too long a time when the solutions contain low initial concentrations of analyte.

We now run into a problem: These equations describe quite complicated behavior. For an assay, we would like to have simpler behavior, because the precision of the assay tends to be better. It turns out that the behavior can be made quite simple if we measure the product concentrations only near the beginning of the reaction (when less than 1% to 2% of the reagent A has reacted). Mathematically, this simplification can be expressed by an approximation given by the series,

$$e^{-kt} \approx 1 - kt + \tfrac{1}{2}k^2t^2 + \cdots$$

For values of kt that are small (this is where the reaction of only 1% to 2% of the reagent comes in), we can write

$$e^{-kt} \approx 1 - kt \tag{16-21}$$

Now, substitute Equation 16-21 into Equation 16-19:

$$\Delta[P]_1 = [A]_{01}\,(1 - kt_0 - 1 + kt_1)$$

which can be simplified to

$$\Delta[P]_1 = [A]_{01}\,k(t_1 - t_0) \tag{16-22}$$

Now, define

$$\Delta t_1 = (t_1 - t_0)$$

and similarly define Δt_2 for the other sample. Thus, for the two samples,

$$\Delta[P]_1 = [A]_{01} \cdot k \cdot \Delta t_1 \tag{16-23}$$

and

$$\Delta[P]_2 = [A]_{02} \cdot k \cdot \Delta t_2 \tag{16-24}$$

Now we see the elegance of the technique (and the analogous algebra). To compute the reaction rates, we choose the time when the *same fixed amount* of product, ΔP, is produced by *both* reactions. Having the same amount of ΔP means that

$$\Delta[P]_1 = \Delta[P]_2 = \Delta P$$

But if $\Delta[P]_1 = \Delta[P]_2$, that means Equations 16-23 and 16-24 can be combined to obtain

$$[A]_{01} \cdot \Delta t_1 = [A]_{02} \cdot \Delta t_2 \tag{16-25}$$

Equation 16-25 says that the initial concentrations are *inversely* related to the time needed to reach the fixed concentration when the reaction is not allowed to proceed too far. Now, let us consider that $[A]_{01}$ is the standard that produces ΔP in time Δt_1. Then, $[A]_{01}\,\Delta t_1$ is constant. So, we can state the analytical relationship algebraically as

$$[A]_{02} = \text{constant} \cdot \frac{1}{\Delta t_2} \tag{16-26}$$

This relation is shown graphically in Figure 16.6.

Catalytic Assay To carry out a variable-time catalytic assay, $[A]_0$ is again held *effectively* constant by keeping its original concentration high. For a catalyst, Equations 16-23 and 16-24 can be restated to describe the assay by using a constant k instead of a constant $[A]_0$. That is,

$$\Delta[P]_1 = [A]_0 k_1\,\Delta t_1 \quad \text{and} \quad \Delta[P]_2 = [A]_0 k_2\,\Delta t_2$$

Table 16.1 Summary of Kinetic Methods Equations

	Direct (k constant)	Catalytic ($[A]_0 \approx$ constant)
Derivative method	$[A]_0 = \text{constant} \times d[P]/dt$	$(d[P]/dt)_{\text{fixed time}} = \text{constant} \times [\text{catalyst}]$
Fixed time method	$[A]_0 = \Delta[P] \times \text{constant}$	$\Delta P = \text{constant} \times [\text{catalyst}]$
Variable time method	$[A]_0 = \text{constant} \times (1/\Delta t)$	$[\text{catalyst}] = \text{constant} \times (1/\Delta t)$

An equation analogous to Equation 16-25 reflects the fact that a catalytic assay is being run: the rate constant changes with the catalyst's concentration.

$$k_1\,\Delta t_1 = k_2\,\Delta t_2$$

That is, an inverse relationship between k_i and t_i is seen. A linear relationship exists between the analyte (catalyst) concentration and the rate. By calibration to Δt_1, the analytical relationship becomes

$$[\text{catalyst analyte}] = \text{constant} \cdot \frac{1}{\Delta t_2}$$

in parallel with Equation 16-26.

A summary of the working equations for these six different kinetic assay forms is shown in Table 16.1.

16A A Deeper Look: Obtaining Simple Kinetic Behavior

In this section we illustrate how reaction conditions are chosen such that reactions of the type

$$\mathrm{A} + \mathrm{R} \underset{k_r}{\overset{k_f}{\rightleftharpoons}} \mathrm{P} \qquad \textbf{(16A-1)}$$

can exhibit simpler kinetic behavior. The change with time of the concentrations of the species present can be expressed by three equations. In this case, all three contain the same information: The right-hand sides are the same.

$$\frac{-d[\mathrm{A}]_t}{dt} = k_f[\mathrm{A}]_t[\mathrm{R}]_t - k_r[\mathrm{P}]_t \qquad \textbf{(16A-2)}$$

$$\frac{-d[\mathrm{R}]_t}{dt} = k_f[\mathrm{A}]_t[\mathrm{R}]_t - k_r[\mathrm{P}]_t \qquad \textbf{(16A-3)}$$

$$\frac{d[\mathrm{P}]_t}{dt} = k_f[\mathrm{A}]_t[\mathrm{R}]_t - k_r[\mathrm{P}]_t \qquad \textbf{(16A-4)}$$

Differential equations such as these are mathematical models for the reaction kinetics. With this in mind, we shall not distinguish between the equation and the chemical reaction from now on. In addition, we drop the subscript t.

The general kinetic behavior of this reaction is complicated and is not particularly useful for straightforward assays. However, the reactions can usually be run under special conditions in which the kinetic behavior is simpler and more useful. One of the greatest simplifications occurs when the reverse reaction does not occur, P $\not\rightarrow$ R + A. Thus, assay reactions that have slow or effectively zero reverse rates are chosen. If this is not true in actual assays, then corrections for the reverse reaction must be made on the assay results.

When no reverse reaction occurs, then $k_r = 0$. Equation 16A-1 becomes

$$A + R \xrightarrow{k_f} P$$

and the behavior of the reaction is described correctly by

$$\frac{-d[A]}{dt} = k_f[A]_t[R]_t \qquad \textbf{(16A-5)}$$

This is Equation 16A-2 written without the second term. Similarly modified, Equations 16A-3 and 16A-4 have the same form.

Even with such simplifications, the behavior of this system over time still can be more complicated than desired for a straightforward analysis, a point that is illustrated in Figure 16A.1. Plotted in Figure 16A.1a are the values of [A], [R], and [P] that would be measured under the specific conditions cited. In Figure 16A.1b are *changes* in concentration over time ($d[A]/dt$, and so on). The data of Figure 16A.1 *can* be analyzed to find the analyte concentration

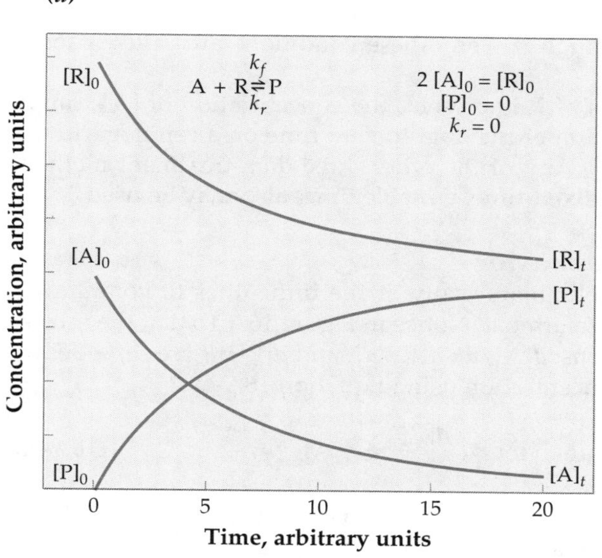

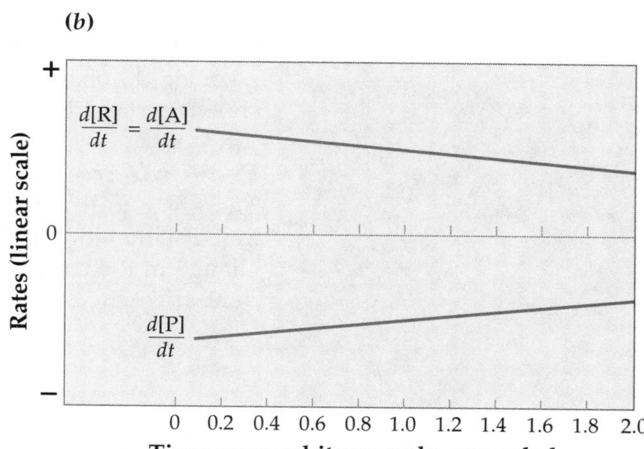

FIGURE 16A.1 ▲
Achieving zero-order conditions.
(a) Graph of the concentrations of A, R, and P for the reaction

$$A + R \rightarrow P$$

with initial concentrations 2 $[A]_0 = [R]_0$ and $[P]_0 = 0$. The units of both time and concentration are arbitrary. This graph illustrates the behavior of a relatively simple reaction system. Pseudo first-order conditions would require a far larger excess of R over A. **(b)** Graph of the slopes of [A], [R], and [P] with time (from figure a) versus an expanded time scale. Only at relatively short times will the system act as a pseudo zero-order system. A true zero-order rate system will exhibit fixed values of the derivatives.

changes. But to derive the desired concentration from these data is far more difficult than necessary even when no reverse reaction occurs. As will be shown, further simplifications arise by adjusting the concentrations of the reagents.

Pseudo First-Order Behavior

Note that the stoichiometry of reaction 16A-1 shows that one A and one R combine to form each product molecule. If we add a large excess of R (say a factor of 20) to the solution, then, while all of the A will be used up, only 5% of the R reacts. The value of [R] changes only slightly during the reaction. We may rewrite the equation for the change of P with time as

$$\frac{d[\mathrm{P}]}{dt} \approx k_{\mathrm{f}}[\mathrm{A}][\mathrm{R}]_0 \tag{16A-6a}$$

or

$$\frac{d[\mathrm{P}]}{dt} \approx k_{\mathrm{f}}'[\mathrm{A}] \tag{16A-6b}$$

where $k_{\mathrm{f}}' = k_{\mathrm{f}}[\mathrm{R}]_0$. These approximations can be made arbitrarily more precise by further increasing [R].

Since $d[\mathrm{A}]/dt = -d[\mathrm{P}]/dt$, the form of Equation 16A-6b is a first-order kinetic reaction like Equation 16-2. A semilogarithmic plot of [P] *vs.* time will be linear. As noted, this simple behavior results only under the special initial conditions: a large excess of the reagent R. Since the reaction is not truly first-order but only acts as though it were, these conditions are called **pseudo first-order** conditions.

When the large excess of R is present, the concentration of $[\mathrm{A}]_0$ can be found from the extrapolation of the data to zero time on a semilogarithmic plot: log[A] or log[P] *vs. t*. Any of the other three data-treatment methods shown earlier (derivative, fixed-time, variable time) also may be used.

Pseudo Zero-Order Behavior

There is another way to eliminate many of the difficulties of complicated kinetic behavior and data treatment. Notice in Figure 16A.1b that the rates of change in the concentrations, $d[\mathrm{A}]/dt$, $d[\mathrm{R}]/dt$, and $d[\mathrm{P}]/dt$ are time-dependent. If we measured a concentration using Equation 16A-6,

$$[\text{analyte}] \propto \frac{d[\mathrm{P}]}{dt} \approx k_{\mathrm{f}}'[\mathrm{A}] \tag{16A-6b}$$

over the course of the reaction, we would find a changing assay value. However, if we look for only a relatively short time, at the early part of the curve, the derivative plots are nearly constant. In the early-time region, the rate alone could be used to determine $[\mathrm{A}]_0$. The mathematical description of these conditions is obtained by modifying Equation 16A-6: substitute $[\mathrm{A}]_0$ for [A].

$$\frac{d[\mathrm{P}]}{dt} \approx k_{\mathrm{f}}[\mathrm{A}]_0[\mathrm{R}]_0 \tag{16A-7}$$

and

$$\frac{-d[\mathrm{R}]}{dt} \approx k_{\mathrm{f}}[\mathrm{A}]_0[\mathrm{R}]_0 \tag{16A-8}$$

Equations 16A-7 and 16A-8 say that the rate of change of [P] or [R] is directly proportional to the concentration $[A]_0$, the value we want to measure. As a result, the initial slope of the plot of *either* $d[P]/dt$ or $d[R]/dt$ can be used to determine $[A]_0$. Therefore, if we use only the initial times of the reaction, the simplified behavior allows a more straightforward approach to the measurement of $[A]_0$.

Note, peripherally, that the rate is proportional to $[R]_0$, not [R]. But $[R]_0$ is a constant; it does not change. As a result, by having a large excess of R and by measuring the generation of product for a short time, the rate of the reaction does not appear to depend on *either* [A] or [R]. These conditions are called **pseudo zero-order** conditions.

Suggestions for Further Reading

VAN EMON, J. M., MUMMA, R. O., eds. 1990. *Immunochemical Methods for Environmental Analysis*, Washington, D. C.: American Chemical Society.

This book was developed from a 1989 symposium. Chapter 11 by Hammock et al. is a good, general introductory chapter. The others are on specific topics full of experimental details.

MARK, H. B., Jr., RECHNITZ, G. A. 1968. *Kinetics in Analytical Chemistry*. New York: Interscience.

A comprehensive but readable work with numerous older examples.

MÜLLER, H. 1982. "Catalymetric Methods of Analysis." *CRC Critical Reviews in Analytical Chemistry*. 13:313–372.

An excellent general review of elemental determinations using contemporary kinetic methods. Highly useful discussions of drawbacks and numerous useful tables as entries to the literature.

GUILBAULT, G. G. 1978. In "Treatise on Analytical Chemistry." Pt. I, Vol. 1, 2nd ed. I. M. KOLTHOFF and P. J. ELVING, eds. New York: Wiley-Interscience. Chap. 11.

A concise review of some kinetic methods with references to examples.

MALMSTADT, H. V., DELANEY, C. J., CORDOS, E. A. 1972. "Reaction-Rate Methods of Chemical Analysis." *Critical Reviews in Analytical Chemistry*. 2:559–619.

MOTTOLA, H. A. 1974. "Catalytic and Differential Reaction-Rate Methods." *Critical Reviews in Analytical Chemistry*. 4:229–280.

Two reviews with numerous references to the literature. Little biochemistry in either.

BERGMEYER, H. U., GRASSL, M. 1983. *Methods of Enzymatic Analysis*. 3rd ed. Weinheim and Deerfield Beach, FL: Verlag Chemie.

An encyclopedic set of books of techniques for analyses of enzymes involving kinetic analyses.

Concept Review

1. What is the function of the analyte in a catalytic, as opposed to a direct, method?

2. Why might you choose to carry out a kinetic assay for some analyte instead of a some other (equilibrium) method?

Exercises

16.1 What method of measurement is used for Clomazone in the ACA in this chapter? (Derivative, variable-time, fixed-time?)

16.2 The enzyme nitrate reductase catalyzes the reduction of nitrate to nitrite. It is assayed by measuring the rate of production of nitrite. The assay reaction is carried out in the presence of a reducing agent, sodium dithionite, to remove oxygen. The reaction is initiated by mixing solutions of nitrate with the enzyme and stopped by bubbling the solution with air at the end of the 10.00-min reaction period. (This also removes the excess sodium dithionite.) The nitrite that has been produced is then reacted with color-producing reagents, and the absorbance at 540 nm is measured *vs.* a blank. The blank consists of an identical enzyme solution that has been treated in the same way except that instead of being mixed with the sodium nitrate solution, it has been mixed with an equal volume of water. The final assay solutions have volumes of 2.5 mL. The following data were found.

Solution	Instrument response
Blank	0.004
Blank + 20 $\mu M\ NO_2^-$	0.269
Blank + 40 $\mu M\ NO_2^-$	0.550
Blank + 60 $\mu M\ NO_2^-$	0.803
Blank + 80 $\mu M\ NO_2^-$	1.087
Blank + 100 $\mu M\ NO_2^-$	1.355
Unknown	0.664

The activity of the enzyme is measured in units. By definition, 1 unit of enzyme reduces 1 μmol of nitrate to nitrite min^{-1} at 30°C and pH 7.0.

(a) Plot the working curve. What is the sensitivity of the assay in instrument response *vs.* enzyme units of activity?

(b) The unknown solution was made from 0.0196 g of dry powder. What is the activity in units per gram dry weight?

(c) Assume that the highest purity enzyme obtained to date contained 32 units per gram dry weight and that the material was "pure." How pure is this preparation? That is, what fraction (w/w) of the unknown is "pure" enzyme relative to that purest material?

16.3 RNA polymerase II(B) cleaves DNA and RNA and is involved in RNA transcription processes. One unit of enzymatic activity causes incorporation of 10 pmol of uridine-5′-triphosphate (UTP) into a precipitatable product of denatured DNA in 15 min at 25°C. This incorporation is found by using radioactive UTP, [(^{3}H)UTP]. The standard [(^{3}H)UTP] produces 3.7×10^4 nuclear decays per second in 100 pmol of the compound. Thus, the number of radioactive decays counted is directly related to the amount of UTP that has become incorporated into the precipitated DNA up to the end of the reaction time. The progress of an enzyme purification was monitored with this kinetic method. Complete Table 16.3.1.

16.4 This exercise involves finding the concentration of nitrite in water. The method consists of injecting solutions of reagents and a sample together into a flowing stream in a carefully controlled manner. After injection, the two solutions automatically are mixed rapidly and thoroughly and pass into a detector. The flow is then stopped, and the progress of the reaction of nitrite in the sample is monitored. First, the nitrite reacts with an aromatic amine. The product of this first reaction then couples to a second aromatic compound to form a dye. In this case, the dye is reddish-purple. The resulting data—the time-dependence of the dye development—are shown in Figure 16.4.1. The curve labels are ppm nitrite in the samples for each. The absorbance scale is linear and directly proportional to the

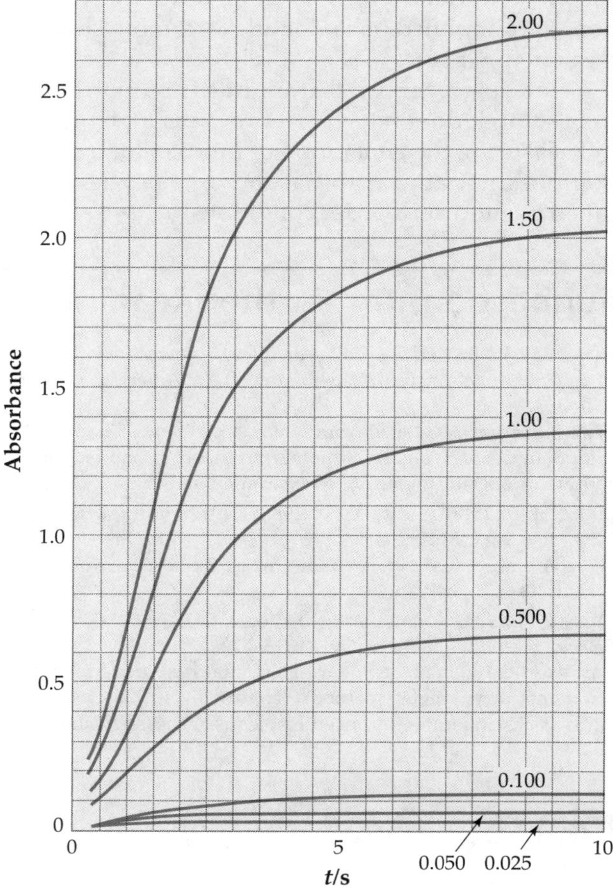

FIGURE 16.4.1 ▲

Table 16.3.1

	Purification Step				
	1	2	3	4	5
Total protein (mg)	191,000	6,830	1,480	92	30
Sample taken (mg)	20	0.20	0.10	0.05	0.01
Radioactivity counts/s above background	4,650	1,430	2,550	18,100	9,300
Total enzyme activity (units)	12,000	13,200	—	—	—
Enzyme activity (units/mg-protein)	0.063	—	—	—	—
n-fold purification	1	31	—	—	—
Extraction yield (%)	100	110	—	—	—

dye (and nitrite) concentration. [Figure 16.4.1 and method reproduced from Koupparis, M. A., et al. 1982. *Analyst* 107:1309.]

(a) Determine the calibration curve for nitrite with the derivative method. There appear to be some irregularities in the curves at early times. How do you overcome this problem?

(b) Determine the calibration curve for nitrite with the fixed-time method.

(c) For the fixed-time method, will your results tend to be more or less precise or have the same level of precision if the time chosen is 2 rather than 5 s?

(d) Determine the calibration curve for nitrite with the variable-time method. Consider carefully where you will fix the concentration line in this case.

(e) Is the fixed-time method or the variable-time method superior to the other?

(f) The following data in the table below were obtained from an unknown. Calculate the concentration (in ppm NO_2^-) using each of your three calibration plots.

Time (s)	Absorbance	Time (s)	Absorbance
0	0.0	4.0	1.36
0.5	0.31	5.0	1.50
1.0	0.43	6.0	1.57
1.5	0.66	7.0	1.60
2.0	0.90	8.0	1.62
2.5	1.07	9.0	1.63
3.0	1.20	10.0	1.64

16.5 What is $t_{1/2}$ for the nitrite reaction in Exercise 16.4?

Additional Exercises

16.6 Extensive myocardial infarction causes an increase in the creatine kinase activity in serum (normal levels 15–150 units L^{-1} in adult males). The standard variable-time catalytic assay for two patients was carried out with the following results.

Sample	Time to Reach Given Response (s)
Standard (120 unit L^{-1})	25
Patient 1	15
Patient 2	63

Calculate the creatine kinase activity for each patient.

16.7 [Combination kinetics–NMR exercise] The conversion of a fatty acid to the corresponding aldehyde was followed by monitoring the ^{13}C NMR of the reaction mixture. As the COOH carbon is converted to a CHO carbon the CHO resonance grows in and the COOH resonance decreases. The following data were obtained.

Time (min)	Integrated Intensity	
	COOH	CHO
0	100	0
1.0	80	36
2.0	64	65
4.0	41	106
8.0	17	149
16.0	3	175
32.0	1	178
64.0	0	180

(a) Plot the integrated intensities on the same graph (intensity vs. time).

(b) Plot the integrated intensities on the same graph (log intensity vs. time).

(c) Which of the two graphs gives a straight line? What does this tell you about how to treat the data?

16.8 Based on the first and last points acquired in Problem 16.7,

(a) What are relative responses for the COOH and the CHO carbons?

(b) If the original fatty acid concentration was 10 mM, what was the reaction rate in mM s^{-1}?

16.9 What is $t_{1/2}$ for the reaction in Exercise 16.7?

Appendices

Appendix I

Solubility Products

Unless noted, all values are extrapolated to ionic strength zero at 25°C. As the ionic strength approaches 0, $K_{thermodynamic}$ and $K_{concentration}$ values are equal.

Formula	pK_{sp}	K_{sp}	Temperature (°C)	Ionic Strength (M)
Bromides				
AgBr	12.30	5.0×10^{-13}		
CuBr	8.3	5×10^{-9}		
$HgBr_2$	18.9	1.3×10^{-19}		0.5
Hg_2Br_2*	22.25	5.6×10^{-23}		
TlBr	5.44	3.6×10^{-6}		
Carbonates				
Ag_2CO_3	11.09	8.1×10^{-12}		
$BaCO_3$	8.30	5.0×10^{-9}		
$CaCO_3$ (calcite)†	8.35	4.5×10^{-9}		
$CdCO_3$	13.74	1.8×10^{-14}		
$CoCO_3$	9.98	1.0×10^{-10}		
$CuCO_3$	9.63	2.3×10^{-10}		
$FeCO_3$	10.68	2.1×10^{-11}		
Hg_2CO_3*	16.05	8.9×10^{-17}		
$La_2(CO_3)_3$	33.4	4.0×10^{-34}		
$MgCO_3$	7.46	3.5×10^{-8}		
$MnCO_3$	9.30	5.0×10^{-10}		
$NiCO_3$	6.87	1.3×10^{-7}		
$PbCO_3$	13.13	7.4×10^{-14}		
$SrCO_3$	9.03	9.3×10^{-10}		
$Y_2(CO_3)_3$	30.6	2.5×10^{-31}		
$ZnCO_3$	10.00	1.0×10^{-10}		
Chlorides				
AgCl	9.77	1.7×10^{-10}		
CuCl	6.73	1.9×10^{-7}		
Hg_2Cl_2*	17.91	1.2×10^{-18}		
$PbCl_2$	4.78	1.7×10^{-5}		
TlCl	3.74	1.8×10^{-4}		
Chromates				
Ag_2CrO_4	11.89	1.3×10^{-12}		
$BaCrO_4$	9.67	2.1×10^{-10}		
$CuCrO_4$	5.44	3.6×10^{-6}		
Hg_2CrO_4	8.70	2.0×10^{-9}		
Tl_2CrO_4	12.01	9.8×10^{-13}		

*Ionizes or dissociates into Hg_2^{2+}.
†This is a specific solid-phase structure.

Formula	pK_{sp}	K_{sp}	Temperature (°C)	Ionic Strength (M)
Cyanides				
AgCN ($2AgCN \rightleftharpoons Ag^+ + Ag(CN)_2^-$)§	15.66	2.2×10^{-16}		
$Hg_2(CN)_2$*	39.3	5×10^{-40}		
$Zn(CN)_2$	15.5	3×10^{-16}		3
Fluorides				
BaF_2	5.76	1.7×10^{-6}		
CaF_2	10.41	3.9×10^{-11}		
MgF_2	8.18	6.6×10^{-9}		
SrF_2	8.54	2.9×10^{-9}		
PbF_2	7.44	3.6×10^{-8}		
ThF_4	28.3	5×10^{-29}		3
Hydroxides				
AgOH	8.17	6.8×10^{-9}		1
Ag_2O ($\rightleftharpoons 2\,Ag^+ + 2\,OH^-$)§	15.42	3.8×10^{-16}		
$Al(OH)_3(\alpha)$†	33.5	3×10^{-34}		
$Au(OH)_3$	5.5	3×10^{-6}		
$Ba(OH)_2 \cdot 8\,H_2O$	3.6	3×10^{-4}		
$Ca(OH)_2$	5.19	6.5×10^{-6}		
$Cd(OH)_2(\beta)$†	14.35	4.5×10^{-15}		
$Ce(OH)_3$	21.2	6×10^{-22}		
$Co(OH)_2$	14.9	1.3×10^{-15}		
$Co(OH)_3$	44.5	3×10^{-45}	19	
$Cr(OH)_3$	29.8	1.6×10^{-30}		0.1
$Cu(OH)_2$	19.32	4.8×10^{-20}		
Cu_2O ($\rightleftharpoons 2\,Cu^+ + 2\,OH^-$)§	29.4	4×10^{-30}		
$Fe(OH)_2$	15.1	7.9×10^{-16}		
$Fe(OH)_3$	38.8	1.6×10^{-39}		
HgO (red) ($\rightleftharpoons Hg^{2+} + 2\,OH^-$)†§	25.44	3.6×10^{-26}		
$In(OH)_3$	36.9	1.3×10^{-37}		
$La(OH)_3$	20.7	2×10^{-21}		
$Mg(OH)_2$	11.15	7.1×10^{-12}		
$Mn(OH)_2$	12.8	1.6×10^{-13}		
$Ni(OH)_2$	15.2	6×10^{-16}		
$Pd(OH)_2$	31	1×10^{-31}		
PbO (red) ($\rightleftharpoons Pb^{2+} + 2\,OH^-$)†§	15.3	5×10^{-16}		
PbO (yellow) ($\rightleftharpoons Pb^{2+} + 2\,OH^-$)†§	15.1	8×10^{-16}		
SnO ($\rightleftharpoons Sn^{2+} + 2\,OH^-$)§	26.2	6×10^{-27}		
UO_2 ($\rightleftharpoons U^{4+} + 4\,OH^-$)§	56.2	6×10^{-57}		
$UO_2(OH)_2$ ($\rightleftharpoons UO_2^{2+} + 2\,OH^-$)§	22.4	4×10^{-23}		
$V(OH)_3$	34.4	4.0×10^{-35}		
$VO(OH)_2$ ($\rightleftharpoons VO^{2+} + 2\,OH^-$)§	23.5	3×10^{-24}		
$Y(OH)_3$	23.2	6×10^{-24}		
$Zn(OH)_2$ amorphous†	15.52	3.0×10^{-16}		

*Ionizes or dissociates into Hg_2^{2+}.
§For the dissolution reaction written in parentheses.
†This is a specific solid-phase structure.

Formula	pK_{sp}	K_{sp}	Temperature (°C)	Ionic Strength (M)
Iodates				
$AgIO_3$	7.51	3.1×10^{-8}		
$Ba(IO_3)_2$	8.81	1.5×10^{-9}		
$Ca(IO_3)_2$	6.15	7.1×10^{-7}		
$Cd(IO_3)_2$	7.64	2.3×10^{-8}		
$Ce(IO_3)_3$	10.86	1.4×10^{-11}		
$Hg_2(IO_3)_2$*	17.89	1.3×10^{-18}		
$La(IO_3)_3$	10.92	1.2×10^{-11}		
$Pb(IO_3)_2$	12.61	2.5×10^{-13}		
$Sr(IO_3)_2$	6.48	3.3×10^{-7}		
$Th(IO_3)_4$	14.62	2.4×10^{-15}		0.5
$TlIO_3$	5.51	3.1×10^{-6}		
$UO_2(IO_3)_2$ ($\rightleftharpoons UO_2^{2+} + 2\,IO_3^-$)§	7.01	9.8×10^{-8}		0.2
$Y(IO_3)_3$	10.15	7.1×10^{-11}		
Iodides				
AgI	16.08	8.3×10^{-17}		
CuI	12.0	1×10^{-12}		
Hg_2I_2	27.95	1.1×10^{-28}		0.5
PbI_2	8.10	7.9×10^{-9}		
TlI	7.23	5.9×10^{-8}		
Oxalates				
BaC_2O_4	6.0	1×10^{-6}	20	0.1
CaC_2O_4	7.9	1.3×10^{-8}	20	0.1
$La_2(C_2O_4)_3$	25.0	1×10^{-25}	20	0.1
SrC_2O_4	6.4	4×10^{-7}	20	0.1
$Th(C_2O_4)_2$	21.38	4.2×10^{-22}		1
$UO_2C_2O_4$ ($\rightleftharpoons UO_2^{2+} + C_2O_4^{2-}$)§	8.66	2.2×10^{-9}	20	0.1
Phosphates				
Ag_3PO_4	17.55	2.8×10^{-18}		
$BaHPO_4$ ($\rightleftharpoons Ba^{2+} + HPO_4^{2-}$)§	7.40	4.0×10^{-8}	20	
$CaHPO_4 \cdot 2\,H_2O$ ($\rightleftharpoons Ca^{2+} + HPO_4^{2-}$)§	6.58	2.6×10^{-7}		
$FePO_4 \cdot 2\,H_2O$	26.4	4×10^{-27}		
$Fe_3(PO_4)_2 \cdot 8\,H_2O$	36.0	1×10^{-36}		
$GaPO_4$	21.0	1×10^{-21}		1
Hg_2HPO_4 ($\rightleftharpoons Hg_2^{2+} + HPO_4^{2-}$)*§	12.40	4.0×10^{-13}		
$InPO_4$	21.63	2.3×10^{-22}		1
$LaPO_4$	22.43	3.7×10^{-23}		0.5
$Pb_3(PO_4)_2$	43.53	3.0×10^{-44}	38	
$SrHPO_4$ ($\rightleftharpoons Sr^{2+} + HPO_4^{2-}$)§	6.92	1.2×10^{-7}	20	
$(VO)_3(PO_4)_2$ ($\rightleftharpoons 3\,VO^{2+} + 2\,PO_4^{3-}$)§	25.1	8×10^{-26}		
$Zn_3(PO_4)_2 \cdot 4\,H_2O$	35.3	5×10^{-36}		

*Ionizes or dissociates into Hg_2^{2+}.

§For the dissolution reaction written in parentheses.

Formula	pK_{sp}	K_{sp}	Temperature (°C)	Ionic Strength (M)
Sulfates				
Ag_2SO_4	4.83	1.5×10^{-5}		
$BaSO_4$	10.04	9.1×10^{-11}		
$CaSO_4$	4.62	2.4×10^{-5}		
Hg_2SO_4*	6.13	7.4×10^{-7}		
$PbSO_4$	6.20	6.3×10^{-7}		
$RaSO_4$	10.37	4.3×10^{-11}	20	
$SrSO_4$	6.50	3.2×10^{-7}		
Sulfides				
Ag_2S	50.1	8×10^{-51}		
CdS	27.0	1×10^{-27}		
CoS(α)†	21.3	5×10^{-22}		
CoS(β)†	25.6	3×10^{-26}		
CuS	36.1	8×10^{-37}		
Cu_2S	48.5	3×10^{-49}		
FeS	18.1	8×10^{-19}		
HgS (black)†	52.7	2×10^{-53}		
HgS (red)†	53.3	5×10^{-54}		
In_2S_3	69.4	4×10^{-70}		
MnS (green)†	13.5	3×10^{-14}		
MnS (pink)†	10.5	3×10^{-11}		
NiS(α)†	19.4	4×10^{-20}		
NiS(β)†	24.9	1.3×10^{-25}		
NiS(γ)†	26.6	3×10^{-27}		
PbS	27.5	3×10^{-28}		
SnS	25.9	1.3×10^{-26}		
Tl_2S	21.2	6×10^{-22}		
ZnS(α)†	24.7	2×10^{-25}		
ZnS(β)†	22.5	3×10^{-23}		
Thiocyanates				
AgSCN	11.97	1.1×10^{-12}		
CuSCN	13.40	4.0×10^{-14}		5
$Hg(SCN)_2$	19.56	2.8×10^{-20}		1
$Hg_2(SCN)_2$*	19.52	3.0×10^{-20}		
TlSCN	3.79	1.6×10^{-4}		

*Ionizes or dissociates into Hg_2^{2+}.

†This is a specific solid-phase structure.

Sources: Martell, A. E., Smith, R. M. 1976. *Critical Stability Constants*, Vol. 4. New York: Plenum Press. Sillén, L.G., Martell, A. E.. 1971. *Stability Constants of Metal-Ion Complexes*. Suppl. No. 1, Special Publication No. 25. London: The Chemical Society.

APPENDIX II

Acid Dissociation Constants

At 25°C and zero ionic strength unless noted otherwise. For bases $K_b = K_w/K_a$. Each acid is written in its protonated form. The acid protons are indicated by boldface.

Name	Structure	pK_a	K_a
Acetic acid	$CH_3COO\mathbf{H}$	4.757	1.75×10^{-5}
Acrylic acid (propenoic acid)	$H_2C{=}CHCOO\mathbf{H}$	4.258	5.52×10^{-5}
Alanine	$\mathbf{NH_3^+}$–$CHCH_3$–$COO\mathbf{H}$	2.348 (COOH) 9.867 (NH_3)	4.49×10^{-3} 1.36×10^{-10}
Aminobenzene (aniline)	C_6H_5–$\mathbf{NH_3^+}$	4.601	2.51×10^{-5}
4-Aminobenzenesulfonic acid (sulfanilic acid)	^-O_3S–C_6H_4–$\mathbf{NH_3^+}$	3.232	5.86×10^{-4}
2-Aminobenzoic acid (anthranilic acid)	C_6H_4($\overset{+}{\mathbf{NH_3}}$)(COO**H**)	2.08 (COOH) 4.96 (NH_3)	8.3×10^{-3} 1.10×10^{-5}
2-Aminoethanol (ethanolamine)	$HOCH_2CH_2\mathbf{NH_3^+}$	9.498	3.18×10^{-10}
2-Aminophenol	C_6H_4(O**H**)($\mathbf{NH_3^+}$)	4.78 (NH_3) (20°) 9.97 (OH) (20°)	1.66×10^{-5} 1.05×10^{-10}
Ammonia	$\mathbf{NH_4^+}$	9.244	5.70×10^{-10}
Arginine	$\mathbf{NH_3^+}$–$CHCH_2CH_2CH_2NHC$(=$\mathbf{NH_2^+}$)(NH_2)–$COO\mathbf{H}$	1.823 (COOH) 8.991 (NH_3) (12.48) (NH_2)	1.50×10^{-2} 1.02×10^{-9} 3.3×10^{-13}

Name	Structure	pK_a	K_a
Arsenic acid	$O{=}As(OH)_3$	2.24 6.96 11.50	5.8×10^{-3} 1.10×10^{-7} 3.2×10^{-12}
Arsenious acid	$As(OH)_3$	9.29 (13.5)	5.1×10^{-10} 3×10^{-14}
Asparagine	NH_3^+ (on C), O (double-bonded) $CHCHOCNH_2$ $COOH$	2.14 (COOH) (μ=0.1) 8.72 (NH_3) (μ=0.1)	7.2×10^{-3} 1.9×10^{-9}
Aspartic acid	NH_3^+ $CHCH_2COOH$ $COOH$	1.990 (α-COOH) 3.900 (β-COOH) 10.002 (NH_3)	1.02×10^{-2} 1.26×10^{-4} 9.95×10^{-11}
Benzoic acid	C_6H_5—COOH	4.202	6.28×10^{-5}
Benzylamine	C_6H_5—$CH_2NH_3^+$	9.35	4.5×10^{-10}
2,2′-Bipyridine	(pyridine ring with N^+–H)—(pyridine ring with N)	4.35	4.5×10^{-5}
Boric acid	$B(OH)_3$	9.236 (12.74) (20°) (13.80) (20°)	5.81×10^{-10} 1.82×10^{-13} 1.58×10^{-14}
Butane-2,3-dione dioxime (dimethylglyoxime)	HON NOH CH_3 CH_3	10.66 12.0	2.2×10^{-11} 1×10^{-12}
[Carbonic acid] = $[H_2CO_3] + [CO_2(aq)]$	$HO{-}C({=}O){-}OH$	6.352 10.329	4.45×10^{-7} 4.69×10^{-11}
Chloroacetic acid	$ClCH_2COOH$	2.865	1.36×10^{-3}
Chlorous acid	$HOCl{=}O$	1.95	1.12×10^{-2}
Chromic acid	$HO{-}Cr({=}O)_2{-}OH$	–0.2 (20°) 6.51	1.6 3.1×10^{-7}

Name	Structure	pK_a	K_a
Citric acid (2-hydroxypropane-1,2,3-tricarboxylic acid)	COOH | $HOOCCH_2CCH_2COOH$ | OH	3.128 4.761 6.396	7.44×10^{-4} 1.73×10^{-5} 4.02×10^{-7}
Cysteine	NH_3^+ | $CHCH_2SH$ | COOH	(1.71) (COOH) 8.36 (SH) 10.77 (NH_3)	1.95×10^{-2} 4.4×10^{-9} 1.70×10^{-11}
Dichloroacetic acid	$Cl_2CHCOOH$	1.30	5.0×10^{-2}
Diethylamine	$(CH_3CH_2)_2NH_2^+$	10.933	4.7×10^{-10}
1,2-Dihydroxybenzene (catechol)	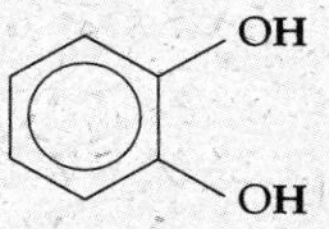 OH OH	9.40 12.8	4.0×10^{-10} 1.6×10^{-13}
1,3-Dihydroxybenzene (resorcinol)	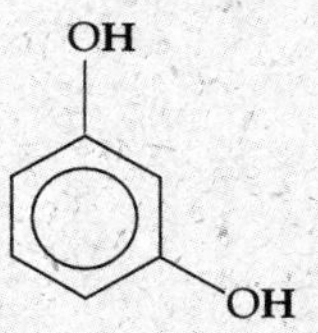 OH OH	9.30 11.06	5.0×10^{-10} 8.7×10^{-12}
2,3-Dimercaptopropanol	$HOCH_2CHCH_2SH$ | SH	8.58 ($\mu = 0.1$) 10.68 ($\mu = 0.1$)	2.6×10^{-9} 2.1×10^{-11}
Dimethylamine	$(CH_3)_2NH_2^+$	10.774	1.68×10^{-11}
Ethane- 1,2-dithiol	$HSCH_2CH_2SH$	8.85 (30°, $\mu = 0.1$) 10.43 (30°, $\mu = 0.1$)	1.4×10^{-9} 3.7×10^{-11}
Ethylamine	$CH_3CH_2NH_3^+$	10.636	2.31×10^{-11}
Ethylenediamine (1,2-diaminoethane)	$H_3\overset{+}{N}CH_2CH_2\overset{+}{N}H_3$	6.848 9.928	1.42×10^{-7} 1.18×10^{-10}
Ethylenedinitrilotetra-acetic acid (edta)	$(HOOCCH_2)_2\overset{+}{N}HCH_2CH_2\overset{+}{N}H(CH_2COOH)_2$	0.0 (COOH) ($\mu = 1.0$) 1.5 (COOH) ($\mu = 0.1$) 2.0 (COOH) ($\mu = 0.1$) 2.68 (COOH) ($\mu = 0.1$) 6.11 (NH) ($\mu = 0.1$) 10.17 (NH) ($\mu = 0.1$)	1.0 0.032 0.010 0.0021 7.8×10^{-7} 6.8×10^{-11}

Name	Structure	pK_a	K_a
Formic acid	HCOOH	3.745	1.80×10^{-4}
Fumaric acid (*trans*-butenedioic acid)	HOOC–CH=CH–COOH (trans)	3.053 4.494	8.85×10^{-4} 3.21×10^{-5}
Glutamic acid	NH_3^+–CH(CH_2CH_2COOH)–COOH	2.23 (α-COOH) 4.42 (γ-COOH) 9.95 (NH_3)	5.9×10^{-3} 3.8×10^{-5} 1.12×10^{-10}
Glutamine	NH_3^+–CH($CH_2CH_2C(=O)NH_2$)–COOH	2.17 (COOH) ($\mu = 0.1$) 9.01 (NH_3) ($\mu = 0.1$)	6.8×10^{-3} 9.8×10^{-10}
Glutaric acid (1,5-pentanedioic acid)	$HOOCCH_2CH_2CH_2COOH$	4.34 5.43	4.6×10^{-5} 3.7×10^{-6}
Glycine	NH_3^+–CH_2–COOH	2.350 (COOH) 9.778 (NH_3)	4.47×10^{-3} 1.67×10^{-10}
Glycolic acid	$HOCH_2COOH$	3.831	1.48×10^{-4}
Guanidine	H_2N–C(=$^+NH_2$)–NH_2	13.54 (27°, $\mu = 1.0$)	2.9×10^{-14}
Histidine	NH_3^+–CH(CH_2–imidazolium, NH, NH^+)–COOH	1.7 (COOH) ($\mu = 0.1$) 6.02 (NH) ($\mu = 0.1$) 9.08 (NH_3) ($\mu = 0.1$)	2×10^{-2} 9.5×10^{-7} 8.3×10^{-10}
Hydrogen azide (hydrazoic acid)	$HN=N^+=N^-$	4.65	2.2×10^{-5}
Hydrogen cyanate	$HOC\equiv N$	3.48	3.3×10^{-4}
Hydrogen cyanide	$HC\equiv N$	9.21	6.2×10^{-10}
Hydrogen fluoride	HF	3.17	6.5×10^{-4}
Hydrogen peroxide	HOOH	11.65	2.2×10^{-12}
Hydrogen sulfide*	H_2S	7.02 11.96	9.1×10^{-8} 1.1×10^{-12}

*Values measured at 18°C.

Name	Structure	pK_a	K_a
Hydrogen thiocyanate	$HSC\equiv N$	0.9	0.13
Hydroxylamine	HON^+H_3	5.96	1.10×10^{-6}
8-Hydroxyquinoline (oxine)	(quinoline ring with HO at C-8 and ring NH^+)	4.91 (NH) 9.81 (OH)	1.23×10^{-5} 1.55×10^{-10}
Hypobromous acid	HOBr	8.63	2.3×10^{-9}
Hypochlorous acid	HOCl	7.53	3.0×10^{-8}
Hypoiodous acid	HOI	10.64	2.3×10^{-11}
Hypophosphorus acid	$H_2P(=O)OH$	1.23	5.9×10^{-2}
Imidazole	(imidazole ring with NH^+ and NH)	6.993	1.02×10^{-7}
Iodic acid	HIO_3	0.77	0.17
Isoleucine	NH_3^+–$CHCH(CH_3)CH_2CH_3$–COOH	2.319 (COOH) 9.754 (NH_3)	4.80×10^{-3} 1.76×10^{-10}
Leucine	NH_3^+–$CHCH_2CH(CH_3)_2$–COOH	2.329 (COOH) 9.747 (NH_3)	4.69×10^{-3} 1.79×10^{-10}
Lysine	NH_3^+–$CHCH_2CH_2CH_2CH_2NH_3^+$–COOH	2.04 (COOH) ($\mu = 0.1$) 9.08 (α-NH_3) ($\mu = 0.1$) 10.69 (ϵ-NH_3) ($\mu = 0.1$)	9.1×10^{-3} 8.3×10^{-10} 2.0×10^{-11}
Maleic acid (*cis*-butenedioic acid)	cis-HOOC–CH=CH–COOH	1.910 6.332	1.23×10^{-2} 4.66×10^{-7}
Malonic acid	$HOOCCH_2COOH$	2.847 5.696	1.42×10^{-3} 2.01×10^{-6}
Mercaptoacetic acid (thioglycolic acid)	$HSCH_2COOH$	(3.60) (COOH) 10.55 (SH)	2.5×10^{-4} 2.82×10^{-11}

Name	Structure	pK_a	K_a
2-Mercaptoethanol	$HSCH_2CH_2OH$	9.72	1.91×10^{-10}
Methionine	NH_3^+ / $CHCH_2CH_2SCH_3$ / COOH	2.20 ($\mu = 0.1$) 9.05 ($\mu = 0.1$)	6.3×10^{-3} 8.9×10^{-10}
Methylamine	$CH_3NH_3^+$	10.64	2.3×10^{-11}
2-Methylaniline (*o*-toluidine)	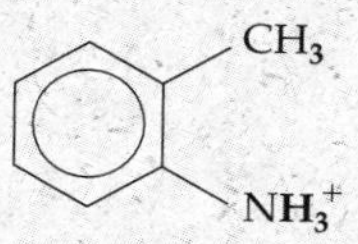 CH_3, NH_3^+	4.447	3.57×10^{-5}
2-Methylphenol (*o*-cresol)	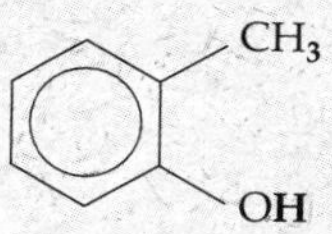 CH_3, OH	10.09	8.1×10^{-11}
Morpholine (perhydro-1,4-oxazine)	 O, NH_2^+	8.492	3.22×10^{-9}
Nitrilotriacetic acid	$H\overset{+}{N}(CH_2COOH)_3$	1.1 (COOH) (20°, $\mu = 1.0$) 1.650 (COOH) (20°) 2.940 (COOH) (20°) 10.334 (NH) (20°)	8×10^{-2} 2.24×10^{-2} 1.15×10^{-3} 4.63×10^{-11}
Nitroethane	$CH_3CH_2NO_2$	8.57	2.7×10^{-9}
N-Nitrosophenylhydroxyl-amine (cupferron)	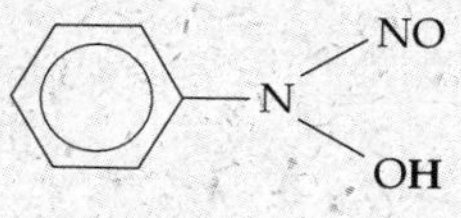 N, NO, OH	4.16 ($\mu = 0.1$)	6.9×10^{-5}
Nitrous acid	HON=O	3.15	7.1×10^{-4}
Oxalic acid	HOOCCOOH	1.252 4.266	5.60×10^{-2} 5.42×10^{-5}
Oxoacetic acid (glyoxylic acid)	O / ‖ / HCCOOH	3.46	3.5×10^{-4}
1,10-Phenanthroline	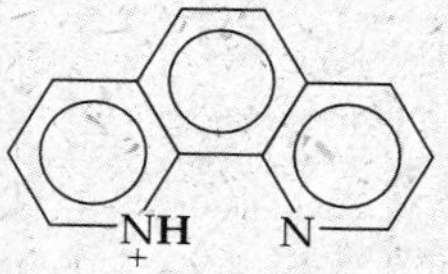 $\overset{+}{N}H$, N	4.86	1.38×10^{-5}
Phenol	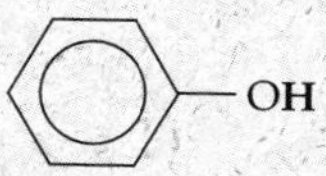 OH	9.98	1.05×10^{-10}

Name	Structure	pK_a	K_a
Phenylacetic acid	$C_6H_5CH_2COOH$	4.310	4.90×10^{-5}
Phenylalanine	NH_3^+–$CHCH_2C_6H_5$–COOH	2.20 (COOH) 9.31 (NH_3)	6.3×10^{-3} 4.9×10^{-10}
Phosphoric acid	$O{=}P(OH)_3$	2.148 7.199 12.35	7.11×10^{-3} 6.32×10^{-8} 4.5×10^{-13}
Phosphorous acid	$HP(=O)(OH)_2$	1.5 6.79	3×10^{-2} 1.62×10^{-7}
Phthalic acid (benzene-1,2-dicarboxylic acid)	$C_6H_4(COOH)_2$	2.950 5.408	1.12×10^{-3} 3.90×10^{-6}
Picric acid	2,4,6-$(NO_2)_3C_6H_2OH$	0.292	5.1×10^{-1}
Piperazine (perhydro-1,4-diazine)	$H_2\overset{+}{N}$ / $\overset{+}{N}H_2$ ring	5.333 9.731	4.65×10^{-6} 1.86×10^{-10}
Piperidine	$\overset{+}{N}H_2$ ring	11.123	7.53×10^{-12}
Proline	ring $\overset{+}{N}H_2$, COOH	1.952 (COOH) 10.640 (NH_2)	1.12×10^{-2} 2.29×10^{-11}
Propanoic acid	CH_3CH_2COOH	4.874	1.34×10^{-5}
Pyridine (azine)	$C_5H_5NH^+$	5.229	5.90×10^{-6}
Pyrophosphoric acid	$(HO)_2P(=O)OP(=O)(OH)_2$	0.8 2.2 6.70 9.40	0.16 6×10^{-3} 2.0×10^{-7} 4.0×10^{-10}

Name	Structure	pK_a	K_a
Pyrrolidine	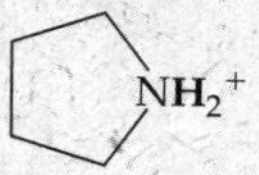 (five-membered ring with NH_2^+)	11.305	4.95×10^{-12}
Pyruvic acid (2-oxopropanoic acid)	$CH_3C(=O)COOH$	2.55	2.8×10^{-3}
Salicylic acid (2-hydroxybenzoic acid)	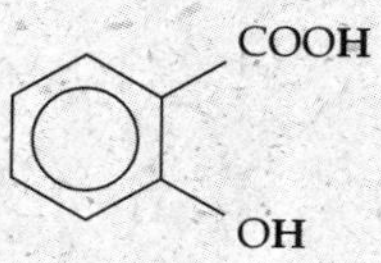 (benzene ring with COOH and OH)	2.97 (COOH) 13.74 (OH)	1.07×10^{-3} 1.82×10^{-14}
Serine	NH_3^+–$CHCH_2OH$–CO_2H	2.187 9.209	6.50×10^{-3} 6.18×10^{-10}
Succinic acid (butanedioic acid)	$HOOCCH_2COOH$	4.207 5.636	6.21×10^{-5} 2.31×10^{-6}
Sulfuric acid	H_2SO_4	1.99 (pK_2)	1.02×10^{-2}
Sulfurous acid	H_2SO_3	1.91 7.18	1.23×10^{-2} 6.6×10^{-8}
L-Tartaric acid L-2,3-Dihydroxybutanedioic acid	HOOCCH(OH)CH(OH)COOH	3.036 4.366	9.20×10^{-4} 4.31×10^{-5}
Thiosulfuric acid	HOS(=O)(=S)OH	0.6 1.6	0.3 3×10^{-2}
Threonine	NH_3^+–$CHCHOHCH_3$–COOH	2.088 (COOH) 9.100 (NH_3)	8.17×10^{-3} 7.94×10^{-10}
Trichloroacetic acid	Cl_3CCOOH	0.66 ($\mu = 0.1$)	0.22
Triethanolamine	$(HOCH_2CH_2)_3NH^+$	7.762	1.73×10^{-8}
Triethylamine	$(CH_3CH_2)_3NH^+$	10.715	1.93×10^{-11}

Name	Structure	pK_a	K_a
1,2,3-Trihydroxybenzene (pyrogallol)	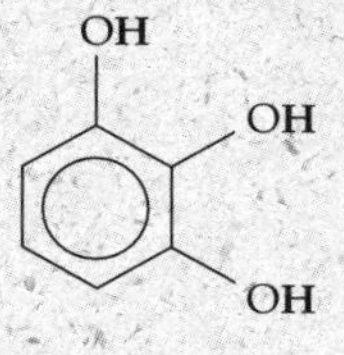 1,2,3-trihydroxybenzene (ring with OH, OH, OH)	8.94 11.08 (14)	1.15×10^{-9} 8.3×10^{-12} 10^{-14}
Trimethylamine	$(CH_3)_3NH^+$	9.800	1.58×10^{-10}
tris(hydroxymethyl)-aminomethane (TRIS or THAM)	$(HOCH_2)_3CNH_3^+$	8.075	8.41×10^{-9}
Tryptophan	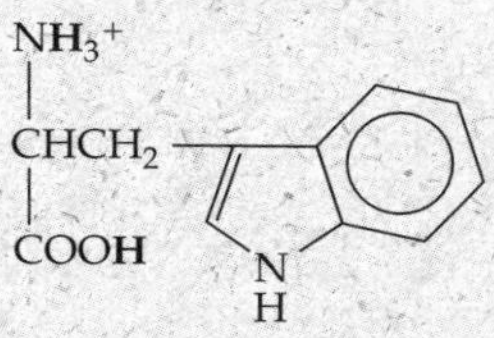 NH_3^+–$CHCH_2$(indolyl)–COOH	2.35 (COOH) ($\mu = 0.1$) 9.33 (NH_3) ($\mu = 0.1$)	4.5×10^{-3} 4.7×10^{-10}
Tyrosine	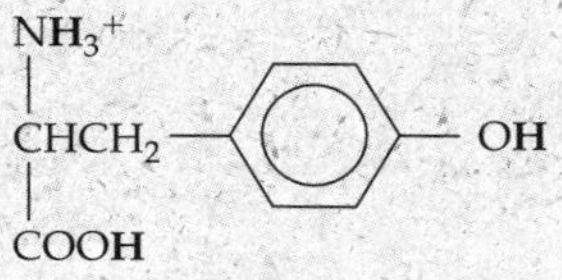 NH_3^+–$CHCH_2$(C_6H_4–OH)–COOH	2.17 (COOH) ($\mu = 0.1$) 9.19 (NH_3) 10.47 (OH)	6.8×10^{-3} 6.5×10^{-10} 3.4×10^{-11}
Valine	NH_3^+–$CHCH(CH_3)_2$–CO_2H	2.286 (COOH) 9.718 (NH_3)	5.18×10^{-3} 1.91×10^{-10}
Water	H_2O	13.997	1.01×10^{-14}

Source: Martell, A. E., Smith, R. M. 1974. *Critical Stability Constants.* New York: Plenum Press.

Appendix III

Selected Standard and Formal Electrode Potentials

The formal potential is the measured potential of a half cell (*vs.* NHE) when the logarithmic term in the Nernst equation is zero and the following conditions hold: the *concentrations* of oxidized and reduced species are expressed overtly, and all other species (such as H^+, ligands, salts) are present at the stated concentrations. Thus, for the general reaction

$$\text{Oxidized form} + e^- \rightleftharpoons \text{Reduced form}$$

when the value of [Red]/[Ox] = 1, the formal potential will be measured. For a reaction with stoichiometry

$$\text{Oxidized form} + e^- \rightleftharpoons 2\ \text{Reduced form}$$

when the value of $[\text{Red}]^2/[\text{Ox}] = 1$, the formal potential will be measured.

In effect, the activity coefficients are incorporated into the potential as are any chemical reactions that occur with either member of the redox couple. The formal potential will vary from medium to medium. One other convention keeps this formalism consistent. In writing the Nernst equation, solids involved in the redox equilibrium are still treated as having unit activity.

All values are for aqueous solutions at 25°C unless otherwise specified.

Half-reaction	$E°$ (V)	Formal potential (V)
Aluminum		
$Al^{3+} + 3\,e^- \rightleftharpoons Al(s)$	−1.662	
Antimony		
$Sb_2O_5(s) + 6\,H^+ + 4\,e^- \rightleftharpoons 2\,SbO^+ + 3\,H_2O$	+0.581	
Arsenic		
$H_3AsO_4 + 2\,H^+ + 2\,e^- \rightleftharpoons H_3AsO_3 + H_2O$	+0.559	0.577, 1 M HCl, $HClO_4$
Barium		
$Ba^{2+} + 2\,e^- \rightleftharpoons Ba(s)$	−2.906	
Beryllium		
$Be^{2+} + 2\,e^- \rightleftharpoons Be(s)$	−1.85	
Bismuth		
$BiO^+ + 2\,H^+ + 3\,e^- \rightleftharpoons Bi(s) + H_2O$	+0.320	
$BiCl_4^- + 3\,e^- \rightleftharpoons Bi(s) + 4\,Cl^-$	+0.16	
Bromine		
$Br_2(l) + 2\,e^- \rightleftharpoons 2\,Br^-$	+1.065	1.05, 4 M HCl

Half-reaction	$E°$ (V)	Formal potential (V)
Bromine (continued)		
$Br_2(aq) + 2e^- \rightleftharpoons 2Br^-$	+1.087*	
$BrO_3^- + 6H^+ + 5e^- \rightleftharpoons \frac{1}{2}Br_2(l) + 3H_2O$	+1.52	
$BrO_3^- + 6H^+ + 6e^- \rightleftharpoons Br^- + 3H_2O$	+1.44	
Cadmium		
$Cd^2 + 2e^- \rightleftharpoons Cd(s)$	−0.403	
Calcium		
$Ca^{2+} + 2e^- \rightleftharpoons Ca(s)$	−2.866	
Carbon		
$C_6H_4O_2$ (quinone) $+ 2H^+ + 2e^- \rightleftharpoons C_6H_4(OH)_2$	+0.699	0.696, 1 M HCl, $HClO_4$, H_2SO_4
$2CO_2(g) + 2H^+ + 2e^- \rightleftharpoons H_2C_2O_4$	−0.49	
Cerium		
$Ce^{4+} + e^- \rightleftharpoons Ce^{3+}$		1.70, 1 M $HClO_4$; 1.61, 1 M HNO_3; 1.44, 1 M H_2SO_4; 1.28, 1 M HCl
Chlorine		
$Cl_2(g) + 2e^- \rightleftharpoons 2Cl^-$	+1.359	
$HClO + H^+ + e^- \rightleftharpoons \frac{1}{2}Cl_2(g) + H_2O$	+1.63	
$ClO_3^- + 6H^+ + 5e^- \rightleftharpoons \frac{1}{2}Cl_2(g) + 3H_2O$	+1.47	
$ClO_4^- + 2H^+ + 2e^- \rightleftharpoons ClO_3^- + H_2O$	+1.19	
Chromium		
$Cr^{3+} + e^- \rightleftharpoons Cr^{2+}$	−0.408	−0.37, 0.1−0.5 M H_2SO_4
$Cr^{3+} + 3e^- \rightleftharpoons Cr(s)$	−0.744	
$Cr_2O_7^{2-} + 14H^+ + 6e^- \rightleftharpoons 2Cr^{3+} + 7H_2O$	+1.33	+1.00, 1 M HCl; 1.025, 1 M $HClO_4$
Cobalt		
$Co^{2+} + 2e^- \rightleftharpoons Co(s)$	−0.277	
$Co^{3+} + e^- \rightleftharpoons Co^{2+}$	+1.808	
Copper		
$Cu^{2+} + 2e^- \rightleftharpoons Cu(s)$	+0.337	
$Cu^{2+} + e^- \rightleftharpoons Cu^+$	+0.153	
$Cu^+ + e^- \rightleftharpoons Cu(s)$	+0.521	
$Cu^{2+} + I^- + e^- \rightleftharpoons CuI(s)$	+0.86	
$CuI(s) + e^- \rightleftharpoons Cu(s) + I^-$	−0.185	

*The standard potentials of these solutions should be those with 1.00-M Br_2 or I_2, respectively. However, at 25°C, the solubilities in water of bromine and iodine are Br_2 (0.18 M) and I_2 (0.0020 M). Therefore, these are not measurable potentials. However, these extrapolated potentials can be used as any other standard potential as a starting point for calculations for runs not under standard conditions.

Half-reaction	$E°$ (V)	Formal potential (V)
Fluorine		
$F_2 + 2\,H^+ + 2\,e^- \rightleftharpoons 2\,HF(aq)$	+3.06	
Gold		
$Au^{3+} + 2\,e^- \rightleftharpoons Au^+$	+1.41	
$Au^{3+} + 3\,e^- \rightleftharpoons Au(s)$	+1.50	
Hydrogen		
$2\,H^+ + 2\,e^- \rightleftharpoons H_2(g)$ (NHE)	0.000	−0.005, 1 M HCl, $HClO_4$
$2\,H_2O + 2\,e^- \rightleftharpoons H_2(g) + 2\,OH^-$	−0.828	
Indium		
$In^{3+} + 2\,e^- \rightleftharpoons In(s)$	−0.33	
Iodine		
$I_2(s) + 2\,e^- \rightleftharpoons 2I^-$	+0.5355	
$I_2(aq) + 2\,e^- \rightleftharpoons 2I^-$	+0.615*	
$I_3^- + 2\,e^- \rightleftharpoons 3\,I^-$	+0.536	
$ICl_2^- + e^- \rightleftharpoons \frac{1}{2}\,I_2(s) + 2Cl^-$	+1.056	
$IO_3^- + 6\,H^+ + 5\,e^- \rightleftharpoons \frac{1}{2}\,I_2(s) + 3\,H_2O$	+1.196	
$IO_3^- + 6\,H^+ + 5\,e^- \rightleftharpoons \frac{1}{2}\,I_2(aq) + 3\,H_2O$	+1.178*	
$IO_3^- + 2\,Cl^- + 6\,H^+ + 4e^- \rightleftharpoons ICl_2^- + 3\,H_2O$	+1.24	
$H_5IO_6 + H^+ + 2\,e^- \rightleftharpoons IO_3^- + 3\,H_2O$	+1.601	
Iron		
$Fe^{2+} + 2\,e^- \rightleftharpoons Fe(s)$	−0.440	
$Fe^{3+} + e^- \rightleftharpoons Fe^{2+}$	+0.771	0.700, 1 M HCl; 0.732, 1 M $HClO_4$; 0.68, 1 M H_2SO_4; 0.46, 2 M H_3PO_4
$Fe(CN)_6^{3-} + e^- \rightleftharpoons Fe(CN)_6^{4-}$	+0.36	0.71, 1 M HCl; 0.72, 1 M $HClO_4$, H_2SO_4
Lead		
$Pb^{2+} + 2\,e^- \rightleftharpoons Pb(s)$	−0.126	−0.14, 1 M $HClO_4$; −0.29, 1 M H_2SO_4
$PbO_2(s) + 4\,H^+ + 2\,e^- \rightleftharpoons Pb^2 + 2H_2O$	+1.455	
$PbSO_4(s) + 2\,e^- \rightleftharpoons Pb(s) + SO_4^{2-}$	−0.350	
Lithium		
$Li^+ + e^- \rightleftharpoons Li(s)$	−3.045	
Magnesium		
$Mg^{2+} + 2\,e^- \rightleftharpoons Mg(s)$	−2.363	

*The standard potentials of these solutions should be those with 1.00-M Br_2 or I_2, respectively. However, at 25°C, the solubilities in water of bromine and iodine are Br_2 (0.18 M) and I_2 (0.0020 M). Therefore, these are not measurable potentials. However, these extrapolated potentials can be used as any other standard potential as a starting point for calculations for runs not under standard conditions.

Half-reaction	$E°$ (V)	Formal potential (V)
Manganese		
$Mn^{2+} + 2e^- \rightleftharpoons Mn(s)$	−1.180	
$Mn^{3+} + e^- \rightleftharpoons Mn^{2+}$		1.51, 7.5 M H_2SO_4
$MnO_2(s) + 4H^+ + 2e^- \rightleftharpoons Mn^{2+} + 2H_2O$	+1.23	1.24, 1 M $HClO_4$
$MnO_4^- + 8H^+ + 5e^- \rightleftharpoons Mn^{2+} + 4H_2O$	+1.51	
$MnO_4^- + 4H^+ + 3e^- \rightleftharpoons MnO_2(s) + 2H_2O$	+1.695	
$MnO_4^- + e^- \rightleftharpoons MnO_4^{2-}$	+0.564	
Mercury		
$Hg_2^{2+} + 2e^- \rightleftharpoons 2Hg(l)$	+0.788	0.274, 1 M HCl; 0.776, 1 M $HClO_4$; 0.674, 1 M H_2SO_4
$2Hg^{2+} + 2e^- \rightleftharpoons Hg_2^{2+}$	+0.920	0.907, 1 M $HClO_4$
$Hg^{2+} + 2e^- \rightleftharpoons Hg(l)$	+0.854	
$Hg_2Cl_2(s) + 2e^- \rightleftharpoons 2Hg(l) + 2Cl^-$ (Calomel)	+0.268	SCE 0.242, sat'd KCl; 0.282, 1 M KCl; 0.334, 0.1 M KCl
$Hg_2SO_4(s) + 2e^- \rightleftharpoons 2Hg(l) + SO_4^{2-}$	+0.615	
Molybdenum		
$Mo^{6+} + e^- \rightleftharpoons Mo^{5+}$	+0.48	+0.53, 2 M HCl
$Mo^{4+} + e^- \rightleftharpoons Mo^{3+}$		+0.1, 4.5 M H_2SO_4
Nickel		
$Ni^{2+} + 2e^- \rightleftharpoons Ni(s)$	−0.250	
Nitrogen		
$N_2(g) + 5H^+ + 4e^- \rightleftharpoons N_2H_5^+$	−0.23	
$HNO_2 + H^+ + e^- \rightleftharpoons NO(g) + H_2O$	+1.00	
$NO_3^- + 3H^+ + 2e^- \rightleftharpoons HNO_2 + H_2O$	+0.94	0.92, 1 M HNO_3
Oxygen		
$H_2O_2 + 2H^+ + 2e^- \rightleftharpoons 2H_2O$	+1.776	
$HO_2^- + H_2O + 2e^- \rightleftharpoons 3OH^-$	+0.88	
$O_2(g) + 4H^+ + 4e^- \rightleftharpoons 2H_2O$	+1.229	
$O_2(g) + 2H^+ + 2e^- \rightleftharpoons H_2O_2$	+0.682	
$O_3(g) + 2H^+ + 2e^- \rightleftharpoons O_2(g) + H_2O$	+2.07	
Palladium		
$Pd^{2+} + 2e^- \rightleftharpoons Pd(s)$		0.987, 4 M $HClO_4$
Phosphorous		
$H_3PO_4(aq) + 2H^+ + 2e^- \rightleftharpoons H_3PO_3(aq) + H_2O$	−0.276	
$H_3PO_3(aq) + 2H^+ + 2e^- \rightleftharpoons H_3PO_2(aq) + H_2O$	−0.50	
Platinum		
$Pt^{2+} + 2e^- \rightleftharpoons Pt(s)$	+1.2	
$PtCl_4^{2-} + 2e^- \rightleftharpoons Pt(s) + 4Cl^-$	+0.73	
$PtCl_6^{2-} + 2e^- \rightleftharpoons PtCl_4^{2-} + 2Cl^-$	+0.68	

Half-reaction	$E°$ (V)	Formal potential (V)
Potassium		
$K^+ + e^- \rightleftharpoons K(s)$	−2.925	
Selenium		
$Se + 2\,H^+ + 2\,e^- \rightleftharpoons H_2Se$	−0.40	
$H_2SeO_3 + 4\,H^+ + 4\,e^- \rightleftharpoons Se(s) + 3\,H_2O$	+0.740	
$SeO_4^{2-} + 4\,H^+ + 2\,e^- \rightleftharpoons H_2SeO_3 + H_2O$	+1.15	
Silver		
$Ag^+ + e^- \rightleftharpoons Ag(s)$	+0.799	0.228, 1 M HCl; 0.792, 1 M $HClO_4$; 0.77, 1 M H_2SO_4
$AgBr(s) + e^- \rightleftharpoons Ag(s) + Br^-$	+0.073	
$AgCl(s) + e^- \rightleftharpoons Ag(s) + Cl^-$	+0.222	0.228, 1 M KCl
$Ag(CN)_2^- + e^- \rightleftharpoons Ag(s) + 2\,CN^-$	−0.31	
$Ag_2CrO_4(s) + 2\,e^- \rightleftharpoons 2\,Ag(s) + CrO_4^{2-}$	+0.446	
$AgI(s) + e^- \rightleftharpoons Ag(s) + I^-$	−0.151	
$Ag(S_2O_3)_4^{3-} + e^- \rightleftharpoons Ag(s) + 2\,S_2O_3^{2-}$	+0.017	
Sodium		
$Na^+ + e^- \rightleftharpoons Na(s)$	−2.713	
Strontium		
$Sr^{2+} + 2\,e^- \rightleftharpoons Sr(s)$	−2.89	
Sulfur		
$S(s) + 2\,H^+ + 2\,e^- \rightleftharpoons H_2S(g)$	+0.141	
$H_2SO_3 + 4\,H^+ + 4\,e^- \rightleftharpoons S(s) + 3\,H_2O$	+0.450	
$S_4O_6^{2-} + 2\,e^- \rightleftharpoons 2\,S_2O_3^{2-}$	+0.08	
$SO_4^{2-} + 4\,H^+ + 2\,e^- \rightleftharpoons H_2SO_3 + H_2O$	+0.172	+0.07, 1 M H_2SO_4
$S_2O_8^{2-} + 2\,e^- \rightleftharpoons 2\,SO_4^{2-}$	+2.01	
Thallium		
$Tl^+ + e^- \rightleftharpoons Tl(s)$	−0.336	−0.551, 1 M HCl; −0.33, 1 M $HClO_4$, H_2SO_4
$Tl^{3+} + 2\,e^- \rightleftharpoons Tl^+$	+1.25	0.77, 1 M HCl
Tin		
$Sn^{2+} + 2\,e^- \rightleftharpoons Sn(s)$	−0.136	−0.16, 1 M $HClO_4$
$Sn^{4+} + 2\,e- \rightleftharpoons Sn^{2+}$	+0.154	0.14, 1 M HCl
Titanium		
$Ti^{3+} + e^- \rightleftharpoons Ti^{2+}$	−0.369	
$TiO^{2+} + 2\,H^+ + e^- \rightleftharpoons Ti^{3+} + H_2O$	+0.099	0.04, 1 M H_2SO_4
Tungsten		
$2\,WO_3(s) + 2\,H^+ + 2\,e^- \rightleftharpoons W_2O_5 + H_2O$	−0.03	
$W^{6+} + e^- \rightleftharpoons W^{5+}$		+0.26, 12 M HCl
$W^{5+} + e^- \rightleftharpoons W^{4+}$		−0.03, 12 M HCl

Half-reaction	$E°$ (V)	Formal potential (V)
Uranium		
$UO_2^{2+} + 4\,H^+ + 2\,e^- \rightleftharpoons U^{4+} + 2\,H_2O$	+0.334	
$U^{4+} + e^- \rightleftharpoons U^{3+}$	−0.61	−0.64, 1 M HCl
$UO_2^{2+} + e^- \rightleftharpoons UO_2^+$	+0.05	
Vanadium		
$V^{3+} + e^- \rightleftharpoons V^{2+}$	−0.256	−0.21, 1 M $HClO_4$
$VO^{2+} + 2\,H^+ + e^- \rightleftharpoons V^{3+} + H_2O$	+0.359	
$V(OH)_4^+ + 2\,H^+ + e^- \rightleftharpoons VO^{2+} + 3\,H_2O$	+1.00	1.02, 1 M HCl, $HClO_4$
Zinc		
$Zn^{2+} + 2\,e^- \rightleftharpoons Zn(s)$	−0.763	
$Zn(NH_3)_4^{2+} + 2\,e^- \rightleftharpoons Zn(s) + 4\,NH_3$	−1.04	

Sources for E° values:

deBethune, A. J., Loud, N. A. S. 1964. *Standard Aqueous Electrode Potentials and Temperature Coefficients at 25°C.* Skokie, IL: Clifford A. Hampel.

Milazzo, G., Caroli, S., Sharma, V. K. 1978. *Tables of Standard Electrode Potentials.* New York: Wiley.

Sources for formal potentials:

Swift, E. H., Butler, E. A. 1972. *Quantitative Measurements and Chemical Equilibria.* San Francisco: W. H. Freeman.

Meites, L., ed. 1963. *Handbook of Analytical Chemistry.* New York: McGraw-Hill.

Appendix IV

Representative Stepwise Formation Constants

For K_a values of the ligands, see Appendix II.

Cation	Conditions*	log K_1	log K_2	log K_3	log K_4	log K_5	log K_6
Ammonia: NH_3							
Ag^+	*a*	3.26	3.94	—	—	—	—
Cd^{2+}	*b*	2.62	2.17	1.37	0.94	−0.2	
Co^{2+}	*c*	2.10	1.57	1.11	0.75	0.2	−0.6
Cu^{2+}	*b*	4.12	3.51	2.88	2.1	—	—
Ni^{2+}	*a*	2.8	2.3	1.8	1.3	0.8	0.2
Zn^{2+}	*b*	2.32	2.49	2.30	2.21	—	—
Cyanide: CN^-							
Ag^+	*b*	log $K_1K_2 = 20.0$		—	—		
Cd^{2+}	*d*	5.55	5.2	4.8	3.5		
Hg^{2+}	*e*	18.00	16.71	3.83	3.0		
Zn^{2+}	*d*	5.3	6.4	5.0	4.9		
Ethylenediamine: $H_2NCH_2CH_2NH_2$							
Ag^+	*e*	4.70	3.00	—			
Cd^{2+}	*f*	5.47	4.55	2.07			
Co^{2+}	*b*	5.93	4.73	3.30			
Cu^{2+}	*g*	11.12	9.61	—			
Hg^{2+}	*h*	14.3	9.0	—			
Mn^{2+}	*i*	2.73	2.06	0.88			
Ni^{2+}	*h*	7.52	6.28	4.26			
Zn^{2+}	*h*	4.71	4.66	1.72			
Nitrilotriacetic acid (NTA)							
Ag^+	*a*	5.16	—				
Al^{3+}	*c*	11.37	—				
Ba^{2+}	*a*	4.82	—				
Ca^{2+}	*a*	6.41	—				
Cd^{2+}	*a*	9.83	3.5				
Co^{2+}	*a*	8.1	—				
Cu^{2+}	*a*	13.05	—				
Mg^{2+}	*a*	5.46	—				
Mn^{2+}	*a*	7.36	—				
Ni^{2+}	*a*	11.54	—				
Pb^{2+}	*a*	11.47	—				
Zn^{2+}	*a*	10.44	—				

*Values for °C, μ are as follows: (*a*) 20°, 0.1; (*b*) 25°, 0.1; (*c*) 25°, 0.2; (*d*) 25°, 1.0; (*e*) 25°, 2.0; (*f*) 25°, 3.0; (*g*) 30°, 0.1; (*h*) 30°, 0.5; (*i*) 30°, 1.0; (*j*) 30°, 2.0. Ionic strength provided by assumed noncoordinating ions.

Cation	Conditions*	log K_1	log K_2	log K_3	log K_4	log K_5	log K_6
Ethylenediaminetetraacetic acid: edta							
Ag^+	*a*	7.11					
Al^{3+}	*c*	16.01					
Ca^{2+}	*a*	10.6					
Cd^{2+}	*a*	16.46					
Cu^{2+}	*a*	18.80					
Mg^{2+}	*a*	8.69					
Mn^{2+}	*a*	14.04					
Ni^{2+}	*a*	18.62					
Pb^{2+}	*a*	18.32					
Zn^{2+}	*a*	16.26					

*Values for °C, μ are as follows: (*a*) 20°, 0.1; (*b*) 25°, 0.1; (*c*) 25°, 0.2; (*d*) 25°, 1.0; (*e*) 25°, 2.0; (*f*) 25°, 3.0; (*g*) 30°, 0.1; (*h*) 30°, 0.5; (*i*) 30°, 1.0; (*j*) 30°, 2.0. Ionic strength provided by assumed noncoordinating ions.

Sources: Sillén, L. G., Martell, A. E. 1964. *Stability Constants of Metal-Ion Complexes.* Special Publication No. 17. London: The Chemical Society.

Sillén, L. G., Martell, A. E. 1961. *Stability Constants of Metal-Ion Complexes.* Suppl. No. 1. Special Publication No. 17. London: The Chemical Society.

Smith, R. M., Martell, A. E. *Critical Stability Constants.* Vol. 4. New York and London: Plenum.

APPENDIX V

Compounds for Elemental Analysis Standard Solutions

Element	Compound	Formula weight (g)	1000 ppm (g/L)	Solvent	Comments
Aluminum	Al metal	26.982	1.0000	Hot, dil. HCl	*b*
Antimony	$KSbOC_4H_4O_6 \cdot \frac{1}{2} H_2O$	333.92	2.7427	Water	*l*
Arsenic	As_2O_3	197.84	2.6406	dil. HCl	*a, c, n*
Barium	$BaCO_3$	197.35	1.4369	dil. HCl	
Beryllium	Be metal	9.0122	1.0000	HCl	*c*
Bismuth	Bi_2O_3	465.96	1.1148	HNO_3	
Boron	H_3BO_3	61.84	5.7200	Water	*o*
Bromine	KBr	119.01	1.4894	Water	*b*
Cadmium	CdO	128.40	1.1423	HNO_3	
Calcium	$CaCO_3$	100.09	2.4972	dil. HCl	*a*
Cerium	$(NH_4)_2Ce(NO_3)_6$	548.23	3.9126	Water	
Cesium	Cs_2SO_4	361.87	1.3614	Water	
Chromium	$K_2Cr_2O_7$	294.19	2.8290	Water	*a*
Cobalt	Co metal	58.933	1.0000	HNO_3	*b*
Copper	Cu metal	63.546	1.0000	dil. HNO_3	*b*
	CuO	79.545	1.2517	HCl, hot	*b*
Dysprosium	Dy_2O_3	373.00	1.1477	HCl, hot	*e*
Erbium	Er_2O_3	382.56	1.1435	HCl, hot	*e*
Europium	Eu_2O_3	351.92	1.1579	HCl, hot	*e*
Fluorine	NaF	41.988	2.2101	Water	*g*
Gadolinium	Gd_2O_3	362.50	1.1526	HCl, hot	*e*
Gallium	Ga metal	69.72	1.0000	HNO_3, hot	*f*
Germanium	GeO_2	104.60	1.4410	1 *m* NaOH, hot	
Gold	Au metal	196.97	1.0000	Aqua Regia, hot	*b*
Hafnium	Hf metal	178.49	1.0000	HF, fusion	*i*
Holmium	Ho_2O_3	377.86	1.1455	HCl, hot	*e*
Indium	In_2O_3	277.64	1.2090	HCl, hot	
Iodine	KIO_3	214.00	1.6863	Water	*a*
Iridium	No suitable compound was found				
Iron	Fe metal	55.47	1.0000	HCl, hot	*b*
Lanthanum	La_2O_3	325.82	1.1728	HCl, hot	*e*
Lead	$Pb(NO_3)_2$	331.20	1.5985	Water	*b*
Lithium	Li_2CO_3	73.890	5.3243	HCl	*b*
Lutetium	Lu_2O_3	397.94	1.1372	HCl, hot	*e*
Magnesium	MgO	40.31	1.658	HCl	
Manganese	$MnSO_4 \cdot H_2O$	169.01	3.0764	Water	*j*
Mercury	$HgCl_2$	271.50	1.3535	Water	*c*
Molybdenum	MoO_3	143.94	1.5003	1 *m* NaOH	
Neodymium	Nd_2O_3	336.48	1.1664	HCl	*e*
Nickel	Ni metal	58.71	1.0000	HNO_3, hot	*b*
Niobium	Nb_2O_5	265.81	1.4305	HF, fusion	*k*
Osmium	Os metal	190.20	1.0000	H_2SO_4, hot	*d*
Palladium	Pd metal	106.40	1.0000	HNO_3, hot	

Element	Compound	Formula weight (g)	1000 ppm (g/L)	Solvent	Comments
Phosphorus	KH_2PO_4	136.09	4.3937	Water	
Platinum	K_2PtCl_4	415.12	2.1278	Water	
Potassium	KCl	74.55	1.9065	Water	*b*
	$KHC_8H_4O_4$	204.22	5.2228	Water	*a, n*
	$K_2Cr_2O_7$	294.19	3.7618	Water	*a, n*
Praseodymium	Pr_6O_{11}	1021.43	1.2082	HCl	*e*
Rhenium	Re metal	186.2	1.0000	HNO_3	
	$KReO_4$	289.30	1.5537	Water	
Rhodium	Rh metal	102.91	1.0000	H_2SO_4, hot	
Rubidium	Rb_2SO_4	267.00	1.5628	Water	
Ruthenium	RuO_4	165.07	1.6332	Water	
Samarium	Sm_2O_3	348.70	2.3193	HCl, hot	*e*
Scandium	Sc_2O_3	137.91	1.5339	HCl, hot	
Selenium	Se metal	78.96	1.0000	HNO_3, hot	
Silicon	Si metal	28.086	1.0000	NaOH, Concd.	
	SiO_2	60.08	2.1391	HF	
Silver	$AgNO_3$	169.87	1.5748	Water	*b, m*
Sodium	NaCl	58.442	2.5428	Water	*a*
	$Na_2C_2O_4$	134.01	2.9146	Water	*a, n*
Strontium	$SrCO_3$	147.63	1.6849	HCl	*b*
Sulfur	K_2SO_4	174.27	5.4351	Water	
Tantalum	Ta_2O_5	441.89	1.2210	HF, fusion	*k*
Tellurium	TeO_2	159.60	1.2507	HCl	
Terbium	Tb_2O_3	365.85	1.1512	HCl, hot	*e*
Thallium	Tl_2CO_3	468.75	1.1468	Water	*b, c*
Thulium	Tm_2O_3	385.87	1.1421	HCl, hot	*e*
Tin	Sn metal	118.69	1.0000	HCl	
	SnO	134.69	1.1348	HCl	
Titanium	Ti metal	47.90	1.0000	H_2SO_4, 1:1	*b*
Tungsten	$Na_2WO_4 \cdot 2\,H_2O$	329.86	1.7942	Water	*h*
Uranium	UO_2	270.03	1.1344	HNO_3	
	U_3O_8	842.09	1.1792	HNO_3	*a, n*
Vanadium	V_2O_5	181.88	1.7852	HCl, hot	
Ytterbium	Yb_2O_3	394.08	1.1386	HCl, hot	*e*
Yttrium	Y_2O_3	225.81	1.2700	HCl, hot	*e*
Zinc	ZnO	81.37	1.2448	HCl	*b*
Zirconium	Zr metal	91.22	1.0000	HF, fusion	*i*

a = Primary standard.
b = These compounds conform very well to the criteria and approach primary standard quality.
c = Highly toxic.
d = Very highly toxic.
e = The rare earth oxides, because they absorb CO_2 and water vapor from the air, should be freshly ignited prior to weighing.
f = mp = 29.6°C. The metal may be warmed and weighed as the liquid.
g = Sodium fluoride solutions will etch glass and should be freshly prepared.
h = Sodium tungstate loses both water molecules at 110°C. After drying, f.w. = 293.83, 1000 ppm = 1.5982 g L^{-1}, 0.1 m = 29.383 g L^{-1}. The water is not rapidly regained, but the compound should be kept in a desiccator after drying and should be weighed quickly once it is removed.
i = Zirconium and hafnium compounds were not investigated in the laboratory. The following methods have been recommended for dissolution of zirconium and hafnium.

(1) 1.0000 g of the powdered metal is placed in a platinum dish with 5–10 mL of water and 1–2 mL of HF (1:5) and covered with a platinum lid or a paraffined watch glass. Once dissolved, the fluorine may be removed by adding 1–2 mL of sulfuric acid (cold) and evaporating to dense fumes or to dryness if required.

(2) A fusion method may also be used. A 5–10 fold excess of $K_2S_2O_7$ is placed in a platinum or quartz crucible along with the sample. After melting to a homogeneous molten mass in a muffle furnace or burner, the fusion product is dissolved in 2*N* sulfuric acid.

(3) A third method avoids the use of platinum ware. The sample of the metal is finely ground and placed in a small heat-resistant beaker. Two to four grams of ammonium sulfate and 3–6 mL of sulfuric acid are then added. A homogeneous melt is obtained on a hot plate and dissolved in 2*N* sulfuric acid.

j = $MnSO_4 \cdot H_2O$ may be dried at 110°C without losing the water of hydration.

k = Niobium and tantalum pentoxides are slowly soluble in 40% HF. The addition of H_2SO_4 accelerates the solution process. They may also be dissolved by a fusion technique. $K_2S_2O_7$ is an often used flux. The pentoxides are fully decomposed at 650–800°C in the presence of an 8- to 10-fold amount of potassium pyrosulfate. A quartz or porcelain crucible is suitable, and the resulting melt may be dissolved in sulfuric acid. Cold HF/H_2SO_4 has been used successfully in plastic beakers with a 10-h solution time.

l = Antimony potassium tartrate loses the $\frac{1}{2}H_2O$ with drying at 110°C. After drying, f.w. = 324.92, 1000 ppm = 2.6687 g L^{-1}, 0.1 *m* = 32.492 g L^{-1}. The water is not rapidly regained, but the compound should be kept in a desiccator after drying and should be weighed quickly once it is removed. The dried compound is water soluble.

m = When kept dry, silver nitrate crystals are not affected by light. Solutions of silver nitrate should be stored in brown bottles.

n = These compounds are sold as primary standards by the National Institute of Standards and Technology, Office of Standard Reference Materials, Washington, D.C. 20234.

o = Boric acid may be weighed accurately directly from the bottle. It will loose one H_2O molecule at 100°C and a second H_2O molecule at approximately 130–140°C and is difficult to dry to a constant weight.

Source: Smith, B. W., Parsons, M. L. 1973. *J. Chem. Educ.* 50:679–681.

Appendix VI

Values of t and $t/\sqrt{N}$ to Calculate the Confidence Limit at Various Levels of Probability

For the equation $\mu = \overline{X} \pm t/\sqrt{N} \cdot s$. See Equation 4.10 for the definition of t.

	Probability level (%)							
	80		90		95		99	
N	t	$\frac{t}{\sqrt{N}}$	t	$\frac{t}{\sqrt{N}}$	t	$\frac{t}{\sqrt{N}}$	t	$\frac{t}{\sqrt{N}}$
2	3.08	2.18	6.31	4.46	12.71	8.99	63.7	4.50
3	1.89	1.09	2.92	1.69	4.30	2.48	9.92	5.73
4	1.64	0.82	2.35	1.18	3.18	1.59	5.84	2.92
5	1.53	0.68	2.13	0.95	2.78	1.24	4.60	2.06
6	1.48	0.60	2.02	0.82	2.57	1.05	4.03	1.65
7	1.44	0.54	1.94	0.73	2.45	0.94	3.71	1.40
8	1.42	0.50	1.90	0.67	2.36	0.83	3.50	1.24
9	1.40	0.47	1.86	0.62	2.31	0.77	3.36	1.12
10	1.38	0.44	1.83	0.58	2.26	0.71	3.25	1.03
20	1.33	0.30	1.73	0.39	2.10	0.47	2.88	0.64
60	1.30	0.17	1.67	0.22	2.00	0.26	2.66	0.34
∞	1.28	—	1.64	—	1.96	—	2.58	—

APPENDIX VII

Colorimetric Indicators for Neutralization Titrations

Color abbreviations: b = blue, c = colorless, g = green, o = orange, p = purple, r = red, y = yellow

Chemical name	Common name	Approximate pH range	Colors: Acid form	Colors: Base form	Common stock solutions*
2,4,6-Trinitrophenol	Picric acid	0.6–1.3	c	y	—
Thymolsulfonphthalein	Thymol blue	1.2–2.8	r	y	0.04% aq.
2,4-Dinitrophenol	α-Dinitrophenol	2.4–4.0	c	y	0.1% alc.
Tetrabromophenol-sulfonphthalein	Bromophenol blue	3.0–4.6	y	b	0.04% aq.
Dimethylaminoazobenzene-*p*-sulfonate	Methyl orange	3.1–4.4	r	o	0.1% aq.
Tetrabromo-*m*-cresol-sulfonphthalein	Bromocresol green	3.8–5.4	y	b	0.1% aq.
Dimethylaminoazobenzene-*o*-carboxylic acid	Methyl red	4.2–6.3	r	y	0.1% in 60% alc.
Dibromo-*o*-cresolsulfon-phthalein	Bromocresol purple	5.2–6.8	y	p	0.04% aq.
Dibromothymolsulfon-phthalein	Bromothymol blue	6.2–7.6	y	b	0.05% aq.
Phenolsulfonphthalein	Phenol red	6.8–8.4	y	r	0.05% aq.
o-Cresolsulfonphthalein	Cresol red	7.2–8.8	y	r	0.05% aq.
Thymolsulfonphthalein	Thymol blue	8.0–9.6	y	b	0.04% aq.
Di-*p*-dioxydiphenyl-phthalide	Phenolphthalein	8.3–10.0	c	p	0.05% in 50% alc.
Dithymolphthalide	Thymolphthalein	9.3–10.5	c	b	0.04% in 50% alc.
m-Nitrobenzeneazo-salicylic acid	Alizarin yellow GG	10.0–12.0	c	y	0.1% alc.
2,4,6-Trinitrophenol-methylnitramine	Nitramine	10.8–13.0	c	o	0.01% aq.

*Solution abbreviations: aq., aqueous; alc., lower alcohols.

Source: Streuli, C.A., Meites, L. 1963. *Handbook of Analytical Chemistry*. L. Meites, ed. Section 3. New York: McGraw-Hill.

APPENDIX VIII

Characteristic Concentrations* and Detection Limits* of Analytical Techniques for Elements in Parts per Billion (Parts per 10^9)

(Abbreviations and conditions are explained in the key. The comments there are important in evaluating the LODs.)

Element	Char. Conc. MAS	Det. Lim. DPP	Det. Lim. ASV	Char. Conc. AAS	Char. Conc. ETA-AAS	Det. Lim. ICP-ES	Det. Lim. ICP/MS	Det. Lim. NAA	Det. Lim. XRF	Det. Lim. IC
Ag	15	60	0.005	20	0.2	0.7	0.04	0.1	1300	100
Al	3	0.03	0.5	30	2	0.7	0.1	1	1600	50
As	15	1	0.5	100	2	2	0.4	0.01	300	—
Au	10	2	0.5	100	0.5	1	0.08	0.0001	600	100
B	1	—	—	15000	100	0.8	0.08	—	—	—
Ba	—	—	—	200	10	0.01	0.02	1	100	100
Be	3	—	—	10	0.2	0.05	0.1	80	300	—
Bi	20	2	0.005	300	2	2	0.06	10	600	—
Br	20	0.03	—	—	—	—	—	0.2	600	8
Ca	20	15	—	20	1	0.005	5	100	100	0.5
Cd	15	0.5	0.005	15	0.05	0.09	0.07	0.5	200	10
Ce	60	—	—	—	—	3	0.01	0.1	200	—
Cl	—	0.3	—	—	—	—	—	2	200	1
Co	15	1	0.05	70	1	0.3	0.01	0.1	30	2
Cr	15	7	—	60	0.5	0.3	0.02	3	200	50
Cs	—	—	—	200	—	—	0.02	0.1	200	100
Cu	10	0.5	0.01	60	1	0.3	0.03	0.01	300	5
Dy	60	—	—	250	—	5	0.04	0.001	—	100
Er	60	—	—	300	—	0.2	0.02	0.1	—	100
Eu	60	—	—	200	—	0.05	0.02	0.01	—	100
F	—	—	—	—	—	—	—	20	1000	0.6
Fe	20	1	0.1	60	1	0.2	0.2	100	300	3
Ga	15	—	—	1000	5	0.5	0.08	0.05	—	—
Gd	60	—	—	—	—	0.4	0.04	0.1	—	100
Ge	3	—	—	1000	—	0.6	0.08	10	—	—
Hg	25	—	0.2	2200	100	2	0.08	0.1	300	—
Ho	60	—	—	300	—	0.6	0.01	0.2	—	100
I	—	0.03	—	—	—	—	0.01	0.2	100	—
In	15	0.5	0.01	300	10	2	0.01	0.1	—	—
Ir	30	—	—	—	60	2	0.06	0.001	—	—
K	—	—	—	20	0.5	6	—	0.2	30	50

*As a result of historic practices, the detection limits and characteristic concentrations are calculated in a variety of ways depending on the assay method involved. The detection limits and characteristic concentrations apply to ideal samples. This means that the samples have no untoward matrix effects and no interferences. The values are for samples directly assayed by the various methods without any preconcentration. If preconcentration treatments could be done without adding interferents, then, in theory, even lower concentrations of analytes in the samples could be determined. For realistic samples, detection limits are about 100 times higher than those noted here.

Element	Char. Conc. MAS	Det. Lim. DPP	Det. Lim. ASV	Char. Conc. AAS	Char. Conc. ETA-AAS	Det. Lim. ICP-ES	Det. Lim. ICP/MS	Det. Lim. NAA	Det. Lim. XRF	Det. Lim. IC
La	60	—	—	—	—	0.3	0.01	0.01	200	—
Li	—	—	—	20	1	0.06	0.06	2	—	0.05
Lu	60	—	—	—	—	0.6	0.01	0.01	—	100
Mg	15	15	—	3	0.03	0.02	0.1	10	600	0.3
Mn	1	3	—	30	0.2	0.04	0.04	0.01	100	3
Mo	30	6	—	300	5	1	0.08	1	600	50
Na	—	—	—	6	2	0.3	0.06	0.1	—	0.08
Nb	15	—	—	—	—	0.2	0.02	3	—	—
Nd	60	—	—	—	—	0.4	0.01	0.5	—	—
Ni	5	1	7	70	0.1	0.5	0.03	10	100	25
Os	60	—	—	300	—	0.2	0.01	1	—	—
P	1	—	—	—	—	2	—	1	600	—
Pb	25	1	0.01	200	0.5	0.8	0.02	1000	600	15
Pd	6	—	—	100	10	0.6	0.06	0.05	—	10
Pr	60	—	—	—	—	1	0.01	0.01	—	—
Pt	200	—	—	600	20	1	0.08	0.5	600	10
Rb	—	—	—	100	—	—	0.02	0.1	—	100
Re	30	—	—	—	—	2	0.06	0.01	—	—
Rh	200	—	—	200	100	0.4	0.02	0.1	—	—
Ru	30	—	—	300	—	2	0.05	1	—	—
S	3	—	—	—	—	8	—	—	300	—
Sb	25	1	0.02	300	1	2	0.02	0.05	30	—
Sc	7	—	—	200	—	0.05	0.08	0.01	—	—
Se	25	0.01	—	100	2	2	1	5	300	—
Si	10	—	—	2000	7	0.6	10	10	100	—
Sn	10	1	0.02	1000	0.4	2	0.03	1	20	100
Sr	—	—	—	100	1	0.02	0.02	10	300	100
Ta	15	—	—	—	—	3	0.02	0.05	—	—
Tc	15	—	—	—	—	—	—	—	—	—
Te	30	—	—	200	10	3	0.04	10	—	—
Ti	10	3	—	100	—	0.1	0.06	3	100	—
Tl	20	3	0.01	200	3	2	0.05	10	—	—
U	30	3	—	—	1	1	0.02	0.01	100	5
V	15	—	1	700	1	0.3	0.03	0.01	100	—
W	60	—	—	—	—	0.4	0.06	0.01	—	50
Yb	60	—	—	100	3	0.2	0.03	0.01	—	100
Zn	7	1	0.01	7	0.02	0.06	0.08	10	300	10
Zr	200	—	—	—	—	0.7	0.03	100	200	—

Key

—: Means either not a method for trace (less than ppm) analysis of the element or not reported.

MAS: Molecular Absorption Spectrometry in conjunction with colorimetric reagents. The values correspond to an absorbance of 0.025 for a 1-cm pathlength with a sample consumption of 1 mL.

DPP: Differential Pulse Polarography, assuming a sample consumption of 10 mL.

ASV: Anodic Stripping Voltammetry, with a sample size of 10 mL and a plating time of 10 min. Various media are used for electrodes. Ni and Co are determined by cathodic stripping of their oxides on Pt.

AAS: Atomic Absorption Spectrophotometry, with flame atomization and pneumatic nebulizer. The values correspond to an absorbance of 0.0025 with a 10-cm path.

ETA-AAS: Electrothermal Atomization—Atomic Absorption Spectrometry. The values correspond to an absorbance of 0.025 with a 50-μL sample.

ICP-ES: Inductively Coupled Plasma-Emission Spectrophotometry. Values are for a pneumatic nebulizer. These are the lowest values reported by any manufacturer for each element and interpolations from that data. Ultrasonic nebulization allows LODs a factor of 2 to 10 lower.

ICP/MS: Inductively Coupled Plasma/Mass Spectrometry. These are values for a simple quadrupole spectrometer with pneumatic nebulization. The limit of detection is an order of magnitude lower with ultrasonic nebulization. High-resolution mass spectrometric detection has LODs for most elements in the parts-per-trillion range. Instrumental LODs for ICP/MS can be as low as 10^{-6} ppb at long integration times for easily ionized elements that do not form oxides.

NAA: Neutron Activation Analysis. The irradiation and counting conditions are optimized for each element.

XRF: X-ray Fluorescence Spectrometry, with a dispersive instrument for a sample of 1 mL spread over a surface area of 7 to 9 cm^2.

IC: Ion Chromatography, without preconcentration, 50 μL sample loop.

Other Sources: Commission on Microchemical Techniques and Trace Analysis. 1982. *Pure & Appl. Chem.* 54:1565–577. Morrison, G. H. 1979. CRC *Crit. Rev. Anal. Chem.* 33:287–320. Manufacturer's literature for ICP/MS and ICP/AES. Neutron Activation Reference: Revel, G. 1987. "Present and future prospects for neutron activation analysis compared to other methods available," in IAEA. Comparison of Nuclear Analytical Methods with Competitive Methods, IAEA-TECDOC-135. Vienna: International Atomic Energy Agency. pp. 147–162.

APPENDIX IX

Partial List of Incompatible Chemicals

Substances that may react violently are incompatible. Not only should they not be mixed, it is now suggested they be stored apart.

Substance	Incompatible with
Acetic acid	Chromic acid, nitric acid, hydroxyl compounds, ethylene glycol, perchloric acid, peroxides, permanganates
Acetylene	Chlorine, bromine, fluorine, copper, silver, mercury
Activated carbon	Calcium hypochlorite, oxidizing agents
Alkali metals	Water, carbon tetrachloride and other chlorinated hydrocarbons, carbon dioxide, halogens
Aluminum alkyls	Water
Ammonia (anhydrous)	Mercury (e.g., in pressure gauges), chlorine, calcium hypochlorite, iodine, bromine, hydrofluoric acid (anhydrous)
Ammonium nitrate	Acids, powdered metals, flammable liquids, chlorates, nitrites, sulfur, fine-particulate organic or combustible materials
Aniline	Nitric acid, hydrogen peroxide
Bromine	See chlorine
Chlorates	Ammonium salts, acids, powdered metals, sulfur, fine-particulate organic or combustible substances
Chlorine	Ammonia, acetylene, butadiene, butane, methane, propane (and other petroleum gases), hydrogen, sodium carbide, turpentine, benzene, powdered metals
Chlorine dioxide	Ammonia, methane, phosphine, hydrogen sulfide
Chromic acid	Acetic acid, naphthalene, camphor, glycerin, turpentine, alcohols, flammable liquids
Copper	Acetylene, hydrogen peroxide
Cumene hydroperoxide	Acids, both organic and inorganic
Cyanides	Acids
Flammable liquids	Ammonium nitrate, chromic acid, hydrogen peroxide, nitric acid, sodium peroxide, halogens
Fluorine	Store separately
Hydrocarbons	Fluorine, chlorine, bromine, chromic acid, sodium peroxide
Hydrocyanic acid	Nitric acid, alkalis
Hydrofluoric acid (anhydrous)	Ammonia (aqueous or anhydrous); ammoniacal compounds (solid or liquid)
Hydrogen peroxide	Copper, chromium, iron, metals and metal salts, alcohols, acetone, organic substances, aniline, nitromethane, combustible substances
Hydrogen sulfide	Fuming nitric acid, oxidizing gases
Iodine	Acetylene, ammonia (aqueous or anhydrous), hydrogen
Mercury	Acetylene, fulminic acid, ammonia
Nitric acid (conc.)	Acetic acid, aniline, chromic acid, hydrocyanic acid, hydrogen sulfide, flammable liquids and gases
Oxalic acid	Silver, mercury
Perchloric acid	Acetic anhydride, bismuth and its alloys, alcohols, paper, wood
Phosphorus	Sulfur, oxygen-containing compounds (such as chlorates)
Potassium	Carbon tetrachloride, carbon dioxide, water
Potassium chlorate	Sulfuric and other acids

Substance	Incompatible with
Potassium perchlorate	See chlorates
Potassium permanganate	Glycerin, ethylene glycol, benzaldehyde, sulfuric acid
Silver	Acetylene, oxalic acid, tartaric acid, ammonium compounds
Sodium	Carbon tetrachloride, carbon dioxide, water
Sodium peroxide	Methanol, ethanol, glacial acetic acid, acetic anhydride, benzaldehyde, carbon disulfide, glycerin, ethylene glycol, ethyl acetate, methyl acetate, furfural
Sulfuric acid	Potassium chlorate, potassium perchlorate, Group IA permanganates

APPENDIX X

Properties of Solvents for Liquid Chromatography Including Miscibilities

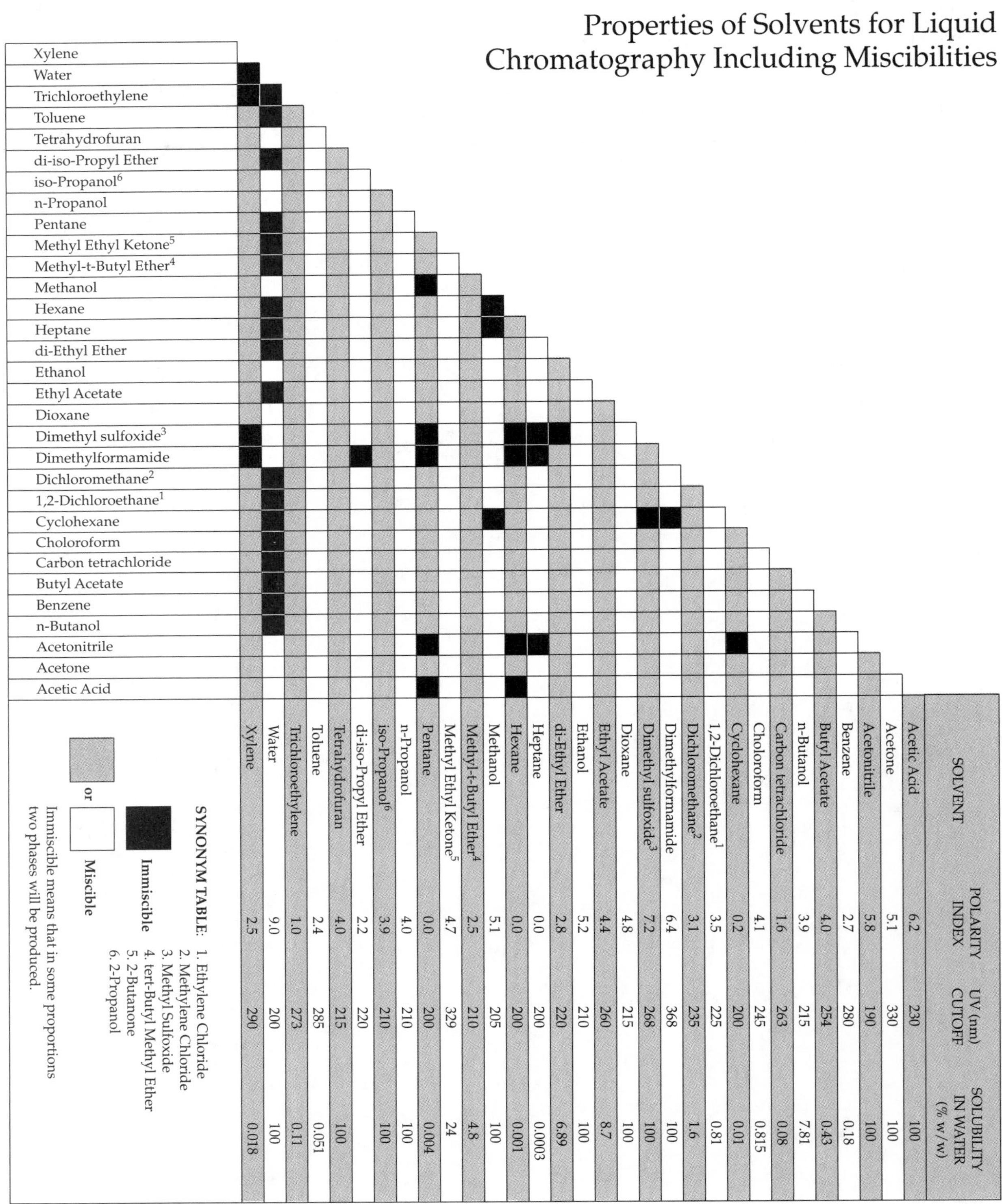

SOLVENT	POLARITY INDEX	UV (nm) CUTOFF	SOLUBILITY IN WATER (% w/w)
Acetic Acid	6.2	230	100
Acetone	5.1	330	100
Acetonitrile	5.8	190	100
Benzene	2.7	280	0.18
Butyl Acetate	4.0	254	0.43
n-Butanol	3.9	215	7.81
Carbon tetrachloride	1.6	263	0.08
Choloroform	4.1	245	0.815
Cyclohexane	0.2	200	0.01
1,2-Dichloroethane[1]	3.5	225	0.81
Dichloromethane[2]	3.1	235	1.6
Dimethylformamide	6.4	368	100
Dimethyl sulfoxide[3]	7.2	268	100
Dioxane	4.8	215	100
Ethyl Acetate	4.4	260	8.7
Ethanol	5.2	210	100
di-Ethyl Ether	2.8	220	6.89
Heptane	0.0	200	0.0003
Hexane	0.0	200	0.001
Methanol	5.1	205	100
Methyl-t-Butyl Ether[4]	2.5	210	4.8
Methyl Ethyl Ketone[5]	4.7	329	24
Pentane	0.0	200	0.004
n-Propanol	4.0	210	100
iso-Propanol[6]	3.9	210	100
di-iso-Propyl Ether	2.2	220	
Tetrahydrofuran	4.0	215	100
Toluene	2.4	285	0.051
Trichloroethylene	1.0	273	0.11
Water	9.0	200	100
Xylene	2.5	290	0.018

SYNONYM TABLE:
1. Ethylene Chloride
2. Methylene Chloride
3. Methyl Sulfoxide
4. tert-Butyl Methyl Ether
5. 2-Butanone
6. 2-Propanol

Immiscible

or **Miscible**

Immiscible means that in some proportions two phases will be produced.

Appendix XI

List of Acronyms

AA	Atomic absorption (spectrometry)
AAS	Atomic absorption spectrometry
ACS	American Chemical Society
ADC	Analog-to-digital converter
AE	Atomic emission (spectrometry)
AEM	Analytical electron microscope
AES	Atomic emission spectrometry
AES	Auger electron spectrometry
AF	Atomic fluorescence
AFM	Atomic force microscopy
AGE	Agarose gel electrophoresis
AOAC	Association of Official Analytical Chemists
APCI	Atmospheric pressure chemical ionization
AR	Analytical reagent grade
ASE	Accelerated solvent extraction™
ASTM	American Society for the Testing of Materials
ASV	Anodic stripping voltammetry
ATR	Attenuated total reflection
B	magnetic sector (in mass spectrometry)
BNA	Bases, neutrals, and acids
BSA	Bovine serum albumin
CAD	Collisionally activated dissociation
CARS	Coherent anti-Stokes Raman scattering (spectrometry)
CCD	Charge-coupled device
CD	Circular dichroism (spectrometry)
CDL	Concentration detection limit
CE	Capillary electrophoresis
CEEC	Capillary electrophoresis electrochemical detection
CEM	Continuous emission (as in stack emission) monitors
CGE	Capillary gel electrophoresis
CI	Chemical ionization
CID	Charge-injection device; Collisionally induced dissociation
CIDNP	Chemically induced dynamic nuclear polarization (NMR)
CIEF	Capillary isoelectric focusing
CITP	Capillary isotachophoresis
CLS	Classical least squares
CMC	Critical micellar concentration
CMRR	Common mode rejection ratio
COSY	Correlated spectroscopy (NMR)
CP-MAS-NMR	Cross polarization–magic angle spinning–NMR
CTD	Charge-transfer device (CCD & CID)
CV	Cold vapor
CVAAS	Cold vapor atomic absorption spectrometry
CVG	Chemical vapor generation; Chemical vapor generator
CW	Continuous wave
CZE	Capillary zone electrophoresis
DAC	Digital-to-analog converter
DAD	Diode array detector
dB	Decibel
DBM	Diazobenzyloxymethyl
DCP	Direct current plasma
DL	Detection limit
DME	Dropping mercury electrode
DOE	Diffractive optical element
DPP	Differential pulse polarography
DSC	Differential scanning calorimetry
DTA	Differential thermal analysis
DTG	Differential thermogravimetry
DTGS	Deuterated triglycine sulfate (IR detector)
DTT	Dithiothreitol
E	Electric sector (in mass spectrometry)
EAA	Electrothermal atomic absorption
EC	Electron capture; Electrochemistry
ECC	Electrokinetic capillary chromatography
ECD	Electron capture detection; Electron capture detector; Electrochemical detection
EDTA	Ethylenediamine tetra-acetate; Ethylenediamine tetra-acetic acid
EDX	Energy-dispersive X-ray (detector)
EDXF	Energy-dispersive X-ray fluorescence
EELS	Electron energy loss spectroscopy
EGA	Evolved gas analysis
EI	Electron ionization; Electron impact

EKC	Electrokinetic chromatography
ELCD	Electrolytic conductivity detector
ELISA	Enzyme linked immunoabsorbent assay
ELS	Evaporative light scattering
ELSD	Evaporative light scattering detector
EMF	Electromotive force (= voltage)
EMP	Electron microprobe analysis
ENDOR	Electron nuclear double resonance
EO	Electroosmosis
EOF	Electroosmotic flow; Electroendosmotic flow
EPR	Electron paramagnetic resonance (spectrometry)
ES	Electrospray; Emission spectrometry
ESCA	Electron scattering for chemical analysis; Electron spectrometry for chemical analysis
ESI	Electrospray ionization
ESR	Electron spin resonance (spectrometry) (also EPR)
ESSA	Extractable semivolatile strong acids
ETA	Electrothermal atomization (electric furnace)
ETV	Electrothermal volatilization
EVA	Ethylene vinyl acetate copolymer
EXAFS	X-ray absorption fine structure
FAA	Flame atomic absorption
FAB	Fast atom bombardment
FAE	Flame atomic emission
FC	Faraday cup
FD	Field desorption
FFF	Field-flow fractionation
FFT	Fast Fourier transform
FIA	Flow injection analysis; Flow injection analyzer
FID	Flame ionization detector; Flame ionization detection; Free induction decay (in magnetic resonance)
FS	Full scale
FT	Fourier transform
FTIR	Fourier transform infrared (spectrometry)
FZCE	Free zone capillary electrophoresis
GC	Gas chromatography
GD	Glow discharge
GDMS	Glow discharge mass spectrometry
GFAA	Graphite furnace atomic absorption
GFC	Gel filtration chromatography
GM	Special reference materials
GPC	Gel permeation chromatography
HCL	Hollow cathode lamp
HDC	Hydrodynamic chromatography
HETP	Height equivalent to a theoretical plate
HFS	Hyperfine splitting
HID	Helium ionization detector
HMQC	Heteronuclear correlation through multiple quantum coherence (NMR)
HOE	Holographic optical element
HPCE	High-performance capillary electrophoresis
HPLC	High-performance liquid chromatography
HPTLC	High-performance thin layer chromatography
HRMS	High-resolution mass spectrometry
IC	Ion chromatography
ICAP	Inductively coupled argon plasma
ICP	Inductively coupled plasma
ICP-MS	Inductively coupled plasma mass spectrometry
ICR	Ion cyclotron resonance
ID	Inner diameter
IEF	Isoelectric focusing
IgG	Immunoglobulin
IMS	Ion mobility spectrometry
INAA	Instrumental neutron activation analysis
IPC	Ion-pair chromatography; Ion-pairing chromatography
IR	Infrared
ISE	Ion-selective electrode
ISFET	Ion-selective field-effect transistor
ISS	Ion-scattering spectroscopy
IUPAC	International Union of Pure and Applied Chemistry
KHP	Potassium hydrogen phthalate
LAAS	Laser atomic absorption spectroscopy
LAMMS	Laser ablation microprobe mass spectrometry
LC	Liquid chromatography
LCEC	Liquid chromatography electrochemical detection
LEED	Low-energy electron diffraction
LEI	Laser-enhanced ionization
LID	Laser-induced dissociation
LIF	Laser-induced fluorescence
LIMS	Laboratory information management system
LLE	Liquid-liquid extraction

LLOD	Lower limit of detection (= LOD)
LOD	Limit of detection
LOQ	Limit of quantitation; Limit of quantification
LSB	Least significant bit
m/z	Mass-to-charge ratio
MAE	Microwave-assisted extraction
MALDI	Matrix-assisted laser desorption and ionization
MAS	Magic angle spinning (NMR)
MCD	Microchannel plate detector; Magnetic circular dichroism (spectrometry)
MCL	Maximum contaminant level
MDC	Minimal detectable concentration
MDL	Method detection limit
MECC	Micellar electrokinetic capillary chromatography
MI	Matrix isolation
MIKES	Mass-analyzed ion kinetic energy spectrometry
MIP	Microwave-induced plasma
MS	Mass spectrometry
MSB	Most significant bit
MSD	Mass spectral detection; Mass spectroscopic detection; Mass spectrometric detection
NAA	Neutron activation analysis
NBS	National Bureau of Standards (U.S.) (now N.I.S.T.)
NC	Nitrocellulose
NEWS	Neutral water-soluble organics
NHE	Normal hydrogen electrode
NIR	Near infrared
NIST	National Institute of Standards and Technology (U.S.)
NMR	Nuclear magnetic resonance
NOESY	Nuclear Overhauser Effect spectrometry (NMR)
NOVA	Nonvolatile acids
NP	Normal phase (in liquid chromatography)
NPD	Nitrogen-phosphorus detector
NQR	Nuclear quadrupole resonance
OCP	Organochlorine pesticides
OD	Outer diameter
ODS	Octyldecylsilyl
OES	Optical emission spectrometry
OPP	Organophosphorous pesticides
ORD	Optical rotatory dispersion
PAGE	Polyacrylamide gel electrophoresis
PAH	Polyaromatic hydrocarbons
PAS	Photoacoustic spectrometry
PC	Plasma chromatography (see ion mobility spectrometry)
PCA	Principal component analysis
PCB	Polychlorinated biphenyls
PCR	Principal component regression; Polymerase chain reaction; Post column reactor
PD	Photodissociation; Plasma desorption
PDA	Photodiode array
PEEK	Polyetheretherketone
PFPD	Pulsed flame photometric detection
PGC	Porous graphitic carbon
pH	Negative of the logarithm (base 10) of the hydrogen-ion effective concentration
PID	Photoionization detector
PIXE	Particle-induced X-ray emission
PLS	Partial least squares
PMT	Photomultiplier tube
ppb	Parts per billion (per 10^9)
ppm	Parts per million
ppq	Parts per quadrillion
ppt	Parts per thousand; Parts per trillion
pptr	Parts per trillion
PSD	Post-source decomposition; Post-source decay; Phase-sensitive detection
PSRR	Power supply rejection ratio
PTFE	Polytetrafluoroethylene (Teflon)
PVDF	Polyvinylidene fluoride; polyvinyldifluoride
Q	Quadrupole sector (in mass spectrometry)
QA/QC	Quality assurance/Quality control
RBS	Rutherford backscattering spectrometry
RCRA	Resource Conservation and Recovery Act
rDNA	Recombinant DNA
RFI	Rediofrequency interference
RGA	Residual gas analyzer
RI	Refractive index
RID	Refractive index detector
RIMS	Resonance ionization mass spectrometry
RIS	Resonance ionization spectrometry
RM	Research materials
RP	Reversed phase (in liquid chromatography)
RPLC	Reversed-phase liquid chromatography

RSD	Relative standard deviation	TG	Thermogravimetry (see TGA)
S/N	Signal-to-noise ratio	TGA	Thermogravimetric analysis
SAM	Scanning Auger microscopy (spectrometry); self-assembled monolayer	TGS	Triglycine sulfate
SAM(-PT)	Strong amines (-primary & tertiary, -secondary)	TIC	Total ion chromatogram
SANS	Small angle neutron scattering	TIMS	Thermal ionization mass spectrometry
SAXS	Small angle X-ray scattering	TISAB	Total ionic strength adjustment buffer
SCE	Saturated calomel electrode	TLC	Thin-layer chromatography; Thick-layer chromatography
SDS	Sodium dodecyl sulfate	TMS	Tetramethylsilane (NMR); Trimethylsilyl- (GC)
SEC	Size exclusion chromatography	TOF-MS	Time-of-flight mass spectrometry
SEM	Scanning electron microscope; Scanning electron microprobe; Secondary electron multiplier; scanning electrochemical microscopy	TR	Transfer ratio
SERS	Surface-enhanced Raman scattering (spectrometry)	TRIKES	Time-resolved ion kinetic energy spectrometry
SFC	Supercritical fluid chromatography	TS	Thermospray
SFE	Supercritical fluid extraction	TSM	Thickness sear mode (piezoelectric oscillators)
SHE	Standard hydrogen electrode	TSMS	Thermospray mass spectrometry
SIM	Selective ion monitoring	TSP	Thermospray
SIMS	Secondary ion mass spectrometry	UHP	Ultra-high purity
SNMS	Secondary neutral(s) mass spectrometry	UV	Ultraviolet (light)
SPE	Solid-phase extraction	V/F	Voltage-to-frequency converter
SPM	Scanning probe microscopy	vis	Visible (light)
SPME	Solid-phase micro extraction	VO	Volatile (purgeable) organics
SRM	Standard reference material(s)	VOSA	Volatile strong acids
STM	Scanning tunnelling microscopy	WABN	Weak acids, bases, and neutrals
SWIFT	Stored waveform inverse Fourier transform	WDX	Wavelength-dispersive X-ray (spectrometers)
TC	To contain	WEDX	Windowless energy-dispersive X-ray analysis
TCD	Thermal conductivity detector	XANES	X-ray absorption near-edge spectrometry
TCLP	Toxicity characteristic leaching procedure	XPS	X-ray photoelectron spectrometry (ESCA)
TD	To deliver	XRD	X-ray diffraction
TDS	Total dissolved solids	XRF	X-ray fluorescence
		YAG	Yttrium aluminum garnet

APPENDIX XII

Concentrated Acids and Bases

	Approximate Weight Percent in Reagent Grade	Molarity Corresponding to to wt %	mL of Reagent Needed to Prepare 1.0 L of ~1.0-M Solution
Acids			
Acetic	99.8	17.4	57.5
Hydrochloric	37.2	12.1	82.6
Hydrofluoric	49.0	28.9	34.6
Nitric	70.4	15.9	62.9
Perchloric	70.5	11.7	85.5
Phosphoric	85.5	14.8	67.6
Sulfuric	96.0	18.0	55.6
Bases			
Ammonia*	28.0	14.5	69.0
Sodium hydroxide	50.5†	19.3	51.8
Potassium hydroxide	52.0	14.2	70.4

*28.0% ammonia is the same as 56.6% ammonium hydroxide.
†Saturated solution at 20°C.

APPENDIX XIII

Formula Weights of Some Common Compounds and Reagents

Compound	Formula weight
$AgBr$	187.78
$AgCl$	143.32
Ag_2CrO_4	331.73
AgI	234.77
$AgNO_3$	169.87
$AgSCN$	165.95
Al_2O_3	101.96
$Al_2(SO_4)_3$	342.15
AS_2O_3	197.84
B_2O_3	69.62
$BaCO_3$	197.35
$BaCl_2 \cdot 2\,H_2O$	244.28
$BaCrO_4$	253.33
$Ba(IO_3)_2$	487.15
$Ba(OH)_2$	171.35
$BaSO_4$	233.40
$BaS_2O_3 \cdot H_2O$	267.48
Bi_2O_3	465.96
CO_2	44.01
$CaCO_3$	100.09
CaC_2O_4	128.10
CaF_2	78.08
CaO	56.08
$CaSO_4$	136.14
$Ca(HSO_4)_4$	528.40
CeO_2	172.12
$Ce(SO_4)_2$	332.24
$(NH_4)_2Ce(NO_3)_6$	548.23
$(NH_4)_4Ce(SO_4)_4 \cdot 2\,H_2O$	632.55
Cr_2O_3	151.99
CuO	79.54
Cu_2O	143.08
$CuSO_4$	159.60
$Fe(NH_4)_2(SO_4)_2 \cdot 6\,H_2O$	392.14
FeO	71.85
Fe_2O_3	159.69
Fe_3O_4	231.54
$HC_2H_3O_2$ (acetic acid)	60.05
$HC_7H_5O_2$ (benzoic acid)	122.12
$HClO_4$	100.46
$H_2C_2O_4 \cdot 2\,H_2O$	126.07
H_5IO_6	227.94
HNO_3	63.01
H_2O	18.015
H_2O_2	34.01
H_3PO_4	98.00
H_2S	34.08
H_2SO_3	82.08
H_2SO_4	98.08
HgO	216.59
Hg_2Cl_2	472.09
$HgCl_2$	271.50
KBr	119.01
$KBrO_3$	167.01
KCl	74.56
$KClO_3$	122.55
KCN	65.12
K_2CrO_4	194.20
$K_2Cr_2O_7$	294.19
$K_3Fe(CN)_6$	329.26
$K_4Fe(CN)_6$	368.36
$KHC_8H_4O_4$ (potassium hydrogen phthalate)	204.23
$KH(IO_3)_2$	389.92
K_2HPO_4	174.18
KH_2PO_4	136.09
$KHSO_4$	136.17
KI	166.01
KIO_3	214.00
KIO_4	230.00
$KMnO_4$	158.04
KNO_3	101.11
KOH	56.11
$KSCN$	97.18
K_2SO_4	174.27
$La(IO_3)_3$	663.62
$MgCO_3$	84.32
$MgNH_4PO_4$	137.32
MgO	40.31
$Mg_2P_2O_7$	222.57
$MgSO_4$	120.37
MnO_2	86.94
Mn_2O_3	157.87
Mn_3O_4	228.81
$Na_2B_4O_7 \cdot 10\,H_2O$	381.37
$NaBr$	102.90
$NaC_2H_3O_2$ (sodium acetate)	82.03
$Na_2C_2O_4$ (sodium oxalate)	134.00
$NaCl$	58.44
$NaCN$	49.01
Na_2CO_3	105.99
$NaHCO_3$	84.01
Na_2H_2edta $\cdot 2\,H_2O$	372.24
Na_2O_2	77.98
$NaOH$	40.00
$NaSCN$	81.07
Na_2SO_4	142.04
$Na_2S_2O_3 \cdot 5\,H_2O$	248.18
NH_4Cl	53.49
$(NH_4)_2C_2O_4 \cdot H_2O$	142.11
NH_4NO_3	80.04
$(NH_4)_2SO_4$	132.14
$(NH_4)_2S_2O_8$	228.20
$PbCrO_4$	323.18
$Pb(NO_3)_2$	331.20
PbO	223.19
PbO_2	239.19
$PbSO_4$	303.25
P_2O_5	141.94
Sb_2S_3	339.69
SiO_2	60.08
$SnCl_2$	189.60
SnO_2	150.69
SO_2	64.06
SO_3	80.06
$Zn_2P_2O_7$	304.68

Appendix XIV

Physical Constants

Term	Symbol	Value*	Multiplication Factor and Units
Avogadro's number	N_A	6.022 136 7(36)	$\times 10^{23}$ mol^{-1}
Elementary charge	e	1.602 177 33(49)	$\times 10^{-19}$ Coulomb
Electron rest mass	m_e	9.109 389 7(54)	$\times 10^{-28}$ g
		5.485 799 03(13)	$\times 10^{-4}$ u
Atomic mass unit†	u	1.660 540 2(10)	$\times 10^{-24}$ g
Gas constant	R	8.314 510(70)	J mol^{-1} K^{-1}
			V C mol^{-1} K^{-1}
			$\times 10^{7}$ erg mol^{-1} K^{-1}
		8.205 784(68)	$\times 10^{-5}$ m^3 atm mol^{-1} K^{-1}
			$\times 10^{-2}$ L atm mol^{-1} K^{-1}
		1.987 21 (2)	cal mol^{-1} K^{-1}
Faraday constant ($=N_Ae$)	$\mathcal{F}$	9.648 530 9(29)	$\times 10^{4}$ C mol^{-1}
Boltzmann's constant ($=R/N_A$)	k_B	1.380 658(12)	$\times 10^{-23}$ J K^{-1}
			$\times 10^{-16}$ erg K^{-1}
Planck's constant	h	6.626 075 5(40)	$\times 10^{-34}$ J s
			$\times 10^{-27}$ erg s
Proton rest mass	m_p	1.672 623 1(10)	$\times 10^{-24}$ g
		1.007 276 470(12)	u
Bohr magneton	μ_B	9.274 015 4(31)	$\times 10^{-24}$ J T^{-1}
			$\times 10^{-21}$ erg G^{-1}
Speed of light in vacuum	c	2.997 924 58	$\times 10^{8}$ m s^{-1}
			$\times 10^{10}$ cm s^{-1}

*Numbers in parentheses are the one-standard-deviation uncertainties in the last digits.
†Calculated from N and $^{12}C = 12\ u$ exactly.

Source: Cohen, E. R., Taylor, B. N. 1987. *Rev. Mod. Phys.* 57:1121. Taylor, B. N., Cohen, E. R. 1990. *J. Res. Natl. Inst. Stand. Technol.* 95:497.

APPENDIX XV

Energy Conversion Factors*

	ergs/molecule	kJ/mol	kcal/mol	electron volts/molecule
ergs/molecule	1	6.0221×10^{13}	1.4393×10^{13}	6.2415×10^{11}
kJ/mol	1.6605×10^{-14}	1	0.23900	1.0364×10^{-2}
kcal/mol	6.9478×10^{-14}	4.184	1	4.336×10^{-2}
electron volts/molecule	1.6022×10^{-12}	9.6485×10^{1}	23.060	1

*See Table 11.2 for other conversion factors.

Answers to Exercises

Chapter 1

Concept Review

1. "Analysis of iron" refers to finding the content of the various other components as well as iron in a sample that is assumed to be all (or at least predominately) iron while "assay for iron" refers to finding the iron content of a sample that may contain any level of iron.
2. "Validation" of a method involves proving that it does indeed give an accurate and precise measure of the content of some analyte; involves use of known samples and/or comparison of the results of the method with some method known to give accurate and precise results; method generally validated for certain analyte in certain type of sample.
3. Define problem; obtain samples; prepare sample; perform assays and data reduction; evaluate results using statistics to determine reliability of method; refine method if necessary
4. Weight/weight measures and weight/volume measures nearly equivalent when density of sample close to 1; for example, for dilute aqueous solutions 1 mg/L and 1 mg/kg both correspond to 1 ppm.

Exercises

1.1 (a) 2.8×10^{-8}; (b) 5.7×10^{-4} 1.2 (a) 0.100%; (b) 1000 ppm; (c) 7.62×10^{-3} 1.3 (a) 2.83×10^{-2}; (b) 0.585%; (c) 9.96×10^{-2} 1.4 18.0 M 1.5 100,000 ppm 1.6 (a) 86 ppm K^+, 2750 ppm Cl^-; (b) 85.6 ppm K^+, 2740 ppm Cl^- 1.7 10% 1.8 47.73 mL 1.9 (a) 20 mg Ca^{2+}, 30 mg Co_3^{2-}; (b) 100×10^{-3} M 1.10 (a) 94.3 mg 1.11 5.45 mL 1.12 504 g 1.13 (a) Na^+: 0.0786 g/L, 0.078g mg/mL; Cl^-: 0.120 g/L, 0.120 mg/mL; (b) all are 3.42 mM 1.14 6.0 ppm 1.15 (a) 9.21 g; (b) 1.27 mg; (c) 1.93 μg 1.16 4.69×10^{-4} 1.17 (a) 0.01 N; (b) 0.1 N; (c) 0.04 N; (d) 0.04 N 1.18 (a) 0.024; (b) 0.12 N; (c) 0.024 1.19 0.0980 N 1.20 1.24 L 1.21 The overall dilution factor is 10^5, so we need five 10-fold dilutions 1.22 (a) 2.00; (b) 5.000; (c) 2.60; (d) 0.344 1.23 (a) 1.0×10^{-6}; (b) 3.14×10^{-4}; (c) 5.62×10^{-4}; (d) 8.9×10^{-12} 1.24 0.08142 N 1.25 1.1 M 1.26 747 mL 1.27 0.1108 M 1.28 0.2572 N 1.29 Dissolve 21.0 mg to give 100 mL of solution. Carry out five 10-fold dilutions. 1.30 (a) 4.000; (b) 5.000; (c) 2.82 1.31 (a) 2.1×10^{-3}; (b) 2.1×10^4; (c) 0.9966 1.32 5.6×10^{-6} 1.33 5.25 1.34 2.0×10^{-5}

Chapter 2

Concept Review

1. (a) Determinate errors are always biased in one direction either high or low (can be constant or proportional); affect closeness of analytical results to the "true" value" (accuracy). Indeterminate errors are random errors and thus are distributed in what approaches a Gaussian distribution around the "true" value; reflected in the precision of the result. (b) Precision reflects the degree of scatter of the data around the mean value, while accuracy reflects the closeness of the mean value to the "true" value. (c) The mean error is the difference between the mean and the "true" value, while the error of the mean is another name for the standard deviation of the mean.
2. Yes. The protocol precision is better than that required by the regulators.
3. Since σ_m is inversely proportional to the square root of N, a *decrease* of a factor of two in σ_m would be assured by an *increase* of a factor of $(2)^2 = 4$ in N.
4. The median is a better indicator of an experimental value when there is a large scatter in the data.

Problems

2.1 100 2.2 (a) 40.4 ppm; 0.7 ppm; (b) 0.7 2.3 0.22% 2.4 21.25%; 0.04%; 0.0016; 0.04% 2.5 4.981 mL; 0.007 mL; 0.003 mL; 0.023 mL 2.6 4% 2.7 (a) Sample 1: 57.78%, RSD = 0.0019; Sample 2: 57.69%, RSD = 0.0029; (b) 95% C.L./mean = 0.0020; about two-thirds the value of the RSD 2.8 (a)

Error in Analysis	
in g	in %
0.0022	—
0.0023	2.3
0.0033	0.84
0.0016	0.22

(b) constant, 0.0026 g; (c) random, 0.0018; (d) at least 0.5 g 2.9 (a)

Beer A	0.53%
Beer B	0.69%
Beer C	−0.28%
Wine A	1.1%
Wine B	−1.2%
Wine C	1.1%

(does not matter which is considered to be "true" in this case, as long as we are consistent); (b) 0.003, no relative bias within limits given 2.10 from the impurities 2.11 (a) 0.60%; (b) close to zero 2.12 13.6 ± 0.4 2.13 25.509 ± 0.004 2.14 −25.88 ± 0.01 2.15 (a) 0.1 to 0.2 % error; (b) 0.04 mg 2.16 (a) 2.84 ± 0.12 ppm; (b) No. Difference would have to have been 0.46 to be significant at the 95% level 2.17 (a) 0.004 ppm; (b) 0.0014 ppm; (c) no 2.18 end up with same number of significant figures by either method 2.19 13.051 2.20 (a) 80; (b) $\bar{X} = 58$; $s = 30$; $\mu = 58 \pm 28$ 2.21 (b) $Y = 125\,X - 1.5$; $s_{slope} = 0.4$; $s_{intercept} = 3.5$; $r = 1.000$; (c) 8.14 2.22 $m = 125$; $b = -1.5$ 2.23 $r = a + b + c$ 2.24 relative error: $r/R = a/A +$

$b/B + c/C$; absolute error: $r = 1/C^2\,(aBC + bAC + cAB)$
2.25 (a) a/A; (b) $a(Ae^A)$; (c) $0.879a/A$; (d) 2.193
2.26 $s_R = \frac{s_A}{A} + \frac{s_B}{B} + \frac{s_C}{C}$

Chapter 3

Concept Review

1. Size of the bulk to be sampled, physical state of the fraction to be analyzed, and chemistry of material to assayed
2. Representative means that content of analytical sample reflects content of bulk sample while homogeneous means that the analytical sample has the same content throughout.
3. Sampling should be carried out at regularly spaced times or positions.
4. Variation is minimized between individual samples from the larger sample, and individual samples are more representative of the larger sample.
5. (a) purge the water sample and trap the gas in a trap filled with BaO; (b) pull a known volume of air through a trap containing finely divided glass wool; (c) extract the smudged portion of the sweater with several small portions of hexane; (d) distillation

Exercises

3.1 (a) 5.2%; (b) both about 6.0%; (c) two regularly spaced: very similar; regular with random: 13% relative error **3.2** (a) 120 s; (b) no; (c) 1.43 **3.3** 39.1% **3.4** (a) 9.92 L; (b) 1.60 g m^{-3}; 3.25 g m^{-3}; (c) 4.04 g m^{-3}
3.5 (a) manual: 1.7; automatic: 0.84; (b) automatic; (c) manual: 10.6%; automatic: 10.8%; (d) better sampling method **3.6** (a) 127.7 L; 77.0 L; (b) 0.230 g; 0.00182 g; (c) 9.93×10^3 g; (d) 1.79 g; (e) yes, 24.3% of total
3.7 (a) 9.0×10^{-10} particles; (b) 0.141 g; 14.1 g; (c) either one as long as sample not limited **3.8** yes, since relationship of volume to dimension given is different (however, *relative* size of samples needed will be same).
3.9 9–10 samples **3.10** (a) 100%; (b) 3.80×10^{-7} moles; (c) 9.3 μL; (d) 0.95 L; (e) 7.3 min

Chapter 4

Concept Review

1. Minimize loss of analyte, bring analyte into proper chemical form for assay method, remove interferents from sample matrix, avoid adding new interferents, include dilution or preconcentration steps to obtain optimum concentration range for analyte based on assay method
2. Specificity implies that it responds *only* to calmodulin; selective implies that its response to calmodulin is much greater (based on concentration or mass) than to other species.
3. Flux digestion: Higher temperatures can be used; concentration of flux reagents much higher; not limited by solvent chemistry. Microwave digestion: since in sealed container minimizes sample loss and results in higher boiling points for solvents; also better control of sample temperature and faster heating.
4. External standards are analyzed separately from the sample while added standards are added to the sample itself before or after analysis.
5. Matrix unknown or very complex; chemistry of method complex and/or recoveries variable; highly precise control of conditions necessary but difficult to obtain

Exercises

4.1 383 ppm; ± 3 ppm; ± 4 ppm **4.2** 328 ppm
4.3 251 ppm **4.4** 4% **4.5** 0.32% **4.6** 6 extractions
4.7 250 ppm **4.8** $C_{sample} = \frac{A_{sample} C_{spike}}{A_{sample+spike} - A_{sample}}$
4.9 (a) 0.0999 ppm; (b) 5 **4.10** (a) extrapolate through the plateau; (b) 197 ± 4 mV; (c) no; (d) 320 mg dL^{-1}
4.11 (a) 10.0 μg; (b) 90.5%; (c) 2.1%; (d) 1.7%; (e) 2.1%; (f) 4.7% **4.12** 0.88 ng mL^{-1}; 8.61 ng mL^{-1}; 1.26 ng mL^{-1}
4.13 49.7% **4.14** 3.67% **4.15** (a) CrO_4^{2-} and OCl^-; both positive; (b) 0.29 for CrO_4^{2-}; varies from 0.14 to 0.043 for OCl^-; (c) zero **4.16** (a) 2.29 mg; (b) 3.01%
4.17 (a) 31.4%; (b) 314 kg **4.18** (a) all positive; (b) pyridine (others within ±5%); (c) no. not enough information
4.19

	chrysene	benzofluoranthenes
(a)	70%	63%
(b)	18%	49%
	benzo(a) pyrene	**dibenzo(a,h)anthracene**
(a)	63%	64%
(b)	23%	27%

(c) chrysene is only one where $\overline{X} \pm s$ assures that minimum error due to sampling will be less than 50%.
4.20 (a) 3–4 minutes (point where response levels off); (b) no, level does not go back down; (c) 1.6 **4.21** Linear gives 21.0 ppm; log gives 24.0 ppm; assuming log solution is "true value," 12.5% error in linear result
4.22 Standard was twice concentration he thought, so recovery actually twice as high. 90% recovery. Corrosion on tray in hood supports that some of sample was lost.
4.23 (a) Chrysene extraction with dichloromethane; (b) 35.0% **4.24** The soxhlet method shows not only a very poor recovery for parathion, but the relative standard deviation for the recovery is 92%. Its recoveries are about the same for the other pesticides, but once again with tetrachlorvos, the sohlet method has a large RSD in the recovery.

Chapter 5

Concept Review

1. Sensitivity refers only to the change in response with a given change in content of analyte, while detection limit refers to the minimum amount that can be reliably (95% CL) detected above background.
2. The detection limit is defined based on the standard deviation in the background signal and is usually calculated from data on samples of optimum form. The quantitation limit is a much more realistic limit since it refers to the amount that can actually be quantitated with a reasonable degree of precision. The instrumental detection limit applies to the instrument only, without errors from any other part of the procedure.
3. The relationship between the two reflects the relationship between the noise voltages measured in two ways; i.e., $V_{rms} = 0.35\ V_{p\text{-}p}$.

Exercises

5.1 (a) macro sample, major component; (b) macro sample, trace component; (c) macro sample, major component; (d) macro sample, minor component 5.2 about 3 ppm 5.3 (a) 2 ng; (b) 6.6 μm 5.4 (a) no; (b) radius = 290 μm; (c) no (impurity would be 5 times expected amount of Mn) 5.5 (a) 50 times; (b) Zn 0.01 ppb; Cu 0.18 ppb; Ni 0.19 ppb; Mn 1.5 ppb; Ag 0.002 ppb; Cd 0.001 ppb; Pb 0.026 ppb; (c) yes 5.6 (a) 12.8%; (b) 4.4% 5.7 (a) hexane; (b) ether; (c) hexane; (d) same answers, since even 1/10 of Zn and Na values still high for other solvents
5.8 (a)

Element	ng/g
Br	253
Co	232
Fe	241,000
Mn	4,340
Hg	25
Cr	183
K	19,700
Na	—
Zn	1,060

(b) Na, K, Zn; (c) No, all could have come from solvent

Chapter 6

Concept Review

1. $A + B \rightleftharpoons C$: $K = \dfrac{[C]}{[A][B]}$; $C \rightleftharpoons A + B$: $K = \dfrac{[A][B]}{[C]}$;
$2\,A + 2\,B \rightleftharpoons 2\,C$: $K = \dfrac{[C]^2}{[A]^2[B]^2}$
second is reciprocal of first (since reaction reversed); third is square of first (since coefficients are doubled)
2. (a) $K_{a,HA} > K_{a,HB}$; (b) A^- is a stronger base that B^-
3. (a) $HOOC\text{-}CH_2\text{-}CH_2\text{-}COOH$, $^-OOC\text{-}CH_2\text{-}CH_2\text{-}COOH$, $^-OOC\text{-}CH_2\text{-}CH_2\text{-}COO^-$; (b) $H_2N\text{-}CH_2\text{-}CH_2\text{-}NH_2$, $^+H_3N\text{-}CH_2\text{-}CH_2\text{-}NH_2$, $^+H_3N\text{-}CH_2\text{-}CH_2\text{-}NH_3^+$
4. Below pH 2.5, predominately $H_2C_4H_4O_6$; pH 2.5 to 4.5, $H_2C_4H_4O_6$ concentration decreases and $HC_4H_4O_6^-$ concentration increases briefly, and then $C_4H_4O_6^{2-}$ concentrations rises and $HC_4H_4O_6^-$ decreases; above pH 4.5, $C_4H_4O_6^{-2}$ predominates.
5. They indicate the fractions of the overall concentration that are present in each of the individual forms at a given pH.
6. Since the ratio in the log term of the Henderson-Hasselbalch equation must change by a factor of 10 in order to change one pH unit, addition of small amounts of acid or base do not produce large changes in pH.
7. The buffer is most effective at pH values near its pK_b, since there are appreciable amounts of both the base and its protonated form at pH = pK_b.
8. The solution that is 0.5 M in each of the forms would have a higher buffer capacity. In the more concentrated buffer the reaction of a small amount of the acid or base form with the added H^+ or OH^- will not produce a large change in the ratio of the concentrations of the two forms. This means, in turn, that there will not be as large a change in the pH of the solution.
9. The interaction between the ions present.

Exercises

6.1 (a) $K = \dfrac{[CO][H_2]^2}{[CH_3OH]}$; (b) $K = \dfrac{[CO_2][H_2]}{[CO][H_2O]}$;
(c) $\dfrac{[\text{gluconate}^-][I^-]^3[H_2O]^2}{[\text{glucose}][OH^-]^3[I_3^-]}$
6.2 (a) $K = \dfrac{[CH_3OH]}{[CO][H_2]^2}$;
mass action expression is reciprocal of that for 6.1a (since reaction reversed) 6.3 (a) shifts right; (b) shifts right
6.4 $K = \dfrac{[D]}{[A]^2[B]^2} = 10^{-6}$ 6.5 $K = \dfrac{[D]^2}{[A][B]^2} = 1$
6.6 hydrolysis of Bar^- 6.7 3.04 6.8 5.92 6.9 (a) 13.3%; (b) 63.6%; (c) 85.6% 6.10 (a) 1.0; (b) 6.97 6.11 (a) 11.13; (b) a little less than 1% 6.12 (a) 3.50; (b) 7.50; (c) 11.00 6.13 decrease 6.14 (a) right; (b) increase 6.15 (a) A = 135, B = 5.75×10^3, C = 2.51×10^5; (b) A = 2.34×10^{-2}, B = 1.00, C = 43.6; (c) A = 1.35×10^{-4}, B = 5.75×10^{-3}, C = 0.251 6.16 (a,b,c) decrease; (d) slightly lower in all cases 6.17 $[H_2CO_3] = 1.59 \times 10^{-5}$ M, $[HCO_3^-] = 0.0676$ M, $[CO_3^{2-}] = 0.0324$ M 6.18 $\alpha_1 = 0.983$ 6.19 (c) 5.66 g H_3BO_3 and 0.704 g NaH_2BO_3 per liter of solution based on α values for each at pH = 8.2 6.20 (a) 3.9×10^{-3} M; (b) 13% difference in $[H^+]$, 2% difference in pH 6.21 (a) 11; (b) $H_2PO_4^-$ and HPO_4^{2-} 6.22 (a) 0.1 M; (b) 0.6 M; (c) 1.0 M 6.23 (a) 0.40; (b) 0.24 6.24 (a) 0.02; (b) 0.12; (c) in H_2O, $\gamma = 0.80$; in NaCl solution, $\gamma = 0.67$; (d) 1.48

Chapter 7

Concept Review

1. Need titrant solution of known concentration, technique to precisely and accurately measure volume of titrant, technique to precisely and accurately measure sample weight or volume, pretreatment method to remove interferents (if present), and technique to detect equivalence point of titration.
2. HA (lowest pK_a means it is strongest acid and therefore is deprotonated first)
4. Leveling effect of water means there would be almost no jump between plateau of buffer region in titration and plateau after equivalence. Need a solvent which is less easily deprotonated than water.
5. Bromcresol green undergoes change from yellow to blue between pH = 3.8 and pH = 5.4, while phenolphthalein is colorless until pH = 8.3, then turns purple by pH = 10.0. "Adding" colors means at first end point, color changes from yellow to blue and then at second endpoint, color changes from blue to purple.

Exercises

7.1 2.72 7.2 (a) 0.01237 M; (b) 0.02852 M; (c) 0.01116 M 7.4 21 mg 7.5 (a) 245 mL; (b) 574 mL; 805 mL 7.6 (a) excess OH^-; (b) 75 mL, pH = 1.48; (c) 100 mL, pH = 7.00; (d) only change is at $\frac{1}{2}V_e$. pH will be 3.48 7.7 4.06 mEq/g 7.8 pH = 11.07; pOH = 2.93; $[CO_3^{2-}] = 8.5 \times 10^{-3}$ M; $[HCO_3^-] = 1.5 \times 10^{-3}$ M 7.9 (a) $V_{e1} = 7.81$ mL, $V_{e2} = 21.48$ mL; (b) same volumes, no difference in mL titrant 7.10 (a) [HOAc] = 0.019 M; $[H^+] = [OAc^-] = 5.9 \times 10^{-4}$ M; $[OH^-] = 1.7 \times 10^{-11}$ M; (b) $[H^+] = 1.75 \times 10^{-5}$ M; [HOAc] = $[OAc^-]$ = 0.01 M; $[OH^-] = 5.7 \times 10^{-10}$ M; (c) $[OAc^-]$ = 0.02 M; $[OH^-] = 3.4 \times 10^{-6}$ M; $[H^+] = 3.0 \times 10^{-9}$ M; $[K^+]$ = 0.020 M 7.11 (a) 5.92×10^{-4} M; (b) 4.95×10^{-3} M; (c) part a 0.956; part b 0.880 7.12 (a) methyl red, bromcresol green or phenolphthalein; (b) methyl red or phenophthalein 7.13 (a) only one; (b) 25.0 mL 7.14 $[HCO_3^-]$ = 0.433 M; $[CO_3^{2-}]$ = 0.582 M 7.15 (b) endpoint is where curve crosses zero 7.17 (a) 1.659 M; (b) no; (c) yes 7.18 12.5 units/mL

Chapter 8

Concept Review

1. (a) electron transfer; (b) formation (or dissociation) of a complex between a metal and ligand(s); (c) dissolution of a sparingly soluble salt to produce aqueous ions
2. If direction in which a half-reaction is written is reversed, the sign changes on the $E°$. Even if must multiply half-reaction by a constant to balance electrons lost and electrons gained, $E°$ remains unchanged.
3. (a) the value of n; (b) the value of T and the ratio of concentrations (or activities) in the log term
4. For a standard potential, T = 25°C, and all reactants are in their standard states. The formal potential for the half reaction:

$$a\,Ox + n\,e^- \rightleftharpoons b\,Red$$

is defined as the potential under conditions where $[Red]^b/[Ox]^a = 1$ and all other conditions are explicitly defined.
5. The trichloride salt (due its charge and also to the fact that any effects of any activity coefficients will be exaggerated).
6. Dissociation equilibria for H_4edta, formation equilibria for the complex, hydrolysis reactions of the metal, and any redox equilibria that affect the concentration of the metal or ligand.
7. (a) a single line |; (b) two parallel lines ||; (c) their symbols or formulas, separated by a comma; (d) the half-cell on the right side

Exercises

8.1 (a) $6\,H^+ + 6\,I^- + ClO_3^- \rightleftharpoons 3\,I_2 + Cl^- + 3\,H_2O$; (b) $10\,H^+ + 4\,Zn(s) + NO_3^- \rightleftharpoons NH_4^+ + 3\,H_2O + 4\,Zn^{2+}$ 8.2 (a) $2\,OH^- + 2\,Al(s) + 6\,H_2O \rightleftharpoons 2\,Al(OH)_4^- + 3\,H_2$; (b) $6\,H_2O + 3\,NiO_2(s) + 2\,Fe(s) \rightleftharpoons 3\,Ni(OH)_2 + 2\,Fe(OH)_3$ 8.3 (a) −0.124 V; (b) +0.250 V; +0.513 V 8.4 (a) 59.2 mV, more negative; (b) 118.4 mV, more positive; (c) 118.4 mV, more negative; (d) no change 8.5 (a) $Mg(s) \rightleftharpoons Mg^{2+} + 2\,e^-$; $Ag^+ + e^- \rightleftharpoons Ag(s)$; (b) $Mg(s) + 2\,Ag^+ \rightleftharpoons Mg^{2+} + 2\,Ag(s)$; (c) 3.162 V; (d) yes; (e) 1220 kJ/mole 8.6 (a) 1.54 V; (b) 1.42 V; (c) $H_2O_2 \rightleftharpoons 2\,OH^-$; no 8.7 (a) zero; (b) more positive; (c) 0.059 V; (d) more negative; (e) more negative 8.8 (a) 0.67 V; (b) 0.72 V; (c) 0.62 V 8.9 $[Co^{2+}]$ = 1.783 M; $[Ni^{2+}]$ = 0.217 M 8.10 (a) 0.06 V; (b) −0.03 V 8.11 0.44 g 8.12 (a) 1.6×10^{-20} M; (b) 1×10^{-97} 8.13 −0.577 V 8.14 1.7×10^{-10} 8.15 (a) 1.08×10^{-18} M; (b) 0.723 V 8.16 7.4×10^{-7} M

Chapter 9

Concept Review

1. The analyte is reacted with an excess of another solution species. The *product from the reaction* is then titrated.
2. All of pX values have a plateau near the bottom left hand corner (values depend on initial concentration of analyte) and slowly increase until just before the endpoint, where there is a sharp increase before another plateau (where values depend on the K for the equilibrium governing the reaction and the concentration of the titrant).

Exercises

9.1 0.0273 M 9.2 0.0674 N 9.3 0.0897 N; 0.0179 M
9.4 (a) $N_{I_2} = \frac{N_{S_2O_3^{2-}}}{V_{I_2}} V_{S_2O_3^{2-}}$;
(b) $\frac{mgI_2}{L} = 1.294 \times 10^5 \left(\frac{N_{S_2O_3^{2-}}}{V_{I_2}}\right) V_{S_2O_3^{2-}}$
9.5 (a) $H_2O + CH_3OH \rightleftharpoons CO_2 + 6\,H^+ + 6\,e^-$;
$H_2O + CH_3CH_2OH \rightleftharpoons CH_3COOH + 4\,H^+ + 4\,e^-$;
(b) 0.256% (w/w) 9.6 (a) 0.05 M; (b) 0.05 M; (c) 1.06 V; (d) 1.57×10^{14}; (e) same; (f) same; (g) $[Fe^{2+}] = [Ce^{4+}] = 1.8 \times 10^{-8}$ M 9.7 1.01% (w/v) 9.8 $[Fe^{3+}] = [Ce^{3+}] = 0.0444$ M; $[Fe^{2+}] = 0.0111$ M; $[Ce^{4+}] = 2.1 \times 10^{-13}$ 9.9 3.1 ppm 9.10 (a) 1.2 mg/mL; (b) 3.9 mg/mL 9.11 1.39×10^{-4} M 9.12 (a) sodium diphenylamine sulfate; (b) Fe^{3+}; (c) No 9.13 (a) 1.36 V; (b) binds to the $Cr_2O_7^{2-}$ 9.14 861 mg/L $CaCO_3$ 9.15 (a) 21.3% (w/w); (b) the Fe^{3+} is more stable, the Fe^{2+} less stable; (c) Zr complex; (d) same stability as with indirect titration; (e) same stability as with indirect titration;
(f) $H_5edta^+ + Zr^{4+} \rightleftharpoons Zr(edta) + 5\,H^+$;
$Zr^{4+} + H_2In^- \rightleftharpoons ZrIn_2^{2-} \rightleftharpoons ZrIn_2^{2-} + 4\,H^+$;
$Zr^{4+} + H_2In^- \rightleftharpoons ZrIn_2^{2-} \rightleftharpoons ZrIn_2^{2-} + 4\,H^+$

Chapter 10

Concept Review

1. (a) reduction of a metal ion in solution, resulting in the deposition of the metal onto the surface of an electrode; (b) loss of mass during heating
2. (a) high; (b) low
3. Low solubility; easy to filter; stable, definite composition in its final form
4. Minimized by any conditions that result in a slow approach to supersaturation: dilute solution for precipation, add precipitating agent slowly with stirring, homogeneous precipitation, heating solution so it approaches supersaturation slowly
5. *Occlusion* refers to impurities being trapped in voids in a crystal while *inclusion* refers to substitution of one ion for another in the crystal lattice
6. Heating speeds rates in equilibrium process of dissolution and reprecipitation, and slow cooling allows impurities to be excluded from the crystal lattice since approaching supersaturation more slowly

Exercises

10.1 (a) low; (b) remains constant 10.2 (a) high; (b) low 10.3 (a) 0.4770; (b) 0.5646; (c) 0.1893 10.4 (a) 103 ppm; (b) 3.24×10^{-4} mol; (c) 312 s 10.5 (a) 0.68, 0.49, and 0.60% (w/w), respectively; (b) 0.6 (±0.1)% (w/w) 10.6 .017 M 10.7 (a) Cu first; Mn, Ca, Ba; (b) 6.10; no 10.8 (a) $[Pb^{2+}] = 0.00495$ M; $[Ba^{2+}] = 2.91 \times 10^{-5}$ M; $[SO_4^{2-}] = 3.43 \times 10^{-6}$; (b) $X_{PbSO_4} = 0.010$; $X_{BaSO_4} = 0.990$

Chapter 11

Concept Review

2. $A = 2 - \log \%T$
3. Spectral interference—absorption or emission due to matrix components; chemical interference—chemical form varies (or analyte decomposes) giving decreased or enhanced absorbance or emission; instrument interference—detection of excess illumination at the transducer due to imperfection in the instrument
4. 0.4 to 0.7 absorbance units
5. Energy proportional to frequency and inversely proportional to wavelength ($E = h\nu = hc/\lambda$)
6. The incident and transmitted radiation is in the visible range for Raman which means that the solvents and sample cells must be transparent there instead of in the IR.
7. Monochromator bandwidth < (1/10) × linewidth of the spectral feature
8. Auger most sensitive for elements lighter than zinc; Rutherford backscattering most sensitive for light elements like carbon and oxygen; X-ray fluorescence most sensitive to heavier elements (heavier than sodium)

Exercises

11.1 (a) decreases; (b) stays same (although if concentrations different, no way to predict)
11.2 (a) 3.42×10^{-19} J; (b) 532 nm = 18,800 cm^{-1}; 632 nm = 15,800 cm^{-1}; (c) 3000 cm^{-1}; (d) transition between outer electronic energy levels; (e) 5.15×10^{14} s^{-1}
11.3 (a) 60 M^{-1} cm^{-1}; (b) 1.87×10^{-7} L μg^{-1} cm^{-1}; (c) 106 M^{-1} 11.4 (a) 2 cm^{-1}; (b) 6.25 μm; (c) IR, transition between vibrational levels 11.5 4.71 L μg^{-1}
11.6 75.9%, 59.6%, 36.3%, 22.4% (in order of increasing concentration) 11.7 0.50 11.8 3.3%
11.9 (b) 144.3 ppm sodium, 5.6 ppm potassium
11.10 (a) TMS, cyclohexane, *t*-butanol, acetone, dioxane, benzene (in order of increasing Hz downfield); (b) 0 ppm, 138 ppm, 163 ppm, 187 ppm, 380 ppm, 715 ppm (in order increasing Hz downfield); c) acetone 20 ppm, benzene, 20 ppm, cyclohexane 10 ppm, *t*-butanol 13 ppm, dioxane 15 ppm 11.11 400 11.12 excitation 367 nm; emission 483 nm; pH 9.35

Chapter 12

Concept Review

1. x axis is m/z; y axis is relative abundance (or counts)
2. (a) the peak corresponding to the m/z of highest abundance; (b) the peak corresponding to the unfragmented molecule (or molecule + H for some ionization methods)
3. Identify the molecular ion; study the isotope distribution pattern; explain the fragmentation pattern

4. (a) MALDI or ESI; (b) ESI; (c) ICP; (d) volatile, EI; nonvolatile, ESI
5. (a) vaporize matrix with laser pulse, simultaneously ejecting analyte into gas phase and ionizing them; (b) pass them through chanber where electrons flowing between two plates of opposite change strike them, remove an electron, and fragment them; (c) spray them from the tip of a very small diameter, electrically charged needle where they pick up charge; (d) ionize them by exciting them with an inductively coupled plasma
6. The first fragment is fragmented once again and those smaller fragments can, in turn, be indentified to confirm the identity of the first fragment.

Exercises

12.1 I—vinyl chloride; II—nitrobenzene; III—carbon tetrachloride; IV—1,2-dichloroethane 12.2 (a) $MW_{(a)} =$ 194.0804; $MW_{(b)} = 194.1055$; $MW_{(c)} = 194.0943$; $MW_{(d)} =$ 194.0579; (b) 0.0112 u 12.3 (a) 15 u; (b) 77 u; (c) 57 u 12.4 fluoro-, chloro-, bromo-, and iodobenzene

Chapter 13

Concept Review

1. (a) $V_R' = V_R - V_0$; (b) $k = (t_R - t_0)/t_0$; (c) $\alpha_{1,2} = k_2/k_1$; (d) $N = 16\left(\frac{t_R}{W}\right)^2$; (e) $HETP = L/N$
2. Using the peak area means that if the retention time of a peak varies the resulting change in peak width and height does not have any effect
3. (a) charge/size is larger for A; (b) A is smaller than B; (c) A is less polar than B
5. Derivatization can increase volatility or, in some cases, provide selective response for given detector
6. An 8-foot long stainless steel column with an outer diameter of 1/8 inch; column packed with particles that pass through a standard 100-mesh sieve, but are trapped by a 120-mesh sieve; the particles have been coated with 5 grams of SE-30 per gram of particles before packing them into the column
7. A corrects for eddy diffusion from relatively larger channels between the stationary phase, B corrects for longtitudinal diffusion, and C corrects for spreading due to transverse mass transfer both between various spots in the mobile phase and between the mobile and stationary phase.
8. In gas chromatography, refers to a increase in temperature over the time required for the separation; in liquid chromatography, refers to an increase in eluting power of the mobile phase over the time required for the separation.

Exercises

13.1 (b) remain same; (c) increase 13.2 (a) increase; (b) decrease; (c) remain same 13.3 (a) 16.3 mL, 18.7 mL; (b) 4.7, 5.6; (c) 3.6 min, 4.1 min; (d) 1.61; (e) 1.24; (f) 2400; (g) 0.104 mm; (h) same as in (b) 13.4 (a) yes; (b) no; (c) use peak areas for quantitation 13.5 270 μg C_{16}; 170 μg C_{14}; 140 μg C_{12} 13.6 (using FWHM) 13.7 (a) *microbore areas:* 1.04×10^{-3}; 1.37×10^{-3}; 4.82×10^{-3} (a.u. $\times$ min); FWHM: 0.0132 mL; 0.0132 mL; 0.0176 mL; *standard areas:* 5.66×10^{-5}; 9.44×10^{-5}; 3.10×10^{-4} (a.u. $\times$ min); FWHM: 0.204 mL; 0.273 mL; 0.477 mL; (b) response factor based on benzene: *microbore:* naphthalene 13; biphenyl 180; *standard:* naphthalene 17; biphenyl 200; (c) greater dilution in standard column accounts for most of difference in response
13.8 100,000 daltons 13.9 Trp, Pro, Ala, Asn
13.11 3.4 s

Chapter 14

Concept Review

1. Negative ions migrate toward higher (positive) voltage and positive toward lower (negative) voltage end of column; average rate of migration of analyte proportional to average charge on ion; average rate of migration of analyte proportional to average potential applied; average rate of motion of analyte inversely proportional to its cross section; quality of separation improves when convection suppressed; quality of separation improves when temperature held more nearly constant.
2. Separation on gel does not result in irreversible unfolding of the protein and loss of protein activity.
3. In pore gradient gel, change in porosity is gradual; in discontinuous gel, abrupt jumps in porosity.
4. Good temperature control minimizes convection; maintains constant pK_a of analyte(s) and buffer(s); keeps macromolecular structure in same form throughout separation (and thus maintains shape and cross sectional area of analyte); maintains constant viscosity of gel
5. Hydrostatic injection avoids problems due to the fact that electrosmotic flow increases volume injected over the amount expected and to discrimination against low mobility species.
6. "Normal" refers to anode (+) at injection end of column, "reversed" indicates that the cathode (−) is at injection end of column.

Exercises

14.1 2.5 cm hr^{-1} 14.2 At pH 3.0: all are positive overall except pepsin. At pH 7.2: ribonuclease, chymotrypsin, cytochrome *c* and lysozyme are still positive, myoglobin is slightly positive and others positive. At pH 10, lysozyme positive, cytochrome *c* is slightly positive, chymotrypsin and ribonuclease are slightly negative, and the rest are negative. 14.3 31.0 kD 14.4 stacking, $N = 13{,}400$; no stacking, $N = 39{,}600$

14.5 (a)

	α_0	α_1	α_2	α_3	(b)z	(c)$\frac{z}{MW^{2/3}}$
oxalic	—	0.0375	0.963		−2	0.100
tartaric	—	0.0521	0.948		−2	0.064
malic	0.0015	0.244	0.754		−1.76	0.070
citric	—	—	0.875	0.125	−2.12	0.074

(c) no

Chapter 15

Concept Review

1. (a) total current passed as function of time; (b) potential at indicator electrode; (c) current; (d) conductivity or specific conductance
2. (a) $Hg_2Cl_2\ (s) + 2\ e^- \rightleftharpoons Hg\ (s) + 2\ Cl^-\ (aq)$; (b) same as in (a); (c) $AgCl\ (s) + e^- \rightleftharpoons Ag\ (s) + Cl^-\ (aq)$; (d) $2\ H^+\ (aq) + 2\ e^- \rightleftharpoons H_2(g)$
3. Semilog (E on y axis and log [X] on x axis)
4. IR drop, concentration polarization, and kinetic overpotential
5. Potential measured between reference and working electrodes; current flow measured between auxiliary and working electrode
6. H^+ and OH^-
7. No difference
8. Area under curve
9. Implies electron transfer proceeds extremely quickly in both directions so system described by Nernst equation; and, for techniques such as cyclic voltammetry, also implies chemical reversibility (oxidized and reduced species both chemically stable under conditions of experiment)

Exercises

15.1 (a) E = constant − 0.059 log [F^-]; (b) $E = -194 - 59$ log [F^-]; 0.956 ppm; (c) yes 15.2 (a) low; (b) high; (c) 4.7; (d) 12.5 15.3 (a) 1.93×10^{-3} M; (b) 0.0441 M; (c) 57.3 mM 15.5 11 ppm 15.6 0.071 M 15.7 (a) cobalt; (b) hydrogen 15.8 (a) $Ag(s) \rightleftharpoons Ag^+ + e^-$; $2\ Cl^- \rightleftharpoons Cl_2 + 2\ e^-$; $2\ I^- \rightleftharpoons I_2$; $2\ H_2O + 2\ e^- \rightleftharpoons H_2(g) + 2\ OH^-$; $2\ H_2O \rightleftharpoons O_2(g) + 2\ H^+ + 2\ e^-$; $HgNH_3edta^{2-} + 2\ e^- \rightleftharpoons Hg(s) + edta^{4-} + NH_3$; $2\ Br^- \rightleftharpoons Br_2 + 2\ e^-$ 15.9 1.47 ppm 15.10 (a) 6-hydroxydopa, L-dopa, tyrosine; (b) No 15.11 (a) −0.375 V, −0.250 V, −0.100 V, 0.0 V; (b) −0.476 V 15.12 $E = E_{measured} + 0.118$ 15.13 Cd is leftmost peak, 0.10 $\mu g\ L^{-1}$; Pb is middle peak, 0.55 $\mu g\ L^{-1}$; Cu is rightmost peak, 1.2 $\mu g\ L^{-1}$

Chapter 16

Concept Review

1. It functions as a catalyst in the reaction (affects k itself)
2. Advantages include: specificity, sensitivity (often many moles product for small amount catalytic analyte), avoid waiting for completion of very slow reaction, can use (chemically) irreversible reactions for analysis

Exercises

16.1 catalytic, fixed-time 16.2 (a) 0.14 unit^{-1}; (b) 0.625 units g^{-1}; (c) 1.95% (w/w) 16.3 Step 2: enzyme activity 1.9 units/mg; Step 3: total activity 10,200 units; enzyme activity 6.9 units/mg; 110-fold purification; 85% yield; Step 4: total activity 9,000 units; enzyme activity 98.0 units/mg; 1550-fold purification; 75% yield; Step 5: total activity 7,560 units; enzyme activity 250 units/mg; 4000-fold purification; 63% yield
16.4 (b) less precise; (e) fixed; (f) 1.24 ppm 16.5 1.8 s

PHOTO CREDITS

Chapter 4
Page 122: Milestone Inc.
Page 123: CEM Corporation.

Chapter 6
Page 174: Science Photo Library/Photo Researchers, Inc.

Chapter 8
Page 239: UPI/Corbis-Bettmann.

Chapter 11
Page 354: Corbis-Bettmann.

Chapter 12
Page 381: Galileo Corporation.

Chapter 13
Page 405: Science Photo Library/Photo Researchers, Inc.
Page 436: Conrad Yu, LLNL.

Chapter 14
Page 458: Cellmark Diagnostics.
Page 468: Hoefer Pharmacia Biotech, Inc.

Chapter 15
Page 518: AP/Wide World Photos.

Chapter 16
Page 546: Corbis-Bettmann.

Endpapers
Figure EP.1: Hach Company. Figure EP.2: Affymetrix. Figure EP.3: C. Falco/Photo Researchers, Inc. Figure EP.4: Michael Heron/Simon & Schuster/PH College. Figure EP.5: Los Alamos National Laboratory/Science Photo Library/Photo Researchers, Inc. Figure EP.6: Photron Pty. Ltd. Figure EP.7: Varian Associates, Inc. Figure EP.8: Varian Associates, Inc. Figure EP.9: Aurora Instruments. Figure EP.10: Geoff Tompkinson/ Science Photo Library/Photo Researchers, Inc. Figure EP.11: Dr. John B. Fenn. Figure EP.12: John Bragagnolo/SGE, Incorporated. Figure EP.13: CMSP/NIH/Custom Medical Stock Photo, Inc Figure EP.14: Bioanalytical Systems.

INDEX

INTERNATIONAL ATOMIC MASS BASED ON ^{12}C = 12 EXACTLY*

Element	*Symbol*	*Atomic Number*	*Atomic Weight*†
Actinium	Ac	89	227.028
Aluminum	Al	13	26.9815
Americium	Am	95	(243)
Antimony	Sb	51	121.75
Argon	Ar	18	39.948
Arsenic	As	33	74.9216
Astatine	At	85	(210)
Barium	Ba	56	137.33
Berkelium	Bk	97	(247)
Beryllium	Be	4	9.0122
Bismuth	Bi	83	208.980
Bohrium	Bh	107	(262)
Boron	B	5	10.81
Bromine	Br	35	79.904
Cadmium	Cd	48	112.41
Calcium	Ca	20	40.08
Californium	Cf	98	(251)
Carbon	C	6	12.011
Cerium	Ce	58	140.12
Cesium	Cs	55	132.905
Chlorine	Cl	17	35.453
Chromium	Cr	24	51.996
Cobalt	Co	27	58.9332
Copper	Cu	29	63.54
Curium	Cm	96	(247)
Dubnium	Db	105	(262)
Dysprosium	Dy	66	162.50
Einsteinium	Es	99	(252)
Erbium	Er	68	167.26
Europium	Eu	63	151.96
Fermium	Fm	100	(257)
Fluorine	F	9	18.9984
Francium	Fr	87	(223)
Gadolinium	Gd	64	157.25
Gallium	Ga	31	69.72
Germanium	Ge	32	72.59
Gold	Au	79	196.9665
Hafnium	Hf	72	178.49
Hassium	Hs	108	(265)
Helium	He	2	4.0026
Holmium	Ho	67	164.930
Hydrogen	H	1	1.0079
Indium	In	49	114.82
Iodine	I	53	126.9045
Iridium	Ir	77	192.2
Iron	Fe	26	55.847
Krypton	Kr	36	83.80
Lanthanum	La	57	138.905
Lawrencium	Lr	103	(260)
Lead	Pb	82	207.2
Lithium	Li	3	6.941
Lutetium	Lu	71	174.97
Magnesium	Mg	12	24.305
Manganese	Mn	25	54.9380
Meitnerium	Mt	109	(266)
Mendelevium	Md	101	(258)
Mercury	Hg	80	200.59
Molybdenum	Mo	42	95.94
Neodymium	Nd	60	144.24
Neon	Ne	10	20.18
Neptunium	Np	93	237.048
Nickel	Ni	28	58.70
Niobium	Nb	41	92.906
Nitrogen	N	7	14.0067
Nobelium	No	102	(259)
Osmium	Os	76	190.2
Oxygen	O	8	15.9994
Palladium	Pd	46	106.4
Phosphorus	P	15	30.9738
Platinum	Pt	78	195.08
Plutonium	Pu	94	(244)
Polonium	Po	84	(209)
Potassium	K	19	39.0983
Praseodymium	Pr	59	140.907
Promethium	Pm	61	(145)
Protactinium	Pa	91	231.036
Radium	Ra	88	226.025
Radon	Rn	86	(222)
Rhenium	Re	75	186.21
Rhodium	Rh	45	102.905
Rubidium	Rb	37	85.468
Ruthenium	Ru	44	101.07
Rutherfordium	Rf	104	(261)
Samarium	Sm	62	150.35
Scandium	Sc	21	44.956
Seaborgium	Sg	106	(263)
Selenium	Se	34	78.96
Silicon	Si	14	28.086
Silver	Ag	47	107.868
Sodium	Na	11	22.9898
Strontium	Sr	38	87.62
Sulfur	S	16	32.06
Tantalum	Ta	73	180.948
Technetium	Tc	43	(98)
Tellurium	Te	52	127.60
Terbium	Tb	65	158.925
Thallium	Tl	81	204.38
Thorium	Th	90	232.038
Thulium	Tm	69	168.934
Tin	Sn	50	118.69
Titanium	Ti	22	47.88
Tungsten	W	74	183.85
Uranium	U	92	238.03
Vanadium	V	23	50.942
Xenon	Xe	54	131.29
Ytterbium	Yb	70	173.04
Yttrium	Y	39	88.906
Zinc	Zn	30	65.38
Zirconium	Zr	40	91.22

Masses are accurate to ± 1 in the last digit.

Numbers in parentheses indicate mass number of the most stable isotope.

◀ FIGURE EP.6

A hollow cathode lamp for atomic absorption spectrometry
The hollow cathode lamp produces a narrow bandwidth emission at the wavelengths characteristic of the metal that makes up the cathode. The open end of the hollow cathode can be seen where the light emission is the most intense. The anode is the ring that projects in front of it. The ring is approximately a centimeter in diameter. (Photron Pty., Ltd.)

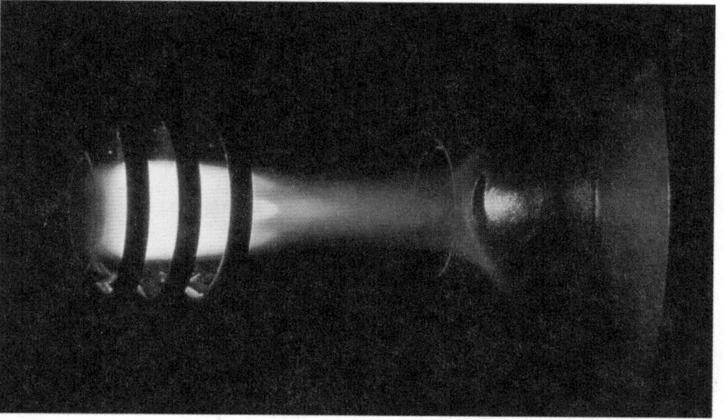

FIGURE EP.7 ▲

An inductively coupled plasma (ICP) torch
The ICP torch's high temperature decomposes the materials passing through it into atoms and ions. The rings obscuring the light are tubes of a copper coil that is the antenna for the radiofrequency emission that furnishes the energy to heat the plasma, which is a few centimeters in diameter. Figure EP.8 shows this part of the spectrometer when the torch is off.

FIGURE EP.8 ▲

Interface of an ICP emission spectrometer (torch off)
This spectrometer collects light from along the axis of an ICP torch. The hot plasma is deflected at the cooled head at the right, and the light emitted from the analytes enters the spectrometer through the hole in the middle of the cone. The three tubes surrounding the silica tube comprise the coil antenna for the radiofrequency power source.

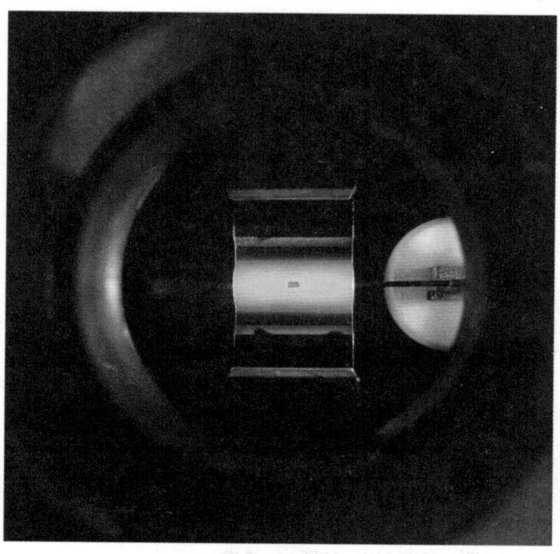

FIGURE EP.9 ▶

Graphite furnace/plasma source for atomic absorption spectrometry
A sample is injected through the rectangular hole (dark) into the graphite furnace tube. After heating the sample to dryness, the hollow graphite tube is heated to incandescence in seconds by passing current from one side to the other (top/bottom). The sample is vaporized and atomized. An inductively coupled plasma is then generated between the graphite bar (anchored at the right) and the furnace wall, and the emission is measured. Without the bar present, atomic absorption can be done.